SURFACE AND NANOMOLECULAR CATALYSIS

edited by
Ryan Richards

CRC Press
Taylor & Francis Group
Boca Raton London New York

CRC Press is an imprint of the
Taylor & Francis Group, an **informa** business

A TAYLOR & FRANCIS BOOK

CRC Press
Taylor & Francis Group
6000 Broken Sound Parkway NW, Suite 300
Boca Raton, FL 33487-2742

First issued in paperback 2019

© 2005 by Taylor and Francis Group, LLC
CRC Press is an imprint of Taylor & Francis Group, an Informa business

No claim to original U.S. Government works

ISBN-13: 978-1-57444-481-0 (hbk)
ISBN-13: 978-0-367-39081-5 (pbk)
Library of Congress Card Number 2005055955

Library of Congress Cataloging-in-Publication Data

Surface and nanomolecular catalysis / edited by Ryan Richards.
    p. cm.
    Includes bibliographical references and index.
    ISBN 1-57444-481-6 (US-HC) -- ISBN 3-13-140291-1 (GTV) 1. Catalysis. 2. Surface chemistry. 3. Nanoscience. I. Richards, Ryan.

QD505.S784 2006
541'.395--dc22
    2005055955

Visit the Taylor & Francis Web site at
http://www.taylorandfrancis.com

and the CRC Press Web site at
http://www.crcpress.com

Dedication

to Sarah

Preface

Few terms have been more commonly used and abused in the scientific literature than *nano*. However, if one is able to sift through the vast amounts of nano literature, there are also numerous reports that are of both academic and commercial importance. This is particularly true for the field of catalysis in which rapid progress is being made that has transformed this once black art into a science, which is understood on a molecular and even atomic level. These gains have been particularly driven by the fields of surface and nanomolecular science with improvements in instrumentation and experimental techniques that have facilitated scientists' observations on the nano-size scale.

While the field of catalysis has a dramatic impact on our daily lives, it does not receive a proportional coverage in the typical undergraduate and graduate educations. This is possibly due to the broad range of expertise involved in the field, which includes physics, chemical engineering, and all subdisciplines of chemistry. The impact of catalysis in our current everyday lives cannot be understated. It was recently estimated that 35% of global GDP depends on catalysis. In addition, there are major hurdles for mankind that may be overcome with developments in catalysis. In particular, the goal of sustainability with regard to energy and environmental concerns will most certainly require significant contributions from catalysis.

Catalysts are materials that change the rate at which chemical equilibrium is reached without themselves undergoing any change. Through the phenomenon of catalysis, very small quantities of a catalytic material can facilitate several thousand transformations. In addition to the remarkable increases in activity observed in the presence of a catalyst, an additional attribute of catalysts is that there is often a selectivity toward certain reaction products. Often, this selectivity is of greater importance than activity since a highly selective process eliminates the generation of wasteful by-products.

The field of nanotechnology has generated a great deal of interest primarily because on this size scale, numerous new and potentially useful properties have been observed. These size-dependent properties include melting point, specific heat, surface reactivities, optical, magnetic, and catalytic properties. In addition to the numerous proposed applications, there are also concerns regarding the environmental and health implications associated with the use of these materials. These concerns are, however, particularly difficult to address because the properties of nanoscale materials are different from both the molecular and bulk forms and can even change as a result of small differences in size and shape. A general understanding of the chemical and physical properties of nanoscale materials as a function of size and shape is necessary to address the concerns about nanomaterials and their applications. A significant contribution to this understanding will be generated through studies of *Surface and Nanomolecular Catalysis*.

Surface and Nanomolecular Catalysis contains an overview of the field as given by several leading international scientists. Chapter 1 by Ma and Zaera provides an excellent overview of characterization methods. This is followed by four chapters highlighting preparation, characterization, and application of traditional subclasses of heterogeneous catalysts. These include metal oxides by Ranjit and Klabunde (Chapter 2), colloids by Bönnemann and Nagabhushana (Chapter 3), microporous and mesoporous materials by Schmidt (Chapter 4), and skeletal catalysts by Smith (Chapter 5). Regalbuto (Chapter 6) follows with an insightful chapter on the preparation of supported metal catalysts. The engineering of catalytic processes is presented by Hočevar (Chapter 7) followed by structure and reaction control by Tada and Iwasawa (Chapter 8). The chapter covering the texturological properties of catalytic systems by Fenelonov and Melgunov (Chapter 9) presents an indepth examination of this critical area. Wallace and Goodman (Chapter 10) then provide an excellent chapter demonstrating how surface science can elucidate reaction mechanisms. The emerging field of combinatorial approaches in catalysis is given by Schunk, Busch, Demuth, Gerlach, Haas, Klein, and Zech (Chapter 11). The final three chapters cover important specialized areas of catalysis. Pârvulescu and Marcu (Chapter 12) present an overview of heterogeneous photocatalysis.

Liquid-phase oxidations catalyzed by polyoxometalates are covered by Mizuno, Kamata, and Yamaguchi (Chapter 13). Finally, the developing field of enantioselective heterogeneous catalysis is presented by Coman, Poncelet, and Pârvulescu (Chapter 14).

Overall, each chapter is designed to be able to stand alone as a short course. However, when taken together, the contents form a comprehensive overview of *Surface and Nanomolecular Catalysis*, appropriate for both a graduate course and as a reference text. In addition, each chapter includes several questions appropriate for a graduate course, which should be particularly helpful to instructors.

Other important aspects of modern catalysis including bio- and homogeneous catalysis are beyond the scope of the current book and are themselves the themes of several excellent books. Further, the emerging areas of computational catalysis and immobilized catalysts are not included here but are covered dedicated texts in the literature.

It is the hope of the editor that this book forms the foundation of graduate-level courses in *Surface and Nanomolecular Catalysis* and aids students in the understanding of this multidisciplinary subject. Further, the editor thanks the contributors for their hard work.

Ryan M. Richards

The Editor

Ryan M. Richards was raised near Flint, Michigan. In 1994, he completed both B.A. in chemistry and B.S. in forensic science at Michigan State University. He then spent 2 years as an M.S. student at Central Michigan University working on organometallic chemistry with Professor Bob Howell. He was awarded a Ph.D. in 2000 for investigating the properties of metal oxide nanoparticles in the laboratory of Professor Kenneth Klabunde at Kansas State University. In 1999, he was an invited scientist at the Boreskov Institute of Catalysis in Novosibirsk, Russia where he began investigating the catalytic properties of nanoscale metal oxides. In 2000, he joined the research group of Professor Helmut Bönnemann investigating colloidal catalysts and heterogeneous catalysis as a research fellow at the Max Planck Institute für
Kohlenforschung, Mülheim an der Ruhr, Germany. He joined the engineering and science faculty at the International University Bremen in 2002 and is leading a research group focusing on the preparation of novel nanoscale materials and catalysis.

Contributors

Helmut Bönnemann
MPI fuer Kohlenforschung
Heterogene Katalyse
Mülheim an der Ruhr, Germany

Oliver Busch
Hte Aktiengesellschaft
Heidelberg, Germany

Simona M. Coman
Department of Chemical Technology
   and Catalysis
University of Bucharest
Bucharest, Romania

Dirk G. Demuth
Hte Aktiengesellschaft
Heidelberg, Germany

Vladimir B. Fenelonov
Boreskov Institute of Catalysis
Novosibirsk, Russia

Olga Gerlach
Hte Aktiengesellschaft
Heidelberg, Germany

D. Wayne Goodman
Department of Chemistry
Texas A&M University
College Station, Texas

Alfred Haas
Hte Aktiengesellschaft
Heidelberg, Germany

Stanko Hočevar
Laboratory of Catalysis and
   Chemical Reaction Engineering
National Institute of Chemistry
Ljubljana, Slovenia

Yasuhiro Iwasawa
Department of Chemistry
Graduate School of Science
The University of Tokyo
Tokyo, Japan

Keigo Kamata
Core Research for Evolutional Science
   and Technology
Japan Science and Technology Agency
Saitama, Japan

Kenneth J. Klabunde
Department of Chemistry
Kansas State University
Manhattan, Kansas

Jens Klein
Hte Aktiengesellschaft
Heidelberg, Germany

Zhen Ma
Department of Chemistry
University of California
Riverside, California

Victor Marcu
Israel Electric Corporation
Orot Rabin Power Station
Hadera, Israel

Maxim S. Mel'gunov
Boreskov Institute of Catalysis
Novosibirsk, Russia

Noritaka Mizuno
Department of Applied Chemistry
School of Engineering
The University of Tokyo
Tokyo, Japan
and Core Research for Evolutional Science
   and Technology
Japan Science and Technology
   Agency
Saitama, Japan

K. S. Nagabhushana
Forschungszentrum Karlsruhe
ITC-CPV
Eggenstein Leopoldshafen
Germany

Vasile I. Pârvulescu
Department of Chemical Technology
   and Catalysis
University of Bucharest
Bucharest, Romania

Georges Poncelet
Unité de Catalyse et Chimie des
   Matériaux Divisée
Universite Catolique de Louvain
Louvain-la-Neuve, Belgium

Ranjit T. Koodali
Department of Chemistry
University of South Dakota
Vermillion, South Dakota

John R. Regalbuto
Department of Chemical Engineering
University of Illinois
Chicago, Illinois

Wolfgang Schmidt
MPI Fuer Kohlenforschung
Heterogene Katalyse
Mülheim an der Ruhr, Germany

Stephan Andreas Schunk
Hte Aktiengesellschaft
Heidelberg, Germany

Andrew J. Smith
School of Chemical Engineering and
   Industrial Chemistry
The University of New South Wales
Sydney, Australia

Mizuki Tada
Department of Chemistry
Graduate School of Science
The University of Tokyo
Tokyo, Japan

W. T. Wallace
Department of Chemistry
Texas A&M University
College Station, Texas

Kazuya Yamaguchi
Department of Applied Chemistry
School of Engineering
The University of Tokyo
Tokyo, Japan
and Core Research for Evolutional Science
   and Technology
Japan Science and Technology
   Agency
Saitama, Japan

Francisco Zaera
Department of Chemistry
University of California
Riverside, California

Torsten Zech
Hte Aktiengesellschaft
Heidelberg, Germany

Contents

Chapter 1
Characterization of Heterogeneous Catalysts 1
Zhen Ma and Francisco Zaera

Chapter 2
Catalysis by Metal Oxides 39
Ranjit T. Koodali and Kenneth J. Klabunde

Chapter 3
Colloidal Nanoparticles in Catalysis...63
Helmut Bönnemann and K.S. Nagabhushana

Chapter 4
Microporous and Mesoporous Catalysts..95
Wolfgang Schmidt

Chapter 5
Skeletal Catalysts 141
Andrew J. Smith

Chapter 6
A Scientific Method to Prepare Supported Metal Catalysts.........................161
John R. Regalbuto

Chapter 7
Catalysis and Chemical Reaction Engineering...195
Stanko Hočevar

Chapter 8
Structure and Reaction Control at Catalyst Surfaces...................................229
Mizuki Tada and Yasuhiro Iwasawa

Chapter 9
Texturology 257
Vladimir B. Fenelonov and Maxim S. Mel'gunov

Chapter 10
Understanding Catalytic Reaction Mechanisms: Surface Science Studies
of Heterogeneous Catalysts...337
W.T. Wallace and D. Wayne Goodman

Chapter 11
High-Throughput Experimentation and Combinatorial
Approaches in Catalysis..373
**Stephan Andreas Schunk, Oliver Busch, Dirk G. Demuth, Olga Gerlach,
Alfred Haas, Jens Klein, and Torsten Zech**

Chapter 12
Heterogeneous Photocatalysis...427
Vasile I. Pârvulescu and Victor Marcu

Chapter 13
Liquid-Phase Oxidations Catalyzed by Polyoxometalates 463
Noritaka Mizuno, Keigo Kamata, and Kazuya Yamaguchi

Chapter 14
Asymmetric Catalysis by Heterogeneous Catalysts 493
Simona M. Coman, Georges Poncelet, and Vasile I. Pârvulescu

Index...533

Characterization of Heterogeneous Catalysts

Zhen Ma and Francisco Zaera

CONTENTS

1.1   Introduction 2
1.2   Structural Techniques 2
    1.2.1   X-Ray Diffraction 2
    1.2.2   X-Ray Absorption Spectroscopy 3
    1.2.3   Electron Microscopy 5
1.3   Adsorption–Desorption and Thermal Techniques 7
    1.3.1   Surface Area and Pore Structure 7
    1.3.2   Temperature-Programmed Desorption and Reaction 8
    1.3.3   Thermogravimetry and Thermal Analysis 9
    1.3.4   Microcalorimetry 10
1.4   Optical Spectroscopies 12
    1.4.1   Infrared Spectroscopy 12
    1.4.2   Raman Spectroscopy 13
    1.4.3   Ultraviolet–Visible Spectroscopy 15
    1.4.4   Nuclear Magnetic Resonance 16
    1.4.5   Electron Spin Resonance 18
1.5   Surface-Sensitive Spectroscopies 19
    1.5.1   X-Ray and Ultraviolet Photoelectron Spectroscopies 19
    1.5.2   Auger Electron Spectroscopy 20
    1.5.3   Low-Energy Ion Scattering 21
    1.5.4   Secondary-Ion Mass Spectroscopy 21
1.6   Model Catalysts 22
1.7   Concluding Remarks 25
References 26
Chapter 1 Questions 32

1.1 INTRODUCTION

Characterization is a central aspect of catalyst development [1,2]. The elucidation of the structures, compositions, and chemical properties of both the solids used in heterogeneous catalysis and the adsorbates and intermediates present on the surfaces of the catalysts during reaction is vital for a better understanding of the relationship between catalyst properties and catalytic performance. This knowledge is essential to develop more active, selective, and durable catalysts, and also to optimize reaction conditions.

In this chapter, we introduce some of the most common spectroscopies and methods available for the characterization of heterogeneous catalysts [3–13]. These techniques can be broadly grouped according to the nature of the probes employed for excitation, including photons, electrons, ions, and neutrons, or, alternatively, according to the type of information they provide. Here we have chosen to group the main catalyst characterization techniques by using a combination of both criteria into structural, thermal, optical, and surface-sensitive techniques. We also focus on the characterization of real catalysts, and toward the end make brief reference to studies with model systems. Only the basics of each technique and a few examples of applications to catalyst characterization are provided, but more specialized references are included for those interested in a more in-depth discussion.

1.2 STRUCTURAL TECHNIQUES

1.2.1 X-Ray Diffraction

X-ray diffraction (XRD) is commonly used to determine the bulk structure and composition of heterogeneous catalysts with crystalline structures [14–16]. XRD analysis is typically limited to the identification of specific lattice planes that produce peaks at their corresponding angular positions 2θ, determined by Bragg's law, $2d \sin\theta = n\lambda$. In spite of this limitation, the characteristic patterns associated with individual solids make XRD quite useful for the identification of the bulk crystalline components of solid catalysts. This is illustrated by the example in Figure 1.1, which displays XRD patterns obtained *ex situ* for a number of manganese oxide catalysts before and after reduction [17]. These data indicate that, regardless of the starting point (MnO_2, Mn_2O_3, or Mn_3O_4), the structure of the catalyst changes after pretreatment with H_2 to the same reduced MnO phase, allegedly the one active for selective hydrogenation. *In situ* XRD is particularly suited to follow these types of structural changes in the catalysts during pretreatments or catalytic reactions [18,19].

X-ray diffraction can also be used to estimate the average crystallite or grain size of catalysts [14,20]. The XRD peaks are intense and sharp only if the sample has sufficient long-range order, and become broader for crystallite sizes below about 100 nm. Average particle sizes below about 60 nm can be roughly estimated by applying the Debye–Scherrer equation, $D = 0.89\lambda/(B_0^2 - B_e^2)^{1/2} \cos\theta$, where B_0 is the measured width (in radians) of a diffraction line at half-maximum, and B_e the corresponding width at half-maximum of a well-crystallized reference sample [14,20]. Figure 1.2 displays an example of the application of this method for the characterization of anatase TiO_2 photocatalysts [21]. In that case, the line width of the (101) diffraction peak at 25.4° was used to calculate the average grain sizes of samples prepared using different procedures: a significant growth in particle size was clearly observed upon high-temperature calcination.

In spite of the large success of XRD in routine structural analysis of solids, this technique does present some limitations when applied to catalysis [1,9]. First, it can only detect crystalline phases, and fails to provide useful information on the amorphous or highly dispersed solid phases so common in catalysts [22]. Second, due to its low sensitivity, the concentration of the crystalline phase in the sample needs to be reasonably high in order to be detected. Third, XRD probes bulk phases,

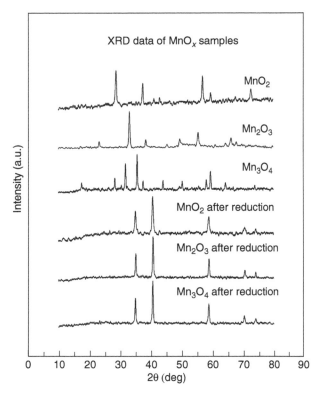

Figure 1.1 XRD patterns for different manganese oxides before and after pretreatment in H₂ at 420°C [17]. The top three traces correspond to the original MnO₂, Mn₂O₃, and Mn₃O₄ solids used in these experiments, while the bottom three were obtained after H₂ treatment. It can be seen here that the catalysts are all reduced to the same MnO phase regardless of the nature of the starting material. It was inferred that MnO is the actual working catalyst in all cases, hence the similarity in methyl benzoate hydrogenation activity obtained with all these MnOₓ solids. (Reproduced with permission from Elsevier.)

and is not able to selectively identify the surface structures where catalytic reactions take place. Finally, XRD is not useful for the detection of reaction intermediates on catalytic surfaces.

1.2.2 X-Ray Absorption Spectroscopy

X-ray absorption can also be used for both structural and compositional analysis of solid catalysts [23–25]. In these experiments, the absorption of x-rays is recorded as a function of photon energy in the region around the value needed for excitation of a core electron of the element of interest. The region near the absorption edge shows features associated with electronic transitions to the valence and conduction bands of the solid. Accordingly, the x-ray absorption near-edge structure (XANES, also called NEXAFS) spectra, which are derived from these excitations, provide information about the chemical environment surrounding the atom probed [26–28]. Farther away from the absorption edge, the extended x-ray absorption fine structure (EXAFS) spectra show oscillatory behavior due to the interference of the wave of the outgoing photoelectron with those reflected from the neighboring atoms. In EXAFS, a Fourier transform of the spectra is used to determine the local geometry of the neighborhood around the atom being excited [25,29,30].

The power of x-ray absorption spectroscopy for the characterization of catalysts is illustrated in Figure 1.3, where both XANES and EXAFS spectra are shown for a pyridine salt of niobium-exchanged molybdo(vanado)phosphoric acid (NbPMo₁₁(V)pyr) active for light alkane oxidation [31]. Specifically, the left panel of the figure displays an enlarged view of the Nb near-edge electronic spectra of the NbPMo₁₁(V)pyr catalyst at different temperatures and under the conditions used for

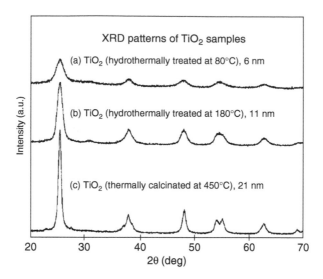

Figure 1.2 XRD patterns for three TiO$_2$ samples obtained by hydrothermal treatments at 80°C (a) and 180°C (b) and after calcination at 450°C (c) [21]. From the six XRD peaks identified with the anatase phase, the broadening of the (101) peak at 25.4° was chosen to estimate the average grain size of these samples. Generally, the sharper the peaks, the larger the particle size. The differences in grain size identified in these experiments were correlated with photocatalytic activity. (Reproduced with permission from The American Chemical Society.)

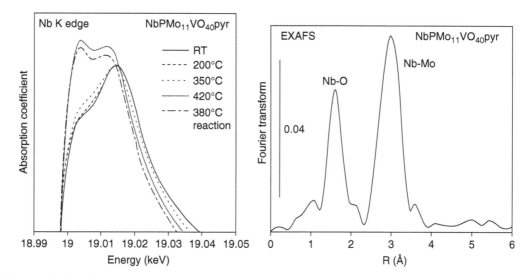

Figure 1.3 *Left:* Detailed view of the Nb K-edge XANES data of a pyridine salt of niobium-exchanged molybdo(vanado)phosphoric acid (NbPMo$_{11}$(V)pry) as a function of temperature [31]. A change in niobium oxidation state, from Nb^{5+} to Nb^{4+}, is identified between 350 and 420°C by a relative increase in absorption about 19.002 keV, and can be connected with the activation of the catalyst for light alkane oxidation. *Right:* Radial Fourier-transform EXAFS function for the NbPMo$_{11}$(V)pyr sample heated to 420°C [31]. The two peaks correspond to the Nb–O (1.5 Å) and Nb–Mo (3 Å) distances in the heteropolymolybdate fragments presumed to be the active phase for alkane oxidation. (Reproduced with permission from Elsevier.)

butane oxidation. The data indicate that below 350°C, the predominant species is Nb^{5+}, as determined by comparison with the spectrum from a reference Nb$_2$O$_5$ sample. At higher temperatures, however, the data resemble that of NbO$_2$, indicating the predominance of Nb^{4+} ions . This change in niobium oxidation state is directly related to the activation of the catalyst for alkane oxidation.

The right panel of Figure 1.3 displays the radial function obtained by Fourier transformation of the k-weighed background-subtracted EXAFS data from the solid heated to 420°C [31]. This spectrum shows two major peaks, one at about 1.5 Å associated with backscattering from O neighbors, and a second at 3 Å related to the Nb–Mo pairs. The measured distances are consistent with a combination of niobium oxo species and heteropolymolybdate fragments, presumably the catalytically active phase.

Several advantages and limitations are associated with the use of x-ray absorption spectroscopy for catalyst characterization. On the positive side, no long-range order is needed in the samples under study, since only the local environment is probed. Also, this technique works well in nonvacuum environments, and can be employed *in situ* during catalysis [19]. However, XANES is not very sensitive to variations in electronic structure, and the interpretation of the spectra is difficult, often requiring the use of reference samples and high-level theory. EXAFS only provides average values for the interatomic distances; it cannot be used to directly identify the chemical nature of the neighboring atoms, and is not very sensitive to the coordination number. Finally, x-ray absorption experiments typically require the use of expensive synchrotron facilities.

1.2.3 Electron Microscopy

Electron microscopy (EM) is a straightforward technique useful for the determination of the morphology and size of solid catalysts [32,33]. Electron microscopy can be performed in one of two modes — by scanning of a well-focused electron beam over the surface of the sample, or in a transmission arrangement. In scanning electron microscopy (SEM), the yield of either secondary or back-scattered electrons is recorded as a function of the position of the primary electron beam, and the contrast of the signal used to determine the morphology of the surface: the parts facing the detector appear brighter than those pointing away from the detector [34]. Dedicated SEM instruments can have resolutions down to 5 nm, but in most cases, SEM is only good for imaging catalyst particles and surfaces of micrometer dimensions. Additional elemental analysis can be added to SEM via energy-dispersive analysis of the x-rays (EDX) emitted by the sample [34].

Figure 1.4 shows SEM and EDX data for a $Mo_9V_3W_{1.2}O_x$ catalyst used in the selective oxidation of acrolein to acylic acid [35]. Although SEM analysis of the fresh sample failed to reveal any crystalline structure (data not shown), the images in Figure 1.4 clearly indicate the formation of well-resolved crystals after activation of the catalyst in the reaction mixture. In addition, the EDX spectra obtained from these samples point to variations in composition among the different crystallites of the catalyst. This analysis helped pin down the crystalline $(MoVW)_5O_{14}$-type structure as the catalytically active phase. The EM images in this example were taken *ex situ*, that is, after transferring the used catalysts from the reactor to the microscope, but *in situ* imaging of working catalysts is also possible [36,37].

Transmission electron microscopy (TEM) resembles optical microscopy, except that electromagnetic instead of optical lenses are used to focus an electron beam on the sample. Two modes are available in TEM, a bright-field mode where the intensity of the transmitted beam provides a two-dimensional image of the density or thickness of the sample, and a dark-field mode where the electron diffraction pattern is recorded. A combination of topographic and crystallographic information, including particle size distributions, can be obtained in this way [32].

Since TEM has a higher resolution than SEM (down to 0.1 nm), it is often used to image nano-sized catalysts such as metal oxide particles, supported metals, and catalysts with nanopores [38–41]. As an example, Figure 1.5 shows a TEM image of Au nanoparticles supported on a TiO_2 solid (A), and the particle size distribution estimated from statistical analysis of a number of similar pictures (B) [42]. Spherical Au particles, well dispersed on the surface of the round TiO_2 grains, are clearly seen in the picture, with sizes ranging from 2 to 8 nm and averaging 4.7 nm. A good correlation was obtained in this study between particle size and catalytic activity for CO oxidation and acetylene hydrogenation reactions. High-resolution TEM (HRTEM), being capable of

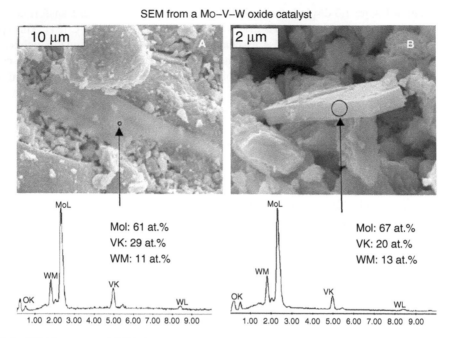

Figure 1.4   SEM images and EDX data from a $Mo_9V_3W_{1.2}O_x$ catalyst after activation during the oxidation of acrolein [35]. The pictures indicate that needle-like (A), platelet-like (B), and spherical (not shown) particles are formed during exposure to the reaction mixture. EDX analysis at different spots, two of which are exemplified here, point to V, Mo, and W contents that vary from 19 to 29, 60 to 69, and 11 to 13 atom%, respectively. It was determined that the *in situ* formation of a $(MoVW)_5O_{14}$-type phase accounts for the increase in acrolein conversion observed during the initial reaction stages. (Reproduced with permission from Elsevier.)

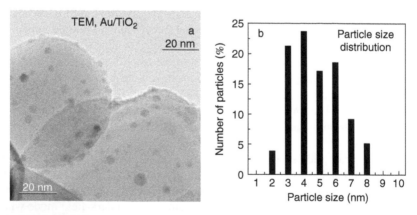

Figure 1.5   Representative TEM image (a) and particle size distribution (b) obtained for a Au/TiO_2 catalyst prepared by grafting of a $[Au_6(PPh_3)_6](BF_4)_2$ complex onto TiO_2 particles followed by appropriate reduction and oxidation treatments [42]. The gold particles exhibit approximately spherical shapes and an average particle size of 4.7 nm. The measured Au particle sizes could be well correlated with the activity of the catalyst for carbon monoxide oxidation and acetylene hydrogenation. (Reproduced with permission from Springer.)

imaging individual planes in crystalline particles, can provide even more detailed structural information on the surface of the catalysts [40,43].

    Electron microscopy does have some limitations. For example, this technique usually requires special sample preparations. Caution also needs to be exercised to minimize any electron beam-induced effects, such as changes in the specimen due to local heating, electronic excitations, or deposition of contaminants during observation [40]. In addition, SEM and TEM work best for sturdy solids, and are not well suited to detect reaction intermediates on catalyst surfaces. Finally, and importantly, statistical analysis of a large number of images is needed to get meaningful information on particle size distributions. It is best to correlate such results with information obtained by other characterization methods [38].

1.3 ADSORPTION–DESORPTION AND THERMAL TECHNIQUES

1.3.1 Surface Area and Pore Structure

Most heterogeneous catalysts, including metal oxides, supported metal catalysts, and zeolites, are porous materials with specific surface areas ranging from 1 to 1000 m^2/g [1]. These pores can display fairly complex size distributions, and can be broadly grouped into three types, namely, micropores (average pore diameter $d < 2$ nm), mesopores ($2 < d < 50$ nm), and macropores ($d > 50$ nm). The surface area, pore volume, and average pore size of such porous catalysts often play a pivotal role in determining the number of active sites available for catalysis, the diffusion rates of reactants and products in and out of these pores, and the deposition of coke and other contaminants. The most common method used to characterize the structural parameters associated with pores in solids is via the measurement of adsorption–desorption isotherms, that is, of the adsorption volume of a gas, typically nitrogen, as a function of its partial pressure [44–48].

Given the complexity of the pore structure in high-surface-area catalysts, six types of adsorption isotherms have been identified according to a classification advanced by IUPAC [45–48]. Out of these six, only four are usually found in catalysis:

- Type II, typical of macroporous solids where the prevailing adsorption processes are the formation of a monolayer at low relative pressures, followed by gradual and overlapping multilayer condensation as the pressure is increased.
- Type IV, often seen in mesoporous solids, where condensation occurs sharply at a pressure determined by Kelvin-type rules.
- Type I, characteristic of microporous solids, where pore filling takes place without capillary condensation, and is indistinguishable from the monolayer formation process.
- Type VI, corresponding to uniform ultramicroporous solids, where the pressure at which adsorption takes place depends on surface–adsorbate interactions, and shows isotherms with various steps each corresponding to adsorption on one group of energetically uniform sites.

A number of models have been developed for the analysis of the adsorption data, including the most common Langmuir [49] and BET (Brunauer, Emmet, and Teller) [50] equations, and others such as t-plot [51], H–K (Horvath–Kawazoe) [52], and BJH (Barrett, Joyner, and Halenda) [53] methods. The BET model is often the method of choice, and is usually used for the measurement of total surface areas. In contrast, t-plots and the BJH method are best employed to calculate total micropore and mesopore volume, respectively [46]. A combination of isothermal adsorption measurements can provide a fairly complete picture of the pore size distribution in solid catalysts. Many surface area analyzers and software based on this methodology are commercially available nowadays.

A recent example of the type of data that can be obtained with such instrumentation is presented in Figure 1.6 [54]. This corresponds to the nitrogen adsorption–desorption isotherm obtained for a mesoporous silica, SBA-15, used as support for many high-surface-area catalysts. The isotherm, identified as type IV according to the IUPAC classification, is typical of mesoporous materials. Three regions are clearly seen with increasing nitrogen pressure, corresponding to monolayer–multilayer adsorption, capillary condensation, and multilayer adsorption on the outer particle surfaces, respectively. A clear H1-type hysteresis loop, characterized by almost vertical and parallel

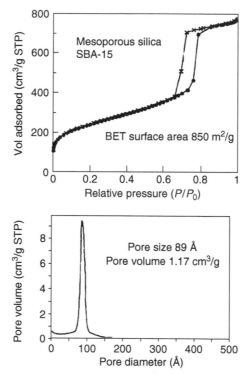

Figure 1.6   *Top*: Low-temperature nitrogen adsorption (•) and desorption (×) isotherms measured on a calcined SBA-15 mesoporous silica solid prepared using an $EO_{20}PO_{70}EO_{20}$ block copolymer [54]. *Bottom*: Pore size distribution derived from the adsorption isotherm reported at the top [54]. A high surface area (850 m^2/g), a uniform distribution of cylindrical nanopores (diameter ~90 Å), and a large pore volume (1.17 cm^3/g) were all estimated from these data. These properties make this material suitable for use as support in the preparation of high-surface-area solid catalysts. (Reproduced with permission from The American Chemical Society.)

but displaced lines in the adsorption and desorption branches, is also observed in the adsorption–desorption isotherm, indicating the presence of uniform cylindrical pore channels.

Aside from N_2 adsorption, Kr or Ar adsorption can be used at low temperatures to determine low (<1 m^2/g) surface areas [46]. Chemically sensitive probes such as H_2, O_2, or CO can also be employed to selectively measure surface areas of specific components of the catalyst (see below). Finally, mercury-based porosimeters, where the volume of the mercury incorporated into the pores is measured as a function of increasing (well above atmospheric) pressures, are sometimes used to determine the size of meso- and macropores [1]. By and large, the limitations of all of the above methods are that they only provide information on average pore volumes, and that they usually lack chemical sensitivity.

1.3.2   Temperature-Programmed Desorption and Reaction

As stated above, when probes with specific adsorption characteristics are used, additional chemical information can be extracted from adsorption–desorption experiments. Temperature-programmed desorption (TPD) in particular is often employed to obtain information about specific sites in catalysts [55,56]. The temperature at which desorption occurs indicates the strength of adsorption, whereas either the amount of gas consumed in the uptake or the amount of desorption upon heating attests to the concentration of the surface sites. The most common molecules used in TPD are NH_3 and CO_2, which probe acidic and basic sites, respectively, but experiments with pyridine, O_2, H_2, CO, H_2O, and other molecules are often performed as well [57–59]. As an example, the ammonia

TPD data in Figure 1.7 show how special treatments of a V_2O_5/TiO_2 catalyst can influence its properties in terms of the strength and distribution of acid sites. These treatments can be used to tune selectivity in partial oxidation reactions [59].

Some solid samples may decompose or react with the probe molecules at elevated temperatures, causing artifacts in the TPD profiles [58]. However, this conversion can in some instances be used to better understand the reduction, oxidation, and reactivity of the catalyst. In this mode, the technique is often called temperature-programmed reduction (TPR), temperature-programmed oxidation (TPO), or, in general, temperature-programmed surface reaction (TPSR or TPR) [55,56,60]. The principles of TPR, TPO, and TPSR are similar to those of TPD, in the sense that either the uptake of the reactants or the yields of desorption are recorded as a function of temperature. Nevertheless, there can be subtle differences in either the way the experiments are carried out or the scope of the application. TPSR in particular often requires the use of mass spectrometry or some other analytical technique to identify and monitor the various species that desorb from the surface. Figure 1.8 shows an example of such application for the case of methanol adsorbed on a MoO_3/Al_2O_3 catalyst. There, the production of water, formaldehyde, and dimethyl ether was detected above 100°C, around 250°C, and about 200°C, respectively [61]. Such information is key for the elucidation of reaction mechanisms.

These TPD techniques reflect the kinetics (not thermodynamics) of adsorption, and are quite useful for determining trends across series of catalysts, but are often not suitable for the derivation of quantitative information on surface kinetics or energetics, in particular on ill-defined real catalysts. Besides averaging the results from desorption from different sites, TPD detection is also complicated in porous catalysts by simultaneous diffusion and readsorption processes [58].

1.3.3 Thermogravimetry and Thermal Analysis

Changes in catalysts during preparation, which often involves thermal calcination, oxidation, and reduction, can also be followed by recording the associated variations in sample weight, as in normal thermogravimetry (TG) or differential thermogravimetry (DTG); or in sample temperature,

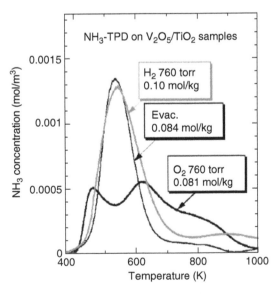

Figure 1.7 Ammonia TPD from a V_2O_5/TiO_2 catalyst after different pretreatments [59]. Two TPD peaks at 460 and 610 K are seen in the data for the oxidized sample, whereas only one is observed at 520 K for the catalyst obtained after either evacuation or reduction. This indicates that the type of treatment used during the preparation of the catalyst influences both the amount and the distribution of acidic sites on the V_2O_5/TiO_2 surface. (Reproduced with permission from Elsevier.)

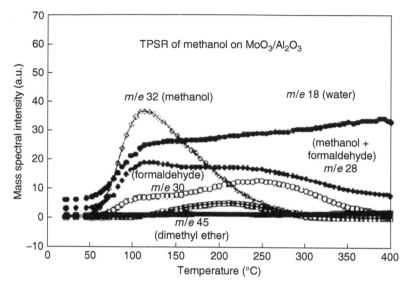

Figure 1.8 TPSR spectra obtained after saturation of a MoO_3/Al_2O_3 catalyst with methanol at room temperature [61]. Seen here are mass spectrometry traces corresponding to methanol ($m/e = 28$ and 32), formaldehyde ($m/e = 28$ and 30), water ($m/e = 18$), and dimethyl ether ($m/e = 45$). These data were used to propose a mechanism for the selective oxidation of methanol on MoO_3-based catalysts. (Reproduced with permission from Elsevier.)

as in differential thermal analysis (DTA) [62–64]. Although these thermal methods are quite traditional, they are still used often in catalysis research. In Figure 1.9, an example is provided on how TG, DTG, and DTA techniques can be used to better understand and design procedures for catalyst preparation [65]. In this case, a $MgFe_2O_4$ spinel, used for the selective oxidation of styrene, was prepared by co-precipitation from a solution containing $Fe(NO_3)_3$ and $Mg(NO_3)_2$, followed by thermal calcination. The data show that the initial amorphous precursor undergoes a number of transformations upon calcination, including the losses of adsorbed and crystal water around 110 and 220°C, respectively, its decomposition and dehydroxylation into a mixed oxide at 390°C, and the formation of the $MgFe_2O_4$ spinel at 640°C.

Besides the prediction of calcination temperatures during catalyst preparation, thermal analysis is also used to determine the composition of catalysts based on weight changes and thermal behavior during thermal decomposition and reduction, to characterize the aging and deactivation mechanisms of catalysts, and to investigate the acid–base properties of solid catalysts using probe molecules. However, these techniques lack chemical specificity, and require corroboration by other characterization methods.

1.3.4 Microcalorimetry

Another thermal analysis method available for catalyst characterization is microcalorimetry, which is based on the measurement of the heat generated or consumed when a gas adsorbs and reacts on the surface of a solid [66–68]. This information can be used, for instance, to determine the relative stability among different phases of a solid [69]. Microcalorimetry is also applicable in the measurement of the strengths and distribution of acidic or basic sites as well as for the characterization of metal-based catalysts [66–68]. For instance, Figure 1.10 presents microcalorimetry data for ammonia adsorption on H-ZSM-5 and H-mordenite zeolites [70], clearly illustrating the differences in both acid strength (indicated by the different initial adsorption heats) and total number of acidic sites (measured by the total ammonia uptake) between the two catalysts.

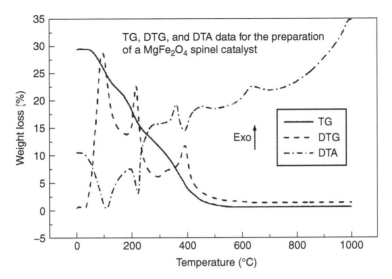

Figure 1.9 TG, DTG, and DTA profiles for an amorphous catalyst precursor obtained by coprecipitation of $Fe(NO_3)_3$ and $Mg(NO_3)_2$ in solution [65]. This precursor is heated at high temperatures to produce a $MgFe_2O_4$ spinel, used for the selective oxidation of styrene. The thermal analysis reported here points to four stages in this transformation, namely, the losses of adsorbed and crystal water at 110 and 220°C, respectively, the decomposition and dehydroxylation of the precursor into a mixed oxide at 390°C, and the formation of the $MgFe_2O_4$ spinel at 640°C. Information such as this is central in the design of preparation procedures for catalysts. (Reproduced with permission from Elsevier.)

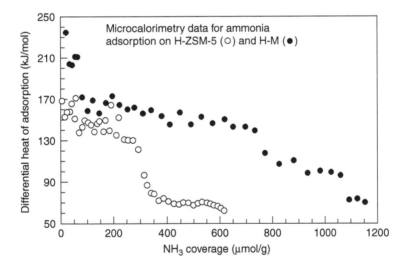

Figure 1.10 Differential heats of adsorption as a function of coverage for ammonia on H-ZSM-5 (o) and H-mordenite (•) zeolites [70]. In both cases, the heats decrease with the extent of NH_3 uptake, indicating that the strengths of the individual acidic sites on each catalyst are not uniform. On the other hand, the H-ZSM-5 sample has a smaller total number of acidic sites. Also, the H-mordenite sample has a few very strong sites, as manifested by the high initial adsorption heat at low ammonia coverage. These data point to a significant difference in acidity between the two zeolites. That may account for their different catalytic performance. (Reproduced with permission from Elsevier.)

Recent advances have led to the development of microcalorimeters sensitive enough for low-surface-area ($\sim$1 cm^2) solids [71]. This instrumentation has already been used in model systems to determine the energetics of bonding of catalytic particles to the support, and also in adsorption and reaction processes [72,73].

1.4   OPTICAL SPECTROSCOPIES

1.4.1   Infrared Spectroscopy

In catalysis, infrared (IR) spectroscopy is commonly used to characterize specific adsorbates. Because of the localized nature and particular chemical specificity of molecular vibrations, IR spectra are quite rich in information, and can be used to extract or infer both structural and compositional information on the adsorbate itself as well as on its coordination on the surface of the catalyst. In some instances, IR spectroscopy is also suitable for the direct characterization of solids, especially if they can be probed in the far-IR region (10–200 cm^{-1}) [74–76].

Several working modes are available for IR spectroscopy studies [74–76]. The most common arrangement is transmission, where a thin solid sample is placed between the IR beam and the detector; this mode works best with weakly absorbing samples. Diffuse reflectance IR (DRIFTS) offers an alternative for the study of loose powders, strong scatters, or absorbing particles. Attenuated total reflection (ATR) IR is based on the use of the evanescent wave from the surface of an optical element with trapezoidal or semispherical shape, and works best with samples in thin films. Reflection–absorption IR spectroscopy (RAIRS) is employed to probe adsorbed species on flat reflecting surfaces, typically metals. In the emission mode, the IR signal emanating from the heated sample is recorded. Finally, both photoacoustic (PAS) and photothermal IR spectroscopies rely on temperature fluctuations caused by radiation of the sample with a modulated monochromatic beam. The availability of all these arrangements makes IR spectroscopy quite versatile for the characterization of catalytic systems.

The applications of IR spectroscopy in catalysis are many. For example, IR can be used to directly characterize the catalysts themselves. This is often done in the study of zeolites, metal oxides, and heteropolyacids, among other catalysts [77,78]. To exemplify this type of application, Figure 1.11 displays transmission IR spectra for a number of $Co_xMo_{1-x}O_y$ ($0 \leq x \leq 1$) mixed metal oxides with various compositions [79]. In this study, a clear distinction could be made between pure MoO_3, with its characteristic IR peaks at 993, 863, 820, and 563 cm^{-1}, and the MoO_4 tetrahedral units in the $CoMoO_4$ solid solutions formed upon Co_3O_4 incorporation, with its new bands at 946 and 662 cm^{-1}. These properties could be correlated with the activity of the catalysts toward carburization and hydrodenitrogenation reactions.

Further catalyst characterization can be carried out by appropriate use of selected adsorbing probes [80–83]. For instance, the acid–base properties of specific surface sites can be tested by recording the ensuing vibrational perturbations and molecular symmetry lowering of either acidic (CO and CO_2) or basic (pyridine and ammonia) adsorbates. Oxidation states can also be probed by using carbon monoxide [84,85]. For instance, our recent study of Pd/Al_2O_3 and Pd/Al_2O_3–25% ZrO_2 catalysts used for nitrogen oxide reduction indicated that the Pd component can be extensively reduced in both samples, with and without the ZrO_2 additive, but oxidized fully to PdO only in the presence of the zirconia [86,87].

Another common application of IR is to characterize reaction intermediates on the catalytic surfaces, often *in situ* during the course of the reaction [76,78,88,89]. Figure 1.12 provides an example in the form of a set of transmission IR spectra obtained as a function of temperature during the oxidation of 2-propanol on Ni/Al_2O_3 [90]. A clear dehydrogenation reaction is identified in these data above 440 K by the appearance of new acetone absorption bands around 1378, 1472, and 1590 cm^{-1}.

New directions have been recently advanced in the use of IR spectroscopy for the characterization of adsorbates, including the investigation of liquid–solid interfaces *in situ* during catalysis. Both ATR [91,92] and RAIRS [86,93] have been recently implemented for that purpose. RAIRS has also been used for the detection of intermediates on model surfaces *in situ* during catalytic reactions [94–96]. The ability to detect monolayers *in situ* under catalytic environments on small-area samples promises to advance the fundamental understanding of surface catalytic reactions.

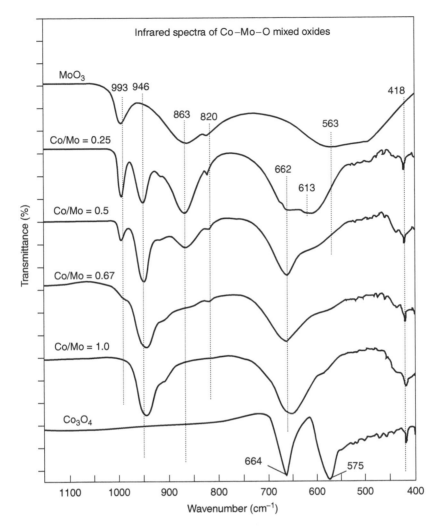

Figure 1.11 Transmission IR spectra from $Co_xMo_{1-x}O_y$ ($0 \leq x \leq 1$) samples obtained by addition of different amount of Co_3O_4 to pure MoO_3 [79]. As the Co/Mo ratio is increased from 0.25 to 1, the IR peaks due to tetrahedral MoO_4 units (at 662 and 946 cm^{-1}) grow at the expense of those associated with the MoO_3 phase (at 563, 820, 863, and 993 cm^{-1}), a trend that indicates the formation of $CoMoO_4$. This example shows how IR can be used to directly characterize solid catalyst samples. (Reproduced with permission from Elsevier.)

Owing to its great molecular specificity, good sensitivity, and high versatility, IR spectroscopy is one of the most widely used techniques for catalyst characterization. Nevertheless, IR catalytic studies do suffer from a few limitations. In particular, strong absorption of radiation by the solid often limits the vibrational energy window available for analysis. For instance, spectra of catalysts dispersed on silica or alumina supports display sharp cutoffs below 1300 and 1050 cm^{-1}, respectively [75]. Also, the intensities of IR absorption bands are difficult to use for quantitative analysis. Finally, it is not always straightforward to interpret IR spectra, especially in cases involving complex molecules with a large number of vibrational modes.

1.4.2 Raman Spectroscopy

Raman spectroscopy offers an alternative for the vibrational characterization of catalysts, and has been used for the study of the structure of many solids, in particular of oxides such as MoO_3, V_2O_5,

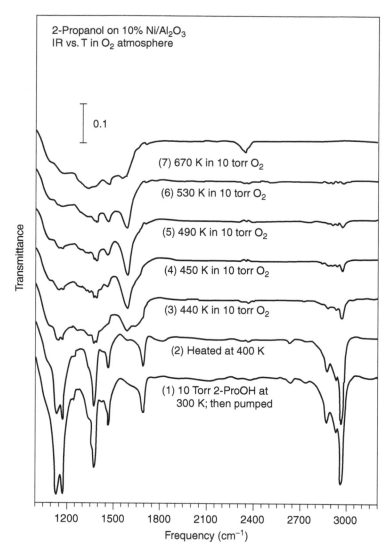

Figure 1.12 Transmission IR spectra obtained during the oxidation of 2-propanol on a Ni/Al$_2$O$_3$ catalyst as a function of reaction temperature [90]. A change in the nature of the adsorbed species from molecular 2-propanol to acetone is seen above 440 K. Experiments such as these allow for the identification of potential reaction intermediates during catalysis. (Reproduced with permission from Elsevier.)

and WO$_3$ [97–99], as well as for the investigation of a number of adsorbates [100,101]. Whereas oxides such as SiO$_2$, Al$_2$O$_3$, and zeolites give low Raman signals, this technique is ideal for the identification of oxygen species in covalent metal oxides. As an example, Figure 1.13 shows the Raman spectra of a series of transition metal oxides dispersed on high-surface-area alumina supports [75,102]. A clear distinction can be made with the help of these data between terminal and bridging oxygen atoms, and with that a correlation can be drawn between the coordination and bond type of these oxygen sites and their catalytic activity. Data such as these can also be used to determine the nature and geometry of supported oxides as a function of loading and subsequent treatment.

Surface-enhanced Raman spectroscopy (SERS) has also been employed to characterize metal catalyst surfaces [103]. The low sensitivity and severe conditions required for the signal enhancement have limited the use of this technique [104], but some interesting work has been published over the years in this area, including studies on model liquid–solid interfaces [105].

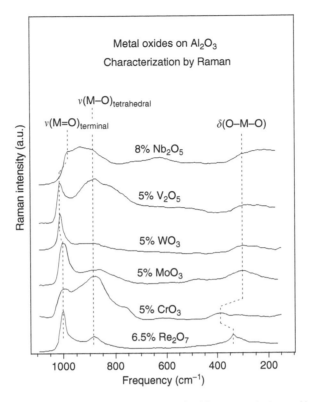

Figure 1.13　Raman spectra for a number of transition metal oxides supported on γ-Al$_2$O$_3$ [75,102]. Three distinct regions can be differentiated in these spectra, namely, the peaks around 1000 cm^{-1} assigned to the stretching frequency of terminal metal–oxygen double bonds, the features about 900 cm^{-1} corresponding to metal–oxygen stretches in tetrahedral coordination sites, and the low-frequency (<400 cm^{-1}) range associated with oxygen–metal–oxygen deformation modes. Raman spectroscopy can clearly complement IR data for the characterization of solid catalysts. (Reproduced with permission from The American Chemical Society.)

Raman spectroscopy does suffer from some severe limitations. For example, Raman intensities of surface species are often quite low. Also, the high laser powers needed for Raman characterization tend to heat the sample, and often cause changes in the physical properties of the solid. Finally, strong sample fluorescence typically masks the weaker Raman signals [8]. Fortunately, some of these difficulties have been recently minimized via the implementation of Fourier transform [106,107] and UV [108,109] Raman spectroscopy arrangements. Figure 1.14 demonstrates the advantages of UV–Raman spectroscopy for catalyst characterization [108]. In this example involving a MoO$_3$/Al$_2$O$_3$ catalyst, no signal other than a sloping background due to fluorescence is seen when using 488 nm radiation, but clear peaks assignable to molybdenum oxide are seen with the 244 nm laser excitation in spite of the low (0.1 wt%) metal oxide loading. There are also new efforts made on the use of Raman spectroscopy *in situ* and under *operando* (in conjunction with activity measurement) conditions [109,110].

1.4.3　Ultraviolet–Visible Spectroscopy

Compared with IR and Raman spectroscopies, ultraviolet–visible (UV–Vis) spectroscopy has had only limited use in heterogeneous catalysis. Nevertheless, this spectroscopy can provide information on concentration changes of organic compounds dissolved in a liquid phase in contact with a solid catalyst, be used to characterize adsorbates on catalytic surfaces, provide information on the

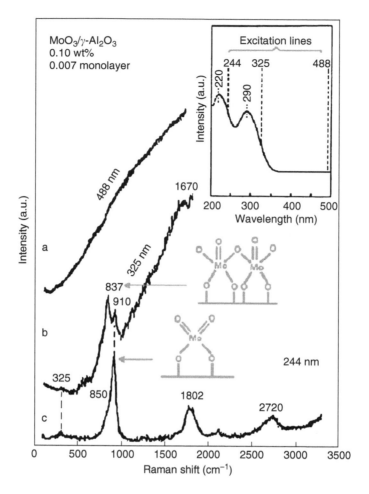

Figure 1.14 Raman spectra from a 0.1 wt% MoO₃/γ-Al₂O₃ catalyst obtained by using different (488, 325, and 244 nm) laser excitation energies [108]. The UV–Vis absorbance spectrum is reported in the inset to indicate that while the catalyst does not absorb light in the visible region, it does show two UV absorption peaks at 290 and 220 nm. The data clearly illustrate the advantage of using ultraviolet (244 nm) light for Raman excitation, since the spectrum obtained with visible (488 nm) radiation is dominated by the fluorescence of the solid. (Reproduced with permission from Elsevier.)

absorption spectra and band gap of photocatalysts, or map the electronic structure of transition metal cations in inorganic materials [111–114]. Figure 1.15 displays an example where Cr³⁺ and Cr⁶⁺ species in calcined, hydrated, and reduced chromia/alumina catalysts are differentiated by UV–Vis [115]. This information was used to optimize the preparation method for Cr⁶⁺-based catalysts for alkane dehydrogenation.

The main drawback of the use of UV–Vis spectroscopy for catalyst characterization is that the data commonly show broad and overlapping absorption bands with little chemical specificity. Also, it is often quite difficult to properly interpret the resulting spectra. Lastly, quantitative analysis is only possible at low metal oxide loadings [114].

1.4.4 Nuclear Magnetic Resonance

Nuclear magnetic resonance (NMR) spectroscopy is most frequently used to analyze liquid samples, but in the magic angle spinning (MAS) mode, this spectroscopy can also be employed to characterize solid catalysts, zeolites in particular [116–120]. For example, the ²⁹Si NMR signal can be used to determine the coordination environment of Si in the framework of the zeolite, taking advantage of the fact that the coordination of each additional Al atom to a given Si center results in a shift of about 5 to 6 ppm from the original peak position in Si(OSi)$_4$ at -102 to -110 ppm. This is illustrated in Figure 1.16 for the case of ruthenium supported on NaY zeolites [121,122]. In addition, the relative population of the Si(xAl) NMR peaks can be used to determine Si/Al ratios in a

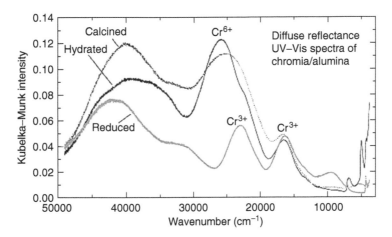

Figure 1.15 Diffuse reflectance UV–Vis spectra from a series of chromia/alumina catalysts after various treatments [115]. All these spectra display a shoulder at about 16,700 cm^{-1} corresponding to the first d–d transition of Cr^{3+}, but the main feature seen in the hydrated and calcined samples at about 26,000 cm^{-1} due to a Cr^{6+} charge transition is absent in the data for the reduced sample. This points to a loss of the catalytically active Cr^{6+} phase upon reduction. (Reproduced with permission from Elsevier.)

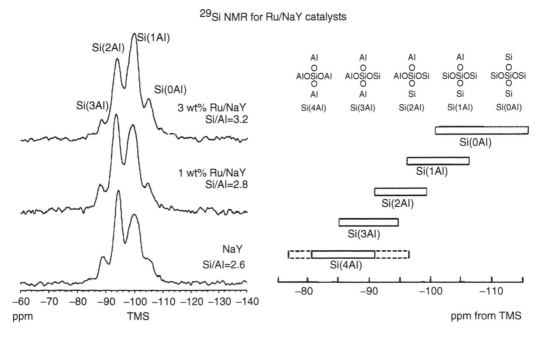

Figure 1.16 ^{29}Si MAS NMR spectra for NaY zeolites with three different (0, 1, and 3 wt%) Ru loading [121]. The slight changes in relative intensities among the different peaks seen in these data are interpreted in terms of changes in Al coordination around the individual silicon atoms, as indicated by the diagram on the right [122]. (Reproduced with permission from Elsevier and The American Chemical Society.)

more reliable fashion than by using other analytical methods, in particular because the NMR data provide information about the framework atoms rather than about the bulk phase of the catalyst, which also contains extra-framework Al species. Caution should be exercised when dealing with dealuminated zeolites because ^{29}Si NMR signals with local $Si(OSi)_{4-x}(OAl)_x$ and $Si(OSi)_{4-x}(OAl)_{x-1}(OH)$ environments often overlap [120], but, fortunately, special $^1H/^{29}Si$ cross-polarization double-resonance experiments can help make this distinction. ^{27}Al MAS NMR can also be used to obtain a picture of the coordination environment around the Al atoms in the solid catalyst by taking advantage of the distinct chemical shifts observed for tetrahedral (60 to 50 ppm), pentacoordinated (about 25 ppm), and octahedral (13 to -17 ppm) environments.

Besides the ^{29}Si and ^{27}Al NMR studies of zeolites mentioned above, other nuclei such as ^1H, ^{13}C, ^{17}O, ^{23}Na, ^{31}P, and ^{51}V have been used to study physical chemistry properties such as solid acidity and defect sites in specific catalysts [123,124]. ^{129}Xe NMR has also been applied for the characterization of pore sizes, pore shapes, and cation distributions in zeolites [125,126]. Finally, less common but also possible is the study of adsorbates with NMR. For instance, the interactions between solid acid surfaces and probe molecules such as pyridine, ammonia, and $P(CH_3)_3$ have been investigated by ^{13}C, ^{15}N, and ^{31}P NMR [124]. *In situ* ^{13}C MAS NMR has also been adopted to follow the chemistry of reactants, intermediates, and products on solid catalysts [127,128].

Nuclear magnetic resonance is certainly a versatile analytical tool with wide applicability to catalysis. Nevertheless, it does have some notable shortcomings. For example, NMR is not a very sensitive spectroscopic technique, and requires catalytic samples with high surface areas. This is often not a big problem, given that most catalytic phases are highly dispersed, but these too have a large number of types of sites, which get averaged in the NMR spectra. In addition, different NMR peaks may overlap in complex mixtures of reactants, intermediates, and products, making the analysis of catalytic systems difficult [10].

1.4.5 Electron Spin Resonance

Electron spin resonance (ESR), also called electron paramagnetic resonance (EPR), is used in heterogeneous catalysis to study paramagnetic species containing one or more unpaired electrons, either catalytic active sites or reaction intermediates [113,129,130]. For instance, a number of ESR studies have been dedicated to the detection and characterization of oxygen ionic surface species such as O_2^-, O^-, O_2^{2-}, and O^{2-}, key intermediates in catalytic oxidation processes [131–135]. Another important use of ESR in catalysis is for the study of the coordination chemistry of transition metal cations incorporated into zeolites or metal oxides [136,137]. As an illustration of this latter application, Figure 1.17 shows the results from ESR studies on the incorporation of vanadium into a silicate-based zeolite for use in selective oxidation catalysis [138]. The calcined catalyst exhibits no ESR signal because of the exclusive presence of the ESR-silent V^{5+} species. However, a strong and complex ESR spectrum develops after photoreduction of the catalyst, indicative of the existence of V^{4+} in tetrahedral coordination; further addition of a small amount of water leads to yet another ESR trace assignable to distorted octahedral VO^{2+} ions. This information could be correlated with both the accessibility and photocatalytic activity of the vanadium centers after different catalyst pretreatments.

Special spin-trapping techniques are also available for the detection of short-lived radicals in both homogeneous and heterogeneous systems. For instance, α-phenyl *N-tert*-butyl nitrone (PBN), *tert*-nitrosobutane (*t*-NB), α-(4-pyridyl *N*-oxide) *N-tert*-butyl nitrone (4-POBN), or 5,5-dimethyl-1-pyrroline *N*-oxide (DMPO) can be made to react with catalytic intermediates to form stable paramagnetic adducts detectable by ESR [135]. Radicals evolving into the gas phase can also be trapped directly by condensation or by using matrix isolation techniques [139].

Although ESR has the advantage over NMR of its high sensitivity toward low concentrations of active sites and intermediates, this method is only applicable to the characterization of paramagnetic substances. In addition, the widths of the ESR signals increase dramatically with increasing

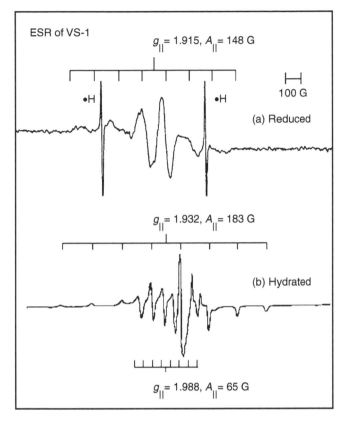

Figure 1.17 (a) ESR spectrum from a vanadium silicate catalyst after photoreduction with H_2 at 77 K [138]. The ESR data obtained indicate the existence of V^{4+} ions in tetrahedral coordination. (b) Addition of a small amount of water leads to a new ESR trace identified with distorted octahedral VO^{2+} ions, indicating the easy accessibility of the vanadium surface species. (Reproduced with permission from Elsevier.)

temperature, making the *in situ* characterization of catalytic systems at reaction temperatures difficult. Finally, ESR methods cannot distinguish surface and bulk species [135].

1.5 SURFACE-SENSITIVE SPECTROSCOPIES

1.5.1 X-Ray and Ultraviolet Photoelectron Spectroscopies

X-ray photoelectron spectroscopy (XPS) is a useful technique to probe both the elemental composition of the surface of catalysts and the oxidation state and electronic environment of each component [140–144]. Qualitative information is derived from the chemical shifts of the binding energies of given photoelectrons originating from a specific element on the surface: in general, binding energies increase with increasing oxidation state, and to a lesser extent with increasing electronegativity of the neighboring atoms. Quantitative information on elemental composition is obtained from the signal intensities. The principles of ultraviolet photoelectron spectroscopy (UPS) are similar to those of XPS, except that ultraviolet radiation (10 to 45 eV) is used instead of soft x-rays (200 to 2000 eV), and what is examined is valence rather than core electronic levels [140].

As an example of the use of XPS for catalyst characterization, Figure 1.18 presents data obtained for a Mo–V–Sb–Nb mixed oxide catalyst after calcination under different conditions (in air vs. nitrogen) [145]. In spite of the fact that each catalyst displays different activity and selectivity

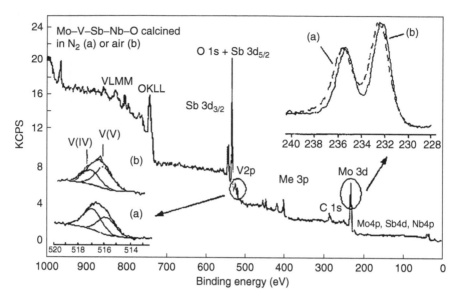

Figure 1.18 Survey and expanded V 2p and Mo 3d XPS spectra form a Mo–V–Sb–Nb mixed oxide catalyst after calcination in nitrogen (a) and air (b) atmospheres [145]. The data indicate a lesser degree of oxidation in nitrogen, a result that was correlated with the promotion of reactions leading to the production of propene and acrylic acid rather than acetic acid, the main product obtained with the fully oxidized sample. (Reproduced with permission from Elsevier.)

for the selective oxidation of propane to acrylic acid, the survey spectra of the two catalysts look quite similar, both showing peaks for Mo, V, Sb, Nb, O, and C on their surfaces. However, a closer inspection of the data indicates that the metal ions, the Sb and V ions in particular, are oxidized to a lesser extent in N_2. Further quantitative analysis shows that there is more Sb but less Nb on the surface of the catalyst calcined in air. A good correlation could be derived between the physical properties determined by XPS and the catalytic behavior of these samples.

X-ray photoelectron spectroscopy is indeed quite informative, but requires the use of expensive instrumentation. Also, the detection of photoelectrons requires the use of ultrahigh vacuum, and therefore can mostly be used for *ex situ* characterization of catalytic samples (although new designs are now available for *in situ* studies [146,147]). Finally, XPS probes the upper 10 to 100 Å of the solid sample, and is only sensitive to the outer surfaces of the catalysts. This may yield misleading results when analyzing porous materials.

1.5.2 Auger Electron Spectroscopy

Auger electron spectroscopy (AES) is based on the ejection of the so-called Auger electrons after relaxation of photoionized atoms. This technique is quite complementary to XPS, and also provides surface-sensitive information on surface compositions and chemical bonding [143]. Figure 1.19 shows AES data obtained during the characterization of a Ru/Al_2O_3 catalyst used for CO hydrogenation [148]. These data were recorded after poisoning with H_2S, and show that the sulfur detected in this catalyst sample is present only on the surface and not on the subsurface; mild sputtering leads to the easy removal of all the sulfur signal. Moreover, the lack of any carbon either before or after sputtering indicates the absence of carbon in the used catalyst.

As opposed to XPS, AES signals typically exhibit complex structure, and sometimes require elaborate data treatment. Also, AES does not easily provide information on oxidation states, as XPS does. On the other hand, AES is often acquired by using easy-to-focus electron beams as the excitation source, and can therefore be used in a rastering mode for the microanalysis of nanosized spots

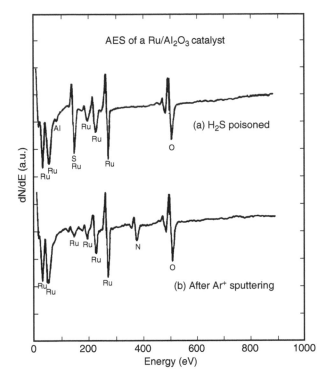

Figure 1.19 AES data from a Ru/Al$_2$O$_3$ catalyst aged in a reaction (CO+H$_2$) mixture containing trace amounts of H$_2$S [148]. Spectra are shown for the sample before (a) and after (b) sputtering with an Ar$^+$ beam for 2 min. The difference between the two spectra indicates the presence of S on the surface but not the subsurface of the poisoned catalyst. (Reproduced with permission from Elsevier.)

within the surface of the catalyst. Given their different sampling depths, XPS and AES can also be combined to obtain a better picture of the profile of the different elements in the solid as a function of distance from the surface. The latter task can be aided by adding sputtering capabilities to the experimental setup, as illustrated in the example in Figure 1.19 [148].

1.5.3 Low-Energy Ion Scattering

Low-energy ion scattering (LEIS), also called ion scattering spectroscopy (ISS), is based on the determination of the energy losses associated with the elastic scattering of monochromatic ions impinging on the surface [149,150]. Like AES and XPS, it is used to determine the atomic composition of surfaces, though, unlike them, LEIS is sensitive only to the outermost atomic layer of the solid. The power of this unique surface sensitivity is illustrated by the example in Figure 1.20, which shows LEIS data for a series of WO$_3$/Al$_2$O$_3$ catalysts with different WO$_3$ loading [151]. The three peaks at $E/E_0 = 0.41$, 0.59, and 0.93 are easily assigned to O, Al, and W, respectively. The almost linear decrease in the Al/O peak intensity ratio and the concomitant increase in the W/O ratio seen with WO$_3$ loading indicate the blocking of the Al sites by the tungsten species, which appear to deposit in two-dimensional monolayers. The surface coverage of WO$_3$ could be determined quantitatively in each case using these data.

1.5.4 Secondary-Ion Mass Spectroscopy

Secondary-ion mass spectroscopy (SIMS) is based on the mass spectrometric detection of the secondary ions emitted upon bombardment of the sample with a primary ion beam. The composition

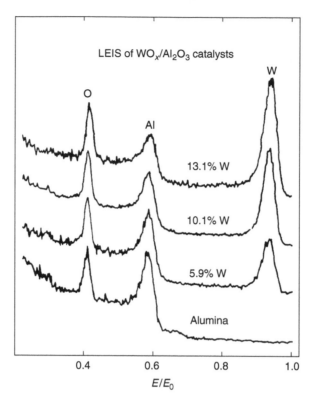

Figure 1.20 LEIS data for an Al_2O_3 support covered with different amounts of WO_3 [151]. It is seen here that as the tungsten loading is increased, the O LEIS signal remains unchanged, whereas the W peak increases at the expense of the Al signal, indicating the direct growth of two-dimensional WO_3 islands on top of the aluminum sites. The Al/O intensity ratios in these data were also used to calculate the surface coverage of WO_3. This technique has proven successful for the study of surface coverages in supported catalysts. (Reproduced with permission from Springer.)

of the ion clusters detected provides an indication of the molecular arrangement of the atoms on the surface [152]. SIMS experiments may be performed in one of two modes— static, where a low sputtering rate is used in order to analyze the topmost surface, and dynamic, in which case the primary ion current density is sufficient to erode the surface for depth profile analysis.

Figure 1.21 shows time-of-flight negative-ion SIMS data from a 0.6% Pt/Al_2O_3 catalyst before and after the reforming of an *n*-heptane reaction mixture [153]. These spectra highlight the high sensitivity of SIMS, in particular given the low metal loading used in the catalyst. Pt^-, PtO^-, $PtCl^-$, $PtClO^-$, and $PtCl_2^-$ clusters are clearly identified in these spectra, proving the pivotal role of residual chlorine in the active catalyst. Also, a substantial decrease in the intensity of most of the Pt-containing clusters after reaction is indicative of the build-up of significant amounts of carbonaceous deposits on the surface.

Although SIMS can provide quite valuable information on the molecular (rather than atomic) composition of the surface, this is a difficult technique to use. Moreover, the resulting spectra are complex, and quantification of the data is almost impossible. To date, SIMS remains a special and seldom-used technique for catalyst characterization.

1.6 MODEL CATALYSTS

The fields of surface science and catalysis have benefited greatly from advances in ultrahigh-vacuum technology during the space race. As a consequence, a large number of surface-sensitive

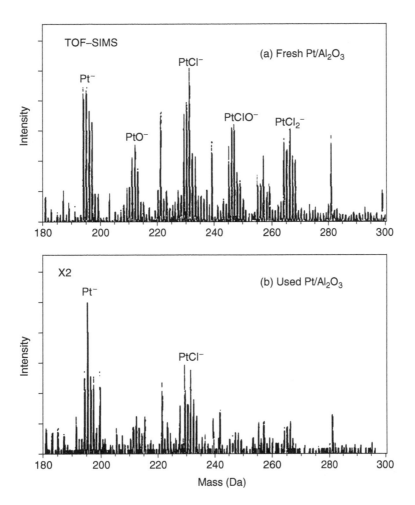

Figure 1.21 (a) Time-of-flight (TOF) negative-ion SIMS data from a fresh 0.6% Pt/Al$_2$O$_3$ catalyst prepared by using a H$_2$PtCl$_6$ solution [153]. Clusters in the 180 to 300 amu mass range arise from Pt$^-$, PtO$^-$, PtCl$^-$, PtClO$^-$, and PtCl$_2^-$ ions. (b) TOF–SIMS data for the same catalyst after having been used for heptane reforming. The total intensity of the Pt$^-$ signal has been attenuated by 70%, but PtCl$^-$ clusters are still observable in the spectra. These data provide direct evidence for the role of residual Cl atoms in the performance of the Pt catalyst. (Reproduced with permission from Elsevier.)

spectroscopies were developed, with capabilities for probing structural, electronic, and chemical properties of both the substrate itself and the molecular adsorbates. A detailed description of these techniques is beyond the scope of this chapter, but can be found in a number of excellent reviews and books [13,154,155].

As mentioned above, most modern surface-sensitive techniques operate under vacuum, and are often used for studies in model systems. Nevertheless, there have been recent attempts to extend that work to more relevant catalytic problems. Great advances have already been made to bridge the so-called pressure and materials gaps, that is, to address the issues related to the differences in catalytic behavior between small simple samples (often single crystals) in vacuum, and supported catalysts under higher (atmospheric) pressures [155–157]. Nevertheless, more work is still needed.

High-pressure surface science experiments with model samples have proven quite useful in advancing the molecular-level understanding of catalytic processes. The studies on ethylene hydrogenation over Pt(111) model catalysts summarized by the data in Figure 1.22 are used here to illustrate this point [158–162]. A number of spectroscopies, including TPD, low-energy electron diffraction (LEED), and high-resolution energy loss spectroscopy (HREELS), were initially used to characterize the surface of the catalyst after ethylene hydrogenation at atmospheric pressures [158,163]. The results from that work led to the inference that ethylidyne species form and remain on the Pt(111) surface during reaction, a conclusion later confirmed by *in situ* IR and sum-frequency generation (SFG) spectroscopies (Figure 1.22) [162,164,165]. Isotope labeling [166] and other experiments [167] have since been used to determine that this ethylidyne layer acts as a spectator, passivating in part the high activity of the metal and helping store hydrogen on the surface, and that a weakly π-bonded species is the active species during catalysis. These observations have many implications for catalysis, since the deposition of carbonaceous deposits is fairly common in hydrocarbon conversion processes [168,169].

The characterization of model-supported catalysts provides another venue for the molecular-level study of catalysis. In particular, metal particles deposited on oxide supports can be emulated by the sequential physical deposition of thin oxide films and metal particles on well-defined refractive substrates [170–172]. Figure 1.23 summarizes some results from an investigation using this approach, in this case for a model Au/TiO_2 catalyst [173]. Gold clusters of 1 to 6 nm diameter were deposited on TiO_2 single-crystal surfaces in a controlled fashion, and the samples characterized under ultrahigh vacuum in order to correlate physical properties with activity. STM and reaction

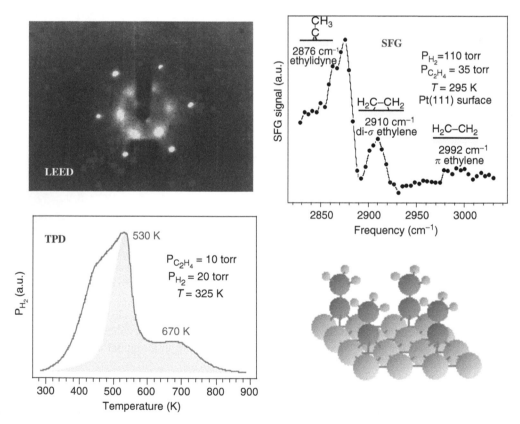

Figure 1.22 *Left*: Low-energy electron diffraction (LEED; *top*) and hydrogen temperature programmed desorption (TPD; *bottom*) data obtained after the catalytic hydrogenation of ethylene on a Pt(111) single-crystal surface [158,159]. The order of the overlayer formed on the surface, as indicated by the (2×2) diffraction pattern in LEED, together with the main H_2 desorption features seen at 530 and 670 K in the TPD data, suggests the formation of an ethylidyne overlayer on the surface during reaction, as shown schematically in the lower right corner. *Right, top*: An SFG spectrum taken *in situ* during ethylene hydrogenation, corroborating the presence of the ethylidyne layer as well as di-σ and π bonding forms of ethylene on the platinum surface [162]. Additional experiments have shown that the π-bonded species is the direct intermediate in the catalytic hydrogenation process [162]. (Reproduced with permission from The American Chemical Society.)

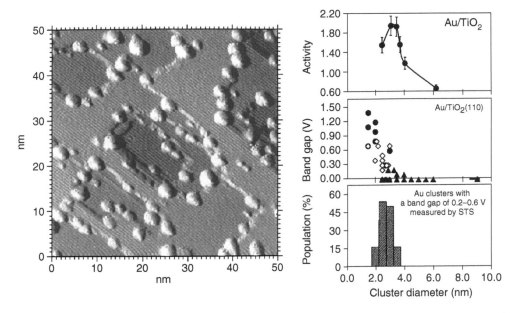

Figure 1.23   *Left*: STM image of a model Au/TiO$_2$ sample used to emulate carbon monoxide oxidation catalysts [173]. This sample was prepared by physical evaporation of gold atoms on a TiO$_2$ (110)-(1×2) single-crystal surface under ultrahigh vacuum, and corresponds to a metal coverage of approximately a quarter of a monolayer. *Right, top*: CO oxidation activity, in turnover frequency, on Au/TiO$_2$ model systems such as that imaged on the left as a function of Au cluster size. A 1:5 CO/O$_2$ mixture was converted at 350 K and a total pressure of 40 torr. *Right, middle*: Cluster band gap, measured by scanning tunneling spectroscopy (STS), again as a function of Au cluster size. *Right, bottom*: Size distribution of two-atom-thick Au clusters with a band gap of 0.2 to 0.6 V. A combination of surface characterization and catalytic measurements, as illustrated in this figure, can be used to establish structure–reactivity correlations and to understand the physical properties responsible for changes in the behavior of catalysts with changing particle size. In this example, the activity of supported gold particle is ascribed to the semiconductor properties of the small (2 to 4 nm) particles. (Reproduced with permission from The American Association for the Advancement of Science.)

kinetics measurements showed that the structure sensitivity of the carbon monoxide oxidation reaction over gold catalysts is related to a quantum size effect, with two-layer thick gold islands being the most active for oxidation reactions. Studies where this methodology is applied to more demanding reactions promise to provide a great insight into the chemistry involved.

The strength of the surface science approach is that it can address the molecular details of catalytic issues by pooling information from a battery of specific analytical spectroscopies and techniques [174]. As more complex model systems are developed, the wealth of characterization techniques available in vacuum environments can be used to better understand catalysis.

1.7 CONCLUDING REMARKS

In this chapter, we have briefly introduced a selection of techniques used to characterize heterogeneous catalysts. Only the most common and useful techniques have been reviewed since a comprehensive list of all the characterization methods available would be never ending. Other spectroscopies, including field emission microscopy (FEM) [175], field ion microscopy (FIM) [176], scanning tunneling and atomic force microscopies (STM and AFM, respectively) [177,178], photoemission electron microscopy (PEEM) [179,180], electron tomography [181], ellipsometry [182], luminescence spectroscopy [183], SFG [184,185], and Mössbauer spectroscopy [186,187], among others, have also been used for the characterization of specific heterogeneous catalysts and model systems.

Chemical probes such as titrations using Hammett indicators [188,189] and test reactions [190] have been often employed as well. Given that each method has its own strengths and limitations, a rational combination of specific techniques is often the best approach to the study of a given catalytic system.

Most of the techniques discussed above are typically used *ex situ* for catalyst characterization before and after reaction. This is normally the easiest way to carry out the experiments, and is often sufficient to acquire the required information. However, it is known that the reaction environment plays an important role in determining the structure and properties of working catalysts. Consequently, it is desirable to also try to perform catalytic studies under realistic conditions, either *in situ* [113,114,157, 191–193] or in the so-called *operando* mode, with simultaneous kinetics measurements [194–196]. In addition, advances in high-throughput (also known as combinatorial) catalysis call for the fast and simultaneous analysis of a large number of catalytic samples [197,198]. This represents a new direction for further research.

The examples introduced above refer to the characterization of the most common types of catalysts, usually supported metals or single, mixed, or supported metal oxides. Many other materials such as alloys [199,200], carbides [201–203], nitrides [204,205], and sulfides [206] are also frequently used in catalysis. Moreover, although modern surface science studies with model catalysts were only mentioned briefly toward the end of the review, this in no way suggests that these are of less significance. In fact, as the ultimate goal of catalyst characterization is to understand catalytic processes at a molecular level, surface studies on well-defined model catalysts is poised to be central in the future of the field [155,174]. The reader is referred to the Chapter 10 in this book for more details on this topic.

REFERENCES

1. J.M. Thomas, W.J. Thomas, *Principles and Practice of Heterogeneous Catalysis*, VCH, Weinheim, 1997.
2. R.A. van Santen, P.W.N.M. van Leeuwen, J.A. Moulijn, B.A. Averill (Eds.), *Catalysis: An Integrated Approach*, (2nd ed.), Elsevier, Amsterdam, 1999.
3. R.B. Anderson, P.T. Dawson (Eds.), *Experimental Methods in Catalytic Research*, Vols. II & III, Academic Press, New York, 1976.
4. W.N. Delgass, G.L. Haller, R. Kellerman, J.H. Lunsford, *Spectroscopy in Heterogeneous Catalysis*, Academic Press, New York, 1979.
5. J.M. Thomas, R.M. Lambert (Eds.), *Characterization of Catalysts*, Wiley, Chichester, 1980.
6. F. Delannay (Ed.), *Characterization of Heterogeneous Catalysts*, Marcel Dekker, New York, 1984.
7. J.L.G. Fierro (Ed.), *Spectroscopic Characterization of Heterogeneous Catalysts*, Elsevier, Amsterdam, 1990.
8. B. Imelik, J.C. Vedrine (Eds.), *Catalyst Characterization: Physical Techniques for Solid Materials*, Plenum Press, New York, 1994.
9. J.W. Niemantsverdriet, *Spectroscopy in Catalysis: An Introduction*, (2nd ed.), Wiley-VCH, Weinheim, 2000.
10. J.F. Haw (Ed.), *In-Situ Spectroscopy in Heterogeneous Catalysis*, Wiley-VCH, Weinheim, 2002.
11. B.M. Weckhuysen (Ed.), *In-Situ Spectroscopy of Catalysts*, American Scientific Publishers, Stevenson Ranch, 2004.
12. J. M. Walls (Ed.), *Methods of Surface Analysis: Techniques and Applications*, Cambridge University Press, Cambridge, 1992.
13. D.P. Woodruff, T.A. Delchar, *Modern Techniques of Surface Science*, (2nd ed.), Cambridge University Press, Cambridge, 1994.
14. M.H. Jellinek, I. Fankuchen, *Adv. Catal.* 1 (1948) 257.
15. H.P. Klug, L.E. Alexander, *X-Ray Diffraction Procedures for Polycrystalline and Amorphous Materials*, Wiley, New York, 1954.
16. E.F. Paulus, A. Gieren, in *Handbook of Analytical Techniques*, Vol. 1, H. Gunzler, A. Williams (Eds.), Wiley-VCH, Weinheim, 2001, p. 373.

17. A.M. Chen, H.L. Xu, Y.H. Yue, W.M. Hua, W. Shen, Z. Gao, *J. Mol. Catal. A* 203 (2003) 299.
18. G. Perego, *Catal. Today* 41 (1998) 251.
19. B.S. Clausen, H. TopsØe, R. Frahm, *Adv. Catal.* 42 (1998) 315.
20. R.J. Matyi, L.H. Schwartz, J.B. Butt, *Catal. Rev.-Sci. Eng.* 29 (1986) 41.
21. Z.B. Zhang, C.-C. Wang, R. Zakaria, J.Y. Ying, *J. Phys. Chem. B* 102 (1998) 10871.
22. Y.-C. Xie, Y.-Q. Tang, *Adv. Catal.* 37 (1990) 1.
23. R.A. van Nordstrand, *Adv. Catal.* 12 (1960) 149.
24. J.H. Sinfelt, G.H. Via, F.W. Lytle, *Catal. Rev.-Sci. Eng.* 26 (1984) 81.
25. Y. Iwasawa (Ed.), *X-Ray Absorption Fine Structure for Catalysts and Surfaces*, World Scientific, Singapore, 1996.
26. J.C.J. Bart, *Adv. Catal.* 34 (1986) 203.
27. J. Stöhr, *NEXAFS Spectroscopy*, Springer, Berlin, 1992.
28. M. Fernández-García, *Catal. Rev.* 44 (2002) 59.
29. B.K. Teo, D.C. Joy (Eds.), *EXAFS Spectroscopy: Techniques and Applications*, Plenum Press, New York, 1981.
30. J.C.J. Bart, G. Vlaic, *Adv. Catal.* 35 (1987) 1.
31. C.J. Dillon, J.H. Holles, R.J. Davis, J.A. Labinger, M.E. Davis, *J. Catal.* 218 (2003) 54.
32. S. Amelinckx, D. van Dyck, J. van Landuyt, G. van Tendeloo (Eds.), *Handbook of Microscopy: Applications in Materials Science, Solid-State Physics and Chemistry*, VCH, Weinheim, 1997.
33. S. Amelinckx, D. van Dyck, J. van Landuyt, G. van Tendeloo, *Electron Microscopy: Principles and Fundamentals*, VCH, Weinheim, 1999.
34. J. Goldstein, D.E. Newbury, D.C. Joy, C.E. Lyman, P. Echlin, E. Lifshin, L.C. Sawyer, J.R. Michael, *Scanning Electron Microscopy and X-Ray Microanalysis* (3rd ed.), Kluwer Academic Publishers, New York, 2003.
35. O. Ovsitser, Y. Uchida, G. Mestl, G. Weinberg, A. Blume, J. Jäger, M. Dieterle, H. Hibst, R. Schlögl, *J. Mol. Catal. A* 185 (2002) 291.
36. G.J. Millar, M.L. Nelson, P.J.R. Uwins, *J. Chem. Soc., Faraday Trans.* 94 (1998) 2015.
37. C. Park, R.T.K. Baker, *Chem. Mater.* 14 (2002) 273.
38. L.D. Schmidt, T. Wang, A. Vacquez, *Ultramicroscopy* 8 (1982) 175.
39. M. J. Yacamán, M. Avalos-Borja, *Catal. Rev.-Sci. Eng.* 34 (1992) 55.
40. A.K. Datye, D.J. Smith, *Catal. Rev.-Sci. Eng.* 34 (1992) 129.
41. J.M. Thomas, O. Terasaki, P.L. Gai, W.Z. Zhou, J. Gonzalez-Calbet, *Acc. Chem. Res.* 34 (2001) 583.
42. T.V. Choudhary, S. Sivadinarayana, A.K. Datye, D. Kumar, D.W. Goodman, *Catal. Lett.* 86 (2003) 1.
43. S. Bernal, J.J. Calvino, M.A. Cauqui, J.M. Gatica, C.L. Cartes, J.A.P. Omil, J.M. Pintado, *Catal. Today* 77 (2003) 385.
44. P.H. Emmett, *Adv. Catal.* 1 (1948) 65.
45. S.J. Gregg, K.S.W. Sing, *Adsorption, Surface Area, and Porosity* (2nd ed.), Academic Press, London, 1982.
46. G. Leofanti, M. Padovan, G. Tozzola, B. Venturelli, *Catal. Today* 41 (1998) 207.
47. J.A. Lercher, in *Catalysis: An Integrated Approach* (2nd ed.), R.A. van Santen, R.W.N.M. van Leeuwen, J.A. Moulijn, B.A. Averill (Eds.), Elsevier, Amsterdam, 1999, p. 543.
48. F. Rouquerol, J. Rouquerol, K.S.W. Sing, in *Handbook of Porous Solids*, F. Schüth, K.S.W. Sing, J. Weitkamp (Eds.), Wiley-VCH, Weinheim, 2002, p. 236.
49. I. Langmuir, *J. Am. Chem. Soc.* 40 (1918) 1361.
50. S. Brunauer, P.H. Emmett, E. Teller, *J. Am. Chem. Soc.* 60 (1938) 309.
51. B.C. Lippens, J.H. de Boer, *J. Catal.* 4 (1965) 319.
52. G. Horvath, K. Kawazoe, *J. Chem. Eng. Jpn.* 16 (1983) 470.
53. E.P. Barrett, L.G. Joyner, P.P. Halenda, *J. Am. Chem. Soc.* 73 (1951) 373.
54. D.Y. Zhao, Q.S. Huo, J.L. Feng, B.F. Chmelka, G.D. Stucky, *J. Am. Chem. Soc.* 120 (1998) 6024.
55. R.J. Cvetanović, Y. Amenomiya, *Catal. Rev.* 6 (1972) 21.
56. J.L. Falconer, J.A. Schwarz, *Catal. Rev.-Sci. Eng.* 25 (1983) 141.
57. S. Bhatia, J. Beltramini, D.D. Do, *Catal. Today* 7 (1990) 309.
58. R.J. Gorte, *Catal. Today* 28 (1996) 405.
59. M. Niwa, K. Habuta, K. Okumura, N. Katada, *Catal. Today* 87 (2003) 213.
60. N.W. Hurst, S.J. Gentry, A. Jones, B.D. McNicol, *Catal. Rev.-Sci. Eng.* 24 (1982) 233.

61. L.E. Briand, W.E. Farneth, I.E. Wachs, *Catal. Today* 62 (2000) 219.
62. W.M. Wendlandt, *Thermal Methods of Analysis* (2nd ed.), Wiley, New York, 1974.
63. J.W. Dodd, K.H. Tonge, *Thermal Methods: Analytical Chemistry by Open Learning*, Wiley, Chichester, 1987.
64. M. Maciejewski, A. Baiker, *J. Therm. Anal.* 48 (1997) 611.
65. N. Ma, Y.H. Yue, W.M. Hua, Z. Gao, *Appl. Catal. A* 251 (2003) 39.
66. P.C. Gravelle, *Adv. Catal.* 22 (1972) 191.
67. N. Cardona-Martinez, J.A. Dumesic, *Adv. Catal.* 38 (1992) 149.
68. A. Auroux, *Top. Catal.* 4 (1997) 71.
69. J.M. McHale, A. Auroux, A.J. Perrotta, A. Navrotsky, *Science* 277 (1997) 788.
70. S.B. Sharma, B.L. Meyers, D.T. Chen, J. Miller, J.A. Dumesic, *Appl. Catal. A* 102 (1993) 253.
71. N. Al-Sarraf, J.T. Stuckless, C.E. Wartnaby, D.A. King, *Surf. Sci.* 283 (1993) 427.
72. Q.F. Ge, R. Kose, D.A. King, *Adv. Catal.* 45 (2000) 207.
73. C.T. Campbell, A.W. Grant, D.E. Starr, S.C. Parker, V.A. Bondzie, *Top. Catal.* 14 (2001) 43.
74. P.R. Griffiths, J.A. de Haseth, *Fourier Transform Infrared Spectroscopy*, Wiley, New York, 1986.
75. F. Zaera, in *Encyclopedia of Chemical Physics and Physical Chemistry*, Vol. 2, J.H. Moore, N.D. Spencer (Eds.), IOP Publishing, Philadelphia, PA, 2001, p. 1563.
76. J. Ryczkowski, *Catal. Today* 68 (2001) 263.
77. J.B. Peri, R.B. Hannan, *J. Phys. Chem.* 64 (1960) 1526.
78. L.M. Kustov, *Top. Catal.* 4 (1997) 131.
79. T.-C. Xiao, A.P.E. York, H. Al-Megren, C.V. Williams, H.-T. Wang, M.L.H. Green, *J. Catal.* 202 (2001) 100.
80. J.A. Lercher, C. Gründling, G. Eder-Mirth, *Catal. Today* 27 (1996) 353.
81. J.C. Lavalley, *Catal. Today* 27 (1996) 377.
82. G. Busca, *Catal. Today* 41 (1998) 191.
83. H. Knözinger, S. Huber, *J. Chem. Soc., Faraday Trans.* 94 (1998) 2047.
84. A. Zecchina, D. Scarano, S. Bordiga, G. Ricchiardi, G. Spoto, F. Geobaldo, *Catal. Today* 27 (1996) 403.
85. K.I. Hadjiivanov, G.N. Vayssilov, *Adv. Catal.* 47 (2002) 307.
86. F. Zaera, *Int. Rev. Phys. Chem.* 21 (2002) 433.
87. H. Tiznado, S. Fuentes, F. Zaera, *Langmuir* 20 (2004) 10490.
88. V.A. Matyshak, O.V. Krylov, *Catal. Today* 25 (1995) 1.
89. G. Busca, *Catal. Today* 27 (1996) 457.
90. F. Zaera, *Catal. Today* 81 (2003) 149.
91. I. Ortiz-Hernandez, C.T. Williams, *Langmuir* 19 (2003) 2956.
92. M.S. Schneider, J.-D. Grunwaldt, T. Bürgi, A. Baiker, *Rev. Sci. Instrum.* 74 (2003) 4121.
93. J. Kubota, F. Zaera, *J. Am. Chem. Soc.* 123 (2001) 11115.
94. M.D. Weisel, F.M. Hoffman, C.A. Mims, *J. Electron Spectrosc. Relat. Phenomena* 64/65 (1993) 435.
95. M. Endo, T. Matsumoto, J. Kubota, K. Domen, C. Hirose, *J. Phys. Chem. B* 105 (2001) 1573.
96. E. Ozensoy, C. Hess, D.W. Goodman, *Top. Catal.* 28 (2004) 13.
97. I.E. Wachs, *Catal. Today* 27 (1996) 437.
98. G. Mestl, T.K.K. Srinivasan, *Catal. Rev.-Sci. Eng.* 40 (1998) 451.
99. M.A. Bañares, I.E. Wachs, *J. Raman Spectrosc.* 33 (2002) 359.
100. R.P. Cooney, G. Curthoys, N.T. Tam, *Adv. Catal.* 24 (1975) 293.
101. T.A. Egerton, A.H. Hardin, *Catal. Rev.-Sci. Eng.* 11 (1975) 71.
102. M.A. Vuurman, I.E. Wachs, *J. Phys. Chem.* 96 (1992) 5008.
103. M.J. Weaver, *J. Raman Spectrosc.* 33 (2002) 309.
104. A. Campion, P. Kambhampati, *Chem. Soc. Rev.* 27 (1998) 241.
105. R.J. LeBlanc, W. Chu, C.T. Williams, *J. Mol. Catal. A* 212 (2004) 277.
106. R. Burch, C. Possingham, G.M. Warnes, D.J. Rawlence, *Spectrochim. Acta A* 46 (1990) 243.
107. J.C. Vedrine, E.G. Derouane, in *Combinatorial Catalysis and High Throughput Catalyst Design and Testing*, E.G. Derouane, F. Lemos, A. Corma, F.R. Ribeiro (Eds.), Kluwer Academic Publishers, 2000, p. 125.
108. C. Li, *J. Catal.* 216 (2003) 203.
109. P.C. Stair, *Curr. Opin. Solid State Mater. Sci.* 5 (2001) 365.
110. M.O. Guerrero-Pérez, M.A. Bañares, *Chem. Commun.* (2002) 1292.

111. H.P. Leftin, M.C. Hobson, Jr., *Adv. Catal.* 14 (1963) 115.
112. M. Che, F. Bozon-Verduraz, in *Handbook of Heterogeneous Catalysis*, Vol. 2, G. Ertl, H. Knözinger, J. Weitkamp (Eds.), Wiley, Weinheim, 1997, p. 641.
113. M. Hunger, J. Weitkamp, *Angew. Chem., Int. Ed.* 40 (2001) 2954.
114. B.M. Weckhuysen, *Chem. Commun.* (2002) 97.
115. P.L. Puurunen, B.M. Weckhuysen, *J. Catal.* 210 (2002) 418.
116. C.A. Fyfe, Y. Feng, H. Grondey, G.T. Kokotailo, H. Gies, *Chem. Rev.* 91 (1991) 1525.
117. J. Klinowski, *Anal. Chim. Acta* 283 (1993) 929.
118. G. Engelhardt, in *Handbook of Heterogeneous Catalysts*, Vol. 2, G. Ertl, H. Knözinger, J. Weitkamp (Eds.), VCH, Weinheim, 1997, p. 525.
119. A.T. Bell, *Colloids Surf. A* 158 (1999) 221.
120. C.P. Grey, in *Handbook of Zeolite Science and Technology*, S.M. Auerbach, K.A. Carrado, P.K. Dutta (Eds.), Marcel Dekker, New York, 2003, p. 205.
121. U.L. Portugal, Jr., C.M.P. Marques, E.C.C. Araujo, E.V. Morales, M.V. Giotto, J.M.C. Bueno, *Appl. Catal. A* 193 (2000) 173.
122. J. Klinowski, T.L. Barr, *Acc. Chem. Res.* 32 (1999) 633.
123. M. Hunger, *Catal. Rev.-Sci. Eng.* 39 (1997) 345.
124. J.F. Haw, T. Xu, *Adv. Catal.* 42 (1998) 115.
125. C.I. Ratcliffe, *Annu. Rep. NMR Spectrosc.* 36 (1998) 123.
126. J.-L. Bonardet, J. Fraissard, A. Gédéon, M.-A. Springuel-Huet, *Catal. Rev.-Sci. Eng.* 41 (1999) 115.
127. J.-Ph. Ansermet, C.P. Slichter, J.H. Sinfelt, *Prog. NMR Spectrosc.* 22 (1990) 401.
128. E.G. Derouane, H.Y. He, S.B.D. Hamid, D. Lambert, I. Ivanova, *J. Mol. Catal. A* 158 (2000) 5.
129. D.E. O'Reilly, *Adv. Catal.* 12 (1960) 31.
130. C.L. Gardner, E.J. Casey, *Catal. Rev.-Sci. Eng.* 9 (1974) 1.
131. J.H. Lunsford, *Adv. Catal.* 22 (1972) 265.
132. M. Che, A.J. Tench, *Adv. Catal.* 31 (1982) 77.
133. M. Che, A.J. Tench, *Adv. Catal.* 32 (1983) 1.
134. E. Giamello, *Catal. Today* 41 (1998) 239.
135. Z. Sojka, *Catal. Rev.-Sci. Eng.* 37 (1995) 461.
136. K. Dyrek, M. Che, *Chem. Rev.* 97 (1997) 305.
137. M. Labanowska, *Chem. Phys. Chem.* 2 (2001) 712.
138. M. Anpo, S. Higashimoto, M. Matsuoka, N. Zhanpeisov, Y. Shioya, S. Dzwigaj, M. Che, *Catal. Today* 78 (2003) 211.
139. D.J. Driscoll, K.D. Campbell, J.H. Lunsford, *Adv. Catal.* 35 (1987) 139.
140. D. Briggs (Ed.), *Handbook of X-Ray and Ultraviolet Photoelectron Spectroscopy*, Heyden, London, 1978.
141. P.K. Ghosh, *Introduction to Photoelectron Spectroscopy*, Wiley, New York, 1983.
142. T.L. Barr, *Modern ESCA: The Principles and Practice of X-Ray Photoelectron Spectroscopy*, CRC Press, Boca Raton, FL, 1994.
143. J.F. Watts, J. Wolstenholme, *An Introduction to Surface Analysis by XPS and AES*, Wiley, Chichester, 2003.
144. A.M. Venezia, *Catal. Today* 77 (2003) 359.
145. E.K. Novakova, J.C. Védrine, E.G. Derouane, *J. Catal.* 211 (2002) 235.
146. D. Teschner, A. Pestryakov, E. Kleimenov, M. Hävecker, H. Bluhm, H. Sauer, A. Knop-Gericke, R. Schlögl, *J. Catal.* 230 (2005) 195.
147. D.F. Ogletree, H. Bluhm, G. Lebedev, C.S. Fadley, Z. Hussain, M. Salmeron, *Rev. Sci. Instrum.* 73 (2002) 3872.
148. P.K. Agrawal, J.R. Katzer, W.H. Manogue, *J. Catal.* 74 (1982) 332.
149. B.A. Horrell, D.L. Cocke, *Catal. Rev.-Sci. Eng.* 29 (1987) 447.
150. E. Taglauer, in *Surface Characterization: A User's Sourcebook*, D. Brune, R. Hellborg, H.J. Whitlow, O. Hunderi (Eds.), Wiley-VCH, Weinheim, 1997, p. 190.
151. C. Pfaff, M.J.P. Zurita, C. Scott, P. Patiño, M.R. Goldwasser, J. Goldwasser, F.M. Mulcahy, M. Houalla, D.M. Hercules, *Catal. Lett.* 49 (1997) 13.
152. A. Benninghoven, F.G. Rüdenauer, H.W. Werner, *Secondary Ion Mass Spectrometry: Basic Concepts, Instrumental Aspects, Applications, and Trends*, Wiley, New York, 1987.

153. Y. Zhou, M.C. Wood, N. Winograd, *J. Catal.* 146 (1994) 82.
154. G.A. Somorjai, *Introduction to Surface Chemistry and Catalysis*, Wiley, New York, 1994.
155. F. Zaera, *Prog. Surf. Sci.* 69 (2001) 1.
156. D.W. Goodman, *Chem. Rev.* 95 (1995) 523.
157. G.A. Somorjai, *CATTECH* 3 (1999) 84.
158. F. Zaera, G.A. Somorjai, *J. Am. Chem. Soc.* 106 (1984) 2288.
159. F. Zaera, A.J. Gellman, G.A. Somorjai, *Acc. Chem. Res.* 19 (1986) 24.
160. F. Zaera, *Langmuir* 12 (1996) 88.
161. F. Zaera, *Israel J. Chem.* 38 (1998) 293.
162. P.S. Cremer, X.C. Su, Y.R. Shen, G.A. Somorjai, *J. Am. Chem. Soc.* 118 (1996) 2942.
163. A. Wieckowski, S.D. Rosasco, G.N. Salaita, A. Hubbard, B.E. Bent, F. Zaera, G.A. Somorjai, *J. Am. Chem. Soc.* 107 (1985) 5910.
164. T.P. Beebe, Jr., J.T. Yates, Jr., *J. Phys. Chem.* 91 (1987) 254.
165. T. Ohtani, J. Kubota, J.N. Kondo, C. Hirose, K. Domen, *J. Phys. Chem. B* 103 (1999) 4562.
166. S.M. Davis, F. Zaera, B.E. Gordon, G.A. Somorjai, *J. Catal.* 92 (1985) 240.
167. H. Öfner, F. Zaera, *J. Phys. Chem. B* 101 (1997) 396.
168. S.M. Davis, F. Zaera, G.A. Somorjai, *J. Catal.* 77 (1982) 439.
169. F. Zaera, *Catal. Lett.* 91 (2003) 1.
170. D.R. Rainer, D.W. Goodman, *J. Mol. Catal. A* 131 (1998) 259.
171. C.R. Henry, *Surf. Sci. Rep.* 31 (1998) 231.
172. H.-J. Freund, M. Bäumer, J. Libuda, T. Risse, G. Rupprechter, S. Shaikhutdinov, *J. Catal.* 216 (2003) 223.
173. M. Valden, X. Lai, D.W. Goodman, *Science* 281 (1998) 1647.
174. F. Zaera, *Surf. Sci.* 500 (2002) 947.
175. R. Gomer, *Adv. Catal.* 7 (1955) 93.
176. E.W. Müller, T.T. Tsong, *Field Ion Microscopy*, American Elsevier, New York, 1969.
177. D.A. Bonnell, *Prog. Surf. Sci.* 57 (1998) 187.
178. R. Wiesendanger, *Scanning Probe Microscopy and Spectroscopy: Methods and Applications*, Cambridge University Press, Cambridge, 2003.
179. W. Weiss, D. Zscherpel, R. Schlögl, *Catal. Lett.* 52 (1998) 215.
180. Y. Yamaguchi, S. Takakusagi, Y. Sakai, M. Kato, K. Asakura, Y. Iwasawa, *J. Mol. Catal. A* 141 (1999) 129.
181. U. Ziese, K.P. de Jong, A.J. Koster, *Appl. Catal. A* 260 (2004) 71.
182. G. Ertl, H.-H. Rotermund, *Curr. Opin. Solid State Mater. Sci.* 1 (1996) 617.
183. G.T. Pott, W.H.J. Stork, *Catal. Rev.-Sci. Eng.* 12 (1976) 163.
184. Y.R. Shen, *Surf. Sci.* 299–300 (1994) 551.
185. G.A. Somorjai, K.R. McCrea, *Adv. Catal.* 45 (2000) 386.
186. H.M. Gager, M.C. Hobson, Jr., *Catal. Rev.-Sci. Eng.* 11 (1975) 117.
187. J.A. Dumesic, H. TopsØe, *Adv. Catal.* 26 (1977) 121.
188. K. Tanabe, M. Misono, Y. Ono, H. Hattori, *New Solid Acids and Bases*, Elsevier, Amsterdam, 1989, p. 6.
189. A. Corma, *Chem. Rev.* 95 (1996) 559.
190. Y. Traa, J. Weitkamp, in *Handbook of Porous Solids*, Vol. 2, F. Schüth, K.S.W. Sing, J. Weitkamp (Eds.), Wiley-VCH, Weinheim, 2002, p. 1015.
191. T. Shido, R. Prins, *Curr. Opin. Solid State Mater. Sci.* 3 (1998) 330.
192. H. TopsØe, *J. Catal.* 216 (2003) 155.
193. A. Brückner, *Catal. Rev.* 45 (2003) 97.
194. M.A. Bañares, M.O. Guerrero-Pérez, J.L.G. Fierro, G.G. Cortez, *J. Mater. Chem.* 12 (2002) 3337.
195. I.E. Wachs, *Catal. Commun.* 4 (2003) 567.
196. B.M. Weckhuysen, *Phys. Chem. Chem. Phys.* 5 (2003) 4351.
197. A. Hagemeyer, B. Jandeleit, Y.M. Liu, D.M. Poojary, H.W. Turner, A.F. Volpe, Jr., W.H. Weinberg, *Appl. Catal. A* 221 (2001) 23.
198. J. Scheidtmann, P.A. Weiβ, W.F. Maier, *Appl. Catal. A* 222 (2001) 79.
199. J.H. Sinfelt, *Acc. Chem. Res.* 10 (1977) 15.
200. V. Ponec, *Appl. Catal. A* 222 (2001) 31.
201. S.T. Oyama, G.L. Haller, *Catalysis* 5 (1982) 333.

202. M.J. Ledoux, C. Pham-Huu, R.R. Chianelli, *Curr. Opin. Solid State Mater. Sci.* 1 (1996) 96.
203. J.G. Chen, B. Frühberger, J. Eng, Jr., B.E. Bent, *J. Mol. Catal. A* 131 (1998) 285.
204. R. Rrins, *Adv. Catal.* 46 (2001) 399.
205. E. Furimsky, *Appl. Catal. A* 240 (2003) 1.
206. R.R. Chianelli, M. Daage, M.J. Ledoux, *Adv. Catal.* 40 (1994) 177.
207. J. Stöhr, E. B. Kollin, D. A. Fischer, J. B. Hastings, F. Zaera, F. Sette, *Phys. Rev. Lett.* 55 (1985) 1468.
208. X. Deng, Y. Yue, Z. Gao, *Appl. Catal. B* 39 (2002) 135.
209. K.V.R. Chary, T. Bhaskar, G. Kishan, K. R. Reddy, *J. Phys. Chem. B* 105 (2001) 4392.
210. F. Zaera, N.R. Gleason, B. Klingenberg, A.H. Ali, *J. Mol. Catal. A* 146 (1999) 13.
211. L. Cao, Z. Gao, S.L. Suib, T.N. Obee, S.O. Hay, J.D. Freihaut, *J. Catal.* 196 (2000) 253.
212. J. Matta, D. Courcot, E. Abi-Aad, A. Aboukaïs, *Chem. Mater.* 14 (2002) 4118.
213. W.-H. Zhang, J. Lu, B. Han, M. Li, J. Xiu, P. Ying, C. Li, *Chem. Mater.* 14 (2002) 3413.
214. Z. Yan, D. Ma, J. Zhuang, X. Liu, X. Liu, X. Han, X. Bao, F. Chang, L. Xu, Z. Liu, *J. Mol. Catal. A* 194 (2003) 153.
215. N.R. Gleason, F. Zaera, *J. Catal.* 169 (1997) 365.
216. R.M. Navarro, M.C. Álvarez-Galván, M. Cruz Sánchez-Sánchez, F. Rosa, J.L.G. Fierro, *Appl. Catal. B* 55 (2005) 229.
217. H. Gnaser, W. Bock, E. Rowlett, Y. Men, C. Ziegler, R. Zapf, V. Hessel, *Nucl. Instrum. Meth. Phys. Res. B* 219–220 (2004) 880.

CHAPTER 1 QUESTIONS

Question 1

Figure 2 in Ref. 65, reproduced on the right, compares the XRD patterns of the precursor and calcined (at 700, 800, 900, 1000, and 1100°C) forms of $MgFe_2O_4$ catalysts. Assign the observed XRD peaks. What can be learned from these XRD patterns about the changes that occur in the sample when the calcination temperature is increased all the way to 1100°C? How can the crystallite size of the $MgFe_2O_4$ spinel calcined at 900°C be determined from the data? Can the appearance of spinel peaks at 700°C be approximately correlated to the results from thermogravimetric (DTA) analysis shown in Figure 1 of Ref. 65?

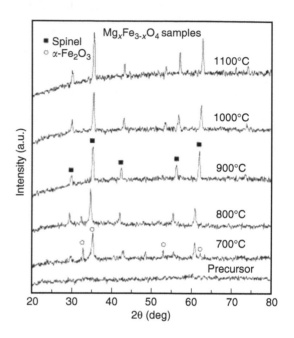

Question 2

A part of Figure 3 in Ref. 207, reproduced on the right, reports radial EXAFS data around the S 1s absorption edge for sulfur adsorbed on the (100) plane of a nickel single-crystal surface. The top trace corresponds to the deposition of atomic sulfur by dehydrogenation of H_2S, while the bottom data were obtained by adsorbing thiophene on the clean surface at 100 K. Based on these data, what can be learned about the adsorption geometry of thiophene? Propose a local structure for the sulfur atoms in reference to the neighboring nickel surface.

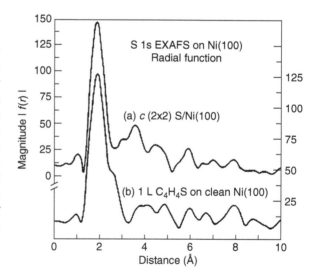

Question 3

Figure 4 in Ref. 208, reproduced on the right, compares the TEM images of four nanosized TiO_2 samples. Based on these images, what can you say, in terms of shape and grain sizes, about the four samples? In Table 2 of the same article, the average grain size of the A-HT-TiO_2-450 sample is reported, based on the broadening of the XRD peaks, at about 12 nm. Is this value roughly consistent with the grain size observed by TEM?

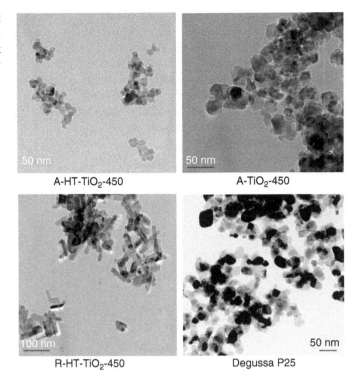

Question 4

Figure 8 in Ref. 209, reproduced on the right, compares the NH_3-TPD data obtained from pure Nb_2O_5 against those from a series of MoO_3/Nb_2O_5 samples. What can be learned from these experiments in terms of the acidity of MoO_3/Nb_2O_5 catalysts? What inferences can be drawn between these NH_3-TPD results and the catalytic data reported in this paper? How did the authors determine the NH_3 uptakes?

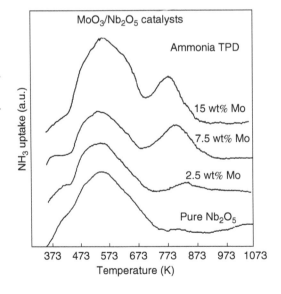

Question 5

Figure 3 in Ref. 210, reproduced on the right, shows TPSR traces obtained from 2-iodopropane adsorbed on a Ni(100) single-crystal surface precovered with oxygen. Describe briefly how these

TPSR experiments were carried out. Based on the products detected, propose a mechanism for the reaction(s) that take place on the surface. What is the main purpose of the XPS data reported in Figure 2 of the same article within the context of these studies? Justify the assignments provided for the I $3d_{5/2}$ XPS peaks at 620.0 and 619.5 eV to molecular iodopropane and atomic adsorbed iodine, respectively.

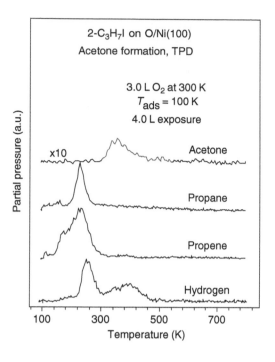

Question 6

Figure 7 in Ref. 211, shown on the right, compares the diffusion reflectance FTIR data of a number of samples based on nanosized TiO_2 catalysts. Based on regular vibrational mode analysis, interpret the features seen in these spectra in terms of possible surface species. What general conclusions can be reached concerning surface reaction intermediates seen during the conversion of toluene on these catalysts, and about the process devised for catalyst regeneration?

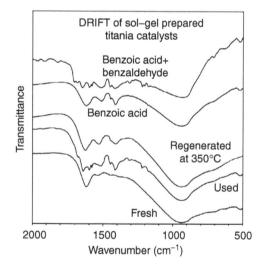

Question 7

Figure 1 in Ref. 212, displayed on the right, shows the Raman spectra obtained from a series of V–Ce–O catalysts and related reference samples. Summarize the key experimental findings from these data in the context of the presence of different solid species as the V loading and calcination temperature are changed.

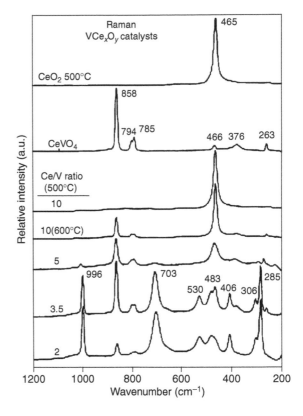

Question 8

Figure 8 in Ref. 213, reproduced on the right, displays diffuse-reflectance UV–Vis spectroscopic data obtained for titanium-substituted mesoporous (Ti-SBA-15) catalysts as a function of titanium content. Assign the main absorption feature observed at 200 to 220 nm and the shoulder seen at about 300 nm. What can we learn from this figure in terms of the different titanium species present on the solid?

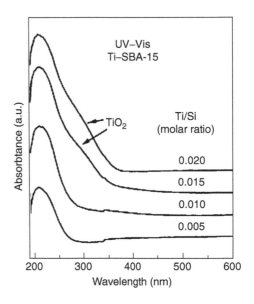

Question 9

The left panel of Figure 4 in Ref. 214, shown on the right, displays ^{27}Al MAS NMR data from an ultra-stable Y (USY) zeolite sample after several treatments with nitric acid solutions.

Summarize the main findings of these experiments in terms of the different Al species present in the solid.

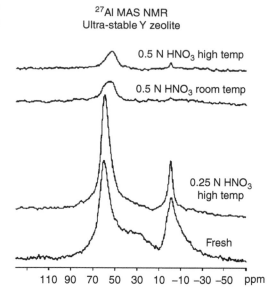

Question 10

Figure 7 in Ref. 215, reproduced on the right, reports ISS spectra from a Ni(100) single-crystal surface partially covered with oxygen as a function of 2-C$_3$H$_7$I exposure. Assign the different peaks seen in these data to the corresponding atoms present on the surface. What was the objective of this study? How do the areas of the two main ISS peaks change as the 2-C$_3$H$_7$I exposure is increased? What is the main lesson from these results?

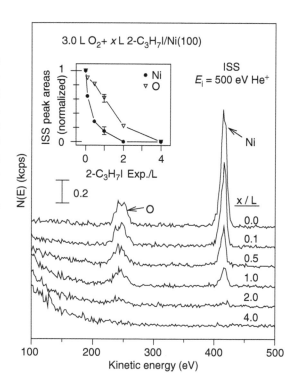

Question 11

Figure 4 in Ref. 216, reproduced on the right, displays Pt $4d_{5/2}$ XPS spectra from calcined alumina-supported platinum catalysts, pure (Pt/A) and doped with lanthanum (Pt/A–L), cerium

(Pt/A–C), and a mixture of both elements (Pt/A–L–C). Provide an interpretation for the results.

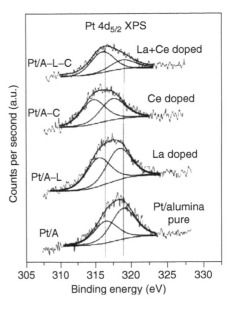

Question 12

Figure 3 in Ref. 217, reproduced on the right, shows SIMS spectra from a CuO/CeO$_2$/ γ-Al$_2$O$_3$ catalyst before and after a methanol steam reforming reaction. Assign the main peaks in the spectra, and provide an interpretation for the changes seen in the catalyst after reaction.

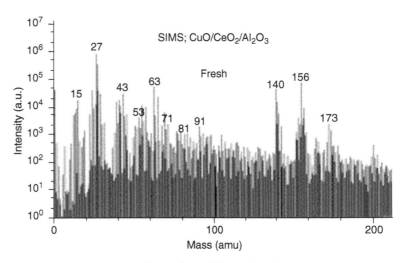

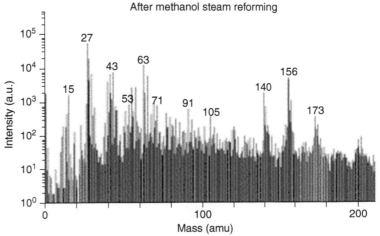

Catalysis by Metal Oxides

Ranjit T. Koodali and Kenneth J. Klabunde

CONTENTS

2.1 Introduction...40
2.2 Properties of Metal Oxides..40
    2.2.1 The Periodic Table as Metal Oxides..41
    2.2.2 Insulators (Highly Ionic) 41
    2.2.3 Semiconductors (Ionic–Covalent Bonding) 42
    2.2.4 Crystal Structures 42
2.3 Surface Structures 44
    2.3.1 Surface Reconstruction...44
    2.3.2 Defect Sites..46
    2.3.3 Hydroxyl Groups 48
2.4 Solid Acids and Bases 48
    2.4.1 Metal Cations as Lewis Acids 49
    2.4.2 Oxygen Anions as Lewis Bases...49
    2.4.3 Hydroxyl Groups as Bronsted Acids and Lewis Bases.................................50
    2.4.4 Spectroscopic Methods of Detecting Lewis Acidity/Basicity and Bronsted
           Acidity/Basicity 50
2.5 Examples of Catalysis 51
    2.5.1 Examples of Solid Acids and Bases 51
        2.5.1.1 Hydrogen–Deuterium Exchange Reaction..51
        2.5.1.2 Hydrogenation (1,3-Butadiene + D_2) 52
        2.5.1.3 Dehydration and Dehydrogenation 53
        2.5.1.4 Dehydrochlorination...53
        2.5.1.5 Benzylation...53
        2.5.1.6 Claisen–Schmidt Condensation-Asymmetric Epoxidation.............54
    2.5.2 Oxidation Catalysis...54
        2.5.2.1 Oxidative Dehydrogenation — Vanadia..54
        2.5.2.2 Oxidative Coupling of Methane 55
    2.5.3 Chloride–Oxide Exchange Catalysis 56
        2.5.3.1 Chlorocarbons 56
        2.5.3.2 Freons 56
2.6 Conclusions...57

References 57
Chapter 2 Questions 61

2.1 INTRODUCTION

Metal oxides constitute an important class of materials covering the entire range from metals to semiconductors to insulators, which exhibit diverse and fascinating properties. They are ubiquitous in heterogeneous catalysis. Indeed, they are the key components for a variety of catalytic reactions, functioning directly as reactive components or as supports (having high-specific-surface areas) to disperse active metal species, or as additives or promoters to enhance the rate of catalytic reactions. The importance of metal oxides in the field of catalysis is thus profound. However, their widespread use as supports has incorrectly led to the popular misconception that they are benign or inert in many catalytic reactions. Although stoichiometric low-index faces of oxides are often rather unreactive, it is important not to assume that this is always true for single-crystal faces, and it certainly is not for step-edges, corners, and other defects.[1]

The surface science of metal oxides has received tremendous attention during recent years. The advent of several advanced techniques to probe their surfaces has led to a new understanding of their surface properties and structures. This in turn has helped catalyst scientists to tailor the properties of the oxide materials for appropriate applications.

Oxides in heterogeneous metal catalysts facilitate well-observed phenomena such as hydrogen spillover and transport of organic reagents between the metal and the oxide support. Many commercial catalysts consist of metal particles supported on high-surface-area oxide. The most commonly used supports are silica and alumina. The use of metal oxides as supports is very extensive and inclusion of them herein is well beyond the scope of this chapter.

An increasing application of transition metal oxides is their use as photocatalysts. A major landmark in the study of transition metal oxides occurred with the report by Fujishima and Honda[2] that TiO_2 could be used as a photoelectrode in a photoelectrolysis cell, to decompose water into hydrogen and oxygen without the application of an external voltage. This report coupled with the energy crisis of the 1970s led to a widespread interest in the synthesis of semiconductor metal oxides for the splitting of water. Although no practical prototype has been developed in spite of 30 years of intense research, the hope is that the results derived from the past studies would help to design a catalyst capable of achieving acceptable efficiency in the future. The subject of photocatalysis is discussed in Chapter 12 and therefore will not be included in the present chapter.

Clays and zeolites too come under the classification of oxide materials. Zeolites are microporous aluminosilicates extensively employed as catalysts and as detergents. They are used in the petroleum industry as cracking catalysts, or as adsorbents to eliminate odor, or as molecular sieves to separate oxygen, argon, nitrogen, and other components of air.[3] The preparation, characterization, and applications of zeolites as catalysts are exhaustive and have been described in detail in Chapter 4 and hence will also not find mention in the present chapter.

2.2 PROPERTIES OF METAL OXIDES

Metal oxides, in particular transition metal oxides, exhibit a wide-ranging array of properties and phenomena. The diverse structures adopted by the metal oxides are helpful to demonstrate the relationship between structures and properties. Correlation of structure and physical properties of transition metal oxides requires an understanding of the valence electrons that bind the atoms in the solid state.[4] The band theory and the ligand field theory have been invoked to explain the electronic properties of transition metal oxides. The important properties of oxides that are of interest are magnetic, electrical, dielectric, optical, Lewis acid/base, and redox.

2.2.1 The Periodic Table as Metal Oxides

Oxide materials can be classified into simple metal oxides and their nonstoichiometric variants, and complex metal oxides such as spinels, perovskites, pyrochlores (exhibiting intersecting tunnel structures), and polyoxometalates (POMs). POMs are discrete molecular species and hence different from bulk structures such as spinels or perovskites. Metal oxides crystallize in a variety of structures, and bonding in the materials can range from ionic (MgO, CaO, BaO, and $Fe_{1-x}O$) to semimetallic (VO) to semiconductor (TiO_2, MnO, NiO, FeO, and CoO) to metallic (TiO and ReO_3). The insulating oxides are made up of elements from both the right and the left sides of the periodic table. Typical examples include SiO_2, Al_2O_3, MgO, and CaO. The oxides of the metals in the middle of the periodic table (Sc to Zn) comprise the semiconducting or metallic oxides. Other oxides of Ru, Mo, W, Pt, V, and Sn find applications in catalysis and sensors. Table 2.1 shows the electrical and magnetic properties of some of the transition metal oxides.

2.2.2 Insulators (Highly Ionic)

We shall briefly discuss the electrical properties of the metal oxides. Thermal conductivity, electrical conductivity, the Seebeck effect, and the Hall effect are some of the electron transport properties of solids that characterize the nature of the charge carriers. On the basis of electrical properties, the solid materials may be classified into metals, semiconductors, and insulators as shown in Figure 2.1. The range of electronic structures of oxides is very wide and hence they can be classified into two categories, nontransition metal oxides and transition metal oxides. In nontransition metal oxides, the cation valence orbitals are of s or p type, whereas the cation valence orbitals are of d type in transition metal oxides. A useful starting point in describing the structures of the metal oxides is the ionic model.[5] Ionic crystals are formed between highly electropositive

Table 2.1   Electrical and Magnetic Properties of Binary Transition Metal Oxides

Metal Oxide	Properties
(d^0) TiO_2, V_2O_5, ZrO_2, Nb_2O_5, MoO_3, WO_3	Diamagnetic semiconductors or insulators
(d^n) TiO, CrO_2, MoO_2, WO_2, IrO_2, ReO_3, RhO_2	Metallic and paramagnetic
(d^n) Ti_2O_3, Ti_3O_5, VO_2, V_2O_3, NbO_2	Exhibit temperature-induced metal to nonmetal transition
(d^n) Cr_2O_3, MnO, Mn_3O_4, FeO, CoO, NiO	Insulators
(f^n) Ln_2O_3 (Ln = rare earth)	Insulators and paramagnetic

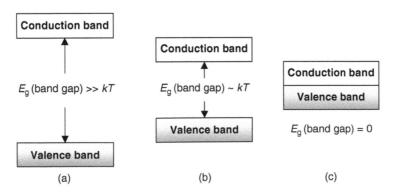

Figure 2.1   Schematic representation of band structure of (a) insulator, (b) semiconductor, and (c) metal.

and highly electronegative elements. Bonding is considered to be ionic if the electronegativity difference between the cation and the anion is sufficiently large, e.g., greater than 2. The crystal structures of the compounds are essentially determined by the nature of bonding and the relative sizes of the cation and the anion. Metal oxides can be nearly ionic as in MgO, but in reality, in many transition metal oxides, the bonding is only partly ionic. Since in metal oxides the anions are generally larger than the cations, the structures can be thought of as consisting of a close-packed array of the anions with the voids or interstices being occupied by the cations. The opposite would be the case when the cations are larger than the anions. Generally, ionic radii correspond actually to free ions and not to ions in crystals. However, the anions in the metal oxide do not exist as O^{2-} in free space but are subject to a Madelung potential, giving rise to a contraction of the charge cloud, while cations are subject to a negative potential causing an opposing effect. Let us, for example, consider MgO, which is a highly ionic compound. The crystal structure of MgO indicates that the distance between the nuclei of O^{2-} and Mg^{2+} is 2.08 Å (0.208 nm). However, the size of the free ions would predict a distance of 2.12 Å. Hence, the effect of lattice potentials can be seen in this difference.

The ionic model suggests that the filled orbitals of highest energy are the filled O 2p on O^{2-} and that the lowest unoccupied ones are the empty Mg 3s orbitals on Mg^{2+}. The ionic model is at best a semiempirical approach to describe the properties of the ionic compound. The optical band gaps (E_g) of insulators are generally over 5 eV. For example, the band gaps of MgO, CaO, SrO, and BaO are 7.7, 6.9, 5.3, and 4.4 eV, respectively. It is a popular misconception that insulators do not conduct electricity because they do not have electrons which can move around. The fact is that the difference in energy levels between the conduction band edge and the valence band edge is very large compared to the thermal energy of the electron, i.e., $E_g \gg kT$, and so it is not possible for the material to conduct at room temperature.

2.2.3 Semiconductors (Ionic–Covalent Bonding)

In compound semiconductors, the difference in electronegativity leads to a combination of both covalent and ionic bonding. In semiconductor metal oxides, the band gap is of almost similar magnitude to that of the thermal energy of the electron, i.e., $E_g \sim kT$. Hence, it is possible to excite the electrons from the valence band to the conduction band with relatively less energy (heating or UV and visible light excitation). The optical band gaps of semiconductor metal oxides are ~3 eV. The semiconductors that are most popularly studied are TiO_2 and ZnO. Other examples include WO_3 and $SrTiO_3$. All of these semiconductors have a band gap of ~3.2 eV and hence require ultraviolet radiation to promote an electron from the conduction band to the valence band. Figure 2.1 shows an illustration of the band structure in insulators and in semiconductors. Titanium dioxide is a popular semiconductor oxide used as a pigment in paints and increasingly as a photocatalyst for the degradation of pollutants.[6] In TiO_2, the valence band is composed mainly of O 2p orbitals and the conduction band of Ti 3d orbitals. However, there is some covalent mixing or hybridization of these levels, which is greater in the lower bonding part of the valence band than in the upper nonbonding part. It is difficult to prepare many transition metal oxide semiconductors without large defect concentrations, and hence their intrinsic properties are difficult to be accurately determined. Even in TiO_2, where reasonably pure and stoichiometric materials can be made, defects can be very easily introduced. The presence of such defects in various oxides such as ZnO and SnO_2 leads to semiconducting properties.

2.2.4 Crystal Structures

The chemistry of metal oxides can be understood only when their crystal structure is understood. Knowledge of the geometric structure is thus a prerequisite to understanding the properties of metal oxides. The bulk structure of polycrystalline solids can usually be determined by x-ray crystallography. In recent years, it has been possible to obtain the detailed structure of the oxides in powder form by employing x-ray and neutron diffraction methods. In fact, these two techniques are used because of the availability of synchrotron x-rays and pulsed neutron sources. In this subsection, we shall describe some of the bulk structures of the metal oxides. At the moment though, we shall restrict ourselves to ideal crystals, in which the bulk atomic arrangement is maintained up to and including the surface planes. In the subsequent section, we shall consider the local or microstructure of the oxides, often arising from defects or compositional changes.

The simplest of structures is the rock salt structure, depicted in Figure 2.2a. Magnesium oxide is considered to be the simplest oxide for a number of reasons. It is an ionic oxide with a 6:6 octahedral coordination and it has a very simple structure — the cubic NaCl structure. The structure is generally described as a cubic close packing (ABC-type packing) of oxygen atoms in the $\langle 111 \rangle$ direction forming octahedral cavities. This structure is exhibited by other alkaline earth metal oxides such as BaO, CaO, and monoxides of 3d transition metals as well as lanthanides and actinides such as TiO, NiO, EuO, and NpO.

The wurtzite structure with hexagonal symmetry depicted in Figure 2.2b has a 4:4 tetrahedral coordination arising from HCP arrangement of anions with half the tetrahedral sites occupied by the cations. Examples include ZnO and BeO.

The rutile structure shown in Figure 2.2c is tetragonal and consists of an infinite array of TiO_6 octahedra linked through opposite edges along the c-axis. Many tetravalent oxides (Ti, V, Cr, Mn, Nb, Sn, Pb, Ir, and W) crystallize in this structure. Other oxides such as VO_2, MoO_2, TcO_2, NbO_2, and WO_2 have a distorted rutile structure. Further, it is possible to substitute stoichiometrically equivalent amounts of appropriate cations to obtain ternary compounds such as $CrNbO_4$, $FeNbO_4$, etc.

Another important structure is the fluorite structure exhibited by oxides with the formula MO_2. The coordination of the cation/anion is 8:4. The fluorite structure shown in Figure 2.2d consists of a cubic close-packed array of cations in which all the tetrahedral sites are occupied by anions. The fluorite structure corresponding to the mineral CaF_2 is exhibited in oxides such as ZrO_2, HfO_2, UO_2, and TbO_2.

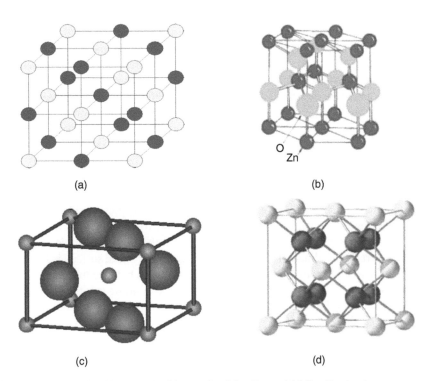

(a)

(b)

O Zn

(c)

(d)

Figure 2.2 Metal oxides with (a) rock salt, (b) wurtzite, (c) rutile, and (d) fluorite structures.

The structures of ternary oxides such as spinels, perovskites, pyrochlores, layered cuprates (high-T_c superconductors), and other lamellar oxides are fascinating subjects by themselves and are beyond the scope of the present discussion.

2.3 SURFACE STRUCTURES

As described in the introduction, metal oxides are key components of a wide range of industrial catalysts. Since catalysis occurs mainly on the surface, the surface properties of the metal oxides assume great importance. The study of the field of metal-oxide surface structures is quite young compared to that of metals or semiconductors. The applications of several ultra-high- vacuum (UHV) techniques in the last 20 years have led to a surge of interest in the study of metal oxide surfaces. There are several reasons for the lag in the growth of knowledge of metal oxide surfaces. The electronic structure of metal oxides is much more complex than that of metals or semiconductors. The number of atoms per unit cell along with the reduced symmetry at the surface and the necessity to separate the surface and bulk effects makes the task of evaluating the surface properties of the metal oxides a daunting one. Another difficulty is a practical one related to experimental techniques used by surface scientists. Many of the oxides such as MgO and alumina are very good insulators and are of great importance to chemical industries. However, many of the powerful surface science techniques such as x-ray photoelectron spectroscopy (XPS) and ultraviolet photoelectron spectroscopy (UPS) involve emission or absorption of charged particles, whether they are ions or electrons. Samples that have negligible conductivity often cannot be studied by these techniques due to surface charging. However, such problems have been circumvented to some degree during the last decade. However, in spite of all the progress made, the fact remains that the range of experimental techniques to probe metal oxide surfaces is smaller than with metals or semiconductors. A comprehensive work devoted to the surface science of metal oxides may be found in the book by Heinrich and Cox.[1] Heinrich and Cox have made an extensive compilation of the literature pertaining to the surface of metal oxides and the book has nearly 1000 references.

According to Idriss and Barteau,[7] the surface of a metal oxide resembles the metal center of mononuclear metal complexes in solution, more than it resembles metal atoms on an extended metal surface. The absence of metal–metal bonds due to the presence of bridging oxygen anions makes the metal cations isolated from each other. Also, the metal centers are not in their zero-valent state unlike the metal centers on a bulk metal surface. Thus, these authors strongly contend that metal oxides find strong analogies with organometallic chemistry and homogeneous catalysis by transition metal complexes. Thus, analogous to organometallic complexes, the key concepts applicable to oxide surfaces are their coordination environment, oxidation state, and redox properties.

2.3.1 Surface Reconstruction

An important factor affecting the property of the surface of the metal oxide is their ability to undergo thermal rearrangement, which is called surface reconstruction or faceting. This arises from two factors, the first being the need for charge balancing at the surface of ionic solids. The second factor is that the average coordination of the metal center on the oxide surface is lower than in bulk metals and this leads to the creation of coordination vacancies to balance the charge. The forces experienced by the surface atoms are different compared to the bonding environment in the bulk, and hence each atom tries to minimize the energy and maximize the coordination; this leads to a reconstructed surface as shown in Figure 2.3. These effects tend to cause reconstruction to form structures that minimize surface polarity and the number of dangling bonds at the surface. An important consequence of surface reconstruction is that polar surfaces are never truly stable. For example, when Mg is burned in air or oxygen, the MgO particles that are formed are almost perfect cubes having (100) faces. Thus, nearly all work on the surface properties of rocksalt oxides have been

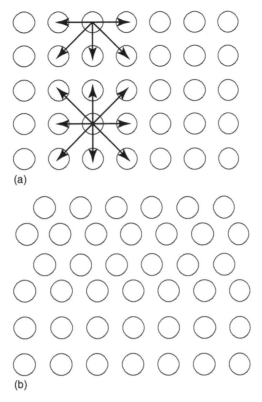

(a)

(b)

Figure 2.3 A simple model depicting reconstruction on the surface of a material: (a) ideal surface and (b) reconstructed surface.

done on (100) surfaces. Attempts to prepare (110) and (111) surfaces have been partially successful, although they tend to reconstruct to surfaces containing (100) planes in order to minimize the surface energy. However, in some cases, the (110) and (111) planes may be stabilized by impurities. The net result is that even low Miller index planes of surfaces undergo reconstruction and faceting. The tendency of surfaces to undergo reconstruction makes it more difficult to vary surface structure of any given oxide. However, it makes it possible to examine the various reconstructed surface structures using only one single-crystal sample. Although the surface reconstruction produces new and interesting reactivity patterns, it also produces a greater challenge to define surface structure and relate it to surface reactivity, since the ultimate goal of a surface scientist is to understand the chemical reactivity in relation to the geometric structure of the oxide surface.

A set of simple physical and chemical principles can be used to understand and predict the surface reconstruction or faceting, which occurs on oxide surfaces.[8,9] These include: (1) autocompensation of stable surfaces, (2) rehybridization of the dangling bond charge density, (3) formation of an insulating surface, (4) conservation of near-neighbor bond lengths (which minimize the formation of surface strain fields), and (5) surface kinetic factors. We shall briefly discuss all of these factors governing the surface reconstruction of metal oxide.

The autocompensation model states that the energetically most stable surfaces are those for which all the cation-derived dangling bonds are completely empty and all anion-derived dangling bonds are completely full. Thus, this model predicts which rearrangement of atoms and which surface terminations will be stable and exist.[10] Surface autocompensation is a necessary but insufficient condition for a stable structure. This means that there may be several autocompensated surfaces that are stable but not observed, presumably because some other autocompensated surfaces are more stable energetically. However, the main drawback of this model is that it cannot predict interlayer relaxations. Based on the autocompensation principles, for example for α-Fe$_2$O$_3$ (0001) and Fe$_3$O$_4$ (0001) surfaces, the most stable surface structures contain Fe atoms terminated at the surface.

Once a particular surface structure has been determined to be stable (i.e., autocompensated), the primary factor determining the nature of the surface reconstruction is the energy that can be gained by rehybridizing the surface dangling bond charge density in response to the reduced coordination at the surface. As a consequence, a charge transfer between the atoms at the surface takes place and this results in the formation of new bonds between surface atoms adsorbed to the surface (also known as adatoms). The formation of new bonds on the surface leads to different chemical and physical properties at the surface.[11]

A third factor responsible for surface reconstruction is the formation of insulating surfaces. Surface dimerization or adsorption leads to the pairing of electrons in the surface dangling bonds. As a result, there is an opening up of a gap between the occupied and unoccupied surface states and hence an insulating surface is formed. The formation of such an insulating surface results in the stabilization of the occupied surface states while destabilizing the unoccupied surface states. This results in a net lowering of the surface energy.

The surface atoms change their positions in order to minimize the energy of the dangling bond charge density, as discussed earlier. As a result, there are changes in the local bonding environment (near-neighbor bond lengths and bond angles). This leads to the formation of local strain fields that are energetically unfavorable. Changes in near-neighbor bond lengths cost the most energy and at times even prevent the surface atoms from moving. Hence, the reconstruction of a surface results from a balance of energy lowering due to rehybridization of the dangling bond charge density, formation of insulating surface, and the elastic energy cost in conservation of near-neighbor bond lengths.

The last factor affecting the surface reconstruction is kinetics. This is important since the experimentally observed surface is that which is kinetically accessible under the experimental conditions. This no longer guarantees that the observed surface is the thermodynamically most stable surface. For example, for cleaved surfaces, the experimentally observed surface is activationless, i.e., the activation barrier to reconstruction is less than the energy supplied by the cleavage process. Thus the surface geometry that is exhibited is an energy balance between formation of local strain fields and rehybridization of surface dangling bond charge density.

It is also worth mentioning briefly the various techniques used for surface determination. The bulk geometrical structures of all crystalline solids are known accurately, but the geometric arrangement of atoms and ions on the surface of solids is known only for a relatively small number of materials. Low-energy electron diffraction (LEED), ion scattering spectroscopy (ISS), and extended x-ray absorption fine structure (EXAFS) are some of the techniques used to understand the geometric arrangement of atoms.[12] Also of considerable importance are scanning tunneling microscopy (STM) and atomic force microscopy (AFM) to probe the surface structures of oxides.[13,14] These two surface structural techniques are truly local in that they image individual atoms and do not necessarily rely on long-range order to produce a signal. Higher spatial resolution has been achieved with STM (~1 Å lateral and ~0.1 Å vertical), but the STM technique requires a conducting surface.

2.3.2 Defect Sites

Thermodynamic considerations imply that all crystals must contain a certain number of defects at nonzero temperatures (0 K). Defects are important because they are much more abundant at surfaces than in bulk, and in oxides they are usually responsible for many of the catalytic and chemical properties.[15] Bulk defects may be classified either as point defects or as extended defects such as line defects and planar defects. Examples of point defects in crystals are Frenkel (vacancy plus interstitial of the same type) and Schottky (balancing pairs of vacancies) types of defects. On oxide surfaces, the point defects can be cation or anion vacancies or adatoms. Measurements of the electronic structure of a variety of oxide surfaces have shown that the predominant type of defect formed when samples are heated are oxygen vacancies.[16] Hence, most of the surface models of oxides have oxygen vacancies. Cation vacancies also exist on oxide surfaces but are considered to be less important when compared with oxygen vacancies. Planar defects include the surface itself, grain boundaries, and crystallographic shear (CS) plane. Point defects may cluster and aggregate in numerous ways: the formation of a crystallographic shear plane can be regarded as the association of oxygen vacancies along a lattice plane accompanied by a shear of the structure in such a way as to eliminate the vacancies completely.

A block model of defects on a single-crystal surface is depicted in Figure 2.4.[17] The surface itself in reality is a two-dimensional defect of the bulk material. In addition, one-dimensional defects in the form of steps which have zero-dimensional defects in the form of kink sites. Terraces, which are also shown in the figure, have a variety of surface sites and may also exhibit vacancies, adatoms, and point defects. Surface boundaries may be formed as a result of surface reconstruction of several equivalent orientations on terraces.

It is worthwhile to look at the geometry of a simple oxide surface in order to fully appreciate the importance of defects in surface chemistry. The MgO structure as mentioned previously is the simplest oxide not only from the point of view of arrangement of atoms, but more importantly the surface energy is far lower for the (100) surface for MgO than for any other oxide. It is interesting to note that periclase (MgO) occurs naturally as octahedral crystals having predominantly (111) faces, but the ideal MgO (111) surface is polar and hence highly unstable. As mentioned previously, the (111) planes tend to reconstruct to (100) planes to minimize the surface energy. Figure 2.5 shows the representation of the surface (100) plane of MgO.[18] The cations and

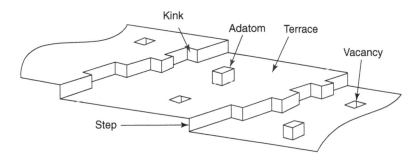

Figure 2.4 A simple model depicting the various types of defects formed on a single-crystal oxide surface. (Reproduced from Brown, G. E. et al., *Chem. Rev.* 1999, *99*, 77–174. Copyright 1999, American Chemical Society. With permission.)

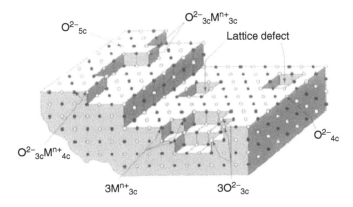

Figure 2.5 Representation of a surface (100) plane of MgO showing steps, kinks, and terraces, which provide sites for M^{n+} and low-coordination oxygen anions. (Reproduced from Dyrek, K. and Che, M., *Chem. Rev.* 1997, *97*, 305–331. Copyright 1997, American Chemical Society. With permission.)

anions on the MgO (100) surface have five nearest-neighbor atoms, four in the surface plane, and one directly below it in the second plane.

2.3.3 Hydroxyl Groups

When a metal oxide surface is exposed to water, adsorption of water molecules takes place as shown in Equation 2.1. Cation sites can be considered as Lewis acids and interact with donor molecules like water through a combination of ion–dipole attraction and orbital overlap. Subsequent protonation and deprotonation of the surface hydroxyls produce charged oxide surfaces as shown in Equation 2.2 and Equation 2.3, respectively:

$$MO_{surf} + H_2O \rightarrow MO.H_2O_{surf} \rightarrow MOH.OH_{surf} \tag{2.1}$$

$$MOH_{surf} + H^+ \rightarrow MOH_{2surf}^+ \tag{2.2}$$

$$MOH_{surf} \rightarrow MO_{surf} + H^+ \tag{2.3}$$

Hence the surface hydroxyl groups can act as either a basic site (Equation 2.2) or as an acidic site (Equation 2.3).

Adsorption of water is thought to occur mainly at steps and defects and is very common on polycrystalline surfaces, and hence the metal oxides are frequently covered with hydroxyl groups. On prolonged exposure, hydroxide formation may proceed into the bulk of the solid in certain cases as with very basic oxides such as BaO. The adsorption of water may either be a dissociative or nondissociative process and has been investigated on surfaces such as MgO, CaO, TiO_2, and $SrTiO_3$.[16] These studies illustrate the fact that water molecules react dissociatively with defect sites at very low water-vapor pressures ($\leq 10^{-9}$ torr) and then with terrace sites at water-vapor pressures that exceed a threshold pressure. Hydroxyl groups will be further discussed in the context of Bronsted acids and Lewis bases.

2.4 SOLID ACIDS AND BASES

Before we discuss the acidity and basicity of metal oxides, the definitions of acidity and basicity as originally proposed by Bronsted and Lewis is worth mentioning. Lowry and Bronsted independently proposed that an acid is any hydrogen-containing species that is capable of releasing a proton, while a base is any species that is capable of accepting a proton.[19,20] According to this view, an acid HA is in equilibrium with a base B to give rise to the conjugated base of HA, A^- and the conjugated acid of B, HB^+ as follows:

$$HA + B = A + HB^+ \tag{2.4}$$

In the same year, Lewis proposed an alternative definition. According to Lewis, an acid is any species that can accept a pair of electrons because of the presence of an incomplete electronic grouping. Hence, a Lewis base is any species that possesses a nonbonding electron pair that can the donated to form a coordination or dative bond. The Lewis acid–base interaction is given by

$$B + A = B^{\delta+} A^{\delta-} \tag{2.5}$$

Thus, according to the above definitions, Lewis bases are also Bronsted bases but Lewis acids do not correspond to Bronsted acids.

As discussed in the previous sections, the surface of a metal oxide consists of unsaturated metal cations and oxide anions. In the case of highly ionic oxides, the ions can act as acids and bases according to the Lewis definition. Therefore, Lewis acid sites (coordinatively unsaturated cations) and basic sites (oxide anions) are normally associated at the surface of ionic metal oxides. The interaction of water (invariably present) converts the surface of the oxide into hydroxyl groups and these groups behave as Bronsted acids, although they mostly behave as basic hydroxyl groups because the bond with which they are held to a metal cation is ionic. In the case of covalent oxides such as SiO_2, water irreversibly coordinates to the surface of these oxides and hence Lewis acidity disappears completely and the formation of surface hydroxyl groups partially is responsible for Bronsted acidity. The oxides of metals in high oxidation state (such as V_2O_5, WO_3, and MoO_3) are also characterized by high covalency. Thus, they exhibit weak or no basicity. However, the presence of metal–oxygen double bonds leads to the presence of strong Lewis acidity in some cases. Further, the dissociative adsorption of water on the surface results in delocalization of the anion charge, which results in medium-to-strong Bronsted acidity.

2.4.1 Metal Cations as Lewis Acids

For clean metal oxide surfaces, without any hydroxyl groups or surface protons, the catalytic behavior of the oxide surfaces is mainly explained in terms of Lewis acidity and basicity. The coordinatively unsaturated cations exposed at the surface of the metal oxide gives rise to surface Lewis acid sites. Thus, basic molecules can react to either form a coordinative bond or increase the overall coordination of the surface cation. The interaction of the surface metal cation and the basic molecule depends on the polarizing power of the unsaturated cation and the basic strength of the adsorbate. Different strengths of Lewis acid sites can be found on oxide surfaces pertaining to different coordination of the surface metal ions. For example, on TiO_2 surfaces, the presence of two types of Lewis acid sites have been observed and they have been attributed to five- and fourfold coordinated titanium cations situated on different faces (edges, corners, and steps).[21] The adsorption of probe molecules such as pyridine, piperidine, ammonia, CO, and NO has been used for characterization of acid/base properties of surface sites on oxides. These characterizations are in turn helpful to explain the catalytic behavior of oxide surfaces.

2.4.2 Oxygen Anions as Lewis Bases

The electron-rich oxygen anions exhibit basic electron donor capacity. Basic metal oxides are commonly used for neutralizing or scrubbing acidic gases. Alkaline earth metal oxides have been used for the removal of NO_x. The surfaces of cubic alkaline metal oxide such as MgO, CaO, and BaO are dominated by the Lewis basicity of surface oxide anions. The basicity increases down the alkaline earth family as the metal ion radii become larger and the charge on the metal ion becomes more positive.

Adsorption of Lewis acids such as CO_2, SO_2, or SO_3 can thus be expected to bond to the oxide surfaces at the anion sites forming surface carbonate (CO_3^{2-}), sulfite (SO_3^{2-}), and sulfate (SO_4^{2-}), respectively. However, the resulting species formed at the surface may have different coordination geometries. The interaction of CO_2 can be regarded as an acid–base reaction between the basic oxide and acidic CO_2 to form surface carbonates. But this simplistic view ignores the variations in the coordination of CO_2 with the oxide surface. One can have a monodentate or a bidentate carbonate species formed on the surface. The formation of a monodentate carbonate species requires only direct interaction with the surface oxide anion, whereas the formation of a bidentate species most likely requires the interaction of CO_2 with oxide anions as well as neighboring cations. In addition to these species, one can have bent and linear unperturbed physisorbed species. A variety of different coordination environments of oxygen anions may be encountered on different oxide materials or different surface orientations of the same material itself. Thus, the metal oxide surfaces can exhibit different strengths of basicity, i.e., surface heterogeneity.

2.4.3 Hydroxyl Groups as Bronsted Acids and Lewis Bases

The surfaces of oxide materials are heterogeneous due to the presence of surface defects, different crystallite faces that terminate the surface, and alsothe presence of multiple hydroxyl groups. The surface properties of a metal oxide thus depend on the distribution, concentration, and nature of the hydroxyl groups. The number of surface hydroxyl groups on an oxide surface has been estimated to be as high as 5 OH/nm2 in the case of SiO_2.[22] For example, on a silica surface, two main types of surface hydroxyl groups (silanol) have been identified. The first type is a single hydroxyl group attached to silicon (single silanols or Q_3 sites) and the second type are those with two hydroxyl groups attached to each silicon (geminal silanols or Q_2 sites). Surface hydroxyl those refers to silicon types groups play an important role in catalytic reactions actingas Bronsted acid sites and are active sites for anchoring or immobilizing chemically active chemical species. The hydroxyl groups that are bonded to metal ions constitute Bronsted acid sites due to their ability to release protons. The stability of the surface hydroxyl groups is temperature-dependent. Complete dehydroxylation is not complete even at temperatures as high as 1100°C. However, during dehydroxylation, two neighboring hydroxyl groups can be dehydrated (resulting in the loss of a water molecule) leading to the formation of a Lewis base site. Various spectroscopic techniques such as nuclear magnetic resonance (NMR),[23] infra-red (IR),[24] Raman,[25] chemical probes, and other analytic tools have been used to investigate surface hydroxyl groups. The concentration of hydroxyl groups and therefore the concentration of Bronsted acid sites could be obtained from these spectroscopic techniques. This will be discussed in detail in the following section.

2.4.4 Spectroscopic Methods of Detecting Lewis Acidity/Basicity and Bronsted Acidity/Basicity

Many methods have been developed to detect the Lewis acidity/basicity and Bronsted acidity/basicity. Liquid- or gas-phase titration, UV spectroscopy, and calorimetry are some of the non-spectroscopic methods of detecting acidity and basicity of an oxide material.[26] However, we shall confine our discussion to the spectroscopic methods of detecting the acidity and basicity of an oxide material. IR, Raman, and NMR are the common spectroscopic techniques used to quantify the acidity or basicity of an oxide material.

The Raman and IR detection of surface acid–base centers is based on the observation of vibrational changes when probe molecules (such as pyridine, CO_2, and CO) are adsorbed on them. As discussed earlier, the presence of coordinatively unsaturated metal ions on the surface give rise to Lewis acidity in an ionic oxide. Consequently, basic molecules can interact with these sites to increase the coordination of the surface metal ion. The Lewis interaction is dependent on the basic strength of the adsorbate and on the polarizing power of the metal cation (charge-to-ionic radius ratio). Upon adsorption of a basic molecule, electrons flow from the basic molecules to the catalyst surface. This results in an electronic perturbation and the lowering of the symmetry. As a consequence, a vibrational perturbation of the adsorbate molecules is observed. In most cases, the vibrational perturbation consists of shifts in vibrational frequencies. The shift in the vibrational frequency is a measure of the Lewis strength of the surface site. The more pronounced the shift in the vibrational frequency, the greater is the Lewis strength of the surface site. Thus, the shift of the position of the bands of the adsorbate molecule can be taken as a measure of the Lewis acid strength of the surface sites. IR spectroscopy in particular has proven to be a powerful technique since it allows one to directly probe the hydroxyls present on a solid surface. In principle, it is also possible to quantitatively estimate the strength of the different types of hydroxyl groups.

Two types of probe molecules have been used for the detection of Lewis and Bronsted acid sites. The first involves the adsorption of relatively strong basic molecules such as pyridine, ammonia, quinoline, and diazines. The second kind involves the adsorption of weak base molecules such as CO, NO, acetone, acetonitrile, and olefins. The pioneering works of Parry[27] and Hughes and

White[28] indicated that the adsorption of pyridine molecule can be used to determine the concentration of Bronsted and Lewis acid sites. When IR is used in conjunction with thermal desorption, an estimation of the acid strength distribution can be obtained.

Pyridine is toxic and also has low volatility. In spite of these drawbacks, pyridine is still the most widely used basic probe molecule for surface acidity characterization. The characteristic bands of pyridine protonated by Bronsted acid sites (pyridinium ions) appear at ~1640 and 1540 cm^{-1}. The bands from pyridine coordinated to Lewis acid sites appear at ~1620 and 1450 cm^{-1}. From a measurement of the intensity of the bands and the values of the extinction coefficients reported in literature,[28–30] it is thus possible to calculate the number of Bronsted and Lewis acid sites. When different hydroxyl groups are present on a metal oxide surface, it is possible to look at the stretching frequency of the different hydroxyl groups before and after adsorption. Heating the oxide progressively to higher temperatures causes desorption of some of the pyridine molecules and thus it is possible to determine which hydroxyl groups are acidic, and their relative acid strength. Laser Raman spectroscopy can also be used to observe the adsorption of pyridine on Bronsted and Lewis acid sites. However, Raman spectroscopy has a lower sensitivity for the detection of Bronsted acidity. Ammonia can also be used to distinguish between Bronsted and Lewis acid sites since it gives IR bands due to NH_4^+ and NH_3 coordinated to Lewis acid sites. Bronsted acid sites can be selectively measured by IR using substituted pyridines such as 2,6-dimethylpyridine as the probe molecule.

Basic molecules such as pyridine and NH_3 have been the popular choice as the basic probe molecules since they are stable and one can differentiate and quantify the Bronsted and Lewis sites. Their main drawback is that they are very strong bases and hence adsorb nonspecifically even on the weakest acid sites. Therefore, weaker bases such as CO, NO, and acetonitrile have been used as probe molecules for solid acid catalysts. Adsorption of CO at low temperatures (77 K) is commonly used because CO is a weak base, has a small molecular size, a very intense $v_{C=O}$ band that is quite sensitive to perturbations, is unreactive at low temperature, and interacts specifically with hydroxyl groups and metal cationic Lewis acid sites.[26]

Thus far, we have discussed the properties of the metal oxide, the surface structures, and their acid–base behavior. In the concluding section, we shall discuss some important catalytic reactions occurring on oxide surfaces.

2.5 EXAMPLES OF CATALYSIS

2.5.1 Examples of Solid Acids and Bases

As discussed in the previous section, metal oxides have both acidic and basic properties. The acid–base properties of metal oxides have led to many interesting catalytic reactions. Catalytic reactions such as H_2–D_2 exchange, hydrogenation, isomerization, dehydrogenation, dehydrohalogenation, and benzylation can be considered as examples of acid–base catalysis reactions.[31–36] These reactions will be briefly discussed in the following section. The remarkable properties of MgO as a catalyst have been well documented in the literature and we shall discuss some of these unique catalytic properties.

2.5.1.1 Hydrogen–Deuterium Exchange Reaction

Metal oxides catalyze the exchange between elemental hydrogen and deuterium (through exchange with the OH group) and the H–D exchange of saturated hydrocarbons. MgO is an important base catalyst because the electron-rich oxygen anions on the MgO surface act as electron-donating sites (basic), while the electron-deficient magnesium ions act as weak electron-accepting sites. Many catalytic reactions such as isomerization, hydrogenation of conjugated dienes, amination, and H–D exchange have been shown to be base-catalyzed.[37] Tanabe and coworkers[38,39] have shown that several basic oxides such as MgO, CaO, SrO, BaO, and La_2O_3 are active for H–D exchange reactions. In these catalytic reactions, the first step involves the heterolytic dissolution of the adsorbed molecules (R-H → R^-_{ads} + H^+_{ads}) through an acid–base-type interaction on the surface of MgO. This results in the formation of protons adsorbed on oxygen anions and conjugate base anions (carbanions) adsorbed on magnesium cations. The formation of carbanions has been verified by Stone and coworkers,[40] who measured the formation of carbanions by allowing them to react with O_2 and measuring the relative amount of O_2^- by electron spin resonance (ESR) studies. The reactions occurring on the surface of MgO can be summarized by the following equations:

$$R\text{-}H \rightarrow R^-_{surf} + H^+_{surf} \tag{2.6}$$

$$R^-_{ads} + O_2(g) \rightarrow R^{\bullet} + O_2^{-}{}_{surf} \tag{2.7}$$

The ESR signal due to O_2^- demonstrates that the heterolytic activation of R–H has occurred. An interesting feature of the H–D exchange reactions over the MgO surface is the low activation energy, i.e., $E_a \sim 2$ kcal/mol. This is lower than the gas phase E_a for the reaction $H^{\bullet} + D_2 \rightarrow 2 D^{\bullet} + HD$. The high activity of the ionic oxides has been attributed to the presence of basic sites and in particular to defect sites that are formed during the oxide preparation that persist even at elevated temperatures.[41,42]

2.5.1.2 Hydrogenation (1,3-Butadiene + D₂)

Another novel catalytic property of MgO is its high activity and selectivity for the hydrogenation (deuteriogenation) of 1,3-butadiene to *cis*-2-butene.[43] Preferential formation of *cis*-2-butene containing 2 D atoms was observed. During the course of the hydrogenation reaction, the molecular identity of hydrogen (or D_2) is maintained, i.e., both H (or D) atoms in a H_2 (or D_2) molecule are incorporated into one hydrogenated molecule. A similar behavior was exhibited by ZnO for the hydrogenation of 1,3-butadiene. Once again, the products formed were those that resulted from an addition of two H (or D) atoms to the original double bond. The mechanism for the deuteriogenation of 1,3-butadiene as suggested by Tanabe et al.[43] is as follows. 1,3-Butadiene consists of 93% *trans* conformer and 7% *cis* conformer in the gas phase. Deuterium adsorbs heterolytically to form D^+ and D^-. A π-allyl carbanion of the *trans* form is formed by the attack of D^- on to a terminal carbon atom. This can either undergo interconversion to form *cis* allyl anion or react with D^+ to form *cis*-2-butene. The electron density in a π-allyl carbanion is highest at the other terminal carbon atom and hence D^+ selectively attacks the other carbon atom to form *cis*-2-butene-d_2. On alkaline earth oxides such as MgO, BaO, and SrO, the interconversion between the *trans* allyl anion and the *cis* allyl anion is faster than the addition of D^+. Thus *cis*-2-butene-d_2 is preferentially formed.[43] However, in the cases of ThO_2, ZrO_2, and rare earth oxides, the addition is faster than the interconversion and hence *trans*-2-butene-d_2 is the main product.

The hydrogenation reactions occurring on heterogeneous basic catalysts have characteristic features, which distinguish heterogeneous basic catalysts from transition metal catalysts.[44] The key features of the base-catalyzed hydrogenation reactions are as follows:

1. There is a preferential formation of 1,4- addition of H atoms in contrast to 1,2- addition which is commonly observed for conventional hydrogenation catalysts. For example, as discussed above in the case of MgO-catalyzed hydrogenation, 2-butene is preferentially formed while 1-butene is the main product over conventional hydrogenation catalysts.
2. There is retention of the molecular identity of H atoms during reaction. The two H atoms used for hydrogenation of a reactant molecule originate from the same hydrogen molecule.
3. The dissociation of H_2 (or D_2) is heterolytic from H^+ and H^- (or D^+ and D^-) on these metal oxides. This is different from conventional metal-based hydrogenation catalysts where H_2 dissociates homolytically and $2H^{\bullet}$ molecules are formed.

The active sites for hydrogenation on alkaline earth oxides such as MgO are believed to be Mg^{2+}_{3c}–O^{2-}_{3c} pairs of low coordination.

2.5.1.3 Dehydration and Dehydrogenation

In general, alcohols undergo dehydrogenation to aldehydes or ketones, and dehydration to alkenes over basic catalysts. The catalytic decomposition of alcohols on oxide catalysts via dehydrogenation and dehydration is considered as a simple route for the production of valuable chemical compounds such as ethane, propylene, and hence has been widely studied in the past decades by several catalysis scientists. Different reaction mechanisms, such as E1, E2, or E1CB have been proposed to explain the product distribution. The surface acid or basic sites participating in the dehydration reaction lead to the formation of alkenes whereas basic or redox sites in dehydrogenation lead to the formation of aldehydes or ketones. In fact, the decomposition of alcohols has become a well-established probe for the study of the surface acid–base and redox properties of heterogeneous oxide catalysts.[45] Basic catalysts such as ZnO, MgO, Cr_2O_3, and CuO are active for alcohol dehydrogenation. In general, dehydration is thought to be catalyzed by surface Bronsted acid sites, while the dehydrogenation sites are surface base sites or acid–base site pairs. However, single-crystal oxides investigated in UHV experiments can dehydrate alcohols in spite of the fact that they are devoid of Bronsted acid sites.

2.5.1.4 Dehydrochlorination

The disposal and destruction of chlorinated compounds is a subject of great importance. In fact, in 1993, some environmental groups had proposed the need for a chlorine-free economy. The cost of complete elimination of chlorinated compounds is quite staggering with the latest estimate as high as \$160 billion/year.[46] The most common method to destroy chlorocarbons is by high-temperature thermal oxidation (incineration).[47] The toxic chlorinated compounds seem to be completely destroyed at high temperatures; however, there is concern about the formation of toxic by-products such as dioxins and furans.[48]

The dehydrochlorination can be expressed by the equation

$$R\text{-}CH_2\text{-}CH_2\text{-}Cl \rightarrow R\text{-}CH = CH_2 + HCl \qquad (2.8)$$

Oxidative catalysis over metal oxides yields mainly HCl and CO_2. Catalysts such as V_2O_3 and Cr_2O_3 have been used with some success.[49,50] In recent years, nanoscale MgO and CaO prepared by a modified aerogel/hypercritical drying procedure (abbreviated as AP-CaO) and AP-MgO, were found to be superior to conventionally prepared (henceforth denoted as CP) CP-CaO, CP-MgO, and commercial CaO/MgO catalysts for the dehydrochlorination of several toxic chlorinated substances.[51,52] The interaction of 1-chlorobutane with nanocrystalline MgO at 200 to 350°C results in both stoichiometric and catalytic dehydrochlorination of 1-chlorobutane to isomers of butene and simultaneous topochemical conversion of MgO to $MgCl_2$.[53-55] The crystallite sizes in these nanoscale materials are of the order of nanometers (~4 nm). These oxides are efficient due to the presence of high concentration of low coordinated sites, structural defects on their surface, and high-specific-surface area.

2.5.1.5 Benzylation

Friedel–Crafts reactions involving electrophilic substitution of aromatic compounds have been reported on solid base catalysts such as thallium oxide and MgO. The rates of benzylation of toluene by benzyl chloride over MgO nanocrystals were found to be of the order CP-MgO > CM-MgO > AP-MgO.[56] An important observation in the study was that x-ray diffraction of the spent catalyst indicated that the CM-MgO bulk phase was unchanged, while CP and AP samples were converted into $MgCl_2$–MgO. Another interesting feature of the reaction was that the AP-MgO sample exhibited an induction period whereas the CP-MgO did not. Thermogravimetric analyzer-mass spectrometer (TGA–MS) studies indicate that the AP-MgO samples strongly adsorbed the organic species, meaning that the catalytic sites were blocked. This is manifested in the observation of an induction period of 80 min for AP-MgO. The unusual trends observed in the rates were explained in terms of the crystal shapes of these oxide crystals. The CP-MgO samples are thin hexagonal platelets about 150 nm in length and 10 nm in thickness, the CM samples are large cubic crystals, and the AP-MgO samples are small stacks of plates with numerous edges, corners, and crystal faces. The MgO (100) face is perhaps the most important crystal face since the CP-MgO hexagonal crystals primarily expose the (100) face surface and allow appropriate adsorption and desorption of the reactant molecules.

2.5.1.6 *Claisen–Schmidt Condensation-Asymmetric Epoxidation*

Chiral epoxides are important precursors to natural products and drug molecules. Thus, asymmetric epoxidation (AE) of α,β-unsaturated ketones to yield chiral epoxides is an important organic reaction in the pharmaceutical industry. The chiral epoxides are generally prepared by homogeneous catalytic processes; heterogeneous catalytic processes are preferred by the industry because of the ease of work-up, handling and separation of reactants and products, and regenerability. Nanocrystalline MgO was found to be a bifunctional heterogeneous catalyst for the Claisen–Schmidt condensation (CSC) of benzaldehyde with acetophenone to yield chiral epoxy ketones with 58% yield and 90% enantioselectivities (ee's) in a two-pot reaction.[57] Different MgO samples were evaluated for the CSC–AE reactions to find the best catalyst. The samples that were studied were CM-MgO, CP-MgO, and AP-MgO. However, AP-MgO was found to the most active catalyst in the condensation and epoxidation reactions. The ee of the product was found to be as high as 90% in the case of AP-MgO, as compared to 30% for CM-MgO and 0% for CP-MgO. The CSC and epoxidation reactions are base-catalyzed reactions and hence the surface hydroxyl groups and O^{2-} of AP-MgO are believed to be the active species responsible for the reaction. The –OH groups present on the edge and corner sites of the AP-MgO surface are more isolated and hence accessible for epoxidation, whereas on CP-MgO, the –OHs are situated on flat planes in close proximity with each other, thus hindering the adsorption of reactant molecules. Figure 2.6 shows the various reactions occurring over nanocrystalline AP-MgO catalyst.

2.5.2 Oxidation Catalysis

The Mars and van Krevelen[58] mechanism is one of the most general mechanisms for oxidation reactions on metallic oxides. The cooperation of redox sites and acid–base is a key feature in this mechanism. The mechanism can be described by a series of elementary steps as follows: (1) activation of the substrate on a metallic cation, (2) insertion of oxygen from lattice oxygen, and (3) transfer of one or more electrons and redox reaction at the metallic site. In the first step, metallic cations act as Lewis acid sites and the surface O^{2-} and OH groups constitute the basic sites. The substrate then undergoes hydrogen abstraction, oxygen insertion, and finally electron transfer. Some of the examples of oxidation catalysis believed to occur through this mechanism include oxidation of butane to maleic anhydride, isobutyric acid oxidative dehydrogenation to methacrylic acid, and propane oxidative dehydrogenation to propene.[59–61]

2.5.2.1 *Oxidative Dehydrogenation — Vanadia*

Vanadia catalysts exhibit high activity and selectivity for numerous oxidation reactions. The reactions are partial oxidation of methane and methanol to formaldehyde, and oxidative dehydrogenation of propane to propene and ethane to ethene.[62–65] The catalytic activity and selectivity of

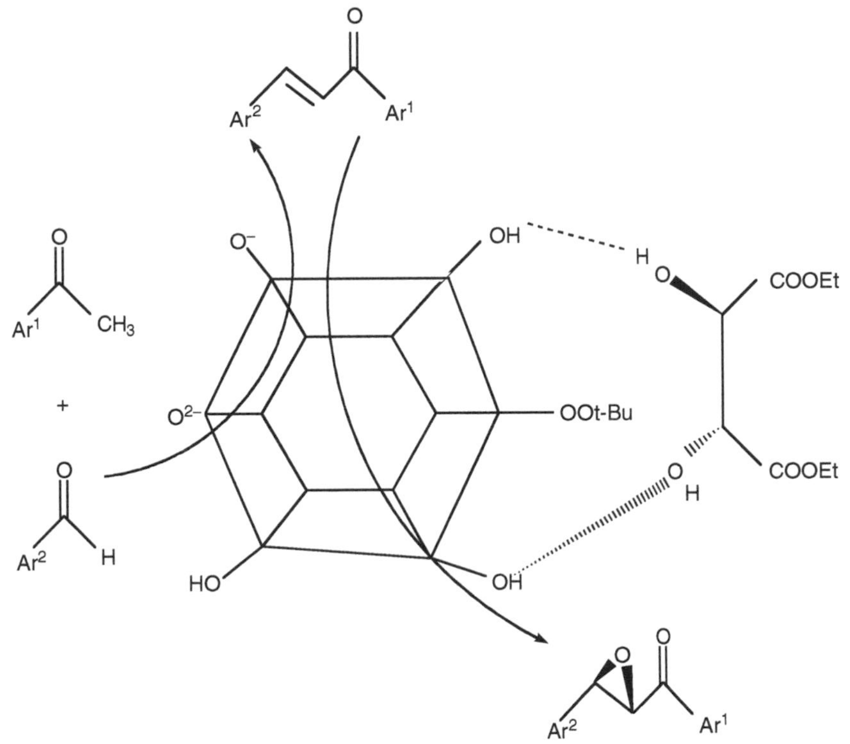

Figure 2.6   Claisen–Schmidt condensation-asymmetric epoxidation reactions over nanocrystalline aerogelprepared AP-MgO catalysts.

vanadia have been found to depend on the structure of the vanadia species.[63] Several techniques such as Raman, IR, UV–visible, solid-state NMR and EXAFS have been used to elucidate the structure and geometry of vanadium species on the surface. Three types of vanadium species have been identified: isolated VO_4 species, polyvanadate species (ribbons and chains of VO_4 units), and V_2O_5 crystallites. The distribution of vanadium species depends on a variety of factors such as vanadium concentration on the surface and sample pretreatment conditions. At low vanadia concentrations, monovanadate species are predominant. With increasing concentrations of vanadia, polyvanadate species are formed. The VO_x species undergoes structural transformations upon oxidation at high temperatures and hence the active site responsible for the ODH reaction is still a matter of intense controversy. However, it appears that tetrahedral V^{5+} are the active sites for the oxidative dehydrogenation of C_2–C_4 compounds.[63]

High-surface-area nanocrystalline MgO, Al_2O_3, and MgO–Al_2O_3 have been examined for the oxidative dehydrogenation of butane to butadiene in the presence of small amounts of iodine.[66] Molecular iodine shifts the equilibrium of the dehydrogenation reactions to the right and makes it possible to achieve high butane conversion. Butadiene selectivity as high as 64% has been achieved in the presence of small amounts of iodine (0.25 vol%) over a vanadia–magnesia catalyst at 82% butane conversion.

2.5.2.2 Oxidative Coupling of Methane

There are large reserves of natural gas throughout the world; in America alone, the Energy Information Administration (EIA) estimates that there are 1190.62 Tcf of technically recoverable natural gas. Natural gas is a combustible mixture of hydrocarbon gases, formed primarily of methane, but also including ethane, propane, butane, and pentane. It is proposed to be one of the cleanest and safest of all the nonrenewable energy sources.[67] Natural gas is a vital component of the world's supply of energy. New processes and technologies are necessary to convert this valuable resource into liquid fuels and chemical feedstocks. One such reaction that has received a great deal of attention is the oxidative coupling of methane (OCM). Methane is passed separately or together with oxygen over a metal oxide catalyst and converted into ethane, ethylene, CO, and CO_2. Ethylene, a key intermediate in the chemical industry, can be oligomerized into liquid fuel or used for manufacture of other important chemicals such as alcohols. A large number of alkali, alkaline and rare earth oxides were found to be active and selective for the oxidative coupling of methane. The basic metal oxides such as MgO, SrO, CaO, and BaO were found to be selective for the OCM reaction.[68] It has been suggested that low-coordinated anion and cation sites, electron-deficient oxide ions (O^-), and anion vacancies are involved in this reaction. Evidence also suggests that the gas phase coupling of methyl radicals is a major pathway for the formation of ethane and ethylene.[69]

2.5.3　Chloride–Oxide Exchange Catalysis

The carcinogenic and toxic properties of chlorocarbons have urged scientists to find efficient ways to destroy them. The desired reaction is the complete oxidation of chlorocarbons to water, CO_2, and HCl, without the formation of any toxic by-products. High surface area and high surface reactivity are desired properties of such destructive adsorbents, and nanoparticles fall into this category. Nanoparticles will react to a greater extent than normal metal oxides because there are more molecules of metal oxide available for reaction with chemicals adsorbed on the surface of the particles. Also, nanocrystals exhibit intrinsically higher chemical reactivities owing to the presence of a large number of defect sites such as edges/corners, and high concentration of coordinatively unsaturated ions. The mechanism to describe the oxidation of chlorocarbons has been developed and involves chemisorption followed by surface to adsorbate oxygen transfer and adsorbate to surface chlorine transfer, i.e., a chloride–oxide exchange.

2.5.3.1　Chlorocarbons

CaO and MgO destroy chlorocarbons such as CCl_4, $CHCl_3$, and C_2Cl_4 at temperatures around 400 to 500°C in the absence of an oxidant, yielding mainly CO_2 and the corresponding metal chlorides.[70–72] In this temperature range, the solid nanocrystalline oxides react nearly stoichiometrically. Nanoparticles of MgO and CaO react with CCl_4 to yield $CaCl_2$ and CO_2. The reaction with $CHCl_3$ yields $CaCl_2$, CO, and H_2O, while C_2Cl_4 yields $CaCl_2$, carbon, and $CaCO_3$. All these reactions are thermodynamically favorable; however, kinetic parameters demand that high-surface-area metal oxides to be used. The mobility of oxygen and chlorine atoms in the bulk of the material becomes important and kinetic parameters involving the migration of these atoms are rate-limiting. The reaction efficiencies can be improved by the presence of small amounts of transition metal oxide as catalyst; for example, Fe_2O_3 on CaO. Morphological studies indicate that iron chloride intermediates help shuttle chloride and oxide anions in and out of the base metal oxide (CaO or MgO) as shown in Figure 2.7.

Nanocrystalline MgO and CaO also allow the destruction of chlorinated benzenes (mono-, di-, and trichlorobenzenes) at lower temperatures (700 to 900°C) than incineration.[73] The presence of hydrogen as a carrier gas allows still lower temperatures to be used (e.g., 500°C). MgO was found to be more reactive than CaO as the latter induces the formation of more carbon.

2.5.3.2　Freons

Freons are a family of compounds containing C–F bonds. The most common of the freons, freon-12, CF_2Cl_2, was developed by Dupont in the 1930s as a substitute for ammonia that was used

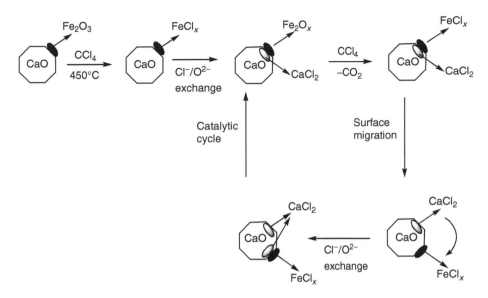

Figure 2.7 Illustration of Fe_2O_3 catalyst on nanocrystalline CaO for the destructive adsorption of CCl_4.

at that time as the gas in refrigerators. However, in recent years, these chemicals are thought to be responsible for the destruction of the ozone layer, allowing potentially dangerous levels of UV light to reach the earth. These compounds are very difficult to decompose. Compounds such as C_2F_6, C_3F_6, C_2ClF, and CHF_3 can be effectively mineralized over CaO and MgO at higher temperatures ($>500°C$).[74] The use of an overlayer of Fe_2O_3 on nanoscale MgO crystallites has proved to be promising for the one-step decomposition of chlorofluorocarbons at temperatures near 400°C. Similarly, the use of $[MgV_xO_y]MgO$ shell/core-like particles led to the decomposition of CCl_3F at temperatures as low as 280°C. No chloride/fluoride was found on the surface of the sample, suggesting the formation of volatile vanadium–halogen products.[75]

2.6 CONCLUSIONS

In this chapter, we have discussed the application of metal oxides as catalysts. Metal oxides display a wide range of properties, from metallic to semiconductor to insulator. Because of the compositional variability and more localized electronic structures than metals, the presence of defects (such as corners, kinks, steps, and coordinatively unsaturated sites) play a very important role in oxide surface chemistry and hence in catalysis. As described, the catalytic reactions also depend on the surface crystallographic structure. The catalytic properties of the oxide surfaces can be explained in terms of Lewis acidity and basicity. The electronegative oxygen atoms accumulate electrons and act as Lewis bases while the metal cations act as Lewis acids. The important applications of metal oxides as catalysts are in processes such as selective oxidation, hydrogenation, oxidative dehydrogenation, and dehydrochlorination and destructive adsorption of chlorocarbons.

REFERENCES

1. Henrich, V.E.; Cox, P.A. *The Surface Science of Metal Oxides*; Cambridge University Press: Cambridge, U.K., 1994.
2. Fujishima, A.; Honda, K. Electrochemical photolysis of water at a semiconductor electrode. *Nature* 1972, *238*, 37–38.

3. Glaesar, R.; Weitkamp, J. The application of zeolites in catalysis. *Springer Ser. Chem. Phys.* 2004, *75*, 161–212.

4. Rao, C.N.R.; Gopalakrishnan, J. *New Directions in Solid State Chemistry*; Cambridge University Press: Cambridge, U.K., 1986.

5. West, A.R. *Solid State Chemistry and its Applications*; Wiley: Chichester, 1984.

6. Anpo, M. Utilization of TiO_2 photocatalysts in green chemistry. *Pure Appl. Chem.* 2000, *72*, 1265–1270.

7. Idriss, H.; Barteau, M.A. Active sites on oxides: From single crystals to catalysts. *Adv. Catal.* 2000, *45*, 261–331.

8. Duke, C.B. Determination of the atomic geometries of solid surfaces. *Appl. Surf. Sci.* 1982, *11–12*, 1–19.

9. LaFemina, J.P. Total energy calculations of semiconductor surface reconstructions. *Surf. Sci. Rep.* 1992, *16*, 133–260.

10. Street, R.A.; Biegelsen, D.K.; Knights, J.C. Defect states in doped and compensated hydrogenated amorphous silicon. *Phys. Rev. B: Condens. Matter Mater. Phys.* 1981, *24*, 969–984.

11. Decker, S.P. Characterization and reactivity of calcium oxide and iron oxide coated — calcium oxide nanocrystals. Ph.D. thesis. Kansas State University, Manhattan, KS, 1998.

12. Woodruff, D.P.; Delchar, T.A. *Modern Techniques of Surface Science*; Cambridge University Press: Cambridge, U.K., 1986.

13. Binnig, G.; Rohrer, H. Scanning tunneling microscopy. *Surf. Sci.* 1983, *126*, 236–244.

14. Hansma, P.K.; Elings, V.B.; Marti, O.; Bracker, C.E. Scanning tunneling microscopy and atomic force microscopy: Application to biology and technology. *Science* 1988, *242*, 209–216.

15. Tilley, R.J.D. *Defect Crystal Chemistry*; Blackie: Glasgow, 1987.

16. Barteau, M.A. Organic reactions at well-defined oxide surfaces. *Chem. Rev.* 1996, *96*, 1413–1430.

17. Brown, G.E., Jr.; Henrich, V.E.; Casey, W.H.; Clark, D.L.; Eggleston, C.; Felmy, A.; Goodman, D.W.; Gratzel, M.; Maciel, G.; McCarthy, M.I.; Nealson, K.H.; Sverjensky, D.A.; Toney, M.F.; Zachara, J.M. Metal oxide surfaces and their interaction with aqueous solutions and microbial organisms. *Chem. Rev.* 1999, *99*, 77–174.

18. Dyrek, K.; Che, M. EPR as a tool to investigate the transition metal chemistry on oxide surfaces. *Chem. Rev.* 1997, *97*, 305–331.

19. Lowry, T.M. The electronic theory of valency. Part IV. The origin of acidity. *Trans. Faraday Soc.* 1924, 13–15.

20. Brönsted, J.N. Acid and basic catalysis. *Chem. Rev.* 1928, *5*, 231–338.

21. Hadjiivanov, K.; Lamotte, J.; Lavalley, J.-C. FTIR study of low temperature CO adsorption on pure and ammonia precovered TiO_2 (anatase). *Langmuir* 1997, *13*, 3374–3381.

22. Zhuravlev, L.T. Concentration of hydroxyl groups on the surface of amorphous silicas. *Langmuir* 1987, *3*, 316–318.

23. Maciel, G.E.; Ellis, P.D. In *NMR Techniques in Catalysis*; Bell, A.T.; Pines, A., Eds.; Marcel Dekker: New York, 1994, p. 231.

24. Hair, M.L. *Infrared Spectroscopy in Surface Chemistry*; Marcel Dekker: New York, 1967.

25. Brinker, C.J.; Tallant, D.R.; Roth, E.P.; Ashley, C.S. Sol-gel transition in simple silicates. III. Structural studies during densification. *J. Non-Crystalline Solid* 1986, *82*, 117–126.

26. Busca, G. Spectroscopic characterization of the acid properties of metal oxide catalysts. *Catal. Today* 1998, *41*, 191–206.

27. Parry, E.P. An infrared study of pyridine adsorbed on acidic solids. Characterization of surface acidity. *J. Catal.* 1963, *2*, 371–379.

28. Hughes, T.R.; White, H.M. A study of the surface structure of decationized Y zeolite by quantitative infrared spectroscopy. *J. Phys. Chem.* 1967, *71*, 2192–2201.

29. Datka, J.; Turek, A.M.; Jehng, J.M.; Wachs, I.E. Acidic properties of supported niobium oxide catalysts: An infrared spectroscopy investigation. *J. Catal.* 1992, *35*, 186–199.

30. Emeis, C.A. Determination of integrated molar extinction coefficients for infrared absorption bands of pyridine adsorbed on solid acid catalysts. *J. Catal.* 1993, *141*, 347–354.

31. Boudart, M.; Delbouille, A.; Derouane, E.G.; Indovina, V.; Walters, A.B. Activation of hydrogen at 78 K on paramagnetic centers of magnesium oxide. *J. Am. Chem. Soc.* 1972, *94*, 6622–6630.

32. Komarewsky, V.I.; Miller, D. Hydrogenation with metal oxide catalysts. *Adv. Catal.* 1957, *9*, 707–715.

33. Kania, W.; Jurczyk, K. Acid-base properties of modified γ-alumina. *Appl. Catal.* 1987, *34*, 1–12.

34. Tetenyi, P. Comparison of metals and metal-oxides from the viewpoint of their effect in effect in some reactions of hydrocarbons. *Catal. Today* 1993, *17*, 439–447.

35. Walker, F.H.; Pavlath, A.E. Dehydrohalogenation of 1,1,1-trihaloethanes. *J. Org. Chem.* 1965, *30*, 3284–3285.

36. Koyande, S.N.; Jaiswal, R.G.; Jayaram, R.V. Reaction kinetics of benzylation of benzene with benzyl chloride on sulfate-treated metal oxide catalysts. *Ind. Eng. Chem. Res.* 1998, *37*, 908–913.

37. Ono, Y.; Baba, T. Selective reactions over solid base catalysts. *Catal. Today* 1997, *38*, 321–337.

38. Tanabe, L. *Solid Acids and Bases*; Academic Press: New York, 1970.

39. Utiyama, M.; Hattori, H.; Tanabe, K. Exchange reaction of methane with deuterium over solid base catalysts. *J. Catal.* 1978, *53*, 237–242.

40. Garrone, E.; Zecchina, A.; Stone, F.S. Anionic intermediates in surface processing leading to O_2^- formation on magnesium oxide. *J. Catal.* 1980, *62*, 396–400.

41. Klabunde, K.J.; Matsuhashi, H.; A comparison of electron donor and proton abstraction of thermally activated pure magnesium oxide and doped magnesium oxides. *J. Am. Chem. Soc.* 1987, *109*, 1111–1114.

42. Hoq, M.F.; Klabunde, K.J. Thermally activated oxide as a selective deuteration catalyst under mild conditions. *J. Am. Chem. Soc.* 1986, *108*, 2114–2116.

43. Hattori, H.; Tanaka, Y.; Tanabe, K. A novel catalytic property of magnesium oxide for hydrogenation of 1,3-butadiene. *J. Am. Chem. Soc.* 1976, *98*, 4652–4653.

44. Hattori, H. Heterogeneous basic catalysis. *Chem. Rev.* 1995, *95*, 537–558.

45. Haffad, D.; Chambellan, A.; Lavalley, J.C. Propan-2-ol transformation on simple metal oxides TiO_2, ZrO_2 and CeO_2. *J. Mol. Catal. A: Chem.* 2001, *168*, 153–164.

46. Amato, I. The crusade against chlorine. *Science* 1993, *261*, 152–154.

47. Exner, J.H. *Detoxification of Hazardous Waste*; Ann Arbor Science: Michigan, 1982.

48. Seeker, W.R. *Incineration of Hazardous Waste*; Koshland, C.P., Ed.; Gordon and Breach Science: Philadelphia, 1992.

49. Mochida, I.; Yoneda, Y. Dehydrochlorination and dechlorination of chloroethanes on chromia catalyst. *J. Org. Chem.* 1968, *33*, 2163–2165.

50. Pistarino, C.; Brichese, F.; Finocchio, E.; Romezzano, G.; Di Felice, R.; Baldi, M.; Busca, G. Catalytic conversion of 2-chloropropane in oxidizing conditions: A FT-IR and flow reactor study. *Stud. Surf. Sci. Catal.* 2000, *130*, 1613–1618.

51. Wagner, G.W.; Koper, O.B.; Lucas, E.; Decker, S.; Klabunde, K.J. Reactions of VX, GD and HD with nanosize CaO: Autocatalytic dehydrohalogenation of HD. *J. Phys. Chem. B* 2000, *104*, 5118–5123.

52. Wagner, G.W.; Bartram, P.W.; Koper, O.; Klabunde, K.J. Reactions of VX, GD and HD with nanosize MgO. *J. Phys. Chem. B* 1999, *103*, 3225–3228.

53. Fenelonov, V.B.; Mel'gunov, M.S.; Mishakov, I.V.; Richards, R.M.; Chesnokov, V.V.; Volodin, A.M.; Klabunde, K.J. Changes in texture and catalytic activity of nanocrystalline MgO during its transformation to $MgCl_2$ in the reaction with 1-chlorobutane. *J. Phys. Chem. B* 2001, *105*, 3937–3941.

54. Mishakov, I.V.; Bedilo, A.F.; Richards, R.M.; Chesnokov, V.V.; Volodin, A.M.; Zaikovskii, V.I.; Buyanov, R.A.; Klabunde, K.J. Nanocrystalline MgO as a dehydrohalogenation catalyst. *J. Catal.* 2002, *206*, 40–48.

55. Gupta, P.P.; Hohn, K.L.; Erickson, L.E.; Klabunde, K.J.; Bedilo, A.F. Transformation of nanocrystalline MgO pellets in reaction with 1-chlorobutane. *AIChE* 2004, *50*, 3195–3205.

56. Choudary, B.M.; Mulukutla, R.S.; Klabunde, K.J. Benzylation of aromatic compounds with different crystallites of MgO. *J. Am. Chem. Soc.* 2003, *125*, 2020–2021.

57. Choudary, B.M.; Kantam, M.L.; Ranganath, K.V.S.; Mahendar, K.; Sreedhar, B. Bifunctional nanocrystalline MgO for chiral epoxy ketones via Claisen–Schmidt condensation-asymmetric epoxidation reactions. *J. Am. Chem. Soc.* 2004, *126*, 3396–3397.

58. Mars, P.; van Krevelen, D.W. Oxidations carried out by means of vanadium catalysts. *Chem. Eng. Sci.* 1954, *3*, 41–59.

59. Zhang-Lin, Y.; Forissier, M.; Sneeden, R.P.; Vedrine, J.C.; Volta, J.C. On the mechanism of *n*-butane oxidation to maleic anhydride on VPO catalysts. I. A kinetics study on a VPO catalyst as compared to VPO reference phases. *J. Catal.* 1994, *145*, 256–266.

60. Marchal-Roch, C.; Bayer, R.; Moisan, J.F.; Teze, A.; Herve, G. Oxidative dehydrogenation of isobu-
     tyric acid: Characterization and modeling of vanadium containing polyoxometalate catalysts. *Top.
     Catal.* 1996, *3*, 407–419.
61. Jibril, B.Y. Propane oxidative dehydrogenation over chromium oxide-based catalysts. *Appl. Catal. A:
     General* 2004, *264*, 193–202.
62. Kung, H.H. Oxidative dehydrogenation of light (C_2 to C_4) alkanes. *Adv. Catal.* 1994, *40*, 1–38.
63. Wachs, I.E. Molecular engineering of supported metal oxide catalysts: Oxidation reactions over sup-
     ported vanadia catalysts. *Catalysis* 1997, *13*, 37–54.
64. Watling, T.C.; Deo, G.; Seshan, K.; Wachs, I.E.; Lercher, J.A. Oxidative dehydrogenation of propane
     over niobia supported vanadium oxide catalysts. *Catal. Today* 1996, *28*, 139–145.
65. Gao, X.; Banares, M.A.; Wachs, I.E. Ethane and *n*-butane oxidation over supported vanadium oxide
     catalysts: An *in situ* UV-visible diffuse reflectance spectroscopic investigation. *J. Catal.* 1999, *188*,
     325–331.
66. Chesnokov, V.V.; Bedilo, A.F.; Heroux, D.S.; Mishakov, I.V.; Klabunde, K.J. Oxidative dehydrogena-
     tion of butane over nanocrystalline MgO, Al_2O_3, and VO_x/MgO catalysts in the presence of small
     amounts of iodine. *J. Catal.* 2003, *218*, 438–446.
67. Hefner, R.A. Toward sustainable economic growth: The age of energy gases. *Int. J. Hydrogen Energy*
     1995, *20*, 945–948.
68. Ito, T.; Lunsford, J.H. Synthesis of ethylene and ethane by partial oxidation of methane over lithium-
     doped magnesium oxide. *Nature* 1985, *314*, 721–722.
69. Lunsford, J.H. The catalytic oxidative coupling of methane. *Angew. Chem. Int. Ed. Eng.* 1995, *34*,
     970–980.
70. Koper, O.; Li, Y.-X.; Klabunde, K.J. Destructive adsorption of chlorinated hydrocarbons on ultrafine
     (nanoscale) particles of calcium oxide. *Chem. Mater.* 1993, *5*, 500–505.
71. Koper, O.; Lagadic, I.; Klabunde, K.J. Destructive adsorption of chlorinated hydrocarbons on ultrafine
     (nanoscale) particles of calcium oxide 2. *Chem. Mater.* 1997, *9*, 838–848.
72. Jiang, Y.; Decker, S.; Mohs, C.; Klabunde, K.J. Catalytic solid state reactions on the surface of
     nanoscale metal oxide particles. *J. Catal.* 1998, *180*, 24–35.
73. Li, Y.-X.; Li, H.; Klabunde, K.J. Destructive adsorption of chlorinated benzenes on ultrafine
     (nanoscale) particles of magnesium oxide and calcium oxide. *Environ. Sci. Technol.* 1994, *28*,
     1248–1253.
74. Decker, S.P.; Klabunde, J.S.; Khaleel, A.; Klabunde, K.J. Catalyzed destructive adsorption of envi-
     ronmental toxins with nanocrystalline metal oxides. Fluoro-, chloro-, bromocarbons, sulfur, and
     organophosphorus compounds. *Environ. Sci. Technol.* 2002, *36*, 762–768.
75. Martyanov, I.N.; Klabunde, K.J. Decomposition of CCl_3F over vanadium oxides and [MgV_xO_y]MgO
     shell/core-like particles. *J. Catal.* 2004, *224*, 340–346.

CHAPTER 2 QUESTIONS

Question 1

Explain why MgO crystallizes in the rock salt (NaCl) structure whereas BeO crystallizes in the wurtzite structure.

Question 2

Explain why TiO (rock salt structure) is a metal whereas TiO_2 (rutile) is a semiconductor.

Question 3

Explain the difference between a semiconductor and an insulator, and why a semiconductor such as TiO_2 cannot be excited by visible light. How can TiO_2 be made to absorb light in the visible region?

Question 4

What is surface reconstruction? and why is it difficult to prepare MgO (110) and (111) surfaces although MgO crystals having (111) faces occur naturally ?

Question 5

Explain the various defect sites (bulk and surface) and their importance in surface chemistry reactions.

Question 6

Describe the acid–base properties of metal oxides.

Question 7

Describe the spectroscopic methods for detection of Lewis acidity/basicity and Bronsted acidity/basicity of metal oxides, and explain why pyridine (in spite of its toxicity and low volatility) is a popular choice as an adsorbate molecule.

Question 8

The hydrogenation of 1,3-butadiene over metal oxide catalysts with D_2 or H_2 gives rise to predominantly 2-butene, whereas conventional metal catalysts give rise to 1-butene as the main product. Explain the differences in the product distribution in light of the reaction mechanisms.

Colloidal Nanoparticles in Catalysis

Helmut Bönnemann and K.S. Nagabhushana

CONTENTS

3.1   Introduction 63
3.2   Mechanism of Stable Particle Formation 64
3.3   Modes of Stabilization 64
3.4   Reduction Methods 66
3.5   Applications in Catalysis 74
       3.5.1   Quasi-Homogeneous Reactions 74
       3.5.2   Heterogeneous Reactions 74
                 3.5.2.1   Precursor Concept 74
                 3.5.2.2   Conditioning: A Key Step in Generating Active Catalysts 75
                 3.5.2.3   Heterogeneous Catalysts in Catalysis 76
                 3.5.2.4   Fuel Cell Catalysts 83
3.6   Conclusion 85
References 86
Chapter 3 Questions 94

3.1   INTRODUCTION

Studies of useful size-dependent properties of nanomaterials are only possible when they are prepared and isolated in a monodisperse form. The synthesis, therefore, should address the need for a great degree of control over the structure, size, and also the composition of the particles. The design of successful synthetic strategies has enabled continuous exploration and exploitation of the unusual properties of nanomaterials that differ both from the single atom (molecule) and the bulk. This also suggests that the intended use of the nanomaterials will dictate the method that can be conveniently applied to obtain them.

Colloidal nanoparticles are generated by using colloidal stabilizers that control the agglomeration of the primarily formed metal particles of uniform sizes, during synthesis. Generally, isolable particles that are less than 50 nm in size are called colloidal nanoparticles. Colloidal metal nanoparticles of controlled size and morphology have been of great scientific interest since their inception. This is due to the access to colloidal metal particles of controlled size, shape, and morphology having unique physical, chemical, and thermodynamical properties that have made them useful in such diverse fields of science as homogeneous and heterogeneous catalysis [1–2], fuel cell catalysis [2–4], electronics [5], optics [6], magnetism [7,8], material sciences [9,10], and, of late, even in biological and medical sciences [11,12]. The newer tunable synthetic approaches have provided chemists with not just compositional variations but also with significant control over size and inner structures [13–16]. Here we provide a detailed report on the preparation of colloidal nanometal particles and their applications in catalysis.

3.2 MECHANISM OF STABLE PARTICLE FORMATION

There are two general approaches for the production of nanostructured materials, namely the top-down and the bottom-up approaches. The top-down approach relies on breaking down of the bulk material into nanosized material with subsequent stabilization by appropriate stabilizers [17,18], while the bottom-up approach works on the principle of building nanoclusters by generating individual atoms that group to form a stable nucleus that are then stabilized. For this reason, it is very difficult to obtain monodispersed particles by the top-down approach, whereas this is the hallmark of the bottom-up approach. The versatile wet chemical approach, electrochemical approach, and thermal decomposition of lower-valent metal complexes are all based on the bottom-up synthetic method and there exists a large volume of information on this subject [19–30]. A plethora of stabilizers like donor ligands, polymers and surfactants are used to control the growth of the primarily formed nanoparticles and to prevent them from agglomeration.

After Faraday's seminal report on the preparation of transition metal clusters in the presence of stabilizing agents in 1857 [31], Turkevich [19–21] heralded the first reproducible protocol for the preparation of metal colloids and the mechanism proposed by him for the stepwise formation of nanoclusters based on nucleation, growth, and agglomeration [19] is still valid but for some refinement based on additional information available from modern analytical techniques and data from thermodynamic and kinetic experiments [32–41]. Agglomeration of zero-valent nuclei in the seed or, alternatively, collisions of already formed nuclei with reduced metal atoms are now considered the most plausible mechanism for seed formation. Figure 3.1 illustrates the proposed mechanism [42].

There have been many recent studies in support of this mechanistic approach. Stepwise reductive formation of Ag_3^+ and Ag_4^+ clusters has been followed using spectroscopic methods by Henglein [33]. Reduction of copper (II) to colloidal Cu protected by cationic surfactants (NR_4^+) through the intermediate Cu^+ prior to nucleation of the particles [36] as monitored by *in situ* x-ray absorption spectroscopy is another example. The seed-mediated synthesis also serves as evidence in support of this mechanism [38–41].

The formation of nanometal colloid via reductive stabilization using aluminum organic reagent operates with a different mechanism as is depicted in Figure 3.2. The mechanism has been elucidated based on various physical and analytical data [37].

The current general understanding is that metal salts are reduced to give zero-valent metal atoms in the embryonic stage of nucleation [37], which can collide in solution with other metal ions, metal atoms, or clusters, to form an irreversible seed of stable metal nucleus (as shown in Figure 3.1). The diameter of this seed nucleus can be controlled to well below 1 nm, depending on the difference in redox potentials of the metal salt and the reducing agent used, and the strength of the metal–metal bonds. Further, size of the resulting metal colloid is determined by the relative rates of nucleation and particle growth.

3.3 MODES OF STABILIZATION

The enormous surface area-to-volume ratio of nanoparticles leads to excess surface free energy that is comparable to the lattice energy leading to structural instabilities. The nanoparticles have to

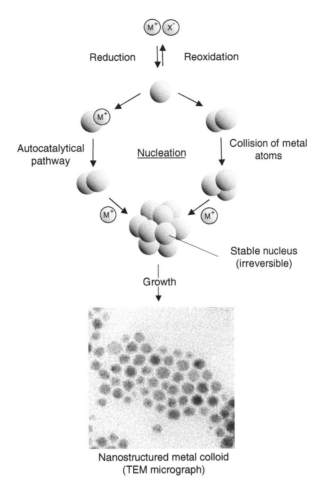

Figure 3.1 Formation of nanostructured metal colloids via the salt reduction method. (Adapted from Maase, M., Neue Methoden zur größen- und formselektiven Darstellung von Metallkolloiden., Ph.D. thesis, Bochum, Germany, 1998.)

be kinetically protected from further condensation reactions to form the thermodynamically favored binary phases. This implies that a ligand sphere coordinating to surface metal centers is necessary to prevent such unwanted agglomerations of nanoparticles in synthesis. The role of the ligand is not just to hold the particles stable against oxidation and agglomeration, but rather also to serve as controller of even crystal growth and shape. This stabilizer should coordinate to nanoparticles strongly enough to prevent them from agglomeration, but should leave the metal surface relatively easy to generate pure metal surfaces as needed in catalysis. Two stabilization modes are well established [43].

The electrical double layer formed by ions adsorbed at the particle surface and the corresponding counter-ions causing a coulombic repulsion between particles provides electrostatic stabilization. Reduction of $[AuCl_4^-]$ with sodium citrate to produce gold sols is one of the examples [19–21].

Steric stabilization is achieved due to the presence of sterically bulky organic molecules on the metal surface acting as protective shields. The main classes of protective groups that have been studied till date include polymers and block copolymers; P, N, and S-donors (e.g., phosphines, amines, thioethers); solvents such as THF, THF/MeOH, or propylene carbonate; long-chain alcohols; surfactants; and organometallics. A recent review on various protective groups [2] provides a broad overview of this subject.

Depending on the nature of these protecting shells, the resulting metal colloid may be taken into organic (organosols) or aqueous (hydrosols) medium. Lipophilic protective agents such as

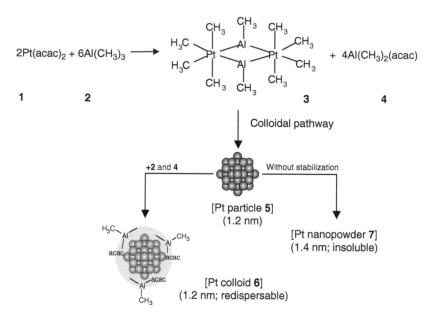

$2Pt(acac)_2 + 6Al(CH_3)_3 \longrightarrow$ [intermediate complex 3] $+ 4Al(CH_3)_2(acac)$

1   2   3   4

Colloidal pathway

+**2** and **4**   Without stabilization

[Pt particle **5**]
(1.2 nm)

[Pt nanopowder **7**]
(1.4 nm; insoluble)

[Pt colloid **6**]
(1.2 nm; redispersable)

Figure 3.2 Schematic representation of platinum seed formation through the intermediate complex (**3**). (Adapted from Angermund, K. et al., *Angew. Chem. Int. Ed.*, 41, 4041, 2002. With permission from VCH.)

tetraalkylammonium or aluminum organics give organosols, while the hydrophilic protective agents e.g., 3-(N,N-dimethyldodecylammonio)-propanesulfonate (SB12), give hydrosols. In metal organosols stabilized by tetraalkylammonium halides (Figure 3.4), the metal core is protected by a surfactant coating [44], in contrast to metal hydrosols that are stabilized by zwitterionic surfactants able to self-aggregate in the form of organic double layers [45].

3.4   REDUCTION METHODS

There are several bottom-up methods for the preparation of nanoparticles and also colloidal nanometals. Amongst these, the salt-reduction method is one of the most powerful in obtaining monodisperse colloidal particles. Electrochemical methods, which gained prominence recently after the days of Faraday, are not used to prepare colloidal nanoparticles on a large scale [26, 46]. The decomposition of lower valent transitional metal complexes is gaining momentum in recent years for the production of uniform particle size nanoparticles in multigram amounts [47,48].

Metal salt reduction is one of the most used wet chemical methods of nanoparticle synthesis because of the simple advantages of having high concentration of metal in solution, reproducibility in multigram amounts, and narrow size distribution of the particles. The classical route of reducing $[AuCl_4]^-$ by sodium citrate is still used to prepare standard 20 nm gold sols [19,43]. An array of chemical reducing agents and methods have successfully been applied to obtain nanostructured colloidal metals. Diboranes as reducing agents with various stabilizers (used by Schmid and coworkers [49–63]), Finke's polyoxoanion- and tetraalkylammonium-stabilizer in conjunction with hydrogen [29,35,64–68], hydrogen in the presence of stabilizers like phenanthroline and bipyridine (Moiseev's giant Pd) [69–74], other stabilizers and polymers [75,76], alcohols containing α-hydrogens (method developed by Hirai and Toshima) [77–83], polyols, CO, formic acid or sodium formate, formaldehyde, benzaldehyde [20,84,85], tetrakis(hydroxymethyl)phosphoniumchloride (THPC) [86–91], silanes [92,93], hydrazine [94–96],

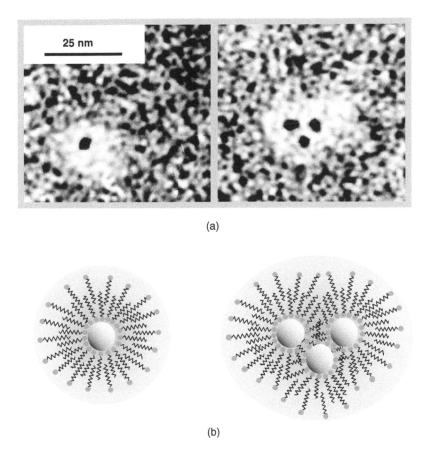

(a)

(b)

Figure 3.3   (a) TEM micrographs of colloidal Pt particles (single and aggregated, average core size=2.8 nm) stabilized by Carboxybetaine 12. (b) Schematic model of hydrosol stabilization by a double layer of the zwitterionic Carboxybetaine 12. (Adapted from Schulze Tilling, A., Bimetallische Edelmetallkolloide als Precursor für Kohlenhydrat-Oxidationskatalysatoren., Ph.D. thesis, RWTH Aachen, 1996.)

hydroxylamine [41,97], even electrons of K^+ $[(crown)_2K]^-$ [98], and borates (though metal boride contaminants are formed) using solvent as stabilizer [99–104] are among the many different reducing agents conveniently used. Alkaline hydrotriorganoborates (MBR_3H) may be regarded as adducts of MH and BR_3, in which the triorganoboron acts as a neutral complexing agent and not as a reducing agent. Use of this reducing agent for the preparation of transition metals (groups 6–12 and also group 14) yielded metal nanopowders without contamination from borides. Early transition metals like Ti, Mn, Zr, Nb, V formed solvent-stabilized nanoparticles of diminished sizes [24]. However, a wide range of applications in the wet chemical reduction of transition metal salts have been performed using hydrotriorganoborate having tetraalkylammonium cations [24,25,27,105–108]. In this case, the reductant (BEt_3H^-) is combined with the stabilizing agent (e.g., NR_4^+). The surface-active NR_4^+ salts are immediately formed at the reduction center in high local concentration and act as powerful protecting agents for the metal particles and prevent their aggregation. By suitable work-up, the resulting colloid can be isolated in solid form and redissolved in various solvents in high molar concentrations of the metal. The other advantages of this process are that trialkylboron is recovered unchanged from the reaction and no borides contaminate the products. Using various advanced physical methods, it has been recently shown in the case of tetrabutyl- and tetraoctyl-ammonium-stabilized Pd colloids [109] that negatively charged chloride is present on the inner side of the protecting shell and is located between the palladium core and N(alkyl)$_4$ groups. Moreover, results also suggest a dependence of the equilibrium position on the chloride between the metal core and the alkyl chain length of the protecting shell (Figure 3.4).

$$MX_v + NR_4(BEt_3H) \rightarrow M_{colloid} + vNR_4X + vBEt_3 + v/2H_2\uparrow \qquad (3.1)$$

where M = metals of the groups 6–11; X = Cl, Br; v = 1,2,3; and R = alkyl, $C_6 - C_{20}$.

From this method of colloidal synthesis, the NR_4^+-stabilized metal raw colloids typically contain 6–12 wt-% of metal. Upon work-up, purified transition metal colloids containing ca. 70–85 wt.-% of metal are obtained [24]. As an alternative, the prepreparation of (NR_4^+ BEt_3H^-) can be avoided when NR_4X is coupled to the metal salt prior to the reduction step. NR_4^+-stabilized transition metal nanoparticles can also be obtained from NR_4X-transition metal double salts. Since the local concentration of the protecting group is sufficiently high, a number of conventional reducing agents may be applied to give protected colloidal particles:

$$(NR_4)_w MX_v Y_w + v[Red] \rightarrow M_{colloid} + vRedX + wNR_4Y \qquad (3.2)$$

where

M = metals; Red = H_2, HCOOH, K, Zn, LiH, LiBEt$_3$H, NaBEt$_3$H, KBEt$_3$H; X, Y = Cl, Br; v, w = 1–3; and R = alkyl, C_6–C_{12}

Figure 3.5 gives a survey of the (BEt$_3$H$^-$) method for the preparation of various transition metal nanoparticles. The average particle size of the resulting nanoparticles is indicated along with the metal atoms.

The advantages of the method (Figure 3.5) may be summarized as follows:

- The method is generally applicable to salts of metals in groups 4–11 in the periodic table.
- It yields extraordinary stable metal colloids that are easy to isolate as dry powders.
- The particle size distribution is nearly monodisperse.
- Bi- and pluri-metallic colloids are easily accessible by co-reduction of different metal salts.
- The synthesis is suitable for multigram preparations and easy to scale-up.

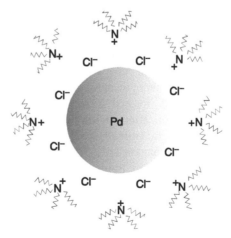

Figure 3.4   Idealized model of NR$_4$Cl stabilized metal core. (Adapted from Bucher, J. et al., *Surf. Sci.*, 497, 321, 2002. With permission from Elsevier Science.)

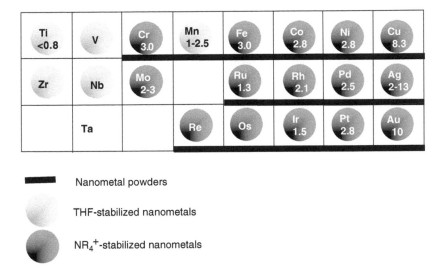

		Cr 3.0	Mn 1-2.5	Fe 3.0	Co 2.8	Ni 2.8	Cu 8.3
Ti <0.8	V						
Zr	Nb	Mo 2-3		Ru 1.3	Rh 2.1	Pd 2.5	Ag 2-13
	Ta			Re	Os	Ir 1.5	Pt 2.8	Au 10

▬▬▬ Nanometal powders

⬤ THF-stabilized nanometals

⬤ NR_4^+-stabilized nanometals

Figure 3.5 Nanopowders and nanostructured metal colloids accessible via the (BEt_3H^-)-reduction method. (Equation 3.1 and Equation 3.2) (including the mean particle sizes obtained) (Adapted from Bönnemann, H. and Brijoux, W., in *Surfactant-Stabilized Nanosized Colloidal Metals and Alloys as Catalyst Precursors/Advanced Catalysts and Nanostructured Materials*, Moser, W., Ed., Academic Press, San Diego, 1996, pp. 165–196, Chap. 7. With permission from Elsevier Science.)

A drawback of this method is that by altering the reaction conditions, the particle size of the resulting sols cannot be varied. This synthetic method has a promising impact in generating precursors for highly effective catalysts, which is dealt in some detail in Section 3.5 [110–112].

Organoaluminum compounds have been used for the reductive stabilization of mono- and bimetallic nanoparticles (see Equation 3.3) to generate organosols [113–116].

$$MX_n + AlR_3 \xrightarrow{\text{Toluene}} \text{[nanoparticle]} + [R_2Alacac]$$

(3.3)

M = Metals of Groups 6-11
X= Halogen, Acetylacetonate
n = 2-4. R = C1-C8-Alkyl,
Particle sizes 1-12nm

From Equation 3.3, it is clear that colloids of zero-valent elements of groups 6–11 of the periodic table (and also of tin) can be prepared in the form of stable, isolable organosols. This method is now used for the production of a wide range of small zero-valent transition metal particles that can be redispersed in organic solvents such as toluene or THF. The key feature of this synthesis is the formation of an organometallic colloidal protecting shell around the particles that can be modified by using modifiers (mono-functional) and also spacer molecules (bifunctional). This modification (Equation 3.4) of the organoaluminum-protecting shell can be used to tailor the dispersion characteristics of the original organosols. A three-dimensional cross-linked nanoparticle network can also be prepared, which finds useful application (Equation 3.4).

$$\qquad\qquad\qquad (3.4)$$

The modifiers are alcohols, carboxylic acids, silanols, sugars, polyalcohols, polyvinylpyrrolidone, surfactants, silica, alumina, etc.

While there are extensive reviews of organosols and the catalysts therefrom in the literature, hydrosols are relatively unknown in spite of the promising electrocatalysts that can emerge from them. Hydrosols of mono-, bi- and multimetallic nanoparticles as isolable precursors for producing supported metal catalyst are an economically beneficial alternative to the traditional wet impregnation of active metal components on carrier surfaces [25].

Organosols with promising catalytic activity are welcome in research, but for economical and ecological reasons, water-based synthesis for the production of catalyst is preferred in practice. Using hydrophilic surfactants as the stabilizer, highly water-soluble mono- and bi- metallic catalyst precursors can be produced in water without appreciable metal precipitations. There are also a few limitations in getting highly soluble colloidal metal nanoparticles in aqueous medium. A few inventions available to date relate to poor metal concentration hydrosols [43,78,79]. Preparation of Lewis base P, N donor-ligand-stabilized hydrosols affect the catalytic performance and are not suitable [43,56]. Moreover, such ligands themselves are prepared by multistep synthetic protocols and thus they are economically nonviable. Auxiliary agents are known to stabilize nanometals in water; however these tensides generally protect the nanometal only at low concentrations, and thus find limited application. A significant breakthrough with regard to water-soluble colloids was achieved by an electrochemical reduction process engineered by Reetz and Helbig [117] using an LiCl salt of the sulfobetaine 3-(dimethyldodecylammonium)propane sulfonate. In one such preparation, a palladium colloid (8 nm) stabilized by sulfobetaine was isolated in solid form. Highly water-soluble hydrosols, particularly those of zero-valent precious metals were synthesized using betaine surfactants instead of NR_4^+ salts (see Equation 3.1). As shown in Equation 3.2, a broad variety of hydrophilic surfactants may be used (see Table 3.1) [25,105,107]. Size- and morphology-selective preparation of metal colloids using tetraalkylammonium carboxylates of the type $NR_4^+R'CO_2^-$ (R = octyl, R' = alkyl, aryl, H) both as the reducing agent and the stabilizer (Equation 3.5) was also reported by Reetz and Maase [42,118,119].

$$M^+ + R_4N^+R'CO_2^- \xrightarrow{\;50-90°C\;} M^0(R_4NR'CO_2)_x + CO + R'\text{-}R' \qquad (3.5)$$

where R = octyl, R' = Alkyl, Aryl, H

The electronic nature of the R' group in the carboxylate had a profound influence on the size of the resulting particles. Electron donors produced small nanoclusters, while electron-withdrawing substituents R', in contrast, yielded larger particles. For example, Pd particles of 2.2 nm size were found when $Pd(NO_3)_2$ was treated with an excess of tetra(n-octyl)ammonium-carboxylate bearing R' = $(CH_3)_3CCO_2^-$ as the substituent. With R' = $Cl_2CHCO_2^-$ (an electron-withdrawing substituent), the particle size was found to be 5.4 nm. The following bimetallic colloids were obtained with tetra(n-octyl) ammonium formate as the reductant: Pd/Pt (2.2 nm), Pd/Sn (4.4 nm), Pd/Au (3.3 nm), Pd/Rh (1.8 nm), Pt/Ru (1.7 nm), and Pd/Cu (2.2 nm). The shape of the particles also depends on the reductant: with tetra(n-octyl) ammonium glycolate reduction of $Pd(NO_3)_2$, a significant amount of trigonal particles in the resulting Pd colloid was detected.

Table 3.1 Various Surfactants Used for the Preparation of Hydrosols

Hydrophilic Type of Tenside	Name	Tenside	Common Name/Formula, Commercial Name
A	Amphiphilic betaines	A1	3-(*N*,*N*-dimethyldodecylammonio) propane sulfonate(SB12)
		A2	Lauryldimethyl carboxymethyl-ammonium betaine, REWO
		A3	Cocoamidopropyl betaine, DEHYTON K,
		A4	Cocoamidopropyl betaine, AMPHOLYT JB130,
B	Cationic tensides	B1	

$$-Cl\ CH_3-\!\!\!\overset{\displaystyle C_{18}H_{37}}{\underset{\displaystyle CH_3}{\overset{|}{\underset{|}{N^+}}}}\!\!\!-CH_2-\overset{\displaystyle OH}{\overset{|}{CH}}-\overset{\displaystyle Cl}{\overset{|}{CH_2}}$$

QUAB 426

| | | B2 | |

$$(R\overset{\displaystyle O}{\overset{\|}{C}}OCH_2CH_2)_n\overset{+}{\underset{\displaystyle CH_3}{\overset{|}{N}}}(CH_2CH_2OH)_{3-n}$$

R = Alkyl radical of partially hydrogenated palm grease ESTERQUAT AU35

C	Fatty alcohol-Polyglycolether	C1	Polyoxyethylene laurylether, BRIJ 35
D	Polyoxyethylene carbohydrate-20 Fatty alkylester	D1	Polyoxyethylene sorbitan monolaurate, TWEEN
E	Anionic tensides	E1	Na-cocoamidoethyl-N-hydroxyethyl glycinate, DEHYTON G
F	Amphiphilic sugar tensides	F1	Alkylpolyglycoside, APG 600

Although this method gives hydrosols in isolable form and redispersible in water in high concentration of metal, a scale-up of this is not possible.

A synthetic alternative to this is the chemical reduction of metal salts in the presence of extremely hydrophilic surfactants have yielded isolable nanometal colloids having at least 100 mg of metal per litre of water [105]. The wide range of surfactants conveniently used to prepare hydrosols with very good redispersibility properties include amphiphilic betaines A1–A4, cationic, anionic, nonionic and even environmentally benign sugar soaps. Table 3.1 presents the list of hydrophilic stabilizers used for the preparation of nanostructured colloidal metal particles, and Table 3.2 shows the wide variety of transition metal mono- and bi-metallic hydrosols formed by this method [105,120].

It is possible to extract the nanocolloids from aqueous solution into an organic phase or to support them onto inorganic supports by what is called the precursor method (described in Section 3.5) to generate heterogeneous catalysts. Such catalysts find application in chemical catalysis, e.g., in selective hydrogenation of fatty acids.

Generating organosols and hydrosols is only part of the complete synthesis of a heterogeneous catalyst. The deposition of these organosols onto a support and then activating them while retaining most of the advantages achieved during their synthesis forms a more important and challenging part of an active catalyst preparation protocol. These preparation details are discussed in Section 3.5.

Table 3.2 Tenside-Stabilized Colloid of Metals of Groups 8–11 of the Periodic Table by Reduction with Hydrogen in H_2O

No.	Metal Salt (g/mmol)	Tenside (g/mmol)	Metal Salt/Tenside Molar Ratio	Reducing Agent	Solvent (mL)	Reaction Conditions			Amount of Product Isolated (g)	Metal Content(%)	Water Solubility in mg Atom Metal/L Water
						T (°C)	T (h)	P (bar)			
1	$RuCl_3 \cdot 3H_2O$ 0.25/0.95	A1 0.96/2.89	1:3	H_2	H_2O 100	60	10	1	0.43	10.98	370
2	$RuCl_3$ 0.16/0.78	A1 0.79/2.34	1:3	H_2	H_2O 100	60	16	50	0.22	9.41	375
3	$RuCl_3 \cdot 3H_2O$ 0.22/0.84	$A2-K_2CO_3$ 0.68/2.53–0.23/1.67	1:3	H_2	H_2O 80	60	3	1	0.38	7.1	160
4	$Ru_2(OAc)_4$ 0.27/0.62	A1 1.25/3.7	1:6	H_2	H_2O 100	60	16	50	0.41	8.78	210
5	$Ru(Acac)_3$ 0.39/0.97	A1 0.98/2.91	1:3	H_2	H_2O 100	60	16	50	0.35	9.37	340
6	$RhCl_3 \cdot 3H_2O$ 0.42/1.6	A1 1.62/4.81	1:3	H_2	H_2O 60	20	2	1	1.02	11.40	330
7	$RhCl_3 \cdot 3H_2O$ 0.40/1.52	A2 1.24/4.57	1:3	H_2	H_2O 100	20	2	1	0.85	10.73	150
8	$PdBr_2$ 1.4/5.3	$A1-Li_2CO_3$ 7.2/21.2–0.4/5.3	1:4	H_2	H_2O 100	20	3	1	9.6	5.3	290
9	$H_2PtCl_6 \cdot 6H_2O$ 2.7/5.3	$A1-Na_2CO_3$ 7.2/21.2–3.4/31.8	1:4	H_2	H_2O 100	20	4	1	12.6	7.1	360
10	$PtCl_2$ 1.4/5.3	$A1-K_2CO_3$ 7.2/21.2–0.73/5.3	1:4	H_2	H_2O 100	20	1.5	1	8.79	9.7	330
11	$PtCl_2$ 1.4/5.3	$A1-K_2CO_3$ 7.2/21.2–0.73/5.3	1:4	H_2	H_2O 100	20	3	1	8.5	10.1	310
12	$PtCl_2$ 1.4/5.3	$A1-Na_2CO_3$ 7.2/21.2–0.56/5.3	1:4	H_2	H_2O 100	20	1.5	1	9.0	10.8	350

13	$PtCl_2$ 21/79.5	A1–Li_2CO_3 108/318–6/79.5	1:4	H_2	H_2O 1500	20	2	1	140	10.3	370
14	$PtCl_2$ 21/79.5	A1–Li_2CO_3 54/159–3/39.75	1:2	H_2	H_2O 1500	20	2	1	73.7	17.9	450
15	$PtCl_2$ 1.4/5.3	A1–Li_2CO_3 7.2/21.2–0.4/5.3	1:4	H_2	H_2O 100	20	3	1	8.4	10.7	380
16	$PtCl_2$ 1.4/5.3	A1–Li_2CO_3 3.6/10.6–0.2/2.65	1:2	H_2	H_2O 100	20	2	1	6.1	16.4	440
17	$PtCl_2$ 1.4/5.3	A1–Li_2CO_3 7.2/21.2–0.4/5.3	1:4	H_2	H_2O 100	20	1.5	1	9.5	10.4	380
18	$PtCl_2$ 1.4/5.3	A1·LiBr 7.2/21.2–0.5/5.3	1:4	H_2	H_2O 100	20	3	1	9.7	9.8	360
19	$PtCl_2$ 1.4/5.3	A1·LiBr 3.6/10.6–0.25/2.65	1:2	H_2	H_2O 100	20	3	1	6.1	16.9	370
20	$PtCl_2$ 1.4/5.3	A2–Li_2CO_3 11.5/42.4–0.8/10.6	1:8	H_2	H_2O 100	20	3	1	14.1	7.1	160
21	$PtCl_2$ 1.4/5.3	A4–Li_2CO_3 14.6/42.4–0.8/10.6	1:8	H_2	H_2O 100	20	3	1	17.3	5.7	110
22	$PtCl_2$ 1.4/5.3	B1–Li_2CO_3 21.7/21.2–0.4/5.3	1:4	H_2	H_2O 100	20	3	1	25.1	4.1	105
23	$PtCl_2$ 1.4/5.3	D1–Li_2CO_3 7/–0.4/5.3	—	H_2	H_2O 100	20	4	1	9.5	10.6	130
24	$PtCl_2$ 1.4/5.3	C1 –Li_2CO_3 7/–0.4/5.3	—	H_2	H_2O 100	20	2	1	9.7	10.6	145
25	$PtCl_2$ 1.4/5.3	E1–Li_2CO_3 8.3/21.2–0.4/5.3	1:4	H_2	H_2O 100	20	4	1	11.5	8.9	110
26	$PtCl_2$ 1.4/5.3	F1–Li_2CO_3 7.2/–0.4/5.3	—	H_2	H_2O 200	20	3	1	13.7	6.8	110
27	$PtCl_2$ 1.4/5.3	B2–Li_2CO_3 7.2/21.2–0.4/5.3	1:4	H_2	H_2O 100	20	3	1	10.8	9.1	105

3.5 APPLICATIONS IN CATALYSIS

3.5.1 Quasi-Homogeneous Reactions

Lipophilic or hydrophilic nanostructured metal colloids dissolved in the form of organosols or hydrosols can serve as catalysts either in organic solution or in the aqueous phase. Schmid [121] has called these quasi-homogeneous catalytic reactions, which include solvated metal atom dispersions (SMAD) [122], heterogeneous catalysis in solution. Rhodium hydrosols dissolved in water were proven to be effective as hydrogenation catalysts in two-phase systems with the olefin in the organic phase [123]. Solutions of Moiseev's giant Pd colloids [69–74] were shown to catalyze a number of reactions in the quasi-homogeneous phase, namely oxidative acetoxylation reactions [69], oxidative carbonylation of phenol to diphenyl carbonate [74], hydrogen-transfer reduction of multiple bonds by formic acid [124], reduction of nitriles and nitroarenes, and acetal formation [125] with high turnover frequencies and significant lifetimes [29]. Giant Pd clusters protected by anionic ligands were prepared and used for catalyzing oxidative acetoxylation of toluene in the presence of molecular oxygen [125]. Finke has reported remarkable catalytic lifetimes for the polyoxoanion- and tetrabutylammonium-stabilized transition metal nanoclusters [69–74]. For example, in the catalytic hydrogenation of cyclohexene, a common test for structure-insensitive reactions, the Ir(0) nanocluster [35] showed up to 18,000 total turnovers with turnover frequencies of 3,200 h^{-1} [126]. As many as 190,000 turnovers were reported in the case of the Rh(0) analog reported recently [127]. These results are quite unprecedented. Obviously, the polyoxoanion component prevents the precious metal nanoparticles from aggregating and hence the active metals exhibit a high surface area [127]. A review [25] compares this new approach with the catalytic properties of other nanometallic systems active in the quasi-homogeneous phase. The formation of olefins from aldehydes and ketones via McMurry-type coupling reactions was reported when nanostructured Ti colloids (3 nm) stabilized by Bu_4NBr were used [128]. THF-protected Ti_{13}-nanoclusters [129] were found to hydrogenate Ti and Zr sponges in the quasi-homogeneous phase [130,131]. The enantioselective hydrogenation of ethylpyruvate in HOAc/MeOH solution was performed using cinchonidine-stabilized Pt colloids. Heck- and Suzuki C-C –bond-coupling reactions were catalyzed with $NR_4^+X^-$ -stabilized Pd and Pd/Ni colloids in dimethylacetamide. The same reactions have been observed when solvent-stabilized Pd particles in propylene carbonate were applied as the catalyst [133,134]. Nanosized Pd colloids generated *in situ* by reduction of Pd(II) to Pd(0) are involved in the catalysis of phosphine-free Heck- and Suzuki reactions [135]. Micelles in which nanosized metals are incorporated are very stable; they undergo no significant change of colloidal properties such as size and polydispersity and thus are effective hydrogenation catalysts. Depending on the strength of the reducing agent, the morphology of the metal core can be varied between a cherry- and a raspberry-type architecture. Block copolymer stabilized Pd-raspberry-colloids have an extraordinary high metal surface and no additional support is needed for catalytic applications in solution. This type of colloid catalyst combines the advantages of homogeneous and heterogeneous catalysis, i.e., the high selectivity and reactivity of homogeneous hydrogenation catalysts coupled with long-term stability of the heterogeneous systems. The scope of transition metal nanoparticles in catalysis has been recently summarized by Schmid and Finke [59,136,137].

3.5.2 Heterogeneous Reactions

3.5.2.1 Precursor Concept

Heterogeneous catalysts are readily obtained when pre-prepared nanometal colloids are deposited on supports [20]. The so-called precursor concept for manufacture of heterogeneous metal colloid catalysts was developed on this basis in the 1990s [24, 25, 27, 105–108]. The catalyst precursor is manufactured by dipping the supports into organic or aqueous media containing the dispersed precursor at ambient temperature to adsorb the pre-prepared particles. This has been demonstrated for supports such as charcoal, various oxidic support materials, and even low-surface materials such as quartz, sapphire, and highly oriented pyrolitic graphite (HOPG). The feasibility of preparing catalysts in this way for industrial purposes has been demonstrated by Degussa.

Using this method, homogeneous alloys, segregated alloys, layered bi-metallics, and decorated particles are all readily accessible. An obvious advantage of the precursor concept over the conventional salt-impregnation method is that both the size and the composition of the colloidal metal precursors may be tailored independent of the support. Further, the metal particle surface may be modified by lipophilic or hydrophilic protective shells and coated by intermediate layers, e.g., of oxide. The modification of the precursor by dopants is also possible.

A combination of AFM, STM, and XPS [138,139] has revealed the interaction of platinum hydrosols with oxide (sapphire, quartz) and graphite single-crystal substrates. The metal core is immediately adsorbed onto the support surface when dipped into aqueous Pt colloid solutions at 20°C. The carpet-like coat formed over the particles by the protecting shell and the support surface cannot be removed from the particle surface even by intense washing with solvent. The organic-protecting shell decomposes on annealing at 280°C and above in UHV. The thermal degradation can be monitored by XPS up to 800°C and by STM. It was shown that the Pt particles remain virtually unchanged at up to ca. 800°C. Sintering processes are observed only above this temperature.

3.5.2.2 Conditioning: A Key Step in Generating Active Catalysts

After depositing the colloidal precursor onto the support, when the protecting shell (especially for organosols) cannot be washed without disturbing the metal dispersion on the surface of the support, the colloidal catalyst precursors are heat treated in order to strip off the protecting shell. The process, is called conditioning or reactive annealing [140] and involves heating the colloidal precursor catalyst at a desired temperature using different gases in order to strip off the organic-protecting shell from the colloidal precursor, eventually producing catalysts with clean surface metal particles on the support. Figure 3.6 depicts details of the precursor concept and the conditioning procedure for a tetraoctylammonium-stabilized catalyst precursor. Figure 3.7, in addition, brings out the salient features of a combination of precursor concept and conditioning in generating uniformly distributed

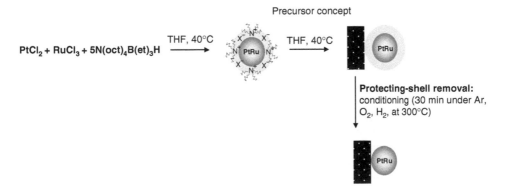

Figure 3.6 Schematic representation of the supported catalyst preparation from its precursor using precursor concept and conditioning. (Adapted from Bönnemann H. et al., *Fuel Cells*, 4, 1, 2004. With permission from Wiley VCH.)

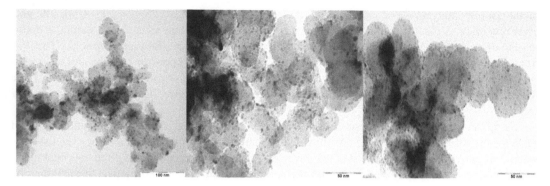

Figure 3.7   TEM micrographs indicating the size and uniform distribution of colloidal nanoparticles on Vulcan XC 72 using $LiBet_3H$, $N(oct)_4Bet_3H$ and aluminumorganic synthetic pathways.

metal nanoparticles on the surface having advantageous catalytic properties. Three different reducing agents, $LiBet_3H$, $N(oct)_4Bet_3H$, and $Al(me)_3$, were used to generate colloidal nanoparticles of different particle sizes and their respective distribution on the support has been shown. It can be observed here that irrespective of the reduction pathway, a uniform distribution of the metal particles on the support surface is seen, though particle size differences are seen owing to the differences in the strength of the reducing agents used and also on the mode of stabilization.

During conditioning at elevated temperatures, some level of sintering of the metal particles occurs, but most of the advantages of the synthetic steps are still retained. In particular, for the tetraoctylammonium-protected catalyst precursors, the three mandatory steps during conditioning involve heating the catalyst precursor in a flow of argon whereby most of the protecting shell is removed; this is then followed by a 3.5% (v/v) oxygen/argon step that ensures oxidative removal of the protective shell and also causes limited surface oxidation of the metal nanoparticles. Therefore, a final step of treating the catalyst with hydrogen is performed to ensure clean and completely reduced metal surfaces. For supported tetraoctylammonium-stabilized nanoparticles, the optimum temperature of 300°C was fixed using results from TGS-MS analysis [140].

3.5.2.3   Heterogeneous Catalysts in Catalysis

$N(octyl)_4Br$-stabilized Pd colloids (typical size 3 nm) have been used as precursors by Reetz and coworkers [141] to generate the so-called cortex-catalyst, where the active metal forms an extremely fine shell of less than 10 nm on the supports (e.g., Al_2O_3). Within the first 1–4 sec, the impregnation of Al_2O_3 pellets by dispersed nanostructured metal colloids leads to the time-dependent penetration of the support, which is complete after 10 sec. As a result, the so-called egg-shell catalysts were obtained, which contain the colloidal metal particles as a thin layer (<250 nm) on the surface of the support [Figure 3.6]. These catalysts exhibited threefold higher activity in olefin hydrogenation than the conventionally prepared catalysts for the same metal loading (5% Pd on Al_2O_3). Schmid [57] has described phosphine-stabilized Rh_{55} nanoclusters, namely $Rh_{55}[P(t-Bu)_3]_{12}Cl_{20}$ and $Rh_{55}[PPh_3]_{12}Cl_{20}$, deposited on TiO_2. These systems were used to catalyze the heterogeneous hydroformylation of propene with high turnover numbers giving equal amounts of *n*- and *i*-butanal. Even though observed turnover frequencies were higher than those for homogeneous complex catalysts, the selectivity was too low for practical applications. The activity of surfactant-stabilized colloidal rhodium (5 wt-% on charcoal) was found to surpass that of conventional salt-impregnation catalysts for butyronitrile hydrogenation test with same metal loading. Similarly, addition of 0.2% of colloidal Ti(0) to the supported noble metal resulted in a significant enhancement in activity [25] (see Figure 3.8).

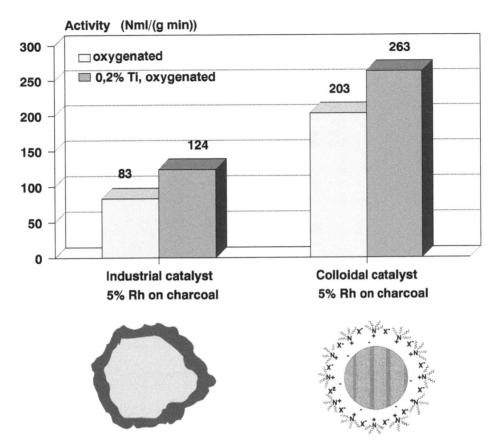

Figure 3.8 Activity of conventional and colloidal Rh/C catalysts, eventually doped with 2% colloidal Ti(0), compared in the butyronitrile hydrogenation test (2d). (Adapted from Bönnemann, H. and Brijoux, W., in *Surfactant-Stabilized Nanosized Colloidal Metals and Alloys as Catalyst Precursors/Advanced Catalysts and Nanostructured Materials*, Moser, W., Ed., Academic Press, San Diego, 1996, pp. 165–196, Chap. 7. With permission from Elsevier Science.)

The addition of dopants is found to have beneficial effects. However, they are not restricted only to transition metals. The hydrogenation of acrylic acid can be promoted significantly by the addition of neodymium ions onto the palladium particles [142]. The selective transformation of 3,4-dichloronitrobenzene to the corresponding aniline has been selected to test pre-prepared Pt hydrosols as heterogeneous catalyst precursors (see Figure 3.9) [143].

The performance of the colloidal catalyst system was evaluated and compared with conventional Pt/C systems in batch and continuous tests. Further, fine-tuning of the process was performed by using the synergistic effect of bimetallic precursors (e.g., Pt/Cu). In summary, the potential of the colloidal heterogeneous catalyst stems from the possibility of tailoring the properties for specific applications by the controlled addition of special dopants or poisons (such as sulfur). The influence of metal ions on the hydrogenation of o-chloronitrobenzene over platinum colloids, and the effect of metal complexes on the catalytic performance of metal clusters have been investigated [144,145]. Furthermore, the doping effect of additional metal on the activity and selectivity of platinum clusters in the homogeneous liquid-phase hydrogenation of cinnamaldehyde was studied [145].

Copper-catalyzed Suzuki cross-coupling reactions using mixed nanocluster catalysts have been studied recently. Copper-based catalysts were shown to be effective as reagents that can present an inexpensive and environmentally friendly alternative to noble metal catalysts. In the hydrogenation of cinnamic acid to corresponding alcohol, the selectivity can be varied by doping Sn with Rh colloid catalysts. A selectivity of 86% was achieved using a colloidal Rh/Sn (Rh/Sn = 1.5:1) catalyst on

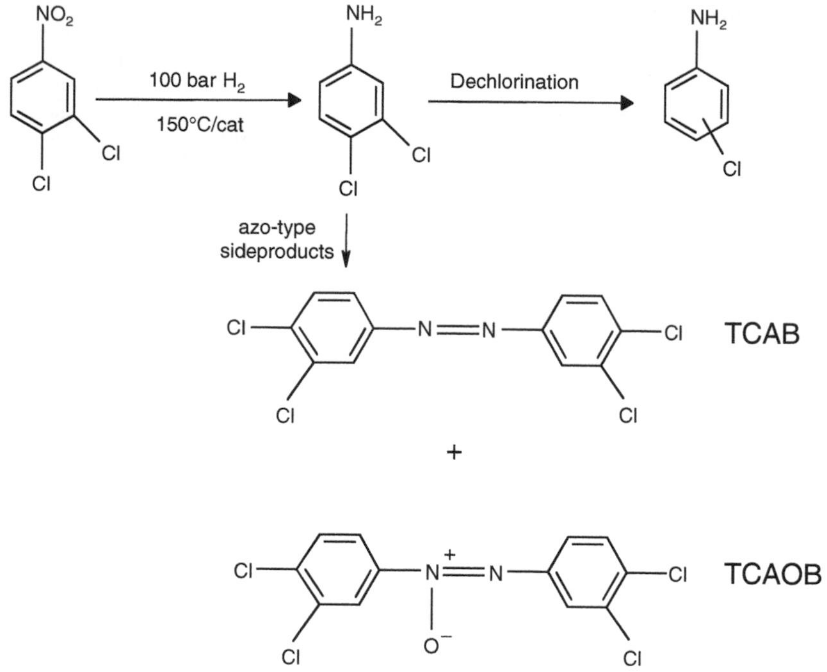

Figure 3.9 Reaction scheme for the hydrogenation of 3,4-dichloronitrobenzene.

carbon [105]. Remarkably, the stabilizing surfactant was shown to modify the colloidal metal surface and hence the catalytic properties [143]. Bulk industrial processes often rely on alloy-like bimetallic catalysts. The mutual influence of two different metals on the catalytic property have opened the possibility of differential studies for nanostructured bimetallics. The controlled co-reduction of two different metal ions has made bimetallic colloids [24,25,78–80,105] readily accessible, while structural characterization and some catalytic aspects of bimetallic colloids have been reviewed [28]. The homogeneous structure of bimetallic particles can be altered to colloidal particles having composition gradients from the core to the shell by the successive reduction of mixed metal ions [146–149]. Bimetallic particles having a gradient metal distribution or a layered structure are most interesting for catalytic applications. In the catalytic hydrogenation of crotonic acid to butanoic acid, a clear synergistic effect of Pt and Rh [24] was observed when bimetallic colloidal precursors that have a gradient core/shell structure were used ($Pt_{20}Rh_{80}$), with an increase in the Rh concentration from the core toward the surface of the particle [146] (see Figure 3.10). Similar effects in the partial hydrogenation of 1,3-cyclooctadiene with $Pt_{80}Pd_{20}$ and $Pd_{80}Au_{20}$ colloid catalysts have been discussed by Toshima [28].

A novel electronic structure evolves because of the electronegativity difference between Pd and Au together with the combination of the partially filled d band of Pd and the completely filled d band of Au. The sequential reduction of gold salts and palladium salts with sodium citrate allows the gold core to be coated with Pd. The layered bimetallic colloid is stabilized by trisulfonated triphenylphosphane and sodium sulfanilate (20–56 nm) to yield isolable (more than 90% metal) particles. Redispersion in water is possible at high concentration. Au/Pd and Pd/Au systems on a TiO_2 support have been used as heterogeneous catalysts for the hydrogenation of hex-2-yne to *cis*-hex-2-ene. In comparison to pure metal catalysts, both palladium-plated gold seeds and gold-plated palladium particles showed considerably increased activities. Herein, the protecting ligand shell enhances the lifetime of the ligand-stabilized colloid catalysts considerably. The influence of the electronegativity difference between Au and Pd on the activity and selectivity of the hydrosilylation reaction [150] on colloidal Pt surfaces [92,93] has been carefully investigated. The results

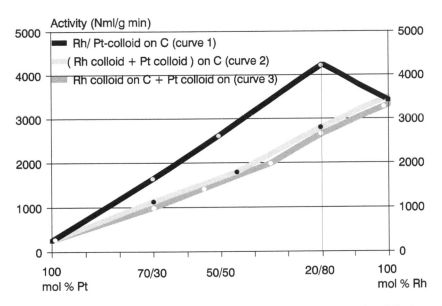

Figure 3.10 Activity plot of alloyed Rh/Pt/C and mixed Rh 1Pt /C catalysts in the crotonic acid hydrogentation test (H_2 consumption). (Adapted from Bönnemann, H. and Brijoux, W., in *Surfactant-Stabilized Nanosized Colloidal Metals and Alloys as Catalyst Precursors/Advanced Catalysts and Nanostructured Materials*, Moser, W., Ed., Academic Press, San Diego, 1996, pp. 165–196, Chap. 7. With permission from Elsevier Science.)

demonstrate that by a defined variation of the core/shell composition in bimetallic precursors, the activity, selectivity, and the lifetime of metal colloid catalysts may be optimized.

Stabilizing agent residues present at the surface of the colloidal metal precursors (1 and 5 wt% metal loading on charcoal, glassy carbon, zeolites, TiO_2, Al_2O_3, $CaCO_3$, SiO_2, single-crystal oxides or HOPG) may act as an effective catalyst modifier for controlling the selectivity and durability of heterogeneous colloid catalysts. Ligands such as phosphines or phenanthrolines [61], surfactants of various types [25,105], or organic "envelopes" such as polymers [151,152] were found to control the regio- and enantioselectivity of heterogeneous metal colloid catalysts. In the case of partial hydrogenation of hex-2-yne to *cis*-hex-2-ene, it is found that even subtle changes in the chain length of the substituents of the alkyl-substituted phenanthrolines used as colloid stabilizers alter the regioselectivity significantly [153]. Seven- and eight-shell palladium clusters on TiO_2, protected by phenanthroline catalyze the semi-hydrogenation of hex-2-yne to *cis*-hex-2-ene with 93% selectivity. A similar selectivity resulted with the 3-*n*-decyl phenanthroline as the stabilizer. On substitution of phenanthroline with *n*-butyl- or *n*-heptyl groups however, the activity drops dramatically and the consecutive isomerization or total hydrogenation of the *cis*-hex-2-ene is completely suppressed. Geometric, i.e., steric factors were proposed to explain the strong influence on the selectivity, which is illustrated by further examples. A surfactant was found to control the selectivity in the synthesis of leaf alcohol, a valuable fragrance, by *cis*-selective partial hydrogenation of 3-hexyn-1-ol (Equation 3.6) [154].

$$\text{HO}\diagup\!\!\!\equiv\!\!\!\diagdown \xrightarrow[\text{Pd}]{H_2} \text{HO}\diagdown\!\!\diagup\!\!=\!\!\diagup\diagdown + \left(\text{HO}\diagdown\!\!\diagup\!\!=\!\!\diagdown\diagup + \text{HO}\diagdown\!\!\diagup\!\!\diagdown\!\!\diagup\right) \qquad (3.6)$$

The performance in this reaction (Equation 3.6) of heterogeneous Pd colloid catalysts on $CaCO_3$ modified by a number of surfactants was compared with conventional Pd/C and Lindlar catalysts under optimized reaction conditions. The selectivity was found to depend on the support and various promoters with the highest activity and the best selectivity (98.1%) toward the desired *cis*-3-hexen-1-ol found when employing a lead-acetate-promoted palladium colloid on CaCO$_3$ modified by the zwitterionic surfactant sulfobetaine-12 (*N*,N-dimethyl-dodecylammoniopropane-sulfonate). Chemisorption measurements have shown that residual amounts of the surfactant are still present on the surface of the immobilized particles. This colloid catalyst was twice as active as a conventional Lindlar catalyst and surpassed its selectivity by 0.5%. The contact of heterogenized metal colloid surfaces with substrates in aqueous media is improved when hydrophilic protecting shells are present. For example, a hydrophilic ruthenium nanocluster catalyst on lanthanum oxide was shown to convert benzene into cyclohexene with 59% selectivity at 50% benzene conversion when suspended in an aqueous solution of sodium hydroxide [155]. Chiral molecules on the surface of the metal colloid can induce enantioselectivity control. Following this concept, a new type of enantioselective platinum sol catalyst stabilized by the alkaloid dihydrocinchonidine was designed (see Figure 3.11) [156,157].

The colloidal catalysts have been prepared in different particle sizes by the reduction of platinum tetrachloride with formic acid in the presence of different amounts of alkaloid. Optical yields of 75–80% *ee* were obtained in the hydrogenation of ethyl pyruvate with chirally modified Pt sols (Equation 3.7). The catalysts were demonstrated to be structure-insensitive since turnover frequencies (ca. 1 sec^{-1}) and enantiomeric excess are independent of the particle size.

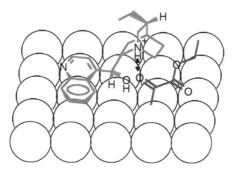

$$ (3.7) $$

In order to evaluate the catalytic characteristics of colloidal platinum, a comparison of the efficiency of Pt nanoparticles in the quasi-homogeneous reaction shown in Equation 3.7, with that of supported colloids of the same charge and of a conventional heterogeneous platinum catalyst was performed. The quasi-homogeneous colloidal system surpassed the conventional catalyst in turnover frequency by a factor of 3 [157]. Enantioselectivity of the reaction (Equation 3.7) in the presence of polyvinyl-pyrrolidone as stabilizer has been studied by Bradley et al. [158,159], who observed that the presence of HCl in as-prepared cinchona alkaloids modified Pt sols had a marked effect on the rate and reproducibility [158]. Removal of HCl by dialysis improved the performance of the catalysts in both rate and reproducibility. These purified colloidal catalysts can serve as reliable

Figure 3.11 Enantioselective hydrogenation of ethyl pyruvate using dihydrocinchonidine stabilized Pt colloids. (Adapted from Bönnemann, H. and Brijoux, W., in *Surfactant-Stabilized Nanosized Colloidal Metals and Alloys as Catalyst Precursors/Advanced Catalysts and Nanostructured Materials*, Moser, W., Ed., Academic Press, San Diego, 1996, pp. 165–196, Chap. 7. With permission from Elsevier Science.)

test systems to screen alternative chiral modifiers to cinchona alkaloids and for precise rate studies. In addition, a systematic variation of the reaction parameters relevant to the catalyst performance is possible without unwanted side effects. It was shown that an excess of polyvinylpyrrolidone present on the colloidal Pt catalyst hinders the access of modifier molecules to the colloidal metal surface. It is most likely that the polymer is adsorbed at the metal surface, reducing the number of modified surface sites available for the enantioselective hydrogenation [159]. Nanostructured metal colloids can be used for the stereoselective and enantioselective transformation of specific prochiral substrates into valuable fine chemicals. The application of a rational ligand control based on molecular modeling, which has been successful in homogeneous metal complex catalysis, promises similar results in heterogeneous metal colloid catalysis.

On the basis of the chemisorption results, it is reasonable to believe that coating on catalytically active nanometal particles, which is permeable for small molecules such as H_2 or O_2, effectively prevents the contact of the metal surface with poisons and enhances the lifetime. In fact, it is evidenced that the lifetime of colloid catalysts is considerably longer than that of conventional precipitation catalysts. The activity of a conventional Pd/C catalyst for example, expires completely in the hydrogenation of cyclooctene to cyclooctane after 38×10^3 catalytic cycles per Pd atom, but the Pd colloid/C catalyst still shows a residual activity after 96×10^3 catalytic turnovers (Figure 3.12) [160].

Superior catalytic oxidation catalysts were obtained when surfactant-stabilized Pd–Pt-precursors were supported on charcoal and promoted by bismuth. By comparison with industrial heterogeneous Pd/Pt catalysts, charcoal-supported Pd_{88}/Pt_{12}-$(Oct)_4NCl$ alloy particles (1.5 to 3 nm) show an excellent activity combined with high selectivity in the glucose oxidation to gluconic acid by molecular oxygen (see Equation 3.8) [161].

$$\tag{3.8}$$

D-(+)-glucose

D-gluconic acid
Na-salt

Greater durability of the colloidal Pd/C catalysts was also observed in this case. The catalytic activity was found to have declined much less than a conventionally manufactured Pd/C catalyst after recycling both catalysts 25 times under similar conditions. Obviously, the lipophilic $(Oct)_4NCl$ surfactant layer prevents the colloid particles from coagulating and being poisoned in the alkaline aqueous reaction medium. Shape-selective hydrocarbon oxidation catalysts have been described, where active Pt colloid particles are present exclusively in the pores of ultramicroscopic tungsten heteropoly compounds [162]. Phosphine-free Suzuki and Heck reactions involving iodo-, bromo- or activated chloroatoms were performed catalytically with ammonium salt- or poly(vinylpyrrolidone)-stabilized palladium or palladium nickel colloids (Equation 3.9) [162, 163].

$$\tag{3.9}$$

(E)-stilbene was the exclusive product of the Pd colloid-catalyzed Heck arylation of styrene with chlorobenzene. Recently, a polymer-mediated self-assembly of functionalized Pd and SiO_2 nanoparticles have been found to be highly active catalysts for hydrogenation and Heck coupling

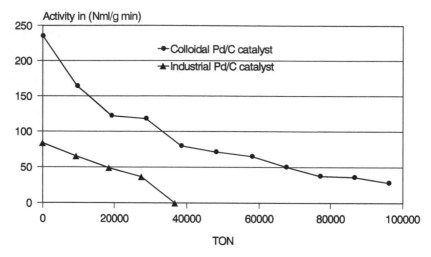

Figure 3.12 Durability of colloidal Pd/C catalysts in the cyclooctene hydrogenation test compared with a conventional Pd/C system. (Adapted from Bönnemann, H. and Brijoux, W., in *Surfactant-Stabilized Nanosized Colloidal Metals and Alloys as Catalyst Precursors/Advanced Catalysts and Nanostructured Materials*, Moser, W., Ed., Academic Press, San Diego, 1996, pp. 165–196, Chap. 7. With permission from Elsevier Science.)

reactions. This type of C–C linkage is valuable for the production of pharmaceutical intermediates and fine chemicals. Turnover frequencies (TOF) of more than 80,000 have been achieved in the coupling of *p*-bromobenzaldehyde with butyl acrylate using poly(vinylpyrrolidone)-stabilized palladium as the catalyst [164]. Reetz [165,166] has applied colloidal, water-soluble PtO_2 (i.e., a colloidal Adams catalyst) in the immobilized form for the reductive amination of benzaldehyde by *n*-propylamine (Equation 3.10).

$$\text{/\textbackslash/NH}_2 \quad + \quad \text{PhCHO} \quad \xrightarrow[\text{cat.}]{\text{H}_2} \quad \text{/\textbackslash/N}\text{/\textbackslash Ph} \qquad (3.10)$$

The selectivity in favor of the desired monobenzylated product was found to be >99% and the immobilized PtO_2 was found to be 4–5 times more active than the commercial Adams catalysts. In solution or in immobilized form, the PtO_2 colloid is effective in the hydrogenation of carbonyl compounds or of olefins. Recently, the heterogeneous catalytic amination of aryl bromides by immobilized Pd(0) particles has been reported [163]. Secondary amines such as piperidine and diethyl amine are used in the amination of aryl bromides and the reaction proceeds with good turnover numbers and regio-control. The catalysts can be reused repeatedly without loss of activity or selectivity after filtration from the reaction mixture.

Noble metal nanoparticles have been used as catalysts for olefin hydrogenation in C–C coupling reactions including Heck and Suzuki coupling reactions. Co nanoparticles are also being used as heterogenous Pauson–Khand catalysts. Fabrication of hollow spheres composed of palladium nanoparticles and their application to heterogenous Suzuki reactions has been recently performed [167–169].

A new class of heterogeneous catalyst has emerged from the incorporation of mono- and bimetallic nanocolloids in the mesopores of MCM-41 or via the entrapment of pre-prepared colloidal metal in sol-gel materials [170–172]. Noble metal nanoparticles containing Mex-MCM-41 were synthesized using surfactant stabilized palladium, iridium, and rhodium nanoparticles in the synthesis gel. The materials were characterized by a number of physical methods, showed that the nanoparticles were present inside the pores of MCM-41. They were found to be active catalysts in the hydrogenation of cyclic olefins such as cyclohexene, cyclooctene, cyclododecene, and norbornene [173]. Pre-prepared Pd colloids stabilized via phenanthroline by the Schmid method [49–63] have successfully been embedded in MCM-41 mesopores (some are deposited on the external surface) to give CO oxidation catalysts [170].

A strategy for the entrapment of monodisperse nanometal colloids, e.g., NR_4X-stabilized Pd particles, has been developed by Reetz and Dugal [171]. The nanosized particles (2 to 3.5nm) were pre-prepared by electrochemical synthesis [171]. The entrapment in hydrophobic sol-gel materials occurs via the fluoride-catalyzed hydrolysis of mixtures of $CH_3Si(OCH_3)$ and $Mg(OC_2H_5)_2$. Finally, the stabilizer is extracted and the "naked" nanometallic particles are left trapped in the pores. Ultramicrotomic techniques in combination with shadow casting TEM has confirmed that the Pd particles are embedded in the oxide matrix and not fixed on the outer surface. Reetz and Dugal [171] have investigated the catalytic properties of the new material. By adjusting the relative amounts of the $CH_3Si(OCH_3)_3$ and $Mg(OC_2H_5)_2$ in the gel they were able to control the hydrophobicity of the matrix. The catalytic activity and selectivity were tested by the partial hydrogenation of COD to give cyclooctene: it was found that the activity of the new systems is governed by the specific pore volume and the hydrophobic properties of the wall. The selectivity for cyclooctene was 93%. The most active of these sol-gel-entrapped Pd catalysts had a considerably higher activity than the commercially available Pd/Al_2O_3 samples. The preparation of a fully alloyed Pd–Au colloid of 3.0 nm particle size, by a modified sol-gel procedure using THF as the solvent, from the co-reduction of Pd and Au salts with tetraalkyl-ammonium-triethylhydroborate [24] and its embedding in a silica has been described [172]. The integrity of the incorporated Pd–Au alloy particles remained virtually untouched. After the removal of the protecting surfactant, a mesoporous texture with a comparatively narrow pore distribution remained. According to the physical characterization by a combination of techniques, the SiO_2-embedded Pd-Au colloid preserves the size and the structural characteristics of the colloidal metal precursor. Tetraalkyl-substituted Ru in either silica or zirconia sol-gel matrix has been used for hydrogenolysis [174,575]. In a similar fashion, colloidal palladium supported on silica gel has been conclusively proved to be involved in the direct reaction of H_2 and O_2 to give H_2O_2 [175]. The material exhibits excellent catalytic properties in selective hydrogenation test reactions. These few examples show that colloidal bimetallic precursors—embedded in matrices or deposited on various supports, and promoted by additives if necessary—may become of practical importance in heterogeneous catalysis. A recent review describes various synthetic strategies applied for the synthesis of metal nanoparticles, which can be used as reusable catalysts with almost all kinds of reactions studied so far. They include hydrosilylation, oxidation, C–C coupling reactions such as carbonylation, Heck, and Suzuki reactions, hydrogenation, chemoselective, regioselective, and stereo and enantioselective reactions [176].

3.5.2.4 Fuel Cell Catalysts

"Fuel cell technology" allows the direct conversion of chemical energy into electricity [177]. The fuel cell is an electrochemical reactor where the catalyst systems are an important component. Among the wide-ranging applications, fuel cells are best suited for low-emission transport systems, stationary power stations, and combined heat and power sources. The classical studies were carried out in the early 1900s and major innovations and improvements have been achieved over the last few years. The first new electric cars are expected to roll on the markets around the year 2010, but further developments are still needed, notably in the catalyst sector. Hydrogen fuel cell catalysts rely on pure Pt, whereas Pt-alloy-electrocatalysts are applied for the conversion of reformer gas or methanol into electricity. The active components in the latter cases are small Pt-containing bi- or trimetallic particles of 1–3 nm size. These systems offer improved efficiency and tolerance against certain contaminants, especially CO in the anode feed [178–183]. It was clear from patents filed in the early 1970s that finely particulated colloidal platinum sols should be the ideal precursors for the manufacture of fuel cell electrodes [178].

This survey focuses on recent developments in catalysts for phosphoric acid fuel cells (PAFC), proton-exchange membrane fuel cells (PEMFC), and the direct methanol fuel cell (DMFC). In PAFC, operating at 160–220°C, orthophosphoric acid is used as the electrolyte, the anode catalyst is Pt and the cathode can be a trimetallic system like Pt/Cr/Co. For this purpose, a trimetallic colloidal precursor of the composition $Pt_{50}Co_{30}Cr_{20}$ (size 3.8 nm) was prepared by the co-reduction of the corresponding metal salts [184–186]. From XRD analysis, the trimetallic particles were found alloyed in an ordered fct-structure. The electrocatalytic performance in a standard half-cell was compared with an industrial standard catalyst (trimetallic crystallites of 5.7 nm size) manufactured by co-precipitation and subsequent annealing to 900°C. The advantage of the trimetallic colloid catalysts lies in its improved durability, which is essential for PAFC applications. After 22 h it was found that the potential had decayed by less than 10 mV [187].

PEM fuel cells use a solid proton-conducting polymer as the electrolyte at 50–125°C. The cathode catalyst is based on Pt alone but because of the required tolerance to CO, a combination of Pt and Ru is preferred for the anode [178–183]. Colloidal Pt/Ru catalysts have been extensively studied for low-temperature (80°C) polymer membrane fuel cells [187,188]. These have also been proposed for use in direct methanol fuel cells (DMFC) or in PEMFC, which are fed with CO-contaminated hydrogen produced in on-board methanol reformers. The ultimate dispersion state of the metals is essential for CO-tolerant PEMFC, and truly alloyed Pt/Ru colloid particles of less than 2 nm size seem to fulfill these requirements [189,190]. Alternatively, bimetallic Pt/Ru PEM catalysts have been developed for the same purpose, where non-alloyed Pt nanoparticles <2 nm and Ru particles <1 nm are dispersed on the carbon support [180]. From the results it can be concluded that a Pt/Ru interface is essential for the CO tolerance of the catalyst regardless of whether the precious metals are alloyed. On the other hand, for DMFC applications, Pt/Ru nanopowders of 3–5 nm size or thin films serve better as catalysts [181–183]. For electrocatalytic methanol oxidation, a Pt metal or a Pt metal alloy catalyst has been developed where an Ru phthalocyanine complex is added as a dopant to reinforce the catalytic effect substantially [186]. For comparison, the electrocatalytic activity was compared with that of a bimetallic Pt_{50}/Ru_{50}-N(Oct$_4$)Cl colloid prepared by the salt co-reduction method [24,107] toward oxidation of CO and a CO/H_2 gas mixture (simulated reformer gas) [189]. According to high-resolution transmission electron microscopy (HRTEM), the mean particle diameter was 1.7 nm. The alloyed state of the particles was verified by point-resolved energy dispersive x-ray (EDX) analysis. CO tolerance and direct methanol oxidation on Pt_3Mo/C have been investigated in PEM fuel cell conditions. This catalyst has shown threefold enhancement for CO tolerance, while for methanol oxidation such enhancements were not observed. It is also emphasized that the mechanism involved in CO oxidation is different from the conventional Pt–Ru catalysts. While ruthenium is converted into its oxide that helps in the oxidation of CO, Mo oxide surface shows no affinity for CO providing a pathway of easy oxidation at lower overpotentials [192].

Glassy carbon-supported Pt_{50}/Ru_{50}-N(Oct$_4$)Cl colloids were examined by CO-stripping voltammetry, and the data were found to be essentially identical with those found in well-characterized bulk alloy electrodes. In a rotating disk electrode, the activity of the colloid toward the continuous oxidation of 2% CO in H_2 was determined at 25°C in 0.5 M H_2SO_4. The results obtained led to the conclusion that these Pt/Ru colloids are very suitable precursors for high-surface-area fuel cell catalysts [190]. Structural information on the precursor was obtained by *in situ* XRD via Debye function analysis. In addition, the XANES data support the bimetallic character of the particles. *In situ* XRD has revealed the catalytic function of the alloyed Ru in the CO oxidation: surface oxide species are formed on the Ru surface at 280°C, which slowly coalesce to RuO_2 particles. After re-reduction, the catalyst shows a pure hcp ruthenium phase and larger platinum-enriched alloy particles [193]. Scanning probe microscopy (SPM) has been applied in order to characterize the real-space morphology of the electrode surfaces of supported nanostructured metal colloids on the nanometer scale[194]. Colloidal Pt_{50}/Ru_{50} precursors (<2nm) raise the tolerance to CO, allowing higher CO concentrations in the H_2 feed of a PEMFC without a significant drop in performance [190]. A selective Pt/Mo oxidation catalyst for the oxidation of H_2 in the presence of CO in fuel cells comprises Pt_xMo_y particles where x is 0.5–0.9 and

y is 0.5–0.1 [195]. The colloid method was found to be a highly suitable exploratory approach to finding improved formulations for binary and ternary anode electrocatalysts. The metals used include Pt, Ru, W, Mo, and Sn. As an alternative to the reductive metal colloid synthesis, the so-called metal oxide concept was developed, which allows the fabrication of binary and ternary colloidal metal oxides as electrocatalyst precursors (Equation 3.11) [165,166].

$$PtCl_4 + RuCl_3 \xrightarrow[\text{stabilizer}]{\text{H}_2\text{O/base}} PtRuO_x \text{colloid} \tag{3.11}$$

There are many recent reports of Pt–Ru being prepared by different methods and supported on different carbon supports to check their catalytic activity [196,197]. Colloidal Pt/RuO$_x$(1.5 $\pm$ 0.4 nm) stabilized by a surfactant was prepared by co-hydrolysis of PtCl$_4$ and RuCl$_3$ under basic conditions. The Pt/Ru ratio in the colloids can be varied between 1:4 and 4:1 by variation of the stochiometry of the transition metal salts. The corresponding zero-valent metal colloids are obtained by the subsequent application of H$_2$ to the colloidal Pt/Ru oxides (optionally in the immobilized form). Additional metals have been included in the metal oxide concept (Equation 3.11) in order to prepare binary and ternary mixed metal oxides in the colloidal form. Pt/Ru/WO$_x$ is regarded as a good precatalyst especially for the application in DMFCs. Main group elements such as Al have been included in multimetallic alloy systems in order to improve the durability of fuel cell catalysts. Pt$_3$AlC$_{0.5}$ alloyed with Cr, Mo or W particles of 4–7 nm size has been prepared by sequential precipitation on conductive carbon supports such as highly dispersed Vulcan XC72® [198]. Alternatively, colloidal precursors composed of Pt/Ru/Al allow the manufacture of multimetallic fuel cell catalysts (1–2 nm) having a metal loading of >20%.

The colloidal Pt/Ru/Al precursor is pre-prepared via the organoaluminum route. Co-reduction of organic Pt and Ru salts using Al(CH$_3$)$_3$ gives halogen-free, multimetallic Pt colloids, e.g., Pt$_{50}$Ru$_{50}$/Al (size 1.2$\pm$0.3 nm). The stochiometric ratio of the metal salts can be adjusted by changing the amount of the second metal. The addition of alcohols or suitable surfactants allows the dispersivity of the colloidal precursor in organic media or water to be tailored without affecting the particle size. In the second step, the Pt/Ru/Al colloid is adsorbed on high-surface-area carbon by treatment at 40°C for 24 h. In the third step (conditioning) the dried Pt/Ru/Al Vulcan catalyst powders are exposed to O$_2$ and H$_2$ for 30 min each at 250–300°C to remove the surfactants completely. The particle size of the Pt/Ru/Al colloid adsorbed on the support was found to be virtually untouched (1.3 $\pm$ 0.4 nm) and after the thermal treatment only a moderate growth was observed (1.5 $\pm$ 0.4nm). A thin sheath of oxidized aluminum was found on the Pt/Ru surface, which accounts for the size stabilization observed in the Pt/Ru particles and for the improved durability of the resulting electrocatalysts. Recently, a fuel cell for generating electric power from a liquid organic fuel (synfuel) was described. It comprises a solid electrolyte membrane directly supporting the anode and cathode layers, which contain 7–10% Pt and Ru, 70–80% of perfluorovinyl ether sulfonic acid and 15–20% polytetrafluoroethylene [199].

3.6 CONCLUSION

This chapter demonstrates how the colloidal precursor route generates novel pathways in nanometal synthesis. The advantages for catalyst preparation are obvious. Using monodisperse colloidal metals is an elegant approach to both organosols and hydrosols, which can be readily re-dispersed with high metal concentrations in a variety of solvents. The colloidal stabilizers can also be washed off the colloidal metal particles to give metal nanopowders. Alternatively, the mono- and bi-metallic colloids can easily be deposited onto any given support to generate supported heterogeneous catalysts. In the latter case, conditioning (reactive annealing) is used as the final step to remove the colloidal protective shell and to activate the metallic surface, yielding catalysts with advantageous performances. Modulations of the particle surface both in terms of the structure and composition are possible. Any bi- or pluri-metal combination having a particular composition and particle structure is easily accessible by suitable manipulations during the preparation steps. The versatility of the precursor method, its tunable approach, and the long-term stability of the nanosized metal particles have been confirmed by testing a series of catalysts and comparing them with commercially available state-of-the-art systems. From the economic point of view it is, however, obvious that multimetallic nanostructured colloids are promising precursors for manufacturing high performance catalysts to be used for the production of either fine chemicals or for efficient nanosized multimetallic fuel cell catalysts that are truly nanosized ($\sim$3 nm) fuel cell catalyts with high metal loadings (30 wt% of metal).

REFERENCES

1. Zhou, B., Hermans, S., and Somarjai, G.A., Eds., *Nanotechnology in Catalysis*, Vols. 1 and 2, Kluwer Academic/Plenum Publishers, New York, 2004.
2. Bönnemann H. and Nagabhushana K.S., Chemical synthesis of nanoparticles, in *Encyclopedia of Nanoscience and Nanotechnology*, Vol. 1, Nalwa, H.S., Ed., American Scientific Publishers, Stevenson Ranch, CA, USA, 2004, pp. 777–813.
3. Bönnemann H. and Nagabhushana K.S., Colloidal nanometals as fuel cell catalyst precursors, in *Encyclopedia of Nanoscience and Nanotechnology*, Vol. 1, Schwarz, J.A., Contescu, C.I., and Putyera, K., Eds., Marcel Dekker, New York, 2004, pp. 739–759.
4. Vielstich, W., Lamm, A., and Gateiger, H.A., *Handbook of Fuel Cells, Fundamentals, Technology and Applications*, Vols. 1–4, Wiley, West Sussex, 2003.
5. Cui, C. and Lieber, C.M., Functional nanoscale electronic devices assembled using silicon nanowire building blocks, *Science*, 291, 851, 2001.
6. Eychmuller, A., Structure and photophysics of semiconductor nanocrystals, *J. Phys. Chem. B*, 104, 6514, 2001.
7. Puntes, V.F., Krishnan, K.M., and Alivisatos, A.P., Colloidal nanocrystal shape and size control: the case of cobalt, *Science*, 291, 2115, 2001.
8. Sun, S. et al., Monodisperse FePt nanoparticles and ferromagnetic FePt nanocrystal superlattices, *Science*, 287, 1989, 2000.
9. Rao, C.N.R., Kulkarni, G.U., and Thomas, P.J., Metal nanoparticles and their assemblies, *Chem. Soc. Rev.*, 29, 27, 2000.
10. Chen, S., Two–dimensional crosslinked nanoparticle networks, *Adv. Mater.*, 12, 186, 2000.
11. Niemeyer, C.M., Biomolecules meet nanoparticles, *Angew. Chem. Int. Ed. Engl.*, 40, 4128, 2001.
12. Kumar C.S.S.R., Hormes, J., and Leuschner, C., Eds., *Nanofabrication Towards Biomedical Applications*, Wiley, Weinheim, 2005.
13. Weller, H., Transistoren und Lichtemitter aus einzelnen Nanoclustern, *Angew. chem.*, 110, 1748, 1998.
14. Alivisatos, A.P., Semiconductor clusters, nanocrystals, and quantum dots, *Science*, 271, 933, 1996.
15. Schmid. G., Ed., *Nanoparticles, from Fundamentals to Applications*, VCH, Weinheim, 2004.
16. Bönnemann, H. and Nagabhushana, K.S., Tunable synthetic approaches for the optimization of nanostructured fuel cell catalysts: an overview, *Chem. Ind. Belgrade J.*, 58, 271, 2004.
17. Gaffet, E., Tachikart, M., El Kedim, O., and Rahouadj, R., Nanostructural materials formation by mechanical alloying-morphologic analysis bases on transmission and scanning electron microscopic observations, *Mater. Charact.* 36, 185, 1996.
18. Amulyavichus, A., Daugvila, A., Davidonis, R., and Sipavichus, C., Chemical composition of nanostructured erosion products produced upon laser cutting of steel, *Fizika Metallov I Metallovedenie*, 85, 84, 1998.
19. Turkevich, J., Stevenson, P.C., and Hillier, J., Nucleation and growth process in the synthesis of colloidal gold, *Disc. Faraday Soc.*, 11, 55, 1951.
20. Turkevich, J. and Kim, G., Palladium- preparation and catalytic properties of particles of uniform size, *Science*, 169, 873, 1970.

21. Turkevich, J., Colloidal gold Part I: historical and preparative aspects, morphology and structure, *Gold Bulletin,* 18, 86, 1985.

22. Schmid, G., in *Aspects of Homogeneous Catalysis,* Ugo, R., Ed., Vol. 7, Kluwer, Dordrecht, 1990, p. 1.

23. Schmid, G., Ed., *Clusters and Colloids,* VCH, Weinheim, Germany, 1994.

24. Bönnemann, H. et al., Preparation, characterization, and application of fine metal particles and metal colloids using hydrotriorganoborates, *J. Mol. Catal.,* 86, 129, 1994.

25. Bönnemann, H. et al., Nanoscale colloidal metals and alloys stabilized by solvents and surfactants. Preparation and use as catalyst precursors, *J. Organomet. Chem.,* 520, 143, 1996.

26. Reetz, M.T., Helbig W., and Quaiser, S.A., Electrochemical methods in the synthesis of nanostructured transition metal clusters, in *Active Metals,* Fürstner, A., Ed., VCH, Weinheim, 1996, p. 279.

27. Bönnemann H. and Brijoux, W., in *Catalytically Active Metal Powders and Colloids/Active Metals,* Fürstner, A., Ed., VCH, Weinheim, 1996, pp. 339–379.

28. Toshima, N. and Yonezawa, T., Bimetallic nanoparticles—novel materials for chemical and physical applications, *New J. Chem.* 1179, 1998.

29. Aiken, III, J.D. and Finke, R.G., A review of modern transition metal nanoclusters: their synthesis, characterization, and applications in catalysis. *J. Mol. Catal.,* A 145, 1, 1999.

30. Johnson, B.F.G., From clusters to nanoparticles and catalysis, *Coord. Chem. Rev.,* 190–192, 1269, 1999.

31. Faraday, M., Experimental relations of gold (and other metals) to light, *Philos. Trans. R. Soc. London,* 147, 145, 1857.

32. Leisner, T. et al., The catalytic role of small coinage metal clusters in photography, *Surf. Rev. Lett.,* 3, 1105, 1996.

33. Tausch-Treml, R., Henglein, A., and Lilie, J., Reactivity of silver atoms in aqueous solution II a pulse radiolysis study, *Ber. Bunsen-Ges. Phys. Chem.,* 82, 1335, 1978.

34. Michaelis, M. and Henglein, A., Reduction of palladium (II) in aqueous solution: stabilization and reactions of an intermediate cluster and palladium colloid formation, *J. Phys. Chem.,* 96, 4719, 1992.

35. Watzky, M.A. and Finke, R.G., Transition metal nanocluster formation kinetic and mechanistic studies. A new mechanism when hydrogen is the reductant: slow, continuous nucleation and fast autocatalytic surface growth, *J. Am. Chem. Soc.,* 119, 10382, 1997.

36. Rothe, J. et al., In situ X-ray absorption spectroscopy investigation during the formation of colloidal copper, *J. Am. Chem. Soc.,* 120, 6019, 1998.

37. Angermund, K. et al., Nanoscopic Pt colloids in the "embryonic state", *Angew. Chem. Int. Ed.,* 41, 4041, 2002.

38. Jana, N.R., Gearheart, L., and Murphy, C.J., Evidence for seed mediated nucleation in the chemical reduction of gold salts to gold nanoparticles, *Chem. Mater.,* 13, 2313, 2002.

39. Brown, K.R., Walter, D.G., and Natan, M.J., Seeding of colloidal Au nanoparticles solutions.2. Improved control of particle size and shape, *Chem. Mater.,* 12, 306, 2000.

40. Heng, H., Gibbons, P.C., Kelton, K.F., and Buhro, W.E., Heterogeneous seeded growth: a potentially general synthesis of monodisperse metallic nanoparticles, *J. Am. Chem. Soc.,* 123, 9198, 2001.

41. Jana, N.R., Gearheart, L., and Murphy, C.J., Seeding growth for size control of 5–40 nm diameter gold nanoparticles, *Langmuir,* 17, 6782, 2001.

42. Maase, M., Neue Methoden zur größen- und formselektiven Darstellung von Metallkolloiden., Ph.D. thesis, Bochum, Germany, 1998.

43. Bradley, J.S., The chemistry of transition metal colloids, in *Clusters and Colloids,* Schmid, G., Ed., VCH, Weinheim, 1994, p. 471.

44. Reetz, M.T. et al., Visualization of surfactants on nanostructured Pd clusters by a combination of STM and high resolution TEM, *Science,* 267, 367, 1995.

45. Schulze Tilling, A., Bimetallische Edelmetallkolloide als Precursor für Kohlenhydrat-Oxidationskatalysatoren., Ph.D. thesis, RWTH Aachen, 1996.

46. Mbindyo, J.K.N., Template synthesis of metal nanowires containing monolayer molecular junctions, *J. Am. Chem. Soc.,* 124, 4020, 2002.

47. Amiens, C. et al., Selective synthesis, characterization, and spectroscopic studies on a novel class of reduced platinum and palladium particles stabilized by carbonyl and phosphine ligands, *J. Am. Chem. Soc.,* 115, 11638, 1993.

48. Suslick, K.S., Hyeon, T., Fang, M., and Cichowlas, A., Sonochemical preparation of nanostructured catalysts, in *Advanced Catalysts and Nanostructured Materials,* Moser, W., Ed., Academic Press, San Diego, 1996, p. 197, chap. 8.

49. Schmid, G. et al., $Au_{55}(P(C_6H_5)_3)_{12}Cl_6$- ein gold cluster ungewöhnlicher größe, *Chem. Ber.*, 114, 3634, 1981.

50. Houbertz, R. et al., STM investigation of compact Au_{55} cluster pallets, *Europhys. Lett.*, 28, 641, 1994.

51. Schmid, G., Metal clusters and cluster metals, *Polyhedron*, 7, 2321, 1988.

52. de Jongh, L.J. et al., Physical properties of high-nuclearity metal cluster compounds, *Z. Phys. D: At., Mol. Clusters*, 12, 445, 1989.

53. Schmid, G., Morum, B., and Malm, J.-O., $Pt_{309}Phen_{36}O_{30\pm10}$, A four-shell platinum cluster, *Angew. Chem. Int. Ed. Engl.*, 28, 778, 1989.

54. Schmid, G., Klein, N., and Korste, L., Large transition metal clusters—VI. ligand exchange reactions on $Au_{55}(PPh_3)_{12}Cl_6$—the formation of a water soluble Au_{55} cluster, *Polyhedron*, 7, 605, 1998.

55. Tominaga, T. et al., Tracer diffusion of a ligand stabilized two shell gold cluster, *Chem. Lett.*, 1033, 1996.

56. Schmid, G., Large clusters and colloids. Metals in the embryonic state, *Chem. Rev.*, 92, 1709, 1992.

57. Wicrenga, H.A. et al., Direct imaging of $Pd_{561}(phen)_{38\pm2}O_n$ and $Au_{55}(pph_3)_{12}Cl_6$ clusters using scanning tunneling microscopy, *Adv. Mater.*, 2, 482, 1990.

58. Schmid, G. and Lehnert, A., Complexation of gold colloid, *Angew. Chem.*, 101, 773–774, 1989; *Angew. Chem. Int. Ed. Engl.* 28, 780, 1989.

59. Schmid, G. Maihack, V. Lantermann, F., and Peschel, S., Ligand-stabilized metal clusters and colloids: properties and applications, *J. Chem. Soc. Dalton Trans.*, 589, 1996.

60. Schmid, G. et al., Catalytic properties of layered gold-palladium colloids, *Chem. Eur. J.*, 2, 1099, 1996.

61. Simon, U. et al., Chemical tailoring of charging energy in metal cluster arrangements by use of bifunctional spacer molecules, *J. Mater. Chem.*, 8, 517,1998.

62. Schmid, G. and Peschel, S., Preparation and scanning probe microscopic characterization of monolayers of ligand-stabilized transition metal clusters and colloids, *New J. Chem.*, 22, 669, 1998.

63. Schmid, G., Pugin, R., Malm, J.-O., and Bovin J.-O., Silsesquioxanes as ligands for gold clusters, *Eur. J. Inorg. Chem.*, 813, 1998.

64. Lin, Y. and Finke, R.G., Novel polyoxoanion- and Bu_4N^+-Stabilized, Isolable, and Redissolvable, 20–30 nm $Ir_{300-900}$ nanoclusters: the kinetically controlled synthesis, characterization, and mechanism of formation of organic solvent-soluble, reproducible size, and reproducible catalytic activity metal nanoclusters, *J. Am. Chem. Soc.*, 116, 8335, 1994.

65. Lin, Y., and Finke, R.G., A more general approach to distinguishing "homogeneous" from "heterogeneous" catalysis: discovery of polyoxoanion- and Bu_4N+-stabilized, isolable and redissolvable, high-reactivity Ir. $apprx._{190-450}$ nanocluster catalysts, *Inorg. Chem.*, 33, 4891, 1994.

66. Nagata, T., Pohl, M., Weiner, H., and Finke, R.G., Polyoxoanion-supported organometallic complexes: carbonyls of Rhenium(I), Iridium(I), and Rhodium(I) that are soluble analogs of solid-oxide-supported $M(CO)n^+$ and that exhibit novel $M(CO)n^+$ mobility, *Inorg. Chem.*, 36, 1366, 1997.

67. Aiken, III, J.D. and Finke, R.G., Nanocluster formation synthetic, kinetic, and mechanistic studies. The detection of, and then methods to avoid, hydrogen mass-transfer limitations in the synthesis of polyoxoanion- and tetrabutylammonium-stabilized, near-monodisperse 40 ± 6 Å Rh(0) nanoclusters, *J. Am. Chem. Soc.*, 120, 9545, 1998.

68. Aiken, III, J.D. and Finke, R.G., Polyoxoanion- and tetrabutylammonium-stabilized, near-monodisperse, 40 ± 6 Å Rh(0)~1500 to Rh(0)~3700 nanoclusters: synthesis, characterization, and hydrogenation catalysis, *Chem. Mater.*, 11, 1035, 1999.

69. Vargaftik, M.N. et al., Giant Pd clusters as catalysts of oxidative reactions of olefins and alcohols, *J. Mol. Catal.*, 53, 315, 1989.

70. Vargaftik, M.N. et al., A novel giant palladium cluster, *J. Chem. Soc. Chem. Commun.*, 937,1985.

71. Volkov, V.V. et al., Long- and short-distance ordering of the metal cores of giant Pd clusters, *J. Cryst. Growth*, 163, 377, 1996.

72. Moiseev, I.I. et al., Palladium-561 giant clusters: chemical aspects of self organization on a nano level, *Mend. Commun.*, 87, 1995.

73. Oleshko, V. et al., High resolution electron microscopy and electron energy-loss spectroscopy of giant palladium clusters, *Z. Phys. D.*, 34, 283, 1995.

74. Vargaftik, M.N. et al., Catalysis with a Pd giant cluster: phenol oxidative carbonylation to diphenul carbonate conjugated with reductive nitrobenzene conversion, *J. Mol. Catal. A: Chem.*, 108, 77, 1996.

75. Fu, X. et al., Shape-selective preparation and properties of oxalate-stabilized Pt colloid, *Langmuir*, 18, 4619, 2002.

76. Del Angel, P. et al., Aggregation state of Pt-Au/C bimetallic catalysts prepared by surface redox reactions, *Langmuir*, 16, 7210, 2000.

77. Toshima, N. and Yonezawa, T., Bimetallic nanoparticles—novel materials for chemical and physical applications, *New J. Chem.*, 1179, 1998.

78. Hirai, H., Nakao, Y., Toshima N., and Adachi, K., Colloidal Rh in polyvinylalcohol as hydrogenation catalysts for olefins, *Chem. Lett.*, 5, 905, 1976.

79. Hirai, H., Nakao Y., and Toshima, N., Colloidal Rh in poly(vinylpyrrolidone) as hydrogenation catalysts for internal olefins *Chem. Lett.*, 7, 545, 1978.

80. Hirai, H., Nakao Y., and Toshima, N., Preparation of colloidal Rhodium in polyvinylalcohol by reduction with methanol, *J. Macromol. Sci.*, *Chem. A*, 12, 1117, 1978.

81. Hirai, H., Nakao Y., and Toshima, N., Preparation of colloidal transition-metals in polymers by reduction with alcohols or ethers, *J. Macromol. Sci. Chem. A*, 13, 727, 1979.

82. Toshima, N. and Hirakawa, K., Polymer protected bimetallic nanocluster catalysts having core/shell structure for accelerated electron transfer in visible-light-induced hydrogen generation, *Polymer J.*, 31, 1127, 1999.

83. Lu, P. et al., Polymer-protected Ni/Pd bimetallic nano-clusters: preparation, characterization and catalysis for Hydrogenation of Nitrobenzene, *J. Phys. Chem. B*, 103, 9673, 1999.

84. Mucalo, M.R. and Cooney, R.P., F.T.I.R spectra of carbon monoxide adsorbed on platinum sols, *J. Chem. Soc. Chem. Commun.*, 94, 1989.

85. Meguro, K., Torizuka, M., and Esumi, K., The preparation of organo colloidal precious metal particles, *Bull. Chem. Soc. Jpn.*, 61, 341, 1988.

86. Curtis, A.C. et al., Morphology and nanostructure of colloidal gold and silver, *Angew. Chem.*, 99, 688–691, 1987; *Angew. Chem. Int. Ed. Engl.*, 26, 676, 1987.

87. Duff, D.G. et al., The microstructure of colloidal silver: evidence for a polytetrahedral growth sequence, *J. Chem. Soc. Chem. Commun.*, 1264, 1987.

88. Curtis, A.C. et al., A copper organosol with well defined morphology, *Angew. Chem.*, 100, 1588–1590, 1988; *Angew. Chem. Int. Ed. Engl.*, 27, 1530, 1988.

89. Duff, D.G. et al., Structural characterization of colloidal platinum by high resolution electron microscopy and EXAFS analysis, *Angew. Chem.* 101, 610, 1989; *Angew. Chem. Int. Ed. Engl.*, 28, 590, 1989.

90. Duff, D.G., Baiker, A., and Edwards, P.P., A new hydrosol of gold clusters. 1. Formation and particle size variation, *Langmuir*, 9, 2301, 1993.

91. Vogel, W., Duff, D.G., and Baiker, A., X-ray structure of a new hydrosol of gold clusters, *Langmuir*, 11, 401, 1995.

92. Lewis, L.N. and Lewis, N., Platinum-catalyzed hydrosilylation—colloid formation as the essential step, *J. Am. Chem. Soc.*, 108, 7228, 1986.

93. Lewis, L.N. and Lewis, N., Preparation and structure of platinum group metal colloids: without solvent, *Chem. Mater.*, 1, 106, 1989.

94. van Rheenen, P.R., McKelvey, M.J., and Glaunsinger, W.S., Synthesis and characterization of small platinum particles formed by the chemical reduction of chloroplatinic acid, *J. Solid State Chem.*, 67, 151, 1987.

95. Xiao, J., Xie Y., and Luo, W., A rational low-temperature approach to the synthesis of gladiate ruthenium nanoparticles, *Chem. Lett.*, 31, 462, 2002.

96. Zheng, H-G., Liang, J-H., Zeng, J-H., and Qian, Y-T., Preparation of nickel nanopowders in ethanol-water system(EWS), *Mater. Rese. Bull.*, 36, 947, 2001.

97. Duff, D.G. and Baiker, A., Preparation and structural properties of ultrafine gold colloids for oxidation catalysis, in *Preparation of Catalysts VI*, Poncelet, G., Martens, J., Delmon, B., Jacobs, P.A., and Grange, P., Eds., Elsevier, Amsterdam, 1995, p. 505.

98. Tsai, K.-L. and Dye, J.L., Synthesis, properties, and characterization of nanometer-size metal particles by homogeneous reduction with alkalides and electrides in aprotic solvents, *Chem. Mater.*, 5, 540, 1993.

99. Cassagneau, T. and Caruso, F., Continuous silver nanoparticle coatings on dielectric spheres, *Adv. Mater.*, 14, 732, 2002.

100. Park, K-W et al., Chemical and electronic effects of Ni in Pt/Ni and Pt/Ru/Ni alloy nanoparticles in methanol electrooxidation, *J. Phys. Chem. B*, 106, 1869, 2002.

101. Kinoshita, K. and Stonehart, P., *Modern Aspect of Electrochemisty*, Plenum Press, New York, 1996, p. 12.

102. Klabunde, K.J. and Mohs, C., Nanoparticles and nanostructural materials, in *Chemistry of Advance Materials,* Interrante, L.V. and Hampden-Smith, L.V., Eds., Wiley-VCH, New York, 1998, Chap. 3.

103. Zhao, S-Y et al., Preparation, phase transfer, and self-assembled monolayers of cubic Pt nanoparticles, *Langmuir,* 18, 3315, 2002.

104. Lee, S-A. et al., Nanoparticle synthesis and electrocatalytic activity of Pt alloys for direct methanol fuel cells, *J. Elect. Chem. Soc.,* 149, A1299, 2002.

105. Bönnemann, H. and Brijoux, W., in *Surfactant-Stabilized Nanosized Colloidal Metals and Alloys as Catalyst Precursors/Advanced Catalysts and Nanostructured Materials,* Moser, W., Ed., Academic Press, San Diego, 1996, pp. 165–196, Chap. 7.

106. Bönnemann, H., Brijoux, W., and Joussen, T., (to Studiengesellschaft Kohle mbH) Microcrystalline-to-amorphous metal and/ or alloy powders dissolved without protective colloid in organic solvents, *U.S. Pat.,* 5, 580, 492, 1993.

107. Bönnemann, H. et al., Erzeugung von kolloiden übergangsmetallen in organischer phase und ihre anwendung in der katalyse, *Angew. Chem.,* 103, 1344, 1991; *Angew. Chem. Int. Ed. Engl.,* 30, 1312, 1991.

108. Bönnemann, H., Brijoux, W., Brinkmann, R., and Richter, J., Process for producing tenside-stabilized colloids of mono- and bimetals of the group VIII and Ib of the periodic system in the form of precursors for catalysts which are isolable and water soluble at high concentration, *U.S. Pat.,* 849,482, 1997 (to Studiengesellschaft Kohle mbH).

109. Bucher, J. et al., Interaction between core and protection shell of N(Butyl)$_4$Cl- and N (Octyl)$_4$Cl-stabilized Pd colloids, *Surf. Sci.,* 497, 321, 2002.

110. Franco, E.G., Neto, A., Linardi, M., and Arico, E., Synthesis of electrocalysts by the Bönnemann method for the oxidation of methanol and the mixture of H$_2$/CO in a proton exchange membrane fuel cell, *J. Braz. Chem. Soc.,* 13, 516, 2002.

111. Neto, A.O. et al., *International Workshop on Ceramic/Metal Interfaces Control at the Atomic Level,* Oviedo, Spain, 2002.

112. Roth, C., Martz, N., and Fuess, H., Characterization of different Pt–Ru catalysts by x-ray diffraction and transmission electron microscopy, *Phys. Chem. Chem. Phys.,* 3, 315, 2001.

113. Bönnemann, H. et al., The reductive stabilization of nanometal colloids by organo-aluminum compounds, *Rev. Roum. Chim.,* 44, 1003, 1999.

114. Sinzig, J. et al., Antiferromagnetism of Colloidal [Mn0.0.3THF] x, *Appl. Organomet. Chem.,* 12, 387, 1998.

115. Bönnemann, H., Brijoux, W., and Brinkmann, R., Method for modifying the dispersion characteristics of metal-organic-prestabilized or pre-treated nanometal colloids, WO 99/59713, (to Studiengesellschaft Kohle), November 25, 1999.

116. Angermund, K. et al., In situ study on the wet chemical synthesis of nanoscopic Pt colloids by Reductive Stabilization, *J. Phys. Chem. B,* 107, 7507, 2003.

117. Reetz, M.T. and Helbig, W., Size-selective synthesis of nanostructured transition metal clusters, *J. Am. Chem. Soc.,* 116, 7401, 1994.

118. Reetz, M.T. and Maase, M., Redox-controlled size-selective fabrication of nanostructured transition metal colloids, *Adv. Mater.,* 11, 773,1999.

119. Bradley, J.S. et al., Surface spectroscopic study of the stabilization mechanism for shape-selectively synthesized nanostructured transition metal colloids, *J. Am. Chem. Soc.,* 122, 4631, 2000.

120. Bönnemann, H. and Nagabhushana, K.S., Advantageous fuel cell catalysts from colloidal nanometals, *J. New Mater. Electrochem. System,* 7(2), 93, 2004.

121. Wicrenga, H.A. et al., Direct imaging of Pd$_{561}$(phen)$_{38\pm2}$O$_n$ and Au$_{55}$(pph$_3$)$_{12}$Cl$_6$ clusters using scanning tunnelling microscopy, *Adv. Mater.,* 2, 482, 1990.

122. Klabunde, K.J., Li, Y.-X., and Tan, B.-J., Solvated metal atom dispersed catalysts, *Chem. Mater.,* 3, 30, 1991.

123. Larpent, C., Brisse-le-Menn, F., and Patin, H., New highly water-soluble surfactants stabilize colloidal rhodium(0) suspensions useful in biphasic catalysis, *J. Mol. Catal.,* 65, L35–L40, 1991.

124. Moiseev, I.I., Tsirkov, G.A., Gekhman, A.E., and Vargaftik, M.N., Facile hydrogen- transfer reduction of multiple bonds by formic acid catalysed with a Pd-561 giant cluster, *Mendeleev Commun.,* 7, 1–3, 1997.

125. Vargaftik, M.N. et al., Catalysis by metal colloids: tragectories for atom assembling in Pd and Pt colloids, *Kinetics Catal.,* 39, 740, 1998.

126. Aiken, III, J.D. Lin, Y., and Finke, R.G., A perspective on nanocluster catalysis: polyoxonation and tetrabutylamonium stabilized Ir_{0-300} nanocluster 'soluble heterogeneous catalysts,' *J. Mol. Catal. A*, 114, 29, 1996.

127. Aiken III, J.D., and Finke, R.G., Polyoxoanion- and tetrabutylammonium-stabilized Rh(0)n nanoclusters: unprecedented nanocluster catalytic lifetime in solution, *J. Am. Chem. Soc.*, 121, 8803, 1999.

128. Reetz, M.T., Quaiser, S.A., and Merk, C., Electrochemical preparation of nanostructured titanium clusters: characterization and use in Mcmurry-type coupling reactions, *Chem. Ber.*, 129, 741, 1996.

129. Franke, R. et al., A study of the electronic and geometric structure of colloidal Ti.0.5THF, *J. Am. Chem. Soc.*, 118, 12090, 1996.

130. Bönnemann, H. and Korall, B., Etherlösliches Ti0 und bis(h6-aren)titan(0)-komplexe durch reduktion von $TiCl_4$ mit triethylhydroborat, *Angew. Chem.*, 104, 1506, 1992, *Angew. Chem. Int. Ed. Engl.*, 31, 1490, 1992.

131. Bönnemann, H. and Brijoux, W., The preparation, characterization and application of organosols of early transition metals, *Nanostruct. Mater.*, 5, 135, 1995.

132. Leff, D.V., Ohara, P.C., Heath, J.R., and Gelbart, W., Thermodynamic control of gold nanocrystal size: experiment and theory, *J. Phys. Chem.*, 99, 7036, 1995.

133. Reetz, M.T., Breinbauer, R., and Wanninger, K., Suzuki and heck reactions catalyzed by performed palladium clusters and palladium/nickel bimetallic clusters, *Tetrahedron Lett.*, 37, 4499, 1996.

134. Beller, M. et al., First palladium-catalyzed heck reactions with efficient colloidal catalyst systems, *J. Organomet. Chem.*, 520, 257, 1996.

135. Reetz, M.T. and Westermann, E., Phosphane-free palladium-catalyzed coupling reactions: the decisive role of Pd nanoparticles, *Angew. Chem. Int. Ed.*, 39, 165, 2000.

136. Schmid, G. et al., Synthesis and catalytic properties of large ligand stabilized palladium clusters, *J. Mol. Catal. A*, 107, 95, 1996.

137. Widegren, J.A. and Finke, R.G., A review of soluble transition metal nanoclusters as arene hydrogenation catalysts, *J. Mol. Cat. A: Chem.*, 191, 187, 2003.

138. Witek, G. et al., Interaction of platinum colloids with single crystalline oxide and graphite substrates: a combined AFM, STM and XPS study, *Catal. Lett.*, 37, 35, 1996.

139. Shaikhutdinov, S.K., Möller, F.A., Mestl, G. and Behm, R.J., Electrochemical deposition of platinum hydrosol on graphite observed by scanning tunneling microscopy, *J. Catal.*, 163, 492, 1996.

140. Bönnemann H. et al., Activation of colloidal PtRu fuel cell catalysts *via* thermal "conditioning process," *Fuel Cells*, 4, 1, 2004.

141. Reetz, M.T., Quaiser, S.A., Breinbauer, R., and Tesche, B., A New Strategy in Heterogeneous catalysis: the design of cortex catalysts/catalysis/clusters/ immobilization/ surface chemistry, *Angew. Chem. Int. Ed. Engl.*, 34, 2728, 1995.

142. Teranishi, T., Nakata, K., Miyake, M., and Toshima, N., Promition effect of polymer-immobilized neodymium ions on catalytic activity of ultra fine palladium particles, *Chem. Lett.*, 277, 1996.

143. Bönnemann, H., Wittholt, W., Jentsch, J.D., and Tilling, A.S., Supported Pt-colloid catalysts for the selective hydrogenation of 3,4-dichloronitrobenzene, *New J. Chem.*, 22, 713, 1998.

144. Yu, W. et al., Modification of metal cations to the supported metal colloid catalysts, *J. Mol. Catal. A*, 147, 73, 1999.

145. Feng H. and Liu, H., The metal complex effect on metal clusters in liquid medium, *J. Mol. Cat. A: Chem.*, 126, L5, 1997.

146. Bönnemann, H. et al., The preparation of colloidal Pt/Rh alloys stabilized by NR_4^+- and PR_4^+-groups and their characterization by X-ray-absorption spectroscopy, *Z. Naturforsch.*, 50b, 333, 1995.

147. Aleandri, L.E. et al., Structural investigation of bimetallic Rh-Pt nanoparticles through X-ray absorption spectroscopy, *J. Mater. Chem.*, 5, 749, 1995.

148. Harada, M., Asakura K., and Toshima, N., Structural analysis of polymer-protected platinum/rhodium bimetallic clusters using extended x-ray absorption fine structure spectroscopy. Importance of microclusters for the formation of bimetallic clusters, *J. Phys. Chem.*, 98, 2653, 1994.

149. Toshima, N., Yonezawa, T., and Kushihashi, K., Polymer protected Pd–Pt bimetallic clusters preparation, catalytic properties and structural considerations, *J. Chem. Soc. Faraday Trans.*, 89, 2537, 1993.

150. Schmid, G., West, H., Mehles, H., and Lehnert, A., Hydrosilation reactions catalyzed by supported bimetallic colloids, *Inorg. Chem.*, 36, 891, 1997.

151. Toshima, N., Takahashi, T., and Hirai, H., Polymerized micelle-protected platinum clusters-preparation and application to catalyst for visible light-induced hydrogen generation, *J. Macromol. Sci. -Chem.*, A25, 669, 1988.

152. Ohtaki, M., Komiyama, M., Hirai, H., and Toshima, N., Effects of polymer support on the substrate selectivity of covalently immobilized ultrafine rhodium particles as a catalyst for olefin hydrogenation, *Macromolecules*, 24, 5567, 1991.

153. Schmid, G., Maihack, V., Lantermann, F., and Peschel, S., Ligand-stabilized metal clusters and colloids: properties and applications, *J. Chem. Soc. Dalton Trans.*, 5, 589, 1996.

154. Bönnemann, H., Brijoux, W., Tilling A.S., and Siepen, K., Application of heterogeneous colloid catalysts for the preparation of fine chemicals., *Top. Catal.*, 4, 217, 1997.

155. Bönnemann, H., Britz P., and Ehwald, H., Herstellung und testung von ruthenium-kolloid-katalysatoren für die selektivhydrierung von benzol zu cyclohexen, *Chem. Tech. (Leipzig)*, 49, 189, 1997.

156. Bönnemann, H. and Braun, G.A., Enantioselektive hydrierung an platinkolloiden, *Angew. Chem.*, 108, 2120, 1996, *Angew. Chem., Int. Ed. Engl.*, 35, 1992, 1996.

157. Bönnemann, H. and Braun, G.A., Enantioselectivity control with metal colloids as catalyst, *Chem. -Eur. J.*, 3, 1200, 1997.

158. Köhler, J.U. and Bradley, J.S., Enantioselective hydrogenation of ethyl pyruvate with colloidal platinum catalysts: the effect of acidity on rate, *Catal. Lett.*, 45, 203, 1997.

159. Köhler, J.U. and Bradley, J.S., A kinetic probe of the effect of a stabilizing polymer on a colloidal catalyst: accelerated enantioselective hydrogenation of ethyl pyruvate catalyzed by poly(vinylpyrrolidone)-stabilized platinum colloids, *Langmuir*, 14, 2730, 1998.

160. Bönnemann, H. Brinkmann, R., and Neiteler, P., Preparation and catalytic properties of NR_4^+-stabilized palladium colloids, *Appl. Organomet. Chem.*, 8, 361, 1994.

161. Bönnemann, H. et al., Selective oxidation of glucose on bismuth-promoted Pd-Pt/C catalysts prepared from $N(Oct)_4Cl$-stabilized Pd–Pt colloids, *Inorg. Chim. Acta.*, 270, 95, 1998.

162. Djakovitch, L., Wagner, M., and Köhler, K., Amination of aryl bromides catalysed by supported palladium, *J. Organomet. Chem.*, 592, 225, 1999.

163. Dhas, A. and Gedanken, A., Sonochemical preparation and properties of nanostructured palladium metallic clusters, *J. Mater. Chem.*, 8, 445, 1998.

164. Bars, J.L., Specht, U., Bradley, J.S., and Blackmond, D.G., A catalytic probe of the surface of colloidal palladium particles using heck coupling reactions, *Langmuir*, 15, 7621, 1999.

165. Reetz, M.T. and Koch, M., PCT/EP 99/08594 (to Studiengesellschaft Kohle m.b.H.), November 9, 1999.

166. Reetz, M.T. and Koch, M., Water-soluble colloidal Adams catalyst: preparation and use in catalysis, *J. Am. Chem. Soc.*, 121, 7933, 1999.

167. Kim, S.W. et al., Fabrication of hollow palladium spheres and their successful application to the recyclable heterogeneous catalyst for suzuki coupling reactions, *J. Am. Chem. Soc.*, 124, 7642, 2002.

168. Crooks, M. et al., Dendrimer-encapsulated metal nanoparticles: synthesis, characterization, and applications to catalysis, *Acc. Chem. Res.*, 34, 181, 2001.

169. Li, Y. and El-Sayed, M.A., The Effect of stabilizers on the catalytic activity and stability of Pd colloidal nanoparticles in the Suzuki reactions in aqueous solution, *J. Phys. Chem. B.*, 105, 8938, 2001.

170. Ihlein, G. et al., Ordered porous materials as media for the organization of matter on the nanoscale, *Appl. Organometallic Chem.*, 12, 305, 1998.

171. Reetz, M.T. and Dugal, M., Entrapment of Nanostructured Palladium Clusters in Hydrophobic Sol-Gel Materials, *Catal. Lett.*, 58, 207, 1999.

172. Bönnemann, H., Endruschat, U., Tesche, B., Rufínska, A., Lehmann, C.W., Wagner, F.E., Filoti, G., Pârvulescu, V., and Pârvulescu, V.I., An SiO_2- embedded nanoscopic Pd/Au alloy colloid, *Eur. J. Inorg. Chem.*, 5, 819, 2000.

173. Niederer, J.P.M. et al., Noble metal nanoparticles incorporated in mesoporous hosts, *Topics Catal.*, 18, 265–269, 2002.

174. Oprescu, C. et al., Hydrogenolysis of 1,1a,6,10b-tetrahydro-1,6-methanodibenzo[a,e]cyclopropa[c]-cycloheptene over silica- and zirconia-embedded Ru-colloids, *J. Mol. Cat. A: Chem.*, 186, 153, 2002.

175. Dissanayake, D., and Lunsford, J. H., Evidence for the role of colloidal palladium in the catalytic formation of H_2O_2 from H_2 and O_2, *J. Catal.*, 206, 173, 2002.

176. Roucoux, A., Schulz, J., and Patin, H., Reduced transition metal colloids: a novel family of reusable catalysts?, *Chem. Rev.*, 102, 3757, 2002.

177. Kordesch, K. and Simader, G., *Fuel Cells and their Applications*, VCH, Weinheim, 1996.
178. Petrow, H.G. and Allen, R.J., US 4,044,193 (to Prototech Comp.), August 23, 1977.
179. Burstein, G.T., Barnett, C. J., Kucernak A.R., and Williams, K.R., Aspects of the anodic oxidation of methanol, *Catal. Today*, 38, 425, 1997.
180. Auer, E., Freund, A., Lehmann, T., Starz, K.A., Schwarz, R., and Stenke, U., EP 0 880 188 A2 (to Degussa AG), November 25, 1998.
181. Wilson, M.S. and Gottesfeld, S., Thin film catalyst layers for polymer electrolyte fuel cell electrodes, *J. Appl. Electrochem.*, 22, 1, 1992.
182. Wilson, M.S., Ren, X., and Gottesfeld, S., High performance direct methanol polymer electrolyte fuel cells, *J. Electrochem. Soc.*, 143, L 12, 1996.
183. Thomas, X., Ren, S., and Gottesfeld, J., Influence of ionomer content in catalyst layers on direct methanol fuel cell performance, *Electrochem. Soc.*, 146, 4354, 1999.
184. Luczak F.J. and Landsman D.A., US 4,613,582 (to United Technologies Corp.), September 23, 1986.
185. Wittholt, W., Ph.D. thesis, RWTH Aachen, 1997.
186. Wendt, H. and Götz, M., EP 0951 084 A2 (to Degussa-Hüls AG), October 20, 1999).
187. Freund, A., Lang, J., Lehmann T., and Starz, K.A., Improved Pt alloy catalysts for fuel cells, *Catal. Today*, 27, 279, 1996.
188. Paulus, U. et al., New PtRu alloy colloids as precursors for fuel cell catalysts, *J. Catal.*, 195, 383, 2000.
189. Schmidt, T.J. et al., Electrocatalytic activity of PtRu alloy colloids for CO and CO/H_2 electrooxidation: stripping voltammetry and rotating disk measurements, *Langmuir*, 13, 2591, 1997.
190. Schmidt, T.J. et al., PtRu alloy colloids as precursors for fuel cell catalysts, *J. Electrochem. Soc.*, 145, 925, 1998.
191. Bönnemann, H. et al., Nanoscopic Pt-bimetal colloids as precursors for PEM fuel cell catalysts., *J. New Mater. Electrochem. Syst.*, 3, 199, 2000.
192. Mukerjee, S. and Urian, R.C., Bifunctionality in Pt alloy nanocluster electrocatalysts for enhanced methanol oxidation and CO tolerance in PEM fuel cells: electrochemical and in situ synchrotron spectroscopy, *Electrochim. Acta*, 47, 3219, 2002.
193. Vogel, P. et al., Structure and chemical composition of surfactant-stabilized PtRu alloy colloids, *J. Phys. Chem. B.*, 101, 11029, 1997.
194. Stimming, U. and Vogel, R., in *Electrochemical Nanotechnology*, Lorenz, W.J. and Plieth, W., Eds., Wiley-VCH, Weinheim, 1998, p.73.
195. Giallombardo, J.R. and De Castro E.S., WO 99/53557 (to De Nora S.P.A.), April 14, 1999.
196. Park, S., Xie, Y. and Weaver, M.J., Electrocatalytic pathways on carbon-supported platinum nanoparticles: comparison of particle-size-dependent rates of methanol, formic acid, and formaldehyde, *Langmuir*, 18, 5792, 2002.
197. Steigerwalt, S.E. et al., A Pt-Ru/graphitic carbon nanofiber nanocomposite exhibiting high relative performance as a direct-methanol fuel cell anode catalyst, *J. Phys. Chem. B.*, 105, 8097, 2001.
198. Freund, T., Lehmann, Starz, K.A., Heinz, G., and Schwarz, R., EP 0 743 092 A1 (to Degussa AG), November 20, 1996.
199. Narayanan, S., Surampudi, S., and Halpert, G., Direct liquid-feed fuel cell with membrane electrolyte and manufacturing thereof, US 5,945,231 (to California Institute of Technology), (August 31, 1999).

CHAPTER 3 QUESTIONS

Question 1

Why is nanoscience interesting?

Question 2

What are nanostructured colloidal particles? What are hydrosols and organosols?

Question 3

What is the role of a protecting agent during nanoparticle synthesis?

Question 4

What are the advantages of Bönnemann's method of colloidal nanoparticle synthesis?

Question 5

Why is heat treatment chosen to better effect the stripping off the protecting shell of a supported colloidal catalyst precursor?

CHAPTER 4

Microporous and Mesoporous Catalysts

Wolfgang Schmidt

CONTENTS

4.1   Setting the Scene 96
4.2   Porous Catalysts 97
4.3   Microporous Catalysts: Zeolites 97
    4.3.1   What Are Zeolites? 97
    4.3.2   Types of Zeolites Used in Catalytic Processes 101
        4.3.2.1   Zeolite X and Zeolite Y 101
        4.3.2.2   ZSM-5 102
        4.3.2.3   Mordenite 102
        4.3.2.4   Zeolite A 103
    4.3.3   Production of Zeolites 103
    4.3.4   Post-Synthesis Treatment of Zeolites and Modification of Zeolites 105
        4.3.4.1   Protonation of Zeolites 105
        4.3.4.2   Dealumination of Zeolites 106
        4.3.4.3   Metals and Metal Complexes in Zeolites 106
    4.3.5   Catalytic Application of Zeolites 107
        4.3.5.1   Zeolite Catalysts in Petrochemical Processes 109
        4.3.5.2   Methanol to Gasoline and Methanol to Olefins 117
4.4   Mesoporous Catalysts 118
    4.4.1   Ordered Mesoporous Silica Materials 118
        4.4.1.1   Surface Modifications of Ordered Mesoporous Silica Materials 122
        4.4.1.2   Catalysis with Ordered Mesoporous Silica Materials 123
    4.4.2   Nonsiliceous Ordered Mesoporous Materials 125
4.5   Characterization of Microporous and Mesoporous Materials 126
    4.5.1   X-Ray Diffraction 127
    4.5.2   Physisorption Analysis 128
    4.5.3   Electron Microscopy 130
    4.5.4   Nuclear Magnetic Resonance Spectroscopy 130
    4.5.5   Infrared Spectroscopy 132
Appendix 134
    Terms and Abbreviations 134
    Further Reading 135

References 135
Chapter 4 Problems 138

4.1 SETTING THE SCENE

This chapter will focus on catalysis using porous catalysts. Since porosity is a feature often found in nature but hardly ever discussed in standard textbooks, this short chapter is intended to set the scene and introduce porous catalysts.

Pores are found in many solids and the term porosity is often used quite arbitrarily to describe many different properties of such materials. Occasionally, it is used to indicate the mere presence of pores in a material, sometimes as a measure for the size of the pores, and often as a measure for the amount of pores present in a material. The latter is closest to its physical definition. The porosity of a material is defined as the ratio between the pore volume of a particle and its total volume (pore volume + volume of solid) [1]. A certain porosity is a common feature of most heterogeneous catalysts. The pores are either formed by voids between small aggregated particles (textural porosity) or they are intrinsic structural features of the materials (structural porosity). According to the IUPAC notation, porous materials are classified with respect to their sizes into three groups: microporous, mesoporous, and macroporous materials [2]. Microporous materials have pores with diameters < 2 nm, mesoporous materials have pore diameters between 2 and 50 nm, and macroporous materials have pore diameters > 50 nm. Nowadays, some authors use the term nanoporosity which, however, has no clear definition but is typically used in combination with nanotechnology and nanochemistry for materials with pore sizes in the nanometer range, i.e., 0.1 to 100 nm. Nanoporous could thus mean everything from microporous to macroporous.

The different pore sizes of specific catalysts have implications not only on their specific surface areas but also on properties concerning diffusion, site accessibility, and catalyst deactivation (e.g., owing to coke formation). In heterogeneous catalysis, conversions proceed on active sites on the surface of solid catalysts. The conversion is thus closely related to the accessible surface area of a given catalyst. If the diffusion of educts and products is not limited, the catalytic conversion scales proportional to the specific surface area, S [m^2 g^{-1}], provided that the active sites are distributed homogeneously on the surface of the catalyst. The specific surface area of a material is defined as the surface area of a material per gram of solid. If the diffusion is limited owing to smaller pores, the catalytic conversion is significantly reduced and does not scale proportional to the specific surface area.

For gas-phase reactions on macroporous catalysts, diffusion limitations are of minor importance, since gas diffusion is typically not restricted in pores of this size. However, for mesoporous and especially for microporous materials, diffusion control of gas molecules plays a major role. If the pore size is of the same dimension as the size of the reacting molecules, a second effect has to be considered. Molecules of sizes smaller than the diameter of a given pore can enter the pore system and react on sites inside the pores. Yet, molecules of sizes larger than the pore diameter are excluded from the pore system and thus from the reactive sites inside the material. Reactions of such larger molecules can only proceed on sites on the external surface of the catalyst particles. Since the external surface area (i.e., outer surface area of particle) is typically two to three magnitudes smaller than the internal surface area (i.e., surface area inside pores), the conversion of these molecules is comparably low. In this way, a selectivity is achieved for smaller molecules. This mechanism is typical for microporous catalysts and is often termed molecular sieving; as a consequence, microporous materials are frequently called molecular sieves. In the following section, typical microporous and mesoporous catalysts will be described. To allow the reader to understand the results of specific characterization methods, a short survey on these methods follows. This survey is intended to provide only some basic information on the different methods. For a more detailed understanding, it is recommended that readers consult Chapter 9 and more specialized textbooks(see Further Reading in the appendix at the end of this chapter).

4.2 POROUS CATALYSTS

Most heterogeneous catalysts are porous to a certain degree. Yet, a closer look shows that the presence of pores in many materials is because of textural porosity, i.e., the pores are voids between small aggregated particles. These type of pores are typically observed in many supported catalysts where oxides or carbon materials are used as supports. Well-known examples are automotive catalysts. They consist of highly dispersed rhodium and platinum particles (as the catalytically active compound) being supported on an alumina washcoat. The alumina washcoat provides a high surface area for the deposition of the noble metal catalysts, together with a textural porosity, which allows the exhaust gas molecules to diffuse to the catalytically active noble metal sites. This type of material will not be considered in the present chapter since their porosity is only because of aggregation of nonporous particles. The main focus will be on materials with an intrinsic porosity due to their structural properties. In the first section, microporous materials will be discussed, followed by a short survey of mesoporous catalysts with structural porosity.

4.3 MICROPOROUS CATALYSTS: ZEOLITES

Microporosity is a feature observed in many different materials (e.g., activated carbons, aerogels, and xerogels). However, with regard to heterogeneous catalysis, zeolites are practically the only microporous catalysts used at present. The following chapter thus only addresses zeolites and their use in catalysis.

4.3.1 What Are Zeolites?

In general, zeolites are crystalline aluminosilicates with microporous channels and/or cages in their structures. The first zeolitic minerals were discovered in 1756 by the Swedish mineralogist Cronstedt [3]. Upon heating of the minerals, he observed the release of steam from the crystals and called this new class of minerals zeolites (Greek: zeos = to boil, lithos = stone). Currently, about 160 different zeolite structure topologies are known [4] and many of them are found in natural zeolites. However, for catalytic applications only a small number of synthetic zeolites are used. Natural zeolites typically have many impurities and are therefore of limited use for catalytic applications. Synthetic zeolites can be obtained with exactly defined compositions, and desired particle sizes and shapes can be obtained by controlling the crystallization process.

The frameworks of zeolites are formed by fully connected SiO_4 and AlO_4 tetrahedra linked by shared oxygen atoms as shown in Figure 4.1 (top) for a Faujasite-type zeolite. Faujasite is a zeolitic mineral, which can be found in nature. Synthetic Faujasite-type zeolites are of particular importance in zeolite catalysis as we will see below.

The silicon and aluminum atoms are located on the centers of the tetrahedra and are frequently denoted as T-atoms (T for tetrahedra). A TO_4 unit could thus represent both SiO_4 and AlO_4 tetrahedra. The Faujasite structure, as shown in Figure 4.1, can be considered as being formed by double rings consisting of six TO_4 units each, denoted as D6R (double six ring) hereafter. Larger cages are formed by connecting these D6R, which build up cavities and open pore windows with a window diameter of about 0.7 nm. Obviously, the structure representation showing all atoms present in a structure is rather confusing and details of the structure are hardly visible. Therefore, a simplified representation is often chosen as shown in Figure 4.1 (middle). The oxygen atoms are omitted and straight lines directly connect the silicon or aluminum atoms in the centers of the SiO_4 and AlO_4 tetrahedra. As a result, substructures such as cages and channels are much easier recognized. Polyhedra representations as shown in Figure 4.1 (bottom), allow an even easier perception of the zeolite cages and pore openings. The large cages of the Faujasite structure are formed by smaller β-cages, which are connected by D6R units. Similar cages are found in many zeolite structures as can be seen in Figure 4.2.

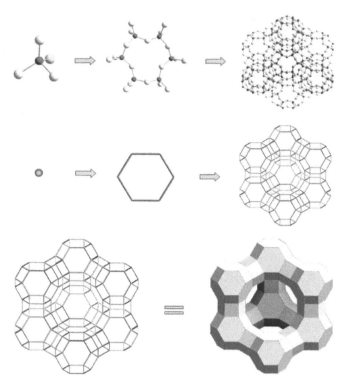

Figure 4.1 The framework structure of a Faujasite-type zeolite and simplified structure representations thereof (top: ball and stick model, middle: simplified stick model, and bottom: comparison stick model and polyhedra model).

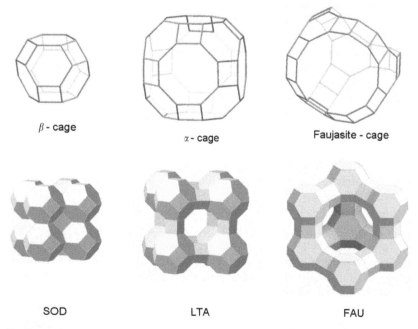

Figure 4.2 Zeolite cages as found in Sodalite (SOD), Zeolite A (LTA), and Faujasites (FAU). (Reprinted from D. Kern, N. Schall, W. Schmidt, R. Schmoll, J. Schürtz, *Winnacker-Küchler: Chemische Technik, Prozesse und Produkte*, Vol. 3, *Siliciumverbindungen*, pp. 849–850. Copyright 2005. With permission from Wiley-VCH.)

The connectivity (topology) of the zeolite framework is characteristic for a given zeolite type, whereas the composition of the framework and the type of extra-framework species can vary. Each zeolite structure type is denoted by a three-letter code [4]. As an example, Faujasite-type zeolites have the structure type FAU. The pores and cages of the different zeolites are thus formed by modifications of the TO_4 connectivity of the zeolite framework.

Besides the more or less stable framework, extra-framework cations, and inorganic or organic guest molecules may be found in the zeolite pores. What is the reason for the presence of extra-framework cations? We know that the zeolite framework is formed by SiO_4 and TiO_4 tetrahedra. Each O^{2-} anion is connected to two Si^{4+} cations, thus each O^{2-} anion provides one negative charge per Si^{4+} cation. Therefore, the four positive charges of a given Si^{4+} cation are compensated by four negative charges of the surrounding oxygen atoms. In the same way, the three positive charges of an Al^{3+} cation are balanced by four negative charges from the four surrounding oxygen atoms, resulting in one negative excess charge. Thus, each Al^{3+} cation in the zeolite framework causes a negative framework charge. These negative charges are balanced by extra-framework cations located in the zeolite pores, as shown schematically in Figure 4.3. Typically, synthetic zeolites are crystallized in alkaline reaction mixtures containing Na^+ cations and the charge compensating extra-framework cations are then Na^+ cations. The composition of such a zeolite can be written as:

$$Na_x(AlO_2)_x(SiO_2)_{y-x}nH_2O \quad \text{or} \quad Na_x[Al_xSi_{y-x}O_{2y+}]nH_2O$$

The chemical composition is often given as the composition of a unit cell of the respective zeolite structure (a unit cell is the smallest repetition unit of a crystal containing all information of the crystal structure). An important feature of a given zeolite material is its silicon to aluminum (Si/Al) ratio. Zeolites with an Si/Al ratio up to ten are called low silica zeolites, those with higher Si/Al ratios high silica zeolites. The framework structures of low silica zeolites contain a large number of aluminum cations. Since all these framework aluminum cations cause negative framework charges, balanced by extra-framework cations, such low silica zeolite have a very high ion exchange capacity. However, the lower limit for the Si/Al ratio is one. This implies that the framework of a normal zeolite never contains more aluminum than silicon.

Zeolite structures typically consist of silicon and aluminum linked by tetrahedrally coordinating oxygen atoms. However, similar structures as found for these aluminosilicates can be formed by substitution of the aluminum by other elements (e.g., Ga in gallosilicates or Ti in titanosilicates). Even the substitution of both Si and Al is possible, as for example in aluminophosphates or

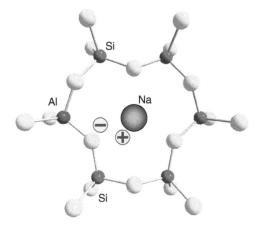

Figure 4.3 Compensation of a negative framework charge by a Na^+ cation.

gallophosphates. These types of materials are also considered to be zeolites but they are of minor importance as catalysts.

The micropores of zeolites typically contain water as guest molecules, which desorbed upon heating. However, most zeolites easily adsorb water from the ambient air. Owing to its specific pore size, the K^+-containing Zeolite A (see Figure 4.2, LTA type) is only capable of adsorbing water from the air. Nitrogen and oxygen molecules are too large to pass the 0.3 nm pore openings of this zeolite, which is thus an excellent water adsorbent and typically used for drying gases or organic liquids (Molsieve 3A).

The negative framework charges can also be balanced by many other cations, e.g., by most alkaline and earth alkaline cations as well as by transition metal cations. The extra-framework cations are usually easy to exchange by other ones, e.g., from aqueous solutions, provided the ions are small enough to fit into the pores. Some cations are surrounded by rather bulky water shells, preventing their migration into the zeolite pores. Stripping the water shell, e.g., by raising the temperature of the exchange solution, may help to overcome this restriction. The easy exchange of cations makes zeolites excellent ion exchangers. For this reason they are widely used as water softeners in detergents; Ca^{2+} and Mg^{2+} cations are removed from the solution by the zeolite and replaced by Na^+ cations.

Apart from metal cations, the negative framework can also be compensated by other positively charged cations such as NH_4^+ and H^+. The protonated forms of zeolites have a high acidity because the proton can be easily removed from the zeolite and replaced by other positively charged species. This feature makes them very efficient solid-state acids, which can be used for acid-catalyzed reactions.

The pore size of a zeolite depends on its structure, but in general, the pore openings are enclosed by a certain number of TO_4 units, forming circular or elliptic rings. Figure 4.4 shows the width of some circular pore openings with respect to the number of TO_4 units. Depending on the shape of the pore window, the width of the pores can vary in the range 0.3 to 0.8 nm for pores formed by rings consisting of six (6MR) to twelve (12MR) TO_4 units. Furthermore, the type and position of specific extra-framework cations can affect the pore opening, since these cations may partially block the pore windows. The diameters of the windows in zeolites determine the size of molecules being able to pass through these windows, making them very efficient as molecular sieves and as selective catalysts.

The pores of zeolites can be regarded as extensions of their surfaces; zeolites have an external surface, i.e., the surface of the zeolite crystallites, and an internal surface, i.e., the surface of their channels and/or cages. In total, the surface areas of zeolites are remarkably large. One gram of a typical Faujasite zeolite expresses a geometric surface area of about 1100 m^2/g (specific surface area). The contribution of the external surface area to this number is almost negligible (about 5 $m^2 g^{-1}$ for 1 μm crystallites), and almost the complete surface area is due to the surface of the micropores.

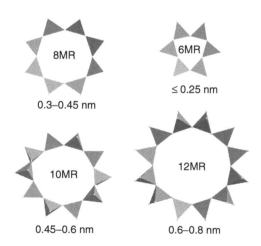

Figure 4.4 Typical pore diameters as observed in zeolites.

The determination of the specific surface area of a zeolite is not trivial. Providers of zeolites typically give surface areas for their products, which were calculated from gas adsorption measurements applying the Brunauer–Emmet–Teller (BET) method. The BET method is based on a model assuming the successive formation of several layers of gas molecules on a given surface (multilayer adsorption). The specific surface area is then calculated from the amount of adsorbed molecules in the first layer. The space occupied by one adsorbed molecule is multiplied by the number of molecules, thus resulting in an area, which is assumed to be the best estimate for the surface area of the solid. The BET method provides a tool to calculate the number of molecules in the first layer. Unfortunately, it is based on a model assuming multilayer formation. Yet, the formation of multilayers is impossible in the narrow pores of zeolites. Specific surface areas of zeolites calculated by the BET method (often termed BET surface area) are therefore erroneous and should not be mistaken as the real surface areas of a material. Such numbers are more related to the pore volume of a zeolite rather than to their surface areas.

Considering all we know up to now, the specific properties of zeolites can be summarized as follows. Zeolites are aluminosilicates with defined microporous channels or cages. They have excellent ion-exchange properties and can thus be used as water softeners and to remove heavy metal cations from solutions. Furthermore, zeolites have molecular sieve properties, making them very useful for gas separation and adsorption processes, e.g., they can be used as desiccants or for separation of product gas streams in chemical processes. Protonated zeolites are efficient solid-state acids, which are used in catalysis and metal-impregnated zeolites are useful catalysts as well.

4.3.2　Types of Zeolites Used in Catalytic Processes

Many types of zeolites are known but only a rather small number of zeolites are used in catalysis. In this section, the most important zeolites will be introduced. We will focus on the most commonly used types which are Zeolite X, Zeolite Y, ZSM-5, and Zeolite Beta. Apart from these, a couple of other zeolites, e.g., Mordenite or Zeolite L, are also used for specific reactions but they are produced on a smaller scale. Most of these zeolites have a remarkable thermal stability and can be heated to a temperature of 600°C without structural damage; some of them resist even temperatures of 800 to 1000°C.

The nomenclature of zeolites is rather arbitrary and follows no obvious rules because every producer of synthetic zeolites uses his/her own acronyms for the materials. However, as mentioned before, at least the structure types of the different zeolites have a unique code. For example, FAU represents Faujasite-type zeolites, LTA Linde Type A zeolites, MFI Mobile Five, and BEA Zeolite Beta. The structure commission of the International Zeolite Association (IZA) is the committee granting the respective three-letter codes [4]. Some typical zeolites, which are of importance as catalysts in petrochemistry, will be described in the following sections.

4.3.2.1　Zeolite X and Zeolite Y

Faujasite is a zeolite which can be found in nature. Unfortunately, natural zeolite deposits typically contain impurities of other mineral phases and the chemical composition and morphology of the zeolite materials can vary widely . For catalytic applications, generally, very well-defined zeolite materials are required. Therefore, rather than natural zeolites, their synthetic homologs are typically used for catalytic conversions. They are obtained in high purity and their properties can be minutely controlled during their formation. Zeolites X and especially Zeolite Y are the most frequently used in petrochemistry. They both have a FAU-type structure (Faujasite, see Figure 4.1) but differ with respect to their chemical composition. Zeolite X has an Si/Al ratio between 1 and 1.4, while Zeolite Y has between 2 and 3. In between these two compositions, mixed forms of both types of zeolites exist. The differentiation into two different zeolites is not arbitrary but due to significant steps in the evolution of the unit cell parameters of the crystal structures with increasing aluminum content. Thus, a typical

Zeolite X has a general composition of $(Ca^{2+}, Mg^{2+}, Na_2^+, K_2^+)_{0.5x}Al_xSi_{192-x}O_{384}\cdot240\ H_2O$ with $x = 96$ to 80 and Zeolite Y a composition of $(Ca^{2+}, Mg^{2+}, Na_2^+)_{0.5x}Al_x\ Si_{192-x}O_{384}\cdot240\ H_2O$ with $x = 64$ to 48). The cations in the parentheses may be found as extra-framework cations in the respective zeolites. The compositions reflect the composition of a unit cell of the respective zeolite. In order to increase the thermal stability and acidity of the zeolite, Zeolite Y is steam-treated and then calcined. During the steaming process, aluminum is released from the zeolite structure. The structure thus damaged is then healed in the calcination process. The resulting material is called dealuminated Zeolite Y (DAY).

The pore windows of Zeolites X and Y consist of 12MR with an approximate window diameter of 0.74 nm. They give access to the larger Faujasite cages (see Figure 4.2) with a diameter of about 1.4 nm. Catalytically active sites, e.g. protons, are located either in these large cages, in the 12MR windows, or in the smaller β-cages and/or D6R cages (see Figure 4.1 and Figure 4.2).

4.3.2.2 ZSM-5

The zeolite ZSM-5 has the MFI-type structure and can be obtained with many different Si/Al ratios typically ranging from about 10 to ∞. If no aluminum is present (Si/Al = ∞), a pure siliceous structure is obtained and the resulting material is then called Silicalite-1. The unit cell composition of an MFI-type zeolite can be written as $(Na^+, H^+)_xAl_xSi_{96-x}O_{192}\cdot n\ H_2O$ with $x \leq 27$ (in most cases <9) and $n \leq 16$ (the more siliceous a zeolite gets, the higher is its hydrophobicity).

The MFI structure has pores formed by channels encircled by ten T-atoms (10MR). Two individual 10MR pore systems are found in the structure, straight channels, and perpendicular to them sinusoidal (zig-zag type) channels. Both channel types intersect and form larger cavities at the intersections. The channel system is open in all three directions and adsorbed molecules can migrate in all three directions. Such a channel system is called three-dimensional. Figure 4.5 shows a schematic view into the straight (left) and sinusoidal (right) channels. The pore diameters of the straight channels are 0.56×0.53 nm and those of the slightly elliptical sinusoidal channels 0.55×0.51 nm.

4.3.2.3 Mordenite

The natural zeolite Mordenite is found as a mineral in large deposits. However, for catalytic applications, only synthetic Mordenite is used. The unit cell composition of a typical MOR-type zeolite can be written as $(Na^+, H^+)_8Al_8Si_{40}O_{96}\cdot nH_2O$ with $n \approx 24$. The Mordenite structure has 12MR channels that are interconnected by smaller 8MR windows, as shown in Figure 4.6. Larger molecules which cannot pass through the 8MR windows can migrate only along the straight 12MR channels. Since larger molecules cannot pass each other inside the channels, they only can diffuse in single file

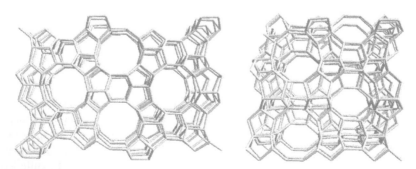

Figure 4.5   View into the pores of the ZSM-5 (MFI) structure along the straight channels (left) and along the sinusoidal channels (right). (Reprinted from D. Kern, N. Schall, W. Schmidt, R. Schmoll, J. Schürtz, *Winnacker-Küchler: Chemische Technik, Prozesse und Produkte*, Vol. 3, *iliciumverbindungen*, pp. 849–850. Copyright 2005. With permission from Wiley-VCH.)

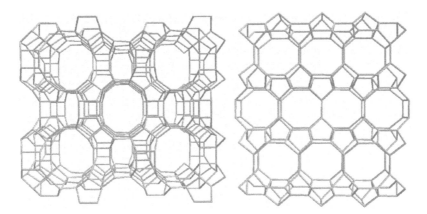

Figure 4.6 View into the 12MR channels (left) and the 8MR windows of the Mordenite (MOR) structure (right).

mode. Figure 4.6 shows a view into the pores of the MOR structure; the straight 12MR channels with a diameter of about 0.71×0.62 nm and 8MR windows with a diameter of about 0.48×0.38 nm.

4.3.2.4 Zeolite A

Zeolite A with the LTA structure type is typically not used as a catalyst but frequently used as a drying agent and as an adsorbent in separation processes. It is indispensable for the separation of product streams, e.g., from the skeletal isomerization process in petrochemistry where it is used for the separation of *n*-alkanes from *iso*-alkanes. A structure model of Zeolite A is shown in Figure 4.2. Eight β-cages located on the corners of a cube are connected by D4R units, thus forming the larger α-cages. The largest pore windows consist of 8MR openings with a diameter of about 0.45 nm. This window diameter may be smaller if cations partially block the windows. The size of the pore openings then depends on the size of the respective cation and is thus tunable. The unit-cell composition of Zeolite A can be given as $Na_{96}Al_{96}Si_{96}O_{384} \cdot 216 \, H_2O$. As for the other zeolites, the sodium cation can be exchanged with various cations. However, a protonated Zeolite A is not stable and decomposes during the protonation process.

The pore types FAU, MFI, MOR, and LTA discussed above contain basically all micropore types possible in zeolites. There are accessible and nonaccessible cages (FAU and LTA) and straight or meandering channels in one, two, or three dimensions, which may be either isolated or connected to each other (MFI and MOR).

4.3.3 Production of Zeolites

The annual consumption of zeolites lies in the range of some million tons (global consumption about 4 Mio t/year in year 2000) and zeolites are produced for many different applications as shown in Figure 4.7. The large amounts of natural zeolites produced in Cuba and China are omitted in this figure. Natural zeolites are basically used for cement production or in agriculture. The largest fraction of synthetic zeolites is used as water softeners in detergents, and catalysts make only about 8% use of all zeolites produced in United States, Western Europe, and Japan. However, about 55 to 60% of the total profits on the zeolite market are made on zeolite catalysts [5].

Typically, for production of zeolites (Figure 4.8), a silicon source such as sodium silicate and an aluminum source such as sodium aluminate, are prepared in solutions containing sodium and water contents as required for the formation of the respective zeolite [6]. These solutions are mixed in a reactor and reacted at temperatures typically in the range between 80 and 200°C. The reaction time may vary from hours to days, and for reactions at temperatures $> 100°C$ the reactions have to be performed in autoclaves. For the production of some zeolites (e.g., ZSM-5), organic additives are added to the reaction mixture. These additives direct the crystallization toward the desired products and are typically ammonium salts, amines, or alcohols. The organic molecules are occluded inside the zeolite pores and have to be removed by calcination in an additional step. After the crystallization, the zeolite is filtered, washed, and, if required, ion-exchanged. After drying of the product, the material can be packed as zeolite powder. For the application as catalyst, the zeolite powder is typically processed to give granules (see Figure 4.8) or extrudates. A binder (e.g., clay slurry) is admixed to the zeolite powder and the resulting mixture is either granulated or extruded. The green bodies are then calcined and sieved to obtain a product with a specific grain size.

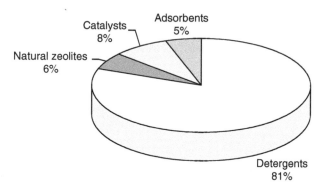

Figure 4.7   Zeolite production in United States, Western Europe and Japan in Year 2000 (total 1.3 Mio t/years). (Reprinted from D. Kern, N. Schall, W. Schmidt, R. Schmoll, J. Schürtz, *Winnacker-Küchler: Chemische Technik, Prozesse und Produkte*, Vol. 3, *Siliciumverbindungen*, pp. 849–850. Copyright 2005. With permission from Wiley-VCH.)

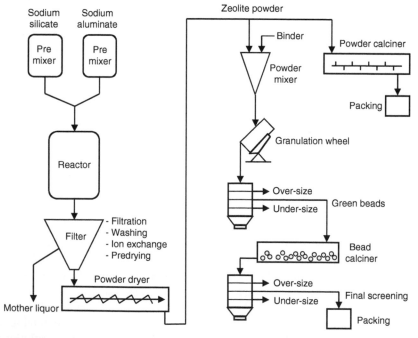

Figure 4.8   Industrial production of zeolites. (Reprinted from *Winnacker-Küchler: Chemische Technik, Prozesse und Produkte*, Vol. 3, D. Kern, N. Schall, W. Schmidt, R. Schmoll, J. Schürtz, *Siliciumverbindungen*, pp. 849–850. Copyright 2005. With permission from Wiley-VCH.)

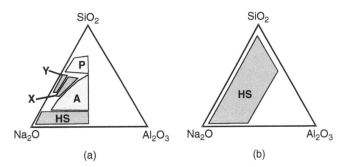

Figure 4.9 Crystallization field in the $Na_2O/Al_2O_3/SiO_2$ system at 100°C and a water content of (a) 90 to 98 mol% H_2O and (b) 60 to 85 mol% H_2O (A,X,Y, Zeolite A, X, Y; P, Zeolite P1 (GIS) and HS, Hydroxysodalite). (Reprinted from D. Kern, N. Schall, W. Schmidt, R. Schmoll, J. Schürtz, *Winnacker-Küchler: Chemische Technik, Prozesse und Produkte,* Vol. 3, *Siliciumverbindungen,* pp. 849–850. Copyright 2005. With permission from Wiley-VCH.)

In addition to these steps, further processing is possible, e.g., impregnation with a noble metal. Zeolites A, X, and Y can be obtained at temperatures below 100°C, while the high-silica zeolites ZSM-5 and Zeolite Beta are typically crystallized at temperatures between 100 and 200°C. The composition of the reaction mixture and the crystallization time are very important parameters that have to be controlled carefully, since even small deviations may result in a product with undesired zeolitic or nonzeolitic by-products or, in the worst case, in complete nonzeolitic products. Figure 4.9a shows the ranges in which Zeolites A, X, and Y can be obtained [7]. Keeping all other parameters constant but reducing the water content results in hydroxysodalite, a non-desired product (Figure 4.9b).

The aging time of the reaction mixture (prior to crystallization) at room temperature or slightly elevated temperature, the heating rate applied to reach the crystallization temperature, the stirring speed (in reactor), and an optional addition of seed crystals (zeolite crystals) are all process parameters influencing the product obtained. While the crystallization of a specific zeolite is the main objective, however, the control of product properties such as the crystallite size, the particle morphology, crystal aggregation, and the chemical composition of the crystallites, are in most cases also of utmost importance. The preparation of zeolite catalysts is a challenge and requires precise control of every process parameter.

4.3.4 Post-Synthesis Treatment of Zeolites and Modification of Zeolites

4.3.4.1 Protonation of Zeolites

Exchanging the extra-framework cations with protons results in an acidic zeolite. These protons are mobile and can participate in chemical reactions inside the zeolite pores. Thus, protonated zeolites can be regarded as solid acids with Bronsted acid sites. They can be obtained by direct exchange of sodium cations by protons in acidic solution, but many zeolites suffer a partial or complete destruction of their frameworks in acidic solutions. This is due to a removal of aluminum from the framework in acidic environment. Framework dealumination can be avoided by exchanging the sodium cations in zeolites with ammonium cations. The resulting NH_4^+-exchanged zeolite is then transformed into its protonated form by simple heat treatment, typically at temperatures of about 450 to 550°C. During this calcination, the ammonium inside the zeolite decomposes into ammonia and protons according to the following equation:

$$Z^-NH_4^+ \rightarrow Z^-H^+ + NH_3\uparrow$$

The ammonia is released and the protons remain in the zeolite, which then can be used as acidic catalysts. Applying this method, all extra-framework cations can be replaced by protons. Protonated zeolites with a low Si/Al ratio are not very stable. Their framework structure decomposes even upon moderate thermal treatment [8–10]. A framework stabilization of Zeolite X or Y can be achieved by introducing rare earth (RE) cations in the Sodalite cages of these zeolites. Acidic sites are obtained by exchanging the zeolites with RE cations and subsequent heat treatment. During the heating, protons are formed due to the autoprotolysis of water molecules in the presence of the RE cations as follows:

$$RE^{n+} + (H_2O)_{n-1} \rightarrow [RE(OH)_{n-1}]^+ + (n-1)H^+$$

In this way, fairly stable acidic REY and REX zeolites are formed containing not only protons but also oxidic RE species. High-silica zeolites are more stable in acidic solutions and can be exchanged directly by protons from acidic solutions. However, a certain structure dealumination is often observed for these materials as well and typically not all cations are replaced by protons.

4.3.4.2 Dealumination of Zeolites

The dealumination may result in unstable zeolites as described above. However, in the case of Zeolite Y, an extremely stable material can be achieved after dealumination under appropriate conditions [10]. Applying either repeated ammonium ion exchanges and successive high-temperature treatments to Na–Y or steaming and elevated temperature to H–Y leads to a so-called Ultrastable Zeolite Y (USY). The successive removal of the aluminum from the zeolite framework according to the reaction shown in Figure 4.10, results in the formation of hydroxyl nests. The interrupted structure is then healed at temperatures around 700 to 850°C. During this step, neighboring silanol groups condense to form new Si–O–Si bonds and silicon atoms migrate to fill up the vacant positions. The extracted aluminum forms oxidic species inside the zeolite pores. The nonframework aluminum can be removed from the zeolite by treatment with sodium hydroxide or EDTA solution. In this way, a Zeolite Y is obtained which is thermally extremely stable and widely used for cracking catalysts in oil refineries.

4.3.4.3 Metals and Metal Complexes in Zeolites

Some metals (e.g., Ti, Fe, Ni, Co) can be incorporated directly into the zeolite framework structure, thus substituting silicon on sites in a given siliceous zeolite. The most prominent material of that type is Titanium Silicalite-1 (TS-1), a zeolite with MFI-type structure. The Ti^{4+} is coordinated tetrahedrally by four framework oxygen atoms. TS-1 is a very efficient catalyst for oxygenation reactions (partial oxidation, e.g., of olefins toward epoxides) using hydrogen peroxide as oxidizing agent [11–13].

In order to combine the catalytic activity of highly dispersed metal species and that of zeolites, metals can be deposited in the pores and on the external surface of zeolite particles. In this way, a catalyst is formed with both a metal functionality, e.g., redox or hydrogenation activity, and an acidic function. The metals can be deposited by different methods. Impregnation of a zeolite with a metal

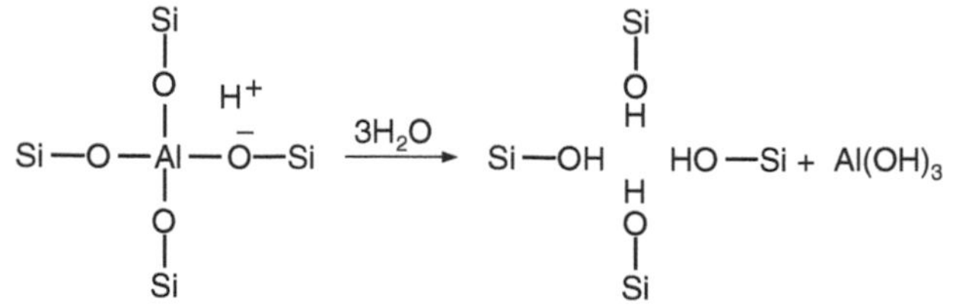

Figure 4.10 Dealumination of a zeolite structure and formation of a hydroxyl nest.

salt solution and successive evaporation of the water results in the deposition of the metal in the zeolite. However, in addition to the metal cations, large amounts of the anionic counterions may also get deposited in the zeolite. A very convenient way to load a zeolite with only the active cationic species is the ion exchange from aqueous metal salt solutions. The anions remain in the aqueous phase and can be removed by washing with water. Cations can also be exchanged by solid-state ion exchange. In this case, the zeolite powder is intimately mixed with a suitable metal salt or oxide. The mixture is then heated to temperatures at which the ion exchange proceeds via a solid-state reaction or from the molten salt covering the zeolite particles. The temperatures for this ion-exchange process typically lie in the range between 200 and 800°C. By exchanging the sodium cations in a given zeolite with a redox-active cation, selective oxidation and/or reduction catalysts can be obtained [11]. Not only hydrated cations but also metal complexes can be ion exchanged. Thus, it is very convenient to exchange Pt^{2+} or Pd^{2+} as $[Pt(NH_3)_4]^{2+}$ or $[Pd(NH_3)_4]^{2+}$ complexes.

Highly dispersed metal particles can be deposited in the cages of zeolites or on their external surface by reducing the metal cations in the exchanged zeolites in a hydrogen stream at elevated temperatures. The reduced metal atoms aggregate and form metal particles. During the reduction of the metal cations by hydrogen, the resulting protons remain in the zeolite and balance the framework charges according to the equation:

$$Z^{n-}Me^{n+} + n/2H_2 \rightarrow Z^{n-}(H^+)_n + Me.$$

Another method to introduce metal particles into zeolites is the deposition of metal carbonyl from a gas stream. The gaseous carbonyls are adsorbed in the micropores of the zeolite and then transformed into the metallic species.

Metal species deposited inside larger cages in zeolite are sometimes transformed into complexes by reacting them with appropriate ligands. In this way, larger complexes that are too large to pass the cage windows can be deposited in a zeolite. This method is frequently called ship-in-the-bottle synthesis.

Finally, one should note that metal species in a zeolite may act as Lewis acid sites. These sites are less acidic than Bronsted acid sites, i.e., protons, but contribute to the total acidity of a given zeolite.

4.3.5 Catalytic Application of Zeolites

The reason for the high selectivity of zeolite catalysts is the fact that the catalytic reaction typically takes place inside the pore systems of the zeolites. The selectivity in zeolite catalysis is therefore closely associated to the unique pore properties of zeolites. Their micropores have a defined pore diameter, which is different from all other porous materials showing generally a more or less broad pore size distribution. Therefore, minute differences in the sizes of molecules are sufficient to exclude one molecule and allow access of another one that is just a little smaller to the pore system. The high selectivity of zeolite catalysts can be explained by three major effects [14]: reactant selectivity, product selectivity, and selectivity owing to restricted size of a transition state (see Figure 4.11).

If reactants are too large to fit into the zeolite pores, they are excluded from the pore system and thus cannot react inside the zeolite pores. Of course, they can react on the external surface area of the zeolite crystals. However, the surface areas of zeolite pores (i.e., the internal surface area of micropores) are several magnitudes larger than the surface area on top of the crystals (i.e., the external surface area) and reactions proceeding on the external surface of a zeolite are in most cases negligible. The high selectivity is thus due to the fact that only a specific fraction of all potential reactants are actually converted by the zeolite catalyst. In a similar way, only products of a distinct size are able to leave the pore system of a zeolite. Larger molecules are retained inside the zeolite pores and are either converted into a product that is small enough to leave the pore system or they are kept inside the zeolite, e.g., in the form of a carbon-rich deposit (coke). The selectivity is thus due to the size of the product molecules. These two types of selectivity are quite obvious; however,

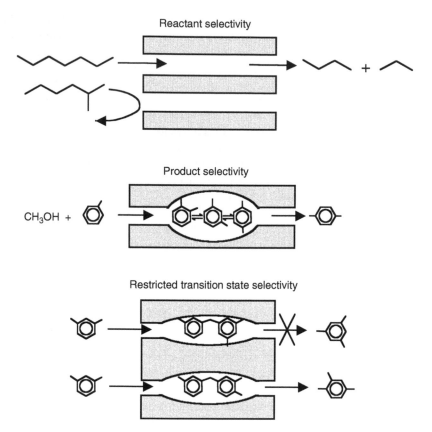

Figure 4.11 Different types of selectivity in zeolite catalysts. (Reprinted from *Introduction to Zeolite Science and Practice, Studies in Surface Science and Catalysis*, Vol. 137, I.E. Maxwell, W.J.H. Stork, *Introduction to Zeolite Science and Practice, Studies in Surface Science and Catalysis*, Vol. 137, *Hydrocarbon Processing with Zeolites*, pp. 747–819. Copyright 2001. With permission from Elsevier.)

there exists a third type of selectivity. For some reactions, rather bulky transition states are required. These transition states can be of different sizes for different products. Only such products can be formed which result from a transition state being small enough to be realized inside the restricted pore geometry of the zeolite. Zeolites are thus highly selective catalysts that find their major catalytic application in petrochemistry. However, they can be also applied profitably in the production of organic base chemicals and fine chemicals.

Zeolites are solids and if they are used as catalysts for the conversion of organics in a batch reactor, they can be easily separated, e.g., by filtration. A distinct separation step is dispensable if gases or vapors are reacted over a zeolite catalyst bed. In this set-up, the gas/vapor passes through a bed of the solid catalyst and the gaseous products leaving the reactor are easily recovered. A complicated separation of the catalyst from the product, as compulsory in homogeneous catalysis, is not necessary. Protonated zeolites are acidic and can replace mineral acids as catalysts in many organic reactions. To describe all possible reactions is beyond the scope of the present chapter. To give a rough impression of the vast number of different reactions possible with zeolite catalysts, a list of possible reactions and the structure type of zeolite used are given in Table 4.1 [15]. For the zeolites used, the table lists only the structure types (three-letter codes). One has to be aware that most of the listed zeolite catalysts are not used in their as-synthesized form. They are typically modified to meet the requirements of a specific reaction (e.g., H, Fe, B, Ti, Cu, K, RE — modified forms of zeolites) and may require an additional metal function (e.g., Pt and Pd, etc.). Table 4.1 gives only a selection of possible reactions and is of course not comprehensive. A review on these and other

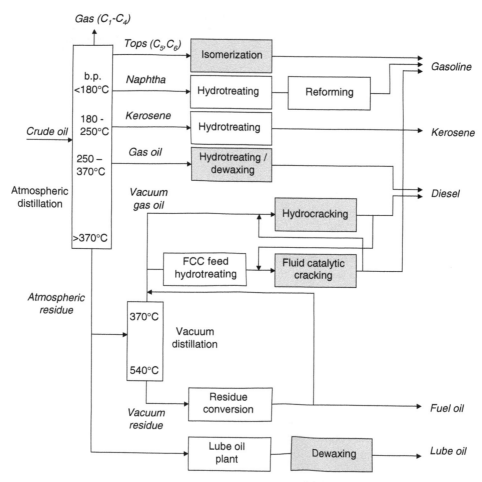

Figure 4.12 Products and simplified product streams of oil refinery processes. The yellow color indicates processes using zeolite catalysts.

processes. It is thus also mandatory to remove nitrogen and sulfur to maintain the catalytic performance of downstream catalysts.

The processes in an oil refinery are very complex and a complete description would exceed the volume of the present chapter. Here we will focus on the main processes involving zeolite catalysts. These processes are indicated by gray boxes in Figure 4.12. One should be aware of the fact that the scheme shown in Figure 4.12 is oversimplified for clarity. Many processes being essential in a petrol plant setup are omitted for simplicity. More comprehensive information can be found in the literature given in the appendix.

4.3.5.1.1 Fluid Catalytic Cracking

Catalytic cracking is a process that is currently performed exclusively over fluidized catalyst beds. The fluid catalytic cracking (FCC) process was introduced in 1942 and at that time replaced the conventional moving bed processes. These early processes were based on acid-treated clays as acidic catalysts. The replacement of the amorphous aluminosilicate catalysts by Faujasite-type zeolites in the early-1960s is regarded as a major improvement in FCC performance. The new acidic catalysts had a remarkable activity and produced substantially higher yields than the old ones.

Fluid catalytic cracking over an acid catalyst converts residual hydrocarbons from the vacuum oil fraction into valuable olefins, gasoline, and diesel products. The catalytic cracking at temperatures of about 500 to 600°C. The long-chain hydrocarbon feed (paraffins), preheated to 200 to 300°C, is mixed with hot fluidized zeolite catalyst having a temperature of about 650 to 750°C. At the temperature resulting in the fluid bed, the long-chain hydrocarbons are cracked upon contact with the acid catalyst. The volume expansion because of the evaporation of the feed and the formation of the smaller cracking products and an injection of a steam flow causes the zeolite particles to lift in the riser. The catalyst moves through the riser within a few seconds (2 to 3 sec) and continues to crack the feed molecules. As a result of the catalytic cracking, not only short-chain products but also a carbonaceous deposit inside the zeolite pores (0.8 to 1.3 wt.%) are formed. This coke leads to a rapid deactivation of the catalyst that has to be regenerated. For this, the coke-loaded zeolite is separated from the product gas stream in a stripper and then transferred to a regenerator by the aid of a steam flow, as illustrated in the scheme in Figure 4.13 [16,17]. The steam also removes residual long-chain hydrocarbons from the zeolite.

In the regenerator, the coke is removed from the zeolite by simply burning it off. For this, air is passed through the zeolite bed in the regenerator operating at about 700°C. During the regeneration, the coke combusts and the zeolite heats up to a temperature of about 650 to 750°C. After this, the hot regenerated zeolite is ready to be transferred again into the riser reactor.

Several reaction pathways for the cracking reaction are discussed in the literature. The commonly accepted mechanisms involve carbocations as intermediates. Reactions probably occur in catalytic cracking are visualized in Figure 4.14 [17,18]. In a first step, carbocations are formed by interaction with acid sites in the zeolite. Carbenium ions may form by interaction of a paraffin molecule with a Lewis acid site abstracting a hydride ion from the alkane molecule (1), while carbonium ions form by direct protonation of paraffin molecules on Bronsted acid sites (2). A carbonium ion then either may eliminate a H_2 molecule (3) or it cracks, releases a short-chain alkane and remains as a carbenium ion (4). The carbenium ion then gets either deprotonated and released as an olefin (5,9) or it isomerizes via a hydride (6) or methyl shift (7) to form more stable isomers. A hydride transfer from a second alkane molecule may then result in a branched alkane chain (8). The

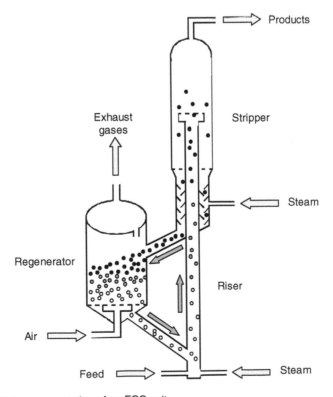

Figure 4.13 Schematic representation of an FCC unit.

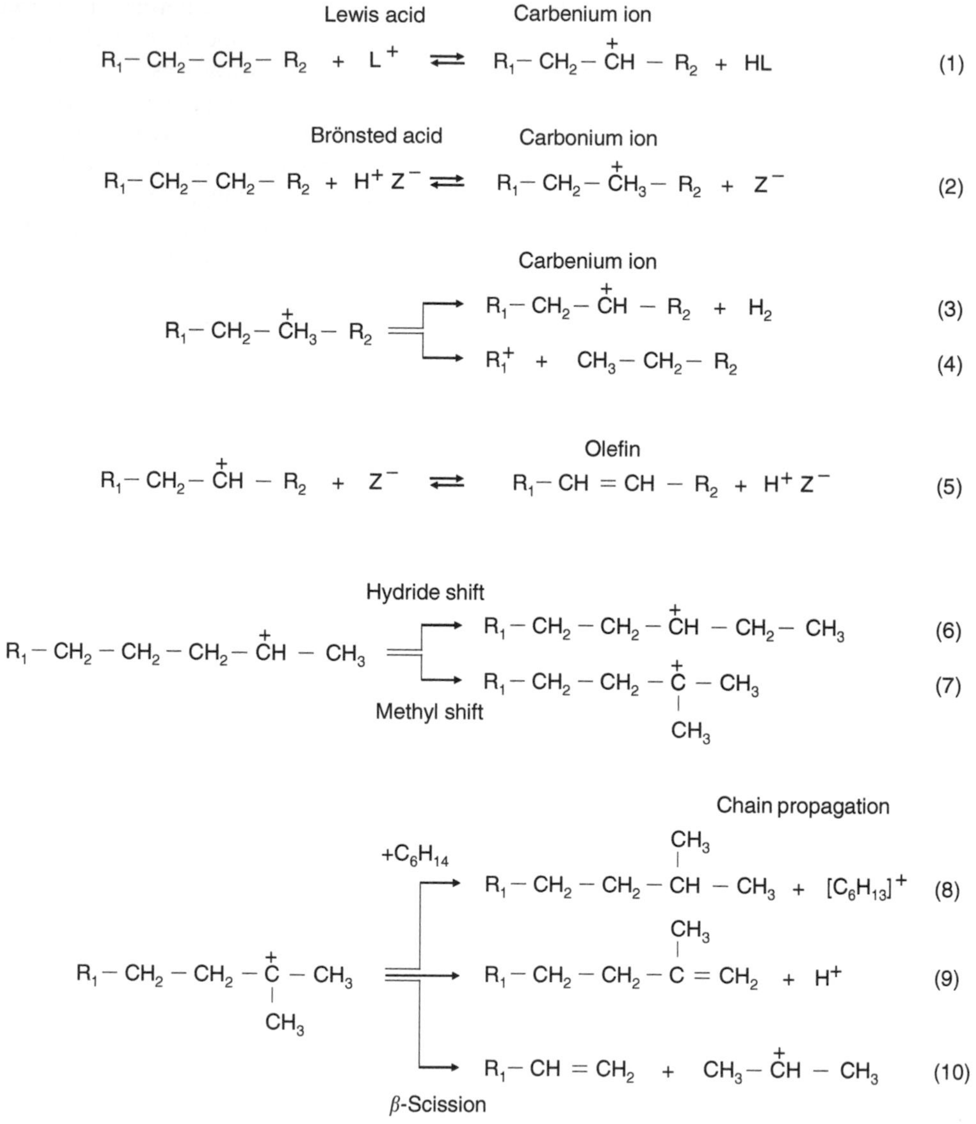

$$\text{Lewis acid} \qquad \text{Carbenium ion}$$

$$R_1 - CH_2 - CH_2 - R_2 \; + \; L^+ \; \rightleftharpoons \; R_1 - CH_2 - \overset{+}{CH} - R_2 \; + \; HL \tag{1}$$

$$\text{Brönsted acid} \qquad \text{Carbonium ion}$$

$$R_1 - CH_2 - CH_2 - R_2 \; + \; H^+ Z^- \; \rightleftharpoons \; R_1 - CH_2 - \overset{+}{CH_3} - R_2 \; + \; Z^- \tag{2}$$

$$\text{Carbenium ion}$$

$$R_1 - CH_2 - \overset{+}{CH_3} - R_2 \left\{ \begin{array}{l} \longrightarrow R_1 - CH_2 - \overset{+}{CH} - R_2 \; + \; H_2 \qquad\qquad (3) \\[2ex] \longrightarrow R_1^+ \; + \; CH_3 - CH_2 - R_2 \qquad\qquad\quad (4) \end{array} \right.$$

$$\text{Olefin}$$

$$R_1 - CH_2 - \overset{+}{CH} - R_2 \; + \; Z^- \; \rightleftharpoons \; R_1 - CH = CH - R_2 \; + \; H^+ Z^- \tag{5}$$

$$\text{Hydride shift}$$

$$R_1 - CH_2 - CH_2 - CH_2 - \overset{+}{CH} - CH_3 \left\{ \begin{array}{l} \longrightarrow R_1 - CH_2 - CH_2 - \overset{+}{CH} - CH_2 - CH_3 \qquad (6) \\[2ex] \longrightarrow R_1 - CH_2 - CH_2 - \overset{+}{\underset{\underset{\textstyle CH_3}{|}}{C}} - CH_3 \qquad\qquad\quad (7) \end{array} \right.$$

$$\text{Methyl shift}$$

$$\text{Chain propagation}$$

$$R_1 - CH_2 - CH_2 - \overset{+}{\underset{\underset{\textstyle CH_3}{|}}{C}} - CH_3 \left\{ \begin{array}{l} \underset{+C_6H_{14}}{\longrightarrow} R_1 - CH_2 - CH_2 - \underset{\underset{\textstyle CH_3}{|}}{\overset{\overset{\textstyle CH_3}{|}}{CH}} - CH_3 \; + \; [C_6H_{13}]^+ \quad (8) \\[3ex] \longrightarrow R_1 - CH_2 - CH_2 - \underset{\underset{\textstyle CH_3}{|}}{\overset{\overset{\textstyle CH_3}{|}}{C}} = CH_2 \; + \; H^+ \qquad\qquad (9) \\[3ex] \longrightarrow R_1 - CH = CH_2 \; + \; CH_3 - \overset{+}{CH} - CH_3 \qquad (10) \end{array} \right.$$

$$\beta\text{-Scission}$$

Figure 4.14   FCC reaction pathways.

second molecule is transferred into a carbenium ion while the first one can leave the zeolite. Finally, the carbenium ion cracks via β-scission into smaller molecules (10). Alkylaromatics from the feed may also get cracked, e.g., forming an alkyl molecule and a smaller aromatic molecule. However, aromatics and coke are also formed by hydrogen-transfer reactions in the zeolite. The extent to which either hydrogen transfer or cracking takes place depends on the reaction parameters, the feed composition, and the catalyst applied. Undesired olefins in the product stream can be hydrogenated with hydrogen in a subsequent step in a hydrotreating or hydrocracking unit.

The main components of FCC catalysts are Zeolite Y, e.g., REY or USY as the major active component (10 to 50%), and a binder that is typically an amorphous alumina, silica-alumina, or clay material. In addition to these main components, other zeolite components, e.g., ZSM-5, and other oxide or salt components are quite frequently used additives in the various FCC catalysts available on the market. The addition of 1 to 5% ZSM-5 increases the octane number of the gasoline. ZSM-5 eliminates feed compounds with low octane numbers because it preferentially center-cracks n-paraffins producing butene and propene [14]. These short-chain olefins are then used as alkylation feedstocks in alkylation processes. Depending on the petrol plant specifications, the feedstock composition, and the desired product composition, the compositions of the catalysts are typically adjusted to the specific requirements. Worldwide, the FCC process is the conversion process applied most widely and FCC catalysts represent the most important single market for refinement catalysts [16].

4.3.5.1.2 Hydrocracking

Hydrocracking is the second large-scale process used to convert heavy fuel oil into more valuable products. It combines two catalytic processes, the hydrogenation of olefins and aromatics, and the cracking of larger molecules. The cracking is performed in the presence of hydrogen (30 to 150 bar) and at temperatures between 350 and 450°C. The products of the hydrocracking process contain less unsaturated compounds than those from the FCC process and the deactivation of the catalyst, e.g., due to coke formation, is extremely reduced as a consequence of the hydrogenation function. The lifetime for hydrocracking catalysts is in the range of several years. The hydrocracking process is very flexible and allows adjustment to various feed compositions and product demands. Hydrocracking units consist either of a single-stage bed reactor containing a catalyst with both a hydrogenation and an acidic (cracking) function, or of two independent bed reactors. In the latter set-up, the first reactor contains the hydrogenation catalyst converting olefins and aromatics into saturated products and eliminating sulfur and nitrogen (HDS and HDN) present in the feed. Sulfur and nitrogen heteroatoms are converted into H_2S and NH_3. In the second reactor, hydrocracking is performed which transforms the long-chain molecules into smaller ones. To achieve complete conversion, the fraction of unconverted feed is recycled to the hydrocracker or further processed in an FCC unit. Various configurations for the hydrocracking process are shown in Figure 4.15 [14]. The reactions taking place in a hydrocracker are summarized in Figure 4.16 and Figure 4.17 [17,18].

Hydrogenation of aromatic compounds can result in cycloalkanes (11) and short-chain fragments from side groups cleavage (12,13). Cycloalkanes can crack into smaller fragments (14) and paraffins are either cracked (15) or isomerized (16). The reaction pathways occurring in hydrocracking are shown schematically in Figure 4.17. In a first step, paraffin molecules are dehydrogenated at the metal site (18). The resulting olefin is then protonated at the acidic site to form a carbenium ion (19). The carbenium ion now may go through hydride (19) or methyl shift (20),

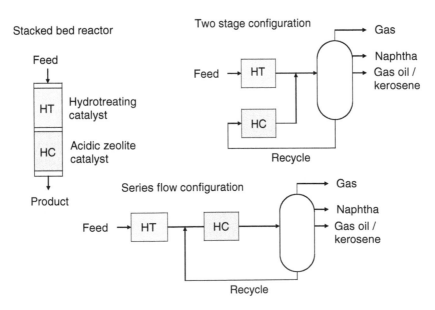

Figure 4.15 Configurations for the hydrocracking process.

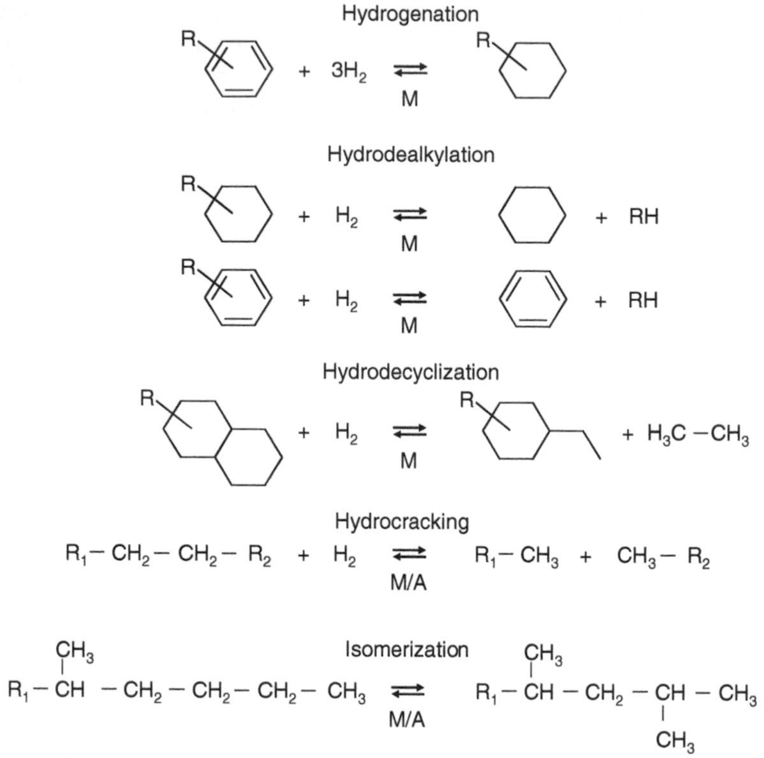

Hydrogenation

$$R\text{—}\langle\text{ring}\rangle \;+\; 3H_2 \;\underset{M}{\rightleftarrows}\; R\text{—}\langle\text{ring}\rangle \tag{11}$$

Hydrodealkylation

$$R\text{—}\langle\text{ring}\rangle \;+\; H_2 \;\underset{M}{\rightleftarrows}\; \langle\text{ring}\rangle \;+\; RH \tag{12}$$

$$R\text{—}\langle\text{ring}\rangle \;+\; H_2 \;\underset{M}{\rightleftarrows}\; \langle\text{ring}\rangle \;+\; RH \tag{13}$$

Hydrodecyclization

$$R\text{—}\langle\text{rings}\rangle \;+\; H_2 \;\underset{M}{\rightleftarrows}\; R\text{—}\langle\text{ring}\rangle\text{—}\!\!\backslash \;+\; H_3C\text{—}CH_3 \tag{14}$$

Hydrocracking

$$R_1\text{—}CH_2\text{—}CH_2\text{—}R_2 \;+\; H_2 \;\underset{M/A}{\rightleftarrows}\; R_1\text{—}CH_3 \;+\; CH_3\text{—}R_2 \tag{15}$$

Isomerization

$$\overset{\displaystyle CH_3}{\underset{\displaystyle |}{R_1\text{—}CH}}\text{—}CH_2\text{—}CH_2\text{—}CH_2\text{—}CH_3 \;\underset{M/A}{\rightleftarrows}\; \overset{\displaystyle CH_3}{\underset{\displaystyle |}{R_1\text{—}CH}}\text{—}CH_2\text{—}\overset{}{\underset{\displaystyle \underset{\displaystyle CH_3}{|}}{CH}}\text{—}CH_3 \tag{16}$$

Figure 4.16 Reactions occurring in hydrocracking process (M, metallic site; A, acidic site).

resulting in isomers (21), or β-scission (22), resulting in cracked products. Finally, olefins are hydrogenated to form n- and iso-paraffins at the metal site (23,24).

The acid component of a hydrocracking catalyst can be an amorphous oxide, e.g., a silica-alumina or a zeolite, e.g., USY. This component usually serves as a support for the metal compound responsible for the hydrogenation function. The metal compound can be a noble metal, e.g., Pt or Pd, or a mixture of sulfides, e.g., of Ni/Mo, Ni/W, or Co/Mo. The relative amounts of the respective compounds have to be thoroughly balanced to achieve an optimum performance.

4.3.5.1.3 Isomerization of n-Paraffins

Gasoline with high octane numbers prevents uncontrolled detonations of air/gasoline mixtures in car engines (knocking). Gasoline with an octane number of 100 has the same knock resistance as pure iso-octane (trimethyl pentane). An octane number of 0 corresponds to the performance of pure n-heptane. A conventional gasoline with an octane number of 96 thus has the knock resistance as a mixture of 96% iso-octane and 4% n-heptane. Linear paraffins have low octane numbers and in the past, conventional gasoline contained octane number boosters, i.e., tetraethyl lead, benzene, and olefins. At the end of the last century, these compounds had to be eliminated from conventional gasoline. Nowadays, to obtain gasoline with high octane numbers, the content of branched paraffins is increased at the cost of linear ones. The isomerization of linear paraffins is thus a process of enormous importance in petrochemistry. Isomerization of n-alkanes (C_5 to C_7 fraction) over an acidic catalyst is thermodynamically favored over cracking at lower temperatures. The process is typically performed in the presence of hydrogen (hydroisomerization) and using a bifunctional catalyst with an acidic and a metallic function [14]. A paraffin molecule is dehydrogenated in a first step at the metal site resulting in an olefin, which is then protonated at the acid site to give a

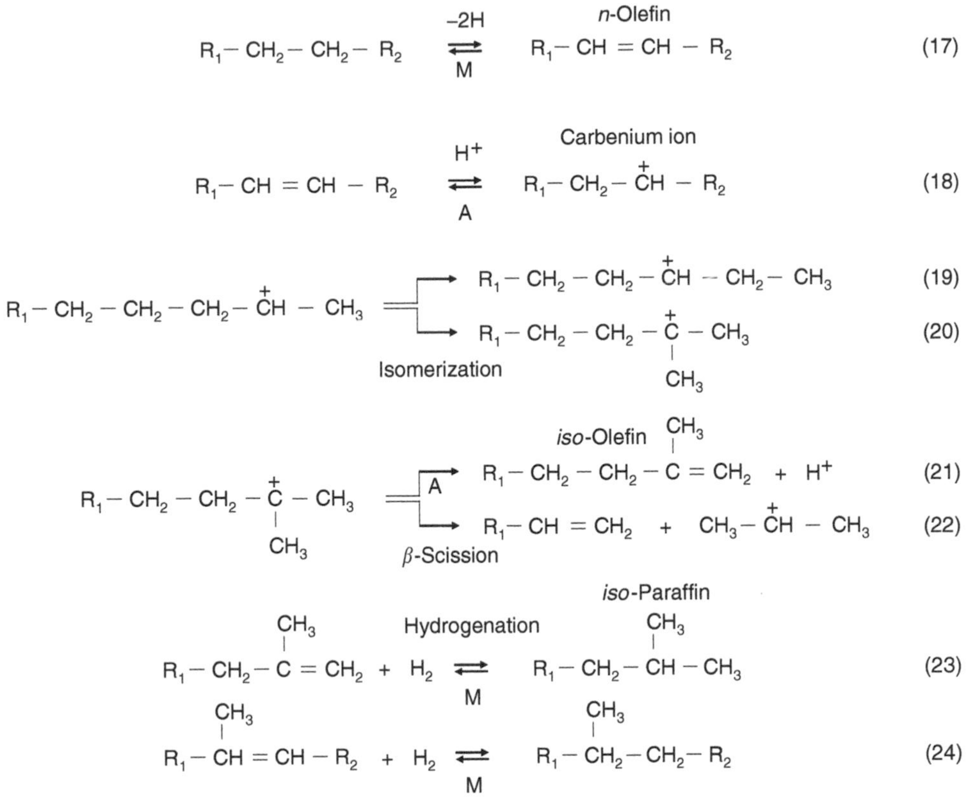

$$R_1 - CH_2 - CH_2 - R_2 \underset{M}{\overset{-2H}{\rightleftharpoons}} \overset{n\text{-Olefin}}{R_1 - CH = CH - R_2} \tag{17}$$

$$R_1 - CH = CH - R_2 \underset{A}{\overset{H^+}{\rightleftharpoons}} \overset{\text{Carbenium ion}}{R_1 - CH_2 - \overset{+}{C}H - R_2} \tag{18}$$

$$R_1 - CH_2 - CH_2 - CH_2 - \overset{+}{C}H - CH_3 \rightleftharpoons$$

$$\longrightarrow R_1 - CH_2 - CH_2 - \overset{+}{C}H \cdots CH_2 - CH_3 \tag{19}$$

$$\longrightarrow R_1 - CH_2 - CH_2 - \overset{+}{\underset{|}{C}} - CH_3 \tag{20}$$

Isomerization CH_3

$$R_1 - CH_2 - CH_2 - \overset{+}{\underset{\underset{CH_3}{|}}{C}} - CH_3 \rightleftharpoons$$

$$\overset{A}{\longrightarrow} \overset{\textit{iso-}\text{Olefin} \quad CH_3}{R_1 - CH_2 - CH_2 - C = CH_2} + H^+ \tag{21}$$

$$\longrightarrow R_1 - CH = CH_2 + CH_3 - \overset{+}{C}H - CH_3 \tag{22}$$

β-Scission

$$\overset{CH_3}{\underset{|}{R_1 - CH_2 - C}} = CH_2 + H_2 \overset{\text{Hydrogenation}}{\underset{M}{\rightleftharpoons}} \overset{\textit{iso-}\text{Paraffin} \quad CH_3}{R_1 - CH_2 - \overset{|}{CH} - CH_3} \tag{23}$$

$$\overset{CH_3}{\underset{|}{R_1 - CH}} = CH - R_2 + H_2 \underset{M}{\rightleftharpoons} \overset{CH_3}{R_1 - CH_2 - \overset{|}{CH_2} - R_2} \tag{24}$$

Figure 4.17 Cracking and isomerization pathways of paraffins in hydrocracking (M, metallic site; A, acidic site).

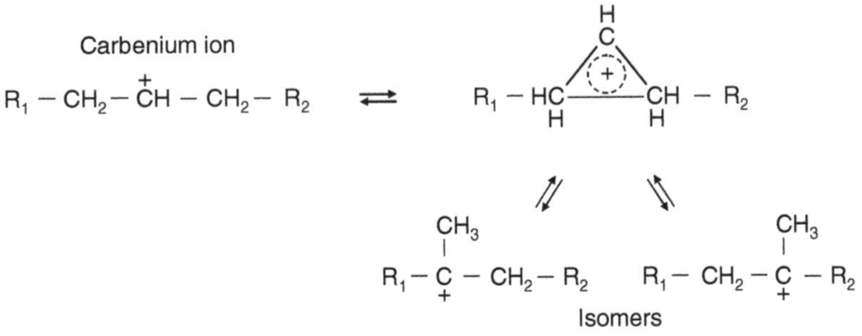

Figure 4.18 Isomerization via protonated cyclopropane intermediate.

carbenium ion. A methyl shift according to reaction (20), as shown in Figure 4.17, can result in a branched isomer of the carbenium ion. This reaction is supposed to proceed via a protonated cyclopropane intermediate as shown in Figure 4.18 [14,18]. Deprotonation of the carbenium ion then restores the Bronsted acid site and releases an *iso*-olefin, which then is hydrogenated at the metallic site, resulting in the *iso*-paraffin (see Figure 4.17, reactions 23 and 24).

Pt supported on an acidic support is a typical catalyst for the skeletal isomerization of light *n*-paraffins. The acidic supports can be acidic oxides, e.g., halogenated (Cl, F) alumina or sulfated zirconia (ZrO_2/SO_4), or an appropriate zeolite, e.g., Mordenite. Pt-(Cl, F)-alumina catalysts have a high performance at low temperatures and efficiently operate at temperatures between 115 and 150°C. Such low temperatures thermodynamically favor isomerization and thus, highly branched products are obtained. Zeolite supports are less active at lower temperatures and have to be operated at about

230 to 285°C and pressures of about 30 bar. This reduces the yield of *iso*-paraffins but Pt-zeolite catalysts are less susceptible toward catalyst poisoning. Pt on H-Mordenite is a prominent catalyst used for the isomerization of light paraffins. The fraction of *n*-alkanes which is not isomerized by the Pt/H-Mordenite can be selectively adsorbed by a bed of Ca–A zeolite (molecular sieve 5A). The branched *iso*-alkanes are not adsorbed by this zeolite and are thus separated from the *n*-alkanes. The *n*-alkanes are finally desorbed from the zeolite in a subsequent step and recycled to the isomerization process. The performance of the catalyst is significantly affected by the acidity of the zeolite, i.e., its framework Si/Al ratio. To obtain optimum performance for a Pt/H-Mordenite, the Si/Al ratio has be adjusted to about 8 to 10 [19,20] and the Pt content should be in the range 0.1 to 0.4%. Acid leaching or steaming of the Mordenite results in both framework dealumination and creation of a certain mesoporosity in the zeolite. In this way, the acidity and the accessibility of the zeolite pore system can be optimized.

4.3.5.1.4   Catalytic Dewaxing

Linear long-chain (*n*-alkanes) and mono-branched paraffins tend to crystallize at low temperature in diesel fuels and lubricants. In order to avoid this flocculation, it is desirable to remove or to transform these paraffins (waxes) from gas oil and lube oil fractions. One solution is a selective cracking of these undesired paraffins. For this purpose, zeolites having smaller pore openings than Faujasites (12MR) are advantageous because strongly branched paraffins cannot enter the smaller pores as easily as linear ones (educt selectivity). Cracking over a ZSM-5 zeolite (10MR) effectively removes wax molecules by transforming them into light paraffins (Mobile Oil process). Another advantage of a ZSM-5 cracking catalyst is its low tendency to form coke as compared to 12MR zeolites. The isomerization of *n*-alkanes and mono-branched alkanes converts them into multibranched molecules. For this purpose, zeolites such as ZSM-22, ZSM-23, and SAPO-11 have proved to be very efficient catalysts (Chevron process). The dewaxing of feeds and product streams can be combined with a hydro-treating process to saturate olefins and remove hetero-atoms from the products [16]. Crystallization of the wax at low temperature and separation of the liquid heavy oil fraction, e.g., using appropriate solvents, represents an alternative method for dewaxing.

4.3.5.1.5   Aromatization of Liquefied Petrol Gases

Aromatization of LPG is used to convert the rather low-value LPG fraction into more valuable products, i.e., mainly benzene, toluene, xylene, and ethylbenzene (BTX fraction) [17]. C_2 to C_4 alkanes can be effectively converted over H-ZSM-5, especially if the zeolite is impregnated with Ga^{3+} [21]. Using a Ga-ZSM-5 catalyst, propane and butane are effectively converted in the Cyclar process developed by BP/UOP [22]. Extra-framework Ga_2O_3 or GaOOH species in the pores of the ZSM-5 is responsible for the catalytic dehydrogenation of the light alkanes. Other efficient catalysts for the aromatization of C_2 to C_4 alkanes are Zn-ZSM-5 and Pt-ZSM-5. Unfortunately, the active ZnO species tend to form metallic Zn-ZSM-5 during the reaction, resulting in a gradual deactivation of the catalyst. Pt in Pt-ZSM-5 tends to aggregate into larger particles on the external surface of the zeolite, also leading to deactivation of the catalyst.

Over an acidic zeolite, e.g., H-ZSM-5, activation of the alkane can proceed via protonation of the alkane (according to reactions 2, 3, and 4) in a similar way as shown for paraffins in Figure 4.14. Since the catalyst is more efficient if Ga sites are present, the gallium species obviously play an important role in the activation of the light alkanes. One possible reaction mechanism on the bifunctional catalysts is sketched in Figure 4.19 [23]. The catalytic reaction proceeds typically at temperatures of about 500 to 600°C. The olefins formed react on the acidic catalyst and form oligomers in subsequent reaction steps. The oligomers then dehydrogenate at the metallic sites and undergo cyclization reactions on the acid sites. Finally, the cyclic products are dehydrogenated at the metallic sites. In this way, aromatics are formed that then desorb from the zeolite.

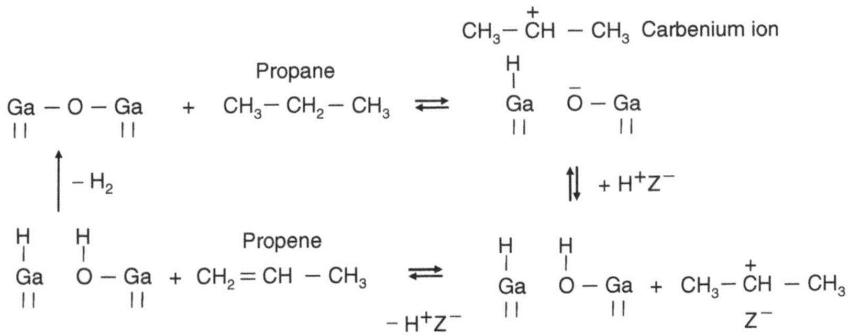

$$CH_3-\overset{+}{C}H-CH_3 \quad \text{Carbenium ion}$$

Propane

$$Ga-O-Ga \quad + \quad CH_3-CH_2-CH_3 \quad \rightleftharpoons \quad \underset{||}{Ga}\ \overset{H}{\underset{|}{}}\ \overset{-}{O}-\underset{||}{Ga}$$

$$\uparrow\ -H_2 \qquad\qquad\qquad\qquad \updownarrow\ +H^+Z^-$$

Propene

$$\underset{||}{Ga}\ \overset{H}{\underset{|}{}}\ O-\underset{||}{Ga} + CH_2=CH-CH_3 \quad \rightleftharpoons \quad \underset{||}{Ga}\ \overset{H}{\underset{|}{}}\ \overset{H}{\underset{|}{}}\ O-\underset{||}{Ga} + CH_3-\overset{+}{C}H-CH_3$$

$$-H^+Z^- \qquad\qquad\qquad\qquad\qquad\qquad Z^-$$

Figure 4.19 Alkane activation on bifunctional Ga/H-ZSM-5.

4.3.5.2 *Methanol to Gasoline and Methanol to Olefins*

All the processes described above are based on raw oil as the starting material. A process, which is related but not based on raw oil, is the conversion of methanol to hydrocarbons. The methanol is converted into gasoline or to olefins that are intermediates in the reaction path toward the paraffins (gasoline). The conversion of methanol to gasoline (MTG) became significant during the raw oil shortage in the 1970s (Arabian Oil Embargo). At present, the MTG and the Fischer–Tropsch process (Sasol-process, conversion of coal to gasoline via syngas) are the only commercial processes toward synthetic fuels. The source of methanol is variable but a major portion of methanol is obtained by the conversion of natural gas into methanol. Methane is converted in a steam reformer into CO and H_2 (syngas) which then reacts to form methanol. The MTG/methanol to olefins (MTO) reaction is a solid Bronsted acid-catalyzed process. Methanol is initially converted into olefins, which then oligomerize and form paraffins and cyclic derivates. A hydrogen transfer completes the reaction toward aromatics. The reaction may be summarized as shown in Figure 4.20 [24]. Dimethyl ether has been found as an intermediate at early stages of the reaction. Its formation from methanol is therefore regarded as the initial step for the reaction [24,25]. The next step, i.e., the formation of the first C–C bond, is still not understood and various models can be found in the literature, including carbene–carbenoid, carbocationic, oxonium ylide, free radical, and concerted reactions. Once the first C–C bond has formed, the further reactions can proceed via carbocations as already discussed in the previous section.

The pore size of the zeolite catalyst has a strong effect on the product distribution of the process. By choosing catalysts with different pore sizes, the amounts of aromatics in the product can be significantly varied. Conversion of methanol at 370°C over the small-pore zeolite Erionite produces no aromatics at all but short-chain olefins and paraffins. Using the medium-pore zeolite ZSM-5 results in about 40% aromatics and using the large-pore zeolite Mordenite gives about 20% organics. Not only does the ratio of paraffins to aromatics vary for zeolites with different pore sizes but also the product distribution of the aromatics as shown in Figure 4.20. The wide-pore zeolites produce larger amounts of heavy aromatics because their pore size allows the formation of larger cyclic molecules. Small-pore zeolites cannot accommodate aromatics in their pores and therefore convert methanol into mainly olefins and paraffins. The use of a zeolite with a medium-pore size is thus advantageous and results in aromatics and light paraffins [24]. For this reason, ZSM-5 is the most widely used catalyst for the MTG process. Process parameters such as the reaction temperature or the reaction pressure are also factors which allow a certain shift of the product composition of the process.

The MTG process can be performed in a fixed-bed reactor or in a fluidized-bed reactor. The fixed bed configuration is technically less demanding. However, for strongly exothermic processes such as the MTG process, the release of heat is a severe problem that results in hot spots and overheating of the reactor. In a fluidized-bed reactor, a gas passes upward through the catalyst bed and

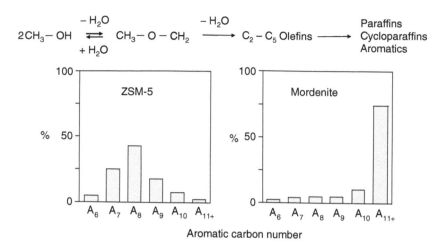

Figure 4.20 MTG/MTO reaction path and aromatics distribution with different zeolites as catalysts. (Reprinted from C.D. Chang, W.H. Lang, W.K. Bell, *Catalysis in Organic Reactions, Molecular Shape-Selective Catalysis in Zeolites*, pp. 93–94. Copyright 1981. With permission from Marcel Dekker.)

the balance between gravity and upward motion due to the gas flow maintains the catalyst particles suspended in the turbulent gas flow. This turbulent flow results in excellent heat transfer and prevents hot spots and a nonuniform catalyst aging.

A higher olefin yield may be obtained by variation of the catalyst (small-pore catalyst) or by variation of the process parameters such as reaction temperature, pressure, and feed flow.

4.4 MESOPOROUS CATALYSTS

Textural mesoporosity is a feature that is quite frequently found in materials consisting of particles with sizes on the nanometer scale. For such materials, the voids in between the particles form a quasi-pore system. The dimensions of the voids are in the nanometer range. However, the particles themselves are typically dense bodies without an intrinsic porosity. This type of material is quite frequently found in catalysis, e.g., oxidic catalyst supports, but will not be dealt with in the present chapter. Here, we will learn that some materials possess a structural porosity with pore sizes in the mesopore range (2 to 50 nm). The pore sizes of these materials are tunable and the pore size distribution of a given material is typically uniform and very narrow. The dimensions of the pores and the easy control of their pore sizes make these materials very promising candidates for catalytic applications. The present chapter will describe these rather novel classes of mesoporous silica and carbon materials, and discuss their structural and catalytic properties.

4.4.1 Ordered Mesoporous Silica Materials

Ordered mesoporous silica (OMS) materials are obtained by condensation of silica around organic micelles via a co-operative or true templating process [26,27]. Surfactant molecules such as alkyl ammonium cations, can form different types of micelles. At low concentrations they build spherical micelles while at higher concentrations extended micelles are forming. Lamellar, hexagonal, and cubic micellar phases are obtained depending on the temperature and surfactant concentration. The terms hexagonal and cubic phases indicate the hexagonal or cubic symmetry of the micellar structures. The polar head-groups of the surfactant molecules are directed toward the water phase. Here, they interact with the negatively charged silica precursors. The silica precursors react with each other and form extended silica domains that enclose the micelles. The formation of the micelles and that of the silica walls typically proceed simultaneously (co-operative process) but the result is also possible via a true templating where the surfactant micelles are formed in a primary step followed by the addition of the silica material. In all cases, the silica forms a rigid-wall structure around the surfactant micelles by condensation of silanol groups. The condensation of the silica is catalyzed either by a base or an acid. The formation of a hexagonally OMS from a hexagonal micelle is schematically illustrated in Figure 4.21. The organic micelles act as templates for the porous silica. After the formation of the silica walls, the surfactant molecules are removed to open the pore system of the silica. Most commonly, the organics are simply combusted at temperatures of about 400 to 600°C. The resulting material has pores that are arranged in exactly the same manner as the parent micelles.

Ordered mesoporous silica materials of this type were developed more or less simultaneously by scientists from Toyota and Waseda University [28,29] and by researchers from Mobil Oil [30,31] using alkyltrimethyl ammonium surfactants (tensides). The researchers from Toyota/Waseda used Kanemite as the silica source and obtained a hexagonally ordered material, which was called FSM-16 (where FSM stands for folded sheet mesoporous material). Mobil used noncrystalline silica precursors and synthesized materials of the so-called M41S type. The different M41S materials were denoted as MCM-41 (hexagonal, P6m symmetry), MCM-48 (cubic, Ia-3d symmetry), and MCM-50 (lamellar) (where MCM stands for mobile composition of matter). MCM-41 and MCM-48 are porous after template removal. The diameters of the channels observed in these materials are typically in the range 2 to 4 nm and the pore walls are about 1 nm thick. The pore systems of these two materials are shown schematically in Figure 4.22. The pores of MCM-41 are arranged in a two-dimensional hexagonal ordering. MCM-48 has a bi-continuous three-dimensional pore system. The complex pore structure of MCM-48 is due to the formation of a minimal surface between the surfactant-water interface and surfactant molecule at a given temperature.

MCM-50 consists of stacks of silica and surfactant layers. Obviously, no pores are formed upon removal of the surfactant layers. The silica layers contact each other resulting in a nonporous silica. It is noteworthy to mention that materials of M41S type were probably already synthesized by Sylvania Electric Products in 1971 [32]. However, at that time the high ordering of the materials was not realized [33]. M41S-type materials are synthesized under basic reaction conditions. Scientists from the University of Santa Barbara developed an alternative synthesis procedure under acidic conditions. They also used alkyltrimethyl ammonium as the surfactant. The porous silica materials obtained (e.g., hexagonal SBA-3; Santa BArbara [SBA]) had thicker pore walls but smaller pore diameters. Furthermore, they developed materials with novel pore topologies, e.g., the cubic SBA-1 with spherical pores.

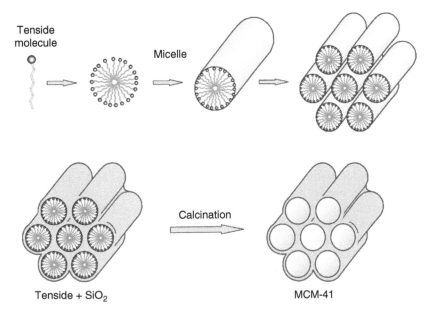

Figure 4.21 Formation of a hexagonal micelle (top) and of MCM-41 silica (bottom).

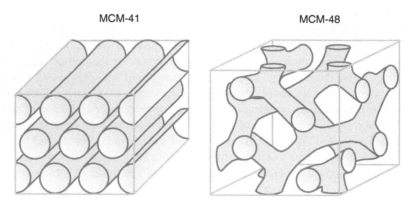

Figure 4.22　Pore systems of MCM-41 and MCM-48.

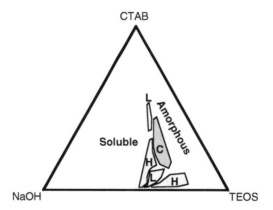

Figure 4.23　Synthesis space diagram for a ternary system composed of tetraethylorthosilicate (TEOS), cetyltrimethylammonium bromide (CTAB), and sodium hydroxide (H, hexagonal phase [MCM-41]; C, cubic phase [MCM-48]; L, lamellar phase [MCM-50]; H_2O/SiO_2 = 100, reaction temperature 100°C, reaction time 10 days). (Reprinted from *Science*, Vol. 267, A. Firouzi, D. Kumar, L.M. Bull, T. Besier, P. Sieger, Q. Huo, S.A. Walker, J.A. Zasadzinski, C. Glinka, J. Nicol, D.I. Margolese, G.D. Stucky, B.F. Chmelka, *Cooperative Organization of Inorganic-Surfactant and Biomimetic Assemblies*, pp. 1138–1143. Copyright 1995. With permission of AAAS.)

The formation of a specific mesophase depends strongly on the silicon-to-surfactant ratio, ionic strength and pH of the reaction mixture as illustrated by the synthesis-space diagram in Figure 4.23 [34]. The pore sizes of such OMS materials can be controlled by using surfactant molecules with different chain lengths. The diameters of the micelles, and thus the pores of the resulting silica, are smaller if short-chain surfactants are used and vice versa. The pore sizes can be further modified by using co-templates, e.g., long-chain alcohols, which result in a swelling of the alkyltrimethyl ammonium surfactant micelles. The acidity or alkalinity of the reaction mixture also has an effect on the pore diameters and pore wall thickness of the resulting material. The shrinkage of the pore walls upon calcination seems to be more pronounced in materials obtained under acidic conditions. This may result in somewhat smaller pores.

The characteristics of a typical MCM-41 sample are shown in Figure 4.24. The transmission electron microscopy (TEM) image allows a view into the hexagonally ordered mesopores of the materials. The x-ray difrraction (XRD) pattern shows four reflections, which is in accordance with the hexagonal symmetry of the material. The XRD pattern is not due to scattering by atoms on defined crystallographic positions but due to periodic variations of the electron density as caused by the pores and pore walls of the material. The walls themselves consist of amorphous silica. For this reason no ordering exists in the *c*-direction and only *hk*0 reflections are observed. The nitrogen adsorption–desorption isotherms were measured at 77 K and show a pronounced step in the relative pressure range (p/p_0) of 0.3 to 0.4 due to condensation of the nitrogen in the mesopores (capillary condensation). The adsorption step at relative pressures above $p/p_0 > 0.9$ is due to adsorption in the voids between the individual MCM-41 particles (textural porosity).

In the late 1990s, Stucky and co-workers [35] from the University of Santa Barbara used block co-polymers instead of ionic surfactant templates (acidic conditions) and obtained large pore materials with similar pore arrangements as from surfactant-templated materials. Block co-polymers with hydrophilic and hydrophobic blocks form micelles in a very similar manner as ionic surfactant molecules. The micelles formed by such block co-polymers are much larger than those formed by surfactant molecules. Consequently, the pores of the block co-polymer — templated silicas (e.g., SBA-15) are significantly larger than those of the materials obtained with alkyltrimethyl ammonium surfactants. Not only are the pores larger (6 to 15 nm) but also the walls between the pores are thicker if block co-polymers are used. For MCM-41 materials, the silica walls have a thickness of about 1 nm, while SBA-15 materials have walls with thicknesses of about 3 to 7 nm [27]. Typical templates for the formation of SBA-15 type silica materials are the triblock co-polymers of the type $(PEO)_x(PPO)_y(PEO)_x$ (where PEO stands for polyethylene oxide and PPO for polypropylene oxide). The more hydrophobic PPO chains form the core of the micelle rods and the more hydrophilic PEO chains are organized at the surface of the rods. The hydrophilic PEO chains can also penetrate the silica walls, thus causing smaller pores after the combustion of the polymer as illustrated schematically in Figure 4.25, which is believed to be the reason for the microporosity of the pore walls of SBA-type materials. Some of these micropores connect neighboring mesopores after the removal

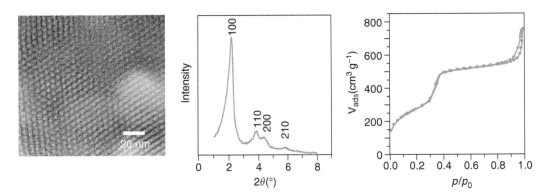

Figure 4.24 TEM image (left), XRD pattern (middle), and nitrogen adsorption–desorption isotherm (right) of a typical MCM-41 material. (Courtesy of F. Kleitz.)

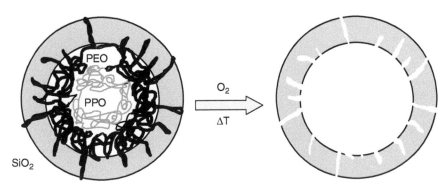

Figure 4.25 Formation of meso- and micropores in SBA-15 after removal of the $(PEO)_x(PPO)_y(PEO)_x$ triblock co-polymer by calcination.

of the triblock co-polymer, e.g., by calcination of the composite in air at temperatures between 400 and 600°C. Block co-polymers can not only be used for the preparation of silicas with hexagonally ordered mesopores but also for many other pore structures including those discussed already for the M41S type materials, i.e., hexagonal (like MCM-41), cubic (like MCM-48), cubic (like SBA-1, cubic arrangement of spherical pores), and others.

The silica networks of mesoporous silicas are terminated at the surfaces of the amorphous pore walls, thus resulting in terminal silanol groups on the walls. The density of the silanol groups of all mesoporous silicas (1 to 3 nm^{-2}) is somewhat lower than usually found for other typical silica materials (4 to 6 nm^{-2}) [27]. The specific surface areas of the different mesoporous silicas vary depending on their pore sizes, the thicknesses of their pore walls, and the density of their silica networks. For some MCM-41- and MCM-48-type materials, surface areas of about 1000 to 1400 m^2 g^{-1} have been reported. The surface areas of SBA-15-type materials can be > 600 m^2 g^{-1} [27].

The very high surface areas of OMS materials and various possibilities to modifying their surface properties make these silicas promising for catalytic applications. In the following sections, a short survey will be given of different methods to modify the properties of the silica materials and of catalysis with OMS-based catalysts.

4.4.1.1 *Surface Modifications of Ordered Mesoporous Silica Materials*

OMS materials are chemically not different from amorphous silicas and thus all modifications which are possible on conventional silica materials are basically also possible on such mesoporous silicas. The most frequently applied modifications are summarized schematically in Figure 4.26 [27]. The acidity of silanol groups is rather low but they can be used as reactive sites for the fixation of organics molecules or inorganic species. Metal organic catalysts can be attached to the surface either directly or via a spacer molecule and silylation of the silica surface results in a more hydrophobic material. These modifications allow the imobilization of a catalyst on the surface and in the pores of the OMS materials and a fine-tuning of the hydrophilic/hydrophobic properties of a given material. A proper immobilization of the organic catalyst to the surface via a chemical bond is mandatory to prevent leaching of the catalyst from the support during the catalytic reaction.

The acidity of the silica material can be enhanced by substitution of a fraction of the silica by alumina. The aluminum is either incorporated into the silica walls, e.g., directly during the synthesis

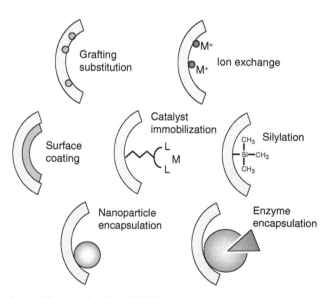

Figure 4.26 Modifications of the functionality of OMS materials.

of the material, or it is deposited on the surface of the pores in a separate step after the removal of the template. The latter method allows the modification of the acidic properties (substitution) and also the modification of the pore wall thickness (grafting). The deposition of alumina in and on the pore walls creates acidic sites, especially when aluminum is tetrahedrally coordinated by oxygen. The ratio between tetrahedrally and octahedrally coordinated aluminum in the OMS materials can be varied by the choice of specific aluminum sources if the aluminum is incorporated directly during the synthesis. Stronger Bronsted acid sites are observed in materials that were prepared with $NaAlO_2$ as the aluminum source rather with aluminum hydroxide $Al(OH)_3$, or aluminum isopropoxide $Al(OPr)_3$ [27,36]. However, since the local environment of the aluminum in OMS materials is more similar to silica-alumina rather than to zeolites, the acidity of Al-substituted OMS is lower than that of zeolites. As a further consequence, the hydrophilicity of the material increases upon decrease of the Si/Al ratio in the pore walls. As in silica-alumina, protons may be exchanged by cations in Al-modified OMS materials. Other metals, such as Fe, Ga, Ti, Sn, V, and B, may also be substituted or grafted on the walls of OMS materials. The acidity of the resulting material is significantly affected by the type of heteroatom present. For MCM-48, the Bronsted acid strength decreases in the order Al > Ga > Fe, while the Lewis acid strength showed an order of Ga ≈ Al >> Fe [27,37]. Complete coating of the silica walls with metal oxide species is also possible as shown for europium-doped Y_2O_3 on SBA-15 [27,38]. Finally, nanoparticles, e.g., Pt, Rh–Pt, Cu–Rh, Pt–Ru, or Au metal particles or clusters, or enzymes can be immobilized in the large mesopores of OMS materials. Such supported materials then can be used as catalysts for metal- or enzyme-catalyzed reactions [27].

4.4.1.2 Catalysis with Ordered Mesoporous Silica Materials

Ordered mesoporous silicas are not as widely applied for catalysis in industrial processes as zeolites but many catalytic processes have been described where OMS-based catalysts have been used (see [27] and references therein). Only some of these reactions will be described here, to illustrate the potential of such catalysts. Directly after the discovery of OMS materials, the use of such silicas as catalysts in petrochemical processes was on the top of the list of their potential applications. The processing of bulky molecules in the large pores seemed to be very attractive. Unfortunately, siliceous and/or aluminum-containing OMS materials have only a very moderate acidity, comparable to that of conventional silicas and silica-aluminas. They proved to be less active and also less stable than zeolite catalysts. However, modifications of the silicas resulted in catalytic performances with activities comparable to conventional catalysts. Sulfated zirconia supported on MCM-41 has a similar catalytic activity as bulk sulfated zirconia for the cracking of cumene and 1,3,5-triisopropyl benzene [39]. Excellent conversions were also reported for the condensation of tert-butanol and methanol to methyl–tert–butyl ether (MTBE) using sulfated zirconia on SBA-15 as the catalyst [40]. The deposition of the heteropoly acid $H_3PW_{12}O_{40}$ on fluorinated MCM-41 gives a catalyst with a high gas-phase MTBE synthesis activity [41].

The weak acidity of the pure OMS materials is disadvantageous for petrochemical processes but for the synthesis of fine chemicals, solids with moderate acidity can be very active catalysts. Table 4.2 gives

Table 4.2 OMS-Catalyzed Reactions toward Organic Fine Chemicals

Reaction	Ordered Mesoporous Silica
Friedel–Crafts alkylation	$Al(OPr^i)_3$-grafted MCM-41, Ga-substituted/impregnated MCM-41, Al-substituted MCM-41, Fe-impregnated MCM-41
Acetalyzation	Siliceous MCM-41, Al-substituted MCM-41
Diels–Alder reaction	$Al(OPr^i)_3$-grafted MCM-41, Zn^{2+}-exchanged MCM-41
Beckmann rearrangement	Al-substituted MCM-41
Aldol condensation	Al-substituted MCM-41
Prins condensation	Sn-substituted MCM-41
Meerwein–Ponndorf–Verley reaction	$Zr(OPr^i)_4$, $Al(OPr^i)_3$-grafted SBA-15, MCM-41, and MCM-48
Metathesis	Siliceous FSM-14 and MCM-41

a summary of reactions for which weakly acidic OMSs have been used successfully as catalysts (see [27] and references therein). The pore size of the silica material can play an important role as shown for the acetalyzation of cyclohexanone on MCM-41 [42]. For this reaction, a pore diameter of 1.9 nm proved to be the optimum as shown in Figure 4.27. To explain for this effect, an optimal arrangement of the silanol groups in the pores of the MCM-41 material is assumed.

Modification of OMS materials with early transition metal elements, such as Ti, V, Nb, Mo, Mn, Fe, Co, and Cu, can result in very active redox catalysts. Unfortunately, leaching of the active compound from the silica support proved to be a severe problem for some systems, especially for oxidation reactions in liquid phase using peroxides as the oxidizing agents [27]. The deposition of metal or metal oxide nanoparticles in the pores of OMS also results in active catalysts. The HDS of feedstock for petrochemical processes is an important process in oil refineries for which supported metal sulfide particles are typically used as catalysts. MoS_2 particles supported on MCM-41 proved to be excellent catalysts for the HDS of dibenzothiophene [43]. Deposition of Ni/WS_2 particles on SBA-15 resulted in a catalyst having up to seven times higher activities for HDS of dibenzothiophene, a commercial HDS catalyst based on Co–Mo/Al_2O_3 [44]. Iron oxide supported on MCM-41 showed to be a catalyst for the oxidation of SO_2 being superior to commercial catalysts at high reaction temperatures as reached in the oxidation of SO_2 from smelters using pure oxygen instead of air [45]. Also metal particles, such as Pt, Pd, Ni, Rh, Ir, and Ru as well as metal alloys, supported on OMS are active catalysts, e.g., for hydrogenation reactions [27]. The immobilization of metal–organic catalysts on the surface and in the pores of OMS materials is an alternative method to prepare catalytically active solids. A crucial point for such catalysts is a real immobilization, i.e., fixation by chemical bonds to prevent leaching of the active species from the silica support. The ligand environment of the immobilized catalysts has a similar effect as in the liquid phase and the catalytic performance of given catalysts can be tuned by variation of the ligand. Such immobilized catalysts can be very active. The catalyst shown in Figure 4.28 was used for the Friedel–Crafts hydroxyalkylation of 1,3-dimethoxybenzene with 3,3,3-trifluoropyruvate [46]. The surface of the MCM-41 support was hydrophilized by silylation of free silanol groups. A conversion of 77% and an enantiomeric excess (ee) of 82% were achieved with the immobilized catalyst. The performance of the immobilized catalyst was significantly better than that of the same catalyst in homogeneous phase (conversion 44%, ee 72%).

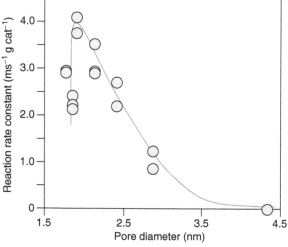

Figure 4.27   Pore-size effect on the acetalyzation of cyclohexanone with methanol. (Reprinted from *J. Am. Chem. Soc.*, Vol. 125, M. Iwamoto, Y. Tanaka, N. Sawamura, S. Namba, *Remarkable Effect of Pore Size on the Catalytic Activity of Mesoporous Silica for the Acetalization of Cyclohexanone with Methanol*, pp. 13032–13033, Copyright 2003. With permission from American Chemical Society.)

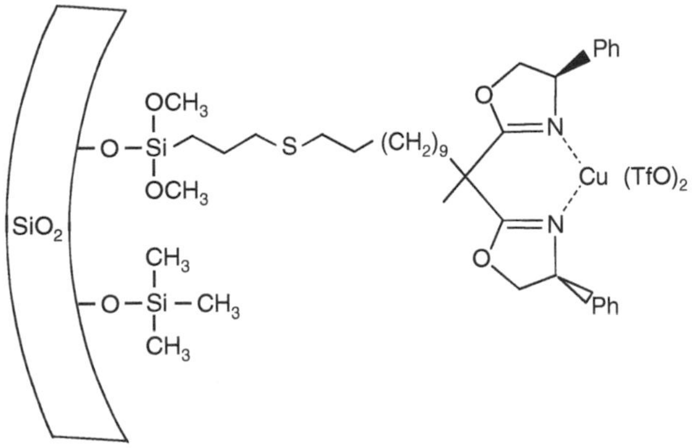

Figure 4.28 Scheme of an immobilized metal-organic catalyst attached to the silica surface via a mercaptoalkyl siloxane spacer.

However, with respect to ee, the same catalyst immobilized on amorphous silica performed even better (conversion 72%, ee 92%) than the one immobilized on MCM-41. This example illustrates an important issue, i.e., OMS-based catalysts have to be compared with those based on amorphous silica or silica-alumina. If the amorphous materials perform as well or even better than the OMS materials, then there is no advantage in using the significantly more expensive OMSs. However, in those cases where the catalytic reaction benefits from the regular and well-defined pore systems of the OMS materials, such materials can be very attractive, e.g., for the conversion of bulkier molecules or to overcome transport limitations in more narrow pores.

4.4.2 Nonsiliceous Ordered Mesoporous Materials

Ordered mesoporous materials of compositions other than silica or silica-alumina are also accessible. Employing the micelle templating route, several oxidic mesostructures have been made. Unfortunately, the pores of many such materials collapse upon template removal by calcination. The oxides in the pore walls are often not very well condensed or suffer from recrystallization of the oxides. In some cases, even changes of the oxidation state of the metals may play a role. Stabilization of the pore walls in post-synthesis results in a material that is rather stable toward calcination. By post-synthetic treatment with phosphoric acid, stable alumina, titania, and zirconia mesophases were obtained (see [27] and references therein). The phosphoric acid results in further condensation of the pore walls and the materials can be calcined with preservation of the pore system. Not only mesoporous oxidic materials but also phosphates, sulfides, and selenides can be obtained by surfactant templating. These materials have pore systems similar to OMS materials.

Ordered mesoporous carbons are accessible via a nanocasting process. A carbon precursor, e.g., furfuryl alcohol, is deposited in the pores of an OMS, such as SBA-15 or MCM-48. Condensation or polymerization of the carbon precursor followed by a carbonization step, e.g., at high temperature under protective atmosphere, results in a carbon–silica composite. The pores of the silica are blocked by the carbon material. Subsequently, the silica is dissolved, e.g., in HF or NaOH solutions, resulting in a pure carbon material consisting of carbon rods. The carbon rods represent a negative replica form of the pores of the parent silica material. The rods are thus arranged in exactly the same way as the pores of the parent silica were oriented before, i.e., hexagonally packed rods for SBA-15 and rods arranged in a cubic symmetry for MCM-48. If SBA-15 is used as the cast (solid template), the rods are fixed on their positions by very narrow carbon bridges, which are due to carbon deposition in the micropores connecting the mesopores of the parent silica. The spaces where the

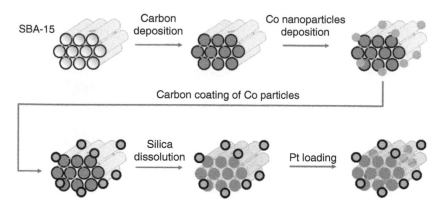

Figure 4.29 Preparation of a magnetically separable carbon-supported catalyst. (Reprinted from *Angew. Chem. Int. Ed.*, Vol. 43, A.-H. Lu, W. Schmidt, N. Matoussevic, H. Bönnemann, B. Spliethoff, B. Tesche, E. Bill, W. Kiefer, F. Schüth, *Nanoengineering of a magnetically separable hydrogenation catalyst*, pp. 4303–4306. Copyright 2004. With permission from Wiley-VCH.)

silica walls had been previously are empty after the silica removal and form a new pore system (CMK-3). Scientists from the Korean Advanced Institute of Science and Technology (KAIST) introduced this type of material in 1999 and denoted them as CMK-n materials [47]. Deposition of a thin layer of carbon on the walls of the OMS materials creates no rods but carbon tubes ordered in exactly the same arrangement as the parent silica pores (CMK-5). Combining successive preparation steps allows the creation of highly advanced catalysts as shown schematically in Figure 4.29. The pores of a mesoporous silica (SBA-15) are filled with a carbon precursor, which is then carbonized inside the pores. After the deposition of magnetic cobalt nanoparticles on the silica–carbon composite, the cobalt particles are coated with a 1 nm carbon layer. The silica is then removed by hydrofluoric acid treatment while the protected cobalt nanoparticles remain on the external surface of mesoporous carbon (CMK-3 type carbon) thus obtained. Finally, the actual catalyst is deposited as Pd nanoparticles inside the mesopores of the resulting carbon. In this way, a highly efficient hydrogenation catalyst is obtained, which is easily separated from the liquid phase by a magnet [48]. OMS as the hard template for nanocasting may also be used for the preparation of metal oxides. By applying this method, some mesoporous metal oxides, such as Co_3O_4, NiO, CeO_2, Cr_2O_3, OsO_4, and InO_2, were accessible (see [27] and references therein). However, this method has its limitations because many oxides simply decompose during the silica-leaching process.

Ordered mesoporous materials, such as described here, have been successfully tested as catalysts or catalyst supports for many different reactions [27]. However, since this class of materials is rather new, the real potential of these materials in catalysis is of course not fully investigated. As for OMS materials, the production of such materials is rather cost intensive, e.g., compared to conventional oxide materials. Therefore, the benefit of a regular mesopore system has to be substantial to justify the use of elaborated but expensive catalyst materials for industrial applications. Nevertheless, many of the materials described above proved to perform very well in many different catalytic reactions and they may of course find applications in this field.

4.5 CHARACTERIZATION OF MICROPOROUS AND MESOPOROUS MATERIALS

The analytical methods for the characterization of catalysts are described extensively in other chapters of this book. Here, only a brief overview on methods of predominant importance for the investigation of micro- and mesoporous materials will be given.

4.5.1 X-Ray Diffraction

If a single crystal is exposed to a monochromatic x-ray beam, a fraction of the radiation interacts with the electrons of the atoms in the crystal and every electron is the source of a new x-ray wave. Interferences of the x-ray waves, caused by the electrons of the different atoms in the crystal, results in complete extinction of the x-ray wave except for specific angles where parallel waves, which are reflected from atoms on certain lattice planes of the structure, have positive interference. The diffracted x-rays thus form a diffraction pattern which can be detected, e.g., on an image plate detector. The intensities and the diffraction angle of the diffracted beams are detected as spots of different intensities at defined positions on the detector. Such diffraction patterns allow the calculation of the position of each atom in the crystal structure. For a polycrystalline material, diffraction by millions of crystallites oriented isotropically in all possible directions overlap. The diffraction pattern on a given plane detector is therefore not a pattern of diffraction points but consists of diffraction rings. The randomly oriented crystallites cause diffraction in all directions, resulting in the diffraction rings. Typically, the complete diffraction rings are not detected but only their intensities along the radii of the diffraction rings (powder pattern). The diffraction intensities are plotted against the diffraction angle 2θ. The calculation of the crystal structure of a given material from a powder pattern is difficult if not impossible in many cases. This is because reflections of different classes may overlap. As a consequence, individual reflections are not discernible. However, even if the crystal structure of a polycrystalline sample may not be solved, a lot of valuable information can be obtained from powder patterns, i.e., the qualitative and quantitative phase composition of a material, the symmetry and lattice constants of a given phase, the exact crystal structure of a phase if its structure topology can be anticipated (Rietveld refinement), and with limitations, the dimensions of the coherently scattering domains (which are often identical with the average size of the crystal particles, i.e., for very small crystallites).

If planes of identical atoms in a crystal structure are considered as lattice planes, a relation exists between the diffraction angle (2θ) and the distance (d) between identical lattice planes. This relation is known as the Bragg equation:

$$n\lambda = 2d \sin(\theta)$$

where n is the multiples of the wavelength and λ the wavelength of the x-rays. Thus, the d values of a crystal structure define the peak positions of the respective powder pattern. For a given structure symmetry, the lattice constants a, b, and c of a crystal structure define the d values resulting from the different lattice planes, defined by the Miller indices hkl. The relations for cubic, orthorhombic, and hexagonal symmetry are shown in Table 4.3. Vice versa, the lattice constants of the structure of a given material can be calculated from the positions of the reflections in the powder pattern.

The expansion of the crystal structure upon substitution of smaller atoms by larger ones is reflected by increasing lattice constants. For a zeolite with cubic symmetry, the lattice constant a decreases with increasing Si/Al ratio. This relation is occasionally used to calculate the Si/Al ratio of the

Table 4.3   Relation between d Value and Lattice Constants for Three Different Crystal Symmetries

Symmetry	Lattice Constants and Angles	d_{hkl}
Cubic	$a = b = c$ $\alpha = \beta = \gamma = 90°$	$[(1/a^2)(h^2+k^2+l^2)]^{-1/2}$
Orthorhombic	$a \neq b \neq c$ $\alpha = \beta = \gamma = 90°$	$[(h^2/a^2)+(k^2/b^2)+(l^2/c^2)]^{-1/2}$
Hexagonal	$a = b \neq c$ $\alpha = \beta = 90°, \gamma = 120$	$[(4/3a^2)(h^2+k^2+hk)+(l^2/c^2)]^{-1/2}$

FAU-type Zeolites X and Y. Several equations have been proposed for this purpose [49]. The number of aluminum atoms per unit cell in the FAU-type structure, N_{Al}, can thus be calculated according to

$$N_{Al} = 101.202(a/f_c - 24.2115)$$

with f_c being a correction factor that is 1 for as-synthesized Faujasite, 0.9957 for Zeolite X, and 0.9966 for Zeolite Y. The proposed method allows the calculation of the number of Al atoms per unit cell (192 T-atoms per unit cell) with an error of a few atoms ($+/-2$ to 5).

4.5.2 Physisorption Analysis

Gas adsorption (physisorption) is one of the most frequently used characterization methods for micro- and mesoporous materials. It provides information on the pore volume, the specific surface area, the pore size distribution, and heat of adsorption of a given material. The basic principle of the methods is simple interaction of molecules in a gas phase (adsorptive) with the surface of a solid phase (adsorbent). Owing to van der Waals (London) forces, a film of adsorbed molecules (adsorbate) forms on the surface of the solid upon incremental increase of the partial pressure of the gas. The amount of gas molecules that are adsorbed by the solid is detected. This allows the analysis of surface and pore properties. Knowing the space occupied by one adsorbed molecule, A_g, and the number of gas molecules in the adsorbed layer next to the surface of the solid, N_m (monolayer capacity of a given mass of adsorbent) allows for the calculation of the specific surface area, A_s, of the solid by simply multiplying the number of the adsorbed molecules per weight unit of solid with the space required by one gas molecule:

$$A_s = N_m A_g$$

The number of gas molecules can be measured either directly with a balance (gravimetric method) or calculated from the pressure difference of the gas in a fixed volume upon adsorption (manometric method). The most frequently applied method to derive the monolayer capacity is a method developed by Brunauer, Emmett, and Teller (BET) [1]. Starting from the Langmuir equation (monolayer adsorption) they developed a multilayer adsorption model that allows the calculation of the specific surface area of a solid. The BET equation is typically expressed in its linear form as

$$\frac{p/p_0}{N(1-p/p_0)} = \frac{1}{CN_m} + \frac{(C-1)}{CN_m}\left(\frac{p}{p_0}\right)$$

where p_0 is the saturation pressure of the gas, N the number of gas molecules adsorbed, N_m the number of molecules in the monolayer, and $C = \exp(q_{ads}/RT)$, where q_{ads} is the net heat of adsorption. A plot of $(p/p_0)/(N(1-p/p_0))$ vs. (p/p_0) should result in a straight line with slope $s = (C-1)/CN_m$ and intercept $i = 1/CN_m$. From the two equations, N_m and C are derived as

$$N_m = \frac{1}{s+i}$$

$$C = \frac{s}{i} + 1$$

The BET surface area is then calculated as

$$A_{BET} = N_m N_L A_g$$

with N_L being the Avogadro constant. For the adsorbed layer, a liquid-like arrangement of the adsorbed molecules is assumed. This allows the calculation of the number of adsorbed molecules, N, from the volume, V, of the adsorbed film and the cross sectional area, A_g, of an adsorbed molecule is considered to be that of the molecule in a liquid at a given temperature. For nitrogen and argon, the most frequently used adsorptives, the cross sectional areas are considered to be 0.162 and 0.138 nm^2, respectively. Nitrogen adsorption measurements are typically performed at liquid nitrogen temperature (77 K) and argon isotherms are either performed at liquid argon (87 K) or liquid nitrogen temperature. In order to evaporate all adsorbed gas molecules (especially water) from the samples, they are usually heated under vacuum for several hours prior to the adsorption measurement.

Typical adsorption isotherms are shown in Figure 4.30. They are typical for microporous, nonporous, and mesoporous materials and are classified according to the IUPAC notation as type I, II, and IV isotherms (type III, V, and VI isotherms are of no interest here). Because of the strong adsorption of gas molecules in micropores as the consequence of strongly overlapping Lennard–Jones potentials in the narrow pores, adsorption takes place at very low relative pressure p/p_0. At higher pressure, only adsorption on the external surface of the particles takes place. Since the number of gas molecules adsorbed on external surface area is typically much smaller than that adsorbed in the micropores, the isotherm runs almost parallel to the p/p_0 axis. This is a unique feature of type I isotherms. Nonporous materials only adsorb on their surface. Multilayers of adsorbed molecules form on the surface and a type II isotherm results, which can be described by the BET equation. At lower pressure, mesoporous materials adsorb gas molecules in the same way as nonporous materials and the isotherms of such materials also can be described by the BET equation in this range. However, at some stage a meniscus forms in the mesopores and the pores are filled in significant step because adsorption is energetically favored over the meniscus in a capillary. This capillary condensation step, which is typical for type IV isotherms, restricts the validity of the BET equation. Only data points at pressures below this step can be taken for the calculation of the specific surface area (typically in the p/p_0 range between 0.05 and 0.2).

Thus, either type I or type IV isotherms are obtained in sorption experiments on microporous or mesoporous materials. Of course, a material may contain both types of pores. In this case, a convolution of a type I and type IV isotherm is observed. From the amount of gas that is adsorbed in the micropores of a material, the micropore volume is directly accessible (e.g., from t plot of α_s plot [1]). The low-pressure part of the isotherm also contains information on the pore size distribution of a given material. Several methods have been proposed for this purpose (e.g., Horvath–Kawazoe method) but most of them give only rough estimates of the real pore sizes. Recently, nonlocal density functional theory (NLDFT) was employed to calculate model isotherms for specific materials with defined pore geometries. From such model isotherms, the calculation of more realistic pore size distributions seems to be feasible provided that appropriate model isotherms are available. The mesopore volume of a mesoporous material is also rather easy accessible. Barrett, Joyner, and Halenda (BJH) developed a method based on the Kelvin equation which allows the calculation of the mesopore size distribution and respective pore volume. Unfortunately, the BJH algorithm underestimates pore diameters, especially at

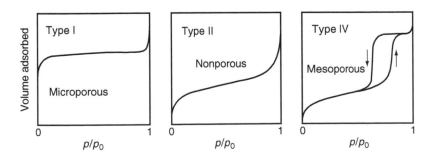

Figure 4.30 Adsorption isotherms according to IUPAC classification.

low relative pressures. The *t*-plot or α_s-plot methods may serve as alternatives for the calculation of the mesopore volume of a material. The latter methods also allow separating the micropore volume from the mesopore volume of a material possessing both types of pores. However, NLDFT proved to also be a more reliable tool for the analysis of mesoporous materials. Applying appropriate models, pore size distributions of model systems with defined pore diameters (e.g., porous glasses, MCM-*n* type materials, or SBA-*n* type materials) can be derived with remarkable accuracy.

4.5.3   Electron Microscopy

Electron microscopy (EM) is a powerful tool for the investigation of porous materials. It provides direct images of the samples under investigation over lengths scales from nanometers up to several hundred micrometers. An electron beam is focused on the sample and images of the morphology of a sample can be obtained by scanning the surface of the material under investigation with the electron beam (scanning electron microscopy [SEM]). Alternatively, the intensity of the electron beam can be imaged after transmission through the sample (TEM). SEM and TEM provide information on different properties of the sample. The shape of crystallites or of other particles as well as twinning, aggregation, and agglomeration of particles can be visualized by SEM as shown for Silicalite-1 crystallites in Figure 4.31. SEM typically provides information on objects with sizes from about hundred nanometers up to some hundred micrometers. Information on features on a smaller size scale is obtained by TEM as shown in Figure 4.24 (left). The image shows a view parallel to the hexagonally ordered channels of the mesoporous silica MCM-41.

4.5.4   Nuclear Magnetic Resonance Spectroscopy

Magic angle spinning nuclear magnetic resonance (MAS NMR) is used to obtain information about the chemical environment of specific elements in solids. Resonance lines are measured with respect to those of standard materials (e.g., tetramethoxy silane [TMS] for ^{29}Si). Deviations from the resonance lines of the standards are called chemical shifts. In a typical NMR spectrum, the intensities of the resonance lines are plotted versus the chemical shifts δ (in ppm). For the characterization of microporous and mesoporous silicates, ^{29}Si and ^{27}Al MAS NMR spectra provide valuable information on different structural features of the materials. For pure siliceous zeolites, silicon atoms on crystallographically nonequivalent sites may be distinguished by different chemical shifts of the signals. For the siliceous MFI-type Silicalite 1 structure, a structure transformation of the monoclinic room-temperature phase (295 K) to the orthorhombic high temperature phase (395 K)

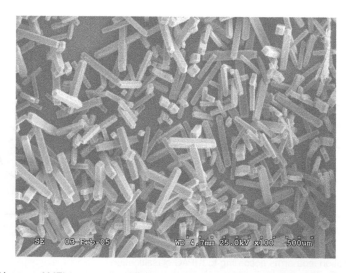

Figure 4.31   SEM image of MFI-type zeolite. (distance between points of size bar = 50 μm; total length 500 μm.)

results in a reduction of 20 lines to only 10 lines [50]. Of the 20 lines, two have threefold intensities. Thus the lines represent the 24 crystallographically nonequivalent sites of the orthorhombic MFI structure. Of the ten lines, two have double intensities. In total, the signals represent the 12 nonequivalent silicon sites of the monoclinic high-temperature MFI structure.

If aluminum is present in the zeolite framework, the ^{29}Si MAS NMR signals become broader and the crystallographic silicon sites cannot be distinguished. However, the different environments due to different numbers of aluminum atoms in the second coordination spheres of the silicon atoms are discernible. Thus, it is possible to distinguish silicon atoms that are connected via oxygen bridges to four, three, two, one, and zero aluminum atoms as shown in Figure 4.32 [51,52]. Determination of the individual intensities, e.g. by deconvolution, allows the calculation of the Si/Al ratio of a given material according to the formula [52]:

$$Si/Al = \frac{\sum_{n=0}^{4} I_n}{\sum_{n=0}^{4} 0.25 n I_n}$$

with $I_{Si(nAl)}$ being the individual intensities of the respective ^{29}Si MAS NMR signals.

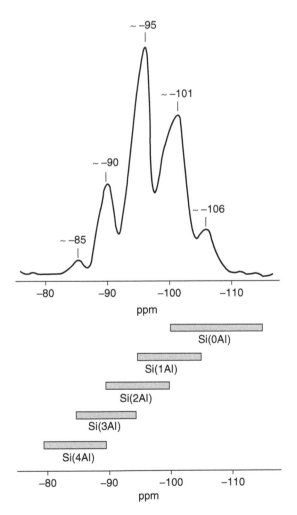

Figure 4.32 ^{29}Si MAS NMR spectrum of a NaY zeolite (Si/Al = 2.6) and the assignment of the different signals. (Reprinted from *Introduction to Zeolite Science and Practice, Studies in Surface Science and Catalysis*, Vol. 137, A. Jentys, J.A. Lercher, *Techniques of Zeolite Characterization*, pp. 345–386. Copyright 2001. With permission from Elsevier.)

Aluminum atoms in zeolites are typically coordinated via oxygen bridges by four silicon atoms. The ^{27}Al NMR spectra are therefore simpler than ^{29}Si NMR spectra. They allow an easy distinction between framework aluminum and extra-framework aluminum in zeolites. Aluminum in the zeolite framework is typically coordinated by four oxygen atoms while extra-framework aluminum is typically coordinated by six oxygen atoms. The signals of tetrahedrally coordinated aluminum are typically observedat about 60 ppm (with respect to aqueous Al(NO$_3$)$_3$ solution) and those of octahedrally coordinated aluminum at about 0 ppm. ^{13}C MAS NMR spectroscopy can be used to investigate organic guest molecules inside micro- and mesoporous materials. It also provides a tool to investigate immobilized organic species in mesoporous silicas (grafting).

4.5.5 Infrared Spectroscopy

Infrared spectroscopy (IR) is a valuable characterization method for studying microporous and mesoporous materials. It allows detection of lattice vibrations of a crystalline (300 to 1300 cm^{-1}) solid as well as investigations on surface species (e.g., OH groups) and adsorbed molecules (e.g., pyridine) (see [53] and references therein). Some lattice vibrations are structure-insensitive while others are structure sensitive. The latter are sometimes taken as an indication for the presence of specific structures or structural sub-units. This kind of evidence is of special importance if other methods such as XRD fail, e.g., for the characterization of zeolite particles sized in the nanometer range. Hydroxyl groups in zeolites and on their surfaces have been investigated intensively. OH stretching vibrations can be used to distinguish different types of acidic hydroxyl groups on zeolites. The higher the acid strength of an OH group, the lower the stretching frequency because the proton is less strongly bound to the oxygen. Figure 4.33 shows three IR bands of protonated Zeolite Y [53]. The band at 3740 cm^{-1} is typically assigned to nonacidic terminal silanol groups, i.e., on the external surface of the zeolite or at structural defects (e.g., silanol nests). The other two bands at 3640 and 3540 cm^{-1} are due to OH groups with lower stretching frequencies and thus are assigned to acidic Si–OH–Al groups (bridging OH groups). The one at 3640 cm^{-1} is supposed to be due to more easily accessible Si–OH–Al groups because it disappears upon adsorption of pyridine.

Other dehydrated zeolites have similar IR spectra with somewhat shifted positions of the bands as shown in Figure 4.34 [53]. This fact was assigned to different acidities of the respective

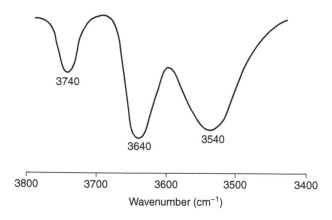

Figure 4.33   IR spectrum of a dehydrated H,Na–Y zeolite in OH stretching region. (Reprinted from *Introduction to Zeolite Science and Practice, Studies in Surface Science and Catalysis*, Vol. 58, J.H.C. van Hooff, J.W. Roelofsen, *Techniques of Zeolite Characterization*, pp. 241–283. Copyright 1991. With permission from Elsevier.)

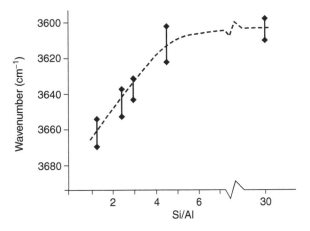

Figure 4.34 Position of IR band (due to acidic OH groups) with respect to varying Si/Al ratios in different zeolites. (Reprinted from *Introduction to Zeolite Science and Practice, Studies in Surface Science and Catalysis*, Vol. 58, J.H.C. van Hooff, J.W. Roelofsen, *Techniques of Zeolite Characterization*, pp. 241–283. Copyright 1991. With permission from Elsevier.)

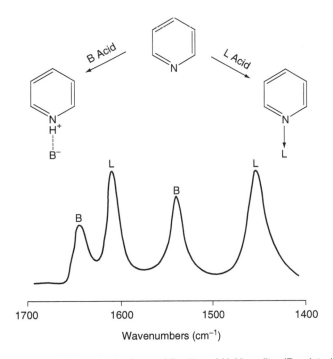

Figure 4.35 IR spectrum of pyridine adsorbed on acidic sites of H–Y zeolite. (Reprinted from *Introduction to Zeolite Science and Practice, Studies in Surface Science and Catalysis*, Vol. 58, J.H.C. van Hooff, J.W. Roelofsen, *Techniques of Zeolite Characterization*, pp. 241–283. Copyright 1991. With permission from Elsevier.)

zeolites. The acid strengths of the bridging OH groups increase up to a Si/Al ratio of about 6 to 7 and then remain more or less constant. This has been explained in terms of a decrease of the acid strength if an acidic OH group is in the direct neighborhood of other acidic OH groups. An aluminum atom is always bound to four silicon atoms via oxygen bridges (Loewenstein rule) but silicon atoms may have 0 to 4 aluminum atoms in their direct neighborhood. Thus, a bridging

OH group becomes more acidic the less aluminum atoms are bound to the respective silicon atom via oxygen bridges.

The type of acid sites in an acidic solid such as a zeolite, can be characterized by using probe molecules which interact with these sites. The basic pyridine interacts with both, Bronsted and Lewis acid sites. A Bronsted acid site transfers the proton to the pyridine molecule while a Lewis acid site interacts with the free electron pair of the nitrogen in the pyridin. The positions of the IR bands of the probe molecule shift with respect to these different interactions and allow a distinction of the sites. Figure 4.35 shows the assignment of bands of an IR spectrum of pyridine adsorbed on a protonted Zeolite Y [53]. Bands due to interaction with Bronsted (B) and Lewis acid sites (L) are clearly discernible. Information upon the acid strength of the sites is available if the temperature of the sample is slowly increased. The less acidic a specific site is, the lower is the temperature required for desorption of the probe molecule. Recording of IR spectra upon heating of a sample with adsorbed probe molecules thus allows a qualitative determination of the acid strengths of the different sites.

APPENDIX

Terms and Abbreviations

porosity	pore volume / total pore volume
specific surface area (*S*)	surface area per gram of solid ($m^2\ g^{-1}$)
micropores	pores with diameters < 2 nm
mesopores	pores with diameters between 2 and 50 nm
macropores	pores with diameters >50 nm
ee	enantiomeric excess (ee $= 100\%(R-S)/(R+S)$, for $R>S$)
MCM-*n*	mobil composition of matter
	(*n* defines a number indicating a specific material)
SBA-*n*	Santa Barbara *n* (*n* defines a number indicating a specific structure)
FSM-*n*	folded sheet mesoporous material *n*
	(*n* defines a number indicating a specific material)
CMK-*n*	carbon material KAIST *n*
	(*n* defines a number indicating a specific material)
	zeolite structure types
AEL	aluminophosphate eleven ($AlPO_4$-11)
AFI	aluminophosphate five ($AlPO_4$-5)
ATO	aluminophosphate thirty one ($AlPO_4$-31)
BEA	zeolite beta
CHA	Chabasite
FAU	Faujasite (e.g., Zeolites X and Y)
HEU	Heulandite
LTA	Linde type A (e.g., Zeolite A)
LTL	Linde type L (e.g., Zeolite L)
MFI	Mobile five (e.g., ZSM-5)
MOR	Mordenite
MWW	Mobil composition of matter twenty two (MCM-22)
MTW	Mobile twelve (e.g., ZSM-12)
OMS	ordered mesoporous silica
RHO	Zeolite rho
SOD	Sodalite

Further Reading

A good deal of information presented in this chapter has been adapted from textbooks and monographs. For further reading, all of them can be recommended. They contain enormous amounts of additional information and give excellent reviews not only on all topics related to catalysis on micro- and mesoporous solids but also on catalysis and on porous solids in general.

1. *Handbook of Heterogeneous Catalysis* edited by Ertl, Knözinger, and Weitkamp [54]
2. *Principles and Practice of Heterogeneous Catalysis* by Thomas and Thomas [55]
3. *Handbook of Porous Materials* edited by Schüth, Sing, and Weitkamp [56]
4. *Introduction to Zeolite Science and Catalysis* edited by van Bekkum, Flanigen, Jacobs, and Jansen [57]
5. *Catalysis and Zeolites: Fundamentals and Applications* edited by Weitkamp and Puppe [58]
6. *Molecular Sieves, Science and Technology* edited by Karge and Weitkamp [59]
7. *Adsorption by Powders and Porous Solids* by Rouquerol, Rouquerol, and Sing [1]

REFERENCES

1. F. Rouquerol, J. Rouquerol, K.S.W. Sing, *Adsorption by Powders and Porous Solids*, Academic Press, London, 1999.
2. K.S.W. Sing, D.H. Everett, R.A.W. Harl, L. Moscou, R.A. Pierotti, J. Rouquerol, T. Siemieniewska, *Pure Appl. Chem.* 1985, *57*, 603–619.
3. A.F. Cronstedt, *Akad. Handl. Stockholm* 1756, *18*, 120–123.
4. Ch. Baerlocher, W.M. Meier, D.H. Olson, *Atlas of Zeolite Framework Types*, 5th revised edition, Elsevier, Amsterdam, 2001.
5. *Zeolites: Industry Trends and Worldwide Markets in 2010*, Frost & Sullivan, 2001.
6. A. Pfenninger, *Molecular Sieves, Science and Technology*, Vol.2 *Structure and Structure Determination*, Eds., H.G. Karge and J. Weitkamp, Springer, Berlin, 1999, p. 163ff.
7. D.W. Breck, *Zeolite Molecular Sieves*, Wiley, New York, 1974.
8. G.H. Kühl, *J. Catal.* 1973, *29*, 270–277.
9. G.H. Kühl, A.E. Schweizer, *J. Catal.* 1975, *38*, 469–476.
10. G.H. Kühl, *Catalysis and Zeolites: Fundamentals and Applications*, Eds. J. Weitkamp, L. Puppe, Springer, Berlin, 1999, pp. 81–197.
11. M. Taramassow, G. Perego, B. Notari, *US Patent* 4410501, 1983.
12. M. Taramasso, G. Manara, V. Fattore, B. Notari, *US Patent* 4666692, 1987.
13. G. Belussi, M.S. Rigutto, *Introduction to Zeolite Science and Practice*, Eds. H. van Bekkum, E.M. Flanigen, P.A. Jacobs, J.C. Jansen, Studies in Surface Science and Catalysis, Vol. 137, Elsevier, Amsterdam, 2001, pp. 911–955.
14. I.E. Maxwell, W.J.H. Stork, *Introduction to Zeolite Science and Practice*, Eds. H. van Bekkum, E.M. Flanigen, P.A. Jacobs, J.C. Jansen, Studies in Surface Science and Catalysis, Vol. 137, Elsevier, Amsterdam (2001), pp. 747–819.
15. W.F. Hölderich, H. van Bekkum, *Introduction to Zeolite Science and Practice*, Eds. H. van Bekkum, E.M. Flanigen, P.A. Jacobs, J.C. Jansen, Studies in Surface Science and Catalysis, Vol. 137, Elsevier, Amsterdam (2001), pp. 821–910.
16. P.M.M. Blauwhoff, J.W. Gosselink, E.P. Kieffer, S.T. Sie, W.H.J. Stork, *Catalysis and Zeolites: Fundamentals and Applications*, Eds. J. Weitkamp, L. Puppe, Springer, Berlin, 1999, pp. 437–538.
17. A. Corma, A. Martínez, *Handbook of Porous Solids*, Eds. F. Schüth, K.S.W. Sing, J. Weitkamp, Wiley-VCH, Weinheim, 2002, pp. 2825–2922.
18. J.A. Martens, P.A. Jacobs, *Introduction to Zeolite Science and Practice*, Eds. H. van Bekkum, E.M. Flanigen, P.A. Jacobs, J.C. Jansen, Studies in Surface Science and Catalysis, Vol. 137, Elsevier, Amsterdam, 2001, pp. 633–671.
19. P.B. Koradia, J.R. Kiovski, M.Y. Asim, *J. Catal.* 1980, *66*, 290–293.
20. M. Guisnet, V. Fouche, M. Belloum, J.P. Bournonville, C. Travers, *Appl. Catal.* 1991, *71*, 283–293.

21. E.E. Davis, A.J. Kolombos, *US Patent* 4175057, 1979.
22. C.T. O'Connor, *Handbook of Heterogeneous Catalysis*, Eds. G. Ertl, H. Knözinger, J. Weitkamp, VCH, Weinheim, 1997, pp. 2069–2122.
23. P. Meriaudeau, C. Naccache, *J. Mol. Catal.* 1990, *59*, L31–L36.
24. C.D. Chang, W.H. Lang, W.K. Bell, *Catalysis in Organic Reactions*, Ed. R. Moser, Marcel Dekker, New York, 1981, pp. 73–94.
25 C.D. Chang, *Handbook of Heterogeneous Catalysis*, Eds. G. Ertl, H. Knözinger, J. Weitkamp, VCH, Weinheim, 1997, pp. 1894–1908.
26. F. Di Renzo, A. Galarneau, P. Trens, F. Fajula, *Handbook of Porous Solids*, Eds. F. Schüth, K.S.W. Sing, J. Weitkamp, Wiley-VCH, Weinheim, 2002, pp. 1311–1395.
27. A. Taguchi, F. Schüth, *Micropor. Mesopor. Mater.* 2005, *77*, 1–45.
28. T. Yanagisawa, K. Kuroda, C. Kato, *Bull. Chem. Soc. Jpn.* 1988, *61*, 3743–3745.
29. T. Yanagisawa, K. Kuroda, C. Kato, *Reactivity of Solids* 1988, *5*, 167–175.
30. J.S. Beck, C.T.W. Chu, I.D. Johnson, C.T. Kresge, M.E. Leonowicz, W.J. Roth, J.C. Vartuli, *WO Patent* 9111390, 1991.
31. C.T. Kresge, M.E. Leonowicz, W.J Roth, J.C. Vartuli, J.S. Beck, *Nature* 1992, *359*, 710–712.
32. V. Chiola, J.E. Ritzko, C.D. Vanderpool, *US Patent* 3556729, 1971.
33. F. Di Renzo, H. Cambon, R. Dutartre, *Micropor. Mater.* 1997, *10*, 283–286.
34. A. Firouzi, D. Kumar, L.M. Bull, T. Besier, P. Sieger, Q. Huo, S.A. Walker, J.A. Zasadzinski, C. Glinka, J. Nicol, D.I. Margolese, G.D. Stucky, B.F. Chmelka, *Science* 1995, *267*, 1138–1143.
35. D. Zhao, Q. Huo, J. Feng, B.F. Chmelka, G.D. Stucky, *J. Am. Chem. Soc.* 1998, *120*, 6024–6036.
36. M.L. Occelli, S. Biz, A. Auroux, G.J. Ray, *Microp. Mesop. Mater.* 1998, *26*, 193–213.
37. H. Landmesser, H. Kosslick, U. Kürschner, R. Fricke, *J. Chem. Soc. Faraday Trans.* 1998, *94*, 971–977.
38. J. Sauer, F. Marlow, B. Spliethoff, F. Schüth, *Chem. Mater.* 2002, *14*, 217–224.
39. Y. Sun, L. Zhu, H. Lu. R. Wang, S. Lin, D. Jiang, F.-S. Xiao, *Appl. Catal. A: Gen.* 2002, *237*, 21–31.
40. M.V. Landau, L. Titelman, L. Vradman, P. Wilson, *Chem. Commun.* 2003, 594–595.
41. Q.-H- Xia, K. Hidajat, S. Kawi, *J. Catal.* 2002, *209*, 433–444.
42. M. Iwamoto, Y. Tanaka, N. Sawamura, S. Namba, *J. Am. Chem. Soc.* 2003, *125*, 13032–13033.
43. E. Rivera-Muños, D. Lardizabal, G. Alonso, A. Aguilar, M.H. Siadati, R.R. Chianelli, *Catal. Lett.* 2003, *85*, 147–151.
44. L. Vradman, M.V. Landau, M. Herskowitz, V. Ezerski, M. Talianker, S. Nikitenko, Y. Koltypin, A. Gedanken, *J. Catal.* 2003, *213*, 163–175.
45. A. Wingen, N. Anastasievíc, A. Hollnagel, D. Werner, F. Schüth, *J. Catal.* 2000, *193*, 248–254.
46. A. Corma, H. Garcia, A. Moussaif, M.J. Sabater, R. Zniber, A. Redouane, *Chem. Commun.* 2002, 1058–1059.
47. R. Ryoo, S.H. Joo, S. Jun, *J. Phys. Chem. B* 1999, *103*, 7743–7746.
48. A.-H. Lu, W. Schmidt, N. Matoussevitch, B. Spliethoff, B. Tesche, E. Bill, W. Kiefer, F. Schüth, *Angew. Chem. Int. Ed.* 2004, *43*, 4303–4306.
49. V. Jorik, *Zeolites* 1993, *13*,187–191.
50. C.A. Fyfe, Y. Feng, H. Grondey, G.T. Kokotailo, H. Gies, *Chem. Rev.* 1991, *91*, 1525–1543.
51. A. Jentys, J.A. Lercher, *Introduction to Zeolite Science and Practice*, Eds. H. van Bekkum, E.M. Flanigen, P.A. Jacobs, J.C. Jansen, Studies in Surface Science and Catalysis, Vol. 137, Elsevier, Amsterdam, 2001, pp. 345–386.
52. G. Engelhardt, *Introduction to Zeolite Science and Practice*, Eds. H. van Bekkum, E.M. Flanigen, P.A. Jacobs, J.C. Jansen, Studies in Surface Science and Catalysis, Vol. 137, Elsevier, Amsterdam, 2001, pp. 387–418.
53. J.H.C. van Hooff, J.W. Roelofsen, *Introduction to Zeolite Science and Practice*, Eds. H. van Bekkum, E.M. Flanigen, J.C. Jansen, Studies in Surface Science and Catalysis, Vol. 58, Elsevier, Amsterdam, 1991, pp. 241–283.
54. *Handbook of Heterogeneous Catalysis*, Eds. G. Ertl, H. Knözinger, J. Weitkamp, VCH, Weinheim, 1997.
55. J.M. Thomas and W.J. Thomas, *Principles and Practice of Heterogeneous Catalysis*, VCH, Weinheim, 1996.
56. *Handbook of Porous Solids*, Eds. F. Schüth, K.S.W. Sing, J. Weitkamp, Wiley-VCH, Weinheim, 2002.

57. *Introduction to Zeolite Science and Practice*, Eds., H. van Bekkum, E.M. Flanigen, P.A. Jacobs, and J.C. Jansen, Studies in Surface Science and Catalysis, Vol. 137, Elsevier, Amsterdam (2001).
58. *Catalysis and Zeolites: Fundamentals and Applications*, Eds., J. Weitkamp and L. Puppe, Springer, Berlin, 1999.
59. *Molecular Sieves, Science and Technology*, Vol. 2, *Structure and Structure Determination*, Eds., H.G. Karge and J. Weitkamp, Springer, Berlin, 1999.

CHAPTER 4 PROBLEMS

Problem 1

Calculate C and the specific surface area A_s of a material from the nitrogen adsorption isotherm according to the BET equation from the data points given in the figure. Use the ideal gas equation to convert the adsorbed volumes into moles (STP indicates that the volumes adsorbed are given for standard temperature and pressure, i.e., 273 K and 101.3 kPa).

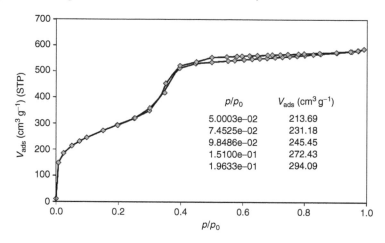

p/p_0	V_{ads} $(cm^3\,g^{-1})$
5.0003e–02	213.69
7.4525e–02	231.18
9.8486e–02	245.45
1.5100e–01	272.43
1.9633e–01	294.09

Problem 2

Calculate the Si/Al ratio of a Na–Y zeolite from a ^{29}Si NMR spectrum with the following intensities:

Signal	Si(4Al)	Si(3Al)	Si(2Al)	Si(1Al)	Si(0Al)
Intensity	1	6.10	15.16	10.68	3.13

Problem 3

A Na–X zeolite (cubic symmetry) has an XRD pattern as shown below. Calculate the lattice constant a of this zeolite from the 2θ positions of the reflections. The wavelength of the x-rays is $\lambda = 0.154056$ nm (Cu K_α radiation).

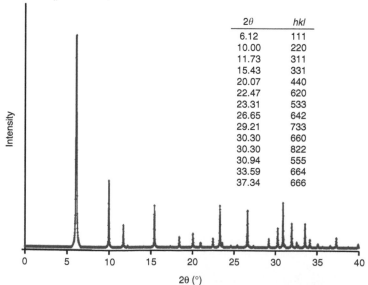

2θ	hkl
6.12	111
10.00	220
11.73	311
15.43	331
20.07	440
22.47	620
23.31	533
26.65	642
29.21	733
30.30	660
30.30	822
30.94	555
33.59	664
37.34	666

Problem 4

Calculate the Si/Al ratio of the Na–X zeolite shown in Problem 3.

Problem 5

Derive simple relations (from Table 4.3) for the calculation of the lattice constant a of a mesoporous material with hexagonal symmetry (MCM-41 or SBA-15 with no ordering along [0 0 1]) from the dhkl values for hkl = (1 0 0), (1 1 0), (2 0 0), and (2 1 0).

Problem 6

Use the formulas derived in Problem 5 and calculate the lattice constant for a material with the XRD pattern shown below.

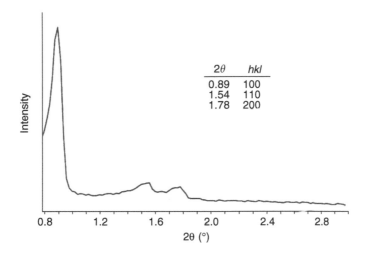

2θ	hkl
0.89	100
1.54	110
1.78	200

Problem 7

Calculate the average thickness th$_p$ of the pore walls of the SBA-15 from Problem 6 by using the calculated lattice constant a and an average mesopore diameter of 8.10 nm (derived from N$_2$ sorption isotherms using the NLDFT method). The scheme illustrating the ordering of the pores given below will help you do this.

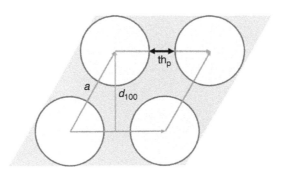

Skeletal Catalysts

Andrew J. Smith

CONTENTS

5.1   Introduction 141
5.2   History 141
5.3   Preparation 142
        5.3.1   Leaching Kinetics 144
        5.3.2   Promoters 145
5.4   Structures 147
5.5   Deactivation/Aging 149
5.6   Applications 151
5.7   Advantages/Disadvantages 153
5.8   Future..154
Acknowledgment 154
References 154
Chapter 5 Questions 159

5.1   INTRODUCTION

Skeletal (Raney®) catalysts are made by a very simple technique. An alloy of two metals in roughly equal proportions, where one metal is the desired catalytic material, and the other is dissolvable in hydroxide, is first made. This alloy is crushed and leached in concentrated hydroxide solution. The soluble metal selectively dissolves, leaving behind a highly porous spongelike structure of the desired catalytic metal. Catalysts formed by this technique show high activity and selectivity, and have found wide use in industry, particularly for hydrogenation and dehydrogenation reactions.

5.2   HISTORY

Skeletal catalysts were first discovered in the 1920s by Murray Raney [1,2]. In recognition of their inventor, the catalysts are often referred to as Raney catalysts, although this trademark is now owned by the Davison division of W.R. Grace & Co., who supply a range of catalysts for industrial use. Another common name is metal sponge, which refers to the porous structure of the catalysts.

Raney's first patent [1] was for an alloy of approximately 50/50 wt% nickel and silicon, subsequently leached in concentrated hydroxide solution. The silicon selectively dissolved, leaving a nickel residue that was quite active in catalyzing the hydrogenation of cottonseed oil, the process in which Raney was engaged at the time. Raney soon discovered that aluminium was a better choice over silicon to be combined with the catalytically active metal [2]. A large number of organic reactions, most of them hydrogenations, have been found to be catalyzed by nickel prepared in this way [3]. The advantages of selective dissolution compared to conventional coprecipitation methods is that the former gives a catalyst active at lower temperatures and pressures and displays a higher degree of selectivity to particular organic groups [4–7]. This latter feature is critical for fields such as pharmaceuticals where complex molecules are involved. Schröter [3] gives an excellent review of the preparation and catalytic uses of skeletal nickel up to the mid-1960s.

Raney predicted that many other metal catalysts could be prepared with this technique, but he did not investigate them [8]. Copper and cobalt catalysts were soon reported by others [4,5]. These catalysts were not nearly as active as Raney's nickel catalyst and therefore have not been as popular industrially; however they offer some advantages such as improved selectivity for some reactions. Skeletal iron, ruthenium and others have also been prepared [9–13]. Wainwright [14,15] provides two brief overviews of skeletal catalysts, in particular skeletal copper, for heterogeneous reactions. Table 5.1 presents a list of different skeletal metal catalysts and some of the reactions that are catalyzed by them.

The activity and stability of skeletal catalysts can be improved with the use of additives, often referred to as promoters. These can be added to the alloy before leaching, or alternatively can be added to the leaching solution [16–19]. An example is the use of zinc to promote skeletal copper for the catalytic synthesis of methanol from synthesis gas [20–22]. Many other promoters have been considered, both inorganic and organic in nature.

5.3 PREPARATION

The desired catalytic metal, or metals, are alloyed with approximately 50 wt% of a reactive metal such as aluminum. Other metals can be used in place of aluminum, such as Raney's original silicon alloy [1], zinc [23], or others; however aluminum gives high activity and remains the most popular choice for most applications.

The exact composition of the precursor alloy depends on which metals are being alloyed. Different intermetallic phases provide different characteristics to the final catalyst in terms of activity and strength [15,20,24]. For example, $NiAl_3$ is readily leached in base, but the residue is friable and disintegrates easily [25]; Ni_2Al_3 is readily leached to form a catalytically active nickel residue [25]; NiAl is only slightly soluble in 20 wt% caustic solution, but it is very tough and can offer strength to the final catalyst [6,26].

Skeletal copper is best made from the $CuAl_2$ intermetallic compound which has very close to 50 wt% aluminum in the alloy and gives an active and selective catalyst [27–29]. Skeletal nickel is also best made from an alloy of about 50 wt% aluminum [25]; however, in this case, the alloy consists of more than one intermetallic phase, the combination of which provides the best activity while maintaining adequate strength in the catalytic residue. The most active skeletal cobalt catalysts are made from an alloy of about 60–65 wt% aluminum, which consists of two intermetallic phases, $Co_2Al_9 + Co_4Al_{13}$ [30].

If the desired catalyst is to consist of two or more catalytic metals after leaching or if a promoter metal is to be included, the precursor alloy becomes even more complicated with respect to phase diagrams. The approximate proportion of reactive metal (aluminum) in these ternary and higher alloys usually remains the same as for the binary metal system for the best results, although the different catalytic activities, leaching behavior and strengths of the various intermetallic phases need to be considered for each alloy system.

Table 5.1 Summary of Skeletal Catalysts and the Types of Reactions Catalyzed by Them (not exhaustive)

Raney Metal	Type of Reaction	Raw Material	Product	Ref.
Ni	Hydrogenation of unsaturates, aromatics, nitro groups, carbonyl groups and nitriles	Linolenate, linoleate (in vegetable oil)	Oleate, stearate	[2]
		Benzene	Cyclohexane	[47]
		p-Nitrotoluene	p-Aminotoluene	[153]
		Cinnamaldehyde	Benzenepropanol	[124]
		Adiponitrile	Hexamethylenediamine	[154]
	Methanation	Syngas $(CO/CO_2/H_2)$	Methane	[68]
	Hydrogen evolution	From biomass		[126]
	Dehydrogenation of alcohols, aldehydes	Cholesterol	Cholest-4-en-3-one	[155]
	Electrocatalyst	Hydrogen oxidation		[156]
	Hydrogenolysis	Benzyl methyl ether	Toluene + methanol	[157]
	Ammonolysis	1,6-Hexanediol	Hexamethylenediamine	[158]
	Reductive dehalogenation	α-Bromoacetophenone	Acetophenone	[159]
	Reductive alkylation	Butyraldehyde + methylamine	N-methylbutylamine	[160]
Cu	Hydrogenation of unsaturates, nitro groups, carbonyl groups and nitriles	Linolenate (in soybean oil)	Linoleate	[161]
		Styrene	Ethylbenzene	[27]
		Acetone	Isopropanol	
		Butyronitrile	Butylamine	
		Nitropropane	Propylamine	
	Methanol synthesis	Syngas $(CO/CO_2/H_2)$	Methanol	[127]
	Water gas shift	$CO + H_2O = CO_2 + H_2$		[162]
	Dehydrogenation of alcohols, aldehydes	Diethanolamine	Iminodiacetic acid	[140]
		Methanol	Methyl formate	[14]
	Hydrolysis	Acrylonitrile	Acrylamide	[114]
	Hydrogenolysis	Ethyl formate	Methanol + ethanol	[14]
	Ammonolysis	1,6-Hexanediol	Hexamethylenediamine	[158]
Co	Hydrogenation of unsaturates, nitro groups, carbonyl groups and nitriles	Styrene	Ethylbenzene	[163]
		Benzaldehyde	Benzyl alcohol	
		Cinnamonitrile	3-Phenylallylamine	[164]
		5-Nitroindole	5 Aminoindole	[165]
	Methanation	Syngas $(CO/CO_2/H_2)$	Methane	[166]
	Reductive dehalogenation	3,4-Dichlorophenol	Phenol	[167]
	Ammonolysis	1,6-Hexanediol	Hexamethylenediamine	[158]
Ru	Hydrogenation of aromatics, carboxyl groups and nitriles	Phenol	Cyclohexanol (Hexanedioic acid)	[168]
		Adipic acid (Hexanedioic acid)	1,6-Hexanediol + 6-hydroxycaproic acid	
		Terephthalonitrile	1,4-bis (Aminomethyl) benzene	
	Methanol synthesis	Syngas $(CO/CO_2/H_2)$	Methanol	[169]
	Ammonia synthesis	N_2	NH_3	[170]
Fe	Fischer–Tropsch	Syngas $(CO/CO_2/H_2)$	Range of hydrocarbons	[171]
Ag	Electrocatalyst	Oxygen reduction reaction		[156]
Pt	Hydrogenation	Acetone	Propane	[172]
		Isopropanol	Ethane + propane	[173]
		Nitroethane	Ethylamine	[174]
Pd	Hydrogenation of unsaturates (selective)	1-(5-Methyl-2-furanyl)-3-phenyl-2-propen-1-one	1-(5-Methyl-2-furanyl)-3-phenyl-1-propanone	[175]
		Linolenate (in soybean oil)	Linoleate, oleate, stearate	[161]
	Electrooxidation	Sequential oxidation of methanol		[145]
Rh	Electroreduction of nitroalkanes	Nitroethane	Ethylamine	[176]
Ir	Electroreduction	Acetone	Isopropanol	[172]
	Electrooxidation	Sequential oxidation of methanol		[177]

Note: Products depend on reaction conditions and on catalyst preparation and treatment. References given as examples only.

The precursor alloy is quenched to form small grains readily attacked by the caustic solution [31]. Quenching can also enable specific intermetallic phases to be obtained, although this is less common. Yamauchi et al. [32–34] have employed a very fast quench to obtain a supersaturation of promoter species in the alloy. It is even possible to obtain an amorphous metal glass of an alloy, and Deng et al. [35] provide a review of this area, particularly with Ni, Ni–P, Ni–B, Ni–Co, and Ni–Co–B systems. The increased catalytic activity observed with these leached amorphous alloy systems can be attributed to either chemical promotion of the catalyzed reaction or an increased surface area of the leached catalyst, depending on the components present in the original alloy. Promotion with additives is considered in more detail later.

The quenched alloy is crushed or ground and screened to a specific particle size range before leaching. The particle size depends on the application envisaged for the catalyst. Particle size affects the catalytic activity [5], most likely due to aging of the leached residue, which occurs even while leaching of the inner particle continues. Aging will be considered in more depth later. Some applications do not use crushed alloy, but rather have the alloy cast in the desired shape and leach it as a large piece, sometimes only on one side. One example of this is fuel cell electrodes, although modern manufacture of these electrodes is done by mixing pelletized or powdered skeletal catalyst with a conducting polymer matrix [36–42].

As an alternative to crushing an alloy into small particles, Ostgard et al. [43] first proposed the manufacture of hollow skeletal catalyst spheres. Precursor alloy is deposited on an organic polymer sphere that is later oxidized completely by heating in air. The hollow alloy spheres that remain are then leached as usual to give the catalyst.

Leaching of the alloy is usually done in strong alkali solution. Acid leaching has been examined; however it generally results in lower activities compared to alkali leaching [44–46]. The general equation for the leaching reaction can be written as

$$2\text{MAl}_{x(s)} + 2x\text{OH}^- + 6x\text{H}_2\text{O} \rightarrow 2\text{M}_{(s)} + 2x\text{Al(OH)}_4^- + 3x\text{H}_{2(g)} \tag{5.1}$$

where M is the desired catalyst metal, such as nickel, copper, and cobalt.

Ultrasonic agitation during leaching has recently been reported to increase the catalytic activity of skeletal nickel for the hydrogenation of benzene to cyclohexane, with the enhanced activity related to changes in the catalyst structure and surface species [47].

Instead of using high-temperature melting to make the precursor alloys, an alternative wet chemistry technique has been proposed where nickel(0) and aluminum coordination compounds are blended together and treated to give nanocrystalline NiAl_x alloys with $1 < x < 3$ [48]. The alloys are leached in the same way as standard skeletal catalysts. Catalysts with higher activity than commercially available Raney® nickel have been prepared by this technique, with the activity attributed to the finer structure and homogeneity of the alloys [48,49].

5.3.1 Leaching Kinetics

The kinetics of leaching for skeletal catalysts are best understood for the copper system. The leaching rate for CuAl_2 is fairly constant and then levels out as particles become fully leached [20], while the kinetics for Cu(Zn)Al_2 alloy were found to be parabolic with time: (leach depth)^2 α time [50]. Leaching kinetics can be followed by measuring evolved hydrogen or dissolved aluminum. Young et al. [20] found good agreement between the two methods. The rate can also be followed by leach depth, although this is more difficult and prone to error [50]. Leach depth simply refers to the distance into a particle that the leach solution has penetrated (Figure 5.1). The alloys are known to have sharp reaction fronts during leaching that are easily identified by optical microscope for both skeletal nickel [51,52] and skeletal copper [20,53,54].

Depending on the way the experiment was set up (e.g., particle size), researchers have found either only reaction control with an activation energy of 69 kJ/mol [55] or a combination of reaction

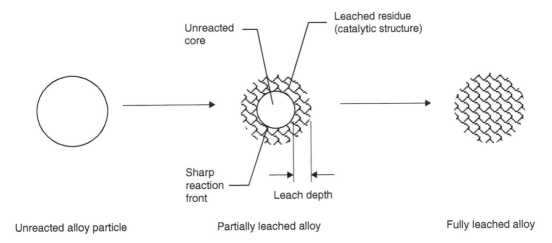

Figure 5.1   Schematic representation of the progress of leaching of an alloy particle. A partially leached alloy shows a sharp reaction front and the extent of leaching is given by the depth of the residual catalytic material.

control at the leach front and diffusion control in the leached residue with an activation energy of 41 kJ/mol [50,56]. The leaching rate increases with increasing temperature for both $CuAl_2$ and $Cu(Zn)Al_2$ alloys [29], while hydroxide concentration has a mixed effect. Addition of zinc to the alloy, or to the lixiviant (leaching solution) slows the reaction rate [17]. Addition of chromate to the lixiviant also reduces the leaching rate [55]. The kinetics have been fitted to Levenspiel's shrinking core model [57], which describes the fractional conversion over time for spherical particles reacting with a sharp reaction front under either reaction- or diffusion-controlled kinetics.

The kinetics of chromium (III) oxide deposition from solution during the leaching of skeletal copper has been studied, and a linear rate was found that is not affected by chromate concentration but decreases with increasing hydroxide concentration [55]. The total amount deposited was greater as the chromate concentration increased.

Choudhary et al. [58] found reaction controlled kinetics with an activation energy of 56.6 kJ/mol for the leaching of skeletal nickel, similar to the leaching of skeletal copper. The kinetics did not fit Levenspiel's shrinking core model [57]; but it should be noted that the leaching solution was agitated with a flat stirrer at 1500 rpm.

The leaching reaction is essentially a dealloying reaction in which one alloy metal is dissolved while the other remains behind. Considerable research has been carried out to understand dealloying reactions, although it is almost entirely in relation to corrosion rather than deliberate dissolution in a hostile solution. Erlebacher et al. [59,60] provide two recent reviews on dealloying and the development of a nanoscale structure in the residue. Parallels between dealloying theory and skeletal copper formation have been drawn that allow a better understanding of the mechanism of formation of the residual catalytic metal during leaching [61]. In general, the mechanism of formation of skeletal copper involves the co-dissolution of both metals and reprecipitation of the copper atoms during the leaching reaction so that the copper forms a coherent, three-dimensional structure. The structure is discussed in a later section of this chapter. This mechanism is also partly responsible for the aging of these catalysts (also discussed later).

5.3.2   Promoters

Promoter species have been mentioned previously. These are additional metals or organic compounds present in either the original alloy or in the lixiviant. They are more than just a second catalytic metal, although bi-metal skeletal catalysts are possible. Promoter species increase the activity of the final catalyst without showing catalytic activity by themselves. The promotion can be chemical in nature, affecting the catalytic metal directly, or it can be structural, if the skeletal metal surface area is increased or stabilized. An excellent example of this promotion effect is zinc in the skeletal copper system [62], which can exhibit both chemical and structural promotion, depending on the conditions of catalyst preparation.

With the addition of zinc to the precursor alloy the Cu_3Al_3Zn intermetallic phase can form, which has higher catalytic activity per unit surface area compared to $CuAl_2$ when leached; however, it only develops low surface areas [63]. This can result in lower overall activity if the ternary phase is present in significant quantities, and emphasizes the need to consider leachability, activity, and strength of different intermetallic phases as previously described for precursor alloys. The right level of zinc addition does provide an increased overall activity to skeletal copper, with the resulting catalyst rivalling conventional coprecipitated catalysts for methanol synthesis from syngas (a hydrogenation reaction) [19,64].

Zinc promotion of the skeletal copper system by addition of sodium zincate to the lixiviant has been compared to addition of zinc metal to the precursor alloy [17]. Both produced higher surface areas than did zinc-free skeletal copper preparation. Dissolving zinc in the lixiviant gave a more uniform distribution of zinc across the leached catalyst particles [16]. The method of adding promoters via the lixiviant opened up the opportunity to investigate a wide range of promoters that could not be alloyed with the metals, or had very low solubilities in the required intermetallic phases (e.g., chromium in $CuAl_2$) [65]. Although creating complex ternary and higher-order alloys with specific intermetallic phases was no longer required, the preparation of solutions with potentially harmful constituents was required.

The range of promoters investigated in various skeletal systems is quite large and includes: calcium [66], gallium[18], titanium[34,67,68], vanadium [34,66,69], chromium [7,18,37,69–75], manganese [18,46,69], zirconium[34,67], niobium [34,66,67], molybdenum [7,18,66, 67,69,72, 76–78], tungsten [66], zinc [18,46], palladium [33,34], platinum [57, 79–81], cadmium[82], indium [66], tantalum[67], boron [35], phosphorus [83], saccharides [84] and glycosides [84]. Modification of skeletal nickel with tartaric acid has also been employed [85,86]. It is difficult to provide a summary of the effects and properties of promoters because their effect depends on the skeletal metal employed, the conditions of catalyst preparation, the reaction catalyzed and the reaction conditions employed. As an example, Table 5.2 presents the effect of several promoting metals on skeletal nickel for the hydrogenation of specific organic groups.

Incorporation of promoters can occur via two distinct mechanisms. A local pH drop at the leach front caused by the aluminum dissolution can cause a solvated promoter to deposit via a shift in the solubility equilibrium. Zincate shows this behavior, depositing as zinc oxide as the pH drops at the leach

Table 5.2   **Effect of Promoters on the Activity of Skeletal Nickel Catalysts for Hydrogenation of Different Functional Groups**

Promoter (metal)	Promoter in alloy (%)	Reactant Undergoing Hydrogenation			
		Butyronitrile (cyano)	Acetone (carbonyl)	Sodium p-nitrophenolate (nitro)	Sodium itaconate (vinyl)
Mo	2.2	6.5	2.9	1.7[b]	1.2
Cr	1.5	3.8	1.5	1.6	—
Fe	6.5	3.3	1.3	2.1	—
Cu	4.0	2.9	1.7	1.3	—
Co	6.0	2.0	1.6[a]	—	—

Note: Activity data reported as reaction rate relative to unpromoted skeletal nickel.

[a]Co promoter = 2.5%.

[b]Mo promoter = 1.5%.

Source: Compiled from S.R. Montgomery, Functional group activity of promoted Raney nickel catalysts, in Catalysis of Organic Reactions, W.R. Moser, Ed., Marcel Dekker, New York, 1981, pp. 383–409.

front (Equation 5.2). An alternative mechanism of incorporation is via electrochemical reduction. Chromate shows this second mechanism, depositing as chromium (III) oxide (Equation 5.3). The electrons for the reduction come from the aluminum oxidation, but can also be supplied from local regions of exposed residual metal if it is more anodic than the promoter species. Ultimately aluminum is the primary supplier, being the most anodic species in the system. Electrochemical reduction can occur anywhere on the leached residue because of electrical conductivity of the structure. The mechanism is partly a displacement reaction with aluminum, but also involves the residual metal to a small extent and occurs in an environment of concentrated caustic solution that actively dissolves the aluminum as well.

$$ZnO_2{}^{2-} + H_2O \rightarrow Zn(O)_{(s)} + 2OH^- \qquad (5.2)$$

$$2CrO_4{}^{2-} + 5H_2O + 6e^- \rightarrow Cr_2O_{3(s)} + 10OH^- \qquad (5.3)$$

Promoter deposition through different mechanisms can account for different catalyst properties. In particular, chromate depositing as chromia does not easily redissolve but, zinc oxide does redissolve once the leach front passes and the pH returns to the bulk level of the lixiviant. Therefore, chromate can provide a more stable catalyst structure against aging, as observed in the skeletal copper system. Of course, promoter involvement in catalyst activity as well as structural promotion must be considered in the selection of promoters. This complexity once again highlights the dependence of the catalytic activity of these materials on the preparation conditions.

Surface modification of skeletal nickel with tartaric acid produced catalysts capable of enantioselective hydrogenation [85–89]. The modification was carried out after the formation of the skeletal nickel catalyst and involved adsorption of tartaric acid on the surface of the nickel. Reaction conditions strongly influenced the enantioselectivity of the catalyst. Both Ni^0 and Ni^{2+} have been detected on the modified surface [89]. This technique has already been expanded to other modified skeletal catalysts; for example, modification with oxazaborolidine compounds for reduction of ketones to chiral alcohols [90].

5.4 STRUCTURES

Consider skeletal copper — the precursor alloy consists essentially of $CuAl_2$. Dissolution of the aluminum causes two-thirds of the atoms in the structure to be removed. What structure can remain behind and how does it form? More importantly, how can its formation be manipulated to provide a better catalyst? Clearly, understanding the preparation conditions of these catalysts and the structures they form is crucial to obtaining optimal catalysts for the industry.

Most research on the structure of skeletal catalysts has focused on nickel and involved methods such as x-ray diffraction (XRD), x-ray absorption spectroscopy (XAS), electron diffraction, Auger spectroscopy, and x-ray photoelectron spectroscopy (XPS), in addition to pore size and surface area measurements. Direct imaging of skeletal catalyst structures was not possible for a long while, and so was inferred from indirect methods such as carbon replicas of surfaces [54]. The problem is that the materials are often pyrophoric and require storage under water. On drying, they oxidize rapidly and can generate sufficient heat to cause ignition.

Skeletal nickel consists of highly-dispersed nickel with a large surface area [68, 91–96], the structure often being likened to a sponge [51,74]. The activity of the catalyst is proportional to the surface area and hence the degree of nickel crystallite dispersion [26,76,91]. The nickel crystallites are about 1–20 nm in size [24,92,94–96], and decrease in size with decreasing temperature

Table 5.3   Summary of Crystallite Size, Surface Area, and Pore Volume of Skeletal Nickel Leached under Different Conditions

Leaching Temperature (°C)	Crystallite Size [97] (Å)	Surface Area [104] (m²/g)	Pore Volume [104] (cm²/g)
10	88[a]	—	—
20	103[a], 52[b]	—	—
50	116[a], 99[b]	110	0.07
103–107	176[a]	80	0.12

[a]20 wt% KOH.

[b]10 wt% KOH.

[26,76,94–98] or caustic concentration [94–97] (Table 5.3). The amount of residual aluminum can also affect the nickel crystallite size [52,99].

In comparison to skeletal nickel, skeletal copper has a significantly larger crystallite size of about 10–100 nm [32,46,92,96,100,101]. Fasman and coworkers [46,100,101] examined the crystal structure more closely and found that it consisted of copper crystals that had agglomerated into granules or precipitated onto oxides. The copper crystal grains and subgrains were of about 10–13 nm in size, while the copper agglomerates were 50–80 nm.

Leaching at higher temperatures produces a coarser skeletal structure. Table 5.3 illustrates this for skeletal nickel at 50°C vs. 107°C. For comparison, skeletal copper shows a pore spacing of 40 nm when leached at 1°C and a pore spacing of 110 nm when leached at 93°C [54]. The narrow pore-size distribution of skeletal copper is actually bimodal and, it is suggested, the larger pores facilitate higher gas diffusivity leading to higher measured catalytic activity [19]. Addition of promoters can significantly alter the skeletal structure, leading to decreased pore and crystallite sizes and increased surface areas [50,76,77,102,103].

The microstructure of the precursor alloy is retained upon leaching, with a sharp interface between the leached and the unleached regions for both skeletal nickel and copper [20,51,53,54,104]. A large amount of cracking exists on the surface probably because of shrinkage on removal of the aluminum [51,92]. During dissolution of the precursor alloy for skeletal nickel, a disordered body-centered cubic (bcc) structure of nickel and aluminum atoms is first formed, while further loss of aluminum results in a face-centered cubic (fcc) structure of nickel atoms [105]. However, the coordination number of the nickel atoms is only 6 instead of the usual 12 for an fcc structure, indicating a very high concentration of lattice defects [49,106,107]. Skeletal cobalt also has an fcc structure, while skeletal iron has a bcc structure [108]. Interestingly, rapidly solidified nickel–aluminum alloys leach with an intermediate fcc tetragonal Ni_3Al_2 phase before proceeding to the fcc nickel in the final skeletal structure [109].

The structure of skeletal catalysts is so fine that electron microscopes are required for sufficient resolution. The use of a focussed ion beam (FIB) miller has enabled a skeletal copper catalyst to be sliced open under vacuum and the internal structure to be imaged directly [61]. Slicing the catalyst enabled viewing beyond the obscuring oxide layer on the surface. A uniform, three-dimensional structure of fine copper ligaments was observed [61], which differed from the leading inferred structure at the time of parallel curved rods [54].

Promoted skeletal copper was also imaged with the FIB. In particular, both zinc- and chromium-promoted skeletal copper have a structure similar to that of un-promoted skeletal copper, but on a much finer scale [110,111]. This observation agrees with the increased measured surface areas for these promoted catalysts. Figure 5.2a shows the fine uniform ligaments in a zinc-promoted skeletal copper catalyst.

Work with the FIB was taken beyond simple imaging by slicing a section of skeletal copper catalyst thinly (~100 nm) before transferring it to a TEM (transmission electron microscope). Very high-resolution imaging of the catalyst in the TEM revealed the agglomerated copper granules making up the fine ligaments observed in the FIB microscope [112], refer Figure 5.2b. In addition, the TEM was fitted with an energy-dispersive x-ray spectrometer, providing an elemental map of the structure and

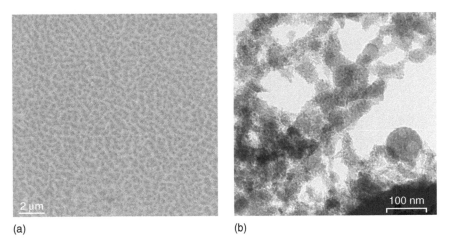

(a)   (b)

Figure 5.2 Promoted skeletal copper showing (a) the uniform network of fine metal ligaments, and (b) the granular structure of those ligaments. (Adapted from A.J. Smith, P. Munroe, T. Tran and M.S. Wainwright, *J. Mater. Sci.*, **30** (2001) 3519.)

revealing critical information about the location of promoter species within the promoted skeletal copper, which were later related to the mechanism of formation of this promoted catalyst.

These techniques for direct structural imaging and analysis utilizing the FIB have not been applied to other skeletal systems. The structures are expected to be similar to copper, albeit on a finer or coarser scale. The FIB has been available for many years, but it has been employed almost exclusively in the semiconductor industry. Rapid growth in the use of the FIB for preparing TEM and SEM (scanning electron microscope) samples of difficult-to-machine materials has resulted in dual-column FIB/SEM microscopes becoming commercially available (e.g., FEI company). This will simplify the technique described earlier for high-resolution imaging and elemental mapping of skeletal systems as well as many other catalytic materials.

5.5 DEACTIVATION / AGING

Skeletal catalysts can lose activity over time. This phenomena has been attributed to several causes depending on the application and that include surface fouling with by-products, surface oxidation, and structure rearrangement.

In studying catalytic activity loss of skeletal copper during hydrolysis of acrylonitrile to form acrylamide, the cause was found to be a combination of surface oxidation (activity was dependent on oxygen concentration in the feed) and fouling of the surface with polymerized acrylamide by-products that blocked access to the pores [113,114]. For hydrogenation of xylose to xylitol over molybdenum-promoted skeletal nickel, deactivation was primarily due to the collapse of the pore structure, with a secondary cause being surface fouling with organic species [115].

Activity loss can occur while the catalysts are still being prepared by leaching [29,116]. Surface areas and activities of leached catalysts pass through a maximum during depth of leaching, particularly for larger particle sizes [29,116]. Figure 5.3 is typical of leaching curves seen in the literature. With activity related to surface area, the problem is the loss of surface area per unit volume of leached residue, despite continued leaching. This is by no means limited to leaching, it also occurs during use of the catalysts, often in parallel with other factors such as fouling of the surface [117].

The loss of surface area and activity is sometimes referred to as sintering or coarsening of the catalyst structure. The term sintering in this case should be considered differently to the common use of the word sinter, where particles fuse together, since for a skeletal catalyst the internal particle struc-

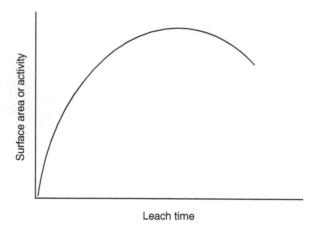

Figure 5.3 Typical shape of the curve for catalytic activity vs. leach time, showing a maximum.

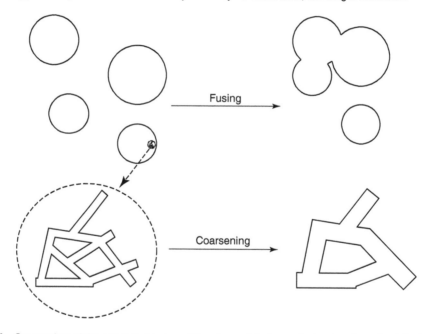

Figure 5.4 Coarsening of the internal structure differs from sintering or fusing together of catalyst particles.

ture is undergoing the change that results in the loss of surface area (Figure 5.4). A somewhat more accurate expression might be 'sintering of ligaments within the structure'. Fusing of individual catalyst particles together can also occur to some extent in certain environments; however, the coarsening of the internal structure is of most relevance to the activity loss. There is a corresponding increase in pore size during the coarsening of the structure, but the total pore volume remains constant.

Several mechanisms have been proposed for the coarsening phenomenon. Most recently, electrochemical measurements during skeletal copper preparation have shown a mechanism of dissolution of copper atoms followed by reprecipitation [61]. This mechanism results in the fine copper structure becoming less fine over time, losing surface area and hence activity. This mechanism was also found by Kalina et al. [100] for skeletal copper using a different technique. The mechanism for other skeletal metal systems has not been concluded but is expected to be similar to that for the copper system. In support of this hypothesis, XANES (x-ray absorption near-edge spectroscopy) analysis of partially leached nickel–aluminum alloys reveals that nickel undergoes an electronic structural change before the geometric structural change, later relaxing into the final fcc crystal structure [118]. The electronic change indicates that nickel is not passive during leaching but is involved chemically. Mikkola et al. have successfully modeled the deactivation semiempirically for skeletal nickel in the hydrogenation of xylose to xylitol [115].

Some promoters are able to stabilize the structure against the rearrangement or coarsening phenomenon. In addition to creating a higher surface area to start with, by facilitating the dispersion of metal crystallites making up the skeletal structure [102,103], promoters can also anchor or protect the structure, thus maintaining the catalytic activity. The protection is not perfect and some surface area will still be lost over time; however, the rate can be significantly reduced. The amount of protection depends on the preparation conditions of the promoted skeletal catalyst and on the environment that the catalyst is exposed to during operation. The exact mechanism of this protection is not known, although theories have been proposed on the basis of electrochemical measurements and the location of the promoters within the structure [111,112]. For example, chromium promotion of skeletal copper results in Cr_2O_3 species on the surface of the copper, hindering dissolution of copper by the coarsening mechanism of dissolution and reprecipitation. In comparison, zinc promotion results in zinc oxide species acting as nucleation sites during the formation of skeletal copper to give a finer structure, but being embedded inside the copper structure, they do not offer any protection against rearrangement by the dissolution and reprecipitation mechanism.

It is possible to age a catalyst before use, significantly reducing the rate of activity loss [119–121]. Aging simply involves exposing the fully leached catalyst to further caustic solution at an elevated temperature. This accelerates the coarsening reaction and will cause an initial drop in activity. After aging, further activity loss is significantly reduced. It is not clear whether this reduced rate of deactivation is because most of the loss has already occurred and the catalyst is simply further down the activity curve, or whether accelerated coarsening is able to impart additional stability to the catalyst, possibly through the strength of larger metallic ligaments making up the structure and providing protection against future deactivation. Interestingly, addition of chromate to skeletal copper increases the initial activity but protects against faster deactivation. Deactivation is at the same rate as that of unpromoted catalysts, or lower [120,121]. Zincate promotion does not impart this same protection to skeletal copper [120].

Applications for this aging method are potentially numerous, particularly for reactions under severe conditions such as oxidative dehydrogenation of alcohols in strong caustic solutions where deactivation by coarsening is a significant problem. Pretreatment or aging of the catalysts before use under these conditions might provide a more stable activity level during use. Accelerated coarsening of the catalysts also serves for accelerated cyclic testing, allowing a better understanding in a shorter time of the long-term stability of these catalysts in hostile environments. Very little work has been published to date on deliberate aging of skeletal catalysts and the effect on activity and deactivation.

Recently, it has been shown that ultrasonic agitation during hydrogenation reactions over skeletal nickel can slow catalyst deactivation [122–124]. Furthermore, ultrasonic waves can also significantly increase the reaction rate and selectivity of these reactions [123,124]. Cavitations form in the liquid reaction medium because of the ultrasonic agitation, and subsequently collapse with intense localized temperature and pressure. It is these extreme conditions that affect the chemical reactions. Various reactions have been tested over skeletal catalysts, including xylose to xylitol, citral to citronellal and citronellol, cinnamaldehyde to benzenepropanol, and the enantioselective hydrogenation of 1-phenyl-1,2-propanedione. Ultrasound supported catalysis has been known for some time and is not peculiar to skeletal catalysts [125]; however, research with skeletal catalysts is relatively recent and an active area.

5.6   APPLICATIONS

Skeletal catalysts are primarily used for hydrogenation and dehydrogenation reactions. The first application of skeletal nickel was hydrogenation of cottonseed oil [1]. Skeletal catalysts have since found use in the manufacture of pharmaceuticals, polymers, surfactants, dyes, resins, agricultural chemicals, and fuel cells.

The types of reactions catalyzed by skeletal catalysts are summarized in Table 5.1. Skeletal nickel remains the most popular of the skeletal catalysts, offering strong hydrogenation/dehydrogenation activity. Other catalyzed reactions include reductive alkylation, ammonolysis, and hydrogenolysis. Skeletal nickel is also particularly attractive for hydrogen evolution from biomass, with promotion by tin improving performance by reducing methane formation [126].

Reductive alkylation:

$$CH_3CH_2CH_2CHO + CH_3NH_2 + H_2 \rightarrow CH_3(CH2)_3NHCH_3 + H_2O \quad (5.4)$$

Ammonolysis:

$$HO(CH_2)_6OH + 2NH_3 \rightarrow H_2N(CH_2)_6NH_2 + 2H_2O \quad (5.5)$$

Hydrogenolysis:

$$CH_3COOCH_2CH_3 + 2H_2 \rightarrow 2CH_3CH_2OH \quad (5.6)$$

Skeletal copper has lower activity toward hydrogenation compared with skeletal nickel, but it offers superior selectivity for certain reactions. Hydrolysis of acrylontrile over skeletal copper yields acrylamide, retaining the unsaturated bond [114,135]:

$$CH_2 = CHCN + H_2O \rightarrow CH_2 = CHCONH_2 \quad (5.7)$$

Methanol synthesis from syngas and related reactions (such as the water gas shift reaction) have been heavily researched by Wainwright's group since 1978 (see, e.g., [22,29,64, 127–134]) resulting in zinc- and chromium-promoted skeletal copper catalysts that rival conventional co-precipitated catalysts for the reaction. Skeletal copper can also be used for the selective oxidation of alcohols to acids through oxidative dehydrogenation. This reaction differs from typical oxidation since it produces hydrogen as a by-product. One application of oxidative dehydrogenation is in the selective conversion of diethanolamine to iminodiacetate (Equation 5.8), a precursor for the herbicide glyphosate. This reaction is performed over skeletal copper with common promoters such as chromium to enhance activity and stability [136–140]. Silver [141] or iron [142] added after catalyst preparation has been claimed to prolong the life of the catalyst for repeated cycles in the hostile reaction conditions.

$$(HOCH_2CH_2)_2NH + 2OH^- \rightarrow (^-OOCCH_2)_2NH + 2H_{2(g)} \quad (5.8)$$

Skeletal nickel and skeletal silver have found application in hydrogen fuel cells [143–147]. Skeletal nickel embedded in a polytetrafluroethylene (PTFE) matrix offers low-temperature operation in hydrogen alkaline fuel cells without the need for precious metal loading. This has become one of the leading electrocatalytic materials for the anode in this type of fuel cell [144, 146–148]. Promoters such as chromium or titanium can enhance the activity of skeletal nickel in this application [36,37]. Direct addition of other metals such as copper or tin as powders to skeletal nickel during the preparation of the PTFE-bonded electrode can improve activity through enhanced conductivity or by providing an alternative mechanism of reaction [148, 149]. Skeletal silver has been successfully used as a cathode in these cells for the oxygen reduction reaction [144,146,147,150,151].

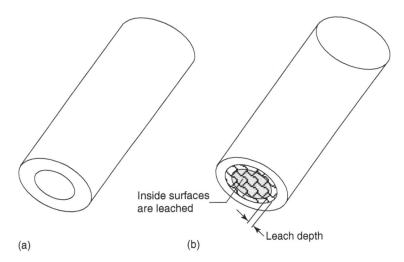

Inside surfaces
are leached

Leach depth

(a)   (b)

Figure 5.5 A large tubular piece of alloy can be partially leached on one side to give a surface-catalyzed reactor.

Enantioselective organic synthesis using modified skeletal catalysts has wide application in areas such as pharmaceutical production; for example, synthesis of chiral alcohols from ketones [90], which is described in detail elsewhere in this book.

Skeletal catalysts are usually employed in slurry-phase reactors or fixed-bed reactors. Hydrogenation of cottonseed oil, oxidative dehydrogenation of alcohols, and several other reactions are performed in slurry phase, where the catalysts are charged into the liquid and optionally stirred (often by action of the gases involved) to achieve intimate mixing. Fixed-bed designs suit methanol synthesis from syngas and catalysis of the water gas shift reaction, and are usually preferred because they obviate the need to separate product from catalyst and are simple in terms of a continuous process.

Skeletal catalysts are usually prepared from an alloy that has been crushed to a defined particle size before leaching. It is also possible to leach large pieces of the alloy. Partially leaching of a large surface can make use of the electrical and heat conductivity of the unleached alloy left behind. This technique can find use in fuel cell electrodes or as surface-catalyzed reactors. Figure 5.1 shows the difference between a partially and fully leached particle while Figure 5.5 shows this applied to a tubular piece of alloy leached only on one side so that it might be used as a surface-catalyzed reactor. This type of application can also be achieved by coating the precursor alloy as a thin layer on another material, and then leaching.

5.7 ADVANTAGES / DISADVANTAGES

The main advantages of skeletal catalysts are high activity at low temperatures, high selectivity, and strength against attrition. Compared with precious metal catalysts, skeletal catalysts are relatively cheap owing to high activity with base metals. Selectivity allows skeletal catalysts to be employed in reactions on complex substrates such as pharmaceuticals. Selectivity also allows high yields of intermediate products, such as methanol from syngas over skeletal copper, where skeletal nickel would give methane (Table 5.1).

Strength against attrition is particularly important for catalysts in slurry-bed reactors, where physical breakage of the catalyst particles, ultimately to fines, can prevent their use for those reactions. The strength of the high surface area skeletal structures can be contrasted against activated carbon, which readily breaks down due to attrition in these types of environments. For the few environments where attrition is still a problem (e.g., oxidative dehydrogenation of alcohols), the skeletal catalytic material can be coated onto a stronger skeletal catalyst (e.g., skeletal copper onto skeletal nickel), or onto a different stronger support [152].

The primary disadvantages of skeletal catalysts include deactivation and waste disposal. Deactivation has already been discussed at some length earlier. The supernatant leach liquor consists of concentrated alkali with dissolved aluminum. In applications involving promoters, small amounts of these will also be present in the leach liquor. This solution presents a disposal issue, particularly if the promoters are hazardous, e.g., chromate ions. In times when industries are being encouraged to reduce waste streams and increase efficiencies of chemical conversions, this waste sodium aluminate liquor presents a problem. Apart from sending the liquor to an alumina refinery should one exist (with interfering promoters not present or removed), there are few other options available for dealing with this material.

5.8 FUTURE

In the 80 years since their first discovery, skeletal catalysts have progressed considerably. From an application for vegetable oil hydrogenation, they have found widespread use across various reactions, mostly hydrogenations but others as well. Different skeletal metals have been identified and investigated, although nickel remains to be the most popular. Gas-phase/fixed-bed and slurry-phase reactions are handled easily with these catalysts, which offer superior selectivity and strength. Deactivation has been an issue, but much work has been devoted to overcoming the problem. Most recently, chromium promotion has been identified as stabilizing the structure against coarsening, allowing sustained activity.

Skeletal catalysts have secured a place in industry and will continue to do so. While skeletal nickel remains popular, it is expected that other metals, particularly skeletal copper with the large amount of research already conducted, will gain popularity. A great deal of work has been published on promotion and stabilization of skeletal catalysts, which has resulted in superior catalysts for industrial applications. This is expected to continue. Further advances in the field of enantioselective skeletal catalysts using tartaric acid modification are inevitable. The parallel application of skeletal nickel and silver in fuel cells will continue to advance as fuel cells begin to reach commercial reality. The unique structure of skeletal catalysts as strong, highly porous, high surface area metals is likely to find other applications in addition to direct catalysis.

ACKNOWLEDGMENT

I would like to thank my mentor Professor Mark Wainwright, who introduced me to these fascinating materials and who kindly edited this chapter.

REFERENCES

1. M. Raney, U.S. Patent 1,563,587, 1925.
2. M. Raney, U.S. Patent 1,628,190, 1927.
3. R. Schröter, Reductions with Raney nickel catalysts, in *Newer Methods of Preparative Organic Chemistry*, Interscience Publishers Inc., New York, 1965, pp. 61–101.
4. L. Faucounau, *Bull. Soc. Chim.*, **4** (1937) 58 (Chem. Abs. 31:3217[1]).
5. B.V. Aller, *J. Appl. Chem.*, **7** (1957) 130.
6. M. Raney, *Ind. Eng. Chem.*, **32** (1940) 1199.
7. S.R. Montgomery, Functional group activity of promoted Raney nickel catalysts, in *Catalysis of Organic Reactions*, W.R. Moser, Ed., Marcel Dekker, New York, 1981, pp. 383–409.
8. M. Raney, U.S. Patent 1,915,473, 1933.

9. P.W. Reynolds and J.W. Donaldson, British Patent 634,097, 1950.

10. P.W. Reynolds, D.A. Dowden and J.A. Mackenzie, British Patent 658,863, 1951.

11. A.A. Vedenyapin, N.D. Zubareva, V.M. Akimov, E.I. Klabunovskii, N.G. Giorgadze and N.E. Barrannikova, *Izv. Akad. Nauk SSSR, Ser. Khim.*, **10** (1976) 2340.

12. A.A. Vedenyapin, N.D. Zubareva, V.M. Akimov, G.K. Areshidze, N.E. Barrannikova and E.I. Klabunovskii, *Izv. Akad. Nauk SSSR, Ser. Khim.*, **8** (1978) 1942.

13. K. Urabe, T. Yoshioka and A. Ozaki, *J. Catal.*, **54** (1978) 52.

14. M.S. Wainwright, *Chem. Ind.*, **68** (Catalysis of Organic Reactions) (1996) 213.

15. M.S. Wainwright, Skeletal metal catalysts, in *Handbook of Heterogeneous Catalysis*, G. Ertl, H. Knözinger and J. Weitkamp, Ed., Wiley-VCH, Weinheim, 1998, pp. 64–72.

16. H.E. Curry-Hyde, M.S. Wainwright and D.J. Young, *Appl. Catal.*, **77** (1991) 75.

17. H.E. Curry-Hyde, M.S. Wainwright and D.J. Young, *Appl. Catal.*, **77** (1991) 89.

18. L. Ma, D.L. Trimm and M.S. Wainwright, Promoted skeletal copper catalysts for methanol synthesis, in *Advances of Alcohols Fuels in the World*, – Proceedings of the XII International Symposium on Alcohol Fuels, Beijing, China, Tsinghua University Press, 1998, pp. 1–7.

19. H.E. Curry-Hyde, G.D. Sizgek, M.S. Wainwright and D.J. Young, *Appl. Catal.*, **95** (1993) 65.

20. D.J. Young, M.S. Wainwright and R.B. Anderson, *J. Catal.*, **64** (1980) 116.

21. M.S. Wainwright and R.B. Anderson, *J. Catal.*, **64** (1980) 124.

22. W.L. Marsden and M.S. Wainwright, *Chem. Eng. Aust.*, **5** (1980) 27.

23. G. Sheela, M. Pushpavanam and S. Pushpavanam, *Int. J. Hydrogen Energy*, **27** (2002) 627.

24. J.R. Anderson, *Structure of Metallic Catalysts*, Academic Press, London, 1975, pp. 228–233.

25. M.L. Bakker, D.J. Young and M.S. Wainwright, *J. Mater. Sci.*, **23** (1988) 3921.

26. J.C. Klein and D.M. Hercules, *Anal. Chem.*, **53** (1981) 754.

27. J.A. Stanfield and P.E. Robbins, Raney Copper Catalysts, *Proc. 2nd Int. Cong. Catal.*, Paris, 1960, pp. 2579–2599.

28. N.K. Nadirov, A.M. Ashirov, A.F. Savel'ev and A. Zhusupova, *Russ. J. Phys. Chem.*, **51** (1977) 82.

29. H.E. Curry-Hyde, D.J. Young and M.S. Wainwright, *Appl. Catal.*, **29** (1987) 31.

30. T.K. Kabiev, D.V. Sokol'skii and G.S. Ospanova, *Dokl. Akad. Nauk. SSSR*, **253** (1980) 1405.

31. H. Hu, M. Qiao, Y. Pei, K. Fan, H. Li, B. Zong and X. Zhang, *Appl. Catal. A*, **252** (2003) 173.

32. I. Ohnaka, I. Yamauchi and M. Itaya, *J. Jpn. Inst. Met.*, **56** (1992) 973.

33. I. Yamauchi, I. Ohnaka and Y. Ohashi, *J. Jpn. Inst. Met.*, **57** (1993) 1064.

34. I. Ohnaka and I. Yamauchi, *Mater. Sci. Eng.*, **A181-182** (1994) 1190.

35. J.-F. Deng, H. Li and W. Wang, *Catal. Today*, **51** (1999) 113.

36. K. Mund, G. Richter and F. Von Sturm, *J. Electrochem. Soc.*, **124** (1977) 1.

37. T. Kenjo, *J. Electrochem. Soc.*, **132** (1985) 383.

38. T. Tomida and I. Nakabayashi, *J. Electrochem. Soc.*, **136**(11) (1989) 3296.

39. W. Jenseit, A. Khalil and H. Wendt, *J. Appl. Electrochem.*, **20** (1990) 893.

40. O. Fuehrer, S. Rieke, C. Schmitz, B. Willer and M. Wollny, *Int. J. Hydrogen Energy*, **19** (1994) 343.

41. M.A. Al-Saleh, Sleem-Ur-Rahman, S.M.M.J. Kareemuddin and A.S. Al-Zakri, *J. Power Sources*, **72** (1998) 159.

42. H.-K. Lee, E.-E. Jung and J.-S. Lee, *Mater. Chem. Phys.*, **55** (1998) 89.

43. D.J. Ostgard, P. Panster, C. Rehren, M. Berweiler, G. Stephani and L. Schneider, German Patent US6573213, 2003.

44. P.W. Reynolds, British Patent 621,749, 1949.

45. J.R. Mellor, R.G. Copperwaithe and N.J. Coville, *Chem. Eng. Comm.*, **167** (1998) 87.

46. M.M. Kalina, S.K. Berdongarova, G.A. Sadykova and A.B. Fasman, Physicochemical characteristics of Raney catalysts prepared from γ-metallides of some binary and ternary systems, in *Mater. Resp. Nauchno-Tekh. Konf. Molodykh Uch. Pererab. Nefti Neftekhim., 3rd*, A. Abdukadyrov, Ed., Sredneaziat. Nauchno-Issled. Inst. Neftepererab. Prom-sti., Tashkent, USSR, 1976, pp. 100–101 (Chem. Abs. 189:204764z).

47. Q. Meng, Y. Wu, Y. Wan and H. Li, *Chin. J. Catal.*, **25** (2004) 529.

48. R. Richards, G. Geibel, W. Hofstadt and H. Bonnemann, *Appl. Organomet. Chem.*, **16** (2002) 377.

49. H. Modrow, M.O. Rahman, R. Richards, J. Hormes and H. Bonnemann, *J. Phys. Chem.*, **107** (2003) 12221.

50. J.B. Friedrich, D.J. Young and M.S. Wainwright, *J. Electrochem. Soc.*, **128** (1981) 1845.

51. P. Colin, S. Hamar-Thibault and J.C. Joud, *J. Mater. Sci.*, **27** (1992) 2326.
52. T. Kubomatsu, *Kagaku to Kôgyô (Osaka)*, **31** (1957) 190.
53. J.B. Friedrich, D.J. Young and M.S. Wainwright, *J. Electrochem. Soc.*, **128** (1981) 1840.
54. J. Szot, D.J. Young, A. Bourdillon and K.E. Easterling, *Philos. Mag. Lett.*, **55** (1987) 109.
55. L. Ma, A.J. Smith, T. Tran and M.S. Wainwright, *Chem. Eng. Proc.*, **40** (2001) 59.
56. N.I. Onuoha, A.D. Tomsett, M.S. Wainwright and D.J. Young, *J. Catal.*, **91** (1985) 25.
57. O. Levenspiel, *Chemical Reaction Engineering*, Wiley , Singapore, 1972, p. 372.
58. V.R. Choudhary, S.K. Chaudhari and A.N. Gokarn, *Ind. Eng. Chem. Res.*, **28** (1989) 33.
59. J. Erlebacher, M.J. Aziz, A. Karma, N. Dimitrov and K. Sieradzki, *Nature*, **410** (2001) 450.
60. J. Erlebacher and K. Sieradzki, *Scr. Materialia*, **49**(10) (2003) 991.
61. A.J. Smith, T. Tran and M.S. Wainwright, *J. Appl. Electrochem.*, **29** (1999) 1085.
62. J.B. Friedrich, *PhD Thesis*, in *School of Chemical Engineering*. 1982, University of New South Wales, Australia: Sydney, Australia.
63. A.J. Bridgewater, M.S. Wainwright, D.J. Young and J.P. Orchard, *Appl. Catal.*, **7** (1983) 369.
64. A.J. Bridgewater, M.S. Wainwright and D.J. Young, *Appl. Catal.*, **28** (1986) 241.
65. G. Ghosh, Aluminium–chromium—copper, in *Ternary Alloys: A Comprehensive Compendium of Evaluated Constitutional Data and Phase Diagrams*, G. Petzow and G. Effenberg, Ed., VCH Publishers, New York, 1992, pp. 311–319.
66. N.K. Nadirov and A.F. Savel'ev, *Kinet. Katal.*, **17** (1976) 1176.
67. A.B. Fasman, G.A. Pushkareva, B.K. Almashev, V.N. Rechkin, Y.F. Klyuchnikov and I.A. Sapukov, *Kinet. Katal.*, **12** (1971) 1271.
68. G.V. Sobolevskii, A.N. Grechenko, I.E. Nemirovskaya, A.I. Nechugovskii and V.V. Lunin, *Kinet. Katal.*, **32** (1991) 714.
69. K. Wimmer and O. Stichnoth, German Patent 875,519, 1953.
70. M. Shimizu and S. Takeoka, *Nippon Kagaku Kaishi*, **6** (1981) 912 (Chem. Abs. 95:68610r).
71. J. Laine, Z. Ferrer, M. Labady, V. Chang and P. Frías, *Appl. Catal.*, **44** (1988) 11.
72. R. Paul, *Bull. Soc. Chim.*, (1946) 208.
73. M. Shimizu and S. Takeoka, *Nippon Kagaku Kaishi*, **6** (1981) 1034 (Chem. Abs. 95:68612t).
74. S. Hamarthibault, T. Koscielski, J.P. Damon and J. Masson, *Appl. Catal.*, **56** (1989) 57.
75. S. Hamarthibault, J. Masson, P. Fouilloux and J. Court, *Applied Catalysis a-General*, **99** (1993) 131.
76. J.C. Klein and D.M. Hercules, *Anal. Chem.*, **56** (1984) 685.
77. A.B. Fasman, G.E. Bedel'baev, N.A. Maksimova, V.N. Ermolaev and A.S. Kuanyshev, *Kinet. Katal.*, **29** (1988) 437.
78. S. Hamarthibault and J. Masson, *J. Chim. Phys. Phys.-Chim. Biol.*, **88** (1991) 219.
79. E. Lieber and G.B.L. Smith, *J. Am. Chem. Soc.*, **58** (1936) 1417.
80. J.R. Reasenberg, E. Lieber and G.B.L. Smith, *J. Am. Chem. Soc.*, **61** (1939) 384.
81. S.S. Scholnik, J.R. Reasenberg, E. Lieber and G.B.L. Smith, *J. Am. Chem. Soc.*, **63** (1941) 1192.
82. Kyowa Hakko Kogyo Co. Ltd., British Patent 1,029,502, 1966.
83. H. Li, W. Wang, B. Zong, E. Min and J.-F. Deng, *Chem. Lett.*, **4** (1998) 371.
84. K.H. Hoffmann, D.R. Anderson and D.L. Werges, US Patent 3,953,367, 1976.
85. P. Kukula and L. Cervený, *Appl. Catal. A*, **210** (2001) 237.
86. N. Haruna, D.E. Acosta, S. Nakagawa, K. Yamaguchi, A. Tai, T. Okuyama and T. Sugimura, *Heterocycles*, **62** (2004) 375.
87. Y. Izumi, *Adv. Catal.*, **32** (1983) 215.
88. P. Kukula and L. Cervený, *J. Mol. Catal.*, **A185** (2002) 195.
89. P. Kukula and L. Cervený, *Appl. Catal. A*, **223**(1–2) (2002) 43.
90. J. Court and M. Lopez, US Patent 6,825,370, 2004.
91. A.S. Kagan, N.M. Kagan, G.D. Ul'yanova and L.G. Mironov, *Russ. J. Phys. Chem.*, **47** (1973) 978.
92. U. Birkenstock, R. Holm, B. Reinfandt and S. Storp, *J. Catal.*, **93** (1985) 55.
93. V.N. Ermolaev, A.B. Fasman, S.A. Semiletov and V.I. Khitrova, *Izv. Akad. Nauk. Kaz. SSR, Ser. Khim.*, **1** (1988) 34.
94. G.G. Urazov, L.M. Kefeli and S.L. Lel'chuk, *Compt. Rend. Acad. Sci. U.R.S.S.*, **55** (1947) 509 (Chem. Abs. 41:7063a).
95. G.G. Urazov, L.M. Kefeli and S.L. Lel'chuk, *Compt. Rend. Acad. Sci.*, **55** (1947) 735.
96. L.M. Kefeli, *Problemy Kinetika i Kataliza, Akad. Nau.k S.S.S.R.*, **6**(Geterogennyi Kataliz) (1949) 210.

97. L.M. Kefeli, *Doklady Acad. Nauk. S.S.S.R.*, **83** (1952) 863 (Chem. Abs. 46:8489g).

98. S.D. Robertson and R.B. Anderson, *J. Catal.*, **23** (1971) 286.

99. S.-I. Tanaka, N. Hirose and T. Tanaka, *Denki Kagaku*, **65**(12) (1997) 1044.

100. M.M. Kalina, A.B. Fasman and V.N. Ermolaev, *Deposited Doc., VINITI*, **1022-80** (1980) 15 (Chem. Abs. 95:66291p).

101. M.M. Kalina, A.B. Fasman and V.N. Ermolaev, *Kinet. Katal.*, **21** (1980) 813 (Chem. Abs. 93:81065v).

102. E.M. Moroz, S.D. Mikhailenko, A.K. Dzhunusov and A.B. Fasman, *React. Kinet. Catal. Lett.*, **43** (1991) 63.

103. M. de Carvalho, C.A. Perez and M. Schmal, *Appl. Catal. a-Gen.*, **264** (2004) 111.

104. J. Freel, W.J.M. Pieters and R.B. Anderson, *J. Catal.*, **14** (1969) 247.

105. S. Knies, G. Miehe, M. Rettenmayr and D.J. Ostgard, *Z. Metallkd.*, **92** (2001) 596.

106. B. Frenzel, J. Rothe, J. Hormes and P. Fornasini, *J. Phys. IV*, **7**(C2) (1997) 273.

107. I. Nakabayashi, E. Nagao, K. Miyata, T. Moriga, T. Ashida, T. Tomida, M. Hyland and J. Metson, *J. Mater. Chem.*, **5** (1995) 737.

108. J. Angeles-Islas, M. Arzola-Mendoza, B.H. Zeifert, J.G. Cabanas-Moreno, H.A. Calderon, H. Yee-Madeira and R. Zamorano, Materials Science Forum, Wiley-VCH, Weinheim, 386–388, 2002, 217–222.

109. R. Wang, Z.L. Lu and T. Ko, *J. Mater. Sci.*, **36** (23) (2001) 5649.

110. A.J. Smith, L. Ma, T. Tran and M.S. Wainwright, *J. Appl. Electrochem.*, **30** (10) (2000) 1097.

111. A.J. Smith, T. Tran and M.S. Wainwright, *J. Appl. Electrochem.*, **30** (10) (2000) 1103.

112. A.J. Smith, P. Munroe, T. Tran and M.S. Wainwright, *J. Mater. Sci.*, **36** (2001) 3519.

113. N.I. Onuoha and M.S. Wainwright, *Chem. Eng. Commn.*, **29** (1984) 13.

114. J.C. Lee, D.L. Trimm, M.S. Wainwright, N.W. Cant, M.A. Kohler and N.I. Onuoha, Copper catalysts for the hydrolysis of acrylontrile to acrylamide: deactivation and cure in *Catalyst Deactivation*, B. Delmon and G. F. Froment, Ed., Stud. Surf. Sci. Catal., Elsevier Science Publishers, Amsterdam, 1987, p. 235.

115. J.-P. Mikkola, H. Vainio, T. Salmi, R. Sjöholm, T. Ollonqvist and J. Väyrynen, *Appl. Catal. A*, **196** (2000) 143.

116. A.D. Tomsett, H.E. Curry-Hyde, M.S. Wainwright, D.J. Young and A.J. Bridgewater, *Appl. Catal.*, **33** (1987) 119.

117. J.P. Mikkola, T. Salmi, A. Villela, H. Vainio, P. Maki-Arvela, A. Kalantar, T. Ollonqvist, J. Vayrynen and R. Sjoholm, *Brazilian J. Chem. Eng.*, **20** (2003) 263.

118. J. Rothe, J. Hormes, C. Schild and B. Pennemann, *J. Catal.*, **191** (2000) 294.

119. M.A. Humbert and A. Sugier, *Revue de l'institut Francais du Petrole*, **35** (1980) 873.

120. L. Ma and M.S. Wainwright, *Chemical Industries (Dekker)*, **89** (Catalysis of Organic Reactions) (2003) 225.

121. D. Liu, N.W. Cant, A.J. Smith, and M.S. Wainwright, *Appl. Catal. A*, **297** (2006) 18.

122. J.-P. Mikkola and T. Salmi, *Catal. Today*, **64** (3–4) (2001) 271.

123. J.-P. Mikkola, B. Toukoniitty, E. Toukoniitty, J. Aumo and T. Salmi, *Ultrasonics Sonochemistry*, **11**(3–4) (2004) 233.

124. R.S. Disselkamp, T.R. Hart, A.M. Williams, J.F. White and C.H.F. Peden, *Ultrasonics Sonochemistry*, **12** (2005) 319.

125. W. Bonrath, *Ultrasonics Sonochemistry*, **12** (2005) 103.

126. G.W. Huber, J.W. Shabaker and J.A. Dumesic, *Science*, **300** (5628) (2003) 2075.

127. W.L. Marsden, M.S. Wainwright and J.B. Friedrich, *Ind. Eng. Chem. Prod. Res. Dev.*, **19** (1980) 551.

128. J.B. Friedrich, M.S. Wainwright and D.J. Young, *Chem. Eng. Commn.*, **14** (1982) 279.

129. J.B. Friedrich, D.J. Young and M.S. Wainwright, *J. Catal.*, **80** (1983) 1.

130. J.B. Friedrich, D.J. Young and M.S. Wainwright, *J. Catal.*, **80** (1983) 14.

131. M.S. Wainwright and D.L. Trimm, *Catal. Today*, **23** (1995) 29.

132. L. Ma and M.S. Wainwright, *Appl. Catal. A*, **187** (1999) 89.

133. L. Ma, T. Tran and M.S. Wainwright, *Top. Catal.*, **22** (3–4) (2003) 295.

134. X. Huang, L. Ma and M.S. Wainwright, *Appl. Catal. A*, **257** (2004) 235.

135. D.L. Werges and L.A. Goretta, U.S. Patent 3,894,084, 1975.

136. X.-J. Zeng, G.-W. Yang, G. Yang and W.-L. Liao, *Jiangxi Shifan Daxue Xuebao, Ziran Kexueban*, **26** (2002) 43.

137. J. Wu, Y.-M. Chen, S.-Q. Qin and Y.-L. Lu, *Guangxi Huagong*, **29** (2000) 5.

138. A. Villanti and G. Conti, European Patent 945,428, 1999.

139. T. Goto, H. Yokoyama and H. Nishibayashi, U.S. Patent 4,782,183, 1988.

140. T.S. Franczyk, U.S. Patent 5,292,936, 1994.

141. J.G. Vigil, U.S. Patent 6,414,188, 2001.

142. D. Ostgard, M. Berweiler and K. Seelbach, U.S. Patent 1,125,634, 2001.

143. E.W. Justi, W. Scheibe and A.W. Winsel, German Patent DE 1,019,361, 1954.

144. G. Sandstede, E.J. Cairns, V.S. Bagotsky and K. Wiesener, History of low temperature fuel cells, in *Handbook of Fuel Cells — Fundamentals, Technology and Applications*, W. Vielstich, H. A. Gasteiger and A. Lamm, Eds., Wiley, New York, 2003 pp. 173–174.

145. H. Binder, A. Koehling and G. Sandstede, Raney catalysts, in *From Electrocatalysis to Fuel Cells*, G. Sandstede, Ed., University of Washington Press, Seattle, 1972, pp. 15–31.

146. G.F. McLean, T. Niet, S. Prince-Richard and N. Djilali, *Int. J. Hydrogen Energy*, **27** (2002) 507.

147. *Fuel Cell Handbook*, National Energy Technology Laboratory, US Department of Energy, Morgantown, West Virginia (2002) Chapter 4.

148. S. Tanaka, N. Hirose and T. Tanaki, *Int. J. Hydrogen Energy*, **25** (2000) 481.

149. M.A. Al-Saleh, S. Gultekin, A.S. Al-Zakri and A.A.A. Khan, *Int. J. Hydrogen Energy*, **21** (1996) 657.

150. M.A. Al-Saleh, S. Gultekin, A.S. Al-Zakri and H. Celiker, *Int. J. Hydrogen Energy*, **19** (1994) 713.

151. S. Gultekin, A.S. Al-Zakri, M.A. Al-Saleh and H. Celiker, *Arabian J. Sci. Eng.*, **20** (1995) 635.

152. D.A. Morgenstern, J.P. Arhancet, H.C. Berk, W.L. Moench and J.C. Peterson, U.S. Patent 2002019564, 2002.

153. V.R. Choudhary and S.K. Chaudhari, *Indian Chem. Engineer (1959–1993)*, **28** (1986) 39.

154. M. Joucla, P. Marion, P. Grenouillet and J. Jenck, *Chemical Industries (Dekker)*, **53** (Catalysis of Organic Reactions) (1994) 127.

155. C.H. Foster, D.R. Nelan and E.B. Cross, *Synth. Commn.*, **13** (12) (1983) 1007.

156. S. Gultekin, M.A. Al-Saleh, A.S. Al-Zakri and H. Celiker, *Energy Environ. Prog. I*, **D** (1991) 179.

157. A. Perosa, P. Tundo and S. Zinovyev, *Green Chemistry*, **4** (2002) 492.

158. T. Horlenko and H.W. Tatum, Application: U.S., U.S. Patent 3215742, 1965.

159. A.F. Barrero, E.J. Alvarez-Manzaneda, R. Chahboun, R. Meneses and J.L. Romera, *Synlett*, (2001) 485.

160. T. Mulligan, *Speciality Chemicals*, **20** (2000) 186.

161. K.D. Mukherjee, I. Kiewitt and M. Kiewitt, *J. Am. Oil Chem. Soc.*, **52** (1975) 282.

162. A. Andreev, V. Kafedjiiski, T. Halachev, B. Kunev and M. Kaltchev, *Appl. Catal.*, **78** (1991) 199.

163. B.V. Aller, *J. Appl. Chem.*, **8** (1958) 492.

164. P. Kukula, M. Studer and H.-U. Blaser, *Adv. Synth. Catal.*, **346** (12) (2004) 1487.

165. J.M. Chapuzet, R. Labrecque, M. Lavoie, E. Martel and J. Lessard, *J. Chim. Phys. Phys.-Chim. Biol.*, **93** (1996) 601.

166. M.D. Schlesinger, J.J. Demeter and M. Greyson, *J. Ind. Eng. Chem. (Washington, D. C.)*, **48** (1956) 68.

167. M. Tashiro, H. Tsuzuki, J. Matsumoto, S. Mataka, K. Nakayama, Y. Tsuruta and T. Yonemitsu, *J. Chem. Res., Synop.*, 12 (1989) 372.

168. K. Morikawa, S. Hirayama, Y. Ishimura, Y. Suyama, T. Nozawa, H. Monzen, M. Miura, K. Marumo and T. Naito, Application: EP, EP Patent 96-100127, 724908, 1996.

169. K. Takeishi and K. Aika, *J. Catal.*, **136** (1992) 252.

170. T. Hikita, K. Aika and T. Onishi, *Catal. Lett.*, **4** (1990) 157.

171. Y.-J. Lu, Z.-X. Zhang and J.-L. Zhou, *J. Nat. Gas Chem.*, **8** (1999) 294.

172. N.V. Kropotova, A.D. Semenova and G.D. Vovchenko, *Vestn. Mosk. Univ., Ser. 2: Khim.*, **16** (1975) 310.

173. A.D. Semenova, N.V. Kropotova and G.D. Vovchenko, *Elektrokhimiya*, **11** (1975) 1051.

174. T.M. Grishina, L.I. Kolganova and G.D. Vovchenko, *Elektrokhimiya*, **13** (1977) 1043.

175. R.A. Karakhanov, T.I. Odintsova, V.B. Yakovlev and A.P. Rodin, *React. Kinet. Catal. Lett.*, **33** (1987) 219.

176. T.M. Grishina, *Vestn. Mosk. Univ., Ser. 2: Khim.*, **37** (1996) 64.

177. R.I. Semkova, E.A. Kolyadko, B.I. Podlovchenko and G.I. Shcherev, *Elektrokhimiya*, **25** (1989) 1129.

CHAPTER 5 QUESTIONS

Question 1

What considerations are required when deciding which composition to make a precursor alloy for a skeletal catalyst? If the alloy is to contain more than one catalytically active metal, does the percentage of aluminum change significantly in proportion?

Question 2

Can skeletal catalysts be prepared by acid leaching instead of alkali leaching? Why is alkali leaching preferred?

Question 3

What are the three mechanisms leading to deactivation of skeletal catalysts during use? Which of these can cause deactivation while the catalyst is still leaching?

Question 4

What are the common types of reactions where skeletal catalysts are employed?

Question 5

What are the main advantages and disadvantages of skeletal catalysts compared to other types of catalysts?

Question 6

The shrinking core models described by Levenspiel cater for both reaction- and diffusion-controlled systems. Referring to the literature, how do these systems differ and which of these models do skeletal catalysts fit during their preparation by leaching?

Question 7

A thin layer of skeletal catalyst precursor alloy can be coated on a supporting material prior to leaching. Referring to the literature, discuss (1) why is a coating used, and (2) what processes are used to achieve this coating, both for a flat surface and for a highly porous support?

A Scientific Method to Prepare Supported Metal Catalysts

John R. Regalbuto

CONTENTS

6.1   Introduction 161
6.2   Early Pioneering Work 162
6.3   Development of the Revised Physical Adsorption Model 166
      6.3.1   Qualitative Discrimination of Mechanisms 166
      6.3.2   pH Shift Modeling 168
      6.3.3   Metal Adsorption Modeling 174
6.4   Case Study: Pt Tetraammine Adsorption over Silica 177
      6.4.1   Survey of Pt/Silica Preparation Methods 177
      6.4.2   Measurement of Oxide Point of Zero Charge 179
      6.4.3   Uptake–pH Survey to Identify Optimal pH 179
      6.4.4   Tuning Finishing Conditions to Retain High Dispersion 182
6.5   The Extension of Strong Electrostatic Adsorption to Alumina and Carbon 185
6.6   Further Applications: Other Oxides, Bimetallics 187
6.7   Summary 190
References 190
Chapter 6 Questions 192

6.1   INTRODUCTION

Supported metal catalysts are used in a large number of commercially important processes for chemical and pharmaceutical production, pollution control and abatement, and energy production. In order to maximize catalytic activity it is necessary in most cases to synthesize small metal crystallites, typically less than about 1 to 10 nm, anchored to a thermally stable, high-surface-area support such as alumina, silica, or carbon. The efficiency of metal utilization is commonly defined as dispersion, which is the fraction of metal atoms at the surface of a metal particle (and thus available to interact with adsorbing reaction intermediates), divided by the total number of metal atoms. Metal dispersion and crystallite size are inversely proportional; nanoparticles about 1 nm in diameter or smaller have dispersions of 100%, that is, every metal atom on the support is available for catalytic reaction, whereas particles of diameter 10 nm have dispersions of about 10%, with 90% of the metal unavailable for the reaction.

The simplest, least expensive, and most prevalent methods to prepare supported metal catalysts begins with the process known as impregnation, whereby the high-surface-area support is contacted with a liquid solution containing dissolved metal ions or coordination complexes such as platinum hexachloride $[PtCl_6]^{2-}$ (derived from chloroplatinic acid [CPA]) or platinum tetraammine $[(NH_3)_4Pt]^{2+}$ (PTA). After impregnation, wet slurries are dried and then heated in various oxidizing or reducing environments in order to remove the ligands and to reduce the metal to its active elemental state.

A simple intuitive picture of an electrostatic adsorption mechanism that might occur during impregnation is given in Figure 6.1. An oxide surface contains terminal hydroxyl groups that protonate or deprotonate depending of the acidity of the impregnating solution. The pH at which the hydroxyl groups are neutral is termed the point of zero charge (PZC). Below this pH, the hydroxyl groups protonate and become positively charged, and the surface can adsorb anionic metal complexes such as CPA. Above the PZC, the hydroxyl groups deprotonate and become negatively charged, and cations such as PTA can be strongly adsorbed. In either case the metal complex might be thought to deposit onto the surface via strong electrostatic adsorption (SEA).

The overriding hypothesis of the SEA approach is that with metals, the simplest and most effective way to synthesize highly dispersed metal particles is to achieve a high dispersion of the metal precursors during impregnation. Once strongly adsorbed, the idea is to perform the pretreatment steps of calcination or reduction, often referred to in industry as catalyst finishing, in such a way that the monolayer morphology of the precursor is maintained as the metal is reduced, such that high metal dispersion is achieved.

In this chapter, the historical development of the above-mentioned electrostatic mechanism will be reviewed and its current application to Pt/silica, Pt/alumina, and Pt/carbon catalysts will be demonstrated. Additionally, its extension to new catalyst systems including bimetallics will be discussed. The method of SEA is in principle applicable to a large number of catalyst systems; it is hoped that readers are able to glean enough from this chapter to be able to employ the SEA approach themselves.

6.2 EARLY PIONEERING WORK

A landmark publication stands out among early efforts to transform the art of catalyst preparation into a science. J.P. Brunelle's widely cited paper [1], published in 1978, was the first in the field of catalysis to postulate a simple, general electrostatic framework for metal complex adsorption over mineral oxides. Figure 6.2, taken from that paper, gives a schematic of the surface polarization of an oxide particle as a function of solution pH (Figure 6.2a). The mechanism of charging is the protonation and deprotonation of surface hydroxyl groups as depicted in Figure 6.1. Figure 6.2 also shows the PZC

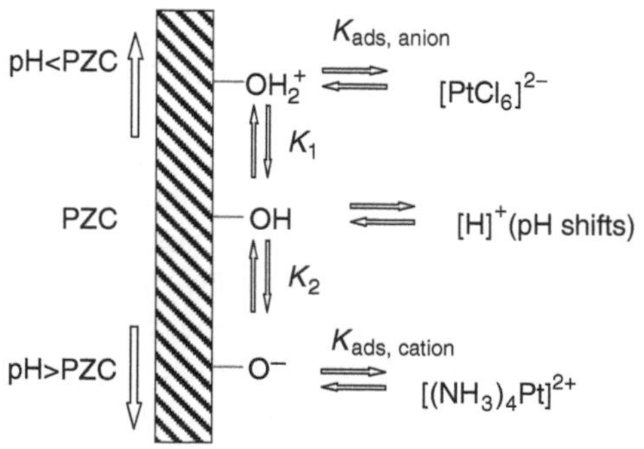

Figure 6.1 A simple electrostatic adsorption mechanism illustrating the protonation–deprotonation chemistry of surface hydroxyl groups on oxide surfaces (which are neutral at the PZC) and the corresponding uptake of anionic or cationic complexes. Proton transfer to or from the surface can significantly affect the solution pH.

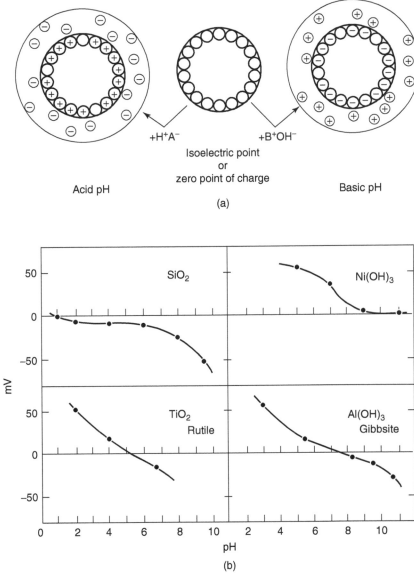

Figure 6.2 Electrostatic adsorption mechanism of Brunelle [1]: (a) surface polarization as a function of pH; (b) measurement of PZC of some oxides (equivalent to isoelectric point) by electrophoresis.

determination of various oxide supports by electrophoresis (Figure 6.2b). Silica possesses the most acidic PZC, titania and alumina are in the midrange, while neodymium hydroxide has a basic PZC.

For a particular metal/oxide catalyst system, anionic or cationic metal complexes are chosen with respect to the PZC of the oxide. For example, silica with an acidic PZC can accrue a strong negative charge as its surface hydroxyl groups deprotonate and so the preferred metal complex is cationic. With their midrange PZC, titania and alumina will adsorb anions at low pH and cations at high pH. Anion adsorption would be preferred to the basic neodymium hydroxide. Brunelle listed a variety of Group 7a, 8, and 1b anionic peroxide and chloride and cationic ammine complexes (Figure 6.3a and Figure 6.3b, respectively) along with a good many supports (Figure 6.3c) with low, midrange, and high PZCs from which suitable complex-oxide systems might be selected.

Another early landmark paper is that of Contescu and Vass [2], in which the electrostatic nature of anionic palladium chloride and cationic Pd ammines was demonstrated over alumina at low and

Mn MnO_4^-	Fe	Co	Ni	Cu
Tc	Ru	Rh $RhCl_6^{3-}$	Pd $PdCl_4^{2-}$	Ag
Re ReO_4^-	Os $OsCl_6^{2-}$	Ir $IrCl_6^{2-}$	Pt $PtCl_6^{2-}$	Au $AuCl_4^-$

(a)

Mn	Fe	Co $Co(NH_3)_x^{2+}$	Ni $Ni(NH_3)_x^{2+}$	Cu $Cu(NH_3)_x^{2+}$
Tc	Ru $Ru(NH_3)_5Cl^{2-}$	Rh $Rh(NH_3)_5Cl^{2+}$	Pd $Pd(NH_3)_4^{2+}$	Ag $Ag(NH_3)_x^{2+}$
Re	Os	Ir $Ir(NH_3)_5Cl^{2+}$	Pt $Pt(NH_3)_4^{2+}$	Au

(b)

Oxide		I.E.P.S.	Adsorption
Sb_2O_5		< 0.4	Cations
WO_3	Hydrous	< 0.5	
SiO_2	Hydrous	1.0 to 2.0	
U_3O_8		~4	Cations
MnO_2		3.9 to 4.5	
SnO_2		~5.5	
TiO_2	Rutile	~6	
	Anatase		
UO_2		5.7 to 6.7	or
γFe_2O_3		6.5 to 6.9	
ZrO_2	Hydrous	~6.7	
CeO_2	Hydrous	~6.75	Anions
Cr_2O_3	Hydrous	6.5 to 7.5	
$\alpha,\gamma Al_2O_3$		7.0 to 9.0	
Y_2O_3	Hydrous	~8.9	
αFe_2O_3		8.4 to 9.0	
ZnO		8.7 to 9.7	Anions
La_2O_3	Hydrous	~10.4	
MgO		12.1 to 12.7	

(c)

Figure 6.3 Electrostatic adsorption mechanism of Brunelle [1]: (a) common peroxide and chloride anionic complexes; (b) common cationic ammine complexes; (c) table of oxides with PZCs and predicted tendencies to adsorb anions, cations, or both.

high pH, respectively. The metal uptake using Pd chlorides at low pH and Pt ammines at high pH is shown in Figure 6.4a. In either case, as the final pH is moved farther from the PZC, higher uptake is observed. This is consistent with a simple model of the protonation–deprotonation chemistry, which is shown in the first plot of their summary figure (Figure 6.4b). As the pH is lowered or raised from the PZC, the surface is increasingly protonated or deprotonated. The authors paid close attention to the speciation of the Pd (second plot). The maximum adsorption capacity as shown in Figure 6.4a is summarized in the third plot, while values for the apparent adsorption equilibrium constants as a function of pH are given in the fourth plot. A Langmuir isotherm could be used at each pH value. In this very comprehensive set of data, shifts in the solution pH, in addition to sketches of the resulting metal profiles in catalyst pellets, were also reported (bottom of the figure).

Very prominent pioneering work in the fundamentals of catalyst impregnation was performed by Schwarz and his group [3–7]. Several seminal results can be excerpted from their four-part series on CPA impregnation of alumina [3–6]. The pH dependence of CPA uptake on alumina was

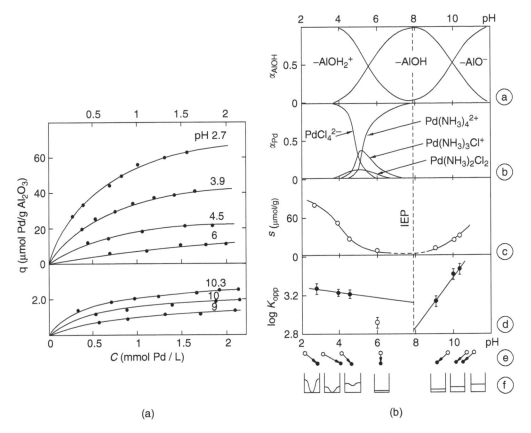

Figure 6.4 Illustrative Pd adsorption over alumina results of Contescu and Vass [2]: (a) uptake of Pd chlorides at low pH and Pd ammines at high pH; (b) summary of analyses.

demonstrated for the first time in their first paper and is reproduced in Figure 6.5. The Pt uptake is plotted vs. pH at constant Pt concentration, as pH was increased using NaOH or decreased with either HCl or HNO$_3$. Uptake exhibits a pronounced maximum at a pH of about 3.

The increase in uptake as pH falls from the PZC at 8 is the same at the increased uptake of Pd anions as pH is lowered from the PZC as shown in Figure 6.3. Uptake falloff at the lowest pH values, however, has not been observed up to this point, and can perhaps be best explained by the work performed in the second part of the Heise and Schwarz series [4]. In that paper, the deleterious effect of high ionic strength on Pt uptake was demonstrated with both 1:1 electrolytes such as NaCl, and 2:1 electrolytes such as CaCl$_2$. These data are shown in Figure 6.6. The similarity of the phenomena is demonstrated when Pt uptake is plotted vs. the activity coefficient of the Pt anion (Figure 6.6c), calculated from the ionic strength of either the 1:1 or 2:1 electrolytes. The same effect may have occurred in the low-pH region of Figure 6.5; as the solution was acidified, ionic strength might be sufficiently high to retard Pt uptake. This effect is central to an electrostatic view of impregnation and will be discussed in detail in the next section.

Another key contribution of the Schwarz group was the recognition of the dramatic influence of oxide surfaces on bulk solution pH. In a landmark 1989 paper, Noh and Schwarz [7] demonstrated the method of mass titration, in which successive additions of oxide cause stepwise shifts in solution pH. This procedure is illustrated in Figure 6.7 [7]. As indicated in Figure 6.1, the protonation–deprotonation chemistry of the surface hydroxyl groups is coupled to the liquid-phase pH. In mass titration, as the mass (or more appropriately, the surface area) of oxide in solution increases, the solution pH is brought to the PZC of the oxide, at which point no driving force for proton transfer exists

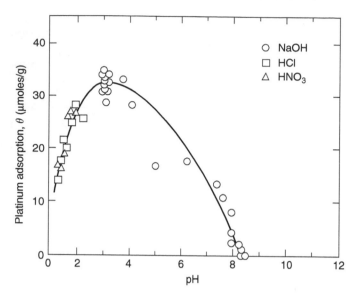

Figure 6.5   First demonstration of pH dependence of Pt uptake on alumnina. (From Heise, M.S., and Schwarz, J.A., *J. Coll. Interf. Sci.* 113, 1986, 55.)

and the solution pH stabilizes. In this manner, oxides might be thought to buffer pH. This method has been applied to determine PZCs of composite oxides [8,9]; the paper has inspired several additional versions of mass titration [10,11].

A comprehensive summary of different types of catalyst preparation appears in Schwarz' excellent review [12].

6.3   DEVELOPMENT OF THE REVISED PHYSICAL ADSORPTION MODEL

Instead of electrostatic (or physical) adsorption, metal uptake onto oxides might be considered chemical in nature. In chemical mechanisms, the metal precursor is envisioned to react with the oxide surface, involving as surface–ligand exchange [13,14] in which OH groups from the surface replace ligands in the adsorbing metal complex. In this section it will be shown that a relatively simple electrostatic interpretation of the adsorption of a number of catalyst precursors is the most reasonable one for a number of noble metal/oxide systems.

6.3.1   Qualitative Discrimination of Mechanisms

The first work of Regalbuto's group in this area sought to explore the reason for the retardation of Pt uptake at low pH over alumina (as seen in Figure 6.5) [15]. At that time it was suspected that dissolution of aluminum and the corresponding loss of adsorption sites at low pH could be the cause [3]. The effects of ionic strength and Al dissolution were determined [15] by comparing CPA uptake over alumina at (1) low pH; (2) the optimal pH (where no Al dissolution occurred), but with ionic strength increased to the level of the low-pH case; and (3) the optimal pH but over an alumina surface that had several monolayers of Al removed by a previous low-pH exposure. The results shown in Figure 6.8 demonstrate that high ionic strength effectively retards Pt adsorption while dissolving a portion of Al from the surface does not. Presumably, fresh hydroxyl groups form on the underlying alumina surface and exhibit the same protonation–deprotonation chemistry as an aged surface. The strong dependence of uptake on ionic strength is an integral part of a purely electrostatic model, as will be detailed later on.

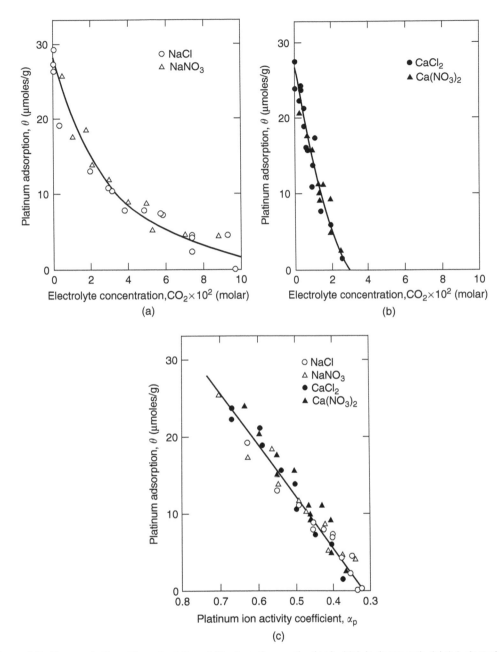

Figure 6.6 Demonstration of the retardation of Pt adsorption on alumina by high ionic strength: (a) 1:1 electrolytes; (b) 2:1 electrolytes; (c) both sets of data plotted on a common graph using Pt activity coefficient as the independent variable. (From Heise, M.S., and Schwarz, J.A., *J. Coll. Interf. Sci.* 123, 1988, 51.)

The next work by Regalbuto's group sought to explore the nature of adsorbed metal complexes [16], with an eye for trying to distinguish electrostatically (or physically) bound adsorbates from chemically bound ones. A key finding, through culling of the literature, was that the maximum surface density of Pt and Pd chlorides corresponds to a close-packed layer of complexes retaining one hydration sheath (Table 6.1, from [16]). This steric maximum amounts to 1.6 μmol/m^2, or about 1 anionic Pt complex/nm^2. The retention of hydration sheaths by adsorbing metal complexes is commonly proposed in colloid science literature [17]. An illustration of Pt complexes retaining a hydration sheath is shown in Figure 6.9 [18].

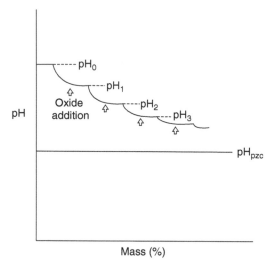

Figure 6.7 Mass titration of Noh and Schwarz. (From Noh, J.S., and Schwarz, J.A., *J. Coll. Interf. Sci.* 130, 1989, 157.)

Uptake measurements were made [16] at several oxide/solution ratios, reported as surface loading (SL) or m^2 oxide surface/liter of solution, as $[PdCl_4]^{-2}$ concentration was increased and pH was held constant at the optimal value (Figure 6.10a). Each SL indeed indicated a plateau near the steric value [16]. For Pt and Pd ammine cations, the maximum surface density over many oxides appears to be a close-packed layer, which retains two hydration sheaths; representative results for PTA uptake over silica from a recent paper [19] are shown in Figure 6.10b. The physical limit of cationic ammine surface density thus appears to be 0.84 $\mu mol/m^2$, or about 1 cationic complex/2 nm^2. Cationic uptake, therefore, is inherently half of anion uptake in many cases.

The retention of hydration sheaths upon adsorption is more consistent with an electrostatic view of adsorption than a chemical one, since by remaining a relatively large distance away from the surface, the metal complexes are less likely to participate in surface–ligand exchange.

Perhaps the most direct proof of physical/electrostatic outer sphere adsorption, as opposed to chemical or inner sphere adsorption, was recently obtained for a model PTA/quartz crystal system using methods of x-ray reflectivity at the Advanced Photon Source at Argonne National Laboratory [20]. These state-of-the-art methods are being employed by geophysicists to obtain electron density profiles normal to the water–oxide interface, that is, to precisely measure how far an adsorbed metal complex resides above a well-structured oxide surface. The result obtained from this study is shown in Figure 6.11; at the terminus of the silica crystal there appears a layer of deprotonated surface hydroxyl groups, and the adsorbing PTA complexes retain either one or two layers of water. This finding confirms the outer sphere nature of adsorbed PTA, and given the difference in oxide substrates (a single crystal of quartz vs. amorphous silica catalyst supports), the agreement with the results in Figure 6.10 is quite reasonable.

6.3.2 pH Shift Modeling

Attention was paid early on to solution pH, and in particular, to a surface — bulk proton balance. Various models of hydroxyl chemistry have been developed in colloid science literature [21]. Perhaps the simplest and most common model assumes a single type of OH group and amphoteric behavior (i.e., one set of K_1 and K_2 from Figure 6.1). More complicated models invoke multiple OH groups and proton affinity distributions [22]. It will be demonstrated below that the simpler type has worked well for the revised physical adsorption (RPA) model.

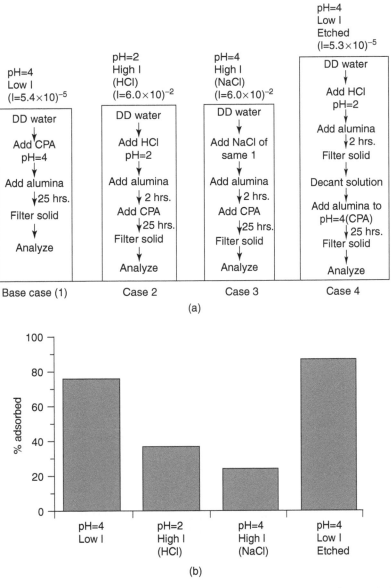

Figure 6.8 Test to isolate the effects of Al dissolution and high ionic strength on Pt retardation at low pH: (a) recipes of the four preparations; (b) Pt uptake resulting from each. (From Shah, A., and Regalbuto, J.R., *Lang.* 10, 1994, 500.)

Park and Regalbuto [11] were the first to simulate the shifts in bulk pH, which occur when aqueous solutions are contacted by various amounts of oxides. A critical parameter in these systems is the total oxide surface area in solution. This parameter, with units of m^2/L, was termed the surface loading (SL) and is illustrated in Figure 6.12 for an oxide support with specific surface area of 105 m^2/g. Low SLs give thin slurries while high SLs give thick slurries. For any particular oxide support, impregnation to incipient wetness (or dry impregnation [DI]) in which the pore volume of the support is just filled with aqueous solution, represents the highest tenable value of SL and is typically of the order of several hundreds of thousands of m^2/L. When comparing oxides with different surface areas, the mass of oxide can be adjusted so as to achieve the same SL.

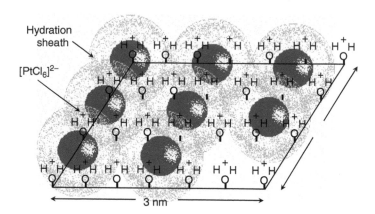

Figure 6.9 A depiction of CPA complexes retaining one hydration sheath adsorbing at a density of about one complex per nm². (From Spieker, W.A., and Regalbuto, J.R., *Chem. Eng. Sci.* 56, 2000, 2365.)

Table 6.1 Maximum Uptake Densities Surveyed from the Literature

Reference	$\Gamma_{experimental}$ (µmol/m²)	Support	pH	SA (mV g)	Metal (wt%)	Method
Species: $PdCl_4^{2-}$; $r_i = 3.12$ Å; $\Gamma_{max\ calculation} = 1.5286$ µmol/m²						
[13]	1.12	Al_2O_3	1.4	281.2	3.24	
	0.87	Al_2O_3	1.4	234.9	2.12	
	0.92	Al_2O_3	1.4	185.6	1.79	
	1.26	Al_2O_3	1.4	151.6	1.99	
	1.21	Al_2O_3	1.4	134.4	1.70	
	1.44	Al_2O_3	1.4	91.3	1.38	
This work	1.25	Al_2O_3	3.5	138	1.73	
Species: $Pd(NH_3)_4^{2+}$; $r_i = 2.55$ Å; $\Gamma_{max\ calculation} = 0.8115$ µmol/m²						
[13]	0.23	Al_2O_3	10.8	281.2	0.67	
	0.26	Al_2O_3	10.8	234.9	0.65	
	0.24	Al_2O_3	10.8	185.6	0.48	
	0.28	Al_2O_3	10.8	151.6	0.44	
	0.22	Al_2O_3	10.8	134.4	0.31	
	0.29	Al_2O_3	10.8	91.3	0.28	
[14]	0.04	Al_2O_3	10.6	174	0.07	A.A.
	0.93	SiO_2	10.6	648	6.0	
This work	0.82	SiO_2	9.5	100	0.87	A.A.
Species: $PtCl_6^{2-}$; $r_i = 2.95$ Å; $\Gamma_{max\ calculation} = 1.6209$ µmol/m²						
[15]	1.59	Al_2O_3	Acidic	78.4	8.3	
[16]	1.33	Al_2O_3		177	1.2	ΔConc.
[17]	2.14	Al_2O_3		100	4.0	Δ Conc.
	1.63	Al_2O_3		133	4.1	
	1.58	Al_2O_3		179	5.2	
	1.55	Al_2O_3		248	7.0	
[18]	0.73	Al_2O_3		150	2.1	Δ Conc.
[24]	1.87	Al_2O_3	4.2	138	8.6	EDXS
Species: $Pt(NH_3)_4^{2+}$; $r_i = 2.41$ Å; $\Gamma_{max\ calculation} = 0.8404$ µmol/m²						
[1]	1.15	SiO_2	Basic	260	5.5	Chemisorption
[20]	≈ 0.5	SiO_2	10	798	≈ 7	Chemisorption
[21]	0.81	SiO_2	9	370	5.5	Chemisorption
	0.20	Al_2O_3	10	204	0.79	
[24]	0.98	SiO_2	9.8	380	13.5	EDXS

Source: From Santhanam, N., Conforti, T.A., Spieker, W.A., and Regalbuto, J.R., *Catal. Today* 21, 1994, 141.

The pH shift model of Park and Regalbuto combined (1) a proton balance between the surface and bulk liquid with (2) the protonation–deprotonation chemistry of the oxide surface (single amphoteric site), and (3) a surface charge–surface potential relationship assumed for an

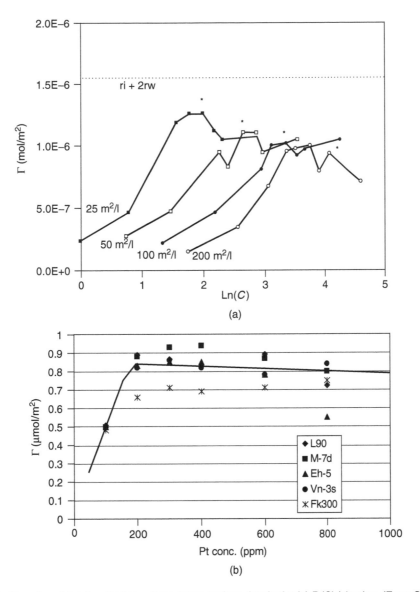

Figure 6.10 Experimental determinations of maximum surface density for (a) PdCl$_4$/alumina. (From Santhanam, N., Conforti, T.A., Spieker, W.A., and Regalbuto, J.R., *Catal. Tod.* 21, 1994, 141). (b) PTA/silica. (From Schreier, M., and Regalbuto, J.R., *J. Catal.* 225, 2004, 190.)

electric double layer. Given the mass and surface area of oxide, the PZC of the oxide, its proto-nation–deprotonation constants K_1 and K_2, and the hydroxyl density are solved simultaneously to give the surface charge, surface potential, and final solution pH. The mass titration experiment of Figure 6.7 can be quantitatively simulated, but perhaps the most powerful simulation is a comprehensive prediction of final pH vs. initial pH, as a function of oxide SL. This relationship, for parameters representative of alumina and silica, is shown in Figure 6.13 and comprises number of key features.

First, the effect of SL is immediately apparent. Low SLs can be employed in the laboratory to minimize pH shifts. At high SLs, oxides exhibit a dramatic effect on pH. In fact, this plot predicts that the final pH of DIs is almost always at the PZC of the oxide, unless the starting solutions are extremely acidic or basic [11]. Alternatively, in DI the hydroxyl groups on the oxide surface far outnumber the protons or hydroxide ions initially in solution, and the surface never becomes significantly charged.

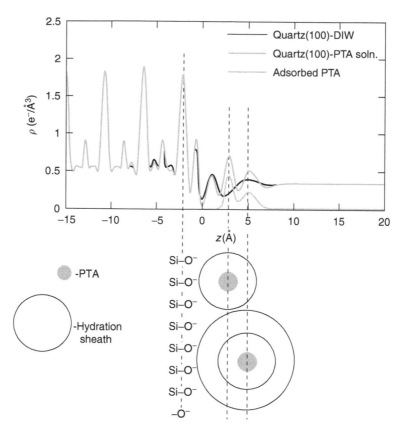

Figure 6.11 Electron density profile of the quartz–PTA interface. (From Park, C., Fenter, P., Sturchio, N., and Regalbuto, J.R., *Phys. Rev. Lett.* 94, 2005, 076104.) A schematic of PTA adsorbed with one or two hydration sheaths.

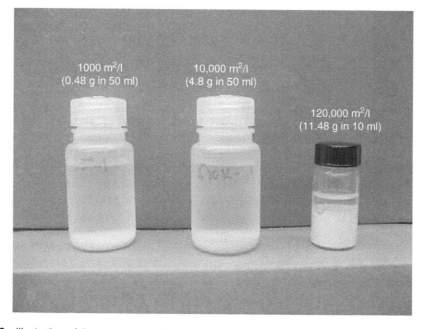

Figure 6.12 Illustration of the parameter surface loading.

Strong electrostatic interactions with ionic metal complexes can only occur when this buffering effect of the oxide is overcome.

The drastic shift in pH, which can be caused even by small amounts of oxide, is illustrated in Figure 6.14. In this series of photographs, 50 g of a 250 m^2/g alumina (SL = 1250 m^2/L) was contacted with 1 L of a solution initially at pH 2.8. The PZC of alumina is normally about 8.5 [23]. After 1 min (Figure 6.14c) the pH increased from 2.8 to 6.6, after 2 min it was 7.8 (not shown) and after 5 min had attained the equilibrium value of 8.2 (Figure 6.14d). This experiment was actually designed with a calculation similar to that shown in Figure 6.13.

Another signficant feature of the pH shift plot of Figure 6.13 are the wide plateaus of final pH seen at the higher SLs. That is, starting from a wide range of initial pH, the final pH is always the same and is in fact the oxide PZC. This suggested that oxide PZC can be measured simply with a pH probe, by measuring the final pH values of a series of oxide–solution slurries at high SL [11]. This method was called EpHL, and measured the equilibrium pH at high loading.

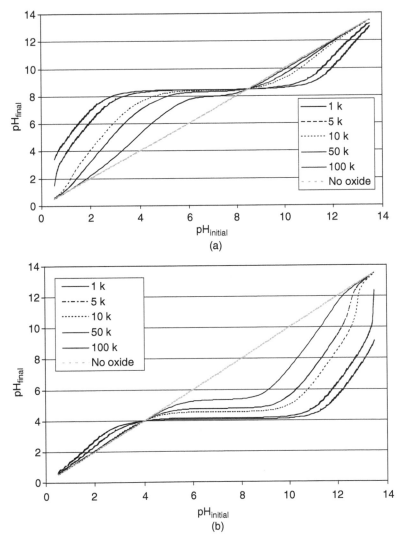

Figure 6.13 pH shift model simulation with varying surface loading. (a) (Alumina) PZC = 8.5, $pK_1 - pK_2 = 5.0$ and $N_s = 5.0$ OH/nm^2. (b) (Silica) PZC = 4.0, $pK_1 - pK_2 = 5.0$ and $N_s = 5.0$ OH/nm^2.

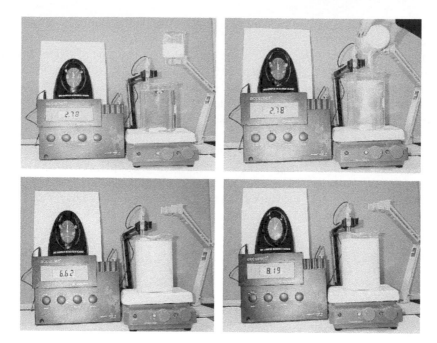

Figure 6.14 Demonstration of the rapid and dramatic shift in solution pH caused by contact with alumina.

In Regalbuto's most recent treatment of surface charging, pH shift data generated at different SLs are fit to the model so as to obtain the best values of K_1 and K_2 [24]. Representative experimental and model results are shown for alumina in Figure 6.15. Having obtained the oxide-charging parameters in the absence of metal adsorption, the parameters can be used with no adjustment in the RPA model to simulate metal uptake. This is described in the next section.

6.3.3 Metal Adsorption Modeling

Here again, a number of models exist in the colloid science literature, ranging from very simple to very complex. Adsorption models employed for catalyst impregnation typically contain a chemical component (see [13,14,25] and references therein); for example, the proposed uptake of CPA by alumina shown in Figure 6.16.

This mechanism has been quantified by employing surface–ligand exchange of OH or O$^-$ from the alumina surface with ligands from the adsorbing CPA complex [13], as follows:

$$\text{AlOH} + [\text{PtCl}_4(\text{OH})_2]^{2-} \rightleftharpoons [(\text{AlOH}) - \text{PtCl}_4(\text{OH})]^- + \text{OH}^-$$

$$\text{AlO}^- + [\text{PtCl}_4(\text{OH})(\text{H}_2\text{O})]^- \rightleftharpoons [(\text{AlO}) - \text{PtCl}_4(\text{H}_2\text{O})]^- + \text{OH}^-$$

$$\text{AlO}^- + [\text{PtCl}_4(\text{OH})_2]^{2-} \rightleftharpoons [(\text{AlO}) - \text{PtCl}_4(\text{OH})]^{2-} + \text{OH}^-$$

On the other hand, a simpler and purely physical (electrostatic) mechanism might go toward describing the uptake of common noble metal precursors onto common oxide supports. The first model of Agashe and Regalbuto [26] was patterned after that of James and Healy [17] from colloid science literature. In the original James and Healy's model, a Langmuir isotherm was employed and

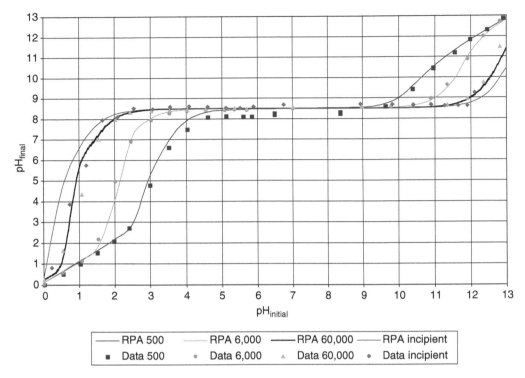

Figure 6.15 pH shift measurements and the pH shift portion of the RPA model at various surface loadings for a typical γ-alumina. (From Regalbuto, J.R., Navada, A., Shadid, S., Bricker, M.L., and Chen, Q., *J. Catal.* 184, 1999, 335.)

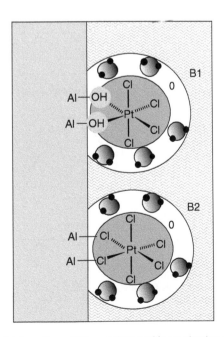

Figure 6.16 Proposed chemical interaction of Pt complexes with an alumina surface, which involves surface–ligand exchange of either surface OH for Cl ligands (model B1) or Cl ligands from the CPA complex for surface hydroxyls (model B2). (From Shelimov, B., Lambert, J.-F., Che, M., and Didillon, B., *J. Mol. Catal.* 158, 2000, 91.)

the free energy of adsorption, on which the adsorption equilibrium constant is based, comprises a coulombic, a solvation, and a nondescript, adjustable chemical energy term:

$$\Delta G_{ads} = \Delta G_{coul} + \Delta G_{solv} + \Delta G_{chem}$$

In the original James and Healy model, the magnitude of the adjustable chemical term usually swamped the other two terms. By using a much smaller value of the solvation energy as suggested by Levine [27], it was found that the adjustable chemical term could be eliminated in the simulation of a number of metal cation/silica systems [26].

The parameters, main equations, and a graphical representation of the RPA model for the CPA/alumina system are shown in Figure 6.17. The surface potential as well as the uptake of Pt is given in the plot. The uptake curve is typically volcano shaped. At the pH of the PZC, there is no surface charge and therefore no uptake. As the pH drops from the PZC, the surface potential builds and uptake increases. In the lowest pH range, the effect of ionic strength is manifested: the high ionic strength inherent to the high acid concentration effectively causes a decrease in the adsorption equilibrium constant.

In later works it was found that the solvation term could be eliminated entirely; the simulation of uptake of CPA on alumina at low pH [19,23,28] and PTA on silica at high pH [19,29] utilizes only the coulombic energy of adsorption. Experimental data fit with the RPA model is shown for these two complementary cases (anions adsorbing over a positively charged alumina surface at low pH and cations adsorbing over a negatively charged silica surface at high pH) in Figure 6.18.

With the RPA model it has been possible to simulate many sets of CPA/alumina data mentioned in the literature [18], with the same set of unadjusted parameters (PZC, K_1 and K_2, OH density). Since pH shifts in the presence and the absence of CPA adsorption on alumina [23] and PTA adsorption on silica [19] are similar, it can be concluded that metal and proton transfer are independent in these systems. Thus the pH shift model can be used in concert with the RPA model not only to predict metal uptake, but also to compute final pH from the initial pH of the contacting solutions [18,28].

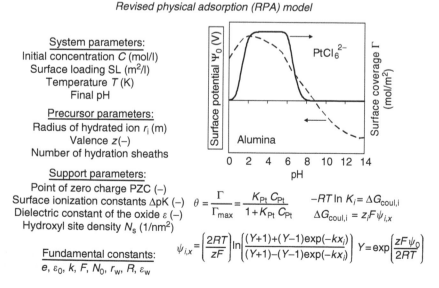

Revised physical adsorption (RPA) model

System parameters:
Initial concentration C (mol/l)
Surface loading SL (m²/l)
Temperature T (K)
Final pH

Precursor parameters:
Radius of hydrated ion r_i (m)
Valence $z(-)$
Number of hydration sheaths

Support parameters:
Point of zero charge PZC (–)
Surface ionization constants ΔpK (–)
Dielectric constant of the oxide ε (–)
Hydroxyl site density N_s (1/nm²)

Fundamental constants:
$e, \varepsilon_0, k, F, N_0, r_w, R, \varepsilon_w$

$$\theta = \frac{\Gamma}{\Gamma_{max}} = \frac{K_{Pt}\,C_{Pt}}{1 + K_{Pt}\,C_{Pt}}$$

$$-RT \ln K_i = \Delta G_{coul,i}$$
$$\Delta G_{coul,i} = z_i F \psi_{i,x}$$

$$\psi_{i,x} = \left(\frac{2RT}{zF}\right)\ln\left(\frac{(Y+1)+(Y-1)\exp(-kx_i)}{(Y+1)-(Y-1)\exp(-kx_i)}\right) \qquad Y = \exp\left(\frac{zF\psi_0}{2RT}\right)$$

Figure 6.17 A summary of the RPA model.

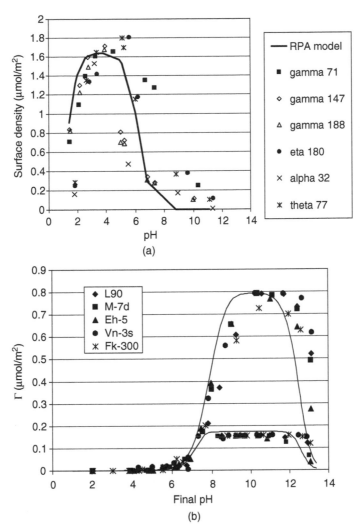

Figure 6.18 Typical uptake results and RPA modeling for (a) CPA adsorption over a number of aluminas. (From Regalbuto, J.R., Navada, A., Shadid, S., Bricker, M.L., and Chen, Q., *J. Catal.* 184, 1990, 335), and (b) PTA adsorption over a number of silicas. (From Schreier, M., and Regalbuto, J.R., *J. Catal.* 225, 2004, 190.)

6.4 CASE STUDY: Pt TETRAAMMINE ADSORPTION OVER SILICA

In this section, the application of the SEA approach for the production of small metal particles on an oxide support will be demonstrated for the case of Pt/silica. Once again, the over-arching hypothesis of the SEA approach is that a correlation exists between strong adsorption of the metal precursor and high dispersion of the reduced metal. The steps of the SEA approach for any particular metal/support system are (1) the measurement of support PZC (which determines which charge of metal ion and which pH range to employ), (2) uptake–pH surveys to determine the pH of the strongest interaction, and (3) tuning the reduction treatment to preserve high dispersion. These steps will be illustrated after a review of Pt/silica preparation methods found in the literature.

6.4.1 Survey of Pt/Silica Preparation Methods

Several steps of preparation, including the support composition, metal salt, method of metal addition, pH, metal loading, calcination, etc., affect the ultimate particle size in the reduced catalyst. The importance of these effects is evident in the survey of the literature on the preparation of Pt/SiO$_2$ catalysts given in Table 6.2 [30–51]. Incipient wetness or DI is often used in the preparation with chloroplatinic acid (H$_2$PtCl$_6$ or CPA) [34,35,38,42,45,48,50,51], but may also be used with platinum tetraammine chloride (or nitrate) (Pt(NH$_3$)$_4$Cl$_2$ or PTA) [35,50]. With this method, the desired amount of metal salt is dissolved in water just sufficient to fill the pore volume of the support. Pt on silica catalysts may also be prepared by several nontraditional methods such as a simultaneous sol–gelsynthesis of a Pt salt and silica [39,41,47] or by DI with organic solvents [41,49]. The method most frequently used, however, is adding PTA to silica from a slurry in excess solution [30–34,36,37,40,41,43–46]. This method is commonly called ion exchange, but can more generally be considered wet impregnation (WI). If the final pH is not controlled in the WI method of preparation, the impregnating solution becomes more acidic as surface hydroxyl groups equilibrate with the bulk solution. If a moderate to high-surface-area silica is used and the ratio of oxide to liquid is high, the solution pH will approach the PZC of silica at about 4 [11], as noted in the previous section. It is only if the impregnating solution is maintained under strong basic conditions that the silica hydroxyl groups will be sufficiently deprotonated and platinum tetraammine cations strongly adsorbed onto the support. It was suggested [19] that it is more correct to describe this interaction as SEA rather than ion exchange, since the attraction occurs only in strong base sufficient to deprotonate the silanol groups.

Table 6.2 lists the methods of preparation, e.g., DI (with PTA or CPA), WI (where the pH moves toward silica's PZC), and SEA (where the pH is controlled to remain high) for Pt/silica catalysts reported in the literature. In addition to the different methods of Pt addition, the metal loading and subsequent calcination and reduction steps are also given. A number of trends affecting the dispersion of the reduced catalysts can be identified in Table 6.2. First, DI with CPA generally gives moderate to poor dispersion, ca. 0.2 to 0.4 [34,35,38,42,45,48,50,51], with the exception at very low loading. Dry impregnation of PTA gives higher dispersions than those prepared from CPA, i.e., about 0.6 to 0.9 with higher dispersions at lower Pt loading [35,50]. Preparation of Pt on silica by WI with PTA also gives dispersions that are moderate to low, generally about 0.3 to 0.5 [30,43,46]. The highest dispersions are achieved by SEA of PTA, i.e., the pH remains high throughout the metal deposition. For catalysts prepared by SEA and WI, Figure 6.19 shows that the Pt dispersion decreases as the metal loading increases. Since the catalysts in Table 6.2 were prepared on silicas with different surface areas, the metal loadings were plotted in Figure 6.19 on the basis of Pt surface density (μmol/m^2). While there are only a few examples of catalysts calcined at higher temperatures, at the

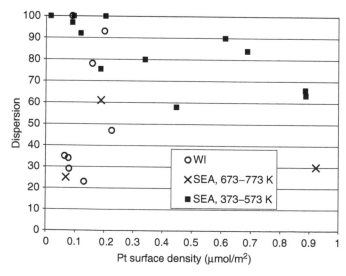

Figure 6.19 Trends in dispersion of Pt/silica catalysts culled from literature. (From Miller, J.T., Schreier, M., Kropf, A.J., and Regalbuto, J.R., *J. Catal.* 225, 2004, 203.)

same Pt surface density, catalysts calcined below 300°C generally have higher dispersions [31,32,34,36,37,40,45,46,48].

6.4.2 Measurement of Oxide Point of Zero Charge

The first step in the SEA approach is to determine the PZC of the support. Low-PZC materials accrue a strong negative charge at high pH and can strongly adsorb cations, while high-PZC materials charge positively at low pH and strongly adsorb anions. Mid-PZC materials might strongly adsorb either in the respective pH range. Point of zero charge is easily determined, as mentioned in the discussion of Figure 6.13 above, by measuring final pH vs. initial pH at high SL. The pH shift results for a number of SLs of a representative silica are shown in Figure 6.20 [19]. As SL increases, the observed pH_{final} plateau falls closer and closer to the PZC of silica; the PZC of this silica sample is seen to be about 4. Correspondingly, a cationic metal complex, PTA, will be chosen as the catalyst precursor and the pH region above pH 4 will be surveyed to determine the pH of the strongest interaction.

A note of caution pertaining to impure oxides should be mentioned at this point. Cationic impurities such as Na^+ and K^+ increase PZC of oxide, while anionic surface impurities such as Cl^- or F^- lower it [52,53]. Surface impurities can be removed by washing [54,55], and the measured PZC values will approach the pure values with repeated washing steps. Attempts to use ion doping to control Pt uptake were unsuccessful [56]; doped, PZC altered oxides behave the same as pure materials with regard to the uptake of CPA and PTA. It is believed that in the pH range of strong metal adsorption, the dopant either redissolves or is forced off the surface by competition from the metal ion [54]. As a consequence, common oxide supports that contain surface impurities can be modeled using the PZC of the pure material.

On the other hand, irreversible changes in the PZC can be effected in some surfaces. The oxidation of carbon surfaces, for example, changes the PZC and does affect the adsorptive properties. Carbon will be discussed in a later section.

6.4.3 Uptake–pH Survey to Identify Optimal pH

Benchtop uptake–pH surveys are conducted with a series of initial pH PTA solutions in 60 mL polypropylene bottles as shown in Figure 6.21 (for illustration, this series of samples is actually CPA, which is colored, over 1000 m^2/L of an alumina support).

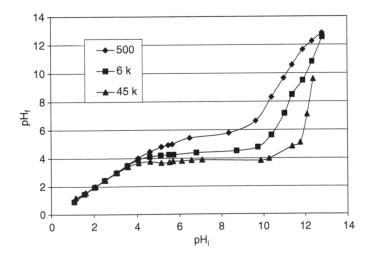

Figure 6.20 The PZC determination of silica. (From Schreier, M., and Regalbuto, J.R., *J. Catal.* 225, 2004, 190.)

Table 6.2   Summary of the Literature on the Methods of Preparation of Pt/SiO$_2$ Catalysts

References	Pt Loading (wt %)	Surface Area (m²/g)	Surface Loading (μmol/m²)	Method	Precursor	pH_{Final}	Calcination (O$_2$) K	Reduction (H$_2$) K	Dispersion
[30]	0.5	n.r.	—	WI	$PTA-Cl_2$	7.0 initial	383	673/603	0.31 to 0.35
[30]	0.5	n.r.	—	WI	CPA	2.1 initial	383	673/603	0.20
[31,32]	6.3	364	0.887	SEA	$PTA-OH$ and Cl_2	8.9	378	673	0.64
[33]	3.6	200	0.923	SEA	$PTA-Cl_2$	NH_4OH	773	—	0.30
[34]	0.8	219	0.187	SEA	CPA to PTA	9 to 11 at 343 K	393	—	0.75
[34]	0.4	219	0.094	DI	CPA in acetone	—	393	—	0.53
[35]	0.05	500	0.005	DI	$PTA-NO_3$	—	573	473	0.99
[35]	0.6	500	0.062	DI	$PTA-NO_3$	—	573	473	0.90
[35]	3.0	500	0.308	DI	$PTA-NO_3$	—	573	473	0.60
[35]	3.1	500	0.318	DI	CPA	—	573	473	0.25
[36]	0.06	198	0.016	SEA	CPA to PTA	9.74	378	533?	1.0
[36]	0.46	198	0.119	SEA	PTA	9.74	378	533C?	0.92
[36]	1.73	198	0.448	SEA	PTA	9.74	378	533 C?	0.58
[37]	0.01 to 7	194	0.003 to 1.85	SEA	$PTA-OH$	8 to 10.5	573	673	0.65
[38]	1.0	260	0.197	DI	CPA	0.2 M HCl	723	773	0.41
[39]	0.83	500 to 800	0.085 to 0.053	Sol–Gel	$PTA-NO_3$	4.3	Dried desiccator	—	0.70
[40]	2.68	200	0.687	SEA	$PTA-Cl_2$	9.0	573	673	0.56
[40]	2.68	200	0.687	SEA	$PTA-Cl_2$	9.0	573 Ar	673	0.84
[40]	2.68	200	0.687	SEA	$PTA-Cl_2$	9.0	673	673	0.85
[41]	1.76	653	0.138	Sol–Gel	CPA	1.4	673	673	0.20
	1.62	480	0.173		$PTA-NO_3$	4.8			0.80
	1.65	632	0.134		$Pt(C_5H_7O_2)_2$	4.0			0.50
[41] (design 2.0%)	0.95	714	0.068	WI	$PTA-NO_3$	8.0	673	673	0.25 to 0.28
[41]	0.91	1097	0.043	Ethanol and acetone	$PTA-NO_3$	2.4	673	673	0.10 to 0.15
[41]	0.91	754	0.062	Sol–Gel	$Pt(AcAc)_2$	1.65	673	673	0.10 to 0.13
	1.65	632	0.134		$Pt(AcAc)_2$	5.0			0.50
	0.8	706	0.058						0.80
	0.65	759	0.044						0.80
	0.40	586	0.035						0.80
[42]	4.9	200	1.26	DI	PtC	—	573	573	0.30
	1.0		0.256						0.50
[43]	1.0	300	0.171	WI	$PTA-Cl_2$	12 initial	293 Vacuum	573	0.30
								673	0.29
								773	0.28
								773 N$_2$	0.98
								773	0.78

Ref	wt%	Support	value	Method	Precursor	pH	Treatment	Temp	Dispersion
[43]	2.0	300	0.342	WI	PTA–Cl$_2$		673 He	1023	0.44
							393	1073	0.20
[44]	0.81	220	0.189	SEA	PTA–Cl$_2$	9.8		673	0.61
[45]	2.65	220	0.617	DI	CPA			673	0.22
	1.44		0.336						0.41
	0.49		0.114						0.45
[45]	3.8	220	0.883	SEA	PTA–Cl$_2$	9.8	Dried	673	0.66
[46]	1.05	596	0.09	SEA	PTA–OH	—	573	573	0.97
[46]	0.90	596	0.077						0.34
[46]	0.87	341	0.131	WI	PTA–OH	—	773	573	0.23
	0.93	238	0.20						0.93
	1.02	580	0.09						1.00
	0.89	202	0.226						0.47
			0.193						
[47]	1.5	399	—	Modified impregnation with Sol–Gel?	Pt(acac)$_2$	—	0.39		
[48]	0.5	377	0.068	DI	Pt(acac)$_2$ CPA	Acidic	823	373	0.24
	1.6		0.218						0.10
	2.5		0.34						1.0
	4.6		0.625						0.50
	0.7		0.095		CPA		393	773	0.30
									0.3
[48]	1.5	377	0.204	SEA	PTA–Cl$_2$	9.0	393	773	1.20
	2.5		0.34						1.00
	4.5		0.612						0.80
	0.5								0.90
[49]	0.5	M-5	—	Organic DH-de hydroxylated support	Pt(Acac)$_2$ Acetyl acetonate	Toluene excess	673	673	0.43
		Solgel						H$_2$/He	0.39
		M-5 (DH)							0.32
		Solgel (DH)							0.31
[49]	0.5	M-5	—	Impreg DH-dehydroxylated support	CPA	Acidic?	673	673	0.12
		Solgel						H$_2$/He	0.12
		M-5 (DH)							0.08
		Solgel (DH)							0.09
[50]	1	270	0.19	DI	PTA[a]	—	773 and above Vacuum dried 383		0.39 to 0.02
[51]	1			DI	CPA	Acidic?		623 to 973	0.84 to 0.13

n.r., not reported; DI, dry impregnation; WI, wet impregnation; SEA, wet impregnation at near neutral pH; SEA, wet impregnation under strongly basic pH.

[a]Text reads Pt(NO$_3$)$_2$(NH$_4$)$_2$; typo assumed.

Source: From Miller, J.T., Schreier, M., Kropf, A.J., and Regalbuto, J.R., *J. Catal.* 225, 2004, 203.

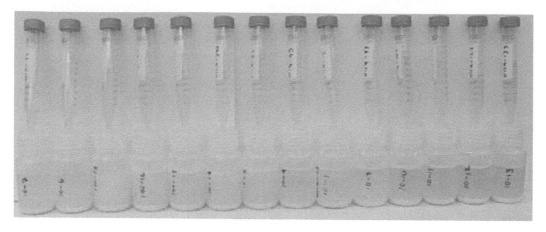

Figure 6.21 A typical adsorption experiment, in this case for CPA on alumina. The alumina powder has been allowed to settle and is visible at the bottom of each bottle. Filtered solutions for ICP analysis are placed in test tubes.

After 1 h on an orbital shaker, pH is recorded and 5 mL aliquots of each thin slurry are filtered for ICP analysis. Over powders, Pt adsorption is complete within minutes [19,23], and 1 h contact is sufficient to ensure equilibration. Samples of the parent solution (before contact) are also filtered and analyzed with ICP, which can be used to determine not only metal uptake, but also dissolution of the support. The metal uptake, in terms of surface density (mol/m^2), is calculated by dividing the concentration difference by the SL:

$$\Gamma_{Pt}\ (\mu mol/m^2) = \frac{C_{Pt,init} - C_{Pt,final}\ (\mu mol/L)}{SL\ (m^2/L)}$$

The amount of Pt adsorbed on silica at an SL of 83,000 m^2/L as a function of the final pH of the solution is shown in Figure 6.22 [29]. At pH values below about 6, there is little adsorption. As the pH increases, the amount of Pt adsorbed increased up to a pH of about 10 with a maximum of about 1.1 μmol/m^2. There was sufficient Pt in solution to give a 1.15 μmol/m^2 had all PTA been adsorbed. The amount of PTA adsorbed is approximately the same from basic solutions adjusted by either NH$_4$OH or NaOH.

The optimal pH for strong adsorption of PTA onto silica is in the range 9.5 to 12. The maximum weight loading, given that the surface area of this silica is 290 m^2/g, is

$$(1.1 \times 10^{-6}\ mol\ Pt/m^2)(195\ g\ Pt/mol\ Pt)(290\ m^2\ silica/g) = 0.062\ g\ Pt/g\ silica$$

or about 6 wt% Pt.

6.4.4 Tuning Finishing Conditions to Retain High Dispersion

Samples were prepared [29] at 1 wt% Pt at pH$_{final}$ 9.5 with ammonium hydroxide to minimize dissolution of the silica and to avoid the use of NaOH, which leaves Na on the catalyst. For comparison, DIs were performed at the same weight percent Pt using PTA dissolved in de-ionized water. Strong electrostatic adsorption and DI samples were then calcined at successively higher temperatures and reduced in flowing hydrogen at 200°C. The dispersion of the reduced SEA and DI-prepared catalysts, determined by hydrogen chemisorption, is shown in Figure 6.23.

At all calcination temperatures except the highest, the dispersion of the SEA-prepared samples is higher than similar catalysts prepared by DI. Notably, the dispersion of the 100°C dried and

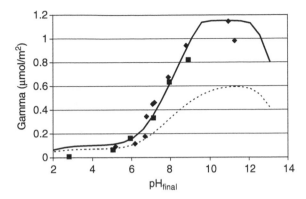

Figure 6.22 A PTA–pH survey for a surface loading of 83,000 m²/L. The solutions were basified with either NaOH (diamonds) or NH₄OH (squares). (From Miller, J.T., Schreier, M., Kropf, A.J., and Regalbuto, J.R., *J. Catal.* 225, 2004, 203.)

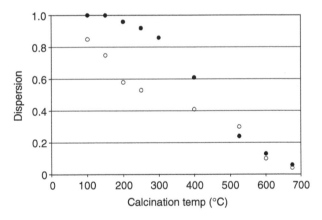

Figure 6.23 Dispersion results of Pt/SiO₂ produced from PTA via SEA (filled circles) or DI (open circles). (From Miller, J.T., Schreier, M., Kropf, A.J., and Regalbuto, J.R., *J. Catal.* 225, 2004, 203.)

150°C calcined catalysts was 100%. This high dispersion corresponds to a particle size of 1 nm. High-angle annular dark-field imaging has been used to verify this ultrasmall particle size; a representative micrograph of the Pt/silica sample prepared by SEA is shown in Figure 6.24. The particles appear homogeneous in both size and spacing. The average particle size is close to 1 nm and confirms the dispersion measurement made from chemisorption.

The effect of calcination temperature on the Pt dispersion for PTA on silica is clearly seen in Figure 6.23. For both methods of preparation, the dispersion is highest when the catalyst is dried at 100°C and reduced directly thereafter. At this temperature, both catalysts are white in color. As the heating/calcination temperature increases, the color first becomes light brown, and at higher temperature turns dark brown. As the calcination temperature increases, the dispersion decreases approximately linearly. Above about 500°C, the dispersion is nearly identical for both methods of preparation.

The SEA results from Figure 6.23 and at several other weight loadings (taken from [29]) are plotted (in Figure 6.25), together with the SEA preparations taken from literature (Table 2). High-temperature calcination had been employed in many of the latter preparations. The results from [29] have been plotted separately as a function of calcination temperature. The trend of decreasing dispersion with increasing calcination temperature is again confirmed; the highest dispersions are obtained when the

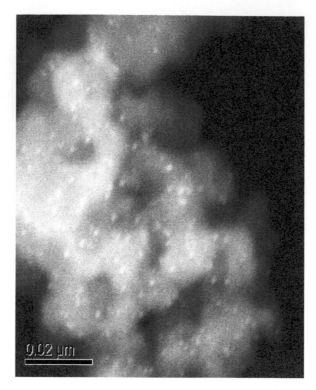

Figure 6.24 Electron microscopy characterization of the SEA-synthesized Pt/silica sample confirming high dispersion.

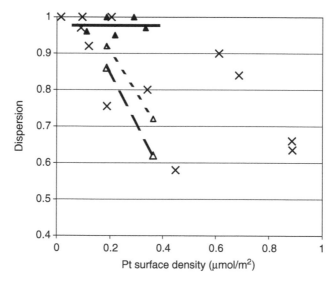

Figure 6.25 Effect of Pt surface density (μmol/m^2) and calcination on the dispersion of reduced Pt/silica catalysts prepared by SEA. $\times$, literature, calcined between 373 and 573 K. This study: solid triangle and solid line — dried at 373 K; small open triangle and dotted line — calcined at 523 K; and large open triangle and dashed line — calcined at 573 K. (From Miller, J.T., Schreier, M., Kropf, A.J., and Regalbuto, J.R., *J. Catal.* 225, 2004, 203.)

catalysts are not calcined, but reduced directly after drying at 100°C. This has been reported previously by several groups [1,48]. Decreases with dispersion at higher calcination temperatures also appear to be a strong function of metal loading. It is also notable that SEA with direct reduction has led to some of the most highly dispersed, highly loaded materials (in terms of surface density) reported in literature.

Finally, the dispersion data in Figure 6.23 and Figure 6.25 indicate that it is possible to control the particle size of Pt. Strong electrostatic adsorption with drying at 100°C and reduction at 200°C can be used to prepare particles with 100% dispersion, that is, where each metal atom is accessible for catalytic reaction. Larger particles can be produced by calcining at higher temperatures.

6.5 THE EXTENSION OF STRONG ELECTROSTATIC ADSORPTION TO ALUMINA AND CARBON

The PZC of carbons can be changed irreversibly by oxidizing the surface; in a series of papers, Regalbuto's group has demonstrated that Pt anion and cation uptake can be manipulated on this basis [55,57–59]. The main characteristics of three types of activated carbon, obtained from Norit, are given in Table 6.3. The main difference is the degree of oxidation of the surface, accomplished by refluxing in concentrated phosphoric acid, which results in a PZC as low as 2.5, compared with a value of 8.1 for the unoxidized material. The PZC determinations of the low- and the high-PZC carbons [57] are shown in Figure 6.26.

The expected trends of CPA and PTA uptake for the three materials are seen in Figure 6.27. The high-PZC material, like alumina in Figure 6.18a, should adsorb the most CPA and the low-PZC material should adsorb the least. Correspondingly, the low-PZC carbon should adsorb the most PTA, like silica in Figure 6.18b, and the high-PZC carbon the least.

The measured uptake of CPA and PTA over the three activated carbons [55] is shown in Figure 6.28, and the trends predicted by the RPA model in Figure 6.27 are at least qualitatively observed. However, at high pH, over the two highest-surface-area carbons (CA and KB), uptake is about half of that predicted by the RPA model. The discrepancy was explained [55] by steric exclusion of the large Pt ammine complexes, believed to retain two hydration sheaths [15,19], from the smallest micropores of the high-surface-area activated carbon.

The value of an SEA preparation for Pt cation uptake over carbon is shown in Figure 6.29, in which a loading of 9 wt% Pt PTA was applied to high-surface-area graphite (TIMREX) at pH 11 [59]. For comparison, DI was conducted with a neutral pH solution of PTA (and a final pH near the PZC of the carbon, as it usually occurs in DI). Both samples were directly reduced at 200°C following drying at 100°C. The DI preparation results in large Pt particles (Figure 6.29a). On the other hand, using SEA yields predominantly 1 to 1.5 nm particles almost as well dispersed as the Pt/SiO$_2$ sample of Figure 6.24.

In the above work with carbon, the same approach as with silica was employed. First, the PZCs were determined. Second, uptake–pH surveys were conducted to determine the pH of the strongest interaction, and third, an SEA sample synthesized at this condition was reduced in a way which retained high dispersion.

A comparison of the Pt dispersion from SEA vs. DI has also been performed with CPA/alumina [60]. In this study, a λ-alumina with surface area of 277 m^2/g and pore volume of 1.0 mL/g was

Table 6.3 Carbon Specifications

Carbon Name	Carbon Type	Abbreviation	Surface Area (m^2/g)	Measured PZC
Norit CA1	Activated Carbon	CA	1400	2.5
Darco KB-B	Activated Carbon	KB	1500	5.0
Norit SX-ULTRA	Activated Carbon	SX	1200	8.1

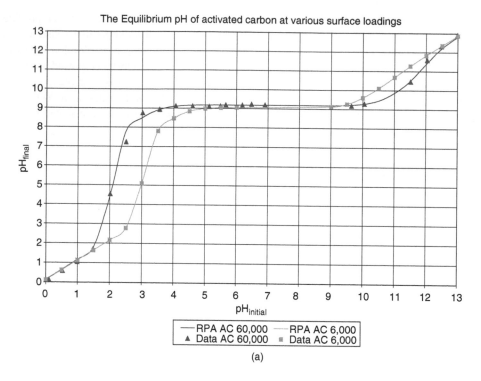

(a)

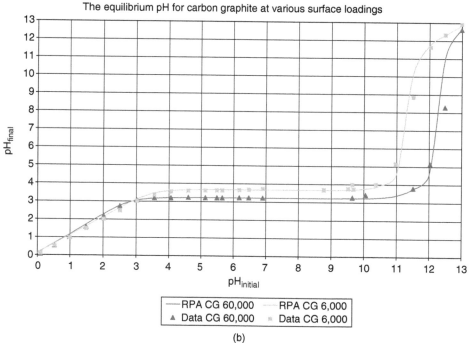

(b)

Figure 6.26 PZC determinations of (a) high and (b) low PZC carbons. (From Hao, X., and Regalbuto, J.R., in preparation.)

employed. As seen in the earlier figures of CPA/alumina (Figure 6.18a and Figure 6.27a), the optimal pH for all aluminas is between 3 and 4. Strong electrostatic adsorption was performed at 1000 m^2/L and at pH 4 with different Pt concentrations to give different Pt loadings. Impregnated samples were reduced in flowing H_2 at 250°C directly after drying, or were calcined at 500°C and then reduced. The

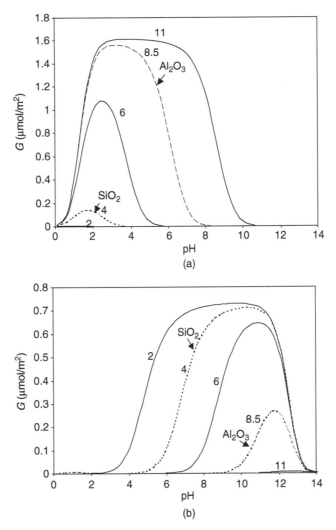

Figure 6.27 RPA prediction of Pt adsorption over materials of various PZCs: (a) CPA, (b) PTA. PZCs representative of silica and alumina are indicated. (From Schreier, M., Ph.D. Dissertation, University of Illinois, Chicago, 2004.)

dispersion was measured by CO chemisorption and is shown in Figure 6.30, along with those for DI samples pretreated in the same way. In every case the SEA preparation yields a higher dispersion than DI, and gives values near 100% even at relatively high Pt loadings. The effect of calcination is much less pronounced than in the case of silica (Figure 6.25).

6.6 FURTHER APPLICATIONS: OTHER OXIDES, BIMETALLICS

The application of the SEA approach to other systems is straightforward, and involves the three steps (PZC determination, uptake–pH survey, and tuned reduction) demonstrated in the earlier sections. It has recently been suggested that electrostatic adsorption over silanol groups is the cause for metal overexchange in low-aluminum zeolites [61]. It is presently being employed to study noble metal uptake on pure oxides of titania, ceria, zirconia, and niobia.

The SEA method can also be applied for the synthesis of bimetallic catalysts. For illustration, the potentially high impact area of bimetallic catalysts for fuel cells will be discussed.

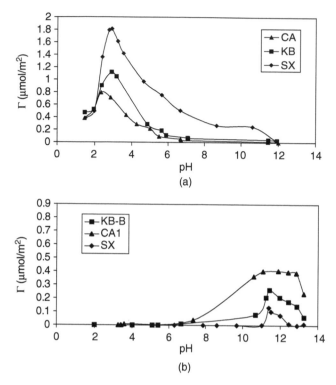

Figure 6.28 Uptake results (1 h contact time) over activated SX, KB, and CA: (a) CPA, (b) PTA. (From Hao, X., Quach, L., Korah, J., and Regalbuto, J.R., *J. Molec. Catal.* 219, 2004, 97.)

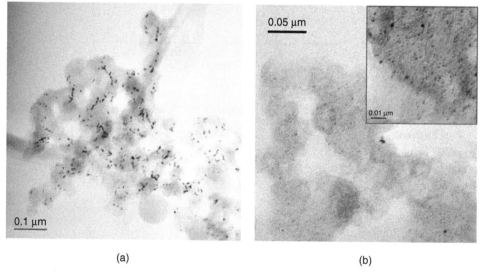

Figure 6.29 Electron microscopy comparison of Pt particle sizes of 9 wt% Pt impregnated as PTA onto TIMREX high-surface-area graphite by (a) DI, diameter = 8 to 10 nm, (b) SEA, near pH 11, diameter =1 to 2 nm.

A promising bimetallic cathode catalyst for proton exchange fuel cells for the automotive industry is Pt/Co [63]. Cobalt is stable as bulk oxide Co_3O_4 and can be made with a surface area of about 200 m²/g and has a PZC of about 9 [62].

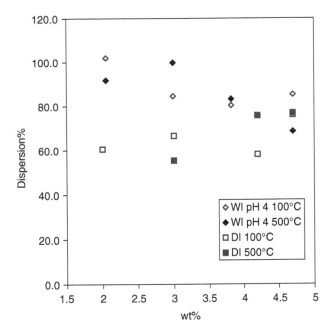

Figure 6.30   Effect of calcination temperature (100°C or 500°C) on dispersion of Pt/γ-Al$_2$O$_3$, reduced at 200°C. (From Liu, J., and Regalbuto, J.R., in preparation.)

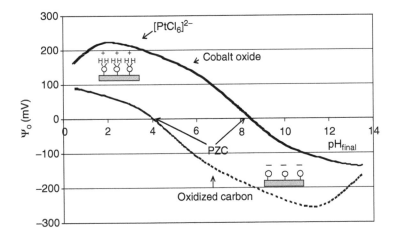

Figure 6.31   Schematic of the use of SEA to achieve selective adsorption of anionic Pt onto carbon-supported cobalt oxide particles.

Currently, Pt/Co bimetallics are prepared by doping copious amounts of cobalt onto a Pt/C catalyst, and then leaching away excess Co not in contact with Pt [63]. A more efficient synthesis might be based on SEA in which cobalt oxide will be formed on a low-PZC (oxidized) carbon and the Pt will be steered onto the cobalt oxide phase. The situation is depicted in Figure 6.31. At a pH midway between the PZC of Co$_3$O$_4$ (9) and of oxidized carbon (4), the cobalt oxide phase will be protonated and positively charged, while the carbon surface will be deprotonated and negatively charged. Pt anions should then be adsorbed selectively onto the cobalt oxide particles. Subsequent reduction in H$_2$ will be used to form the bimetallic particles.

Prior to synthesizing the carbon-supported bimetal, however, an uptake–pH survey of Pt onto the cobalt oxide itself could be performed to give a clear idea on the pH of maximum interaction of CPA with cobalt oxide.

6.7 SUMMARY

A simple electrostatic model of metal adsorption onto oxide surfaces has great utility for the preparation of typical supported metal catalysts. Owing to the chemistry of the surface hydroxide groups, the oxide surfaces become protonated and positively charged at pH values below their PZC, and are capable of strongly adsorbing metal anions. Above the PZC of oxide, the surface is deprotonated and negatively charged and capable of strongly adsorbing cations. In either case, however, the oxide buffering effect must be overcome. This is accomplished in practice by measuring and controlling the final pH of the slurry solution. It is hypothesized that once the metal precursor has been strongly adsorbed, it can be reduced to its active elemental state at conditions which retain its high dispersion.

The SEA approach can be applied to a novel system in three steps: (1) measure the PZC of the oxide (or carbon) and choose a metal cation for low-PZC materials and an anion for high-PZC materials, (2) perform an uptake–pH survey to determine the pH of the strongest interaction in the appropriate pH regime (high pH for low PZC and vice versa), and (3) tune the calcination/reduction steps to maintain high dispersion. Highly dispersed Pt materials have been prepared in this way over silica, alumina, and carbon. Other oxides can be employed similarly. For bimetallics, the idea is to first adsorb a well-dispersed metal that forms an oxide intermediate with a PZC very different to the support. In this way the second metal can be directed onto the first metal oxide by SEA. Reduction may then result in relatively homogeneous bimetallic particles.

REFERENCES

1. Brunelle, J.P., *Pure Appl. Chem.* 50 (1978), 1211.
2. Contescu, C., and Vass, M.I., *Appl. Catal.* 33 (1987), 259.
3. Heise, M.S., and Schwarz, J.A., *J. Coll. Interf. Sci.* 107 (1985), 237.
4. Heise, M.S., and Schwarz, J.A., *J. Coll. Interf. Sci.* 113 (1986), 55.
5. Heise, M.S., and Schwarz, J.A., *J. Coll. Interf. Sci.* 123 (1988), 51.
6. Schwarz, J.A., and Heise, M.S., *J. Coll. Interf. Sci.* 135 (1990), 461.
7. Noh, J.S., and Schwarz, J.A., *J. Coll. Interf. Sci.* 130 (1989), 157.
8. Noh, J.S., and Schwarz, J.A., *J. Coll. Interf. Sci.* 139 (1990), 139.
9. Subramanium, S., Noh, J.S., and Schwarz, J.A., *J. Catal.* 144 (1988), 433.
10. Bourikas, K., and Lycourghiotis, A., *J. Phys. Chem. B.* 107 (2003), 9441.
11. Park, J., and Regalbuto, J.R., *J. Coll. Interf. Sci.* 175 (1995), 239.
12. Schwarz, J.A., *Chemical Rev.* (1995), 477.
13. Mang, T., Breitscheidel, B., Polanek, P., and Knozinger, H., *Appl. Catal. A: Gen.* 106 (1993), 239.
14. Shelimov, B., Lambert, J.-F., Che, M., and Didillon, B., *J. Mol. Catal.* 158 (2000), 91.
15. Shah, A., and Regalbuto, J.R., *Lang.* 10 (1994), 500.
16. Santhanam, N., Conforti, T.A., Spieker, W.A., and Regalbuto, J.R., *Catal. Tod.* 21 (1994), 141.
17. James, R.O., and Healy, T.W., *J. Coll. Interf. Sci.* 40 (1972), 65.
18. Spieker, W.A., and Regalbuto, J.R., *Chem. Eng. Sci.* 56 (2000), 2365.
19. Schreier, M., and Regalbuto, J.R., *J. Catal.* 225 (2004), 190.
20. Park, C., Fenter, P., Sturchio, N., and Regalbuto, J.R., *Phys. Rev. Lett.* 94 (2005), 076104.
21. Healy, T.W., and White, L.R., *Adv. Coll. Interf. Sci.* 9 (1978), 303.
22. Contescu, C., Jagiello, J., and Schwarz, J.A., *Lang.* 9 (1993), 1754.

23. Regalbuto, J.R., Navada, A., Shadid, S., Bricker, M.L., and Chen, Q., *J. Catal.* 184 (1999), 335.
24. Schreier, M., Timmons, M., Feltes, T., and Regalbuto, J.R., submitted.
25. Karakonstantis, L., Bourikas, K., and Lycourghiotis, A., *J. Catal.* 162 (1996), 295.
26. Agashe, K., and Regalbuto, J.R., *J. Coll. Interf. Sci.* 185 (1997), 174.
27. Weise, G.R., James, R.O., and Healy, T.W., *Chem. Soc. London Faraday Disc.* 51 (1971), 302.
28. Hao, X., Spieker, W.A., and Regalbuto, J.R., *J. Coll. Interf. Sci.* 267 (2003), 259.
29. Miller, J.T., Schreier, M., Kropf, A.J., and Regalbuto, J.R., *J. Catal.* 225 (2004), 203.
30. Arai, M, Guo, S.L., Nishiyama, Y., *Appl. Catal.* 77 (1991), 141.
31. Bond, G.C., and Wells, P.B., *Appl. Catal.* 18 (1985), 221.
32. Geus, J.W., and Wells, P.B., *Appl. Catal.* 18 (1985), 237.
33. Freysz, J.L., Saussey, J. et al., *J. Catal.* 197 (2001), 131.
34. Sermon, P.A., and Sivalingam, J., *Coll. Surf.* 63 (1992), 49.
35. Fanson, P.T. et al., *J. Catal.* 204 (2001), 35.
36. Folgado, M.A., Mendioroz, S. et al., *Vac.* 37 (l987), 451.
37. Goguet, A., Candy, J.P. et al., *J. Catal.* 209 (2002), 135.
38. Douidah, A., Marecot, P., and Barbier, J., *Appl. Catal. A: Gen.* 225 (2002), 11.
39. Gonzalez, R.D., Zou, W., Lopez, T. et al., *Mat. Lett.* 24 (1995), 35.
40. Gonzalez, R.D., and Zou, W., *Catal. Tod.* 15 (1992), 443.
41. Gonzalez, R.D., and Zou, W., *J. Catal.* 152 (1995), 291.
42. Hwang, C.P., Yeh, C.T., *J. Catal.* 182 (1999), 48.
43. Zhao, F., Arai, M. et al., *J. Molec. Catal. A: Chem.* 180 (2002), 259.
44. Kim, M.H, Vannice, M. A. et al., J. Catal. 204 (2001), 348.
45. Singh, U.K., Vannice, M.A., *J. Catal.* 191 (2000), 165.
46. Denton, P., Praliaud, H. et al., *J. Catal.* 189 (2000), 410.
47. Jayat, F., Lembacher, C., Schubert, U., Martens, J.A., *Appl. Catal. B: Env.* 21 (1999), 221.
48. Benesi, H.A., Curtis, R.M., Studer, H.P., *J. Catal.* 10 (1968), 328.
49. Dias, E. et al., *Chem. Eng. Sci.* 58 (2003), 621.
50. Burch, R., and Millington, P.J., *Catal. Tod.* 29 (1996), 37.
51. Avai, M., Takahashi, H., Shivai, M., Nishiyama, Y., and Ebina, T., *Appl. Catal. A. Gen* 176 (1999), 229.
52. Mieth, J.A., Schwarz, J.A., Huang, Y.-J., and Fung, S.C., *J. Catal.* 122 (1990), 202.
53. Akratopulu, K.C., Vordonis, L., and Lycourghiotis, A., *J. Catal.* 109 (1988), 41.
54. Korah, J., Spieker, W., and Regalbuto, J.R., *Catal. Lett.* 85 (2003), 123.
55. Hao, X., Quach, L., Korah, J., and Regalbuto, J.R., *J. Molec. Catal.* 219 (2004), 97.
56. Spieker, W., Regalbuto, J., Rende, D., Bricker, M., and Chen, Q., *Stud. Surf. Sci. Catal.* 130, 2000, 203.
57. Hao, X., and Regalbuto, J.R., in preparation.
58. Hao, X., Miller, J.T., Kropf, A.J., and Regalbuto, J.R., in preparation.
59. Castorano, M., Robles, J., and Regalbuto, J.R., in preparation.
60. Liu, J., and Regalbuto, J.R., in preparation.
61. Schreier, M., Terens, S., Belcher, L., and Regalbuto, J.R., *Nanotech.* 16 (2005), S582.
62. Gasteiger, H.A., Kocha, S.S., Sompalli, B., and Wagner, F.T., *Appl. Catal. B: Env.* 56 (2005), 9.

CHAPTER 6 QUESTIONS

Question 1

Calculate the surface loading at incipient wetness for the following cases:
  (a) silica with SA 380 m^2/g and pore volume 2.1 mL/g, assume 5 OH/nm^2,
  (b) alumina with SA 200 m^2/g and pore volume 0.80 mL/g, assume 8 OH/nm^2,
  (c) carbon black with SA 1500 m^2/g and pore volume 9.0 mL/g, assume 0.5 OH/nm^2.

Question 2

How acidic must a solution be to protonate half the OH groups on the above supports using incipient wetness? How must it be basic to deprotonate half the groups?

Question 3

Hydroxyl groups on oxide and carbon surfaces are often modeled as a one-site, two-pK model as shown in Figure 6.1. Defend this choice of model with the pH shift data for alumina (Figure 6.15). Might a different type of site be envoked for silica (Figure 6.20) and unoxidized carbon (Figure 6.26a)? See [21] for more other types of acid–base group models.

Question 4

Using the uptake–pH surveys and the pH shift data presented in this chapter, estimate the initial pH needed for the strongest electrostatic adsorption for:
  (a) CPA uptake over alumina at 500 m^2/L,
  (b) CPA uptake over alumina at 50,000 m^2/L,
  (c) CPA uptake over alumina at incipient wetness (use the parameters of problem 1 above).
  (d) PTA uptake over silica at 1000 m^2/L,
  (e) PTA uptake over silica at 50,000 m^2/L,
  (f) PTA uptake over silica at incipient wetness (use the parameters of problem 1 above).

Question 5

Using the protonation–deprotonation equations of [18], reproduce the alumina OH speciation plot given in Figure 6.4b. (What does the PZC appear to be from the figure?) Choose pK1 and pK2 to get the best fit. Note that pK1 and pK2 must be equidistant from the PZC. For example, if PZC = 8 and pK1 = 6 (i.e., K1 = 1026), then pK2 must equal 10 (K2 = 10210).

Question 6

Extend the model from Problem 5 to simulate the pH shifts for alumina and silica seen in Figure 13. Then model other oxides: titania (PZC 6, 8 OH/nm2), niobia (PZC 2.5, 5 OH/nm2), magnesia (PZC 12, 10 OH/nm2).

Question 7

What would the maximum uptake density of (a) CPA and (b) PTA be if they adsorbed with no hydration sheath?

Question 8

Extend the model from Problem 6 to simulate the Pt uptake vs. pH over alumina and silica seen in Figure 18.

Question 9

Carbon comes in many forms and surfaces areas. Graphite can have surface area below 1 m2/g, while high surface area carbon blacks approach 2000 m2/g. If Pt adsorbs onto carbon at the same surface density, the Pt loading (Pt, wt%) will then be a function of the carbon surface area. Assuming that CPA can be adsorbed onto carbon at 1.6 mmol/m2, and PTA at 0.84 mmol/m2 at the respective SEA conditions, plot Pt (wt%) vs. surface area of carbon.

Question 10

Niobia has a PZC of about 2.5. It is desired to use niobia as a catalyst promoter for Pt, by supporting the niobia onto an oxide support. It will then be attempted to impregnate Pt only onto the niobia phase. Which common support and what Pt complex should be used and why? A sketch of the surface potential vs. pH for niobia and the support will help.

Question 11

An alumina-based catalyst will be bound, for the purpose of mechanical strength, with carbon. The alumina–carbon mixture is essentially a composite support for adsorbing the Pt precursor. If it is desired that all metal go onto the alumina phase, which type of carbon (oxidized or unoxidized) and what type of Pt complex should be used and why? A sketch of the surface potential vs. pH for alumina and the carbon binder will help.

Question 12

Another useful bimetallic for fuel cell electrodes is Pt/Ru. Ruthenium is readily oxidized to RuO2 by calcination after it is impregnated. The PZC of ruthenium oxide is unknown. Propose a comprehensive sequence of experiments with which the SEA method can be applied for the synthesis of a Pt/Ru bimetallic catalyst supported on carbon. The goal is to have intimate contact between the Pt and Ru phases in the final, reduced catalyst.

Catalysis and Chemical Reaction Engineering

Stanko Hočevar

CONTENTS

Annotations 195
    Greek Letters 196
    Subscripts 196
7.1   Introduction 196
7.2   Overview of Heterogeneous Catalysis and Chemical Reaction Engineering 197
7.3   Hydrogen Production and Cleaning: Catalysis and Reaction Engineering 199
    7.3.1   Conventional Processes and Catalysts for Hydrogen Generation and Their
          Limitations in Low-Temperature Fuel Cells Technology 200
          7.3.1.1   Fuel Reforming 200
          7.3.1.2   Water-Gas Shift Reaction 202
          7.3.1.3   Preferential Oxidation of Carbon Monoxide 202
    7.3.2   Fuel Cells and Primary Fuel Processing for Low-Temperature Fuel Cells 204
          7.3.2.1   Catalytic Processes of Hydrogen Production for Proton-Exchange
                  Membrane Fuel Cell 205
7.4   Conclusions 224
Acknowledgments 224
References 225
Chapter 7 Questions 227

ANNOTATIONS

a specific surface area of a catalyst particle ($m^2_{\text{external surface}}/m^3_{\text{particle}}$)

A_i pre-exponential factor of the rate constant i (various units)

A_{td} pre-exponential factor of the reverse oxygen diffusion coefficient (cm^2/sec)

C_{TOT} total molar concentration of gas phase (mol/m_{gas}^3)

D diffusion coefficient of oxygen species in the solid catalyst lattice (m^2/sec)

E_i activation energy of the rate constant i (kJ/mol)

E_{td} activation energy of the reverse oxygen diffusion coefficient (kJ/mol)

$E_{\text{af}}, E_{\text{ab}}$ activation energies of forward and backward reactions, respectively (kJ/mol)

F_{TOT} surface capacity of catalyst inert phase (mol/kg_{cat})

H_{TOT}	volumetric capacity of catalyst (mol/kg$_{cat}$)
$H_{f,I}$	enthalpy of formation of species I
H_r	reaction enthalpy
K_{eq}	thermodynamic equilibrium constant
k_i	rate constant of the component i reaction (mol/g$_{cat}$ sec)
k_1	rate constant of elementary reaction step 1 (m^3/mol sec)
k_2	rate constant of elementary reaction steps 2 and 6 (sec^{-1})
k_3	rate constant of elementary reaction step 3 (m^3/mol sec)
k_4	rate constant of elementary reaction step 5 (kg$_{cat}$/mol sec)
k_5	rate constant of elementary reaction step 7 (m^3/mol sec)
k_f''	rate constant of forward reaction
k_b''	rate constant of backward reaction
L_{TOT}	surface capacity of catalyst active phase (mol/kg$_{cat}$)
m_{cat}	mass of catalyst (kg)
n	partial reaction order
P_i	partial pressure of species i
R_r	universal gas constant (J/mol K)
r''	surface reaction rate
r	pseudo-homogeneous rate
T	temperature (K)
t	contact time, time (sec)
t_d	characteristic diffusion time (sec)
X_i	conversion of species i
x	dimensionless radial position in catalyst crystallite particle
y_i	molar fraction of gas component i
z	dimensionless distance of catalyst bed

Greek Letters

Φ_v	volumetric flow (mL/min)
δ_n	fractional surface coverage of species n adsorbed on catalyst inert phase
ε_B	void fraction of catalyst bed = 0.35
ρ_B	catalyst bulk density (kg$_{cat}$/m$_r$)
θ_j	fractional surface coverage of species j adsorbed on catalyst active phase
τ	space time (sec)
ξ_m	fractional volumetric coverage of species m that diffuses through the catalyst particle

Subscripts

b	oxygen vacancy located in the catalyst bulk lattice
s	oxygen vacancy located on the catalyst surface
b,b	oxygen species located in the catalyst bulk lattice that originated in the bulk
s,b	oxygen species located on the catalyst surface that originated from the bulk catalyst lattice
s,s	oxygen species located on the catalyst surface that originated on the catalyst surface

7.1 INTRODUCTION

A great majority of current industrial chemical processes use catalysts, and most of these processes — in terms of quantity of catalysts, quantity of products, and financial value in chemical industry — are heterogeneously catalyzed processes in petrochemical industry. This sector of the chemical industry is closely related to the energy sector for which it produces various kinds of fuels derived from petroleum oil and natural gas as two nonrenewable (fossil) primary fuel resources. Most of the fuels produced are used to feed internal combustion engines (ICEs), on the use of which the progress of our civilization has been based since the dawn of the 20th century. These industrial chemical processes are the result (and the best illustration) of the very close link between (heterogeneous) catalysis and chemical reaction engineering.

In the absence of aging (deactivation) of the catalyst, a phenomenological definition of catalysis would be enhancement of chemical reactions or a change of their rate under the influence of substances — catalysts — which several times enter into transient chemical interactions with reaction participants and then, after each cycle of transient interactions, regenerate their chemical identity [1].

A catalyst is a substance that increases the rate at which a chemical reaction reaches equilibrium. Catalysis is the word used to describe the action of the catalyst. Heterogeneous catalysis describes the enhancement in the rate of a chemical reaction brought about by the presence of an interface between two phases [2].

Historically, catalysis has developed within the domain of physical chemistry. In this framework the fundamental question to which an answer has been sought has been posed as "why certain reactions in the presence of certain substances (catalysts) proceed faster and more selectively than in the absence of this substance or in the presence of other substances?" Eventually, the answer was found and the nature of catalyst functioning was clarified. The alter ego of man would immediately ask the next question: "how can one use the catalytic act to produce a desired product from certain starting materials in an economically sound way?" The answer to this question concerning the chemical treatment step of a process requires an integrated approach involving the use of information, knowledge and experience from many areas including thermodynamics, chemical kinetics, fluid mechanics, heat transfer, mass transfer, and economics. The combination of these factors to design a suitable chemical reactor is the subject of chemical reaction engineering.

Chemical reaction engineering is that engineering activity concerned with the exploitation of chemical reactions on a commercial scale [3]. Given a particular thermodynamically permissible chemical reaction network, the task of the chemical engineer and applied kineticist is essentially that of engineering the reaction to achieve a specific goal. That goal, or end, is the transformation of given quantities of particular reactants to particular products. This transformation (reaction) ought to be realized in equipment of reasonable, economical size under tolerable conditions of temperature and pressure [4].

Chemical reaction engineering and, particularly, catalysis is still "... an area within which progress is largely realized by art, some science, and a generous portion of serendipity" as Professor James J. Carberry wrote nearly 30 years ago in a preface to his fundamental book *Chemical and Catalytic Reaction Engineering* [4].

The scope of this chapter is to present a concise and fundamental overview of the relationship between modern catalysis and chemical reaction engineering through one of the topics that will certainly be present among the future directions in both fields: production of hydrogen fuel from fossil and renewable energy sources and its use in the fuel cells-based energy conversion technology.

7.2 OVERVIEW OF HETEROGENEOUS CATALYSIS AND CHEMICAL REACTION ENGINEERING

Catalysis by itself is an older discipline than chemical reaction engineering. It was formally initiated by Berzelius [5], who first used this term in 1836. In 1889, Arrhenius [6] laid the foundation of the modern development of the theory of reaction rates by showing that the specific rate of the reaction grows exponentially with inverse temperature. However, it was only in the first decade of the 20th century, owing to the kinetic studies of Bodenstein and Lund [7] and Ostwald [8], and the thermodynamic principles introduced earlier by van't Hoff [9] that the first theoretical foundation of catalysis was laid. In the next decade, Eyring and Polanyi [10] postulated an activated complex theory for predicting the rate of reaction on the basis of fundamental information: configurations, dimensions, and interatomic forces of the reacting molecules. During the 1930s, the application of the methods of quantum mechanics and statistical mechanics enabled chemists to develop further the theory of absolute reaction rates [11]. However, it is only with the development of modern *in situ* microscopy (high-resolution electron microscopy [HREM], atomic force microscopy [AFM], and scanning tunneling microscopy [STM]), diffraction (x-ray diffraction [XRD], neutron diffraction [ND], and electron diffraction [ED]), and spectroscopy (Fourier-transform infrared [FTIR], nuclear magnetic resonance [NMR], x-ray photoelectron spectroscopy [XPS], and extended x-ray absorption fine structure [EXAFS]) techniques that it has been possible in the last 30 years to deal with more and more complex, heterogeneously catalyzed reaction systems. Several numerically based theoretical approaches (quantum mechanics/molecular mechanics [QM/MM], density functional theory [DFT] calculations, and molecular dynamics [MD] simulations), which demand efficient computer applications, were used to investigate bulk metals and metal oxides [12]. However, heterogeneous catalysis proceeds at the interface between two phases. In case one has a catalyzed system with solid catalyst and gaseous or liquid reactants, the reaction proceeds over the surface (outer or inner) of the solid catalyst. Therefore, it is of paramount importance to understand how surface atoms or ions interact with gaseous reactant molecules. Only a small number of MD simulations, however, have been devoted to the study of metal oxide systems including their surfaces. Even fewer experimental and theoretical studies about the motion and vibrations of individual surface ions have been performed [13].

It was recently demonstrated that catalyzed reactions of diatomic molecules have two parts: the dissociation of the reacting molecules and the removal of the dissociation products. The rate of dissociation is determined by its dissociation energy, E_a, while the rate of product removal is mainly driven by the intermediate stability, ΔE. A good catalyst is characterized by a low activation energy and by the weak bonding of the intermediates. It has long been realized that E_a and ΔE are often correlated such that the best catalyst is a compromise having adsorbate–surface interactions of intermediate strength. This is known as the Sabatier principle [14], and was further developed by Balandin [15] to explain the volcano-shaped curves obtained in the relationships between catalytic activity and the various thermodynamic quantities (observables) related to the energy of interaction of the reactant (intermediate) molecule with the catalyst active centers, for instance, heats of formation of compounds between reactant molecule and transition metals, and heats of adsorption of reactant molecules on the surface of transition metals. It was recently shown for a case of diatomic molecules that there is indeed a linear Brønsted–Evans–Polanyi type [16,17] of relationship between the activation energy for dissociation and the bond energy of reactant molecule–active center in several cases [18–21]. Combining the microkinetic model of a given catalyzed reaction with the energy-level diagram obtained from the DFT calculations [22] of reactants and intermediates interactions with the catalyst surface atoms, one can get quite a sound explanation of the elementary reaction steps and of the rate-limiting step(s) governing the rate equation [23]. These efforts have led to new insights into the sequences of elementary reaction steps proceeding during the catalytic cycles of most relevant heterogeneously catalyzed reaction systems.

The first large-scale conscious use of industrial catalysts started in 1750 with the introduction of the lead chamber process for the manufacture of sulfuric acid, in which oxides of nitrogen formed from nitric acid in the presence of water, were used as catalysts [24,25]. Although the usefulness of catalysts in chemical industry was recognized, the scientific basis of their chemical and kinetic action was developed much later. However, the use of catalytic processes has grown almost exponentially from the early 18th century to the present. Early catalytic processes were mostly developed for the production of inorganic chemicals and only in the mid-19th century were catalytic processes involving organic reactions introduced. They have become dominant with the application of catalysis in fuel production (i.e., petrochemical industry). The need for large-capacity catalytic re-actors boosted the theoretical development of chemical reaction engineering soon after World War I.

Modern catalysts have to be very active and very (100%) selective, that is, they have to catalyze the desired reaction in the temperature window, where the equilibrium conversion is the highest possible and the reaction rate is high enough to permit suitable process economics. To engineer the reaction, one has to obtain first the intrinsic reaction rate, free of heat- and mass-transfer limitations. In many cases this is very difficult, because in the core of the catalytic process there are several physical and chemical steps that must occur and which may preclude the reaction running in the kinetic regime. These steps are as follows:

1. *External diffusion of reactants.* This step depends on the fluid dynamic characteristics of the system. Reactants must first diffuse from the bulk gaseous phase to the outer surface of the carrier through a stagnant thin film of gas. Molecular diffusion rates in the bulk have the activation energy $E_1 = 2$ to 4 kcal/mol and they vary with $T^{3/2}$.

2. *Internal diffusion of reactants.* This step depends on the porosity of the catalyst and the size and shape of the catalyst particles, and occurs together with the surface reaction. The active catalyst component is usually highly dispersed within the three-dimensional porous support. The reactant molecules have to diffuse through the network of pores toward the active sites. The activation energy for pore diffusion E_2 may represent a substantial share of the activation energy of the chemical reaction itself.

3. *Adsorption.* This step depends on the possible interaction between molecules and the catalyst surface. When the reactants reach the active sites, they chemisorb on adjacent active sites. The chemisorption may be dissociative and the adjacent active sites may be of the same or different origin. The chemisorbed species react and the kinetics generally follow an exponential dependence on temperature, $\exp(-E_3/RT)$, where E_3 is the activation energy of chemisorption.

4. *Chemical reaction on the surface.* The reaction may proceed through one or more sequential steps in which different intermediates are formed. The intermediate with the highest energy profile represents the rate-limiting step. Once the reaction passes this barrier, the final product is formed. The kinetics of this step also depends exponentially on the temperature and the activation energy E_4 is of the same order of magnitude as in step 3.

5. *Desorption.* The product(s) desorbs from the active site with the kinetics exponentially dependent on temperature and with activation energy E_5, which is of the same order of magnitude as in the previous two steps.

6. *Internal diffusion of products.* The desorbed products then diffuse through the network of pores to reach the outer surface of the catalyst with kinetics and activation energy similar to those in step 2.

7. *External diffusion of products.* The last step is the diffusion of product(s) through the stagnant film into the bulk gas under conditions similar to those in step 1.

Any of the seven steps above may be rate limiting and may control the overall reaction rate. When one of the steps, 1, 2, 6, or 7 is rate limiting, the reactor is mass-transfer-limited. If one of the steps 3 to 5 is rate limiting then the reactor is reaction-rate-limited.

The performance of a chemical reactor can be described, in general, with a system of conservation equations for mass, energy, and momentum. To solve this system we must have a model for the reaction on the basis of which we can derive the intrinsic rate equation on one side, and a model of the reactor in which we want to run the reaction on the other side. Both tasks are, of course, interconnected and difficult to solve without reduction of more general equations to a suitable limiting reactor type to be used for each particular reaction system [4,26].

7.3 HYDROGEN PRODUCTION AND CLEANING: CATALYSIS AND REACTION ENGINEERING

Hydrogen is the most abundant element in the universe, accounting for 90% (by wt) of the universe. Owing to its high reactivity it is not found commonly in its pure form, but rather in compounds such as water (covers 70% of the Earth's surface) and all organic matter. When pure, in the form of the hydrogen molecule H_2, it is among the strongest chemical fuels in combustion processes: its high heating value (HHV) is 2.6 times larger than that of methane. Hydrogen is also the cleanest fuel: its combustion product is water and there are no particulate and noxious gaseous emissions. It is an energy carrier that can be produced from a wide variety of renewable (like water and biomass) and fossil (like coal, oil, and natural gas) primary energy sources. Hydrogen can be produced in a distributed or centralized manner. In the first case, the production is usually small and located at the point of use. In the second case, hydrogen is produced in large quantities and then it has to be stored and transported through diverse transport infrastructures to the point of use. The scheme in Figure 7.1 illustrates in a very concise way some of the hydrogen energy pathways. It is clear that this diversity of options enables hydrogen production almost anywhere in the world.

All hydrogen production processes are based on the extraction of hydrogen from hydrogen-containing feedstock. The extraction method to be used is dictated by the nature of feedstock and by the energy transformation technology at the point of use. In 2002, the amount of hydrogen produced worldwide was about 26 million tons. The percentage of feedstock from which this hydrogen was produced is given in Table 7.1. Over 96% of hydrogen is produced nowadays from fossil primary energy sources. The reason for this lies in available established technology for hydrogen production from hydrocarbon feedstock.

7.3.1 Conventional Processes and Catalysts for Hydrogen Generation and Their Limitations in Low-Temperature Fuel Cells Technology

7.3.1.1 Fuel Reforming

Fuel-reforming process should be understood in a broader sense as including all options such as partial oxidation (POX), steam reforming (SR), and their combination, i.e., autothermal reforming (ATR). In general, the fuel-reforming process can be represented by the following equation:

$$C_xH_yO_z + mO_2 + nH_2O \rightarrow aH_2O + (n + y/2 - a)H_2 + bCO + (x - b)CO_2 \qquad (7.1)$$

The stoichiometry for specific cases such as SR, POX, and ATR is given in Table 7.2.

Partial oxidation is based on extreme rich fuel combustion (low air/fuel ratios). The process is highly exothermal and can be performed in both a catalytic and a noncatalytic manner. If a catalytic system is used, the reformer can be operated at a much lower temperature and the heat can be supplied directly into the catalyst bed. The advantage with this process is that it is rather insensitive to contaminants and that it is rather independent of fuel. The biggest drawback is the risk for carbon

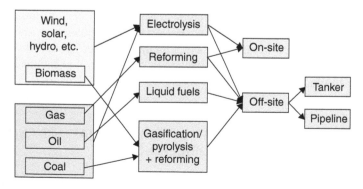

Figure 7.1 Sources for hydrogen and energy pathways.

Table 7.1   World Hydrogen Production Sources Today

H_2 Source	Percentage
Natural gas	48
Oil	30
Coal	18
H_2O splitting	4
Biomass	Low

Table 7.2   The Stoichiometry for Autothermal Reforming (ATR), Partial Oxidation (POX), and Steam Reforming (SR)

Reaction	Condition	Stoichiometry
ATR	$n = 2x + y/2 - z; n > m$	$C_xH_yO_z + mO_2 + nH_2O$ $(2x + y/2 - z - 2m)H_2 + xCO_2$
POX	$n = 0; m = x + y/4 - z/2$	$C_xH_yO_z + mO_2 + nH_2O$ $(y/2)H_2O + xCO_2$
	$n = 0; m = x - z/2$	$C_xH_yO_z + mO_2 + nH_2O$ $(y/2)H_2 + xCO_2$
	$n = 0; m = x/2 - z/2$	$C_xH_yO_z + mO_2 + nH_2O$ $(y/2)H_2 + xCO$
SR	$n = 2x - z; m = 0$	$C_xH_yO_z + mO_2 + nH_2O$ $(2x + y/2 - z)H_2 + xCO_2$

formation if the gas composition exceeds the equilibrium in any of the following carbon-forming reactions.

$$2CO(g) \rightleftharpoons CO_2(g) + C(s) \tag{7.2}$$

$$CH_4(g) \rightleftharpoons 2H_2(g) + C(s) \tag{7.3}$$

$$CO(g) + H_2(g) \rightleftharpoons H_2O(g) + C(s) \tag{7.4}$$

$$CO_2(g) + 2H_2(g) \rightleftharpoons 2H_2O(g) + C(s) \tag{7.5}$$

The first reaction (Boudouard equilibrium) favors carbon formation at lower temperatures compared to POX. The hydrogen concentrations attained depend on the fuel used in POX but it never reaches the theoretical level [27].

Steam reforming is, on the other hand, an endothermal process, which lowers otherwise high system efficiency. The advantage of SR is its high concentration level of hydrogen produced. For instance, with methanol as a feedstock, 75 vol% of hydrogen can be obtained at stoichiometric conditions and total conversion. Copper-based catalysts have been mostly used for ethanol or methanol SR. For methanol, $CuO/ZnO/Al_2O_3$ catalyst is usually used at a temperature between 180 and 250°C. When unconverted methanol is present, the rate of the water-gas shift (WGS) reaction is negligible. As the conversion of methanol approaches 100%, the rate of the WGS reaction becomes significant and the CO concentration in the reformer product gas approaches equilibrium. The reforming of ethanol proceeds better over CuO/ZnO catalyst at temperatures above 300°C [27–29].

Autothermal reforming is a combination of SR and POX, which in theory can be totally heat balanced. When air is used instead of pure oxygen, the reaction can be represented by the following equation:

$$C_xH_yO_z + m(O_2 + 3.76N_2) + (2x - 2m - z)H_2O \rightarrow$$
$$(2x - 2m - z + y/2)H_2 + xCO_2 + 3.76mN_2 \tag{7.6}$$

where m is the oxygen/fuel ratio. The concentration of hydrogen in the product gas (in %) can then be expressed as

$$\{(2x-2m-z+y/2)/(x+(2x-2m-z+y/2)+3.76m)\}\times 100 \qquad (7.7)$$

and the reaction enthalpy is calculated as

$$H_r = xH_{f,CO_2} - (2x-2m-z)H_{f,H_2O} - H_{f,fuel} \qquad (7.8)$$

The calculated thermoneutral oxygen/fuel ratios (m_{O_2}) and theoretical yields for different fuels are presented in Table 7.3.

The CO concentration in the reformate gas, generated by reforming of fuels, is much higher than that permitted for most of the chemical processes in which hydrogen is further used: synthesis of ammonia, hydrogenation of organic compounds, etc. This is true also for the hydrogen used in low-temperature fuel cells. Usually, the CO concentration in the reformate gas must be lowered to the level of parts per million (ppm) because CO is chemisorbed more strongly on the surface of metallic catalysts than H_2, and therefore blocks the active sites for hydrogen activation on the surface of catalysts.

7.3.1.2 Water-Gas Shift Reaction

The primary means of reducing the CO concentration in the reformate fuel gas (synthesis gas) is the WGS reaction:

$$CO+H_2O \rightleftharpoons H_2 + CO_2 \qquad (7.9)$$

Based on the thermodynamic equilibrium, the lower the reaction temperature, the lower is the CO concentration that can be achieved. Figure 7.2 represents the equilibrium data of CO conversion as a function of H_2O/CO ratio and temperature, calculated for a specific case when the CO/H_2 ratio in the inlet gas composition is 1:2. High CO conversion can be achieved only at relatively low temperatures (below 600 K) and relatively high H_2O/CO ratios. Two suitable catalysts are commercially available: an iron–chrome oxide catalyst that operates at 300 to 450°C, and a copper–zinc oxide catalyst that operates at 160 to 270°C. Industrially, the WGS reaction is conducted using two or more reactor stages that operate adiabatically based on the operating temperature regime of the catalyst used. Usually, in the high-temperature (HT) WGS, the concentration of CO in a fuel that is rich in hydrogen is lowered from about 10 to 2–3%. In the low-temperature (LT) WGS step, the CO concentration is further lowered to about 0.5%.

7.3.1.3 Preferential Oxidation of Carbon Monoxide

Various technologies have been investigated to reduce the concentration of CO in fuel gas exiting the shift reactor to 10 ppm or less. Among the candidates are membrane separation, methanation, and

Table 7.3 Calculated Thermoneutral Oxygen/Fuel (m_{O_2}) Ratios and Theoretical Yield

$C_xH_yO_z$	x	y	z	$\Delta H_{r,fuel}$ (kcal/mol)	$y/2z$	m_{O_2} $\Delta H_r = 0$	Efficiency (%)
Methanol (CH_3OH)	1	4	1	−57.1	2	0.230	96.3
Methane (CH_4)	1	4	0	−17.9	2	0.443	93.9
iso-Octane (C_8H_{18})	8	18	0	−62.0	1.125	2.947	91.2
Gasoline ($C_{7.3}H_{14.8}O_{0.1}$)	7.3	14.8	0.1	−52.0	1.014	2.613	90.8

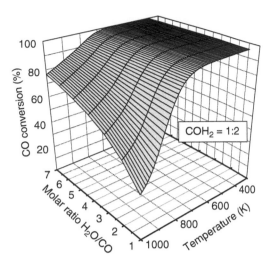

Figure 7.2 Equilibrium conversion of CO in WGS reaction as a function of reaction temperature and H_2O/CO molar ratio at CO/H_2 molar ratio of 1:2.

preferential CO oxidation (PrOX). For membrane separation, Pd alloy membranes can effectively remove CO from the fuel gas, but such membranes require a large pressure difference and a high temperature, which reduce system efficiency. For methanation, CO reacts with H_2 to generate methane and water; however, the amount of H_2 required is three times the amount of CO removed. For PrOX, a small quantity of air is bled into the fuel gas, and CO is selectively oxidized to CO_2 over H_2, using supported and promoted noble metal catalyst such as Pt, Rh, or Ru. For onboard fuel processing, PrOX is the preferred method because of the lower parasitic system load and energy requirement compared to membrane separation and methanation [30]. The selectivity for CO oxidation compared to H_2 oxidation is reduced at higher temperatures. A loss of hydrogen occurs, and as a consequence, a drop in system efficiency. It is also necessary to keep track of the inlet concentration of CO. If the concentration of CO is so high that full surface coverage of CO is attained, the losses of hydrogen are rather small. After most of the CO in the system is consumed, the hydrogen loss increases. The mass of catalyst needed for PrOX process can be minimized with more active catalyst and by reducing the CO concentration in the reformer outlet gas (more efficient WGS steps). However, the increase in reaction temperature decreases the selectivity and consequently increases the loss of hydrogen.

However, in developing new fuel processors for hydrogen production to feed low-temperature fuel cells, one should bear in mind the following limitations of the conventional processes [29]:

- Current SR catalysts based on Ni are extremely sulfur-sensitive and deactivate considerably in the presence of traces of sulfur.
- Hydrodesulfurization (HDS) process operates at a pressure highly exceeding the pressure of natural gas available in the existing infrastructure.
- Ni-based reforming catalysts are pyrophoric; they will sinter if exposed to air and they represent a fire hazard for consumers.
- Steam reforming is an endothermic process that requires complicated heat management of the system.
- High- and low-temperature WGS catalysts based on Fe and Cu respectively, require slow and carefully controlled activation procedures. After reduction they are highly reactive toward air and can be a fire hazard to the consumer.
- Methanation of CO requires removal of CO_2 due to the highly exothermic competitive methanation.
- Ni-based methanation catalysts are also pyrophoric.
- Pressure swing adsorption requires high pressure.

- Industrial H_2 production plants operate at steady state. They were not designed either for numerous start-ups and shutdowns or for the cycling in load. The catalyst and other reformer materials can be chemically and physically damaged.

Catalytic SR process operates at low space velocity gas hour space velocity (GHSV) between (3000 and 8000 h^{-1}) owing to its slow kinetics. Although these conditions are not convenient for transient operation, this process gives the highest yield of hydrogen as compared with POX and ATR processes. Ni-based catalysts are cost effective and commercially available, but in addition to being pyrophoric, they are prone to coke formation at lower H_2O/C ratios. Commercial Cu-based methanol SR catalysts deactivate when exposed to liquid water during the shutdown mode and are also pyrophoric.

For on-board fuel processing, there are two principal concerns: (1) the feasibility of keeping the iron–chrome and copper oxide catalysts in the reduced state, especially during periods of shutdown; and (2) the pyrophoric nature of the copper oxide catalyst in the reduced state [29]. Because of these concerns, considerable research and development is being conducted to develop new WGS catalysts for on-board fuel processing.

7.3.2 Fuel Cells and Primary Fuel Processing for Low-Temperature Fuel Cells

Among the energy converters, fuel cells are unique direct energy conversion devices capable of converting energy of chemical reactions into electricity with the highest maximum-feasible efficiency of 90% [31]. The value of deriving electric current directly from the chemical reactions of fuels was recognized well before electricity became a commodity sold by power utilities. The first investigations go back to 1839 and Sir William Grove. It was not until the 1960s, however, that fuel cells were employed for practical uses. NASA used them first to provide electric power on board the Gemini space mission. Since then, fuel cells have been used on board nearly every space mission, regardless of the country undertaking the mission. Steady progress over the past 40 years has made it possible for fuel cells to start displacing combustion from its central technological role. Nowadays, we see the use of fuel cell technology in demonstration projects from electric cars, buses, and mobile and stationary use. These uses span power generation from few tens of watts to few megawatts. Fuel cell technology is now at the dawn of commercialization [32].

Most fuel cells being developed consume either hydrogen or fuels that have been preprocessed into a suitable hydrogen-rich form. Some fuel cells can directly consume sufficiently reactive fuels such as methane, methanol, carbon monoxide, or ammonia, or can process such fuels internally. Different types of fuel cells are most appropriately characterized by the electrolyte that they use to transport the electric charge and by the temperature at which they operate. This classification is presented in Table 7.4.

Further, we shall concentrate on the low-temperature proton-exchange membrane fuel cells (PEMFCs) as the most representative H_2/O_2 or H_2/air fuel cell. We shall do this deliberately since PEMFCs, working at low temperature, have high thermodynamic equilibrium potential and therefore they can reach high, open-circuit voltage and potentially high efficiency in energy conversion. The low working temperature also poses fewer restrictions on the construction materials and field of application.

Table 7.4   Fuel Cells Ordered According to Operating Temperature and Type of Electrolyte

Type	Electrolyte	Charge Carrier	Temperature (°C)
AFC	Solution of KOH	OH^-	50–90
PEMFC	Polymer membrane	H^+	50–95
PAFC	H_3PO_4	H^+	190–210
MCFC	Li_2CO_3/K_2CO_3	CO_2^{2-}	630–700
SOFC	Y stab. ZrO_2	O^{2-}	900–1000

A PEMFC consists of a negatively charged electrode (cathode), a positively charged electrode (anode), and a thin proton-conducting polymer electrolyte membrane. Hydrogen is oxidized on the anode and oxygen is reduced on the cathode. Protons are transported from the anode to the cathode through the electrolyte membrane and electrons are carried to the cathode over an external circuit. On the cathode, oxygen reacts with protons and electrons forming water and producing heat. Both the anode and the cathode contain a catalyst to speed up the electrochemical processes. The schematic construction and both the half-cell reactions are depicted in Figure 7.3. The electrical and heat energies are produced by the cathode reaction. Theoretically, the Gibbs energy of the reaction is available as electrical energy and the rest of the reaction enthalpy is released as heat. In practice, a part of the Gibbs energy is also converted into heat via the loss mechanisms.

The PEMFC is, in fact, a superb example of a catalytic membrane reactor performing a variety of reactions and separations [33]. We need hydrogen to fuel it. Distributed combined heat and power (CHP) generation based on current common (logistic) fuels demands that hydrogen is produced either on-site (for stationary applications) or on-board (for mobile applications). A general scheme for PEMFC-grade hydrogen production from renewable (biomass, organic waste, and ethanol) or from fossil fuels (coal, oil, and natural gas) is represented in Figure 7.4. As can be noticed, nearly all processes in the train are catalytic and all of them have been practiced in the chemical industry for many years. Therefore, every step forward in the new catalysts formulation and new reactor concept are very demanding because of pressure for time to enable fuel cell technology for market penetration and because of the relatively long and rich history of each and all the catalytic processes involved in this technology [28]. In the following sections, recent results on catalysts development for WGS reaction and for PrOX are briefly discussed.

7.3.2.1 Catalytic Processes of Hydrogen Production for Proton-Exchange Membrane Fuel Cell

For mobile applications, the most suitable reforming technology appears to be ATR because of the adiabatic design that permits a compact and smaller reactor, with low pressure drop. The design combines a highly exothermic POX reaction and endothermic SR. A new generation of natural gas ATR reactor design is based on the overlapped reaction zone concept: the bottom wash-coat layer with Pt/Rh SR catalyst is covered with the Pt/Pd POX catalyst. The heat released in the POX layer is consumed by the SR reactions immediately without going through any heat transfer barriers [29]. Another efficient radial flow ATR reactor, which uses Cu/SiO_2 and Pd/SiO_2 for POX and SR reactions, was

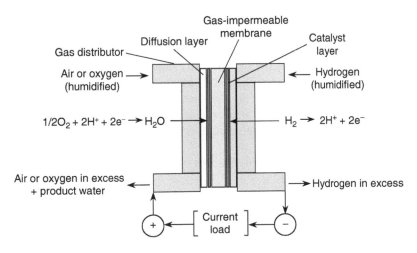

Figure 7.3 Schematic representation of the PEMFC cross section.

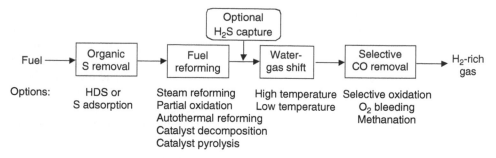

Figure 7.4   Steps and options for on-site and on-board processing of renewable and fossil fuels. (Reprinted from [28]. With permission from Elsevier.)

used for methanol reforming [34]. Recently, researchers from Tsukuba Research Center in Japan have tested a new, non-noble metal, nanostructured cerium oxide-based Cu catalyst in SR of methanol. They have found that this catalyst, containing 3.8 wt% Cu, gives higher methanol conversion (53.9%) than Cu/ZnO (37.9%), Cu/Zn(Al)O (32.3%), and Cu/Al$_2$O$_3$ (11.2%) catalysts containing the same amount of Cu [35,36].

The reformate gas contains up to 12% CO for SR and 6 to 8% CO for ATR, which can be converted to H$_2$ through the WGS reaction. The shift reactions are thermodynamically favored at low temperatures. The equilibrium CO conversion is 100% at temperatures below 200°C. However, the kinetics is very slow, requiring space velocities less than 2000 h^{-1}. The commercial Fe–Cr high-temperature shift (HTS) and Cu–Zn low-temperature shift (LTS) catalysts are pyrophoric and therefore impractical and dangerous for fuel cell applications. A Cu/CeO$_2$ catalyst was demonstrated to have better thermal stability than the commercial Cu–Zn LTS catalyst [37]. However, it had lower activity and had to be operated at higher temperature. New catalysts are needed that will have higher activity and tolerance to flooding and sulfur.

The gas at the outlet of the WGS reactor still contains CO from 0.1 to 1.0% depending on the operating conditions. In the last step of hydrogen production for low-temperature fuel cells, the CO concentration has to be reduced to a minimum. The most effective mechanism for CO removal in PEMFC-grade H$_2$ production is selective oxidation. Because of the high ratio of H$_2$ to CO (>>100:1) at the outlet from LTS reactor, the oxidation catalyst has to be highly selective. The process is therefore called selective oxidation or preferential oxidation (PrOX). The process runs in the temperature window between 80°C (the working temperature of PEMFC) and about 200°C (the working temperature of LTS reactor). The PrOX reactor must run for all varieties of flow. This poses additional demands on the catalyst selectivity. Pt-based PrOX catalysts, for instance, should produce CO by the reverse WGS reaction at longer residence times, since oxygen is consumed in the first part of the catalyst bed. CO oxidation over this catalyst is a multistep process, commonly obeying Langmuir–Hinshelwood kinetics for a single-site-competitive mechanism between CO and O$_2$. An optimum range of O$_2$/CO ratio is required to obtain a proper balance of adsorbed CO and adsorbed O$_2$ on adjacent sites. However, pure precious metals lack the selectivity that is required for PrOX. Recently, we have developed a nanostructured Cu$_x$Ce$_{1-x}$O$_{2-y}$ catalyst, which is highly selective, active, and stable at given reaction conditions [38].

7.3.2.1.1   *Water-Gas Shift Reaction in Excess of H$_2$ Over the Nanostructured Cu$_x$Ce$_{1-x}$O$_{2-y}$ Catalyst*

The product composition from the fuel reformer generally consists of 35 to 40% H$_2$ and 6 to 10% CO balanced with H$_2$O, CO$_2$, and N$_2$ [39]. The CO is further reduced to 2 to 3% by HT WGS and then down to <0.5% CO with LT WGS. It is not possible to reduce the concentration of CO down to a few ppm with LT WGS because of equilibrium constraints. This has to be done with preferential oxidation of CO in the last step. However, owing to the development of high-temperature PEMFC capable of working at temperatures higher than 423 K, the need for PrOX stage of the fuel reformer becomes obsolete because the anode noble metal-based catalysts can tolerate much higher concentrations of CO (usually more than 2%) in the hydrogen fuel. This possibility, however, raises a need to develop a completely new concept of fuel reformer, which would comprise only two stages: a fuel-reforming stage and a WGS stage. To accomplish the development of a compact fuel processing system, the WGS catalyst has to be improved. The WGS reaction

$$CO + H_2O \rightleftharpoons CO_2 + H_2 \quad \Delta H = -41.2 \, kJ/mol \quad (7.10)$$

is thermodynamically favored at low temperature, and the kinetics over existing catalysts is so slow (space velocities below 2000 h^{-1}) that the shift reactor alone would occupy 50% of the entire fuel processor volume. The WGS reaction is a very important industrial reaction [40] increasing the hydrogen content in synthesis gas. The catalysts used in LT WGS process usually consist of different combinations of CuO, ZnO, and Al$_2$O$_3$ components. Unfortunately, these catalysts are extremely pyrophoric in the activated (reduced) state and may explode on exposure to air, which makes them impossible to use in automobile and several other applications [41]. They are also very susceptible to shutdown–start-up cycling because the active catalyst component is leached out by condensed water during quenching or deactivated by the formation of surface carbonates. The desire is to accomplish the WGS reaction at temperatures below 623 K and higher space velocities above 30,000 h^{-1} while maintaining high CO conversion. In this case the product gases could be fed directly on the anode of the HT PEMFC. In short, a WGS catalyst for automobile application has to be cost-effective, more active, nonpyrophoric, stable, and processable on a monolith support surface to achieve the required high flow rates.

Catalyst systems for the WGS reaction that have recently received significant attention are the cerium oxides, mostly loaded with noble metals, especially platinum [42–46]. Jacobs et al. [44] even claim that it is probable that promoted ceria catalysts with the right development should realize higher CO conversions than the commercial CuO–ZnO–Al$_2$O$_3$ catalysts. Ceria doped with transition metals such as Ni, Cu, Fe, and Co are also very interesting catalysts [37,43–47], especially the copper–ceria catalysts that have been found to perform excellently in the WGS reaction, as reported by Li et al. [37]. They have found that the copper–ceria catalysts are more stable than other Cu-based LT WGS catalysts and at least as active as the precious metal–ceria catalysts.

The WGS reaction was examined at low temperatures over the nanostructured Cu$_{0.1}$Ce$_{0.9}$O$_{2-y}$ catalyst with a BET surface area of 22.7 m^2/g, prepared by a sol–gel method [48], and previously studied as a PrOX catalyst having extremely good activity and selectivity for CO oxidation [49,50], and over the Cu$_{0.2}$Ce$_{0.8}$O$_{2-y}$ catalyst with a BET surface area of 2.8 m^2/g, prepared by the coprecipitation method [48]. To compare the efficiency of the two copper–ceria catalysts, a CuO–ZnO–Al$_2$O$_3$ commercial WGS catalyst, G-66 A, from Süd-Chemie AG, Munich, Germany, containing 42 wt% of Cu was tested under the same experimental conditions. The WGS reaction kinetics was studied at atmospheric pressure in a plug-flow micro-reactor at different flow rates and feed compositions. The details of catalyst preparation and experimental set-up are described in Refs. [48,49,51].

While the H$_2$O/CO ratio is crucial for the performance of LT WGS, it was particularly interesting to study the activity of catalysts at stoichiometric ratio and at H$_2$O/CO ratio of 3:1. Both are lower than those used in the commercial LT WGS processing of the gas exiting HT WGS. This was done deliberately for two reasons. The first is that there was no CO$_2$ present in the feed. Hence, the H$_2$O/CO ratio could be lower because there was no need to compensate the CO$_2$ influence on equilibrium with higher H$_2$O concentration (due to reverse WGS reaction). The second reason was the intention to study the behavior of LT WGS catalysts at relatively low inlet CO concentration (0.5 vol%) with respect to the usual inlet CO concentrations used in the industrial process (1.5 to 3 vol%). The feed composition used here was similar to that reported in Refs. [45,46], except that the CO concentration and the H$_2$O/CO ratio were lower.

Figure 7.5 shows CO conversion over G-66 A, Cu$_{0.2}$Ce$_{0.8}$O$_{2-y}$, and Cu$_{0.1}$Ce$_{0.9}$O$_{2-y}$ catalysts, with a feed mixture containing 1.8% CO, 1.8% H$_2$ (CO/H$_2$ = 1:1) diluted in 96.4% He, at a space velocity

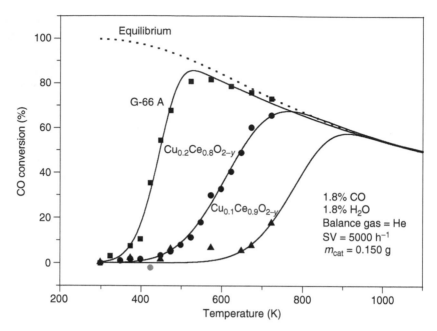

Figure 7.5 CO conversion for the WGS reaction over G-66 A, $Cu_{0.2}Ce_{0.8}O_{2-y}$, and $Cu_{0.1}Ce_{0.9}O_{2-y}$ catalysts. The solid lines are model fits assuming first-order reversible kinetics. The dotted line represents equilibrium conversion for the specific feed gas composition. (Reprinted from [51]. With permission from Elsevier.)

(SV) of 5000 h^{-1}. The G-66 A catalyst is, as shown in the figure, extremely active, but is, as mentioned earlier, pyrophoric and has to be carefully activated by reduction with H_2. Nevertheless, the G-66 A was tested over identical experimental conditions to compare its performance with those of the copper–ceria catalysts. The G-66 A is truly an LT WGS catalyst as can be seen in Figure 7.5. It has $T_{10} = 390$ K (the temperature at which 10% conversion is achieved), and the conversion rapidly increases up to 85% close to the equilibrium curve (dotted) at about 500 K. The $Cu_{0.2}Ce_{0.8}O_{2-y}$ catalyst prepared by coprecipitation method shows good activity. The fitted curve reaches the equilibrium curve at 760 K with a 77% CO conversion. The activity of the $Cu_{0.2}Ce_{0.8}O_{2-y}$ catalyst is in the mid-temperature (MT) WGS range. The nanostructured $Cu_{0.1}Ce_{0.9}O_{2-y}$ catalyst prepared by a sol–gel method is less active: at 673 K it has only reached 17% CO conversion, which confirms the importance of the preparation of the catalyst to form active surface sites. The nanostructured $Cu_{0.1}Ce_{0.9}O_{2-y}$ catalyst prepared by a sol–gel method that has shown extremely good conversion and selectivity [49] for the PrOX reaction is not the first choice as a WGS catalyst. In addition to the preparation technique, the difference in activity could also depend on the metal content. The G-66 A contains 42 wt% copper, the copper–ceria catalysts contain significantly less copper, the $Cu_{0.2}Ce_{0.8}O_{2-y}$ and the $Cu_{0.1}Ce_{0.9}O_{2-y}$ catalysts contain 7.5 and 3.9 wt% copper, respectively. Li et al. [37] have shown that there is no significant difference in the light-off temperature over the WGS reaction over different Cu–Ce(10% La)O$_x$ catalysts in which copper contents range between 5 and 40 at.%. All catalysts have probably enough active sites to sustain the WGS reaction. Our results are different, and they demonstrate that the catalysts prepared in different ways behave in distinctly different manners. The dotted line is the equilibrium conversion for the feed gas composition. The equilibrium CO conversions were calculated using the GASEQ software package [52] and calculated for this reactant composition using the following expression:

$$K_{eq} = \frac{[CO_2][H_2]}{[CO][H_2O]}$$

(7.11)

The solid lines in the figure are model fits of the experimental data. For fitting the experimental data, numerous research groups have proposed more or less complex models [45,47,53,54]. Here we apply a simple rate expression derived by Wheeler et al. [45], and approximating the WGS process as a single reversible surface reaction assuming an elementary reaction with first-order kinetics with respect to all species in the WGS reaction:

$$r'' = k_f'' P_{CO} P_{H_2O} - k_b'' P_{CO_2} P_{H_2} \tag{7.12}$$

The rate expression can be further simplified because in a real system the partial pressures of H_2O and H_2 are much higher than the partial pressures of CO and CO_2, which mean that the partial pressures of H_2O and H_2 are practically constant.

$$r = \frac{\text{area}}{\text{volume}} r'' = k_f P_{CO} - k_b P_{CO_2} \tag{7.13}$$

Here the pseudo-homogeneous rate r is related to the surface reaction rate r'' through the area of active catalyst per unit volume of reactor. Assuming further a plug-flow regime, the integration of the mass balance equation for this simple rate expression gives an expression for CO conversion:

$$X_{CO}(t) = \frac{k_f}{k_f + k_b} \left[1 - e^{-(k_f + k_b) \cdot t} \right] \tag{7.14}$$

where t is the residence time in the reactor, which varies with temperature as predicted by the ideal gas law. This expression was used to fit all experimental data, but instead of varying t, we measured the residence time and it was then set constant at all temperatures giving the following expression:

$$X_{CO}(T) = 100 \times \frac{k_f e^{(-E_{af}/RT)}}{k_f e^{(-E_{af}/RT)} + k_b e^{(-E_{ab}/RT)}} \left[1 - e^{-(k_f e^{(-E_{af}/RT)} + k_b e^{(-E_{ab}/RT)}) \cdot t} \right] \tag{7.15}$$

where $X_{CO}(T)$ is the CO conversion and T the temperature. The activation energy, E_{af}, was calculated by plotting $\ln(K_F)$ vs. $1/T$, which was a straight line. For fitting the experimental data the activation energy was set constant at all temperatures. The fitted values for the pre-exponential coefficients, k_f and k_b, and the backward activation energy, E_{ab}, were obtained by using the Levenberg–Marquardt algorithm. They are presented in Table 7.5 together with various literature values for copper–ceria catalysts. Even though Wheeler et al. [45] used a monolithic reactor system with very high flow rates and short contact times, the model fits of the experimental data obtained in this study show that this simplified model is also capable of fitting data derived from catalyst powder in a packed-bed reactor with lower reaction rates. The curves fit all the experimental points within the accuracy limits of the data.

7.3.2.1.1.1 *Influence of the Oxygen Storage Capacity*

The cerium oxide catalysts are known for their high oxygen storage capacity and it is clear that the cerium oxide has a direct role in the catalytic activity. However, the function of ceria and the metal component promoting the WGS reaction is not clear. Some research groups [37,42,43] have claimed that redox mechanism and oxygen storage capacity have a direct role in the WGS reaction. The other mechanism was proposed [44,55] to proceed through the formation of surface formate intermediates. Nevertheless, in Figure 7.6, the oxygen storage capacity of the copper–ceria catalyst can be observed. Copper–ceria catalysts have oxygen stored in the catalyst lattice, as described by the formula $Cu_{0.2}Ce_{0.8}O_{2-y}$, where the oxygen storage capacity is reported to be $y = 0.17$ [49]. The physisorbed oxygen reacts quickly with CO in the gas flow to form CO_2 as can be seen in the

Table 7.5　WGS Reaction Kinetics, Apparent Activation Energies, E_{af} (Forward), and Modeled Values for the Backward Activation Energy E_{ab} and Pre-Exponential Factors k_{of}, k_{ob}, Assuming an Elementary Reaction with First-Order Kinetics of the WGS Reaction

Catalyst	E_{af} (kJ/mol)	Conditions	E_{ab} (kJ/mol)	K_{of} (sec^{-1})	k_{ob} (sec^{-1})
$Cu_{0.2}Ce_{0.8}O_{2-y}$ [51]		473–623 K	61	1.8×10^3	1.1×10^4
	34	$CO/H_2O = 1/3$			
$Cu_{0.1}Ce_{0.9}O_{2-y}$ [51]		573–673 K	78	4×10^3	2.4×10^4
	51	$CO/H_2O = 1/3$			
8% $CuCeO_2$ [47]		513 K			
	56	$CO/H_2O = 1/3$			
5% Cu–Ce(10%		448–573 K			
La)O$_x$ [37]	30.4	$CO/H_2O = 1/7.5$			
	19.2	$CO/H_2O = 1/5.3$			
42% CuO–ZnO–		396–448 K	71	4.9×10^6	2.2×10^7
Al_2O_3 (G-66 A) [51]	47	$CO/H_2O = 1/3$			
40% CuO–ZnO–		463 K			
Al_2O_3 [47]	79	$CO/H_2O = 1/3$			

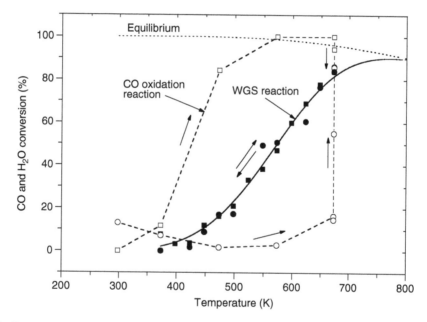

Figure 7.6　Oxygen storage capacity of the $Cu_{0.2}Ce_{0.8}O_{2-y}$ catalyst. Empty squares represent CO oxidation, and empty circles the conversion of H_2O in WGS reaction. Filled squares and filled circles are CO and H_2O conversions respectively, in WGS reaction. The dotted line represents the WGS reaction equilibrium curve. The solid line is model fit assuming first-order reversible kinetics. (Reprinted from [51]. With permission from Elsevier.)

Figure 7.6 (squares). When this takes place, no hydrogen is formed (white circles). But, as soon as all oxygen stored in the catalyst has reacted, the WGS reaction takes over. As can be observed in the figure, the WGS equilibrium line is crossed (white squares), which unfortunately makes it clear that this is CO oxidation and not WGS reaction. Hence the catalyst had to be pretreated (reduced) in the actual gas flow before starting to observe the WGS reaction. In the above experiment, it took more than 70 min to discharge all the oxygen in the catalyst to get stable results for the WGS reaction, which probably also means that oxygen from the crystalline bulk material was liberated by that time at these temperatures (up to 673 K) due to the reducing atmosphere. This was observed for both copper–ceria catalysts and always appeared when the catalyst was left in an oxygen-containing atmosphere. After the initial treatment (see Figure 7.6) the catalysts exhibit stable operation with slight or no deactivation (black squares [CO] and black circles [H_2O]).

7.3.2.1.1.2 Influence of Feed Gas Composition, H_2O/CO Ratio, and Presence of Hydrogen

In Figure 7.7, we observe the importance of utilizing the right feed gas composition over the $Cu_{0.2}Ce_{0.8}O_{2-y}$ catalyst prepared by the coprecipitation method. The dotted lines represent different equilibrium curves calculated for three different feed mixtures: 0.5% CO and 1.5% H_2O; 1.8% CO and 1.8% H_2O; and 50% H_2, 0.5% CO, 1.5% H_2O, all diluted with He. The first mixture represents a suitable feed gas composition for the WGS reaction in which high conversions can be accomplished (over 99% CO conversion at 550 K). It is in this equilibrium conversion region that the WGS reaction has to be carried out. The 1:1 CO/H_2O feed mixture was used to show how fast the WGS equilibrium decreases due to the lower water content in the feed gas composition. The feed mixture containing a large amount of hydrogen (50% H_2, 0.5% CO, and 1.5% H_2O) was also used to show the effect of excess hydrogen on the decrease of equilibrium CO conversion. This feed composition represents conditions that are close to those used in the industry with regard to H_2 content: 473 K, 30 bar, and a steam/dry gas ratio of 0.4 with a dry gas composition of 2% CO, 20% CO_2, and 78% H_2 [40]. It was also intended to examine the behavior of the catalysts in the presence of H_2 in the gas feed with respect to possible onset of methanation reactions on the catalyst surface. Yet the typical industrial gas composition has an equilibrium conversion even higher than the 0.5% CO and 1.5% H_2O gas mixture, since the higher concentrations of both reactants and higher H_2O/CO ratio leads to higher equilibrium conversion of CO.

7.3.2.1.1.3 Methanation Reactions

While carrying out the WGS reaction, methane can be formed in the reactor through the methanation reaction, which is the reverse methane SR reaction and is highly exothermal.

$$CO + 3H_2 \rightleftharpoons CH_4 + H_2O \quad \Delta H = -205.8\,kJ/mol \tag{7.16}$$

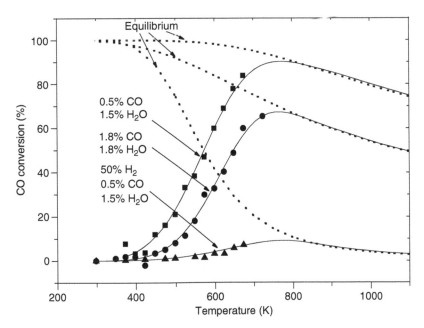

Figure 7.7 Influence of the increasing molar ratio of water and carbon monoxide, and of the addition of 50% H_2 to the feed gas mixture on the CO conversion in WGS reaction over $Cu_{0.2}Ce_{0.8}O_{2-y}$, catalyst at different feed compositions with SV = 5000 h^{-1}. The solid lines are model fits assuming first-order reversible kinetics. The dotted lines represent the equilibrium conversions for the specific feed compositions. (Reprinted from [51]. With permission from Elsevier.)

Methane is an undesired product in the WGS reaction, and for every CH_4 molecule formed, two H_2 molecules are taken away from the product stream, which leads to lower fuel processor efficiency. Figure 7.8 shows two WGS equilibrium curves for the hydrogen-containing conditions. The upper broken line is the WGS equilibrium curve including the methanation reaction, whereas the lower dotted line is the WGS equilibrium curve without methanation reactions. The figure shows that even though the methanation reaction is highly exothermal, this reaction does not occur on the catalyst. This confirms the selectivity of the copper-containing catalysts for the WGS reaction, otherwise, the experimental data would cross the pure WGS equilibrium line (lower dotted) and would eventually reach the (upper dotted) equilibrium line, which includes the methanation reaction. This can also be seen in Figure 7.9, showing equilibrium curves for the WGS reaction with or without methanation over the $Cu_{0.2}Ce_{0.8}O_{2-y}$, catalyst, including the feed gas composition of 0.5% CO and 1.5% H_2O. For this specific gas composition, the difference between the WGS equilibrium with and without methanation reactions is very small and in this case it would not be possible to draw the above conclusion about the selectivity of the catalysts. The reactor outlet gases were analyzed by a gas chromatograph and no methane formation was detected in any experiment.

7.3.2.1.1.4 Influence of Contact Time   As revealed in the Introduction, one important characteristic of the WGS catalyst is that it should be able to operate at high flow rates so as to reduce the total reactor size for the production of hydrogen. To examine the impact of the contact time, we increased the SV over the catalysts for the feed mixture of 0.5% CO and 1.5% H_2O in He from 5000 to 30,000 h^{-1}, as can be seen in Figure 7.10. All three different catalysts responded very similarly: the CO conversion decreases on increasing the SV, i.e., decreasing the contact time. To achieve the same CO conversion at the higher SV, ca. 50 to 100 K higher temperatures are required.

The copper–ceria catalysts in WGS were found to be nonpyrophoric and stable, showing little or no deactivation during the experiments. The $Cu_{0.2}Ce_{0.8}O_{2-y}$ catalyst prepared by coprecipitation method showed good catalytic activity for the WGS reaction. The $Cu_{0.1}Ce_{0.9}O_{2-y}$ catalysts prepared by the sol–gel method were found to be less active, which could be due to the lower number of active

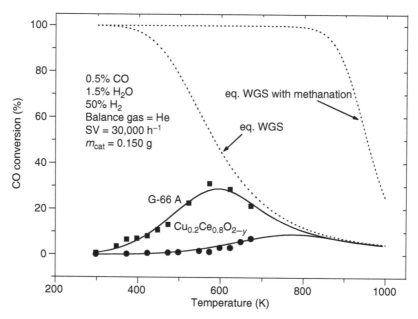

Figure 7.8   Equilibrium curves (dotted lines) for the WGS reaction with or without methanation reaction for feed gas composition with 50% H_2. The filled squares and circles are CO conversions over G-66 A and $Cu_{0.2}Ce_{0.8}O_{2-y}$, catalysts, respectively. The solid lines are model fits assuming first-order reversible kinetics. (Reprinted from [51]. With permission from Elsevier.)

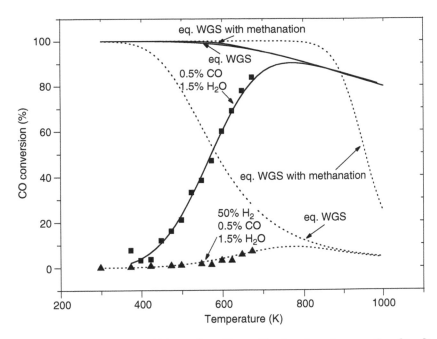

Figure 7.9 Equilibrium curves for the WGS reaction with or without methanation over $Cu_{0.2}Ce_{0.8}O_{2-y}$ catalyst. Filled squares represent feed gas composition of 0.5% CO, and 1.5% H_2O, and the solid line is the model fit assuming first-order reversible kinetics. The other two solid lines are the respective equilibrium curves. Filled triangles represent the feed composition of 50% H_2, 0.5% CO, and 1.5% H_2O and the dotted line is the model fit assuming first-order reversible kinetics. The two other dotted lines are the respective equilibrium curves. (Reprinted from [51]. With permission from Elsevier.)

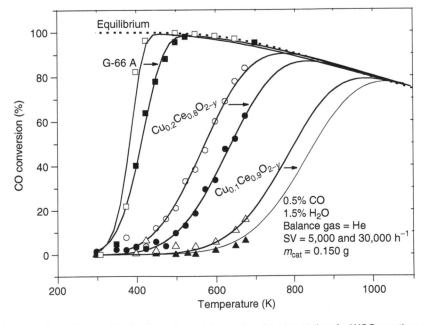

Figure 7.10 The effect of increasing the flow rate and decreasing the contact time for WGS reaction over G-66 A, $Cu_{0.2}Ce_{0.8}O_{2-y}$, and $Cu_{0.1}Ce_{0.9}O_{2-y}$ catalysts. Empty symbols illustrate low flow rate, SV = 5000 h^{-1} and filled symbols high flow rate, SV = 30.000 h^{-1}. The dotted line represents the equilibrium curve for a feed gas composition of 0.5% CO and 1.5% H_2 in He. The solid lines are model fits assuming first-order reversible kinetics. (Reprinted from [51]. With permission from Elsevier.)

copper sites, or due to different CuO crystallite size and structure. The copper–ceria catalysts were shown to be selective for the WGS reaction and no methanation reaction was observed over any catalyst under the experimental conditions used.

Model fits of the experimental data show that it is also possible to use simplified first-order elementary reaction kinetics for these catalysts to approximate the WGS reaction as a single reversible surface reaction. Furthermore, the fitted values for the pre-exponential coefficients and the activation energies have been evaluated and are not much different from other data available in the open literature.

Owing to low copper content, copper–ceria catalysts are nonpyrophoric and stable, showing little or no deactivation during the experiments. The $Cu_{0.2}Ce_{0.8}O_{2-y}$ catalyst prepared by coprecipitation method showed good catalytic activity for the WGS reaction. The $Cu_{0.1}Ce_{0.9}O_{2-y}$ catalyst prepared by sol–gel method was found to be less active, which could be due to lower number of active copper sites, or to different crystallite size and structure of copper-containing species. The copper–ceria catalysts were shown to be selective for the WGS reaction and no methanation reactions were observed over any catalyst under the experimental conditions used.

Model fits of the experimental data show that it is possible to use simplified first-order elementary reaction kinetics also for these catalysts to approximate the WGS reaction as a single reversible surface reaction. Furthermore, the fitted values for the pre-exponential coefficients and the activation energies have been evaluated.

7.3.2.1.2 Selective CO Oxidation in Excess of H_2 (PrOX) over the Nanostructured $Cu_xCe_{1-x}O_{2-y}$ Catalyst

To lower the cost and improve the selectivity of the catalyst, a novel nonstoichiometric nanostructured $Cu_xCe_{1-x}O_{2-y}$ catalyst for the selective low-temperature oxidation of CO in excess of H_2 was synthesized by coprecipitation and by sol–gel methods and patented [38,56]. The sol–gel method of catalyst preparation is particularly convenient for deposition on diverse geometries of support (i.e., honeycomb supports) or reactors, which can be used in PrOX processes. This type of catalyst is also capable of converting methanol directly into hydrogen and CO_2 by SR through the WGS reaction [57,58]. By using this catalyst, the three previously mentioned reactors (reformer, two-stage WGS reactor, and PrOX reactor) could be incorporated into a single unit. The capability of selective CO oxidation in an excess of hydrogen over this catalyst is demonstrated by using a fixed-bed reactor operated at both steady- and unsteady-state conditions. The inlet gaseous mixture composition simulates the real composition at the outlet of the LT WGS reactor with regard to the concentrations of CO, H_2, and O_2, except that no CO_2, H_2O, and unconverted CH_3OH were present.

Figure 7.11 shows the conversion of CO and O_2 as well as the selectivity obtained in the CO oxidation reaction over the nanostructured $Cu_{0.1}Ce_{0.9}O_{2-y}$ catalyst [49]. Regarding the selectivity of the catalyst, it is obvious that it stays at 100% at all temperature ranges in the case when H_2 is not present in the reactor feed. However, when the reactor feed contains H_2, the selectivity starts decreasing at temperatures higher than 90°C. If we examine the effect of hydrogen content on the selectivity, it can be observed that above 90°C, the selectivity is always less in the case where only oxygen, carbon monoxide, and hydrogen are present in the reactor feed, compared to the case when almost 50 vol% of He dilutes the hydrogen (full squares compared to full circles as well as empty squares compared to empty circles). It is normal, because in the previous case the H_2 partial pressure (potential to form water) is higher compared to the latter case.

In addition to the hydrogen partial pressure in the feed, the O_2/CO stoichiometric ratio also influences the selectivity of the catalyst. If excess oxygen is present in the reactor feed ($\lambda = 2.5$), more oxygen is available for the hydrogen oxidation reaction to form water as compared to the case when these two reactants are present in the stoichiometric ratio equal to 1 ($\lambda = 1$), which is clear if we compare full and open circles as well as full and open squares in Figure 7.11a.

The conversion of carbon monoxide depends on both hydrogen and oxygen partial pressures in the reactor feed gas. At temperatures of up to 90°C, where no side reaction of hydrogen oxidation

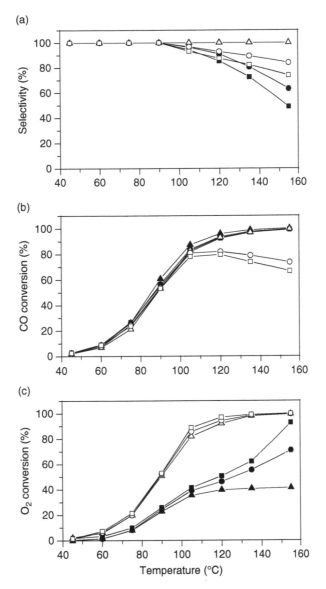

Figure 7.11 Selectivity (a) and conversions of CO (b) and O_2 (c) obtained over $Cu_{0.1}Ce_{0.9}O_{2-y}$ catalyst as a function temperature, λ value ($\lambda=2P_{O_2}/P_{CO}$) and presence of H_2 in the reactor feed. ($\blacksquare$, $\square$) CO, O_2, and H_2 in the reactor feed, no He; ($\bullet$, $\circ$) CO, O_2, 50% H_2, He balance gas; ($\blacktriangle$, $\triangle$) CO, O_2, and He in the reactor feed, no H_2; full catalyst; ($\blacksquare$) $\lambda = 2.5$, $P_{H_2}= 0.5$ bar, He balance; ($\bullet$) $\lambda = 1$, $P_{H_2}= 0.5$ bar, He balance; ($\blacktriangle$, $\triangle$) $\lambda = 1$, $P_{H_2}= 0.985$ bar, no He. Full symbols denote increase in temperature ($\lambda = 2.5$), while empty symbols denote lowering of temperature ($\lambda = 1$). In all the experiments, $P_{CO} = 0.01$ bar, total pressure = 1 bar, $m_{cat} = 100$ mg, and $\Phi_V = 100$ mL/min. (Reprinted from [49]. With permission from Elsevier.)

occurs, the CO conversion is independent of hydrogen partial pressure, while there is a very weak dependence on oxygen partial pressure (empty symbols are slightly lower than full ones in Figure 7.11b) in that temperature region. If the temperature is raised above 90°C, water is also formed. In that case, the CO conversion becomes much more dependent on oxygen and hydrogen partial pressures. This dependence is interconnected and subordinate to water formation reaction. If the stoichiometric ratio of oxygen ($\lambda = 1$) is fed to the reactor in the presence of hydrogen, the CO conversion reaches its maximum value of around 80% at a temperature of 105° C. At higher temperatures, the CO conversion curve lowers again. Because more and more water is formed with increasing temperature, less and less oxygen remains available for the CO oxidation reaction. There is only a small difference in the CO conversion if the hydrogen partial pressure is 0.5 or 0.985 bar, as depicted by open circles and squares, respectively, in Figure 7.11b. If there is no hydrogen present in the reactor feed, CO conversion reaches 100%.

When oxygen is present in excess in the reactor feed ($\lambda = 2.5$), the CO conversion curve does not decrease from 100% even at a reaction temperature of 155°C. At that temperature, there is still around 10 or 30% (full squares and full circles, respectively, in Figure 7.11c) oxygen left in the system. That is enough to attain 100% CO conversion plus the formation of the corresponding amount of water. It seems that the conversion of carbon monoxide is literally independent of the hydrogen partial pressure over all the temperature range as long as there is enough oxygen fed to the reactor. Only the amount of water formation is dependent on hydrogen partial pressure, as seen from closed symbols in Figure 7.11a and Figure 7.11c, respectively, which is a side reaction.

In the case of $\lambda = 2.5$, it is interesting to follow O_2 conversion curves. O_2 conversion is independent of H_2 partial pressure up to the temperature of 90°C as depicted in Figure 7.11c (full symbols). In the case when there is no H_2 present in the system, O_2 conversion curve is very similar to the CO conversion curve (full triangles). It reaches 40% at high temperatures, as predicted from stoichiometry, i.e., $100\%/\lambda$. The O_2 conversion curve has a *single-S shape*. In case hydrogen is present in the reactor feed, it takes off again above 40% because of water formation reaction at the reaction temperature of 105°C and starts to approach 100%. It has a characteristic *double-S shape*. In case of $\lambda = 1$, and when hydrogen is present in the reactor feed the oxygen conversion reaches 100% faster, because in that case it is consumed both for the CO as well as for the H_2 oxidation reaction.

By carrying out selective CO oxidation with some addition of CO_2 and H_2O to the feed gas over a similar nanostructured $Cu_xCe_{1-x}O_{2-y}$ catalyst prepared by a coprecipitation method [56], 15 vol% of CO_2 in the feed gas decreases the activity of the catalyst. Under these conditions the same values of activity and selectivity were obtained at temperatures 15 to 35°C higher. The addition of 10 vol% of H_2O shifted the activity and selectivity curves to temperatures 20 to 40°C higher with respect to the curves where only CO, O_2, H_2, and He were used in the feed.

The comparison of catalytic properties was made under identical reaction conditions, among three important candidate catalysts, namely, the Pt/γ-Al_2O_3, Au/α-Fe_2O_3, and $Cu_xCe_{1-x}O_{2-y}$ systems [50]. The catalytic tests were performed in the reactant feed containing CO, H_2, CO_2, and H_2O — the so-called reformate fuel. The effects of the presence of both CO_2 and H_2O in the reactant feed on the catalytic performance (activity and selectivity) of these catalysts as well as their stability with time under reaction conditions have been studied. The composition of the prepared samples and their BET specific surface areas are presented in Table 7.6. The results obtained with the three catalysts in the presence of 15 vol% CO_2 and of both 15 vol% CO_2 and 10 vol% H_2O in the reactant feed (with contact time m_{cat}/Φ_v = 0.144 g sec/cm³ and $\lambda = 2.5$) are shown in Figure 7.12. For comparison, the corresponding curves obtained under the same conditions but without water vapor in the feed are also shown in Figure 7.12.

The presence of H_2O provokes a significant decrease in the activity of Au/α-Fe_2O_3 and $Cu_xCe_{1-x}O_{2-y}$ catalysts. In fact, for both of these samples, a given CO conversion (obtained in the absence of H_2O) is achieved at about 40 to 45°C higher reaction temperature in the presence of H_2O in the feed. A slightly lower inhibition of the activity is observed for the O_2 conversion in both

Table 7.6   Composition and BET Specific Surface Area of the Catalysts Used in Comparative Study of the Selective Oxidation of CO in Synthetic Reformate Fuel

Catalyst	Composition	S_{BET} (m²/g)
Au/α-Fe_2O_3	2.9 wt% Au	49.8
$Cu_xCe_{1-x}O_{2-y}$	1.9 wt% Cu	19.5
Pt/γ-Al_2O_3	5.0 wt% Pt	224.0

Source: Reprinted from [50]. With permission from Elsevier.

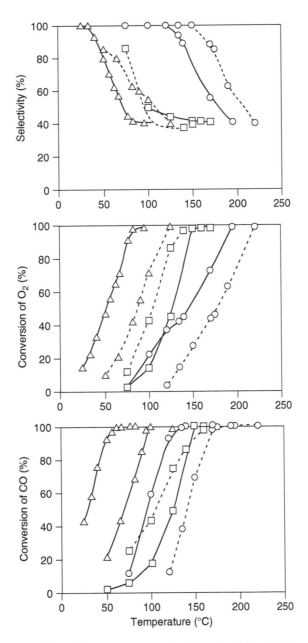

Figure 7.12 Variation of the CO and O_2 conversion and of the selectivity with the reaction temperature for the selective oxidation of CO at $m_{cat}/\Phi_v = 0.144$ g sec/cm³ and $\lambda = 2.5$ over the Au/α-Fe$_2$O$_3$ (△), CuO–CeO$_2$ (○), and Pt/γ-Al$_2$O$_3$ (□) catalysts in the presence of 15 vol% CO$_2$ (solid lines) and in the presence of both 15 vol% CO$_2$ and 10 vol% H$_2$O in the reactant feed (dotted lines). (Reprinted from [50]. With permission from Elsevier.)

samples. In addition to this negative effect on the activity, the presence of H$_2$O also diminishes the selectivity exhibited by these two catalysts. Indeed, a given CO conversion achieved in the absence of H$_2$O is less selectively obtained in its presence. However, this effect is much more pronounced in the case of the Au/α-Fe$_2$O$_3$ sample. For example, in the CO conversion range of 60 to 95%, the selectivity achieved with the Au/α-Fe$_2$O$_3$ sample decreased from ca. 100 to 74% in the absence of H$_2$O and from 76 to 56% in the presence of H$_2$O. For the same CO conversion range (60 to 95%), the selectivity achieved with the $Cu_xCe_{1-x}O_{2-y}$ sample was practically constant at 100% (dropping to 98% only when the CO conversion was 95%) in the absence of H_2O, while in its presence it was gradually decreasing from 100 to 89%.

As compared to the $Au/\alpha\text{-}Fe_2O_3$ and $Cu_xCe_{1-x}O_{2-y}$ catalysts, the effect of H_2O on the $Pt/\gamma\text{-}Al_2O_3$ catalyst was markedly different. For reaction temperatures lower than 145°C (at that temperature range the CO conversion is lower than 90%), the $Pt/\gamma\text{-}Al_2O_3$ sample is more active (giving both higher CO and higher O_2 conversions) when H_2O is present in the reactant feed. It turns out that when both CO_2 and H_2O are present in the feed then this catalyst is more active than when both of these compounds are absent from the feed. However, this beneficial effect does not manifest itself for higher reaction temperatures (when the CO conversion becomes higher than 90%). Indeed, for temperatures higher than 145°C, the CO conversion achieved at a given temperature is significantly lower in the presence of H_2O than in its absence. For example, at 150°C, the CO conversion obtained over the $Pt/\gamma\text{-}Al_2O_3$ sample was 100% (with 41.5% selectivity) in the absence of H_2O but only 92.8% (with 39.8% selectivity) in its presence. The beneficial effect of H_2O that has been observed (namely, the increase in CO conversion at low reaction temperatures) cannot possibly be attributed to a part of the CO being consumed by the WGS reaction, since at this low reaction temperature range (<145°C) the extent of this reaction should be negligible, if any at all [57]. Thus, one may conclude that the observed increase in the CO conversion during selective CO oxidation in the presence of H_2O is due to an enhancement provoked by the presence of H_2O on the CO oxidation rate. This conclusion is corroborated by previous investigations on the CO oxidation over alumina-supported platinum catalysts, which showed that the presence of H_2O in the CO/O_2 reactant feed enhances the rate of CO oxidation [59–61]. Moreover, this enhancement of the CO oxidation rate was reported to decrease as the reaction temperature increased [60].

The influence of H_2O on the selectivity of the $Pt/\gamma\text{-}Al_2O_3$ sample is similar to that for the $Au/\alpha\text{-}Fe_2O_3$ and $Cu_xCe_{1-x}O_{2-y}$ catalysts, namely, a given CO conversion achieved over the $Pt/\gamma\text{-}Al_2O_3$ catalyst in the absence of H_2O is less selectively obtained in its presence, and this was observed for the whole reaction temperature region studied. However, owing to the fact that the selectivity of $Pt/\gamma\text{-}Al_2O_3$ was already low without water in the feed, the decrease of the selectivity provoked by its presence is not as pronounced as it was for the other two catalysts. For example, in the CO conversion range of 60 to 95%, the selectivity achieved with the $Pt/\gamma\text{-}Al_2O_3$ sample was gradually decreasing from 44 to 42% in the absence of H_2O, while in its presence it varied in the region 36 to 39%.

Under these reaction conditions, the $Au/\alpha\text{-}Fe_2O_3$ sample was again superior to the other two at the low reaction temperature range, exhibiting its best catalytic performance (99.5% CO conversion with 54.5% selectivity) at 100°C. The $Pt/\gamma\text{-}Al_2O_3$ sample gave its best results at 160°C, with 97.5% CO conversion and 41.5% selectivity. At practically the same reaction temperature (170°C), the $Cu_xCe_{1-x}O_{2-y}$ sample gave the same CO conversion but with the remarkably higher selectivity of 88%. The $Cu_xCe_{1-x}O_{2-y}$ sample exhibited its best catalytic performance (99.6% CO conversion with 62.3% selectivity) at 190°C. Thus, the $Cu_xCe_{1-x}O_{2-y}$ sample, being remarkably more selective, outperforms the $Pt/\gamma\text{-}Al_2O_3$ at the high reaction temperature range, also in the presence of H_2O in the reactant feed.

The reaction kinetics for the system containing only CO, H_2, and O_2 in the gas feed could be best represented by the redox mechanism [49]. Such a redox reaction can be described by the following two-step reaction:

$$Cat\text{--}O + Red \rightarrow Cat + Red\text{--}O \tag{7.17}$$

$$Cat + Ox\text{--}O \rightarrow Cat\text{--}O + Ox \tag{7.18}$$

The first step in this reaction mechanism is the catalyst reduction. Cat–O represents an oxidized catalyst, which is attacked by a reductant (Red). The catalyst itself undergoes reduction, while the reductant is oxidized. The second step represents reoxidation of the catalyst by the oxidant (Ox–O), which donates an oxygen atom to the catalyst while it reduces itself.

The kinetics of selective CO oxidation over the $Cu_xCe_{1-x}O_{2-y}$ nanostructured catalysts can be well described by employing Mars and van Krevelen type of kinetic equation derived on the basis of a redox mechanism:

$$r_{CO} = \frac{k_{CO}k_{O_2}P_{CO}P_{O_2}^n}{0.5k_{CO}P_{CO} + k_{O_2}P_{O_2}^n} \tag{7.19}$$

$$k_{CO} = A_{CO}\exp(-E_{a,CO}/RT) \tag{7.20}$$

$$k_{O_2} = A_{O_2}\exp(-E_{a,O_2}/RT) \tag{7.21}$$

The parameters k_{CO} and k_{O_2} are taken to be the reaction rate constants for the reduction of surface by CO and reoxidation of it by O_2. The parameters k_{CO}, k_{O_2}, and n at one temperature were obtained by fitting the experimental values of P_{CO}, P_{O_2}, and reaction rate with the above rate equation. The parity plot for calculated vs. experimental values of reaction rate is presented in Figure 7.13. The agreement between experimental and calculated values is very good over three orders of magnitude of reaction rate.

Figure 7.13 represents the calculated vs. experimental values of reaction rates for the Mars and van Krevelen model of the selective CO oxidation in excess of hydrogen over the catalyst used. From the figure one can see that most scatter of data represents the use of eight different catalyst samples; the data obtained over one catalyst sample lie almost on a straight line, within 95% confidence limits.

Unsteady-state oxidation experiments were carried out by employing the step change in CO concentration over the preoxidized catalyst [62]. Figure 7.14 represents the CO and CO_2 responses after a step change from He to 1 vol% CO/He over the fully oxidized $Cu_{0.1}Ce_{0.9}O_{2-y}$ nanostructured catalyst. At low temperatures, CO breakthrough is delayed for a few seconds as can be seen from Figure 7.14a. At a temperature of 250°C, however, 20 sec is needed for the first traces of CO exit

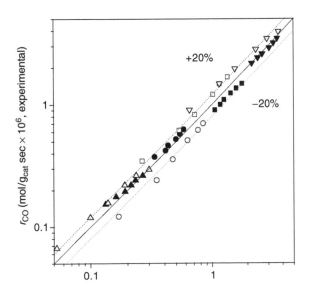

Figure 7.13 Calculated vs. experimental values of reaction rates for selective CO oxidation in excess hydrogen. (Reprinted from [49]. With permission from Elsevier.)

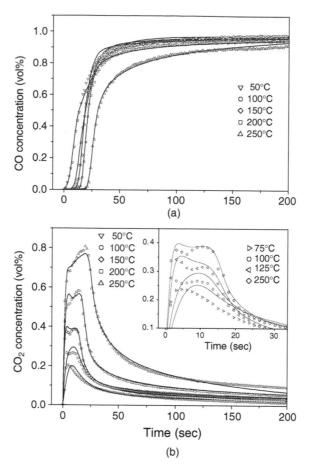

Figure 7.14 Concentration of (a) CO and (b) CO_2 in the reactor effluent stream as a function of temperature. Conditions: $m_{cat} = 200$ mg; $\Phi_v = 200$ mL/min. (Reprinted from [62]. With permission from Elsevier.)

the reactor. On the other hand, the evolution of CO_2 in the reactor effluent stream has no delay as represented in Figure 7.14b. However, the nature of the CO_2 peak as a function of temperature changes significantly. At temperatures lower than 100°C, only one peak is visible for CO_2. At 100°C the CO_2 peak broadens and at 125°C, two separate peaks are clearly visible. The first peak is narrow, followed by a second broader peak. When the temperature is increased, the first peak for CO_2 becomes invisible, because it is covered by the second peak. Only the origin of the first peak is signified by the fast evolution of CO_2 in the reactor effluent stream. It is also important to notice how the maximum of the second peak shifts to the right when the temperature is increased. At 250°C, the catalyst surface responds almost instantaneously to a CO step change by producing CO_2. The concentration of CO_2 in the reactor effluent gas after 3 sec is 0.65 vol% as shown in Figure 7.6b. However, the CO_2 concentration in the reactor effluent gas rises further and reaches 0.80 vol% after 25 sec. This is followed by a sharp decrease in the CO_2 concentration, which stabilizes after 100 sec at 0.2 vol%. Following this, the concentration in CO_2 decreases very slowly and falls to zero after 13 min.

Based on the experimental data and some speculations on detailed elementary steps taking place over the catalyst, one can propose the dynamic model. The model discriminates between adsorption of carbon monoxide on catalyst inert sites as well as on oxidized and reduced catalyst active sites. Apart from that, the diffusion of the subsurface species in the catalyst and the reoxidation of reduced catalyst sites by subsurface lattice oxygen species is considered in the model. The model allows us to calculate activation energies of all elementary steps considered, as well as the bulk diffusion coefficient of oxygen species in the $Cu_{0.1}Ce_{0.9}O_{2-y}$ nanostructured catalyst. The diffusion coefficient obtained by the mathematical modeling of step experiments is shown to be in the range of bulk diffusion coefficients measured over other oxide catalysts. The elementary reaction steps, the mass balance equations, the initial and boundary conditions, and the estimated kinetic parameters are given in Table 7.7 through Table 7.10.

It is generally accepted that CuO and CeO_2 have great synergistic effects when prepared as a composite CuO/CeO_2 catalyst. The reason for that is probably the formation of a solid solution of CuO and CeO_2 phases [63]. The amount of solid solution between those two phases, however, is small [48]. Most probably, the substitutional solid solution forms tiny intergrowths at the interface between XRD amorphous CuO and nanocrystalline CeO_2 phases [49,64]. The substitutional solid solution is most probably sandwiched between the dispersed CuO clusters and CeO_2 crystallites [48], in the interfacial region. This interface might have a thickness of only a few atomic layers. When the Cu^{2+} species reduces to the Cu^{1+} state, it is subsequently reoxidized by reduction of the Ce^{4+} ions in their vicinity into Ce^{3+}, and the following redox equilibrium is established [64]:

$$Ce^{4+} + Cu^{1+} \rightleftharpoons Ce^{3+} + Cu^{2+} \tag{7.22}$$

This equilibrium has a buffer-like effect stabilizing the presence of cationic copper species in the structure even in a highly reductive atmosphere. The above scheme of copper oxide–ceria interactions indicates clearly that the catalyst is mutually promoted, i.e., both copper and ceria cooperate in the redox mechanism.

In our studies we have demonstrated that the redox mechanism that was used to model dynamic behavior of CO oxidation is consistent with a kinetic model of the selective CO oxidation obtained under steady-state mode of operation [62]. We propose the following tentative scheme (Figure 7.15) for the selective CO oxidation over the $Cu_{0.1}Ce_{0.9}O_{2-y}$ catalyst: CO and H_2 adsorb on the

Table 7.7   Elementary Reaction Steps Considered in the Kinetic Modeling of the CO Concentration Step Change Experiments for the Oxidation of CO Over Completely Oxidized $Cu_{0.1}Ce_{0.9}O_{2-y}$ Nanostructured Catalyst in the Absence of Oxygen in the Reactor Feed

Step Number	Elementary Reaction Step
1	$CO + Cu^{2+}O_{s,s} \xrightarrow{k_1} CO \cdots Cu^{2+}O_{s,s}$
2	$CO \cdots Cu^{2+}O_{s,s} \xrightarrow{k_2} CO_2 + Cu^{+}\square_s$
3	$CO + Cu^{+}\square_s \xrightarrow{k_3} CO \cdots Cu^{+}\square_s$
4	$Ce^{4+}O_{b,b} \xrightarrow{D} Ce^{3+}\square_b + O_{s,b}$
5	$CO \cdots Cu^{+}\square_s + O_{s,b} \xrightarrow{k_4} CO \cdots Cu^{2+}O_{s,b}$
6	$CO \cdots Cu^{2+}O_{s,b} \xrightarrow{k_2} CO_2 + Cu^{+}\square_s$
7	$CO + * \xrightarrow{k_5} CO*$
	$2CO + Cu^{2+}O_{s,s} + Ce^{4+}O_{b,b} \rightarrow 2CO_2 + Cu^{+}\square_s + Ce^{3+}\square$

Note: Oxygen vacancy is represented by $\square$. The meaning of subscripts accompanying oxygen species and oxygen vacancies is explained in the text.

Source: Reprinted from [62]. With permission from Elsevier.

Table 7.8   Mass Balance Equations for Gas-Phase, Surface, and Subsurface Species Corresponding to Elementary Reaction Steps Given in Table 7.7.

$$\frac{\partial y_{CO}}{\partial t} + \frac{1}{\tau}\frac{\partial y_{CO}}{\partial z} = \frac{\rho_B}{\varepsilon_B}\left(-k_1 L_{TOT} y_{CO}\theta_{Cu^{2+}O_{s,s}} - k_3 L_{TOT} y_{CO}\theta_{Cu^+\square_s} - k_5 F_{TOT} y_{CO}\delta_*\right)$$

$$\frac{\partial y_{CO}}{\partial t} + \frac{1}{\tau}\frac{\partial y_{CO_2}}{\partial z} = \frac{\rho_B}{\varepsilon_B}k_2\left(L_{TOT}/C_{TOT}\right)\left(\theta_{CO\cdots Cu^{2+}O_{s,s}} + \theta_{CO\cdots Cu^{2+}O_{s,b}}\right)$$

$$\frac{\partial\theta_{Cu^{2+}O_{s,s}}}{\partial t} = -k_1 C_{TOT} y_{CO}\theta_{Cu^{2+}O_{s,s}}$$

$$\frac{\partial\theta_{CO\cdots Cu^{2+}O_{s,s}}}{\partial t} = k_1 C_{TOT} y_{CO}\theta_{Cu^{2+}O_{s,s}} - k_2\theta_{CO\cdots Cu^{2+}O_{s,s}}$$

$$\frac{\partial\theta_{Cu^+\square_s}}{\partial t} = k_2\left(\theta_{CO\cdots Cu^{2+}O_{s,s}} + \theta_{CO\cdots Cu^{2+}O_{s,b}}\right) - k_3 C_{TOT} y_{CO}\theta_{Cu^+\square_s}$$

$$\frac{\partial\theta_{CO\cdots Cu^+\square_s}}{\partial t} = k_3 C_{TOT} y_{CO}\theta_{Cu^+\square_s} - k_4 H_{TOT}\theta_{CO\cdots Cu^+\square_s}\xi_{O_{s,b}}$$

$$\frac{\partial\theta_{CO\cdots Cu^{2+}O_{s,b}}}{\partial t} = k_4 H_{TOT}\theta_{CO\cdots Cu^+\square_s}\xi_{O_{s,b}} - k_2\theta_{CO\cdots Cu^{2+}O_{s,b}}$$

$$\frac{\partial\delta_*}{\partial t} = -k_5 C_{TOT} y_{CO}\delta_*$$

$$\frac{\partial\xi_{O_{b,b}}}{\partial t} = \frac{1}{t_d}\left(\frac{\partial^2\xi_{O_{b,b}}}{\partial x^2} + \frac{1}{x}\frac{\partial\xi_{O_{b,b}}}{\partial x}\right)$$

Note: $\square$ indicates oxygen vacancy.

Source: Reprinted from [62]. With permission from Elsevier.

copper–ceria interfacial region of the catalyst, the most reactive places for both CO and H_2 oxidation reactions. It is further proposed that CO (and H_2) uses mostly copper cations as the adsorption sites, while cerium oxide must also be present in the close vicinity. Copper oxide might also form a solid solution with cerium oxide at least in the form of small intergrowths at the interface, which are XRD-invisible. In this concerted mechanism of copper and cerium oxide, the copper cation has the following role: it is the adsorption site for the CO (and H_2). When either of the two reactants is adsorbed on the copper cation, it extracts oxygen from the surface and copper is reduced from Cu^{2+} to Cu^{1+}. The cerium cation, which lies next to the copper cation, can supply additional oxygen atom from the catalyst lattice while it reduces itself simultaneously from the Ce^{4+} into the Ce^{3+} form. Cerium oxide acts as an oxygen supplier (buffer) when it is needed at the place of reaction. A single copper ion is enough to convert one molecule of CO (or H_2) into CO_2 (or H_2O). When the product molecule is desorbed, the site becomes available for the next reactant molecule, either CO or H_2. Upon extraction of surface oxygen from the catalyst lattice, oxygen vacancy may be refilled directly from the gas phase or by oxygen diffusion through the bulk of the catalyst. The latter mechanism is observed at higher temperatures.

The $Cu_xCe_{1-x}O_{2-y}$ nanostructured catalyst prepared by the sol–gel method is a very efficient and selective CO oxidation catalyst even under the highly reducing conditions which are present in a PrOX reactor. It is energy efficient toward the PEM fuel cell technology, because it oxidizes CO

Table 7.9   Initial and Boundary Conditions Corresponding to Mass Balance Equations Given in Table 7.8

$$y_{CO}\,(z,0)=0, \qquad y_{CO}\,(0,t)=0.01$$

$$y_{CO_2}\,(z,0)=0, \qquad y_{CO_2}\,(0,t)=0$$

$$\theta_{Cu^{2+}O_{s,s}}\,(z,0)=1$$

$$\theta_{CO\cdots Cu^{2+}O_{s,s}}\,(z,0)=0$$

$$\theta_{Cu^{+}\square_{s}}\,(z,0)=0$$

$$\theta_{CO\cdots Cu^{+}\square_{s}}\,(z,0)=0$$

$$\theta_{CO\cdots Cu^{2+}O_{s,b}}\,(z,0)=0$$

$$\delta_*\,(z,0)=1$$

$$\xi_{O_{b,b}}\,(z,x,0)=1$$

$$\frac{\partial \xi_{O_{b,b}}}{\partial x}(z,0,t)=0$$

$$a\,\frac{1}{t_d}\frac{\partial \xi_{O_{b,b}}}{\partial x}(z,1,t)=-k_3 L_{TOT}\theta_{CO\cdots Cu^{+}O_{b,b}}\,\xi_{O_{b,b}}\,(z,1,t)$$

Note: $\square$ indicates oxygen vacancy.

Source: Reprinted from [62]. With permission from Elsevier.

Table 7.10   Estimates of the Kinetic Parameters Obtained by Regression of He → 1 vol% CO/He Concentration Step Experiments for the CO Oxidation Over Fully Oxidized Catalyst in the Absence of Oxygen in the Reactor Feed Gas

Ln (A_1) [m³/mol sec]	14.6 ± 1.2
E_1 [kJ/mol]	39.6 ± 4.6
Ln (A_2) [sec]	1.85 ± 2.1
E_2 [kJ/mol]	9.7 ± 8.1
Ln (A_3) [m³/mol sec]	10.0 ± 1.6
E_3 [kJ/mol]	25.8 ± 5.9
Ln (A_4) [kg$_{cat}$/mol sec]	19.9 ± 5.2
E_4 [kJ/mol]	72.9 ± 19.8
Ln (A_5) [m³/mol sec]	4.07 ± 0.63
E_5 [kJ/mol]	13.9 ± 2.3
Ln (A_{td}) [sec]	2.29 ± 2.74
E_{td} [kJ/mol]	40.0 ± 10.3

Note: The parameters are obtained in the temperature range 125 to 250°C. The mass balance equations are given in Table 7.8, while the corresponding initial and boundary conditions are given in Table 7.9.

Source: Reprinted from [62]. With permission from Elsevier.

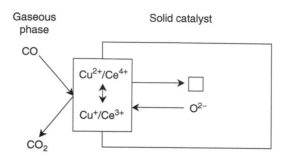

Figure 7.15 Scheme of the Mars–van Krevelen mechanism for selective CO oxidation in excess hydrogen over $Cu_{0.1}Ce_{0.9}O_{2-y}$ nanostructured catalyst. (Reprinted from [62]. With permission from Elsevier.)

with 100% selectivity close to working temperature of the PEM fuel cell. These performances are obtained with catalyst that contains cheap copper and cerium oxides rather than costly noble metals.

7.4 CONCLUSIONS

Despite the very promising and interesting experimental results with regard to catalytic activity of $Cu_xCe_{1-x}O_{2-y}$ catalysts in several important catalytic processes such as SR of methanol, WGS, PrOX, NO reduction, wet oxidation of phenol, etc., only one theoretical study [65] has appeared until now. This is probably the consequence of the difficulties in the modeling and calculation of the 4f elements. The only study mentioned above presents theoretical proof for the Cu–Ce interaction in $Cu_xCe_{1-x}O_{2-y}$. The DFT calculations and the quantum chemical MD simulation on a somewhat peculiar catalyst cluster geometry have indicated possible overlap between Cu 3d, Ce 4f, and O 2p orbitals. This coupling of atomic orbitals between Cu and Ce may make the valence change between Cu^{2+} and Cu^{1+} easier, and hence the catalyst can keep high activity in the redox-based reaction mechanisms.

Further theoretical studies supported by *in situ* spectroscopy and high-resolution microscopy are needed to be able to understand this unusually strong bonding between Cu and Ce. To apply such first-principles quantum chemical MD approach, new computational methods accelerating computational time by several orders of magnitude must be developed.

On the other side, new materials, including nanostructured catalysts like $Cu_xCe_{1-x}O_{2-y}$ open new challenges for reactor engineering. These types of multifunctional catalysts enable engineers to design compact reactor systems such as fuel processors for hydrogen production, which can be integrated dimensionally and functionally with new energy conversion devices like fuel cells. In this respect, a strong accent on the theory of catalytic membrane reactors is expected as well as accent on the computational fluid dynamics methods development, which will be able to treat non-Darcy transport processes in porous media.

ACKNOWLEDGMENTS

My sincere thanks go to Professor Janez Levec, Head of the Laboratory for Catalysis and Chemical Reaction Engineering at the National Institute of Chemistry; to colleagues at this laboratory—Dr. Jurka Batista, Dr. Gorazd Berčič, and Dr. Albin Pintar; to postdoctoral research fellow Dr. Henrik Kušar; and to Dr. Gregor Sedmak, ex Ph.D. student, who all contributed significantly to the research results and fruitful discussions. The work has been performed in part with financial support from the Slovenian Research Agency under Grant No. P2-0152, which I gratefully acknowledge.

REFERENCES

1. G.K. Boreskov, *Catalysis*, Part 1 and 2 (2nd ed., Russian), Nauka (Siberian Branch), Novosibirsk, 1971.
2. J.M. Thomas, W.J. Thomas, *Introduction to the Principles of Heterogeneous Catalysis*, Academic Press, London, 1967.
3. O. Levenspiel, *Chemical Reaction Engineering*, 3rd ed., Wiley, New York, 1999.
4. J.J. Carberry, *Chemical and Catalytic Reaction Engineering*, McGraw-Hill, New York, 1976.
5. J.J. Berzelius, *Fort. Physik. Wissenshaft Tübingen*, **243** (1836).
6. S. Arrhenius, *Z. Physik. Chemie*, **4** (1889) 226.
7. M. Bodenstein, S.C. Lund, *Z. Phys. Chem.*, **57** (1907) 108.
8. W.H. Ostwald, *Grundriss der Algemeinen Chemie*, Verlag von Wilhelm Engelmann, Leipzig, 1902.
9. J.H. van't Hoff, *Ansichten Über die Organische Chemie*, Braunsweig, 1878.
10. H. Eyring, M. Polanyi, *Z. Physik Chem. B*, **12** (1931) 279.
11. S. Glasstone, K.J. Laidler, H. Eyring, *The Theory of Rate Processes*, McGraw-Hill, New York, 1941.
12. R.A. van Santen, *Theoretical Heterogeneous Catalysis*, Lecture and Course Notes in Chemistry, Vol. 5, World Scientific, Singapore, 1991.
13. M. Baudin, M. Wojcik, K. Hermansson, A.E.C. Palmqvist, M. Muhammed, *Chem. Phys. Lett.*, **335** (2001) 517.
14. P. Sabatier, *La catalyse en chimie organique*, B´erange, Paris, 1920.
15. A.A. Balandin, *Adv. Catal.*, **19** (1969) 1.
16. 1. N. Brønsted, *Chem. Rev.*, **5** (1928) 231.
17. M.G. Evans, N.P. Polanyi, *Trans. Faraday Soc.*, **34** (1938) 11.
18. V. Pallassana, M. Neurock, *J. Catal.*, **191** (2000) 301.
19. Z.-P. Liu, P. Hu, *J. Chem. Phys.*, **114** (2001) 8244.
20. A. Logadottir, T.H. Rod, J.K. Nørskov, B. Hammer, S. Dahl, C.J.H. Jacobsen, *J. Catal.*, **197** (2001) 229.
21. J.K. Nørskov, T. Bligaard, A. Logadottir, S. Bahn, L.B. Hansen, M. Bollinger, H. Bengaard, B. Hammer, Z. Sljivancanin, M. Mavrikakis, Y. Xu, S. Dahl, C.J.H. Jacobsen, *J. Catal.*, **209** (2001) 229.
22. R.G. Parr, W. Yang, *Density-Functional Theory of Atoms and Molecules*, International Series of Monographs on Chemistry, Vol. 16, Oxford University Press, Oxford, 1989.
23. C.J.H. Jacobsen, S. Dahl, A. Boisen, B.S. Clausen, H. Topsøe, A. Logadottir, J.K. Nørskov, *J. Catal.*, **205** (2002) 382.
24. H. Heinemann, *Catalysis – Science and Technology*, Vol. 1, Springer, Berlin, 1982, pp. 1–41.
25. J.A. Moulijn, P.W.N.M. van Leeuwen, R.A. van Santen, Eds., An integrated approach to homogenous, hetrogenous and industrial catalyses studies, in *Surface Science and Catalysis*, Vol. 79, Elsevier, Amsterdam, 1993.
26. J.B. Butt, *Reaction Kinetics and Reactor Design*, 2nd ed., Marcel Dekker, New York, 2000.
27. R. Westerholm, L.J. Pettersson, *State of the Art Multi-Fuel Reformers for Automotive Fuel Cell Applications: Problem Identification and Research Needs*, KFB Swedish Transport & Communications Research Board, Stockholm, October 31, 1999.
28. C. Song, *Catal. Today*, **77** (2002) 17.
29. R. Farrauto, S. Hwang, L. Shore, W. Ruettinger, J. Lampert, T. Giroux, Y. Liu, O. Ilinich, *Annu. Rev. Mater. Res.*, **33** (2003) 1.
30. L. Shore, R.J. Farrauto in *Handbook of Fuel Cells — Fundamentals, Technology and Applications* (Eds. W. Vielstich, A. Lamm, H.A. Gasteiger), Wiley, Chichester, 2003, pp. 211–218.
31. J.O'M. Bockris, S. Srinivasan, *Fuel Cells: Their Electrochemistry*, McGraw-Hill, New York, 1969
32. S. Kartha, P. Grimes, *Phys. Today*, November, 1994, 55–61.
33. T. Thampan, S. Malhotra, J. Zhang, R. Datta, *Catal. Today*, **67** (2001) 15.
34. S. Golunski, *Platinum Metals Rev.*, **42** (1998) 2.
35. Y. Liu, T. Hayakawa, K. Suzuki, S. Hamakawa, S. Tsunoda,. T. Ishii, M. Kumagai, *Appl. Catal. A: Gen.*, **223** (2002) 137.
36. Y. Liu, T. Hayakawa, T. Tsunoda, K. Suzuki, S. Hamakawa, K. Murata, R. Shiozaki, T. Ishii, M. Kumagai, *Top. Catal.*, **22** (2003) 205.
37. Y. Li, Q. Fu, M. Flytzani-Stephanopoulos, *Appl. Catal. B: Environ.*, **27** (2000) 179.
38. S. Hočevar, J. Batista, H. Matralis, T. Ioannides, G. Avgouropoulos, EP 1255693, 2002-11-13

39. R.M. Heck, R.J. Farrauto, with S.T. Gulati, *Catalytic Air Pollution Control: Commercial Technology*, 2nd ed., Wiley, Inc., New York, 2002.
40. C.V. Ovesen, B.S.C. Lausen, B.S. Hammershøi, G. Steffensen, T. Askgaard, I. Chorkendorff, J.K. Nørskov, P.B.R. Asmussen, P. Stoltze, P. Taylor, *J. Catal.*, **158** (1996) 170.
41. W. Ruettinger, O. Ilinich, R.J. Farrauto, *J. Power Sources*, **118** (2003) 61.
42. T. Bunluesin, R.J. Gorte, G.W. Graham, *Appl. Catal. B: Environ.*, **15** (1998) 107.
43. S. Hilaire, X. Wang, R.J. Gorte, J. Wagner, *Appl. Catal. A: Gen.*, **258** (2004) 271.
44. G. Jacobs, E. Chenu, P.M. Patterson, L. Williams, D. Sparks, G. Thomas, B.H. Davis, *Appl. Catal. A: Gen.*, **258** (2004) 203.
45. C. Wheeler, A. Jhalani, E.J. Klein, S. Tummala, L.D. Schmidt, *J. Catal.*, **223** (2004) 191.
46. Q. Fu, S. Kudriavtseva, H. Saltsburg, M. Flytzani-Stephanopoulos, *Chem. Eng. J.*, **93** (2003) 41.
47. N.A. Koryabkina, A.A. Phatak, W.F. Ruettinger, R.J. Farrauto, F.H. Ribeiro, *J. Catal.*, **217** (2003) 233.
48. S. Hočevar, U. Opara Krašovec, B. Orel, A.S. Aricó, H. Kim, *Appl. Catal. B: Environ.*, **28** (2000) 113.
49. G. Sedmak, S. Hočevar, J. Levec, *J. Catal.*, **213** (2003) 135.
50. G. Avgouropoulos, T. Ioannides, Ch. Papadopoulou, J. Batista, S. Hocevar, H.K. Matralis, *Catal. Today*, **75** (2002) 157.
51. H. Kušar, S. Hočevar, J. Levec, *Appl. Catal. B: Environ.*, **63** (2006) 194–200.
52. *GASEQ Ver. 0.78*, http://www.gaseq.co.uk
53. C. Callaghan, I. Fishtik, R. Datta, M. Carpenter, M. Chmielewski; A. Lugo, *Surf. Sci.*, **541** (2003) 21.
54. A.B. Mhadeshwar, D.G. Vlachos, *J. Phys. Chem. B*, **108** (2004) 15246.
55. T. Shido, Y. Iwasawa, *J. Catal.*, **136** (1992) 493.
56. G. Avgouropoulos, T. Ioannides, H.K. Matralis, J. Batista, S. Hočevar, *Catal. Lett.*, **73** (2001) 33.
57. G. Kim, *Ind. Eng. Chem. Prod. Res. Dev.*, **21** (1982) 267.
58. B.I. Whittington, C.J. Jiang, D.L. Trimm, *Catal. Today*, **26** (1995) 41.
59. M.J. Kahlich, A. Gasteiger, R.J. Behm, *J. Catal.*, **171** (1997) 93.
60. R.H. Nibbelke, M.A.J. Campman, J.H.B.J. Hoebink, G.B. Marin, *J. Catal.*, **171** (1997) 358.
61. H. Muraki, S.I. Matunaga, H. Shinjoh, M.S. Wainwright, D.L. Trimm, *J. Chem. Technol. Biotechnol.*, **52** (1991) 415.
62. G. Sedmak, S. Hočevar, J. Levec, *J. Catal.*, **222** (2004) 87.
63. C. Y. Ying, A. Tschöpe, *Chem. Eng. J.*, **64** (1996) 225.
64. S. Hočevar, J. Batista, J. Levec, *J. Catal.*, **184** (1999) 39.
65. Y. Luo, Y. Ito, H. Zhong, A. Endou, M. Kubo, S. Manogaran, A. Imamura, A. Miyamoto, *Chem. Phys. Lett.*, **384** (2004) 30–34.

CHAPTER 7 QUESTIONS

Question 1

Heterogeneous catalysts increase the rate at which a chemical reaction reaches equilibrium, but cannot shift this equilibrium. Mention some of the possible ways to increase the CO conversion over the equilibrium value in the case of the water-gas shift reaction.

Question 2

Calculate the equilibrium conversions of CO in water-gas shift reaction at 500 K and 1 atm: at molar ratio $H_2O/CO = 3$; and at molar ratios $H_2O/CO = 3$ and $H_2/CO = 2$ (in excess of H_2 as the reaction product).

How much lower is the equilibrium CO conversion in case (b) as compared to case (a)?

Question 3

Calculate the equilibrium CO concentration for the following reactant gas composition at 1 atm in the temperature interval between 373 and 773 K.

H_2	48 mol
H_2O	26 mol
CO	3 mol
CO_2	12 mol
N_2	37 mol

Plot CO concentration vs. temperature.

a. Below which temperature does the equilibrium CO conversion exceeds 0.99?

b. Above which temperature does the equilibrium CO concentration exceed 3 mol?

Question 4

Does any industrial water-gas shift catalyst exist that can reach equilibrium CO conversion at 410 K for Question 3a?

Question 5

What is the reason for the increase of equilibrium CO concentration above 3 mol in Question 3b?

Question 6

How can the use of modern quantitative theoretical approaches help in designing new and better WGS catalysts?

Question 7

Why is preferential CO oxidation (PrOX) needed to produce H_2 containing less than 10 ppm CO?

Question 8

Why is it so difficult to oxidize CO with high selectivity in the presence of excess H_2?

Question 9

Would the preferential CO oxidation reaction be needed if the proton-exchange membrane fuel cell (PEMFC) with Pt anode catalyst were able to work at temperatures higher than about 403 K?

Question 10

Below which temperature will the equilibrium CO conversion in PrOX be higher than 0.99 when the process runs at 1 atm and the following inlet gas composition?

H_2 50 mol
CO_2 20 mol
CO   0.5 mol
H_2O 15 mol
N_2 30 mol
O_2 0.25 mol

Structure and Reaction Control at Catalyst Surfaces

Mizuki Tada and Yasuhiro Iwasawa

CONTENTS

8.1   Introduction 229
8.2   Regulation of Catalysis by Coadsorbed Molecules 231
    8.2.1   Self-Assisted Dehydrogenation of Ethanol on an Nb/SiO$_2$ Catalyst 231
    8.2.2   Reactant-Promoted Water-Gas-Shift Reactions 233
        8.2.2.1   WGS Reactions on ZnO 234
        8.2.2.2   WGS Reactions on Rh/CeO$_2$ 235
    8.2.3   Regulation of Selective Oxidation of Methanol on a Modified Mo(112) Surface 236
        8.2.3.1   Reaction Aspect of Methanol Oxidation 236
        8.2.3.2   Reaction Scheme of Methanol Oxidation in TPR 240
        8.2.3.3   Reaction Kinetics of the Steady-State Methanol Oxidation 242
        8.2.3.4   Regulation of the Methanol Oxidation by Extra Oxygen Atoms 242
8.3   Regulation of Catalysis by Design of Active Structures 244
    8.3.1   Chemical Tuning of Active Sites 244
    8.3.2   ReO$_x$ Clusters Produced *in Situ* 246
8.4   Design of Reaction Intermediate and Transition-State Analogue for a Target Reaction on Oxide Surfaces 248
    8.4.1   Reaction Regulation by Molecular Imprinting 248
    8.4.2   Design of a Reaction Intermediate on Catalyst Surface 252
8.5   Conclusion 254
References 254
Chapter 8 Problems 256

8.1   INTRODUCTION

Selectivity in catalytic oxidation/reduction and acid–base reactions has been a long-term challenge in the catalysis field. While it has been recognized that the control of molecular activation and reaction intermediates is critical in achieving high selectivity, this issue has not been adequately addressed and is a serious challenge to the field.

The aim of this chapter is to document how catalytic reactions can be promoted and regulated by the presence of coadsorbed species and by the design of active structures at catalyst surfaces. To clarify these issues, this chapter examines two classes of well-defined catalyst surfaces: chemically designed surfaces and single-crystal surfaces.

A simplified form of the usual mechanism for heterogeneous catalytic reactions is shown in Figure 8.1a, where a reaction intermediate (Cat-X) is transformed to a product (P) by surface decomposition/bond rearrangement; that is, a stoichiometric reaction step proceeds without aid of other molecules. In a typical catalyst, the role of the active site is to directly activate the adsorbed intermediate, which makes it possible for the bond rearrangement to occur in a desired manner. A typical example is seen for ethanol oxidation ($C_2H_5OH + 1/2O_2 \rightarrow CH_3CHO + H_2O$) on a supported Mo-oxide catalyst with an active dimer structure, where the Mo-ethoxide intermediate ($Mo-OC_2H_5$) at the surface decomposes to acetaldehyde accompanied with H_2O formation (Figure 8.2) [1]. The behavior of the intermediate is similar under vacuum and under catalytic reaction conditions when the surface coverage (equivalent to concentration) is identical under both atmospheres. This kind of surface reaction requires no additional gas-phase molecules.

In contrast to the simple expectation of no special role of additional gas-phase molecules in a catalytic mechanism, the reaction intermediate of an important catalytic reaction and hence the reaction rate and selectivity can be profoundly influenced by coexisting gas-phase molecules A′ as shown in Figure 8.1b. The transformation of the intermediate (Cat-X) to a product (P) is promoted by the coexisting A′, or alternatively a new reaction path from the intermediate to another product (Q) is opened by the coexisting gas-phase molecules A′, even when they are weakly adsorbed or undetectable at the surface. This aspect is not observed under vacuum but is observed, under catalytic reaction conditions in the presence of gas-phase molecules, though the reaction intermediate is the same species in both cases. This principle is associated with a principle of the genesis of catalysis, where the rate and selectivity can be regulated by the coexisting molecules concerted with the reactivity of catalyst surfaces.

Catalytic activity and selectivity also strongly depend on structures and ensemble sizes of active sites at catalyst surfaces (Figure 8.1c). The requirement and design of molecular structures and quantitative ensemble sizes for efficient catalysis represent important but as yet unaddressed challenges to the field. Although the efforts on the design of excellent catalysts have been acutely difficult challenges, of late, molecular-level catalyst preparation has become realistic on the basis of modern physical techniques and accumulated knowledge of oxide surfaces [2–4].

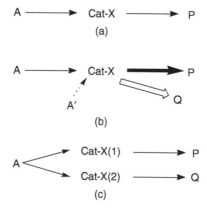

Figure 8.1 (a) A simple catalytic reaction (**A** -------->**P**). **Cat-X** is a reaction intermediate at the surface. (b) Catalytic reaction regulated by coexisting molecules at the catalyst surface, where enhancement of the reaction rate for the formation of **P** or switchover of the reaction path from the **P** formation to the **Q** formation occur by coexisting molecules (**A′**). (c) Catalytic reaction (**A**----->**P**) via intermediate (**Cat-X(1)**) on an active structure, while catalytic reaction (**A**----->**Q**) via intermediate (**Cat-X(2)**) on the other active structure at catalyst surfaces.

8.2   REGULATION OF CATALYSIS BY COADSORBED MOLECULES

8.2.1   Self-Assisted Dehydrogenation of Ethanol on an Nb/SiO$_2$ Catalyst

The SiO$_2$-supported Nb Catalyst (**1**), {SiO}$_2$Nb(=O)$_2$, which is prepared by supporting Nb(η^3-C$_3$H$_5$)$_4$ on SiO$_2$, followed by chemical treatments with H$_2$ and O$_2$ stepwise in a controllable manner, exhibits high activity and selectivity for the dehydrogenation reaction at 423–523 K [5]. Acetaldehyde and hydrogen were stoichiometrically produced during the catalytic reaction (Figure 8.3). Ethanol dissociatively adsorbs on Nb to form OC$_2$H$_5$ (a) and OH (a), {SiO}$_2$Nb(=O)(OH)(OC$_2$H$_5$) (**2**) in Scheme 8.1.

When the gas-phase ethanol was evacuated in the course of the dehydrogenation at 523 K, the reaction completely stopped as shown in Figure 8.3. However, the amount of adsorbed ethanol, {SiO}$_2$Nb(=O)(OH)(OC$_2$H$_5$) species (**2**), remained unchanged by the evacuation, as evidenced by the intensity of the v_{OH} and v_{CH} peaks. In other words, the adsorbed ethanol was converted selectively into acetaldehyde and hydrogen under the ambient ethanol, whereas the adsorbed ethanol did not decompose at all under vacuum in the same temperature range. This might be an unexpected finding in a sense, because ethanol dehydrogenation has been thought to be a surface reaction that proceeds via decomposition of the adsorbed ethanol, the rate of which depends on the coverage of the adsorbed ethanol. To gain insight into the reactivity of adsorbed ethanol, a temperature-programmed desorption (TPD) spectrum for {SiO}$_2$Nb(=O)(OH)(OC$_2$H$_5$) (**2**) formed during the catalytic ethanol dehydrogenation was measured as shown in Figure 8.3 (inset), which revealed that

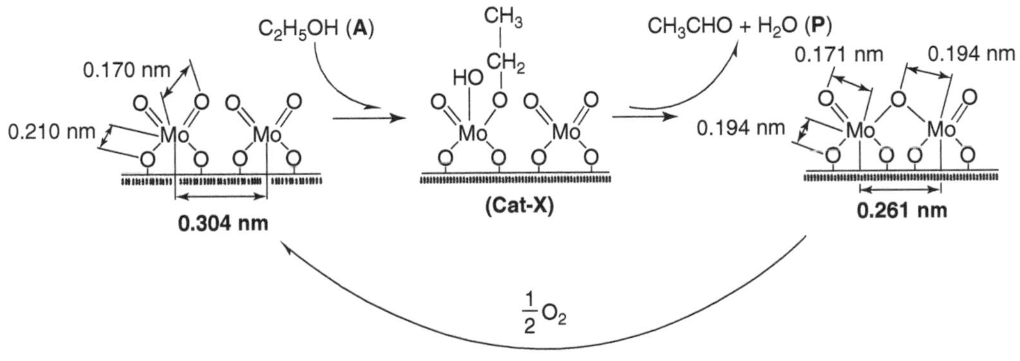

Figure 8.2   Reaction mechanism for ethanol oxidation on an Mo dimer/SiO$_2$ catalyst as an example of the reaction mode (a) in Figure 8.1

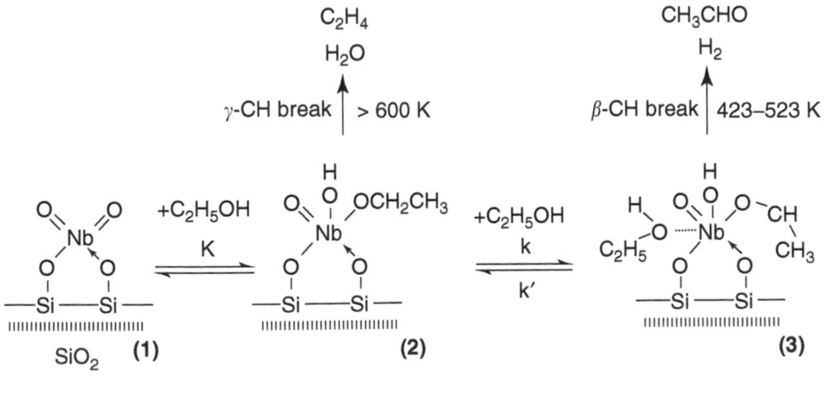

Scheme 8.1   Switchover of the reaction paths by weakly adsorbed ethanol.

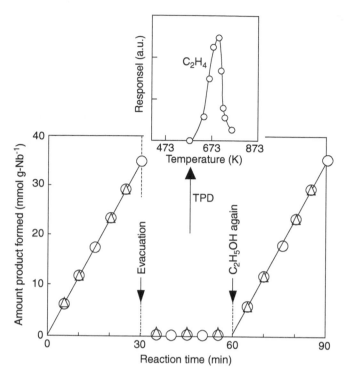

Figure 8.3 Ethanol dehydrogenation on Nb/SiO$_2$ (O, H$_2$; Δ, CH$_3$CHO) and TPD spectrum of adsorbed ethanol (species **2**); heating rate: 4 K/min.

the reaction of adsorbed ethanol (**2**) in vacuum was only possible at temperatures higher than 600 K, with a TPD peak at 700 K. The TPD peak is observed at a much higher temperature range than the 423–523 K for the catalytic reactions, and the TPD products are ethene and water (dehydrated products), in contrast to acetaldehyde and hydrogen (dehydrogenated products) produced in the catalytic reactions. Note that the behavior of adsorbed ethanol (**2**) in vacuum is entirely different from that in the presence of ambient ethanol. The dehydrogenation reaction started again by introducing ethanol vapor onto adsorbed ethanol (**2**) as shown in Figure 8.3. Thus, it seems that the catalytic dehydrogenation reaction is assisted by the ambient ethanol, where the reaction path of adsorbed ethanol (**2**) is switched from dehydration to dehydrogenation by the ambient ethanol. In other words, adsorbed ethanol (**2**) prefers dehydration to form ethene and water by *γ-hydrogen abstraction*, while in the presence of the ambient ethanol the *β-hydrogen abstraction* from absorbed ethanol (**2**) to form acetaldehyde and hydrogen dominates as shown in Scheme 8.1. Note that weakly postadsorbed ethanol promotes the dehydrogenation of strongly preadsorbed ethanol.

To examine how and why the surface ethanol reaction is assisted by the gas-phase ethanol, the following experiments were conducted in a closed circulating reactor. Ethanol vapor was first admitted onto the dioxoniobium monomer catalyst (**1**), {SiO}$_2$Nb(=O)$_2$, to form the niobium ethoxide (**2**), {SiO}$_2$Nb(=O)(OH)(OC$_2$H$_5$), at 373 K, followed by evacuation, and then the system was maintained at 523 K for 10 min, where no H$_2$ evolution was observed because the niobium ethoxide (**2**) was stable up to 600 K in vacuum. After the confirmation of no H$_2$ formation from the preadsorbed ethanol (**2**), *tert*-butyl alcohol was introduced to the system at 523 K, which led to a stoichiometric evolution of H$_2$ and CH$_3$CHO. As the *tert*-butyl alcohol molecule has no extractable α-hydrogen, it is evident that both H$_2$ and CH$_3$CHO were produced from the preadsorbed ethanol by the assistance of the postdosed *tert*-butyl alcohol.

To confirm this mechanistic feature and to examine the interaction between the niobium ethoxide species and the postadsorbed molecule, various electron-donating compounds were postadsorbed on the surface that had been preadsorbed with the same amount of ethanol, {SiO}$_2$Nb(=O)(OH)(OC$_2$H$_5$)

(2). Equimolar amounts of H_2 and CH_3CHO were produced from the preadsorbed ethanol (2) by admission of electron donors. The logarithm of the initial rates of the CH_3CHO (H_2) formation from the preadsorbed ethanol (2) was plotted against the logarithm of the equilibrium constant for the formation of electron donor–acceptor complexes between the nucleophilic molecules and $SbCl_5$ [6]. The logarithm of the equilibrium constants is regarded as the electron donor strength of the postadsorbed molecules. It was found that the logarithm of the initial rates (i.e., activation energy E_a) was proportional to the logarithm of the equilibrium constants (i.e., $\Delta G°$). The linear relationship indicates that the electron donor–acceptor interaction between the postadsorbed molecule and the coordinatively unsaturated Nb d^0 ion is a key issue for the dehydrogenation of the preadsorbed ethanol (2).

Assuming the mechanism in Scheme 8.1, the following equation is derived from a steady-state procedure:

$$\frac{[Nb]_0}{v} = \frac{1}{k_3} + \frac{k' + k_3}{kk_3} + \frac{KP + 1}{KP^2} \tag{8.1}$$

where v, $[Nb]_0$, and P represent the reaction rate, the number of Nb sites, and the ethanol pressure, respectively. Plots of $[Nb]_0/v$ against $(KP+1)/KP^2$ showed a linear relationship, suggesting the validity of the mechanism in Scheme 8.1. Further, the mechanism is also supported by the fact that the rate constant k_3 (1.5 mmol/min [g of Nb]^1) determined from the steady-state equation is almost the same as the value of k_3 determined from the initial rate of the dehydrogenation of species (3) under the condition of ethanol adsorbed at saturation. The equilibrium constant k/k', for weak adsorption of ethanol, was calculated to be 9.0×10^{-4} Pa^{-1}. The weakly adsorbed ethanol is in equilibrium with the gas-phase ethanol, and easily desorbs from the surface in vacuum.

In β-CH elimination on d^8-metal ethoxide complexes, the orbital interactions have to take place in such a manner that the electron donation from σ(CH) to σ*(MO) and the back-donation from σ(MO) to σ*(CH) are required to form the MH σ and CO π bonds and break the CH σ and MO σ bonds. Also in case of d^0-metal–ethoxide complexes, the presence of a weak M---H agostic interaction is predicted by theoretical calculation, but the Ti-β-CH angle is unfavorable for the overlap of the occupied Ti d-orbital and the CH antibonding orbital. Furthermore, there is formally no d-electron available for the promotion of the CH bond scission. The former boundary is satisfied by attaching Nb d^0 ions to the SiO_2 surface through Nb–O–Si bonding. The electronic structure of a distorted tetrahedral dioxo-Nb monomer on SiO_2 calculated by the DV(discrete variable)-Xα cluster method shows that the Nb 4d orbitals is hybridized with the higher occupied O 2p levels, enabling the β-CH breaking. The support electronically modifies the metal oxide species through chemical or ionic bonds and induces the structural change of the surface metal oxides needed for catalysis. It predicts new catalysis involving β-elimination of the CH bond by coordinatively unsaturated tetrahedral Nb monomers chemically attached to the SiO_2 surface. The latter boundary of the orbital overlap seems to be less rigid for the tetrahedral Nb monomer structure [7]. As a result, the Nb monomer (1) catalyzes the dehydrogenation of ethanol with good selectivity of >96%.

8.2.2 Reactant-Promoted Water-Gas-Shift Reactions

The water-gas-shift (WGS) reaction ($H_2O + CO \rightarrow H_2 + CO_2$) on MgO, ZnO, and Rh/CeO_2 is another example of a surface catalytic reaction that is assisted by gas-phase molecules. It is known that the WGS reaction proceeds via surface formate intermediate ($HCOO^-$), which can be monitored by FT-IR. The behavior of the surface intermediates ($HCOO^-$) (Cat-X in Figure 8.1a) is remarkably influenced by weakly coadsorbed water molecules (A' in Figure 8.1b). The characteristic aspect of the WGS reactions on ZnO and Rh/CeO_2 are as follows:

$$H_2O + CO \underset{\text{backward decompostion}}{\overset{\text{formate formation}}{\rightleftharpoons}} HCOO^-(a) + H^+(a) \overset{\text{forward decomposition}}{\longrightarrow} H_2 + CO_2 \tag{8.2}$$

8.2.2.1 WGS Reactions on ZnO

The $v(OD)$ peak at 2706 cm^{-1} on an OD-covered ZnO surface attributable to linear OD groups on two-coordinated Zn ions, decreased by reaction with CO at 473 K and accompanied with the appearance of $v_{as}(OCO)$ and $v_s(OCO)$ peaks for surface bidentate formats (DCOO$^-$) at 1586 and 1342 cm^{-1}, respectively, suggesting that the OD groups react with CO to produce the bidentate formates. The formates (DCOO$^-$) react with the D atoms of bridge (2682 cm^{-1}) or threefold-hollow (2669 cm^{-1}) OD groups at 573 K as monitored by FT-IR, evolving D$_2$, CO$_2$, D$_2$O, and CO in the gas phase.

It was found that the rate constant of the forward decomposition of the surface bidentate formate (DCOO$^-$) to produce D$_2$ and CO$_2$ increased from 0.34×10^{-4} sec^{-1} under vacuum to 5.3×10^{-4} sec^{-1} under ambient water. Electron donors such as NH$_3$, CH$_3$OH, pyridine, and THF also increased the decomposition rate; the rate constants of the forward decomposition of the surface formates at 553 K were determined to be 28.0×10^{-4}, 7.7×10^{-4}, 8.1×10^{-4}, and 6.0×10^{-4} sec^{-1} under NH$_3$, methanol, pyridine, and THF vapors (0.4 kPa), respectively. It is likely that the driving force for the forward decomposition of the formate is electron donation of the adsorbed molecule to the Zn ion on which the bidentate formate adsorbs. The reactant-promoted mechanism for the catalytic WGS reaction on ZnO is illustrated in Scheme 8.2.

It has been proposed that the catalytic WGS reaction on ZnO may proceed as follows. The first water molecule dissociates at the Zn$_{2c}$–O$_{3c}$ pair to form a linear OH group and a bridge or threefold-hollow OH group. The linear OH group on Zn reacts with CO to produce the bidentate formate through unidentate formate [8]. Seventy percent of the formate backwardly decompose to the original H$_2$O and CO under vacuum, while 30% forwardly decompose by reacting with the bridge or threefold-hollow OH groups to produce H$_2$ and CO$_2$. In the presence of gas-phase water, the weakly adsorbed (second) water molecule adsorbs on the Zn atom and enhances the forward decomposition, where almost 100% of the formate decompose to H$_2$ and CO$_2$. The activation energy for the forward decomposition of the formate decreases from 155 kJ/mol under vacuum to 109 kJ/mol under the ambient water. CO$_2$(ad) produced by the decomposition of the bidentate formate in the presence of the second water molecule is the unidentate carbonate as demonstrated by FT-IR [8]. The decomposition of the unidentate carbonate to form CO$_2$ is also markedly promoted by the second water molecule adsorbed on the Zn atom. The accompanying dissociation of the adsorbed water to Zn–OH and O–H may assist the CO$_2$ desorption from the carbonate. In the steady-state reaction, the rates of the two reaction steps (formate--->carbonate+H$_2$ and carbonate--->CO$_2$) are balance in the presence of adsorbed water.

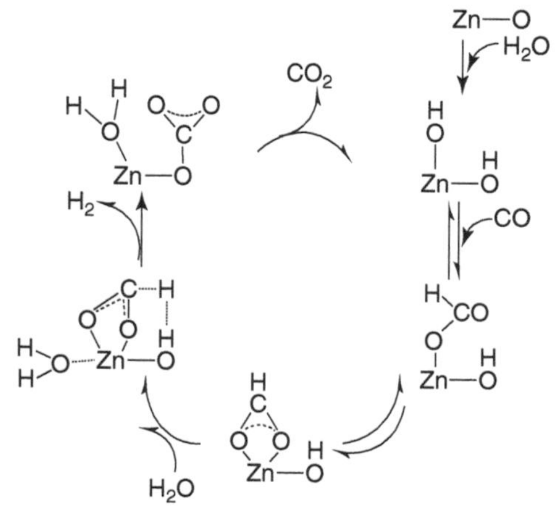

Scheme 8.2 Mechanism of WGS reaction on ZnO.

The WGS reaction is a reversible reaction; that is, the WGS reaction attains equilibrium with the reverse WGS reaction. Thus, the fact that the WGS reaction is promoted by H_2O (a reactant), in turn implies that the reverse WGS reaction may also be promoted by a reactant, H_2 or CO_2. In fact, the decomposition of the surface formates produced from H_2+CO_2 was promoted 8–10 times by gas-phase hydrogen. The WGS and reverse WGS reactions conceivably proceed on different formate sites of the ZnO surface unlike usual catalytic reaction kinetics, while the occurrence of the reactant-promoted reactions does not violate the principle of microscopic reversibility. The activation energy for the decomposition of the formates (produced from H_2O+CO) in vacuum is 155 kJ/mol, and the activation energy for the decomposition of the formates (produced from H_2+CO_2) in vacuum is 171 kJ/mol. The selectivity for the decomposition of the formates produced from H_2O+CO at 533 K is 74% for H_2O+CO and 26% for H_2+CO_2, while the selectivity for the decomposition of the formates produced from H_2+CO_2 at 533 K is 71% for H_2+CO_2 and 29% for H_2O+CO as shown in Scheme 8.3. The drastic difference in selectivity is not presently understood. It is clear, however, that this should not be ascribed to the difference of the bonding feature in the zinc formate species because $v(CH)$, $v_{as}(OCO)$, and $v_s(OCO)$ for both bidentate formates produced from H_2O+CO and H_2+CO_2 show nearly the same frequencies. Note that the origin (H_2O+CO or H_2+CO_2) from which the formate is produced is "remembered" as a main decomposition path under vacuum, while the origin is "forgotten" by coadsorbed H_2O.

In the reverse WGS reaction, hydrogen promoted both decomposition paths of the formate to H_2+CO_2 and H_2O+CO, and the decomposition selectivity did not change. Thus, the mechanism of hydrogen promotion is different from that of electron donors in the WGS reaction. CO_2 not only blocks the adsorption sites of H_2 but also suppresses the decomposition of the formate intermediate. The rate constant for the steady-state reaction is higher than that obtained from the formate decomposition in vacuum, but it is smaller than that for the formate decomposition under the ambient H_2. As a result, the reverse WGS reaction proceeds with a balance of H_2 promotion and CO_2 suppression.

8.2.2.2 WGS Reactions on Rh/CeO₂

CeO_2 is contrasted with ZnO; the surface formate on CeO_2 is stabilized by the coexistence of water vapor, where the selectivity to H_2 and CO_2 only increases as a result of suppression of the backward decomposition of formate more than that of the forward decomposition by water vapor. This property of the CeO_2 surface was modified by doping with a small amount (0.2 wt%) of Rh. The Rh/CeO₂ catalysts have been commonly used as automobile exhaust gas-cleaning catalysts, on which the WGS reaction proceeds.

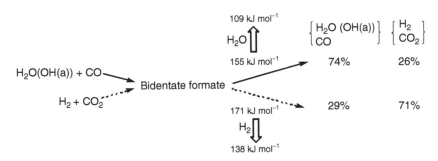

Scheme 8.3 The reactivity of bidentate formates (reaction intermediates) in normal and reverse WGS reactions on ZnO, and the promotion of the reactivity by H_2O and H_2 respectively, accompanied with change in the selectivity.

Surface formates on both Rh/CeO_2 and CeO_2 surfaces have been determined to be of the bidentate type. The bidentate formate is produced and decomposed more easily on Rh/CeO_2 than on CeO_2. The activation energy of the forward decomposition of the formate on Rh/CeO_2 (56 kJ/mol) is much lower than 270kJ/mol for CeO_2.

In the catalytic WGS reaction on Rh/CeO_2, linear OH groups reacted with CO to produce bidentate formates. In vacuum, 65% of the surface formates decomposed backwardly to H_2O+CO, and 35% of them decomposed forwardly to H_2+CO_2. When water vapor coexisted, 100% of the formates decomposed forwardly to H_2+CO_2 as shown in Table 8.1. The activation energy for the forward decomposition of the formate decreased from 56kJ/mol in vacuum to 33 kJ/mol due to the presence of water(D_2O) vapor. By addition of a small amount of Rh (0.2 wt%) to CeO_2, the rate of the WGS reaction increased tremendously, and the value of the forward decomposition rate constant (k_+) was promoted about 100-fold by the coexistence of gas-phase water (Table 8.1).

The isotope effect is observed with the hydrogen atom of the formate and not with the hydrogen atom of the water molecule. The result is similar to that observed on ZnO, where the rate-determining step of the formate decomposition is suggested to be dissociation of the CH bond of the bidentate formate. In summary, the reaction mechanism for the catalytic WGS reaction on Rh/CeO_2 is essentially the same as that on ZnO.

Weakly adsorbed molecules (A′ in Figure 8.1b), which can exist only under catalytic reaction conditions, play an important role in surface catalytic reactions even if the adsorption of the "promoter" is very weak or is undetectable at the surface. Surface intermediates (Cat-X in Figure 8.1b) under the ambient gas molecules behave in a different way from those under vacuum, showing rate enhancement and selectivity change of the surface reaction in the presence of ambient gas.

8.2.3 Regulation of Selective Oxidation of Methanol on a Modified Mo(112) Surface

8.2.3.1 Reaction Aspect of Methanol Oxidation

Control of the reaction path of catalytic reactions by atomic level design of catalyst surfaces is a key issue, which is crucial to success in surface science. The surface designed by optimizing the structural and electronic properties may provide information on the origin of activity and selectivity, and show a new catalytic performance that may overcome the catalytic performance of existing catalysts. Molybdenum is used as a principal and promoting element in many industrial catalysts for various kinds of reactions, partly because of its wide range of chemical reactivity and oxidation states. This section presents a successful modeling of selective catalytic oxidation of methanol by modifying an Mo(112) surface with ordered oxygen atoms to form a one-dimensional reaction field, which is the first example of selective oxidation reaction on Mo single-crystal surfaces [9,10].

Table 8.1 Rate Constants for Forward (k_+) and Backward (k_-) Decompositions of the D-Labeled Formates in Vacuum and Under Ambient D_2O

Catalyst	Reaction Temperature (K)	Gas Phase	$k_+ + k_-$ (sec^{-1})	k_+/k_- (%/%)	E_a (k_+)/ kJ/mol
MgO	600	Vacuum	13.0×10^{-4}	0/100	–
		D_2O	1.9×10^{-4}	74/26	–
ZnO	533	Vacuum	1.3×10^{-4}	26/74	155
		D_2O	5.3×10^{-4}	100/0	109
Rh/CeO_2	443	Vacuum	1.1×10^{-5}	35/65	56
		D_2O	1.1×10^{-3}	100/0	33

Note: D_2O pressure: 0.40 kPa for MgO and ZnO and 0.67 kPa for Rh/CeO_2.

The Mo(112) surface has a ridge-and-trough structure, where the top layer of Mo atoms form close-packed atomic rows along the [111] direction, at a distance of 0.445 nm from each other and adsorbed oxygen atoms produce an Mo(112)-(1×2)-O ordered surface, where oxygen atoms occupy quasi-threefold sites composed of one second-layer and two first-layer Mo atoms. Figure 8.4a shows a model of the (1×2)-O surface ($\theta_O=1.0$), which was proposed on the basis of LEED patterns and CO titration experiments. Every second Mo row is coordinated by oxygen atoms (Mo_{2C}) on both sides, while the other Mo rows have no oxygen atoms directly coordinated (Mo_{NC}). This structure preserves adsorption sites on the Mo_{NC} for methanol. Selective blocking of the second-layer Mo atoms by oxygen atoms can suppress bond breaking of C–O and stabilize $CH_3O(a)$ above 500 K, resulting in the selective oxidation of methanol to formaldehyde ($CH_3OH + 1/2O_2 \rightarrow CH_2O + H_2O$) [9,10]. This trend is remarkably promoted by the presence of extra oxygen on the one-dimensional Mo rows of Mo(112)-(1×2)-O (Figure 8.4b). The extra oxygen atoms greatly increase the selectivity to formaldehyde and lower the activation energy of the selective oxidation. By supplying extra oxygen atoms on the surface, the selective catalytic oxidation of methanol successfully proceeds without deactivation. On single-crystal surfaces of Mo metal such as (100), (110), and (112), major products in methanol reaction are CH_4, CO, and H_2, and a little CH_2O is produced. Also on an oxidized Mo(100)-(1×1)-O ($\theta_O=1.5$), a negligible amount of CH_2O is observed [11].

An Mo(112)-(1×2)-O ($\theta_O=0.1$) surface was exposed to oxygen at 300 K and the coverage of extra oxygen (θ_O') adsorbed on the surface was measured by Auger electron spectroscopy (AES). The extra oxygen was saturated at 0.5 ML (ML, monolayer), which corresponds to the number of Mo atoms in Mo_{NC} rows. The surface after 14 L (1 L: 1.33×10^{-4} Pa sec) exposure showed sharp subspots of (1×2) in LEED (low energy electron diffraction) pattern, indicating that the substrate preserved a (1×2) structure. These results suggest that the extra oxygen species adsorb on Mo_{NC} rows of the Mo(112)-(1×2)-O surface. The (1×2) structure is destroyed by heating to temperatures >800 K.

A temperature-jump method was adopted to measure the amount of catalytic reaction products during the feed of CH_3OH and O_2 (10^{-6} to 10^{-5} Pa) on Mo(112)-(1×2)-O. Constant pressures of CH_3OH and O_2 were introduced to the chamber through two variable leak valves on the (1×2)-O surface at 450 K, where the reaction did not occur; then the sample temperature was ramped to a given reaction temperature for several minutes, and decreased to 450 K again. Therefore, the area of a mass signal over the base line, which is bound between the signals at 450 K, corresponds to the amount of a product in the catalytic reaction at the surface. Data were accumulated by repetition of temperature ramps.

In TPR spectra of methanol from the Mo(112)-(1×2)-O surface after exposure to 4 L of methanol at 200 K, the major product was CH_2O with 50% selectivity, and formation of H_2O was not observed at any temperature. Oxygen atoms in the (1×2)-O structure were not incorporated into the reaction products. Hence, the (1×2)-oxygen atoms do not react with methanol, but work as modifiers on the surface. It is to be noted that CH_2O is not formed on Mo(112) surfaces modified with lower oxygen coverages than the (1×2)-oxygen coverage. Effective blockage of the second-layer Mo atoms, which are supposed to show higher electronic fluctuation leading to higher activity, results in formation of formaldehyde on the Mo(112)-(1×2)-O surface.

The extra oxygen adsorbed on the Mo(112)-(1×2)-O surface drastically changes the selectivity of the reaction. TPR spectra of methanol from the (1×2)-O surface with 0.20 ML of preadsorbed extra oxygen after exposure to 4 L of methanol at 200 K are different from the spectra for the surface without the extra oxygen on the following points: (1) considerable reduction of the peaks of CH_4 and H_2 at 560 K, the second is (2) disappearance of the peak of recombinative desorption of CO at 800 K, and (3) appearance of the peak of H_2O at 580 K. The amounts of desorption products are summarized in Table 8.2. Selectivity to CH_2O increased to 88%. Particularly, reduction of recombinative desorption of CO at 800 K indicates that complete decomposition of methoxy to C(a) and O(a) is considerably suppressed by the presence of extra oxygen. Detection of H_2O and

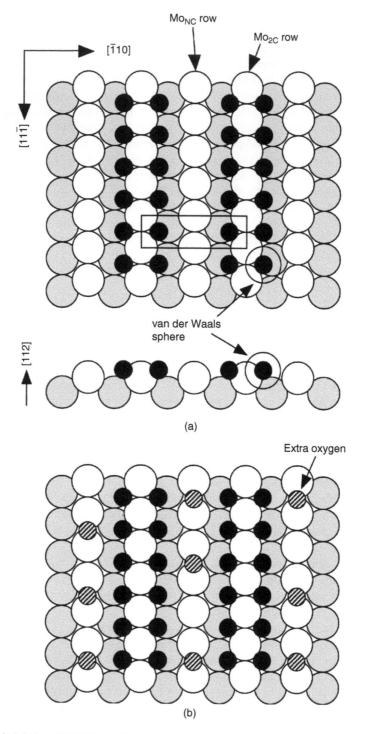

Figure 8.4 (a) Model of an Mo(112)-(1×2)-O surface (θ_O=1.0) with a top view and a plane view. Quasi-three-fold sites with Mo–O distance of 0.21 nm are postulated. (b) Extra oxygen species adsorbed on Mo_{NC} rows of the Mo(112)-(1×2)-O surface($\theta_O \sim 0.22$).

Table 8.2   The Product Distrubution in TPR for Methanol Reaction Around 560 K on the Mo(112) Surfaces Modified with Oxygen

Products	Yield (ML) (Selectivity (%))	
	(1×2)-O Surface (θ_O=1.0)	Extra oxygen (θ_O'=0.20) +(1×2)-O surface (θ_O=1.0)
H_2 (g)	0.10	0
H_2CO (g)	0.09 (50)	0.05 (88)
H_2O (g)	0	0.05
CH_4 (g)	0.04 (22)	<< 0.01 (5)
CO (g)	0.02 (11)	<< 0.01 (7)
C (a)	0.03 (17)	0 (0)
O (a)	0.07	0 [a]

[a] θ_O' after TPR was 0.15.

suppression of H_2 suggest that hydrogen is desorbed effectively as H_2O by reaction with extra oxygen. Thus, more reactive extra oxygen species change the selectivity of the reaction and desorb as H_2O during the reaction.

The simultaneous desorption peaks observed at 560–580 K in TPR are of reaction-limited desorption. The peak temperatures of these peaks do not depend on the coverage of methoxy species, indicating that the desorption rate (reaction rate) on both surfaces has a first-order relation to the coverage of methoxy species. Activation energy (E_a) and the preexponential factor (v) for a first-order process are given by the following Redhead equation [12]:

$$\ln(T_p^2/\beta) = E_a/(RT_p) + \ln[E_a/(Rv)] \tag{8.3}$$

where T_p, β, and R represent peak temperature, heating rate, and the gas constant, respectively. It is found from Equation 8.3 that the extra oxygen atoms lower the activation energy by 20 to 24 kJ/mol for formaldehyde desorption from the Mo(112)-(1×2)-O surface and the (1×2)-O surface with 0.20 ML of extra oxygen. The desorption temperature for CD_3OD as reactant is higher than that for CH_3OH, which suggests that the C–H (C–D) bond cleavage of methoxy is the rate-limiting step of the reaction. The reaction rate itself decreases by the presence of preadsorbed extra oxygen due to the smaller frequency factor.

Catalytic reactions of methanol on an Mo(112)-(1×2)-O surface under a constant flow of CH_3OH and O_2 (10^{-6}–10^{-5} Pa) were monitored as a function of reaction time by the temperature-jump method. Total amounts of the products are summarized in Table 8.3. When only CH_3OH was fed, the reaction rate exponentially decayed with reaction time. After the reaction ceased in both conditions, the surfaces were covered with nearly 1 ML of C(a) (Table 8.3) and the sharp (1×2) LEED subspots of the surface before the reaction almost disappeared due to an increase in background intensity. As shown in Table 8.3, the selectivity of the reaction at 560 K is similar to that obtained by TPR (Table 8.2). The C(a) species formed with 26% selectivity cover the surface, resulting in the exponential decay of the reaction rate. O(a) species are also formed on the surface but they are desorbed as H_2O by reaction with hydrogen atoms. It should be noted that neither C(a) nor a small amount of O(a) change the selectivity in this case.

When CH_3OH was fed with O_2, the selectivity to CH_2O increased in any conditions employed which resulted in longer lifetime of the reaction. As expected from the results of TPR in Table 8.2, extra oxygen atoms formed on the surface enhanced the selectivity and reduced accumulation of C(a). As shown in the bottom of Table 8.3, formaldehyde was formed with 89% selectivity without significant deactivation. At the higher oxygen pressure than the methanol pressure, the selectivity to CH_2O increased as shown in the third line of Table 8.3. The coverages of methoxy and extra oxygen were well balanced, and the reaction proceeded without deactivation. The activation energies for the CH_2O formation were determined to be 54±12 kJ/mol and 51±5 kJ/mol for CH_3OH and CH_3OH+O_2, respectively.

Table 8.3   Yields of the Products and Selectivities in Catalytic Methanol Reactions on Mo(112)-(1×2)-O

P_{CH_3OH} (Pa)	P_{O_2} (Pa)	T_R (K)[a]	Time (s)	Yield (ML)[b] Selectivity (%)					
				$H_2CO(g)$ C(a)	$CH_4(g)$ O(a)		CO(g)	$H_2(g)$	$H_2O(g)$
2.1×10^{-5}	–	560	1110	1.7	0.9 (43)	0.3 (22)	3.2 (8)	1.7	1.05	<0.05 (26)
2.1×10^{-5}	6.5×10^{-6}	560	1650	7.3 (69)	2.1 (20)	0.6 (6)	3.5	5.8	0.55 (5)	0.40
8.1×10^{-6}	1.6×10^{-5}	560	2010	4.5	0.3 (84)	0.4 (5)	0.4 (8)	5.0	0.20	0.30 (4)
2.1×10^{-5}	–	700	2200	5.3	1.4 (59)	1.1 (16)	7.9 (13)	2.6	1.10	⁻0.15 (12)
2.1×10^{-5}	6.5×10^{-6}	700	1870	15.2	<0.1 (89)	1.7 (0.3)	1.3 (10)	15.9	0.05	0.15 (0.3)

[a]T_R = reaction temperatures.
[b]ML = monolayer.

8.2.3.2   Reaction Scheme of Methanol Oxidation in TPR

The first step of the reaction path in TPR on Mo(112)-(1×2)-O without extra oxygen is the dissociation of methanol to form $CH_3O(a) + H(a)$ and the recombinative desorption of the adsorbed hydrogen, which occur above 300 K and at 380 K, respectively.

$$CH_3OH(a) \xrightarrow{\ 300\,K\ } CH_3O(a) + H(a) \qquad (8.4)$$

$$H(a) \xrightarrow{\ 380\,K\ } 1/2 H_2(g) \qquad (8.5)$$

Therefore, above 480 K only methoxy species are left on the surface. The major reaction path with 50% selectivity is formation of formaldehyde:

$$CH_3O(a) \rightarrow CH_2O(a) + H(a) \qquad (8.6)$$

$$CH_2O(a) \rightarrow CH_2O(g) \qquad (8.7)$$

$$H(a) \rightarrow 1/2 H_2(g) \qquad (8.8)$$

where step 8.6 is the rate-limiting step of the reaction.

When extra oxygen species are coadsorbed with methanol, some modifications of the reaction steps are needed. Although a promotion effect of adsorbed oxygen atoms on formation of methoxy from methanol molecule has been reported on some oxygen-modified metal surfaces:

$$CH_3OH(a) + O(a) \rightarrow CH_3O(a) + OH(a) \qquad (8.9)$$

such an effect may not be important on Mo(112)-(1×2)-O where methoxy coverage is low (Table 8.3). However, the hydrogen atom formed by step 8.4 is probably trapped by extra oxygen atom $(O_e(a))$:

$$H(a) + O_e(a) \rightarrow O_eH(a) \qquad (8.10)$$

because desorption of H_2 was not detected below 400 K. As the selectivity to formaldehyde is as high as 88% in the presence of extra oxygen, one should only consider the reaction path from methoxy to formaldehyde. The fact that the activation energy of hydrogen extraction from methoxy is reduced by 20 to 24 kJ/mol in the presence of extra oxygen suggests that hydrogen is extracted at 580 K by an extra oxygen atom (step 8.11) or by an Mo atom which is electronically modified with extra oxygen atoms (step 8.12).

$$CH_3O(a) + O_e(a) \rightarrow CH_2O(a) + O_eH(a) \tag{8.11}$$

$$CH_3O(a) + Mo_m \rightarrow CH_2O(a) + H(a) \tag{8.12}$$

When the hydrogen atoms released from methoxy species in step 8.14 are trapped with extra oxygen atoms (step 8.10), step 8.11, and step 8.12 cannot be discriminated from each other. Alternatively, the hydrogen atoms react with $O_eH(a)$ to produce $H_2O(g)$:

$$O_eH(a) + H(a) \rightarrow H_2O(g) \tag{8.13}$$

$H_2O(g)$ is also produced by dehydration of two $O_eH(a)$:

$$2O_eH(a) \rightarrow H_2O(g) + O_e(a) \tag{8.14}$$

It is also possible that $CH_3O(a)$ reacts with $O_eH(a)$ to produce CH_2O and H_2O at 580 K:

$$CH_3O(a) + O_eH(a) \rightarrow CH_2O(g) + H_2O(g) \tag{8.15}$$

Reaction Steps 8.6 to 8.8 are also relevant to the catalytic reaction of methanol in a flow of CH_3OH, although the steps 8.5 and 8.8 cannot be separated anymore. The net reaction for dissociative adsorption of methanol is expressed by:

$$CH_3OH(g) \rightarrow CH_3O(a) + H(a) \tag{8.16}$$

Recombinative desorption of the methoxy and hydrogen atom,

$$CH_3O(a) + H(a) \rightarrow CH_3OH(g) \tag{8.17}$$

becomes one of major processes because of the higher concentration of hydrogen atoms on the surface in the flow reaction conditions than in TPR. Assuming that the rate-limiting step is C–H bond scission of methoxy (step 8.6), the initial rate of formaldehyde formation (v) can be expressed by:

$$v = \nu S_{CH_2O} \theta_{CH_3O} \exp(-E_a/RT) = A \exp(-E/RT) \tag{8.18}$$

where ν and E_a are the preexponential factor and activation energy of C–H bond scission of methoxy species obtained by TPR ($\nu = 7 \times 10^{15 \pm 1}$ sec^{-1}; $E_a = 175 \pm 13$ kJ/mol), respectively, and S_{CH_2O} is the selectivity to CH_2O. E is an apparent activation energy for the CH_2O formation determined from the Arrhenius plot (54 ± 12 kJ/mol), and A is a constant. S_{CH_2O} is determined by the measured rates of the formation of products (CH_2O, CH_4 CO, and $C(a)$) at each temperature. Then, θ_{CH_3O} can be calculated from Equation 8.18. The θ_{CH_3O} decreased from 0.06 at 560 K to 0.01 at 600 K. These values are much smaller than the observed maximum coverage of methoxy (~ 0.25 ML). The coverage of H(a) indirectly affects the coverage of methoxy species. The desorption energy of hydrogen atoms by step 8.8 can be estimated by TPR at 92 ± 5 kJ/mol, assuming a typical preexponential factor of 1×10^{-2} cm^2/sec. By using these values, θ_H can be calculated from the measured H_2 formation rate. It has an almost constant value of 0.001 from 560 K to 600 K. The rate of recombinative desorption (step 8.18) was estimated to maintain θ_{CH_3O} as a constant value by using E_{ad} as a parameter and assuming a typical preexponential factor of 1×10^{-2} cm^2/sec. The initial rate of formaldehyde formation (v) was simulated by using these values.

8.2.3.3 Reaction Kinetics of the Steady-State Methanol Oxidation

When a mixture of CH$_3$OH and O$_2$ is supplied, extra oxygen atoms produced on the Mo$_{NC}$ rows simplified the reaction paths. Hydrogen atoms produced by dissociative adsorption of methanol (step 8.16) are immediately trapped by the extra oxygen atoms (step 8.10). A minority of them recombine with methoxy to be desorbed as methanol (step 8.17), and only a small amount of H$_2$ was detected during the reaction. Methoxy decomposition to form CH$_2$O by steps 8.11, 8.12 (followed by step 8.7), or 8.15 showed a high selectivity of 89%. With sufficient amounts of extra oxygen and a small coverage of O$_e$H(a) in the catalytic reaction conditions, the contribution of step 8.15 was much lower than that of step 8.11 or step 8.12. The initial rate of formaldehyde formation (v') can be given by Equation 8.19 similar to Equation 8.18.

In the steady state of the catalytic CH$_3$OH+O$_2$ reaction,

$$v' = v'S'_{CH_2O}\,\theta_{CH_3O}\exp(-E'_a/RT) = A'\exp(-E'/RT) \qquad (8.19)$$

where v' and E_a' are the preexponential factor and activation energy of C–H bond scission (step 8.11 or equally step 8.12 + step 8.10) obtained by TPR ($v'=2\times10^{13\pm2}$ sec^{-1}, $E_a'= 155\pm18$ kJ/mol), respectively, and S'_{CH_2O} is the selectivity to CH$_2$O. E' is an apparent activation energy for the CH$_2$O formation determined from the Arrhenius plot (51±5 kJ/mol), and A' is a constant coefficient. The coverage of extra oxygen is not included in the rate because it is a constant in the steady state. S'_{CH2O} is almost unity under these conditions. Thus, θ_{CH_3O} was calculated from Equation 8.19. It decreased from 0.09 at 560 K to 0.03 at 600 K. θ_H was calculated in a similar way and it was found to be less than 10^{-4}. Then, the methoxy species is formed by dissociative adsorption (step 8.16) and consumed by (which?) reaction (step 8.11 or step 8.12). The extra oxygen atoms are adsorbed competitively with methoxy species on one-dimensional Mo$_{NC}$ rows. The coverage of extra oxygen is estimated by the simulation to be 0.28, 0.33, and 0.30 at 560, 580, and 600 K, respectively. It decreases at the higher temperature because of a higher reaction rate to be desorbed as H$_2$O.

8.2.3.4 Regulation of the Methanol Oxidation by Extra Oxygen Atoms

It is suggested that methoxy species exclusively adsorb on Mo$_{NC}$ because of considerable steric blocking of oxygen atoms on Mo$_{2C}$ rows (see the van der Waals sphere of an oxygen atom in Figure 8.5). The saturation coverages of extra oxygen (0.5 ML) and methoxy ($\sim$0.25 ML) also suggest the selective adsorption of these species on the Mo$_{NC}$ rows. These two species competitively adsorb on the one-dimensional Mo$_{NC}$ rows. Figure 8.5 shows a model of the Mo(112)-(1$\times$2)-O surface coadsorbed with extra oxygen atoms ($\theta_o \sim$0.21) and methoxy species ($\theta_{CH_3O} \sim$ 0.07). This model suggests that C–O bond scission of the methoxy is inhibited not due to steric blocking but due to electronic modification of Mo atoms by the extra oxygen atoms. C(a) species is accumulated on the Mo(112)-(1$\times$2)-O surface during the catalytic reaction with CH$_3$OH feed alone (Table 8.3), but it does not inhibit C–O bond scission of methoxy species in contrast to extra oxygen atoms. This result also excludes the possibility of steric blocking as a major reason of the inhibition.

The extra oxygen decreased the activation energy of C–H bond scission of the methoxy, which is the rate-limiting step of the selective methanol oxidation. TPD spectra of CO indicate that extra oxygen species reduce the electron density of Mo atoms in Mo$_{NC}$ rows. This modification causes the decrease of the activation energy for the methoxy dehydrogenation. The extra oxygen is

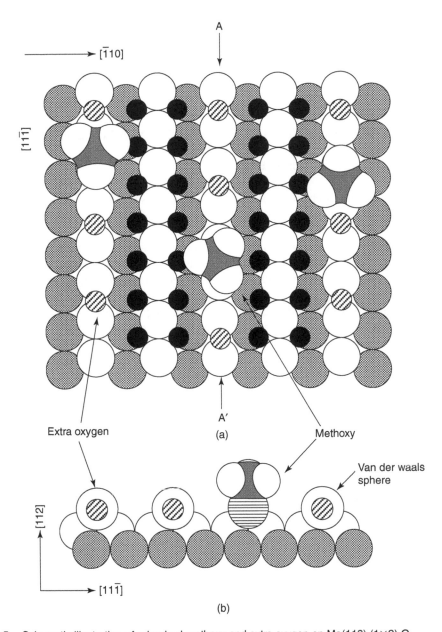

A

[$\bar{1}$10]

[11$\bar{1}$]

Extra oxygen

A'

(a)

Methoxy

Van der waals sphere

[112]

[11$\bar{1}$]

(b)

Figure 8.5 Schematic illustration of adsorbed methoxy and extra oxygen on Mo(112)-(1×2)-O.

desorbed as H_2O during the catalytic reaction. Thus, the selective catalytic oxidation of methanol may involve the direct extraction of a hydrogen atom by an extra oxygen atom (step 8.11), which is similar to the mechanism proposed in oxidative dehydrogenation of methanol on oxide surfaces [13]. One dimensionality of the Mo_{NC} row may increase the probability of the direct interaction between methoxy and extra oxygen species.

A high preexponential factor of $7 \times 10^{15 \pm 1}$ sec^{-1} is observed on the Mo(112)-(1×2)-O surface without extra oxygen. A high preexponential factor has also been reported on CO desorption from Ru(001), which was rationalized by high mobility of transition state species on the surface [14]. Methoxy species in the transition state are probably weakly bound to the surface and may diffuse freely along an Mo_{NC} row. The extra oxygen atoms decreased the preexponential factor of the reaction on the other hand. The extra oxygen atoms may restrict the diffusion of transition state species, resulting in a lower preexponential factor $(2 \times 10^{13 \pm 2} \text{ sec}^{-1})$. As one can consider that dissociation of C–H and C–O bonds of methoxy occur on Mo atoms in an Mo_{NC} row, effective motion of the methoxy for dissociation is hindered rotation along the Mo_{NC} row. Extra oxygen atoms may restrict the diffusion and hindered rotation of methoxy even at the lower coverage due to the one dimensionality of the Mo_{NC} row.

8.3 REGULATION OF CATALYSIS BY DESIGN OF ACTIVE STRUCTURES

8.3.1 Chemical Tuning of Active Sites

In this section a series of chemical designs of Nb structures on SiO_2 is introduced as an example of a one-component tunable catalyst, where the selectivity of ethanol reactions strongly depends on the Nb structures.

The SiO_2-attached Nb monomer catalyst with a four-coordinate structure was prepared by the use of $Nb(\eta^3\text{-}C_3H_5)_4$ as precursor and characterized by extended x-ray absorption fine structure (EXAFS), FT-IR, Raman, ESR, and XPS (x-ray photoelectron spectroscopy). The monomer catalyst (**1** in Scheme 8.1 and Figure 8.6) exhibits high activity and selectivity for the dehydrogenation of ethanol to form acetaldehyde and H_2 as shown in Table 8.4 [7]. The activity is much higher than that of a usual impregnation Nb catalyst and the selectivity is as high as 95 to 100%, whereas the impregnation catalyst is unselective. The dehydrogenation reaction proceeds via the Nb-ethoxide intermediate, but the intermediate is very stable and is not decomposed until 600 K as shown in Scheme 8.1. Even if an adsorbed species at the surface is very stable in vacuum or before catalysis, the catalytic reaction is able to proceed through the same species by activation of the intermediate

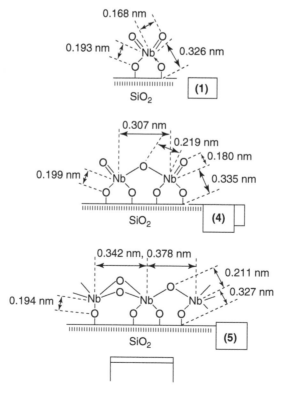

Figure 8.6 Proposed structures of Nb monomer, Nb dimer, and Nb monolayer on SiO_2.

by the reactant [6]. Thus, it may be critical for the dehydrogenation on Nb sites to create a vacant site with an appropriate conformation for the transition state on which electron donation-induced activation of β-hydrogen of C_2H_5O group is favorable.

If a vacant site is occupied by another Nb atom such that it is a dimer, new catalysts may be designed. The Nb dimer catalyst (**4**) in Figure 8.6 was prepared by the reaction of $[Nb(\eta^5\text{-}C_5H_5)H\text{-}\mu\text{-}(\eta^5,\eta^1\text{-}C_5H_4)]_2$ with SiO_2 at 313 K, followed by treatment with O_2 at 773 K. The proposed structure (**4**) was characterized by EXAFS, x-ray absorption near-edge structure (XANES), FT-IR, DR(diffuse reflectance)-UV/VIS, and XPS, which showed Nb–Nb (coordination number, CN: 0.9) and Nb–Si (CN: 2.3) inter-atomic distance at 0.303 and 0.328 nm, respectively, besides Nb–O bonds [7]. The Nb dimer/SiO_2 catalyst (**4**) shows high selectivity for the dehydration of ethanol irrespective of the presence or absence of the ambient ethanol (Table 8.4). The dehydrogenation observed on the Nb monomer catalyst is significantly suppressed to 1/300 and the dehydration is promoted four times on the dimer, indicating that the Nb dimers on SiO_2 have an acidic character. It is to be noted that the change of the number of Nb atoms at the active sites from one to two metal atoms gives rise to a complete reverse of basicity/acidity in the catalytic properties. In dimer (**4**) the access of a second ethanol molecule to the Nb atom coordinated with a first ethanol molecule in a preferable conformation is difficult, unlike in the case of a monomer (**1**). Furthermore, the Lewis acidity of the Nb atoms in the dimer is increased by the oxygen bridge.

A Nb-oxide monolayer (**5**) was successfully prepared by using a $Nb(OC_2H_5)$ precursor. The monolayer growth was monitored mainly by Nb–Si bond (0.327 nm) formation characterized by EXAFS. The monolayer (**5**) showed two different Nb–Nb separations at 0.378 and 0.342 nm, both being considered to have one or two bridging oxygens (Nb–O: 0.211 nm), respectively. The monolayer (**5**) is active and selective for C_2H_4 formation from C_2H_5OH in the temperature range 373 to 573 K as shown in Table 8.4. The monolayer catalyst (**5**) always shows the selective intramolecular dehydration of ethanol, suggesting that the Lewis acid sites in the monolayer niobium oxides may be distributed in an isolated manner. The niobium oxide layer is somewhat distorted by the structural mismatch and the strong Nb–O–Si interaction between the Nb oxide overlayer and the SiO_2 surface, as proven by EXAFS. The distortion and mismatch should be released by the creation of the Lewis-acidic Nb sites. Nb-oxide monolayers on various oxide supports have been extensively investigated and characterized by Raman spectroscopy [15]. These structures are described as being made up of an octahedrally coordinated NbO_6 structure with different degrees of distortion. The Lewis acid sites may be dispersed basically in the NbO_6 overlayer, rendering this type of catalyst applicable to intramolecular dehydration processes. Note that the catalytic property of Nb species can be regulated by the number of Nb atoms, where the structure requirement of catalysis is sensitive to molecular-level arrangement of Nb atoms on the surface.

Table 8.4   Catalytic Performances of the Nb Monomer, Dimer, and Monolayer Supported on SiO_2, and an Impregnation Nb Catalyst for Dehydrogenation and Dehydration (Intra and Inter) of Ethanol

Catalyst	Initial Rate (mmol/min/g-Nb)			Selectivity (%)		
	Total	AA	E+DE	AA	E	DE
Monomer[a]	1.25	1.20	0.049	96.1	2.8	1.1
Dimer[a]	0.18	0.004	0.176	2.1	24.2	73.7
Monolayer[a]	0.11	0.001	0.106	0.9	99.1	0.0
Impreg.[a]	0.17	0.052	0.118	30.5	20.2	49.3
Nb_2O_5 bulk[b]	0.0026	0.0004	0.0022	14.8	46.9	38.3

Note: AA, acetaldehyde; E, ethene; DE, diethyl ether; Ethanol = 3.1 kPa.

[a] 523 K.
[b] 573 K.

8.3.2 ReO$_x$ Clusters Produced *in Situ*

Rhenium is one of the oxophilic atoms effective for oxidation reactions. ReO$_x$ species are likely to have chemical interaction with various oxide supports and exhibit unique catalytic properties that cannot be observed on monomeric rhenium oxides. A new active six-membered octahedral Re cluster in zeolite pores (H-ZSM-5 [HZ]) is produced from inactive [ReO$_4$] monomers *in situ* under selective propene oxidation to acrolein ($C_3H_6+O_2 \rightarrow CH_2=CHCHO+H_2O$) in the presence of ammonia that is not involved in the reaction equation [16]. The cluster is transformed back to the original inactive monomer in the absence ammonia. Note that coexistence of spectator NH$_3$ is indispensable for the selective oxidation.

The ReO$_x$/HZ catalyst was prepared by the following procedure. Methyl trioxorhenium (MTO) was sublimed under vacuum at 333 K and the vapor was allowed to enter the chamber, where the zeolites were pretreated *in situ* at 673 K under vacuum. After the chemical vapor deposition (CVD) into zeolite pores, undeposited MTO was removed by evacuation at RT. The catalyst was treated at 673 K in He before using.

The CVD catalyst exhibits good catalytic performance for the selective oxidation/ammoxidation of propene as shown in Table 8.5. Propene is converted selectively to acrolein (major) and acrylonitrile (minor) in the presence of NH$_3$, whereas cracking to C$_x$H$_y$ and complete oxidation to CO$_2$ proceeds under the propene+O$_2$ reaction conditions without NH$_3$. The difference is obvious. HZ has no catalytic activity for the selective oxidation. A conventional impregnation Re/HZ catalyst and a physically mixed Re/HZ catalyst are not selective for the reaction (Table 8.5). Note that NH$_3$ opened a reaction path to convert propene to acrolein. Catalysts prepared by impregnation and physical mixing methods also catalyzed the reaction but the selectivity was much lower than that for the CVD catalyst. Other zeolites are much less effective as supports for ReO$_x$ species in the selective oxidation because active Re clusters cannot be produced effectively in the pores of those zeolites, probably owing to its inappropriate pore structure and acidity.

Table 8.5 Performance of ReO$_x$/Zeolite Catalysts in Selective Oxidation/Ammoxidation of Propene on a CVD HZ$_{cvd}$ Catalyst, an Impregnated HZ$_{imp}$ Catalyst, and a Physically Mixed HZ$_{phys}$ Catalyst at 673 K

	HZ	HZ$_{cvd9.2}$[a]	HZ$_{cvd9.2}$[b]	HZ$_{cvd4.5}$	HZ$_{imp4.5}$	HZ$_{phys9.6}$
Re (wt%)	0.0	9.2	9.2	4.5	9.5	9.6
Conversion (%)	19.6	15.2	3.8	15.0	20.0	22.0
Cracking	93.2	5.3	31.8	6.5	59.0	48.3
Ethene	21.5	0.0	18.0	0.0	8.5	10.5
Methylamine	15.0	1.7	0.0	1.5	11.2	9.8
Dimethylamine	14.8	1.2	0.0	1.1	12.5	8.0
Ethylamine	22.0	2.4	0.0	3.9	14.0	10.2
Butene	19.9	0.0	13.8	0.0	11.8	9.8
Oxidation/ ammoxidation	0.0	82.5	0.0	81.6	7.5	10.2
Acrolein	0.0	63.0	0.0	63.5	6.0	6.2
Acrylonitrile	0.0	19.5	0.0	18.1	1.5	4.0
Other products	6.8	12.2	68.2	11.9	33.5	41.5
Acetonitrile	0.0	4.2	4.0	3.8	16.5	20.0
CO$_2$	2.0	4.3	60.0	3.5	10.0	12.6
BTX	4.8	3.7	4.2	4.6	7.0	8.9

Note: Reaction conditions: GHSV 19,200 h^{-1}, C$_3$H$_6$/NH$_3$/O$_2$/He = 7.5/7.5/10.0/75.0%.

[a]HZ into sample was treated in NH$_3$/He (50:50%) to form HZ$_{cvd}$.

[b]The reaction was conducted in the absence of NH$_3$ (C$_3$H$_6$/O$_2$/He = 7.5/10.0/82.5%,

To investigate active Re species and the promoting role of NH_3, characterizations of the CVD catalyst have been performed by means of XRD, ^{29}Si and ^{27}Al solid-state MAS NMR, FT-IR, XANES, and EXAFS at the sequential stages for the reaction. Figure 8.7 illustrates the structural changes of ReO_x species in the zeolite pore in successive steps such as the MTO CVD, the He treatment at 673 K, the NH_3 treatment at 673 K, and the treatment with a mixture of propene and O_2 at 673 K. MTO adsorbs on a tetrahedral Al site and near an OH group in zeolite pores at the first step. EXAFS analysis indicates the existence of two shells, Re–O (CN= 3.0, bond distance R = 0.171 nm) and Re–C (CN= 1.1, R= 0.201 nm), indicating retention of its original MTO structure. FT-IR peak intensity at 3610 cm^{-1} attributed to zeolitic protons significantly decreases upon adsorption. The adsorbed MTO in zeolite pore (**6**) decomposes at 673 K to form $[ReO_4]$ species (**7**) (Re–O = 0.172 nm), evolving CH_4 as proved by Re L_1-edge XANES and L_3-edge EXAFS [16]. The bonding of hydrophilic $[ReO_3]$ moiety to the Si–O–Al bridge oxygen may increase the hydrophilic nature of HZ. The chemical shift of tetrahedral Al species in ^{27}Al solid-state NMR changes from 60 ppm for species (**6**) to 50 ppm for species (**7**), indicating the additional bonding between $[ReO_3]$ and tetrahedral Al site in Figure 8.7.

After NH_3 exposure, the XANES and EXAFS spectra change remarkably and Re–Re bonding appears. EXAFS data are well fitted by three shells, Re=O (CN = 2.3, R= 0.172 nm), Re–O (CN = 1.8, R = 0.203 nm), and Re–Re (CN = 3.9, R = 0.276 nm). The CN for Re–Re suggests the formation of a six-membered Re_6 cluster framework in octahedral geometry (**8**). Re atoms occupy six corners of the octahedron and each Re atom has four-coordinated with neighboring Re atoms. Accompanied with the change of Re framework by the NH_3 exposure, the shifted MAS NMR signal for tetrahedral Al returns to the original position at 60 ppm, indicating that a part of

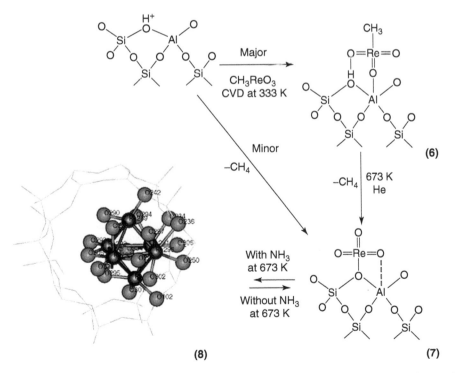

Figure 8.7 Structural changes of ReO_x species in HZcvd catalyst preparation and the catalytic reaction conditions, and a proposed structure of active $[Re_6O_{17}]$ cluster in the ZSM-5 pore channel, where the $[Re_6O_{13}]$ cluster is bound to the pentagonal rings of the zeolite inner wall via three lattice oxygen atoms, and the oxygen atoms are tentatively arranged on the Re_6 octahedron.

Al interacts with the Re species due to Re clusterization and that the additional interaction between the Re cluster and tetrahedral Al is weak.

The proposed Re_6 cluster (**8**) with terminal and bridged-oxygen atoms acts as a catalytic site for selective propene oxidation under a mixture of propene, O_2 and NH_3. When the Re_6 catalyst is treated with propene and O_2 at 673 K, the cluster is transformed back to the inactive $[ReO_4]$ monomers (**7**), reversibly. This is the reason why the catalytic activity is lost in the absence of ammonia (Table 8.5). Note that NH_3, which is not involved in the reaction equation for the acrolein formation $(C_3H_6+O_2 \rightarrow CH_2=CHCHO+H_2O)$ is a prerequisite for the catalytic reaction as it produces the active cluster structure under the catalytic reaction conditions.

8.4 DESIGN OF REACTION INTERMEDIATE AND TRANSITION-STATE ANALOGUE FOR A TARGET REACTION ON OXIDE SURFACES

If it is assumed that enzymes can accommodate the structure of the reactant molecule in the transition state [17], the activation energy can be drastically decreased by the stabilization of the transition state. If one can synthesize a molecule that fits a structure of the transition state, the obtained molecule (transition-state analogue) can be used as a template for an antibody [18]. A stable transition-state analogue (phosphuric ester) for the proposed transition state in the enzyme-catalyzed reaction can be chemically synthesized. By using it, its antibody can be produced in some organisms. This antibody exactly fits the transition state and hence catalyzes the reaction through a preferable interaction with the transition state [19–21]. Preparation of an artificial enzyme catalyst as molecular architecture has generally been a difficult and unaddressed challenge to date.

Molecular design of the active site on catalyst surfaces regarded as a reaction intermediate for a target catalytic reaction is also a way to provide efficient catalysis [22]. In order to achieve the catalyst design, the reaction mechanism including the structural and electronic change of active metal sites must be known at a molecular level.

8.4.1 Reaction Regulation by Molecular Imprinting

Since the discovery of catalysis of metal complexes by Wilkinson et al. [23], many metal complexes have been used as catalytically active species in homogeneous systems, and catalytic reaction mechanisms on metal complexes have been investigated for various reactions. It is well known that ligands coordinated to a metal center significantly modify and regulate not only reactivity of the metal electronically but also the space around the metal geometrically. As a result, tremendous selectivity can be achieved on metal–complex catalysts, which may be difficult to obtain on metal particles and metal single–crystal surfaces. Optimum regulation by ligands in metal complexes finds selective catalysis and asymmetric synthesis.

Homogeneous metal complexes in solution easily gather during catalytic cycles and cause decomposition of metal complexes, resulting in loss of the catalytic activities. Transformation of homogeneous catalysts to heterogeneous catalysts with molecular-level active structures has been accomplished by attaching metal complexes on oxide or polymer supports [4, 6, 7, 24–29]. The new and distinct materials and chemistry prepared stepwise in a controllable manner by using organometallic and inorganic complexes as precursors provide an opportunity for the development of efficient catalytic molecularly organized surfaces. In the development of novel catalysts, new chemical concepts regarding composition or structure are conceived. This section discusses the progress in molecular imprinting as a new tool for design of shape-selective metal complexes at surfaces.

To prepare artificial enzymatic systems possessing molecular recognition ability for particular molecules, molecular imprinting methods that create template-shaped cavities with the memory of the template molecules in polymer matrices, have been developed [22, 30–35] and established in receptor, chromatographical separations, fine-chemical sensing, etc. in the past decade. The molecular design of catalytic sites is necessary for the regulation of catalysis. The regulation of dynamic catalytic processes requires careful strategy of chemical design of the catalytic sites.

The principle of cavity creation with a similar shape to that of a template molecule in an appropriate matrix, by molecular imprinting involves two processes as shown in Figure 8.8, which illustrates an example of metal complexes immobilized on a support surface. An organic or inorganic polymer matrix is produced around a particular molecule used as a template, covering the space around the template. Then the template molecule is removed from the polymer matrix, and as a result, a template-shaped cavity in place of the removed template is obtained in the matrix. The cavity acts as a selective recognition space for molecules with a shape similar to the template. In most cases of metal-complex imprinting, ligands of the complexes are used as template molecules, which aim to create a cavity near the metal site. Molecular imprinting of metal complexes enables realization of several features, e.g., (1) attachment of metal complex on robust supports, (2) surrounding of the metal complex by polymer matrix, (3) formation of new active structure in the matrix, and (4) production of shape-selective space on the metal site. Most of the imprinted metal-complex catalysts have been prepared by imprinting in bulk polymers. However, active sites prepared by bulk imprinting may often be located inside the bulk polymer matrices, where the access of substrate molecules to the sites is difficult. Further, such organic polymers tend to have limited durability in organic solvents or under severe catalytic conditions such as in the presence of oxidants, at high temperatures, etc. On the other hand, oxide supports are inert and robust materials, on which metal complexes can be attached by chemical bondings. Oxide-supported metal complexes often exhibit unique properties and reactivities so that molecular imprinting of metal complexes at oxide surfaces has many possibilities to produce new excellent catalysts similar to enzymatic systems. Recently, for the time,imprinted rhodium complexes on an SiO_2 surface for catalytic shape-selective hydrogenation of simple alkenes without any functional group was designed [36–38]. The preparation steps are presented in Figure 8.9. In heterogeneous systems, generally metal–metal interaction is crucial to dissociate hydrogen molecules, which is most relevant to hydrogenation. The metal dimer is regarded as a minimum active structure for heterogeneous metal catalysis involving metal–metal bonding. In order to create an Rh dimer structure with a template cavity on an Ox.50 (SiO_2 with a low surface area) surface, an $Rh_2Cl_2(CO)_4$ and $P(OCH_3)_3$ were used as an Rh dimer precursor and a template ligand, respectively. The template $P(OCH_3)_3$ is regarded as an analogue to a half-hydrogenated alkyl intermediate of hydrogenation of 3-ethyl-2-pentene.

$Rh_2Cl_2(CO)_4$ is attached on the Ox.50 surface, retaining the bridged-dimer structure monitored by FT-IR. By exposing the surface to $P(OCH_3)_3$, the surface-attached Rh carbonyl dimer is converted into an Rh monomer pair (Rh_{2sup})(**9**) with two $P(OCH_3)_3$ ligands on Rh (Rh–P: 0.224 nm), which is attached on the surface via Rh–O bonding at 0.203 nm in a bidentate form. The first step of molecular imprinting for the attached rhodium complexes is performed by hydrolysis-polymerization of $Si(OCH_3)_4$, which possesses methoxy groups by positive interaction with the template. The second step is the removal of template $P(OCH_3)_3$ by evacuation at 363 K. Note that

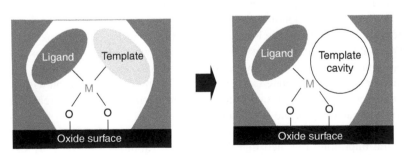

Figure 8.8 Creation of a template-shaped cavity in the matrix for metal complexes supported on an oxide surface.

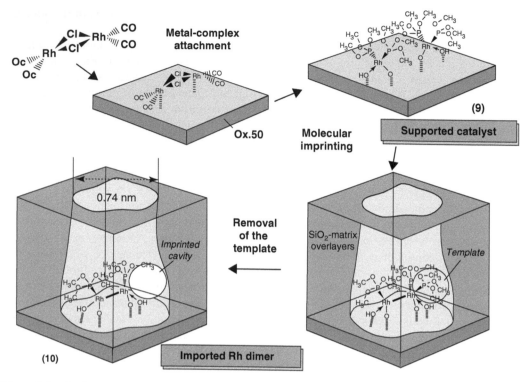

Figure 8.9   Design of a molecular imprinting Rh_{2imp} catalyst with Rh dimer structure on an SiO_2 surface.

the surface imprinting causes dimerization of the Rh monomers to produce highly active Rh dimers (**10**) with a direct Rh–Rh bond at 0.268 nm in the pores of 0.74 nm dimension in the SiO_2-matrix overlayers. By the molecular imprinting procedure, the coordination number of Rh–P bond at 0.221 nm becomes unity. The change in Rh structures in the imprinting processes can be explained by DFT calculation [37]. We assume that the elimination of the two phosphite ligands from the monomer pair to produce the $Rh_2(P(OCH_3)_3)_2$ structure (Rh_{2imp} catalyst (**10**)) occurs stepwise via the stable intermediate $Rh_2(P(OCH_3)_3)_3$ with one $P(OCH_3)_3$ on an Rh atom and with two $P(OCH_3)_3$ on another Rh atom in the dimer at the Rh–Rh bond of 0.271 nm. Therefore, the stable intermediate $Rh_2(P(OCH_3)_3)_3$ is the imprinted Rh species under the CVD process at 348 K. The elimination of a template $P(OCH_3)_3$ ligand from the intermediate at 363 K under vacuum generates the template size cavity in the SiO_2-matrix overlayers.

Hydrogenation reactions are remarkably promoted for all alkenes by surface imprinting. The imprinted Rh_{2imp} catalyst (**10**) is tremendously active for the alkene hydrogenation. For example, hydrogenation of 2-pentene is promoted by factor of 51 as compared to that of the supported Rh catalyst (**9**). The metal–metal bonding and coordinative unsaturation of the Rh dimer are key factors for the remarkable activity of the Rh_{2imp} catalyst (**10**). The Rh_{2imp} catalyst is highly durable and surprisingly air-stable in spite of its unsaturated structure, which is advantageous in practical handling of the system. Further, it can be reused without any loss of catalytic activity. It seems that the chemical attachment in a tetradentate form (Rh–O: 0.211 nm) and the location with stable fitting in the micropore of 1.9 nm thickness prevent the Rh dimers from leaching to the reaction solution, and decomposition and gathering of the Rh dimers.

Ratios of turnover frequencies (TOFs) corresponding to the degree of enhancement of the reaction rates by the imprinting revealed that the imprinted Rh-dimer catalyst (**10**) showed size and shape selectivities for the alkenes as shown in Figure 8.10. Selectivity for the alkene hydrogenation on the Rh_{2imp} catalyst (**10**) depends on the size and shape of the template cavity as reaction site in

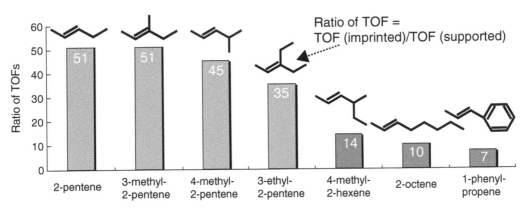

Figure 8.10 Shape and size selectivity of the imprinted Rh-dimer catalyst in alkene hydrogenation.

the micropores of the SiO_2-matrix overlayers on the Ox.50 surface, in addition to the electronic and geometric effects of the ligands. The TOF ratios reduce with gain of alkene size as shown in Figure 8.10. It is to be noted that there is a large decrease between 3-ethyl-2-pentene and 4-methyl-2-hexene (4-ethyl-2-pentene) due to the difference in the shape of the alkenes. The difference in the TOF ratios between 4-methyl-2-pentene and 4-methyl-2-hexene is also large where the difference in the size of a methyl group is discriminated on the Rh_{2imp} catalyst. The TOF ratio for 4-methyl-2-hexene is much smaller than that for 2-pentene, where the difference is attribuited to the ethyl group. There is also a big difference between 2-pentene and 2-octene, where a propyl group can be discriminated. Thus, the molecular imprinting catalyst discriminates the size and shape of the alkenes. However, there was no significant difference in the TOF ratios between 2-pentene and 4-methyl-2-pentene, between 3-methyl-2-pentene and 3-ethyl-2-pentene, and between 2-pentene and 3-ethyl-2-pentene (Figure 8.10). Thus, the molecular imprinting catalyst does not discriminate the existence of methyl and ethyl groups among the alkenes smaller than the template size. It is worth noting that the reaction rates of 2-pentene, 2-octene, and 1-phenylpropene on the supported Rh_{2sup} catalyst (**9**) are similar to each other, whereas thc rate enhancements (TOF ratios) for 2-octene and 1-phenylpropene (10 and 7 times, respectively) are much less than that for 2-pentene (51 times). Because the length of the linear alkene chains cannot be discerned by the ligand-coordinated metal site, it has been suggested that the difference is caused by the wall of the template cavity around the Rh-dimer site in the imprinted catalyst (**10**).

Activation energies for the hydrogenation on the Rh_{2sup} catalyst (**9**) are divided into two values, about 30 and 42 kJ/mol for the linear alkenes and the branched alkenes, respectively, in Table 8.6. Such a large difference in the activation energy has also been observed with the Wilkinson complex. On the contrary, activation entropies for the hydrogenation on the Rh_{2sup} catalyst are similar to each other, which values are in the range -200 to -215 J/mol/ K for all the alkenes.

After the imprinting, significant differences between small and large alkenes is observed in both activation energies and activation entropies. The activation energies for the hydrogenation of 3-ethyl-2-pentene and smaller alkenes (the four alkenes on the left in Figure 8.10) on the imprinted Rh_{2imp} catalyst (**10**) are 26 to 43 kJ/mol, which are similar to the values observed on the Rh_{2sup} catalyst (**9**). The activation entropies for the four alkenes are in the range -170 to -195 J/mol/K, which are larger than those (-200 to -210 J/mol/K) obtained for the Rh_{sup} catalyst (**9**). In contrast, for the larger alkenes such as 4-methyl-2-hexene, 2-octene, and 1-phenylpropene, the activation energies are 10, 7, and 8 kJ/mol, respectively, which are small compared with those for the Rh_{2sup} catalyst (**9**) and other metal complex catalysts. Furthermore, the activation entropies reduce significantly from about -210 J/mol/K to about -260 J/mol/K. The conspicuous change in the kinetic parameters for the larger alkenes is parallel to the change in enhancement of the reaction rates. These dramatic decreases in the activation energy and the TOF ratio can be explained by the

Table 8.6   Activation Energies and Activation Entropies for the Alkene Hydrogenation on the Supported and Imprinted Rh-Dimer Catalysts

Reactant	Supported Catalyst (Rh$_{sup}$)		Imprinted Catalyst (Rh$_{imp}$)	
	E_a (kJ/mol)	$\Delta^{\ddagger}S$ (J/K/ mol)	E_a (kJ/mol)	$\Delta^{\ddagger}S$ (J/K/ mol)
2-Pentene	34	−205	26	−195
3-Methyl-2-pentene	44	−200	43	−170
4-Methyl-2-pentene	40	−207	40	−175
3-Ethyl-2-pentene	42	−210	39	−189
4-Methyl-2-hexene	40	−212	10	−276
2-Octene	28	−215	7	−257
1-Phenylpropene	29	−213	8	−256

shift of the rate-determining step from alkyl formation to the coordination of alkene to the Rh site. The activation entropies for the larger alkene molecules also decreased largely to about −260 from about −180 J/mol/K for the other small molecules. It suggests that the conformation of the coordinated alkene in the template cavity is regulated by the wall of the cavity and the remaining P(OCH$_3$)$_3$ ligands. For the alkenes with the larger sizes and different shapes compared to the template, the coordination to the Rh site through the cavity space becomes the slowest in the reaction sequences. The location of the Rh center to which the alkenes coordinate, the conformation of the remaining P(OCH$_3$)$_3$ ligand, the orientation of template vacant site on Rh, the template cavity shape, the architecture of the cavity wall, and the micropore surrounding the Rh monomer in the SiO$_2$-matrix overlayers provide the active Rh$_{2imp}$ catalyst (**10**) for size- and shape-selective hydrogenation of the alkenes.

The method combining metal-complex attaching and molecular imprinting on the surface demonstrates that the strategy can regulate and design chemical reactions at a molecular level like artificial enzyme catalysts. Achievement of more strict recognition of molecules needs further investigation and molecular recognition of reactants with functional groups is the next stage of the study.

8.4.2   Design of a Reaction Intermediate on Catalyst Surface

Alkene hydroformylation on mononuclear Rh complexes in homogeneous catalytic systems proceeds via an Rh-acyl intermediate. On the other hand, heterogeneous catalysis generally involves more than one metal atom at catalyst surfaces. In fact, ethene hydroformylation on supported Rh catalysts shows a maximum activity at an optimum particle size of about 2 nm, suggesting the contribution of a metal ensemble to the catalytic performance. It may be of importance to know the role of metal–metal bonding in metal catalysis. An example of catalyst design by synthesizing an Rh dimer with an acyl ligand at a SiO$_2$ surface is introduced here.

Trans-[(RhCp*CH$_3$)$_2$(μ-CH$_2$)$_2$] (Cp*=pentamethy-clopentadienyl) has CH$_3$ and μ-CH$_2$ ligands which can react with each other to form an ethyl ligand on the Rh atom. The complex was supported on an SiO$_2$ surface with retention of the Rh–Rh bond, and the CH$_3$ and μ-CH$_2$ were reacted to form C$_2$H$_5$ on an Rh atom (species **11** in Scheme 8.4), which were characterized by EXAFS and FT-IR. The Rh dimers (**11**) were exposed to CO to form *gem*-carbonyls (**12**), which appeared at 2032 and 1969 cm^{-1}. One of the carbonyls is inserted into ethyl ligand to form an acyl ligand (1710 and 1394 cm^{-1}) by heating at 423 K as illustrated in Scheme 8.4. Note that the Rh dimer obtained (**13**) is regarded as a catalytic intermediate for ethene hydroformylation. The Rh dimer (**13**) was used as a catalyst for the catalytic hydroformylation and performed efficiently, as shown in Table 8.7. The performances of Rh dimers supported on TiO$_2$, Al$_2$O$_3$, and MgO are also shown in Table 8.7. The Rh dimer catalysts were all inactive expect for the SiO$_2$-supported Rh dimer catalyst (Rh$_2$/SiO$_2$), which exhibited much higher activity and selectivity for propanal formation than the impregnated Rh/SiO$_2$ catalyst. This result is opposite to the supported Rh cluster catalysts that showed higher activity when Rh$_4$(CO)$_{12}$ and Rh$_6$(CO)$_{16}$ were deposited on basic supports like MgO, ZnO, and La$_2$O$_3$,

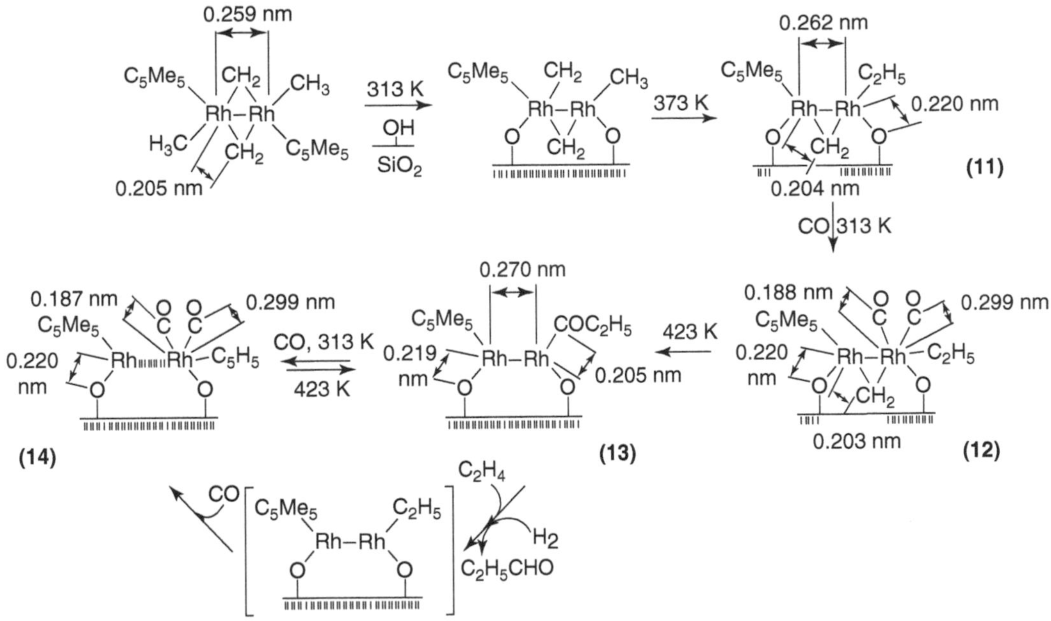

Scheme 8.4 The preparation step for Rh_2/SiO_2 and the structural change during ethene hydroformylation.

Table 8.7 Catalytic Performances of Ethene Hydroformylation on the Rh Dimer Catalysts at 413 K

Catalyst	TOF (Total)	TOF (Ethane)	TOF (Propanal)	Selectivity (%)
Impreg. Rh/SiO$_2$	22.8	21.5	1.3	5.6
Rh$_2$/SiO$_2$	36.9	4.1	32.8	88.9
Rh$_2$/TiO$_2$	6.3	6.3	0	0
Rh/Al$_2$O$_3$	6.8	6.8	0	0
Rh$_x$/MgO	56	56	0	0

Note: TOF = 10^{-4} min^{-1}; CO/H$_2$/C$_2$H$_4$ = 1:1:1 (total pressure = 40.0 kPa).

compared with that supported on acidic supports like SiO$_2$ and Al$_2$O$_3$. Thus, the possibility of another mechanism and support effect that works in the Rh dimer catalyst system is suggested.

When species (**13**) is exposed to CO at 313 K, dicarbonyl species is formed, while Rh–Rh bond is cleaved (species **14**). Transformation between (**13**) and (**14**) reversibly occurs as proved by both FT-IR and EXAFS. CO insertion from (**14**) to (**13**) proceeds under CO-deficient conditions. Generally, in homogeneous systems, CO insertion proceeds more favorably under CO-rich (high-pressure CO) conditions or with the coexistence of cocatalysts like PPh$_3$, because excess CO or PPh$_3$ occupies the vacant site on the Rh atom to stabilize the acyl species produced. On Rh$_2$/SiO$_2$, the regenerated Rh–Rh bond plays a similar role as an additional ligand. The CO insertion is promoted by the formation of metal–metal bonding. This mechanism is also rationalized by kinetic data of CO hydroformylation [39].

On Rh$_2$/TiO$_2$ the Rh–Rh bond is also regenerated by heating the dicarbonyl species, but is accompanied by the formation of a CO-bridged species, where one of the dicarbonyl ligands is merely desorbed. Thus, CO insertion promoted by Rh–Rh bond formation does not take place on Rh$_2$/TiO$_2$. On Rh$_2$/Al$_2$O$_3$, Rh dimer structure is observed as the main species by EXAFS. Rh species on MgO is aggregated to form a small Rh cluster with direct Rh–O bonding. The Rh–Rh bonding in the cluster on MgO is stable and not cleaved by exposure to CO. The metal-assisted CO insertion is not observed with the Rh$_2$/MgO catalyst.

The reason why the activity varies so much according to the kind of supports is that the surface structure properties vary with the kinds of supports, which causes significant difference in reactivity. The Rh dimer complex is supported on SiO_2, keeping its dimer structure. The Rh–Rh bond has such a flexible character that it makes CO insertion promoted by Rh–Rh bond formation possible. On TiO_2, the Rh dimer is supported, keeping its original structure to some extent. Rh–Rh bond is broken at one, but is regenerated by heating at 473 K. Unlike Rh_2/SiO_2, the formation of Rh–Rh bond does not promote CO insertion; instead μ-CO is formed. On Al_2O_3, metal-support interaction becomes stronger and the Rh dimer is supported as monomer, and formation of Rh–Rh bond is prohibited. Again, metal–metal-assisted CO insertion does not occur on Al_2O_3 support. On MgO, the Rh dimer is converted into clusters during the heating of the surface-Rh dimers. No hydroformylation reaction occurs. On Rh_2/TiO_2, Rh_2/Al_2O_3 and Rh_2/MgO, promotion of CO insertion by the adjacent Rh atoms is not observed, possibly because the interaction between Rh dimers and supports are too strong.

8.5 CONCLUSION

In this chapter, the issues of promoting and regulating catalytic reactions by the presence of coadsorbed species and the design of active structures at catalyst surfaces have been addressed for two classes of well-defined catalyst surfaces: chemically designed surfaces and single-crystal surfaces. However, the precise way of controlling molecular activation and reaction intermediates that are critical in achieving high selectivity has not been adequately addressed. The requirement and design of molecular structures and quantitative ensemble sizes for efficient catalysis represent important but as yet unaddressed challenges to the field. This sort of study should be tackled. Understanding and controlling catalyst surfaces are the key issues for development of new catalysts, but conventional supported metal catalysts are generally heterogeneous and complicated. Their characteristics are difficult to ascertain, especially under catalytic reaction conditions, which results in the restriction of the development of new catalytic materials. Molecular-level control of catalyst surfaces is necessary for design of multifunctionalized catalytic sites. New and distinct materials and chemistry prepared stepwise in a controllable manner by using organometallic and inorganic complexes as precursors provide an opportunity for the development of efficient catalytic molecularly organized surfaces. The key factors of chemical design of supported catalyst surfaces include, for example, composition, structure, oxidation state, distribution, morphology, and polarity, which should be organized at the surface. In the development of novel catalysts, new chemical concepts and strategies regarding composition or structure are conceived, which can be associated with *in situ* characterization. Well-defined single-crystal surfaces are model catalyst surfaces that have advantages in finding the key requirements for selective catalysis with the aid of modern physical techniques that can be applied most efficiently to single-crystal surfaces. The establishment of molecular-level rational design of catalyst surfaces with good performances and the development of *in situ* characterization techniques for catalyst surfaces under the working conditions are still important subjects to be solved as an objective of the field.

REFERENCES

1. Y. Iwasawa, K. Asakura, H. Ishii, and H. Kuroda, *Z. Phys. Chem. N. F.* 144 (1985) 105.
2. B. C. Gates, L. Guczi, and H. Knözinger (eds.), *Metal Cluster in Catalysis*, Elsevier, Amsterdam, 1986.
3. Y. Iwasawa, *Adv. Catal.* 35 (1987) 187.
4. Y. Iwasawa, *Catal. Today* 18 (1993) 21.
5. M. Nishimura, K. Asakura, and Y. Iwasawa, *J. Chem. Soc., Chem. Commun.* (1986) 1660.
6. Y. Iwasawa, *Acc. Chem. Res.* 30 (1997) 103.
7. Y. Iwasawa, *Stud. Surf. Sci. Catal.* 101 (1996) 21.

8. Y. Iwasawa, *Elementary Reaction Steps in Heterogeneous Catalysis* (R. W. Joyner and R. A. van Santen Eds.), NATO ASI Series, Kluwer, the Netherlands, 1993 pp. 287–304 .

9. T. Aruga, K. Fukui, and Y. Iwasawa, *J. Am. Chem. Soc.* 114 (1992) 4911.

10. K. Fukui, K. Motoda, and Y. Iwasawa, *J. Phys. Chem. B* 102 (1998) 8825.

11. E. I. Ko, R. J. Madix, *Surf. Sci.* 112 (1981) 373.

12. P. A. Redhead, *Vacuum* 12 (1962) 203.

13. J. Haber, in *Molybdenum: An Outline of its Chemistry and Uses* (E. R. Braithwaite, and J. Haber, Eds.), *Stud. Inorg. Chem.*, vol. 19, Elsevier, Amsterdam, 1994, p. 477.

14. H. Pfnür, P. Feulner, H. A. Engelhardt, and D. Menzel, *Chem. Phys. Lett.* 59 (1978) 481.

15. J. M. Jehng and I. E. Wches, *J. Phys. Chem.* 95 (1991) 7373.

16. N. Viswanadham, T. Shido, T. Sasaki, and Y. Iwasawa, *J. Phys. Chem. B* 106 (2002) 10955.

17. L. Pauling, *Am. Sci.* 36 (1948) 51.

18. W. P. Jencks, *Catalysis in Chemistry and Enzymology*, McGraw-Hill, New York, 1969.

19. S. J. Pollack, J. W. Jacobs, and P. G. Schultz, *Science* 234 (1986) 1570.

20. J. Jacobs, P. G. Schultz, R. Sugasawara, and M. Powell, *J. Am. Chem. Soc.* 109 (1987) 2174.

22. A. Tramontano, K. D. Janda, and R. A. Lerner, *Science* 234 (1986) 1566.

22. M. Tada and Y. Iwasawa, *J. Mol. Catal. A: Chemical* 199 (2003) 115.

23. J. A. Osborn, F. H. Jardine, J. F. Young, and G. Wilkinson, *J. Chem. Soc. (A)* (1966) 1711.

24. Y. Iwasawa, *Tailored Metal Cataysts*, Reidel, Dordrecht, 1986.

25. Y. Iwasawa, *Adv. Catal.* 87 (1987) 187.

26. F. R. Hartley, *Supported Metal Complexes*, D. Reidel, Dordrecht (1985).

27. R. Psaro, S. Recchia, *Catal. Today* 41(1–3) (1998) 139.

28. B. C. Gates, *Topics in Catal.* 14(1–4) (2001) 173.

29. A. Zecchina, C. O. Arean, *Catal. Rev. —Sci. Eng.* 35 (1993) 261.

30. P. A. Brady and J. K. M. Sanders, *Chem. Soc. Rev.* 26 (1997) 327.

31. M. E. Davis, A. Katz, and W. R. Ahmad, *Chem. Mater.* 8 (1996) 1820.

32. B. Shellergren, *Angew. Chem. Int. Ed.* 39 (2000) 1031.

33. M. J. Whitecombe, C. Alexander, and E. N. Vulfson, *Synlett* 6 (2000) 911.

34. K. Mosbach, *Chem. Rev.* 100 (2000) 2495.

35. M. Tada and Y. Iwasawa, *J. Mol. Catal. A: Chemical* 204–205 (2003) 27.

36. M. Tada, T. Sasaki, and Y. Iwasawa, *Phys. Chem. Chem. Phys.* 4 (2002) 4561.

37. M. Tada, T. Sasaki, T. Shido, and Y. Iwasawa, *Phys. Chem. Chem. Phys.* 4 (2002) 5899.

38. M. Tada, T. Sasaki, and Y. Iwasawa, *J. Catal.* 211 (2002) 496.

39. K. K. Bando, K. Asakura, H. Arakawa, K. Isobe, and Y. Iwasawa, *J. Phys. Chem.* 100 (1996) 13636.

CHAPTER 8 PROBLEMS

Problem 1

Describe the promising methods of design of active structures on support surfaces.

Problem 2

In section 3.1 the selectivity of ethanol dehydrogenation and dehydration on supported Nb catalysts is controlled by design of active Nb structures. Describe the important issues to control the selectivity of catalysis.

Problem 3

The rate of ethane formation in ethanol dehyration ($C_2H_5OH \dashrightarrow C_2H_4 + H_2O$) on the Nb dimer/$SiO_2$ catalyst is accelerated 1.7 times when C_2H_5OH is replaced by C_2D_5OH. The substitution of the OH hydrogen with deuterium gives no effect on the rate. Explain the inverse isotope effect.

Problem 4

Considering the catalytic reaction, $C_2H_5OH + 1/2\ O_2 \dashrightarrow CH_3CHO + H_2O$,
When this reaction proceeds via the following two steps:

$$C_2H_5OH + [Mo^{6+}=O] \dashrightarrow CH_3CHO + H_2O + [Mo^{4+}]$$

$$[Mo^{4+}] + 1/2\ O_2 \dashrightarrow [Mo^{6+}=O]$$

give the catalytic reaction rate v for the ethanol oxidation. $[Mo^{6+}=O]$ and $[Mo^{4+}]$ represent an active site with Mo = O bond and its reduced site in a Mo-oxide catalyst.

Texturology

Vladimir B. Fenelonov and Maxim S. Mel'gunov

CONTENTS

9.1   Introduction 258
9.2   Basic Principles of Physical Chemistry of Dispersed Systems 261
    9.2.1   Gibbsian Classical Thermodynamic Theory 261
    9.2.2   Flat Interface 262
    9.2.3   Curved Interface 264
    9.2.4   Surface Curvature 265
    9.2.5   The Limits of the Classical Thermodynamic Theory 266
    9.2.6   Typical Mechanisms and Processes of Texture Genesis Derived
              from the Laws of Surface-Capillary Phenomena 267
           9.2.6.1   Fundamental Mechanisms of Texture Genesis 267
           9.2.6.2   Fundamental Processes of Texture Genesis 269
9.3   Adsorption as a Primary Instrument for Texture Characterization 274
9.4   Morpho-Independent Textural Parameters 280
    9.4.1   Density and Porosity 280
    9.4.2   Experimental Techniques of Measurements of True, Apparent,
              and Bulk Density 283
    9.4.3   The Properties of Porosity 284
    9.4.4   The Specific Surface Area 289
9.5   Morpho-Dependent Textural Parameters: Mean Sizes of Particles and Pores 290
9.6   General Problems of Porous Solids Texture Modeling 293
    9.6.1   Morphology of Porous Solids and Problems with Modeling 293
    9.6.2   Classification of Porous Systems and Texture Modeling 294
    9.6.3   Generalized Models and Systematic Sets of Models 299
9.7   Modeling Particles and Pores in a Local Arrangement 301
    9.7.1   Voronoi–Delaunay Method for Description of Corpuscular
              and Sponge-Like Porous Solids 301
    9.7.2   Ordered Packings 306
    9.7.3   Disordered Packings 311

9.8   Modeling the Ensembles (Clusters) of Particles and Pores on the
       Basis of a Fractal Approach 314
9.9   Lateral and Statistical Models of Pores and Particles Arrangement 320
       9.9.1   Percolation Theory 320
               9.9.1.1   Problem of Bonds 321
               9.9.1.2   Problem of Sites 322
       9.9.2   The Stochastic and other Statistical Models of Long-Range Order 324
9.10   Conclusions 327
References 328
Chapter 9 Problems 335

9.1   INTRODUCTION

A researcher in the field of heterogeneous catalysis, alongside the important studies of catalysts' chemical properties (i.e., properties at a molecular level), inevitably encounters problems determining the catalyst structure at a supramolecular (textural) level. A powerful combination of physical and chemical methods (numerous variants x-ray diffraction (XRD), IR, nuclear magnetic resonance (NMR), XPS, EXAFS, ESR, Raman of Moessbauer spectroscopy, etc. and achievements of modern analytical chemistry) may be used to study the catalysts' chemical and phase molecular structure. At the same time, characterizations of texture as a fairytale Cinderella fulfill the routine and very frequently senseless work, usually limited (obviously in our modern transcription) with electron microscopy, formal estimation of a surface area by a BET method, and eventually with porosimetry without any thorough insight.

The problem of textural studies is not just reduced to the development of their experimental base, but more important is interpretation of the obtained results. The reliable interpretation should be based on the accounting of specific rules and laws that determine the interrelations between textural parameters, their origin and evolution during synthesis, and processing of the catalyst. By the way, that is not a problem limited to heterogeneous catalysis. The same problems arise during studies of porous films in physics, soils in geology, as well as for colloidal and biological systems. Therefore, we believe that nowadays allocation of these problems to an independent interdisciplinary scientific direction becomes imperative. The attempted development of such an approach including a set of theoretical and experimental methods, designated here as *Texturology* (texture, *Latin*: features of a structure of something considered as a whole, caused by an arrangement of its components; and logos, *Greek*: knowledge, doctrine), is the focus of this chapter.

Here, following the works of J.H. De Boer (Delft, The Netherlands, see elsewhere [1,2]), by texture one means the individual geometrical structure of catalysts, supports, and other porous systems (PSs) at the level of pores, particles and their ensembles (i.e., on a supramolecular level scale of 1 nm and larger). In a more complete interpretation, texture includes morphology of porous space and the skeleton of a condensed (solid or sometimes liquid) phase, the shape, size, interconnectivity, and distribution of individual supramolecular elements of the system: particles and pores (or voids) between particles, various phases, etc. In turn, texturology also involves general laws of texture formation and methods for its characterization [3].

Many of the physicochemical laws that determine formation and properties of PS texture have the same origins as J.-M. Lehn's supramolecular chemistry [4]. Those origins highlight a determining role of weak intermolecular interactions, specific influence of geometry and topology, special mechanisms of assembling, including self-assembling, frustration, molecular recognition, etc., that usually are not considered in molecular chemistry.

Figure 9.1 shows some examples of a possible texture (supramolecular structure) variety for PSs of identical molecular composition and structure. All of these materials represent SiO_2 phase

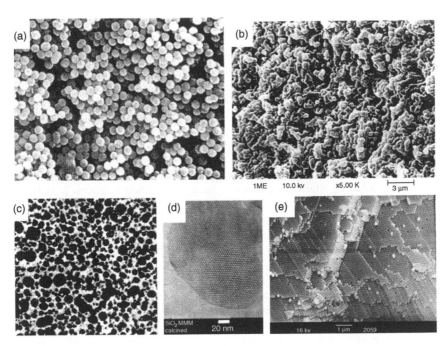

Figure 9.1   Examples of texture of the materials formed on a short range from amorphous SiO_2: (a) silica gel; (b) hydrothermally treated silica gel; (c) porous glass; (d) mesoporous mesophases type of MCM-41; and (e) opal.

amorphous at the molecular level, but their supramolecular structure varies radically due to different synthetic conditions. For example, as-synthesized silica gels (Figure 9.1a) are usually formed of a random packing of spherical particles (globules). However, their sintering under hydrothermal conditions results in accretion of individual globules into conglomerates of various shapes (Figure 9.1b). The texture of a porous glass type of Vycor (Figure 9.1c) is formed during removal of the soluble additives from the amorphous SiO_2 phase. It is similar to sponge, and represents a system of interconnected cavities of various sizes. Using alternative conditions of SiO_2 synthesis one can obtain highly ordered mesophases that combine the structural features of an amorphous phase at a level of short-range order, with the long-range order characteristic of crystals (Figure 9.1d and Figure 9.1e). For example, the structure of mesoporous mesophase type of MCM-41 (Figure 9.1d) is similar to a honeycomb, which is a 2D hexagonal packing of parallel pores of the identical cross-sectional size and shape, whose walls have an apparent amorphous structure. In turn, the structure of opals (Figure 9.1e) is generated by a regular packing of globular particles (size of tens of nanometers), which have amorphous internal structure. Similar textural variations of materials with consistent chemical and phase composition are characteristic of many of PSs [3].

It is essential that all PSs are multiphase. The easiest case to handle is the biphase system consisting of a condensed phase (solid) and a void inside porous particles or between consolidated ensembles of nonporous or porous particles. The void occupies a part of the volume, ε, which is referred to as porosity. The other part of a PS volume is equal to $\eta = (1-\varepsilon)$, and is termed density of packing. It is filled with the condensed phase (see Section 9.4). Generally, PSs can include various condensed phases of different structure, including combinations of solid(s) and liquid(s).

The surface of PSs is determined as an interface between the condensed phase and the space of the pores. For catalysts with supported active components textural characteristics of individual phases are important, including their accessible, surface areas and the interfacial surface areas [5,6].

Here we come to the definition of fundamental textural characteristics of PSs that can be conditionally divided into two groups [3]. We attribute the characteristics that do not require obligatory introduction of any morphological models for a PS or its components to the first group, and designate these characteristics as *morphoindependent*. Porosity and density of packing, whose values are related to unit of volume of a porous solid, are included into this group. For porous powders, tablets, granules, and other PSs, one can allocate three principal values of density [7] as follows:

- *True* or *skeletal density*, ρ, which is the mass, m, divided by the volume, V_ρ, of the solid excluding open and closed pores
- *Apparent* or *envelope density*, $\delta = m/V_\delta$, where V_δ is the volume that includes all the pores inside PS
- *Bulk density*, $\Delta = m/V_\Delta$, which corresponds to the mass of a bed of porous particles, granules, tablets, etc., related to the volume of the whole bulk V_Δ including the pores between granules in a PS bed

The important and commonly used textural characteristics also concerning this group are:

- *Specific surface area* (A) related to a unit of weight or volume (A_V), and also the accessible surface area of the individual components, if some exist in a composite PS
- *Volume of pores*, V_{pore}, related to a unit of weight

The second group of fundamental textural characteristics is closely related to morphology and mutual arrangement of particles and pores. There are, first of all, characteristic sizes of particles, D, and pores, d, for which a definition inevitably requires introduction of morphological models. The distributions of volume, surface, etc., on the appropriate sizes also belong to this group of characteristics. For many tasks the uniformity of distribution of textural parameters on the radius of a grain of the catalyst as well as the degree of connectivity and order are essential. However, such analysis, as a rule, requires introduction of geometrical and topological models [3,8]. Therefore, the characteristics of this group are *morphodependent*.

The atomic structure of a heterogeneous catalyst determines its chemical and phase properties, but texture determines a wide range of additional features that dictate such characteristics as adsorption and capillarity, permeability, mechanical strength, heat and electrical conductivity, etc. For example, the apparent catalytic activity, ϑ, of a grain, taking into account diffusion of reagents, depends on the interrelation between the rates of reaction and diffusion, and the latter is determined by a porous structure.

At this point, we encourage the reader to attempt solving Problem 1 before continuing.

Any heterogeneous catalyst with the ideal chemical composition may be killed if the surface area, porosity, and other textural characteristics are not optimal. Here, we are ready to formulate the first fundamental statement of texturology in application to heterogeneous catalysis: There is no catalysis without chemical activity, but there is no good heterogeneous catalyst without optimization of its texture. This statement is essential since in another form it is applicable to other fields, for example, to such a field being relatively far from catalysis as oil recovery and production. The oil itself can be as perfect as possible, but if the texture of oil reservoir rocks (pattern) is porous and interconnectivity of this porous media is not enough to form an infinite cluster (in a sense of percolation theory; see Section 9.9.1) the production of oil is zero. Thus, there is no money if there is no oil, but there is no oil at all until the pattern is ready to give it.

Optimization of texture is the practical aspect of texturology, and the sections below, along with the derivation of the basic interrelations between textural parameters, illustrate some possible routes for texture optimization.

Let us start with the fundamental laws of capillarity, which from one side determine the catalysts' textural and adsorption properties, and from the other side direct the mechanisms of formation of the catalysts.

9.2 BASIC PRINCIPLES OF PHYSICAL CHEMISTRY OF DISPERSED SYSTEMS

Atoms (or molecules) on the surface of a condensed (solid or liquid) phase have the excessive free energy compared to the atoms in the bulk. This is the key factor that determines the ability of highly dispersed systems for the spontaneous decrease of the excessive free energy any time when possible.

One can estimate this excessive energy using a ratio of the number of surface atoms N_S to the total number, N_V, of atoms in a considered object as $N_S/N_V \approx A/V$, that is, the ratio of the surface area, A, of the object to its volume, V. The ratio A/V is known as *dispersion*, and $(A/V)^{-1}$ expresses the characteristic size of the object [3].

The subnanometer (<1 nm) objects have the majority of the atoms on their surface, which is the case of ultradispersed systems or nanoclusters. This is the transition region from individual atoms to phases. The part of the surface atoms in 1 to 100 nm objects decreases with size, reaching the value of 0.1%. This is the region of highly dispersed systems of nanoparticles (nanodrops if liquid). The combination of the properties related to excessive surface energy and the properties of bulk phase is a characteristic to these systems. This allows the determination of the interface dividing different phases. The objects, whose size exceeds 100 nm have low level of dispersion and the effect of excessive free energy is insignificant, although such objects show the ability to agglomerate, their packing is very loose, they have the increased sedimentation time, and can be easily enticed with a flow.

The genesis of the highly dispersed systems is determined by the specific laws of the thermodynamics of interface phenomena, kinetic, and geometrical factors. One can determine two stages in the development of the general theory of interface phenomena divided with 1878. Before this year, a *classical* variant of this theory was developed basing on molecular mechanics (it was considered as a part of general theory of molecular interactions). This approach is based on the works of Young, Laplace, Dupre, Rayleigh, Grahame, Kelvin,[*] J. Thomson, van der Waals, and other scientists [9]. But, 1878 is marked with publication of the second part of fundamental Gibbs paper On the equilibrium of heterogeneous substance (*Trans. Connect. Acad.*, 1878, pp. 343–524), which gave a general thermodynamic description of equilibrium in heterogeneous systems [10]. Thus, the following development of the theory became impossible without Gibbs's fundamental ideas that lay in the basis of modern theory of interface phenomena including thermodynamic and statistical–mechanical approximations.

9.2.1 Gibbsian Classical Thermodynamic Theory

Gibbs offered the universal method to analyze the events on the interface [10]. Let us consider a simple example of adsorption of guest component, G, on the surface of host phase, H. The dividing region between phases G and H with possible dependence of density, $\rho(h)$, on the distance, h, from the interface is shown in Figure 9.2a. Gibbs introduced a model (Figure 9.2b), according to which the properties of G and H change in steps at some dividing surface. This surface can be placed at any distance from a real surface. The difference between the real density of guest $\int \rho_G(h)dh$ and the density in the model system is the surface excess of guest. The surface excesses of internal energy, U^*, entropy, S^*, and free energy, F^*, are determined by analogy. The surface excess can be positive, negative, or zero depending on the location of dividing surface.

Gibbs postulated the fundamental equation for the surface excess of internal energy as [10,11]

$$U^* = U^*\left(S^*, N_i^*, A, g_1, g_2\right) \tag{9.1}$$

---

[*]William Thomson (1824–1907) became Lord Kelvin in 1892. He has several famous namesakes: J. Thomson proposed the law that described the dependence of phase transition temperature on size, J.J. and J.P. Thomson (father and son) are the Nobel prize awarded physicists, etc. Thus, to avoid misunderstanding we refer W. Thomson as Kelvin by analogy with J.W. Strutt (1842–1919), who got his title (Lord Rayleigh) in 1873 by the inheritance.

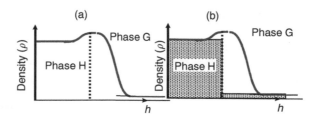

Figure 9.2 The scheme of calculation of surface excessess by Gibbs.

where N_i^* is the molar surface excess of component i, and g_1 and g_2 are the two principal surface curvatures, whose physical meaning is discussed below.

Let us consider the typical mechanisms of spontaneous processes that decrease U^*. The direction and driving force of such mechanisms are determined by the laws of equilibrium thermodynamics, and the rate is proportional to diffusion in gases, viscosity in liquids, and transfer of atoms, vacancies, and other defects in solids.

9.2.2 Flat Interface

In this case, the impact of g_1 and g_2 is negligible, and we can write the equation for the excess of internal energy under isothermal conditions in the following differential form:

$$dU^* = T\,dS^* + \sum_i \left(\mu_i\,dN_i + N_i\,d\mu_i\right) + \sigma\,dA + A\,d\sigma \tag{9.1a}$$

where T is the temperature, μ_i is the chemical potential of component i, and $\sigma=(\partial U^*/\partial A)_{S,N_i}$ is the surface tension, which also can be considered as the 2D equivalent of pressure. One can see that in isothermal regime close to equilibrium, U^* can be decreased by decreasing S^*, A, and σ. The letter expresses the effect of the surface tension on internal energy excess.

The impact of $T dS^*$ is usually small and is frequently taken into account only for the processes of aggregation or desegregation. It is also convenient to consider the systems with $\Sigma_i \mu_i N_i^* =$ const.

The decrease in U^* is caused by small shifts of atoms located in a layer of 3 to 5 atomic diameters near the interface. Such shifts can be clearly observed in monocrystals (reconstruction and relaxation phenomena) [12]. There are mechanisms based on the decrease of A at $V =$ const with the decrease of dispersion A/V. The results of action of these mechanisms are change of particle and pore shape, decrease of the micropore amount and surface roughness, etc. during sintering, coalescence, etc.

The systems with several types of interfaces with different surface tensions σ_i, for example, crystals (or cavities in the crystals), are forced to self-minimization of $\Sigma_i \sigma_i A_i$, which originates Gibbs–Curie–Wulff equation for equilibrium form of crystal of constant volume:

$$\sum_i \sigma_i A_i \Rightarrow \min \tag{9.2}$$

The crystals of nonequilibrium form have the value of $\Sigma_i \sigma_i A_i$, which exceeds the minimum. Thus, under the conditions when the atoms in the crystal become mobile, the spontaneous transition to an equilibrium form becomes probable. In the system of contacting crystals, this frequently results in the decrease of A/V.

When the temperature is close to the melting point, solid particles start acting like liquid drops. The equilibrium form of a liquid drop (G) on a flat surface (H) is determined by Young's equation, which was offered in 1805 before Gibbs. The origin here is also the requirement (9.2) taking into account the intermolecular interaction on the interfaces G/H, P/G, and P/H, where P is the environment (e.g., the vapor of G). The mechanical equilibrium in such systems is controlled by

$$\sigma_{HP} - \sigma_{GH} = \sigma_{GP}\cos\theta \tag{9.3}$$

where the subscript index determines the interface, and θ is the contact angle. In 1869, Dupre found the relation between θ and the work of adhesion W_{GH} (the specific (per unit area) work of separation of phase G and phase H) and cohesion W_{GG} (the specific (per unit area) work that is necessary to divide a single drop of G to two drops):

$$\cos\theta = 2\frac{W_{GH}}{W_{GG}} - 1 = \frac{W_{GH}}{\sigma_{GP}} - 1 \tag{9.4}$$

Equation 9.3 and Equation 9.4 are interrelated and the form of Equation 9.3 is referred to as Young—Dupre equation [9]. Mathematically this equation is incorrect when $W_{GH} > W_{GG}$, but in real situations this condition usually results in the spreading of G over the surface of H. The conditions of wetting/spreading that are taken into account with Equation 9.3 and Equation 9.4 (Figure 9.3) and their derivatives are fundamental for the modern theory of formation, growth, and sintering of supported catalyst particles [13]. There are three possible mechanisms of growth of supported particles: (1) layer by layer, when $W_{GH} - W_{GG} > 0$ (Frank–van der Merwe mechanism), (2) islands at $W_{GH} < W_{GG} \leq 0$ (Volmer–Weber mechanism), and (3) growth of crystallites over a monolayer at $W_{GH} - W_{GG} < 0$ (Stranski–Krastanov mechanism).

Another possible factor for the decrease of U^* is adsorption that is excess of G on a surface of H. According to Gibbs, the isotherm of adsorption (the specific surface excess of ith guest) has a form*

$$\Gamma_i^* = \frac{N_j^*}{A} = -\left(\frac{\partial\sigma}{\partial\mu_i}\right)_{T,S,N_j,A} \tag{9.5}$$

One can transform Equation 9.5 to a more convenient way, substituting μ_i with $\mu_{i,0} + RT\ln a_i$:

$$\Gamma_i^* = -\frac{a_i}{ART}\left(\frac{\partial\sigma}{\partial a_i}\right)_{T,S,N_j,A} \tag{9.5a}$$

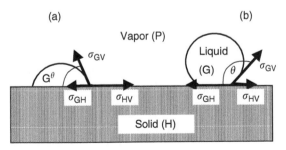

Figure 9.3 A drop of liquid on wetting (a) and not-wetting; (b) surface.

---

*Determination of adsorption of any component is made under the condition of a special location of the dividing surface, which corresponds to $\Sigma_{j,j \neq i}\mu_j N_j = 0$. This surface is referred to as *equimolecular* for component i.

where a_i is concentration, partial pressure, or fugitiveness of guest i. Equation 9.5a means that the positive adsorption of guest i is determined with the condition $\partial\sigma/\partial a_i < 0$, that is, the increase of its concentration results in the decrease of systems' σ, and thus U^*.

9.2.3 Curved Interface

Each individual particle has a closed shape, and thus a curved surface. Obviously, this should be taken into account in the Gibbs equation [11]:

$$dU^* = T\,dS^* + \sum_i \mu_i\,dN_i + \sigma\,dA + \frac{1}{2}\Big[\big(C_{1g} + C_{2g}\big)d\big(g_1 + g_2\big) + \big(C_{1g} - C_{2g}\big)d\big(g_1 - g_2\big)\Big] \qquad (9.1b)$$

where $g_1 = 1/r_1$ and $g_2 = 1/r_2$ are the principal surface curvatures, $C_{1g} = (\partial U^*/\partial g_1)$ and $C_{2g} = (\partial U^*/\partial g_2)$, at other parameters correspondingly fixed. According to Gibbs, we can postulate the special location of a dividing surface having *moderate* curvature ($g_1 - g_2 \approx 0$) when the direct dependence of U^* on surface curvature becomes insignificant. Gibbs called this surface as surface of tension. The following, from Equation 9.1b, is the Young–Laplace equation (law) known from the early XIXs:

$$\Delta P = \frac{2\sigma}{r_m} = 2\sigma H \qquad (9.6)$$

where r_m is the mean curvature radius, and H the mean surface curvature, determined from g_1 and g_2:

$$H = 1/r_m = (g_1 + g_2)/2 \qquad (9.7)$$

One can see that such specific form of Equation 9.1b was used for convenience.

Equation 9.6 determines the conditions of a mechanical equilibrium of the curved interface. This can be illustrated with an example of a spherical bubble of radius r. To compete the surface tension the pressure inside the bubble should exceed the external pressure with ΔP, which is determined from the work, W, for virtual change of r: $dW = \Delta P\,dV - \sigma\,dA$. Under equilibrium, $dW = 0$ and $\Delta P\,dV = \sigma\,dA$, thus

$$\Delta P = \sigma\frac{dA}{dV} = \sigma\frac{d\,(4\pi r^2)}{d\,(4\pi r^3/3)} = 3\sigma\frac{2r\,dr}{3r^2\,dr} = 2\sigma/r \qquad (9.6a)$$

Gauss has shown [14] that Equation 9.6a is true for any surface with constant curvature H for which the reversible fluctuations do not break the relation

$$\frac{dA}{dV} = \frac{2}{r_m} = 2H \qquad (9.8)$$

Equation 9.6 and Equation 9.7 require that any interface between two fluids in equilibrium must have the constant curvature H. Appearance of two points with different curvatures results in appearance of a corresponding ΔP, which forces substance transfer until equalization of H. This gives rise to a number of simple mechanisms of texture formation and transformation. Also, Young–Laplace law (Equation 9.5) is the basis for Kelvin–Gibbs and Gibbs–Freundlich–Ostwald equations.

Kelvin's equation determines the equilibrium vapor pressure over a curved meniscus of liquid:

$$\frac{P}{P_0} = \exp\left[\pm\frac{2\sigma v_M}{RTr_m}\right] \tag{9.9}$$

where P and P_0 are the equilibrium pressures over curved and flat surfaces, respectively, and v_M is the molar volume of a condensate. A + sign in the exponent corresponds to convex and − to concave interfaces. Thus, the pressure exceeds P_0 over convex, and has lower values than P_0 over concave interfaces. Substitution of P/P_0 with C/C_0, where C is the equilibrium solubility of the surfaces with the curvature radius of r_m and C_0 is that of the flat surface (when $r_m \to \infty$), gives the Gibbs–Freundlich–Ostwald equation [15]

$$\frac{C}{C_0} = \exp\left[\pm\frac{2\sigma v_M}{RTr_m}\right] \tag{9.10}$$

Equation 9.9 and Equation 9.10 express the dependence of the chemical potential on the surface curvature:

$$\mu = \mu_0 \pm 2\sigma v_M / r_m = \begin{cases} \mu_0 \pm RT \ln(P/P_0) & \text{vapor−liquid} \\ \mu_0 \pm RT \ln(C/C_0) & \text{solution} \end{cases} \tag{9.11}$$

where μ_0 is the chemical potential of a flat surface. Equation 9.11 shows that μ for convex surfaces exceeds μ_0, which in turn exceeds μ for concave surfaces. This means that change of a curvature results in appearance of a chemical potential gradient, which stimulates spontaneous mass transfer from the places with convex to the places with concave interfaces.

There is another important law that follows from the classical theory of capillarity. This law was formulated by J. Thomson [16], and was based on a Clausius–Clapeyron equation and Gibbs theory, formulating the dependence of the melting point of solids on their size. The first known analytical equation by Rie [17], and Batchelor and Foster [18] (cited according to Refs. [19,20]) is

$$T_{MP} = T_{MP,0}\left(1 - \frac{2\sigma v_M}{\lambda r_m}\right) \tag{9.12}$$

where T_{MP0} is the melting point of a macroscopic object and λ is the heat of phase transformation.

Equation 9.6 and Equation 9.9 through Equation 9.12 are the basis of the classic theory of capillarity [9]. The moderate surface curvature that was assumed for these equations follows the fundamental Gibbs Equation 9.1 and Equation 9.1b. However, there was a problem of application of the classic theory of capillarity to the region of high surface curvatures that corresponds to the nanoparticles (down to ~2 nm).

9.2.4 Surface Curvature

The local surface curvature is determined by construction of a vector normal to the surface and drawing of two orthogonal planes through the normal vector (Figure 9.4). The location of the planes is chosen according to a requirement that the principal radii, r_1 and r_2, of curvature of lines formed by intersection of the planes with the surface have the minimum and the maximum values. In inverse proportion to them are the principal surface curvatures, $g_1 = 1/r_1$ and $g_2 = 1/r_2$.

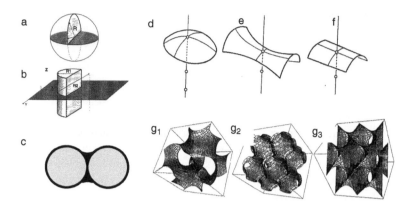

Figure 9.4 The examples of curved surfaces: schematic diagrams showing the principle of surface curvature measurement for sphere (a) and cylinder (b); (c) is the diametric section of meniscus around a zone of contact of two particles; (d)–(f) three basic types of surfaces: (d) is ellipsoidal, (e) hyperbolic; (f) parabolic; (g_1)–(g_3) are the examples of the surfaces of constant mean surface curvature (minimal surface).

The mean local surface curvature is expressed by Equation 9.7, and the full (Gaussian) local surface curvature is determined as

$$K = g_1 g_2 \qquad (9.13)$$

Let us determine that the convex surface has a positive curvature, and the concave surface has a negative curvature. The surfaces that are characterized by $g_1 = -g_2$ have a zero mean surface curvature ($H=0$). There are three characteristic types of surfaces (Figure 9.4d to Figure 9.4f): ellipsoidal (Figure 9.4a and Figure 9.4d) with g_1 and g_2 having the same sign ($K>0$, but $H \neq 0$ and can be positive or negative); parabolic (Figure 9.4b and Figure 9.4f) with one of the principal curvatures being zero ($K=0$, but $H=0.5g_1 \neq 0$); and hyperbolic or saddle-shaped (Figure 9.4c and Figure 9.4e) [21], when g_1 and g_2 have different signs ($K<0$). In the latter case, H can have zero values, $g_1 = -g_2$. Such surfaces are designated as minimal surfaces. For example, a soap film spanned on a nonflat wire contour is a minimal surface, which has the minimum area from all possible cases. The surfaces shown in Figure 9.4g_1–g_3, are other examples of minimal surfaces. Such surfaces are characteristic to SiO_2 mesophases formed in the presence of surfactants. A meniscus of liquid in a region of contact between two particles (Figure 9.4c) can also form a minimal surface, which is called *catenoid*.

9.2.5 The Limits of the Classic Thermodynamic Theory

Gibbs found the solution of the fundamental Equation 9.1 only for the case of moderate surfaces, for which application of the classic capillary laws was not a problem. But, the importance of the world of nanoscale objects was not as pronounced during that period as now. The problem of surface curvature has become very important for the theory of capillary phenomena after Gibbs. R.C. Tolman, F.P. Buff, J.G. Kirkwood, S. Kondo, A.I. Rusanov, P.A. Kralchevski, A.W. Neimann, and many other outstanding researchers devoted their work to this field. This problem is directly related to the development of the general theory of condensed state and molecular interactions in the systems of numerous particles. The methods of statistical mechanics, thermodynamics, and other approaches of modern molecular physics were applied [11,22,23].

The results of application of different mathematical models can be reduced by allocation of some additives to the Young–Laplace equation

$$\Delta P = \sigma H + f(g_1, g_2) \tag{9.14}$$

or in another way by introduction of the dependence of surface tension on curvature, which for spherical particles, is written in the following way:

$$\sigma/\sigma_0 \approx 1 - 2\delta_T/r \tag{9.15}$$

where σ_0 is the surface tension of the flat surface, r the sphere diameter, and δ_T the so-called Tolman parameter, which is proportional to the size of molecules on the surface, but in general can depend on the surface curvature and the properties of molecular interaction. The recent review [11] on this subject has shown the uncertainty in shape and even in the sign of $f(g_1,g_2)$, theoretically derived by different scholars. Also, ambiguous are the sign and value of δ_T.

The situation has dramatically changed during last two decades with synthesis of new materials, appearance of precise experimental methods, and intensive development of numerical methods. The new family of mesoporous mesophase materials type of MCM-41 with highly ordered structure has confirmed a satisfactory applicability of Kelvin equation 9.9 for the pores of 3 to 10 nm size [24,25]. The direct measurements of capillary forces have shown the constancy of σ for a number of liquid hydrocarbons and water until 2 to 4 nm [26,27]. The independent proof of insignificant dependence of σ on surface curvature follow from the recent studies of melting and freezing points for particles of different phases in a nanometer scale [28–32], as well as studies of nucleation of vapors [33]. This problem was studied by Baidakov et al. (see, e.g., Refs. [34,35]), who on the basis of theoretical analysis, computer simulations (molecular dynamics), and physical experiment (nucleation in the mixtures of molecular oxygen and helium) have shown that the value of δ_T in the equations of type (9.15) is of ~0.1 of molecular size, and increase of curvature in a region of very small size (d, less than 3 nm) always results in decrease of σ.[*]

Thus, the set of stated arguments allows one application of the equations of classic theory of capillarity in the whole range of nanoparticle size without accounting the dependence of σ on curvature. This makes it possible to use these equations for the description of basic typical mechanisms of catalysts' texture genesis.

9.2.6 Typical Mechanisms and Processes of Texture Genesis Derived from the Laws of Surface-Capillary Phenomena

9.2.6.1 Fundamental Mechanisms of Texture Genesis

Let us start with the action of Young–Laplace law (Equation 9.6), which determines the equilibrium configuration of the fluids (liquid and liquid-like phases) and the driving force of mass transfer that cause the spontaneous formation of equilibrium configurations.

If one puts vertically a capillary of radius R in a Petri dish filled with water, the latter will rise in the capillary to the height h. The height is determined from gravitation and capillary forces, $h=2\sigma H/\rho g$, where ρ is the true density of liquid and g the acceleration of a free fall.

Figure 9.5 illustrates the properties of the mechanism of water evaporation from a capillary, which has an arbitrary number of narrowings and widenings. Let the capillary be filled with liquid

---

[*]The similar conclusion is stated in a number of other modern experimental and theoretical works. For example, δ_T is estimated as ~0.1 nm in Ref. [34].

(a)

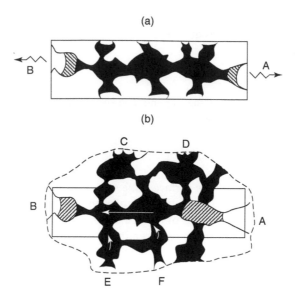

(b)

Figure 9.5 A scheme of liquid redistribution in 1D (a) and 3D (b) capillaries of arbitrary irregular form (by [3]).

with menisci A and B at the liquid–vapor interface. The condition of mechanical equilibrium (Equation 9.6) requires the equality of the menisci curvatures and their coordinated changes ($r_{mA} > r_{mB}$). Let the narrowing at side A be wider than the narrowing at side B ($r_{nA} = r_{nB}$). During evaporation when the meniscus A passes the narrowing with r_{mA}, the condition of mechanical equilibrium breaks and the most plausible possibility to restore it is a spontaneous movement of a part of liquid to the expansion B until r_{mB} becomes equal to r_{mA} at new position of menisci. The zones of such movement are shown in Figure 9.5 with shading. This spontaneous movement is also known as Haines jumps (HJ) [36]. Such HJs accompany any passing through the narrowings of a capillary, after which there are expansions. In the case of a 3D cluster (Figure 9.5b) any expansion of B, C, D, E, or F at the liquid–vapor interface can accept transferring liquid under the condition that corresponding narrowing is smaller than that of donor A.

This fundamental HJ mechanism is of particular interest for supported catalyst preparation at the stage of drying, for example, in the case when the precursor of the supported component does not considerably adsorb on the support [3,37]. When all pores of the support are totally filled with precursor solution, the precursor is evenly distributed over the porous space of support. In absence of HJ mechanism the amount of precipitated precursor in each pore would be proportional to the volume of the pore. However, HJ mechanism provides the significant redistribution of solution during the drying process. For example, in the situation of Figure 9.5a the precursor will be transferred during evaporation from a region of A to a region of B.

Another example of Young-Laplace law action is illustrated in Figure 9.6a and Figure 9.6b, where two spheres (liquid drops in equilibrium with their vapors) of radius R_{sph} are considered at a direct contact. If the external pressure is P_0, according to Equation 9.6, the pressure inside the spheres is $P_2 = P_0 + 2\sigma/R_{sph}$. However, the curvature at the region of contact, $2/r_m$, has a sign opposite to the curvature of the rest surface, and the pressure in this region is determined as $P_2 = P_0 - 2\sigma/r_m$, which is lower than in the bulk of spheres. The difference in pressures results in appearance of a driving force for mass transfer to the region of contact (coalescence), which acts until both spheres merge. This is also accompanied with rapprochement of sphere centers [38].

The mechanism of coalescence of solid spheres (Figure 9.6c and Figure 9.6d) differs from that considered above. In this case the viscous coalescence is impossible, but the mass transfer occurs through the surface diffusion or solubility [39]. According to Equation 9.11, a motive force for this process is

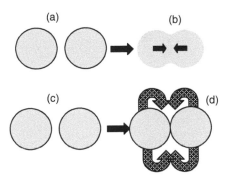

Figure 9.6 Coalescence of fluid drops is accompanied with merging (a)→(b), and coalescence of solid spheres is accompanied with accretion without rapprochement of their centers (c)→(d).

caused by a difference in chemical potentials of convex (donor) and concave (acceptor) regions. There is no change in the positions of the centers of particles during such coalescence. To differentiate the two considered cases, let us designate coalescence of liquid-like particles as CL, and that of solid particles as CS. In the application to texture of heterogeneous catalysts, the mechanism CL is characteristic to high-temperature sintering of dispersed solids, and CS is characteristic to lowtemperature (including hydrothermal) sintering, when only the surface atoms have considerable mobility.

The difference of chemical potentials determines the driving force of mass transfer in the case of separated particles as well. This mechanism was offered by Ostwald in 1901 [15], and called Ostwald ripening (OR). It works during aging of sols (including aerosols) and suspensions (Figure 9.7), and describes sintering of dispersed supported catalysts. The mathematical simulation of OR is applied frequently; however, according to a recent review [40] the most plausible models include the combinations of OR with one of the mechanisms described above since coalescence is inevitable due to contacts between the particles during collisions.

Modeling of coalescence in the ensembles of particles is more complicated (Figure 9.8). We will mention few steps here. Coalescence starts from the regions of contact between pairs of neighboring particles. After the filling of the accepting zones the curvature starts to have an integral character.[*] At this step the regions with the most friable packing of particles start acting as donors, and the densest regions act as acceptors. Mass transfer between these regions can be described by OR mechanism [3]. As a result, the scheme in Figure 9.8 is characteristic to the process that includes several elementary mechanisms.

9.2.6.2 Fundamental Processes of Texture Genesis

Figure 9.9 shows the scheme of the processes that occur during drying of supported catalysts [3,37]. Let us start with the situation when all pores of support are totally filled with precursor solution, and the precursor is evenly distributed over the support. According to HJ mechanism, the drying process results in significant redistribution of the precipitated precursor. There are two limiting cases: the slow drying (the rate of solvent evaporation is lower than the rate of redistribution), and rapid drying (opposite to previous).

In the beginning (Figure 9.9a), the support grain is totally filled with solvent, the degree of filling $U \approx 1.0$, and the starting concentration of the active component precursor is C_0. The liquid phase is continuous and forms a single cluster where the dissolved compound is distributed evenly. The

---

[*]The mean integral curvature is determined as $M = \dfrac{1}{A}\iint_A \dfrac{1}{2}(g_{1,i}+g_{2,i})\mathrm{d}A = \dfrac{1}{A}\iint_A H_i\,\mathrm{d}A$ where $g_{1,i}$ and $g_{2,i}$ are the local principal curvatures, and H_i is the mean curvature in a surface point i.

Figure 9.7 Scheme of recondensation or Ostwald ripening (small particles are donors, large particles are acceptors).

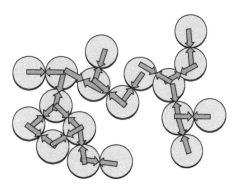

Figure 9.8 Scheme of coalescence in ensemble of particles. Arrows show the direction of mass transfer in the regime of coalescence of liquid (mechanism CL).

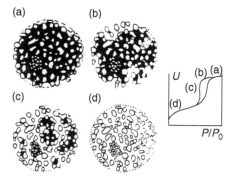

Figure 9.9 A scheme of distribution of liquid phase (black at the successive stages of slow drying: (a), (b), (c), and (d). The insert shows the corresponding points at the desorption branch of isotherm, where U is the degree of pore filling and P/P_0 the partical pressure of solvent vapor [3].

evaporation interface is formed with menisci and adsorbed film at the external surface of a grain. Let us designate as drying stage I the region of movement of the evaporation interface within the grain until any of the meniscus is broken (this happens when the evaporation interface passes the narrowest part of a pore channel). At this stage, the redistribution of liquid in porous space is guided by the diffusion between the pores of different form. The moment of a breakthrough switches on the HJ mechanism, drying stage II starts (Figure 9.9b and Figure 9.9c). The evaporation interface becomes fractal (see Section 9.8) in the process of its movement within the grain. The single cluster exists until some definite degree of filling ($U > U_{cr}$), and this cluster is connected to the external surface with pores that are already free. These pores have narrowings that accept liquid from the donating pores in the cluster. At this stage the rate of drying is constant because it is limited due to evaporation from the accepting zones at the external surface. The single cluster tears at $U = U_{cr}$ to small clusters, whose number consequently increases, and size decreases. The rate of evaporation decreases at $U<U_{cr}$, since the diffusion through the empty pores becomes limiting. The filling degree U_{cr} is a typical percolation threshold (see Section 9.9.1). The tearing of a single cluster brakes the continuity of the liquid phase, and the resulting numerous clusters are connected to each other with a solvent film on the surface of the support, and this film becomes thinner with evaporation. Therefore, the redistribution of liquid proceeds by HJ mechanism within separate clusters only. This process stops when the remaining liquid fills only the places of contacts between the particles and the surface of the support (stage III). This critical point $U_{cr,h}$ approximately corresponds to the lower closure point of the hysteresis loop of the isotherm (see the inset in Figure 9.9).

Precipitation of the active component precursor starts when its concentration in the liquid phase reaches the saturation point C_S. At this point the degree of pore volume filling is U_S. If the precursor is not volatile, one can determine U_S as a ratio of $U_S = U_0 C_0 / C_S$. The redistribution of the precursor is minimal if $U_S < U_{cr,h}$. Aspiration of U_S to U_0 results in a predominant yield of the precursor at the external surface of a support grain.

The regime of rapid drying, when the evaporation interface moves continually and rapidly toward the pellet center, significantly inhibits the impact of the HJ mechanism. The single cluster can exist even until $U \approx U_{cr,h}$ in this case. The impact of HJ can also be neglected by decrease in the surface tension of the solvent, or increase in its viscosity. The distribution of precursor can be affected by its diffusion within the solvent and the necessity of oversaturation for precipitation. In the case of strongly adsorbing precursors, drying only slightly changes the precursor profile established during impregnation stage [3,37].

Drying process is one of the most important steps for formation of texture of many PSs that are formed through sol–gel technique. It is essential that the explanation of the role of different stages that precede drying is usually formulated on the basis of textural studies, during which the sample must be dried before analysis. Let us consider a particular example of texture formation of SiO_2 xerogel (silica gel) in the process of hydrogel drying.

Silica gels are the widespread and mostly explored PSs, and the role of drying for their evolution was discussed seemingly in detail (see, e.g., Ref. [41]). However, even the settling monograph by Iler [42] presents only a simplified first stage from the whole scenario of formation. Let us consider this stage (Figure 9.10).

The evaporation proceeds from the external boarder at this stage. Formation of concave menisci between the primary particles falls into operation of the Young–Laplace law. The hydrostatic pressure is lower than external pressure, hence there is negative pressure in the liquid that forces it to cover the free surface of particles moving them closer to each other. Simultaneously, there are two forces that act on the particles at the external boarder at liquid–vapor interface: (1) the deviation of the inside and outside pressures ΔP and (2) the surface tension force acting along the free surface of the liquid. The location of this particle is determined in terms of wetting perimeter radius r, and the value of total acting force is determined as

$$F_\sigma = \pi r^2 \, \Delta P + 2\pi r \sigma = 2\pi r \sigma \left(\frac{r}{r_m} + 1 \right) \approx 2\pi r \sigma \left(\frac{1-\varepsilon_S}{\alpha_{XY}\varepsilon_S} + 1 \right) \qquad (9.16)$$

where r_m is the curvature radius of a meniscus between the primary particles and σ the surface tension at the liquid–vapor interface. The right-hand side of the equation is obtained after substitution of r and r_m with 2D porosity ε_S and form factor α_{XY} using Equation 9.58 (for details see Section 9.5). F_C takes the values from minimum (at $r = 0$) to maximum ($r = R_0$ or minimal ε_S). The regime of rapid drying can result in a variety of situations [3,42], which we will not consider. Let us consider here the regime of relatively mild drying, when the capillary contraction force results in a uniform contraction of gel globules accompanied with densification (decrease of ε_S) of the external level of particles. Densification of the external level increases its resistance to the following deformations. If the tension forces at the evaporation interface are transferred within the volume

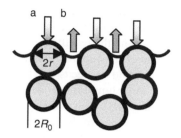

Figure 9.10   Scheme of capillary forces action, which result in contraction of gel. Force acts on the primary particle size of R_0, force b acts on a liquid phase between the particles; r is the wetting perimeter radious.

of hydrogel, the carcass of gel becomes more and more dense. There is some definite density of the external level for which F_C reaches the maximum value, and the evaporation interface breaks into the volume of hydrogel, and stage II of drying begins. The following events are illustrated in Figure 9.11.

Drying process usually results in volumetric shrinkage, which can be determined by the ratio $f_V = 1 - V_\Sigma/V_0 \approx 1 - L/L_0$, where V_Σ and V_0 are the volumes of pores in xerogel and initial hydrogel (per unit of mass of dry product), respectively, and $\Delta L/L_0$ is the relative decrease in a linear size of a granule with initial size L_0. The most significant contraction proceeds at stage I; hence this stage is the most determinative for the total porosity of the final product. Movements of evaporation interface at stage II is analogous to the earlier considered stage of a rigid PS drying. However, the action of capillary forces that are applied to the external surface of single hydrogel clusters can result in their additional contraction. This results in changes of pore volume–size distributions in the final xerogel [3,43]. Stage III is the final stage, when the ionic solvate shells over the globules that did not allow their direct contacts disappear, and all dissolved compounds precipitate. The temperature usually reaches maximum at this stage. The complex of these factors results in accretion of the primary globules, which explains the usually observed decrease in the surface area. This was proved experimentally elsewhere [3].

Sol–gel technology with fast supercritical drying of gel is intensively applied for various processes. Supercritical conditions effectively eliminate the action of capillary forces (see, e.g., Refs. [44]). But, this technology is expensive, and knowledge of texture formation laws on all stages of material formation allows obtaining similar results in different, less expensive ways. A general approach for the description of texture formation for silica gels and other amorphous porous solids is discussed elsewhere [43].

Formation of texture of crystalline systems differs due to the existence of phase transitions. Very important in this case is the parameter $\Delta_{PB} = V_A/V_B$, offered by Pilling and Bedworth in 1923 [45]. This parameter is equal to the ratio of the volume, V_A, of product solid phase to the volume, V_B, of initial solid phase. If $\Delta_{PB} > 1.0$, the phase transformation is accompanied with the increase in solid phase volume, and when $\Delta_{PB} < 1.0$, there is the decrease in solid phase volume. The examples of processes with $\Delta_{PB} > 1.0$ are shown in Figure 9.12. In this case, solid-phase transformations under thermal treatment of dry powders are frequently accompanied with preservation of external morphology and apparent volume of a precursor phase. Such transformations are referred to as pseudomorphous. The ideal pseudomorphous transformation results in formation of internal porosity of the product $\varepsilon = 1 - \Delta_{PB}$. This is schematically illustrated in Figure 9.12b. The case of aging of a precipitate in a liquid-phase transformation of the former is accompanied by dispersion of the product followed by reaggregation, and such processes are referred to as recrystallization [46]. One can see the corresponding scheme in Figure 9.12c and Figure 9.12d. This scheme illustrates the real transformations of aluminum hydroxide precipitated from $Al(NO_3)_3$ and aged at 473 K and pH = 6 [47]. The starting precipitate has amorphous structure with low dispersion and approximate composition of $Al_2O_3 \cdot 3H_2O \cdot 0.5NO_3$ (Figure 9.12c). Aging

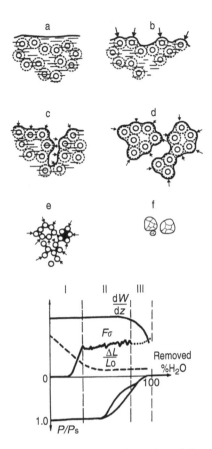

Figure 9.11 Scheme of silica gel texture formation under the regime of slow drying: (a) and (b) represent stage I, (c) and (d) correspond to stage II, (e) is the case of stage III, and (f) represents the stage of sintering. F_σ is the contraction force, $\Delta L/L_0$ the linear contraction, and P/P_0 the pressure of water vapor.

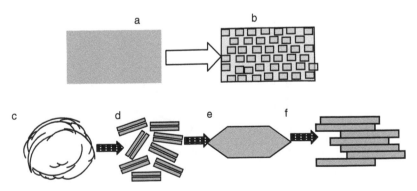

Figure 9.12 Typical transformations during phase transformations at $\Delta_{PB} < 1$: (a) and (b) for powders and granules; (c)–(f) for aging of precipitates in liquid phase (recrystallization).

results in crystallization to pseudo-boemite ($Al_2O_3 \cdot 1.4H_2O$; Figure 9.12d), with $\Delta_{PB} < 1.0$ accompanied with self-dispersion. The next step is the formation of platelets of boemite ($Al_2O_3 \cdot H_2O$; Figure 9.12e) that aggregate into the oriented packings (Figure 9.12f).

One can find the description of other mechanisms and processes of formation of catalysts texture elsewhere [3,46,48].

9.3 ADSORPTION AS A PRIMARY INSTRUMENT FOR TEXTURE CHARACTERIZATION

The surface area of PSs can be measured by means of quantitative electron microscopy (EM) [49–51], knowing the sizes of particles and pores [51,52], wetting heat [7,53] measurements, etc. But, the most universal methods are based on adsorption measurements [51,53–55], corresponding to the traditional methods of pore size distribution measurements in the range 0.3 to 100 nm.

First, one should clearly differentiate physical and chemical adsorption. The first is mainly caused by London–van der Waals and electrostatic (Coulomb, dipole–dipole, etc.) forces. The heat of physical adsorption is close to the heat of condensation (usually it rarely exceeds twice the value of the latter). The origin of chemical adsorption is in the dissociation of adsorbed molecules and formation of new chemical bonds with the interaction energy close to the energy of corresponding chemical reaction. The gap between these two limiting cases of adsorption is filled with a variety of possibilities. The specificity of these types of adsorption is in the reversibility of physical adsorption, as opposed to the selectivity and irreversibility of chemical adsorption. Both cases are important to texturology, since nonspecific physical adsorption can be used for measuring the total surface area of catalysts, distribution of pores on their sizes, etc. Specific adsorption and chemisorption can be used for measuring textural characteristics of different phases in PS [3,53,54].

Basic textural parameters are evaluated by analysis of adsorption isotherms (AI), $q(P/P_0)$, that characterize the equilibrium adsorbed uptake, q, in an accessible range of partial pressure of an adsorptive, P/P_0, at constant temperature T.[*] Figure 9.13 shows the basic types of AIs on PSs [7,53,54].

Type I isotherms are characteristic of strong interactions of the adsorbate and adsorbent. Typical examples are chemisorption and physical adsorption in microporous solids having a correspondingly small amount of mesopores. In this case, adsorbate–adsorbate interaction is significantly weaker than adsorbate–adsorbent (guest–host) interaction (G/G < G/H).

Type II and IV AIs are characteristic of polymolecular adsorption in nonporous (macroporous) (II) and mesoporous (IV) PS. The steep part of such AIs close to the origin corresponds to relatively strong G/H interaction. Type III and V AIs are characteristic of G/G > G/H.

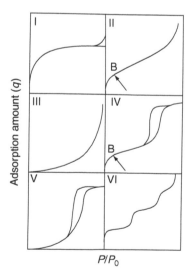

Figure 9.13 Basic types of adsorption isotherms by IUPAC classification [53].

---

[*]Also considered are adsorption isobars, which are the dependences of $q(T)$ at p = const., and adsorption isosters, that is, dependence of $p(T)$ at q = const. [53].

Existence of a large amount of mesopores usually results in the appearance of capillary condensation hysteresis loop. Type II AIs transform to type IV, and type III AIs transform to type V. Type VI AIs are characteristic to low-temperature adsorption of some noble gases over energetically homogeneous surfaces.

The well-known Langmuir AI has the form [56]

$$\frac{q\,(P/P_0)}{q_m} = \theta = \frac{C_L P/P_0}{1 + C_L P/P_0} \tag{9.17}$$

where q_m is an ultimate adsorption uptake of a monolayer, θ the degree of monolayer filling, and C_L indicates the strength of G/H interaction. This equation corresponds to localized monomolecular adsorption over a homogeneous surface without taking into account G/G interactions. The monolayer capacity q_m is related to the surface area, A, according to the equation

$$A = q_m w N_A \tag{9.18}$$

where w is the average area occupied by one adsorbate molecule in a completed monolayer, N_A the Avogadro constant, and q_m is expressed in moles of adsorbate per gram of adsorbent.

Brunauer et al. [57] extended the Langmuir approach to polymolecular adsorption, proposing the famous BET equation

$$\frac{q(P/P_0)}{q_m} = \theta = \frac{C_{BET} P/P_0}{(1 - P/P_0)\left[1 + (C_{BET} - 1)P/P_0\right]} \tag{9.19}$$

where C_{BET} is an equilibrium constant (analogous to C_L in Equation 9.17). For convenience of plotting, it can be rewritten as

$$\frac{P/P_0}{q(P/P_0)(1 - P/P_0)} = \frac{1}{q_m C_{BET}} + \frac{C_{BET} - 1}{q_m C_{BET}} P/P_0 \tag{9.20}$$

According to this equation, a plot of dependence of $[1/(P_0/P - 1)]/q(P/P_0)$ on P/P_0 should have a linear form allowing evaluation of q_m and C_{BET}. The surface area is calculated by Equation 9.18 and the corresponding value is frequently designated by A_{BET}. One can also determine the pressure, P_m/P_0, under which a monolayer is being formed:

$$\frac{P_m}{P_0} = \frac{1}{1 + \sqrt{C_{BET}}} \tag{9.21}$$

The BET equation (or method) is widely used for evaluation of surface areas, A_{BET}. Routine measurements of A_{BET} are based on AIs of N_2 at liquid nitrogen temperature of ~77 K and $w_{N_2} = 0.162\ nm^2$. One usually uses AIs measured in the range $0.05 \le P/P_0 \le 0.25$ [7,53]. In the absence of micropores or surface modification the values of A_{BET} have an agreement with independent measurements [51,53,58]. However, micropores and surface modification strongly decrease the accuracy of BET method.

This is the result of an action of assumptions that were set at the derivation of BET (as well as Langmuir) equations: energetically homogeneous surfaces and the absence of lateral interactions of G/G type (see Figure 9.14). But, the surfaces of real PSs are usually energetically heterogeneous, the

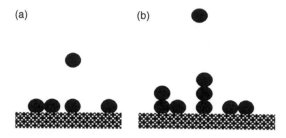

Figure 9.14 Scheme of adsorption by Langmuir (a); BET (b).

existence of micropores is equivalent to the existence of surface species of increased adsorption potential, and surface modification with molecularly dispersed species usually results in a local decrease in adsorption potential. On the other hand, the adsorption on truly homogeneous solids (graphite, type VI AI) can hardly be described by the BET model. There is a real problem of evaluation of w for adsorbates different from nitrogen [59] (a variant of overcoming this difficulty is proposed in Ref. [60]). Nowadays, it is conventional that BET method has a lack of theoretical basis, but Equation 9.19 is a convenient empirical equation, which can generally be used for estimation of surface area, and gives accurate enough values of surface areas with an insignificant distribution of surface sites on adsorption energies without micropores and surface modification [3,7,53].

The deficiencies of measuring the surface areas by the BET method are overcome in comparative methods (CMs) [3,61,62], known in the literature by their modifications: α_S method by Sing [53,63], t method by Lippens–de Boer [2,64], f method by Gregg [65], t/f method by Kadlec–Dubinin [66,67], micropore analysis (MP) method by Brunauer [68], etc. All of these variants are based on comparison of a measured AI with the reference (or standard) AI (RAI). Obviously, both AI and RAI should be measured under similar conditions, that is, usually N_2 at 77 K in some specific range of P/P_0, which is determined by the nature of the studied material. There are numerous variants of RAI, measured on standards, but there are strong requirements to the choice of a standard: it should have no micro- and mesopores, surface modification, and one should be able to measure its surface area by means of independent methods, at least by the BET method.

A pioneer in the search for RAI was Shull [69], who observed the coincidence of overlapping of N_2-AIs on different macroporous materials when the AIs are related to the corresponding values of A_{BET}. This finding was confirmed by others, for example, by authors of [53,61,70–76], who published their own N_2-RAI differing only by the methods of expressing the adsorption units. Kiselev and his coworkers claimed the correct expression of adsorption units as $\alpha(P/P_0)$ ($\mu mol/m^2$) = $q(P/P_0)/A_{BET}$ [71,72], while Shull [69] and de Boer [64,73] gave a form of statistical thickness of the adsorbed film t (nm) = $\left[q(P/P_0)/A_{BET}\right]\sigma_m$, where σ_m is the thickness of a single adsorbed molecular layer. Sing [53] uses a relation $\alpha_S(P/P_0)=q(P/P_0)/q(0.4)$, where $q(0.4)$ is the adsorption uptake at $P/P_0 = 0.4$. A number of statistical monolayers were also used for this purpose [74,75].

Figure 9.15a shows a comparison of N_2-RAIs by different authors normalized to Kiselev's units α_i ($\mu mol/m^2$) vs. P/P_0, and Figure 9.15b shows the linearization of these RAIs in the form of a dependence of $\alpha_i(P/P_0)$ vs. $\alpha_0(P/P_0)$ at similar P/P_0, where $\alpha_0(P/P_0)$ is the RAI, offered in Ref. [76].* One can see that all N_2-RAIs coincide in a range of $0.1<P/P_0<0.4$. The latter range is the region

---

*RAI [76] has a mean of 15 AIs of N_2 at ~77 K, measured for coarsely dispersed systems (carbon black, Al_2O_3, SiO_2, powders of metal, etc) with A_{BET} surface areas in the range 0.3 to 1.5 m^2/g. The independent measurements of A by EM method for these samples shown that the relative deviation of A_{EM} and A_{BET} was less than ±6%. The measurements of these AIs were carried out on DigiSorb-2500 instrument (Micromeritics) at the maximum loading of sample cells with the samples. Normalization to A_{BET} in the range $0.1 \leqslant P/P_0 \leqslant 0.99$ resulted in coincidence of the AIs. According to Ref. [253] this averaged RAI does not deviate from RAI, that was accurately measured in [254] for macroporous silica gel with A_{BET} of 25 m^2/g. Analyzing the literature, one can find that the later the RAI is measured the lower are deviations from the average.

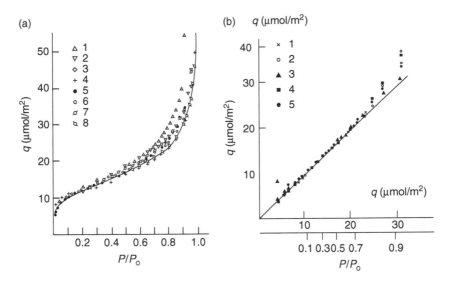

Figure 9.15   (a) References adsorption isotherms (RAIs) by different authors: (a) 1 by [73]; 2 by [79]; 3 by [69]; 4 by [75]; 5 by [80]; 6 by [53]; 7 by [72]; 8 by [76]; (b) comparison of RAIs: 1 by [53]; 2 by [80]; 3 by [81]; 4 by [82]; 5 by [75] with RAI by [76].

of possible capillary condensation in mesopores. If the dates of RAI publications are accurately analyzed one will find that the later the RAI is measured the smaller are the deviations. This is explained by a conflict between two aspirations: an increase in the accuracy of measurements and a decrease in the surface area of a PS. The experimental techniques used in earlier works, for example Refs. [64,69], allowed accurate measurements only for systems with relatively high specific surface areas of $A \sim 100$ to 200 m^2/g. Later publications involve PSs with A in the range 10 to 20 m^2/g, and the samples are subjected to a pretreatment to eliminate micro- and mesopores; hence the obvious reason for the deviations is capillary condensation.

The major property of accurately measured RAI is the independence of its derivative on possible effects of deviations of adsorption potential of a bare surface. This property together with the obvious additivity of adsorption on various parts of a surface allows measuring the textural characteristics of real microporous or modified PSs [3]. Indeed, if the PS has parts with increased (e.g., micropores) or decreased (e.g., modificators) adsorption potential, in the majority of cases the total AI, that is, $q(P/P_0)$, before capillary condensation is expressed as

$$q(P/P_0) = q_\mu(P/P_0) + A_{Me}\alpha(P/P_0) \qquad (9.22)$$

where A_{Me} is the specific surface area of mesopores, $\alpha(P/P_0)$ the specific adsorption on a unit of mesopores surface, that is RAI, and $q_\mu(P/P_0)$ the additive of specific adsorption in micropores ($q_\mu(P/P_0) > 0$) or on the modified partitions ($q_\mu(P/P_0) < 0$). The situation of microporous systems is the most obvious. The increased adsorption potential forces the adsorbate to fill micropores long before the monolayer over the mesopores surface is formed. After volumetric filling of micropores $q_\mu(P/P_0) = $ const. $= V_\mu$, where V_μ is the ultimate adsorption amount in micropores, i.e., the volume of micropores. Hence, the slope of the comparative plot $q(P/P_0)$ vs. $\alpha(P/P_0)$ in a range after the filling of micropores but before capillary condensation is determined by a value of the mesopore surface area, A_{Me}. The intercept of this linear dependence on the ordinate axis (at $P/P_0 \to 0$) determines V_μ (see Equation 9.22).

The advantage of CM over BET is not only in excluding the term w, but the more significant advantage is in its differential character, as compared to the integral character of BET method.

When in some range $q_\mu(P/P_0)$ becomes independent of P/P_0 (i.e., a condition of measurement of the micropore volume) the differentiation of Equation 9.22 withdraws $q_\mu(P/P_0)$ from the consideration (the derivative of a constant parameter is zero). Thus, the result of the measurement (of course under the named assumption) does not depend on the absence or presence of micropores at all. To the contrary, the BET method assumes that the entire adsorbed gas lays on a flat surface, including the portion adsorbed volumetrically (i.e., by different mechanism) in micropores.

The disadvantages of the BET method are clearly observed in the case of modified surfaces. For example, if one deposits small amounts of CH_3OH or any other organic component over graphitized carbon black or quartz and then measures AI of N_2, the formally applied BET method gives a significant (up to a twofold) decrease in A_{BET} (as well as C_{BET}) [3,62,79–83].[*] However, one can hardly believe that the real surface area of a nonporous material would decrease so significantly due to modification with limited amounts of organic guest (if its amount is less than that necessary to form a monolayer). The overall reason of the observed inaccuracy of the BET method is the influence of molecularly dispersed modifier on energetic heterogeneity of the surface in a range, where such heterogeneity significantly affects adsorption, and by which the BET method is taken into account due to its integral nature. Figure 9.16a shows the comparative plots of N_2 AI on modified graphitized carbon black; here the AI on an unmodified carbon black was used as RAI. The linearity of these plots in a range of polymolecular adsorption and constant slope indicate that there was no change in the surface area during modification. The curved parts near the origin are the result of a significant influence of the modifier on N_2 adsorption in a small range of N_2 coverage. Thus, the overall adsorption is decreased, but the additions (derivative) of adsorption in a polymolecular range remain unchanged. The same results were obtained for adsorption of N_2 and Ar on a nonporous quartz modified with CH_3OH, C_6H_6, and H_2O, after deposition of metal–organic complexes over SiO_2, etc. [3,62]. As an additional illustration, Figure 9.16c shows an example from a well-known monograph [53]: adsorption of N_2 on microporous TiO_2 after preadsorption of nonane and various temperatures of vacuum treatment. Plot 1 corresponds to a total desorption of nonane, when all micropores are free; plot 4 shows that nonane fills all micropores and covers some surface of mesopores. Summarizing, we plot three possible types of comparative graphs in Figure 9.16b.

The discussed problems result in a gradual substitution of traditional BET method for the surface area measurements with CM method. The modern adsorption instruments such as ASAP-2020

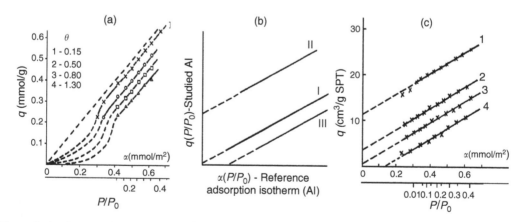

Figure 9.16 The practical examples of comparative plots: (a) N_2 adsorption on a graphitized carbon black, modified by physical adsorption of methanol (the numbers correspond to the amount of CH_3 OH in the fractions of monolayer capacity) [83]; (b) the usual types of comparative plots by [3]; and (c) N_2 isotherms on microporous titanium oxide after various amount of preadsorbed nonane by [53].

---

[*]One can also find numerous examples of modified mesoporoes where the discussed problem arises as well.

(Micromeritics) or Autosorb-1 (Quantachrome) allow measurements of AIs starting from $P/P_0 \sim$ 10^{-7}, and comparative plots in this range discover new opportunities for more detailed studies of micropore structure and surface heterogeneity evaluation [84–87].

Comparative methods may be effectively used for measurements of partial surface areas, A_C, of components in porous composites, for example for active surface area in supported catalysts. The traditional methods of A_C measurements are based on chemisorption of H_2, O_2, CO, NO_x, and some other gases that chemisorb on an active component, and have negligible adsorption on a support [5,54]. The calculation of A_C is fulfilled by an equation similar to Equation 9.18 assuming some values of w and atomic stoichiometry of chemisorption [54]. But, unfortunately chemisorption is extremely sensitive to insignificant variations of chemical composition and structure of surface, which alters the results of the measurements.

There is a one-point modification of a chemisorption method, which is widely used for measurements of A_C. In this case, only one adsorption point of a chemisorption isotherm is measured, and is compared with only one point on a chemisorption isotherm on a reference material (usually, powder [black] or foil). The identity of the chemisorption properties of the active components in supported and pure form is postulated, but very often does not fulfill, making one-point modification an inaccurate procedure, which can hardly be used in scientific studies. For example, studies of supported Rh catalysts by O_2 and CO chemosorption have shown that three different blacks of Rh yield three different results [88]. The multipoint comparison of chemisorption isotherms shown that only one black had a chemisorption isotherm that had affinity to the isotherm on a supported metal.

To avoid chemisorption problems, an alternative and more universal adsorption method was proposed [89,90]. The method is based on the differences in potentials of physical adsorption on the different components. The background of this method is discussed in Problem 8. It advances a CM, and its efficiency was tested on the model mechanical mixtures and PS type of C/SiO_2, C/Al_2O_3, C/MgO, $SiO_2 + Al_2O_3$, etc. by adsorption of CO_2 or hydrocarbons. The main requirement of the adsorbate for such measurements is the existence of some specificity of adsorption on different components.

Finally, let us consider the modern methods of porosity measurements (Table 9.1). Traditional adsorption and mercury intrusion methods are at the top. These methods have significantly progressed through the last decades due to advances in their theory and computer simulation [84–87], the synthesis of model porous solids and catalysts [24,91,92], and combination with independent physical methods, which increase the depth and reliability of traditional measurements. The multiple variants of microscopy have no competitors by the visualization of the results and cannot be overestimated in determining the morphology of pores and particles. The family of these methods includes HRTEM (high-resolution transition electron microscopy), EMT (electron microscopic tomography), STM (scanning–tunneling microscopy), AFM (atomic force microscopy), and others. X-ray projection tomography microscopy, also called high-resolution three-dimensional x-ray

Table 9.1 The Methods of Porous Structure Characterization

Method	Scale: Sizes of Pores/Particles (nm)
Adsorption	10^{-1}–10^2
Mercury intrusion	1–10^6
Electron microscopy	10^{-1}–10^6
NMR (^{129}Xe etc.)	10^{-1}–10^2
XRD	1–10^2
SAXS (SANS)	10^{-1}–10^2
Light scattering	10^3–10^6
Thermoporometry	1–10^2
Permeametry	10^2–10^5

microscopy [93,94], is very new. Unfortunately, the resolution of this method is low (~ 1 µm), although in some cases it was possible to investigate objects of several dozens of nanometer size. The x-ray methods, especially x-ray microscopy, and some other methods are the spectroscopic methods, most of which are based on interaction of electrons, radiation, or probing molecules with the PS, and studies of quantum-mechanical transitions caused by this interaction. This is the case of very popular state-of-the-art methods of the NMR, based on measurements of relaxation times and chemical shifts during the resonance absorption of electromagnetic energy by the molecules with uncoupled spins (^{129}Xe, ^{15}N$_2$, ^{13}CO$_2$, D, etc.). Application of these methods does not eliminate the problem of particle form, and the calibration of these methods is usually provided on the basis of adsorption studies. Among other spectroscopic methods let us mention XRD, small-angle x-ray (SAXS), and neutron (SANS) scattering, as well as light scattering, and laser diffraction.

Some of the methods of analysis of porosity are based on specific properties of porous and disperse materials, namely, thermoporometry method is based on shifts of the temperature of phase transitions and permeametry methods are based on characteristics of mass transfer through porous media. Each method has its advantages, for example low cost of equipment and high performance. Each has its own range of optimal measurements. But, all the methods are really doomed for coexistence, and in many cases they supplement each other.

9.4   MORPHO-INDEPENDENT TEXTURAL PARAMETERS

9.4.1   Density and Porosity

Definition of density for a bulk homogeneous nonporous solid is simple and obvious, that is, the relation of its mass and volume. However, as soon as one introduces some heterogeneity this definition loses its unambiguity. Complications arise immediately when we divide a material into two complementary partitions. For example, such partitions can be solid particles and pores between the particles, porous aggregates of particles and the pores between the aggregates (see Figure 9.17a), or an active supported component and the support. In the general case of a solid of mass M and volume V, which can be divided in the named manner, we deal with partition 1 with mass M_1, volume V_1, and density $\Gamma_1 = M_1/V_1$, and partition 2 of the mass M_2, volume V_2, and density $\Gamma_2 = M_2/V_2$, and

$$V_1 + V_2 = V, \quad M_1 + M_2 = M \tag{9.23}$$

or

$$v_1 + v_2 = 1, \quad m_1 + m_2 = 1 \tag{9.24}$$

if one assumes that $v_1 = V_i/V$ is the partial volume and $m_i = M_i/M$ the partial mass of partition i. It is natural to consider that the apparent density of the PS as a whole is equal to

$$\Gamma = \frac{M}{V} = \frac{M_1 + M_2}{V_1 + V_2} \tag{9.25}$$

According to the definition of v_i and m_i,

$$m_i = \frac{M_i}{M} = \frac{\Gamma_i}{\Gamma} \frac{V_i}{V} = \gamma_i v_i \tag{9.26}$$

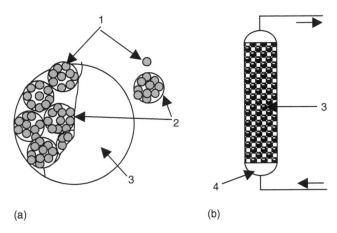

Figure 9.17 The basic schemes of (a) bidisperse (biporous) porous solid structure: 1, nonporous primary particles, 2, aggregates of primary particles (secondary particles), 3, porous solid (granule, grain, pallet, etc.); (b) a bed of granules in a catalytic reactor 4.

where $\gamma_i = \Gamma_i/\Gamma = m_i/v_i$ is the ratio of the apparent density of the partition to the apparent density of a PS as a whole. Noteworthy, parameter $\Gamma_i v_i/m_i = \Gamma$ is invariant for all partitions allocated in a given PS. The combination of Equation 9.24 and Equation 9.6 gives

$$v_2 = 1 - v_1 = 1 - m_1/\gamma_1 \tag{9.27}$$

and

$$\gamma_1 = \frac{m_1}{1 - v_2} = \frac{1 - m_2}{1 - v_2} \tag{9.28}$$

The main advantage of the presented formulation is that the properties of a partition (e.g., its volume, density) can be calculated if one knows the properties of a PS as a whole and the properties of another partition. This is characteristic of not only volume, but also all other morpho-independent textural parameters, for example the surface area. Thus, we will use it as a second fundamental statement of texturology: the morpho-independent parameter of a partition of PS may be expressed through the corresponding parameter of a whole PS and the parameters of the other partitions. Two simple practical examples that illustrate the significance of this property for the textural analysis are formulated as Problems 4 and 5, which are expected to be solved by the reader before continuing.

Let us consider a case when several partitions can be allocated in a porous solid. For example, consider an active component (partition 3) supported over a porous support, assuming that partition 2 is all pores in the supported catalyst and partition 1 is all particles of the support.

By analogy, we can consider an interrelation between the partial volumes of partitions:

$$V = \sum_i V_i = \sum_{i \neq k} V_i + V_k = \bar{V}_k + V_k \tag{9.29}$$

or

$$1 = \bar{v}_k + v_k; \quad \bar{v}_k = 1 - v_k \tag{9.30}$$

where v_k is a partial volume of partition k and $\bar{v}_k$ a sum of partial volumes of all other partitions that we will consider as an addition to v_k, i.e., we use the second statement of texturology again.

Since

$$1 = \sum_i v_i = \sum_i (1 - \bar{v}_i) \tag{9.31}$$

it is obvious that

$$\bar{v}_k = \sum_{i \neq k}(1 - \bar{v}_i), \quad v_k = 1 - \sum_{i \neq k} v_i = 1 - \sum_{i \neq k}(1 - \bar{v}_i) \tag{9.32}$$

In all cases, v_i can be substituted with $\bar{v}_i$ if definition or measurement of the latter is easier. This again means the simple rule that both partial volume of a partition and its addition are expressed using partial volumes of other partitions.

Interesting and interrelated with the previous case is one of enclosed partitions, when one of two partitions can be further divided into two partitions. An illustrative example is shown in Figure 9.17a. A granule of catalyst can be divided into two partitions: porous aggregates (secondary particles—partition 1) and pores between the aggregates (partition 2). Partition 1 can also be divided into two partitions: nonporous particles (primary particles—partition 11) and pores between particles (partition 12), excluding pores between aggregates. Another case of enclosed partitions has already been considered: the case of a porous supported catalyst, which can be divided into pores and a solid phase, while the solid phase can be divided into the support and the active component.

Let us consider a case of one enclosed partition. There are two levels of partitioning in the considered PS. The upper level of partitioning includes partitions 1 and 2 (corresponding subscript indexes are 1 and 2). The lower level of partitioning includes partitions 11 and 12 (corresponding subscript indexes are 11 and 12) that form partition 1. At each level of partitioning, Equation 9.23 through Equation 9.28 are correct, thus

$$v_1 = 1 - v_2 \ (V_1 = V - V_2), \quad v_{11} = 1 - v_{12} \ (V_{11} = V_1 - V_{12}) \tag{9.33}$$

Multiplying,

$$v_1 v_{11} = \frac{V_1}{V}\frac{V_{11}}{V_1} = \frac{V_{11}}{V} = (1 - v_2)(1 - v_{12}) = (1 - \bar{v}_1)(1 - \bar{v}_{11}) \tag{9.34}$$

This means that the volume of the smallest partition in a PS can be determined from the volumes of other partitions. The reader may use this formulation to solve Problems 6 and 7.

It is obvious that in the case of multiple enclosed partitions the following expression is true:

$$\prod v_j = \frac{V_{11.1}}{V} = \prod (1 - \bar{v}_j) \tag{9.35}$$

Where j is the level of partitioning. If $V_{11..1}$ in Equation 9.35 is the volume of entire solid in a PS, the ratio of $V_{11..1/V}$ is the density of solid phase packing, η. In this case the total porosity, ε which is complementary to η is derived by:

$$\eta = 1 - \varepsilon = \prod(1 - \varepsilon_j), \varepsilon = 1 - \prod(1 - \varepsilon_i) \tag{9.36}$$

where ε_i is the porosity at level i. And, of course, using the main principle of the given consideration that the properties of a part of a system can be determined by that of a system as a whole and its other parts we can evaluate the porosity at level i as

$$\varepsilon_i = 1 - \frac{1 - \varepsilon}{\Pi_{k \neq i}(1 - \varepsilon_k)} \tag{9.37}$$

If $V_{11..1}$ in Equation 9.35 is the volume of all pores in a PS, the ratio of $V_{11..1/V}$ is the total porosity, ε, thus:

$$\frac{V_{11...2}}{V} = (1 - v_{11...1}) \cdots (1 - v_{11})(1 - v_1) \tag{9.38}$$

The fundamental difference between Equation 9.36 and Equation 9.38 is in the direction of partitioning. Equation 9.36 corresponds to the system of *substraction*, as if one has a given original nonporous solid (e.g., with volume of a catalyst bed (see, Problem 7)) and starts to allocate pores ("substract" a solid phase from an original volume): first between the granules, then pores in granules between the porous aggregates, and finally the pores inside porous aggregates. The opposite system is one of *addition*, when one has an originally void given volume and starts to fill it with the solid phase. System of addition is described with Equation 9.38. One PS can be treated in both ways, however one should remember, that physical meanings of ε_j in Equation 9.36 and Equation 9.38 differ due to different allocation of corresponding partitions. As in the case of Equation 9.36 one can determine the porosity at level j for the system of addition as:

$$\varepsilon = \Pi \varepsilon_j \tag{9.39}$$

9.4.2 Experimental Techniques of Measurements of True, Apparent, and Bulk Density

Although, the true density of solid phase $\rho = m/V_\rho$ (e.g., g/cm^3) is defined by an atomic–molecular structure (*i*), it has become fundamental to the definition of many texture parameters. In the case of porous solids, the volume of solid phase V_ρ is equal to the volume of all nonporous components (particles, fibers, etc.) of a PS. That is, V_ρ excludes all pores that may be present in the particles and the interparticular space. The PS shown in Figure 9.17a is formed from nonporous particles that form porous aggregates, which, in turn, form a macroscopic granule of a catalyst. In this case, the volume V_ρ is equal to the total volume of all nonporous primary particles, and the free volume between and inside the aggregates (secondary particles) is not included.

The volume of the solid phase V_ρ is usually measured by a pycnometric technique, which measures the excluded volume of a pycnometric fluid, whose molecules cannot penetrate the solid phase of PS. A simple example of a pycnometric fluid is helium [55]. The pycnometric fluid fill in all void space (pores) accessible to it, and presumably do not adsorb on the surface of PS. In the case of microporous PSs, measurement of the volume accessible for guests with various sizes allows the determination of a distribution of micropores volume *vs.* the characteristic size of guest molecules. This approach lays the basis of the method of molecular probes. The essence of this method is in the following: we have a series of probe molecules with different mean sizes ($d_1 > d_2 > d_3 > \cdots$). The pycnometric measurements of the excluded volume will give a series $V_{\rho 1} < V_{\rho 2} < V_{\rho 3} \cdots$. The difference $\Delta V = V_{\rho i} - V_{\rho i}(i > j)$ corresponds to the volume of micropores with pycnometric sizes of d in a range $d_i < d < d_j$. The same can be said in terms of densities, $\Delta V = 1/\rho_i - 1/\rho_i$ (see, e.g., Ref. [95]).

Thus, there are two limitations of the pycnometric technique mentioned: possible adsorption of guest molecules and a molecular sieving effect. It is noteworthy that some PSs, e.g., with a core–shell structure, can include some void volume that can be inaccessible to the guest molecules. In this case, the measured excluded volume will be the sum of the true volume of the solid phase and the volume of inaccessible pores. One should not absolutely equalize the true density and the density measured by a pycnometric technique (the pycnometric density) because of the three factors mentioned earlier. Conventionally, presenting the results of measurements one should define the conditions of a pycnometric experiment (at least the type of guest and temperature). For example, the definition $\rho\,_{298}^{He}$ shows that the density was measured at 298 K using helium as a probe gas. Unfortunately, use of He as a pycnometric fluid is not a panacea since adsorption of He cannot be absolutely excluded by some PSs (e.g., carbons) even at 293 K (see van der Plas in Ref. [2]). Nevertheless, in most practically important cases the values of the true and pycnometric densities are very close [2,7].

The apparent density δ (g/cm^3) is usually measured using mercury as a pycnometric fluid. Mercury does not wet most of the solids and, thus, does not penetrate pores until pressure is applied. Mercury is not the only choice; highly dispersed powders can serve as a guest fluid with the same penetration properties as well [55]. Reciprocal to δ is the specific apparent volume of PS, which is equal to the sum of the volumes of the pores and solid phase (e.g., the total volume of a granule shown in Figure 9.17a), and is obviously related to the mass of a PS. Relation between true and apparent density and porosity was considered in Problem 4.

The bulk density, Δ, is measured by a widely used tap-density technique by weighing of a convenient known macroscopic volume (e.g., a barrel) of PS [55]. A void between granules can be generally considered as macroscopic pores, and its specific volume can be determined as $V_{\Delta}=\Delta^{-1}-\delta^{-1}=(1-\Delta/\delta)/\Delta$.

9.4.3   The Properties of Porosity

We have already considered porosity in the above examples. Summarizing, let us outline that porosity is the ratio of the volume of pores in a PS to the total volume of that PS.

It is essential that porosity does not depend on size of the building blocks that form a PS. Indeed, let porosity of a bed of lead shot be equal to ε. Let all sizes (size of shot, distances between shots, etc.) in such PS increase with a scaling factor of 10^9 (to the size of the Earth). The porosity of such a globular system is still equal to ε and will remain at any isotropic scaling of a system if the form and arrangement of the globules remain.

If we apply another restriction, ρ=const., in addition to porosity, all other volume-related characteristics, such as δ, specific pore volume, V_{pore}, etc. become independent of scaling. Indeed, if the space is isotropic the unit volume is directly proportional to the unit mass. Scaling results in change of sizes and surface areas, but the volume related to the mass remains the same.

It is noteworthy that porosity is remarkably independent of the dimensionality. For a given PS, the volumetric porosity ε_V (void volume/total volume) in 3D space is equal to the interfacial porosity ε_A (void area/total area, measured on a representative cross section) in 2D and the linear porosity ε_L (void interval/total length, measured on a representative traverse line) in 1D [96]:

$$\varepsilon = \varepsilon_V = \varepsilon_A = \varepsilon_L \tag{9.40}$$

This statement is *a priori* obvious for isotropic PSs, but works also for anisotropic PSs as well. Essential here is the term representative, which means that the multiple measurements of 1D or 2D porosities in a number of different directions (traverse lines for 1D case, and cross sections for 2D case) are provided. In the case of anisotropic PSs, dependence of ε_i on a location (direction) of a corresponding cross section or traverse line in a PS gives a picture type of a rose of winds, which shows the properties of the anisotropy; but averaged for all directions, ε corresponds to Equation 9.40 [96]. The latter relation was suggested by Delessy (1848) and Rosival (1898) and was strictly confirmed in Ref. [96]. It is widely used in geology, forestry, ecology, biology, and sociology to determine ε in 1D, 2D, and 3D regions.

Generally, ε varies in the range $0 \leqslant \varepsilon \leqslant 1.0$, where the left limit corresponds to a nonporous system and the right limit corresponds to a void bulk. In the case of PSs formed of rigid monodisperse building blocks, the lower value of $\varepsilon = \varepsilon_{\min}$ corresponds to the densest ordered packing (DOP) and depends on the shape of the building blocks. For example, for the DOP of monospheres $\varepsilon_{\min} = 0.2595$ and for the DOP (hexagonal packing) of cylindrical rods $\varepsilon_{\min} = 0.093$ (see Section 9.7.2). Ordered packings of cubes or rods with polygonal cross sections can have $\varepsilon_{\min} = 0$. For random packings, higher values of $\varepsilon_{\min}$ are characteristic, e.g., for a dense random packing (DRP) of monospheres, $\varepsilon_{\min} \sim 0.36$ to 0.40. In this case, optimization of the conditions of a packing formation and vibrodensification allow decrease of $\varepsilon_{\min}$ to 0.31 to 0.33. However, the minimum porosity of DRP is always higher than that of DOP.

Porosity of random packings increases with an increase in building block anisotropy, roughness of their surface, and the presence of internal porosity. Building blocks of straight rods or cylinders can self-assemble in ordered structures, which results in a decrease in porosity [3]. Onzager [97] has shown a general explanation of this effect using the general idea of dependence of free energy on the concentration and orientation of N hard rods that experience infinitive repulsions. The free energy includes two entropy terms: the first describes the orientation distribution of particles and is proportional to N: the second depends on the number of possible packings and is proportional to N^n, where $n \geqslant 2$ (a decrease of packing entropy with ordering). The effect becomes more pronounced with an increase in the number of building blocks in a unit volume of PS and the ratio L/D_{ef}, where L is the characteristic length of a building block and D_{ef} the diameter of its effective cross section, usually perpendicular to the length. Another reason for spontaneous formation of ordered packings is related to the decrease in PS free energy caused by compensation of intermolecular (interparticle) interaction forces [98,99].

Porosity can decrease in the packings of polydisperse building blocks of various sizes D_i, due to the presence of smaller building blocks between larger ones (Figure 9.18), which is a typical case of a system of addition (see the end of Section 9.4.1).

To illustrate, consider the examples of possible packings of the densest packing of polydisperse circles, formed by limiting filling of a free space with circles (or spheres) of decreasing diameter (Figure 9.18a), and a bidisperse system of spheres (Figure 9.18b to Figure 9.18d). The former is called the Appolonian packing after the ancient Greek mathematician (about 2000 BC), who had considered the problem of the circle inscribed among three given circles in a plane. In both cases, decrease of ε is caused by decrease in the volume excluded from the volume between the larger building blocks (the systems of addition).

Let us consider this effect in detail in the practically important case of a bidispersed PS shown in Figure 9.18b, formed of globules (monospheres) of size $D_1 \gg D_2$. As for any system of addition we can determine the porosity of this PS (Equation 9.38) as

$$\varepsilon = \varepsilon_1 \varepsilon_2 = (1 - \eta_1)(1 - \eta_2) \tag{9.41}$$

where ε_1 and ε_2 are the porosities and η_1 and η_2 the densities of packings of only large and only small globules, respectively. The minimum possible value of ε is achieved when ε_1 and ε_2 are minimal. When $D_1 \gg D_2$, if both small and large globules form DOPs, the resulting minimum $\varepsilon_{\min} = \varepsilon_{\min}$ DRP1 $\varepsilon_{\min}$ DRP2 $= (0.2595)^2 = 0.0673$; if both packings from DPRs, $\varepsilon_{\min} = \varepsilon_{\min}$ DRP1 $\varepsilon_{\min}$ DRP2 $= (0.31)^2 = 0.096$. In all cases, the porosity of bidisperse PS increases when one or both packings begin to demonstrate nondense packings. Typical dependences of ε on partial volume of the large globules, X, and the ratio $K = D_1/D_2$ is shown in Figure 9.19 [61]. The $\varepsilon_{\min}$ values decrease with an increase in K.

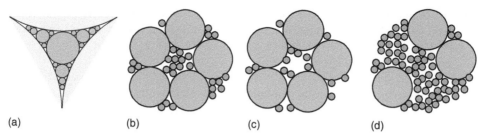

(a)   (b)   (c)   (d)

Figure 9.18 Random packings of (a) densest 2D Appolonian packing; (b)–(b) the building blocks of two sizes: (b) both packings of small and large globules are dense; (c) packing of large globules is dense, but packing of small globules is loose; (d) opposite to (c) surplus of small globules.

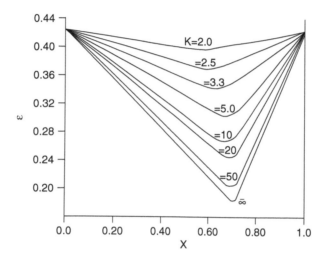

Figure 9.19 Typical dependence of the porosity, ε, of a bidisperse PS with loose random packing of globules size of $D_1 > D_2$ on partial volume of the larger particles X, and $K = D_1/D_2$.

Under constant K, there is a critical value of X ($=X_{CR}$), when both packings have minimal porosities, and the resulting total porosity is minimal. Decrease of ε in the range $0 < X < X_{CR}$ is caused by insufficiency of the quantity of the small globules to fill all voids between the large globules (Figure 9.18c), and increase of ε in a range of $X_{CR} > X > 1$ is explained by a surplus of small globules (Figure 9.18d).

The influence of K on ε is caused by the so-called wall effect and an increase of void volume that is accessible for small particles with K. Both of these effects correspond to the appearance of the excluded volume in the places of contact of rigid particles between themselves (Figure 9.18) or a rigid wall (Figure 9.20).

The problem of investigating such effects is known for decades, but it is far from a general solution up to date (see, e.g., Refs. [100–107]). The best description of the wall effect is fulfilled for DRP of monospheres in beds of the simplest form (cylinder, sphere, or rectangle). The porosity ε of these systems depends on the particle-to-container diameter ratio $K = D/D_C$, where D and D_C are the particles and container diameters, respectively. Dependence of ε on K is considerable at $K > 0.1$, and according to Ref. [102] it has three characteristic regions (Figure 9.21). Porosity increases with K in Region 1 at $0 < K < 0.25$. This region is interpolated with a correlation equation [102]

$$\varepsilon(K) = \varepsilon_0 + A_Z[\exp(-BK) - 1] \tag{9.42a}$$

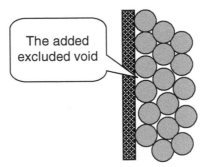

Figure 19.20 A scheme of a possible appearing of an additional void volume in a region of contact of particles with the wall of a container (wall effect).

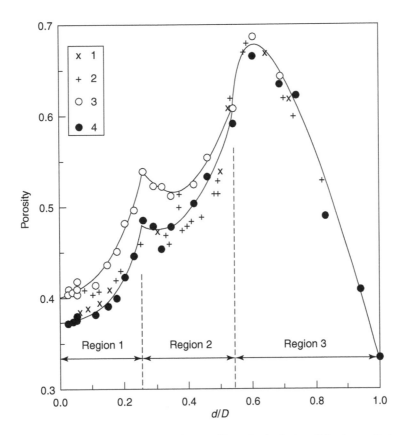

Figure 19.21 Dependence of porosity ε of a random packing of spherical particles in a cylinder container according to 1 [104], 2 [105], (3), (4) for loose and dense packings by [102]. Solid lines correspond to calculations using Equation 9.42a through Equation 9.42c.

where $\varepsilon_0 = 0.40$, $A_z = 0.010$, and $B = 10.686$ for a loose random packing (LRP); and $\varepsilon_0 = 0.372$, $A_z = 0.002$, and $B = 15.306$ for a DRP. These values were calculated for low ratios of D/H, where H is the height of a bed. The increase in D/H ratio in the range 0.1 to 0.2 and higher increases the porosity ε (the top–bottom effect).

Increase in K in Region 2 ($0.25 < K < 0.53$ to 0.54) results in a parabolic decrease or increase of $\varepsilon(K)$, and according to Ref. [102] it is caused by a gradual transition from random packing to an ordered one:

$$\varepsilon(K) = B_1 - B_2 K + B_3 K^2 \tag{9.42b}$$

The values of coefficients for DRP are $B_1 = 0.681$, $B_2 = 1.363$, and $B_3 = 2.241$.

Region 3 ($0.53 - 0.54 < K \leqslant 1.0$) is characterized by a decrease in ε with K, but it is similar for both dense and loose packings and does not depend on the D/H ratio. This peculiarity is caused [102] by the formation of an ordered packing; dependence of $\varepsilon(K)$ in this region is well interpolated by the equation

$$\varepsilon(K) = 1 - \frac{2K^3}{3(2K-1)^{1/2}} \tag{9.42c}$$

The condition of $K > 1.0$ results in inaccessibility of a container for spheres, and $\varepsilon = 0$, looses its physical meaning, correspondingly.

Ayer and Soppet [107] proposed a smoother approximate relationship for DRP in the region of $K < 0.7$:

$$1 - \varepsilon(K) = 0.635 - 0.216 \exp\left(-\frac{0.313}{K}\right) \tag{9.43}$$

As an example of a simpler approximation at $K < 0.3$ one can use the correlation equation proposed in Refs. [3,106]:

$$\varepsilon(K) = \varepsilon_0 + (1 - \varepsilon_0) K^\zeta \tag{9.44}$$

where the exponent $\zeta = 2$ for a cylindrical and $\zeta = 3$ for a spherical container. The relative deviation between $\varepsilon(K)$ values calculated by Equation 9.42 through Equation 9.44 for a cylindrical container does not exceed 10%.

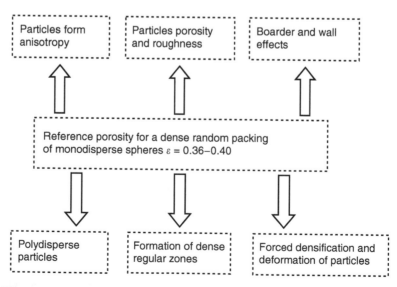

Figure 9.22 The major reasons for change of porosity value comparing to a porosity of a dense random packing of monodispherse. Upward arrows correspond to increase, and downward to decrease of porosity (see text for more comments).

The major directions of changing porosity in DRP are schematically shown in Figure 9.22 [3,61]. As a starting point, one can use the porosity of DRP of monospheres $\varepsilon_0 = 0.36$–0.42. Values of $\varepsilon > \varepsilon_0$ increase with particles' anisotropy, roughness, and internal porosity, and also with the influence of wall effects at $K > 0.1$, and $D/H > 0.05$. Values of $\varepsilon < \varepsilon_0$ are characteristic for polydisperse particles when denser zones with ordered or unidirectional packings are formed, and also under forced densification and deformation of particles, correspondingly.

With the exception of the reasons given in Figure 9.22, porosity significantly depends on particle orientation and packing that can be characterized by a coordination number of packing of particles, n_p, or pores, Z_C, which will be discussed in Section 9.6 and Section 9.7.

9.4.4   The Specific Surface Area

Measuring the specific surface area, A, related to the mass of PS does not require a textural model (a morpho-independent parameter, i.e., one can apply an approach of partitioning and, correspondingly, the second statement of texturology, as we have already done for volume-related parameters). Let us consider the most widespread adsorption method based on proportionality of adsorption, Q, and the specific surface area in the absence of volumetric effects (capillary condensation, micropore filling, etc.):

$$\frac{Q(P)}{A} = q(P) \tag{9.45}$$

where P is adsorbate pressure (concentration) and $q(P)$ some theoretical (Henry, Langmuir, BET, etc.) or experimental (reference) model of adsorption of adsorbate on a unit of surface. Depending on the specificity of adsorbate interaction with the surface, one can allocate methods of total and partial surface area measurements. For example, adsorption of N_2 at 77 K at relatively high pressures ($0.1 < P/P_0 < 0.3$, where P_0 is the saturation pressure) over the majority of PSs is not specific, and one can use the same reference $q(P)$ for the total specific surface area measurements even if the chemical composition of a PS is complex. Adsorption of other gases (CO_2 at 273 K, etc.) may be specific enough to distinguish, for example, $q_1(P)$ and $q_2(P)$ characteristic to different solid phase components of a PS. This principle forms the basis for measuring partial surface areas of components (try to derive that basis by solving Problem 8 by using the second statement of texturology).

Values of A and density allow calculations of the specific surface area related to the volume of PS, A_V:

$$A_V = A\delta = A\rho(1-\varepsilon) \tag{9.46}$$

where $A\rho$ corresponds to the specific surface area related to the volume of solid phase in PS.

In some cases (e.g., textural studies of supported catalysts), it is convenient to use A_V instead of A since the external volume of a granule remains constant, while introduction of an active component results in an increase in mass of the resulting material. In these cases, one can directly compare A_V after deposition of an active component with A_{V0}—the volumetric surface area of initial support. For this purpose it is convenient to use the parameter

$$\psi = \frac{A_V}{A_{V0}} = \frac{A\delta}{A_0\delta_0} = \frac{A}{A_0(1-m_C)} \tag{9.47}$$

where A and A_0 are the specific (related to mass) surface areas as they are measured by means of standard methods and m_C is a partial mass of the active component as before. Parameter ψ is convenient since it is >1.0 when the surface area in a granule increases, and <1.0 when the surface area decreases. For more discussion and practical examples we recommend the literature elsewhere [90,108].

9.5   MORPHO-DEPENDENT TEXTURAL PARAMETERS: MEAN SIZES OF PARTICLES AND PORES

There are a large number of experimental methods to measure particles sizes, D, and distribution functions for D. These methods include sieving, microscopy (optical, electron, etc.), sedimentation, centrifugation, electrical conductometry (the Coulter principle), radiation scattering, dynamic light-scattering, permeability measurements and gas diffusion, adsorption and mercury intrusion, powder XRD, NMR, etc. These methods are thoroughly described in previous literature (see, e.g., [50–52,55]) and we will not focus on them here. More significant is the discussion of the results of these methods, their correlations, and other problems that arise when the sizes of complex particles are evaluated.

Indeed, it is easy to define the size (sizes) of simple particles (e.g., in the case of spherical globules or cylinders). But, for many real PSs, the form of particles and pores is complicated. The sizes of complicated particles or pores are expressed with equivalent diameters (sizes). The particular choice of an equivalent is directed by measuring the technique or other reasons. Following are some frequently used expressions for equivalent diameters [52]:

- $D_V = (6V/\pi)^{1/3}$ is the volumetric diameter, equal to the diameter of sphere of equivalent volume V.
- $D_A = (A/\pi)^{1/2}$ is the interfacial diameter, equal to the diameter of a sphere of the equivalent surface area A.
- $D_{AV} = 6V/A = D_v^3/D_A^2$ is the surface-volumetric diameter, equal to the diameter of sphere or cube with equivalent surface-to-volume ratio.
- D_{Si} is the sieving diameter, equal to the width of the minimum aperture through which the particle will pass.
- $D_{st} \approx (D_v^3/D_A)^{1/2}$ is the Stokes' diameter, equal to the diameter of sphere, which in a laminar region (low Reynolds number $Re < 0.2$), sediments with the same velocity as the considered particle.

In addition, when one considers 2D cross sections or projections:

- $D_a = (4S_A/\pi)^{1/2}$ is the projected area diameter, equal to the diameter of a circle having the same area S_A as the projected area of the particle resting in a stable position. For particles with size anisotropy, S_A corresponds to the mean value derived from all possible orientations.
- $D_p = P_A/\pi$ is the perimeter diameter, equal to the diameter of a circle having the same perimeter P_A as the projected outline of the particle.
- $D_H = 4S_A/P_A$ is the hydraulic diameter, equal to the diameter of a circle with the same projection area to perimeter ratio.

Equivalent diameters D_a, D_p, and D_H are convenient for expression of sizes of windows (narrowings), d_W, between the cavities and contact cross sections, D_{PC}, between the particles. The diameters D_V, D_A, and D_{AV} are convenient for determination of characteristic sizes of particles and cavities.

Feret's diameter, D_F (the mean value of the distance between pairs of parallel tangents to the projected outline of the particle), and Martin's diameter, D_M (the mean chord length of the projected outline of the particle[96] $D_F \approx D_M = V_s/A = S_A/P_A$) are used in automated analysis of microscopic images [49,50].

Interrelation between different equivalent diameters is derived from the relations between the real values of the volume of solid V_s and surface area A of a particle with equivalent diameters D_i and D_j, where i and j are the methods of measurement or types of equivalent sizes. In a general case,

$$V_s = K_{V,i} D_i^3 = K_{V,j} D_j^3 \tag{9.48}$$

$$A = K_{A,i} D_i^2 = K_{A,j} D_j^2 \tag{9.49}$$

where $K_{V,i}$ and $K_{A,i}$ are the form coefficients in a chosen system of equivalent diameters i, and $K_{V,j}$ and $K_{A,j}$ the form coefficients for a system of equivalent diameters j at the same real values of V and A. It is convenient to introduce the surface-to-volume ratio for a particle as

$$\frac{A}{V} = \Lambda_{AV,i} D_i = \Lambda_{AV,j} D_j \tag{9.50}$$

where $\Lambda_{AV,i} = K_{A,i}/K_{V,i}$ and $\Lambda_{AV,i} = K_{A,i}/K_{V,i}$ are the surface-to-volume form coefficients in the systems of equivalent sizes i and j, respectively. Equation 9.50 shows that different methods of particles and pores size expressions are equivalent, and the sizes measured by various methods are strictly related to the appropriate surface-to-volume form coefficients ($\Lambda_{AV,i}$, $\Lambda_{AV,j}$, etc., or in general Λ_{AV}).

One of the earliest defined shape factors is the sphericity, ψ_W, which was defined by Wadell [109] as the surface area of a sphere having the same volume as the particle, related to the surface area of the particle as

$$\psi_W = \left(\frac{D_V}{D_A}\right)^2 = \left(\frac{D_{St}}{D_V}\right)^4 = (36\pi)^{1/3} \frac{V_s^{2/3}}{A} \tag{9.51}$$

For a sphere $\psi_W = 1.0$, and it decreases while the form anisotropy of a particle increases. For example, for cylinders, when $H/D = 0.1$, $\psi_W = 0.51$; for $H/D = 1.0$, $\psi = 0.826$; and for $H/D = 10$, $\psi = 0.169$. Some authors [100] use the sphericity coefficient $\psi_{sph} = (\psi_W)^3$, which is equal to 0.302 for tetrahedron, 0.523 for cube, 0.604 for octahedron, 0.829 for icosahedrons, and 1.0 for a sphere. A number of other methods for characterization of particle form by relating them to a sphere or cylinder are discussed elsewhere [50–52,108,110,111].

Let us use the mentioned relations to establish the relation of the surface area with the size of particles.

The specific surface area A of a porous solid that is formed of the monodisperse particles (solid phase) of volume $V_{s,i}$ and surface area A_i is equal to the multiplication of the surface area of one particle by the number of particles N_m in unit of mass. Using Equation 9.50,

$$A = N_m A_i = \frac{1}{V_{s,i}\rho} A_i = \frac{A_i}{V_{s,i}} \frac{1}{\rho} = \Lambda_{AV} \frac{1}{D_i \rho} \tag{9.52}$$

These relations can be spread to the systems of particles of one form, but of various sizes, D. The specific surface area of such PS is equal to the sum of surface areas of particles related to the entire mass. At known particle size distribution (PSD), $v(D)$, one has

$$A = \frac{\Lambda_{AV}}{\rho} \frac{\int D^2 v(D)dD}{\int D^3 v(D)dD} = \frac{\Lambda_{AV}}{\rho \bar{D}} \tag{9.53}$$

where $\bar{D}$ is a mean surface-to-volume size of particles defined as

$$\bar{D} = \frac{\int D^3 v(D)dD}{\int D^2 v(D)dD} \tag{9.54}$$

We know from statistics the moment of distribution of random values:

$$\langle X^r \rangle = \int X^r v(X)dX \tag{9.55}$$

which is called a zero moment at $r = 0$, first moment at $r = 1$, second moment at $r = 2$, etc. One can see that we used a ratio of the third moment to the second moment when we derived the volume-to-surface ratio in Equation 9.54. Other methods use the ratios of other moments. For example, the mean size averaged from the linear sizes of electron microscopy images is determined by the ratio of the first-to-zero moments of distribution, and that obtained from Roentgen diffraction data is determine by the ratio of the fourth-to-third moments of distribution, etc. [61]. Also, different methods use different methods of approximation of particles formed in the cases of complicated geometry. All this is essential when comparing the results measured by different methods.

The function $v(D)$ can be measured experimentally, or in some cases be simulated as normal, lognormal, etc. distribution. It is also possible to obtain polymodal distributions with several maximums or some special kind of distribution. For example, the distribution of the particles formed by crashing is frequently described by a Rosin–Rammler distribution [51,52] as

$$R = 100\exp\left(-bD^n\right) \tag{9.56}$$

where R is the weight percent of a material, retained on the sieve of aperture D (a plot of R vs. D gives the cumulative percentage oversize curve), n and b are the coefficients (b is assumed to be a measure of a range of particle size present and n a characteristic of the material under consideration). Various other PSD functions have been proposed. These are generally in the form of two-parameter (n and b) potential distribution functions such as by Gates–Gaudin–Schumann: $\phi_{GGS}=(bD)^n$, or by Gaudin–Meloy, $\phi_{GGS}=[1-(1-bD)^n]$, where ϕ is the undersize fraction [51]. One can find a more detailed discussion of PSD elsewhere [51,52,111].

By analogy, one can derivate the formulation to determine the sizes of pores (cavities), d, and establish their relation to volume and surface area. For a system of uniform cavities of constant size and shape with the volume and shape of one cavity of $V_{p,i}=k_v d_i^3$ and $A_i=k_A d_i^2$, and total pore volume, V_P, porosity is ε, specific surface area, A is equal to

$$A = A_i \left(\frac{V_p}{V_{p,i}}\right) = \frac{\lambda_{AV}}{d} V_p = \frac{\lambda_{AV}}{d} \frac{\varepsilon}{\rho(1-\varepsilon)} \tag{9.57}$$

where $\lambda_{AV}=k_A/k_V$ is the surface-to-volume form factor for pores in a sense similar to the form factor for the particles. The values of V_P and porosity ε in this equation are the textural parameters for a given porous solid. For a PS with a variety of pores the mean size of pores $\bar{d}$ is also determined through the relation of the third and second statistical moments of distribution (see Equation 9.54), in the following form:

$$\bar{d} = \frac{\int d^3 v(d)\mathrm{d}(d)}{\int d^2 v(d)\mathrm{d}(d)}$$

(9.54a)

where $v(d)$ is the pore size distribution (psd) function.

Assuming a PS to be a system of cavities with a mean size $\bar{d}$, and particles with that of $\bar{D}$, and taking into account their common interface, one can write [3,61]

$$\frac{\bar{d}}{\bar{D}} = \frac{\lambda_{AV}}{\Lambda_{AV}} \frac{\varepsilon}{1-\varepsilon}$$

(9.58)

Equation 9.58 determines a relation between the mean sizes of pores and particles and is strictly accurate if all pores and particles have constant form. For example, for spherical particles of a mean size $\bar{D}$, simulating the pores between the particles with cylinders of a mean size d_C, introduction of the appropriate form factors $\Lambda_{AV} = 6.0$, and $\lambda_{AV} = 4.0$ gives $\bar{d}/\bar{D} = 0.66[\varepsilon/(1-\varepsilon)]$.

The mean sizes of windows, $\bar{d}_w$, and contacting cross sections, $\bar{D}_{pc}$, can be measured during analysis of the electron microscopy images as the relation of the first statistical moment to the zero one; the sizes of $\bar{d}_w$ can also be measured by adsorption methods (see Section 9.3). The direct interrelation between $\bar{d}_w$ and, for example, $\bar{D}_{pc}$, is determined in view of a used model (e.g., in the framework of a model of isotropic deforming lattice of particles). Besides, also possible are correlations type of $d_{w,i} \leq d_{C,i}$ that relate the possible size of a cavity $d_{C,i}$ to corresponding sizes of windows $d_{w,i}$ from the cavity to the neighboring cavities.

The discussion supports the necessity of accounting for PS morphology when one uses the morpho-dependent textural parameters and relations that include these parameters [112].

9.6 GENERAL PROBLEMS OF POROUS SOLIDS TEXTURE MODELING

9.6.1 Morphology of Porous Solids and Problems with Modeling

The world of porous solids is complicated and full of variety. By the number of possible combinations of particles, pores, and morphologies it seems that it is analogous to the world of biological objects. Majority of biological objects can be described as porous and dispersed. Their voids are usually filled with liquid, whose removal allows obtaining PSs in usual understanding, and these PSs totally or partially keep the morphology of original precursor [3].

We have demonstrated at the beginning (Figure 9.1) the variety of structures that can be obtained from amorphous SiO_2 phase. But, there are many structures of biological origins that are also formed from SiO_2. For example, silicate shells and skeletons of sea sponges, diatomite, corals, radiolarians, etc. (see Figure 9.23). Many of these forms are mesophases (they have long- and no short-range order). It is noteworthy that diatomite, also known as kieselgur, is widely used as an inexpensive silicate support with a specific surface area of more than 100 m^2/g [112,113].

The illustrations shown are just a portion of a variety of textures of real porous solids, also used as adsorbents and catalysts. It is obvious that when one goes from descriptions to quantitative

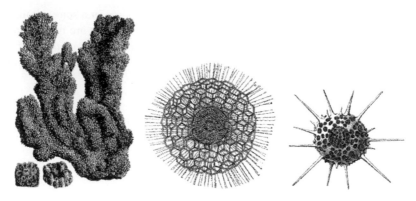

Figure 9.23 Silicate skeletons of coral, radiolarian, and diatomite (left to right).

estimations of morpho-dependable textural characteristics, it is necessary to substitute a complicated reality with simple geometrical models that are convenient for mathematical formulization. One should also note that the model of a porous solid also determines the possible models of practically any process in this porous solid. It frequently predetermines properties of mechanisms and events that occur in PS. This is why development of a geometrical model is an intensive part of many models of physicochemical processes in PS, and its oversimplification results in altering of a real process description in the same order of magnitude [3].

In some cases a model of a porous space is necessary (e.g., a problem of mass transfer), while in other cases a model of a carcass structure (e.g., problems of mechanical strengths, sintering, etc.) is appropriate. This is why full information on PS texture is desirable, including the interrelated characteristics of porous space and carcass structure. Only such an approach allows finding the plausible correlations between, for example, mechanical strength and mass transfer, or evaluating the laws of texture evolution at the different stages of PS synthesis and application. This enforces elaboration of a classification scheme (or a system of classifications) that should be taken into account for accurate modeling of texture and processes that happen inside PS. It can be based on extraction of the major properties of the structure of various PSs, dividing them into groups with common characteristics.

9.6.2 Classification of Porous Systems and Texture Modeling

Table 9.2 illustrates a convenient classification of porous solids by the characteristic sizes of their pores. This classification was originally proposed by Dubinin [114], and in 1972 it was officially adopted by International Union of Pure and Applied Chemistry (IUPAC) [7,53,58].

By size of pore one can mean the diameter of an equivalent cylindrical or the distance between the sides of a slit-shaped pore (i.e., in general a diameter of the largest circle that can be inscribed in a flat cross section of a pore of arbitrary form). The basis of this classification is that each of the size ranges corresponds to characteristic adsorption effect that is manifested in the isotherm of adsorption [53,115]. In micropores, the interaction potential is significantly higher than in wider pores, owing to the proximity of the walls. This explains that such pores become totally full with adsorbate at low relative pressures. In mesopores, one will observe formation of mono- and then multilayer molecular film forming over the walls. After formation of a multilayer molecular film,

Table 9.2 Classification of PS by Pore Sizes

Micropores (or microporous systems)	Less than ~2 nm
Mesopores (or mesoporous systems)	Between ~2 and ~50 nm
Macropores (or macroporous systems)	More than ~50 nm

the volumetric filling by capillary condensation mechanism occurs. Adsorption and desorption inside these pores is usually accompanied with a hysteresis on the AIs. The size of the macropores is too large for capillary condensation effects, and one should expect only reversible mono- and multilayer molecular adsorption to occur over the accessible surface [53].

Recently, the micropore range has been subdivided into two additional categories: very narrow pores or ultramicropores (<0.7 to 1.0 nm), where the enhancement effect is found; and supermicropores (0.7 to 2.0 nm), which fill the gap between the ultramicropores and mesopore ranges [53, 115]. This classification has become widespread, but it only takes into account adsorption effects and ignores the morphology of PS elements and their integrity.

Morphological classifications have been proposed by many authors (see, e.g., Refs. [3,54]). One of the most popular was offered by A.V. Kiselev in 1958 [116]. He divided all PSs into two main classes: corpuscular and sponge-like. Typical examples of corpuscular PS are silica gels and opals (see Figure 9.1a and Figure 9.1e). In such systems, one can easily describe the morphology of particles, i.e., corpuscles, but the description of the porous space between them is a complicated task. Typical examples of sponge-like systems are porous glasses and mesophases of MCM-41 type (see Figure 9.1c and Figure 9.1d), which are systems with a simple form of pores and a complicated form of a solid carcass. This classification is based on the simplicity of description of characteristic element (pores or particles) morphology.

This classification was advanced by Karnaukhov [54], who proposed six types of corpuscular models (monodisperse spheres or ellipsoids, platelets, spindles, filled tubes or rods, and regular polyhedrons) and three types of sponge-like structures (having cylindrical, bottlelike, and spherical cavities [pores]). Additionally, a class of mixed structures was introduced, which combines different elements of corpuscular and sponge-like structures at different structural hierarchy levels. A typical example of the latter structures is the texture of granules formed from zeolite crystals. The porous structure of crystals is considered as sponge like, but the packing of crystals in a granule is corpuscular. However, the analytical description of the considered texture geometries exists only for limited cases. Moreover, this classification does not include some types of PSs, shown, for example, in Figure 9.1 and Figure 9.23, and requires unlimited broadening due to real PS variety.

A set of classifications for quantitative characterization of powder dispersion morphology has been introduced in the literature; these standard scales were offered to evaluate characteristic morphological types of particles [55,110]. de Boer [117] simulated the capillary condensation hysteresis loop form that may occur during adsorption for 15 types of pores that differed morphologically and for their simplest combinations. According to these simulations, he proposed the classification that is based on adsorption measurements in the range of a capillary condensation hysteresis. However, it was shown lately that the form of a hysteresis loop (especially the desorption branch) depends on the interconnection of the pores to the same degree as on morphology and size of the individual pores in the bulk of PS [118–122], primarily due to percolation effects (see Section 9.7.4).

Radushkevich [123] proposed classification of porous solids according to two attributes: (1) the mechanism of formation and (2) general character of texture. With regard to the first attribute, he divided porous solids into two basic groups: systems of addition and systems of growth (subtraction), which we have discussed in Section 9.4. The systems of addition are formed with casual interconnection (summation) of primary particles or their aggregates. The case of nonporous primary particles is analogous to Kiselev's corpuscular systems. Systems of growth are the result of development of the pore system, for example, during burning out, dissolution, etc. or directed growth of a skeleton of a solid phase occurring, for example, at formation of zeolites, corals, etc. (see Figure 9.23). Frequently, the systems of growth have a characteristic individual and almost unique morphology. Therefore, detailed description of their texture should be interconnected with the mechanisms of their formation. The systems of addition can be described by more universal statistical laws. Combinations of the systems of growth and addition describe the complex and combined systems. Classification by the second attribute is the general character of texture. Accordingly to it, the systems are classified as ordered and disordered. Zeolites are systems of

growth of ordered structures, while the majority of active carbons are the systems of growth with disordered structure. Special attention is paid to the systems with practically closed cavities (foamed
plastics, foams, strongly sintered systems, etc.) that require a special approach when modeled.

The second classification by Radushkevich can be widened by introducing the degree of
ordering as summarized in Table 9.3.

Two limiting cases chaos and order are determined here, but in addition one can also consider
chaos with some correlation to particles localization (type 3, Table 9.3); type 4 assumes presence of
some order, for example long-range order in silicate mesophases, or platinum particles in a xerogel,
etc. One can also consider division of these types, which allow or disallow overlapping of particles.

The most comprehensively studied are the models of hard spherical particles that are contiguous or divided by small gaps. Models of types 1 and 3 in Table 9.3 have their origin in crystal
structure characterizations (see, e.g., Refs. [124–126]), while models of types 2 and 4 are widely
used in the theory of nonideal gases, amorphous solids and liquids [49,97,127–135]. The localization of atoms or particles in models of types 2 and 4 is determined by a correlation function
$W(R)$, which gives a probability of finding a particle at a distance R from the center of any
arbitrarily chosen particle. A typical shape and scheme of $W(R)$ evaluation for a random, densely
packed system of monodisperse spheres of radius R_0 is shown in Figure 9.24. One can set the
origin to the center of particle (black particle in Figure 9.24). Then the spheres of increasing radius
R are described having their centers at the origin. The number of particles dN in a spherical layer
of thickness dR at distance R from the origin is $dN = 4\pi R^2 (N/V)W(R)dR$, where N is the total

Table 9.3   Classification of PSs by the Degree of Ordering

No.	Type of Order	Realization
1	Order	Regular packing of elements
2	Chaos	Short- and long-range ordering are absent
3	Chaos in order	Defects in the ordered structures
4	Order in chaos	Presence of only long- or occasionally short-range order

Source: From [124].

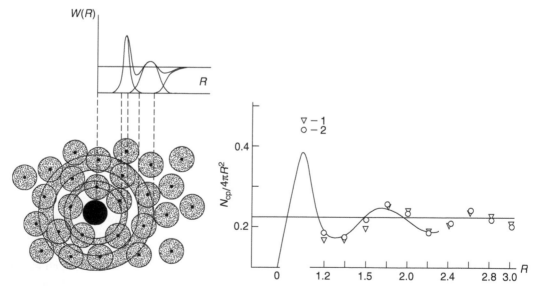

Figure 9.24   A radial distribution function (RDF) for a DRP of monospheres (right) and a scheme for its evaluation (left), $N_{mean}/4\pi R^2$ is an equivalent of $W(R)$, where N_{mean} is the mean number of spheres in the
intervals of $0.2R$. The solid curve illustrates the data obtained from neutron diffraction in liquid
argon; 1, 2 are the experimental data by Scott [128] and Bernal [127] obtained for the models of
steel spheres (cited in [127]).

number of particles in volume V. When $R \leqslant R_0$, $W(R) = 0$, but when $R > R_0$ a growing coordination sphere begins to intersect the neighboring particles and the values of $W(R)$ increase, reach a maximum and then decrease. Further increase of R repeats the situation, but is accompanied with broadening of extremums of $W(R)$. At the limit, value of $W(R)$ is proportional to the mean density of the packing. The number of sharp maxima usually does not exceed 3 and their positions are the average distances from the center to the primary, secondary, etc. neighbors. The zone of the first maximum is called the first coordination sphere, the area under this maximum is proportional to the coordination number of the given structure, that is, the mean number of neighbors; the zone of the second maximum is called as the second coordination sphere, etc.

The models of PSs considered above mainly concern the packings of particles, that is, from the texturological point of view they can be considered as development of a corpuscular model by Kiselev. The morphology of a porous space in corpuscular systems is usually more complex. For simplification, it is traditionally simulated by a group or network of nonintersecting cylindrical channels of varying sizes or flat slits, the models of goffered channels are less often used, etc.

The hard skeleton and porous space complement each other similarly to positive and negative in a photo [123]. Therefore, a corpuscular model can be inverted into a sponge-like one and vice versa. Such possibility to complement each other allows simulation of a porous solid both as a system of pores and as a system of particles; the choice point of readout is defined by conditions of a task or simplicity of the description. The majority of PSs can be modeled by two interconnecting labyrinths, where the labyrinth of pores forms a void that occupies a part ε from the whole volume, and this part is equal to the porosity; and the labyrinth of particles forms a skeleton (carcass) that occupies a part of the whole volume $(1 - \varepsilon)$, which is equal to the density of skeleton packing. Both labyrinths have a common interface and can have the ordered or disorganized structure, form coherent systems, or include discrete clusters as the isolated groups of pores or particles. Many properties of such labyrinths can be simulated by lattices of sites and bonds, if one considers the centers of expansion (particle or cavities) as sites and places of contacts of sites as bonds. Such model lattices are convenient for the definition of integrity in both systems of particles and systems of pores. The integrity essentially depends on the value of the coordination numbers of packing of particles, n_p, and pores, Z_C; therefore, we shall define these terms in the following.

Through coordination number of packing of particles n_p we understand the number of closest neighboring particles in the volume of the first coordination sphere that surround an arbitrary chosen particle (see Figure 9.24 and appropriate discussion). To avoid broader effects, one should select a particle in the bulk sufficiently far from the external surface of PS. In a model of a corpuscular porous solid of ordered or random spherical particles touching each other, one can determine the value of $n_{P,f}$, which is equal to the number of the closest neighbors with which the randomly selected particle has one point contact. We will refer to these simple contacts as full contacts. In a general case, incomplete contacts with small gaps between the particles are possible. These gaps, for example, are less than $D(\sqrt{2} - 1)$, where D is the diameter of the particles [3]. One can also consider the linear (or curved 1D) contacts (e.g., between rod-shaped particles), interface contacts (e.g., between faceted particles or the contacts that form as a result of sintering or deformation of the particles of the arbitrary form). The latter may be termed stain contacts or contact interfaces between the particles. One can also differ the contacts between similar or different particles in the PSs formed of packings of nonuniform particles (different chemical composition, morphology, etc.). Hence, the total number of contacts n_P may include the contacts of various types, which in some cases should be evaluated (for additional information see Section 9.7).

In turn, porous space of many real and model porous materials can be considered as a lattice of expansions cavities (or sites), connected with narrower windows or necks (bonds). With such a definition of sites and bonds it is acceptable to have the whole volume of pores concentrated only in cavities of different sizes and forms. In this case, windows are considered as volumeless figures that correspond to the flat sections in places of the smallest narrowings between the neighbors (as well as bond in a lattice of particles) [3,61]. This approach seems to be the most plausible,* although in some situations it is more convenient to identify bonds with cavities, and sites with windows [8]. The average number of bonds per site corresponds to the average coordination number of a pore lattice, Z_C.

Correspondingly, the values of n_P characterize the connectivity of the lattice of particles or skeleton of a PS and the values of Z_C characterize the interconnectivity of the lattice of pores of the same PS. Connectivity of PS is the major topological attribute, which in the general case does not depend on the shape and size of the PS's individual supramolecular elements, although the latter characterize the major geometrical properties of PS [8]. Appropriately, the classification of PSs by the degree of interconnectivity with allocation of various types of integrity is possible.

Also useful is the classification that takes into account the characteristics of the texture on various scaling (hierarchical) levels. The simplest example of hierarchical organization of PS is shown in Figure 9.17. Figure 9.25 extends the example with another set of simple situations characteristic of the corpuscular systems and formed of monodisperse particles of similar size $D_1 \approx$ const. Let us remind that we refer to pore size distribution as psd and particle size distribution as PSD. The situation in "a" shows a narrow PSD when particles form an isotropic random packing. The pore labyrinth is formed of cavities and windows between the particles; the experimentally measured psd has single maximum. These systems can be designated as homoporous and monodisperse. If all the particles form a regular ordered lattice all windows have the same size, and appropriate psd degenerates to a δ-function (monoporous systems). The model system shown in Figure 9.25b is formed of aggregates of particle size of $D_2 >> D_1$. There are two different labyrinths of pores: internal pores in aggregates and external pores between aggregates. As a result, the psd has two maximums (a model of biporous and simultaneously bidisperse structure). Figure 9.25c shows an example of a heteroporous structure. Characteristic of such structures is overlapping of the sizes of primary particles and aggregates, and correspondingly overlapping of psd's inside aggregates and between them.

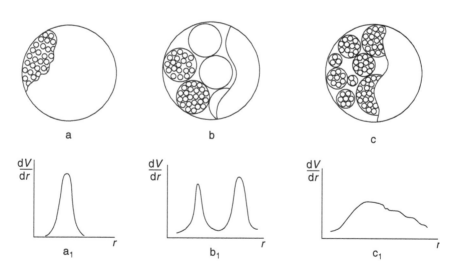

Figure 9.25   Models of granules of monodisperse particles; characteristic psds (pore size distributions) are given below: (a) uniform packing; (b) bidisperse packing of aggregates of particles of similar sizes; (c) same as (b) but the size of aggregates vary in a wide range.

---

*For example, most of us live in the rooms (or *voids*), and doors and windows are used for our connections with the rest of the world. But, there are some guys that prefer торчать in doors or windows and disturb everybody. In the world of PSs the similar effects influence accessibility and mass transfer characteristics of porous space, which is why their control has a significant interest from the practical point of view.

Porous systems of the type shown in Figure 9.25b are of practical interest since they have an interconnected system of large pores that decreases the diffusion resistance in mass transfer processes. The pores between aggregates play the role of transport channels, while the majority of the surface and catalytically active species are located inside the aggregates. This explains the wide use of systems with this type of structure.

Besides the simplest types of hierarchical organization shown in Figure 9.25, structures with three and more scaling levels are possible. In general, for corpuscular PSs, the shape of psd is determined by PSD and particles' space arrangement (PSA). The situations corresponding to the interrelation of PSD and PSA can be classified by a scheme shown in Figure 9.26.

Let us consider the first level of classification, systems with wide and narrow PSDs, which we divide into those having isotropic and anisotropic PSA. The combinations of PSD and PSA shown result in typical psds.

Similar schemes may also be derived for the sponge-like systems that are considered as negative to the system of particles or systems of growth by Radushkevich [123].

Characteristic for many PSs is another type of heterogeneity that can be considered as radial anisotropy (RA). This heterogeneity in composition, density, and porosity can be found in medicine pills formed of layers of different compounds. Less evident is RA in some granules of adsorbents and catalysts formed of pastes or powders. For example, formation by extrusion results in the appearance of RA due to the gradient of the rate of the forming paste movement through the draw plates; adding pressure gives RA due to an inhomogeneous profile of the applied pressure in a press form [61]. Radial anisotropy can appear as a result of nonuniform extraction, gasification, or other types of removal, or vice versa via condensation of additional components, nonuniform densification of various gels and pastes during drying, etc. Thus, classification based on various types of RA can be considered. At last, classifications based on the mechanical properties (PSs can be rigid, elastic, plastic, or fragile), electrical conductivity, chemical properties, etc. are possible.

The given discussion shows that rather universal and simple classification of porous materials equivalent to classification of crystals is absent. However, one can consider a system of interrelating classifications that take into account order, morphology and sizes at different hierarchical levels, degrees of integrity, structure, heterogeneity of a various type, etc. Such a systematic approach can be used as well for adequate modeling of various hierarchical levels of a porous material structure.

9.6.3 Generalized Models and Systematic Sets of Models

While modeling the structure and properties of porous materials one usually is interested in structural properties of a desirable hierarchical level. For example, for chemical properties the molecular structure is major, and the specific adsorption and catalytic properties are guided by the structure and composition of particle surface. Diffusion permeability is determined by the supramolecular

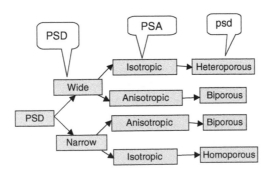

Figure 9.26 The scheme of the simple interrelation of pore size distributions (psd), particle size distributions (PSD), and particles space arrangement (PSA).

structure of the porous space, and the hydrodynamic resistance of a bed of granules depends on the structure of the bed. Hence, the rational modeling of the molecular and supramolecular structure of the porous solid as well as the processes that occur in it may be based on the hierarchical system of models following the wise principle of William Ockam (~1285 to 1349). This principle, known as Ockam's razor, can be formulated as *entia non sunt multiplicanda praeter necessitatem* (Latin; i.e., *the substances should not be added without necessity*). According to this principle, to obtain the necessary property one may allocate only one hierarchical level i, determine its major structural properties, and set the boarder conditions determined by the properties of a lower $(i - 1)$ and upper levels $(i + 1)$.

The corresponding hierarchical system of models may have the following sequence [3,61]:

1. Model of a molecular structure, which determines a mutual arrangement and interaction of atoms that form a molecule, molecular site, or a short-range order in a solid material. This is the area of molecular chemistry, stereochemistry, etc. A characteristic element of this level is the atom with prevailing interatomic interactions.

2. Model of a supramolecular structure of polymolecular ensembles or clusters, determined by interaction and mutual arrangement of the forming molecules. At this level, the specific mechanisms of supramolecular chemistry, including molecular recognition, self-assembly, etc. [4] can be allocated. In most cases, it is possible to limit this area to objects with the sizes under 1 to 2 nm, since further increase in the sizes admits application of statistical concepts like phase and interphase surface.

3. Model of a primary supramolecular structure (texture) of different sites and bonds of a lattice of pores and particles, that is, individual cavities and windows, connecting them, or individual particles and zones of their contacts with nearest neighbors. This model takes into account the form of both sizes of individual sites and bonds, and can be used for analysis of the processes limited to the space of individual sites and bonds (molecular sieving effects, initial stages of capillary condensation, sintering, etc.). The characteristic geometrical size or scale of this level corresponds to the sizes of separate sites and bonds, possibly considering their average sizes or size distributions.

4. The model of clusters or ensembles of sites and bonds (secondary supramolecular structure), whose size and structure are determined on the scale of a process under consideration. At this level, the local values of coordination numbers of the lattices of pores and particles, that is, number of bonds per one site, morphology of clusters, etc. are important. Examples of the problems at this level are capillary condensation or, in a general case, distribution of the condensed phase, entered into the porous space with limited filling of the pore volume, intermediate stages of sintering, drying, etc.

5. Model of a lattice of clusters ensembles, which takes into account local heterogeneity in distribution of sites and bonds, for example, their distribution on the radius of a granule and the presence of more or less obvious agglomeration of particles or pores. Such a model is important when one studies granules with a biporous structure, or aggregated bidisperse structures, where the characteristic scaling size is determined by the size of the appropriate aggregates. The absence of obvious anisotropy eliminates the necessity of this level of modeling; in such cases the model of level 4 at once advances to a model of level 6.

6. Model of a granule of a porous solid as a lattice (labyrinth) of pores and particles, which takes into account the average values of coordination number of bonds and distribution of sites and bonds over the characteristic sizes.

7. The model of an element of a bed of porous granules that describes the processes of heat and mass transfer in a bed.

Further, by analogy, the appropriate models of a bed of granules, chemical reactor as a whole, shop, factory, city, country, etc., up to the Universe can be introduced. But, this systematic set of models is only one among infinite possibilities. While solving particular problems, the considered hierarchical levels can be extended or narrowed with introduction of new sublevels, or a combination of above-mentioned levels if necessary. The parameters of the given level and only those structural features and boundary conditions determined by the neighboring levels are taken into account at each level considered and modeled. The modeling can be based on a structure of porous space and skeleton of a solid phase. One of the advantages of such an approach is the opportunity of independent parallel development of models for different levels (both geometrical models and models of the appropriate processes). Let us consider, using the Ockam's razor, some modern opportunities of realization of this system of models and the approach as a whole.

Limited volume of this chapter allows us to discuss only some of the advanced problems of modern texturology. Let us pay attention to the modern approaches to the modeling ensembles of particles and pores that demonstrate achievements and opportunities of texturology on existing and, obviously, the initial stage of its development [3]. Let us begin from a universal enough approach to the modeling of corpuscular and sponge-like PSs on the basis of Voronoi–Delaunay (VD) tessellations that can be used on the levels of 3 to 7 of the given hierarchical system of models.

9.7 MODELING PARTICLES AND PORES IN A LOCAL ARRANGEMENT

9.7.1 Voronoi–Delaunay Method for Description of Corpuscular and Sponge-Like Porous Solids

The way of the best choice to model PS's structure on both molecular and supramolecular levels begins with allocation of primary building units (PBUs), which without gaps and overlaps would fill a 3D space occupied by a PS. An universal method for allocation of such PBUs in both ordered and randomly arranged PSs, formed of packings of convex particles (or pores), is based on the construction of the assembles of Voronoi polyhedra (V-polyhedra) and Delaunay simplexes (or D-polyhedra), which form Voronoi–Delaunay tessellation [100].

The terms Voronoi polyhedron and Delaunay simplex have English geometry school origin. The first was Rodgers [136], who started using them regarding a great fundamental impact to this field from Russian mathematicians G.F. Voronoi (1868 to 1908) and B.N. Delaunay (1890 to 1980). The term V-polyhedron was used by many mathematicians [131–134]. This makes sense because similar

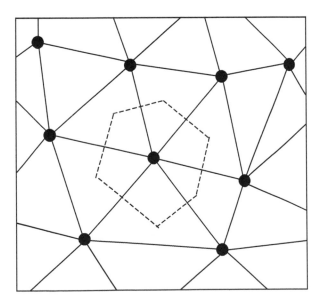

Figure 9.27 Example of Voronoi (V, dotted line) and Delaunay (D, solid line) tessellations for a 2D case.

polyhedra are named Thiessen polyhedra in hydrology [137], Wigner–Seitz cells [138], Dirichlet polyhedra [139], or Brillouin zones [140] in physics, etc.[*]

The VD is applicable to an arbitrary ensemble of points that can be randomly placed in space. These points can be the centers of particles, molecules, atoms, etc. Both V- and D-polyhedra are geometrically interrelated in the following manner. D-polyhedra are built by linking the named points with the closest neighboring points as shown in Figure 9.27 for the 2D case (solid lines). In the 3D case, instead of lines one should use planes that are to be drawn through three neighboring points. V-polyhedra are assumed to be arranged by the planes that are drawn perpendicularly through the centers of the linking lines (Figure 9.27, broken lines). Any V-polyhedron includes one point (particle, molecule, or atom) and a part of the surrounding void that is closer to this point than to others. V-polyhedron determined in such a way can be considered as a PBU since the ensemble of such PBUs fills a 2D as well as a 3D space without gaps, and overlaps forming a tessellation by Voronoi. The presented partition can be used to consider any ensemble of randomly distributed matter; for the case of texturology the Voronoi PBU system for particles (or pores), which we will refer to as PBU/P for brevity, is interesting.

Since drawing of 3D Voronoi tessellations is not as vivid as 2D (Figure 9.27) we will illustrate some typical PBU/P V-polyhedra for PSs formed of nonoverlapping monsphers (Figure 9.28). One can put an entered sphere (not shown) inside a drawn PBU/P. The PBUs of a to d types are characteristic of regular (ordered), and types e and f correspond to irregular packings of monospheres.

The geometry of V-polyhedra and V-tessilations was elaborated by Voronoi and Delaunay [143,144]. They have shown that all PBU/Ps have a convex shape, each facet is common for two neighboring PBU/Ps, each edge is formed by no less than d PBU/Ps (d is the dimension of V-tessellation), and no less than $d + 1$ PBU/Ps intersect at each vertex. We write "no less" but an exact coincidence usually takes place.

Eventually, in regular packings the number of Voronoi planes can be greater than the minimum number of planes necessary to form a PBU. The planes that do not form a PBU facet (have only

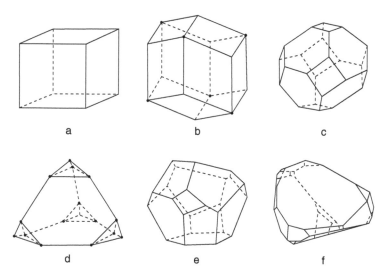

Figure 9.28 Typical V-*polyhedra* formed in regular (a–d) and irregular(e, f) packing of monospheres (regular packings): (a) simple cubic, (b) face-centered cubic, (c) bulk centered cubic, (d) diamond; irregular packings, (e) dense random, and (f) random with tetrahedron coordination.

---

[*]To be on the safe side, we should note that this kind of partitioning of space was used by Descartes in 1644 [139], and its origin can possibly be found in ancient times [141]. The further development was proposed by Gauss [142], Dirichlet [139], and others, but the most detailed and thorough mathematical description was given by, namely, Voronoi and Delaunay [143,144].

one common point or line with PBU) are named as degenerated planes. For example, V-polyhedron for a simple cubic packing has the shape of a cube (Figure 9.28) formed with six planes between the nearest neighbors. However, one can draw (for this particular packing) Voronoi-type planes through each of the 12 edges of the cube taking into account the neighbors from the second coordination sphere (see Figure 9.28), and through each of the eight vertices taking into account the neighbors from the third coordination sphere. The degeneration effect disappears when even a slight random shift (a minimum chaos) of particle centers is introduced [100,143–145]. Thus, degenerated planes are not characteristic of irregular packings, and the majority of PBU/Ps have the form of simple cells contacting $(d + 1)$ neighboring cells at the vertices and d neighboring cells at the edges.

According to the fundamental Euler equation, one can write

$$f = B - S + 2 \qquad (9.59)$$

for any convex polyhedron in a 3D space, where S is the number of vertices (framework bundles), B the number of edges (framework bonds), and f the number of facets. This means that only two of these three parameters are independent. The number of facets f corresponds to the particle packing coordination number n_P ($f=n_P$). Thus, one can rewrite Equation 9.59 as

$$f - 2 = n_p - 2 = B - S \qquad (9.59a)$$

A condition of $n_P = 2$ and $B = S$ corresponds to simple closed contours without ramifications, for example, polyangles. A condition $B<S$ (i.e., $n_P < 2$) corresponds to separate fragments of chains formed from sites and bonds. A condition of $B>S$ is characteristic of a 3D case ($n_P > 2$).

In the case of a primitive polyhedra (in each vertex of which only three edges cross-link) one can use an additional condition $B=(3/2)S$. As a result, Equation 9.59 transforms to

$$f = \frac{B}{3} + 2 = \frac{S}{2} + 2 \qquad (9.60)$$

For convex deltahedra (polyhedra for which all edges are triangular), $B=\frac{3}{2}f$, and

$$f = \tfrac{2}{3}B = 2(S-2) \qquad (9.60a)$$

Thus, for these two simple cases only one parameter is independent, significantly simplifying calculation of coordination numbers.

Now let us consider the possibilities of analysis of the void around particles using a V-tessellation concept. For each V-polyhedron, a particle occupies a part of PBU/P volume $(1-\varepsilon_i)$, where ε_i is the local porosity of a given PBU/P. The distance h between the centers of the two similar neighboring spherical particles (radius R_0) that contact each other at one point is $2R_0$. Thus, if $h > 2R_0$ there is a gap between the particles, and if $h < 2R_0$ the particles overlap (or incorporate each another). The total number of contacts with the neighboring particles is equal to the number of polyhedron edges. The larger cavities that surround the V-polyhedron include the polyhedron vertices (the points of maximum distance from all neighboring particles). Therefore, the number of the neighboring cavities is equal to the number of polyhedron vertices S. The edges of the polyhedra form the axis of the channels (pores) between neighboring cavities. The characteristic size of the cross section of these channels is maximum (r_{max}) at the vertices and minimum (r_{min}) in the direction of channels. The minimum cross section size of a channel can be considered as a window between neighboring cavities.

The sizes of the cavities, channels, and windows can be evaluated by a probing procedure [100] when one moves a probing sphere along a channel axis (Figure 9.29).

This procedure is used for evaluation of the borders and morphology of the pore. The pores are considered as the linked parts of the whole porous space accessible for a probing sphere with a size $r \leqslant r_Z$, where r_Z is a chosen size [100]. For numerous simulations, it is convenient to determine the windows between the cavities as the borders and to assume that the whole porous space is located in the cavities, and windows do not have volume and are only the 2D cross sections between the cavities. In this sense, the windows only determine accessibility of the cavities. This approach allows consideration of the porous space as a network of sites (cavities) and bonds (windows).

The probing procedure allows us to draw a navigational map of a PS. The bonds of a network correspond to a fairway with wide parts at the sites (centers of the cavities) and the narrowings at the windows. Such a navigation map allows all necessary information about all possible movements for the zonds of a given size to be obtained. Recently, a special software has been developed for calculations of such transport in the packings of both nonoverlapping and overlapping spheres of single or various sizes, and also of nonspherical particles [100,141]. Thus, V-tessilation allows fundamental information about the structure of a skeleton of particles and the porous space between the particles to be obtained.

The difference between V- and D-tessilations is as follows: each of V-polyhedra includes one network point (or particle) and a void that is closer to this point than to others, each of D-polyhedra includes one cavity and parts of the particles that are the closest to the center of the cavity and all windows that are on the borders with other neighboring cavities. It is convenient to term the latter as PBU/C, where C means cavity. The local coordination number of cavities Z_C is equal to the number of the faces of PBU/C (D-polyhedra or D-polygons), and their local porosity ε (or ε_A in 2D space) is equal to the unoccupied volume. Typical D-polyhedra are shown in Figure 9.30 and Figure 9.31.

These D-polyhedra are always tetrahedra (simplexes), generally with various edges (Figure 9.30a) in the case of undegenerated systems. However, the form of D-polyhedra differs from tetrahedral for degenerated systems. For example, it is a cubic cavity for a cubic structure (Figure 9.30b). There are D-polyhedra of octahedral form in the regular densest cubic (face-centered) and hexagonal

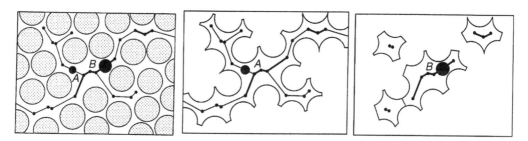

Figure 9.29 Probing of a porous space with the zonds A and B of different size. The thin lines show the cannels which zond A can follow, the thick lines correspond to possible movements of zond B. The dots show the centers of the cavities. The voids for the possible migration for zond A are shown in the center and that for zond B are shown on the right scheme.

Figure 9.30 Typical D-polyhedra (PBU/C) in the case of packings of monospheres: (a) tetrahedral; (b) cubic, and (c) octahedral.

packings (Figure 9.30c). This effect disappears if some chaos is introduced as in the considered case of degenerated V-polyhedra [100,141,144]. Moreover, the degenerated polyhedra can be divided into tetrahedral simplexes.[*]

Comparing to a lattice of edges in a V-tessellation that determines a navigation map for a system of pores, the lattice of edges in a D-tessellation determines a navigation map of the system of particles that form a skeleton of a PS. Migration in a solid phase is obvious if one considers the contact stains between deformed or intergrown particles. These maps are useful for the problems of PS strength, thermo- and electroconductivity, sintering, etc. As a result, a complex VD-approach, which takes into account the analysis of both V- and D-tesselations, is a prospective basis for a all-round analysis of a PS as a labyrinth (lattice) of pores and particles.

It is significant that both V- and D-tessellations are unequivocally interrelated because they form geometrically dual graphs.[†] For a brief expression of a relation between these two lattices let us designate a total number of facets in V- and D-lattices as $f_{V,\Sigma}$ and $f_{D,\Sigma}$, and the total number of edges as $B_{V,\Sigma}$ and $B_{D,\Sigma}$, respectively. For dual graphs the relation is [100,146,147]

$$f_{V,\Sigma} \leftrightarrow B_{D,\Sigma}, \quad f_{D,\Sigma} \leftrightarrow B_{V,\Sigma} \tag{9.61}$$

which is derived from a principle that exactly one bond of a dual lattice passes through each facet (see Figure 9.27). Moreover, each site of one lattice corresponds to the center of a cavity or particle of a dual lattice; thus

$$N_D \leftrightarrow S_V, \quad N_V \leftrightarrow S_D \tag{9.62}$$

where N_D is a number of cavities around PBU/P, and N_V is a number of particles around PBU/C.[‡]

Unfortunately, the relations (9.61) and (9.62) do not allow establishment of an unequivocal interrelation between the coordination numbers of packings of particles (n_P) and pores (Z_C) for corresponding V- and D-lattices, but, for undegenerated D-polyhedra of tetrahedron form $Z_C = 4$. Typical values of n_P for random packings and regular packings of monospheres are discussed in Section 9.7.3 and Section 9.7.2, respectively.

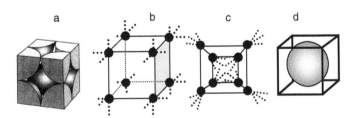

Figure 9.31 Some varients of allocation of PBUs for a primitive cubic packing (Pc): (a) PBU/C; (b) PBU as a 3D graph; and (c) as 2D planar graph; (d) PBU/P; black circles (b) and (c) are the sites of the lattice, that correspond to the centers of particles (or atoms), the solid lines are the bonds between them, the dotted lines are the bonds with the sites from the neighbor cells.

---

[*]However, different particular partitions may result in different number of simplexes. For example, a cube can be divided into five or six tetrahedral simplexes [100].
[†]*Dual*-form Latin dualis in modern mathematics means *interequivocal*. For example, duals are cube, octahedron, dodecahedron, and ikosahedron; tetrahedron is dual to itself [145–147].
[‡]Design of dual graphs in a 2D space is based on inter-cross-sectioning of bonds (edges) of corresponding lattices (see Section 9.7). As a result, the relation (9.61) transforms [146] into $B_{D,\Sigma} = B_{V,\Sigma}; f_{V,\Sigma} = S_{D,\Sigma}; f_{D,\Sigma} = -S_{V,\Sigma}$ (9.61a). Thus, duality works formally for cube and octahedron, which have the same number of edges ($B = 12$), dodecahedron and ikosahedron ($B = 30$), and there is only tetrahedron, which has $B = 6$.

Taking interchangeability of V- and D-lattices that make a uniform VD approach into account one should note that use of D-partitioning allows simplification of mathematical calculations because it is based on the simplest and unequivocal interrelations of the simplest polyhedra.

There is appropriate software based on VD approach for the analysis of the systems of overlapping and nonoverlapping convex mono- and polydispersed particles [100,134,141]. These programs allow studies of permeability of PSs using the navigation maps for movements of probes of various sizes [100]. This approach is used for analyzing results of MP [148]. Conceptually, the same approach is used for modeling the kinetics of topochemical processes with random nucleation [149–152]. The nuclei of a new phase appear randomly and independently in the bulk of initial phase at random times, and their growth continues until they contact with other growing nuclei; when all given space becomes filled with a new phase it stops. As a result, a tessellation similar to V-tessellation is formed. It is natural that the VD-approach may be further applied to the sponge-like systems. Other examples of construction and application of similar tessilations are discussed elsewhere [50,100,134,139–141,145,153,154].

Concluding this section we note that VD-tesselations may have numerous different applications. For example, each lion, which lives in Savanna assumes himself as a host of the whole territory that is closer to his den than to the den of his neighbor [100]. Thus, the boarders of the territory of each lion may be estimated by V-polyhedra. By analogy, each archaeologist attempts to identify the parts of region under the influence of different neolithic clans by VD-tesselation, the urban planner locates public schools in a city, a physiologist examines a capillary supply to muscle tissue, etc. [141].

Let us consider, taking into account VD-approach, some peculiarities of regular packings of spherical particles. These packings are widely used in texturology for many years [3,125,126,129,133,134]. It was proposed by Freundlich in 1928 [155], and proved by Carman [104] and Plank and Drake [156] that the structure of silica gel is formed of packing of spherical globules, and White and Walton [157], Manegold et al. [158], and Karnaukhov [54] proposed the models of regular globule arrangement.

Until nowadays, the regular packings of particles and pores, including the spherical ones, were considered as convenient models of ideal PS. But, the recent decade discovered a variety of PSs with regular ordered texture (see, e.g., Figure 9.1e and Refs. [24,91,158–161]). As a result, the studies of regular packings found a new practical significance.

9.7.2   Ordered Packings

Ordered packings are investigated in the finest details by crystallography for the description of structure of crystal lattices, where the role of particles is given to atoms, ions, molecules (including molecules of a complex form), or their groups, and packings of particles of one or different sizes are considered. Here, for brevity, we shall be limited to packings of monospheres.

Crystallographic classification of ordered packings based on symmetry of their internal structure allows allocation of seven packing systems (syngonias), 32 morphological classes, and 230 space groups [125,126], where the basic five types are 2D and there are 14 types of 3D lattices by Bravais (1885). Symmetry allows allocation of PBUs that can be treated as V-polyhedrons, filling a 3D space without gaps and overlaps.

In principle, there are a number of methods to allocate PBUs. Let us consider some of them using the simplest example of primitive cubic packing (Pc) of rigid spheres of one diameter D (Figure 9.31a). Each sphere (site) in this packing has contact with six neighboring sites. As a result, the packing has the coordination number $n_p = 6$ and the porosity $\varepsilon = 1 - \pi/6 \approx 0.476$. Figure 9.31a shows the structure of a cell with one cavity and six windows that connect the cell with its neighbors. This structure corresponds to a D-polyhedron, that is, PBU/C. The variant in Figure 9.31b determines a PBU in the form of a cell of a 3D lattice of sites and bonds. The relaxed bonds with the sites of the neighboring cells are shown as dotted lines. Figure 9.31c shows this packing in the form of a 2D graph. In a Pc packing $Z_C = n_p$, thus graphs b and c describe both a lattice of pores and a lattice of particles. And finally, the variant in Figure 9.31d corresponds to a V-polyhedron, that is, PBU/P, because it includes exactly one particle and all its contacts with particles in neighboring cells (not shown). A coordination number of this cell is equal to the number of facets of the polyhedron.

Constructing PBUs in the form of graphs allows the analysis of the properties of lattices in a mathematical language of topology, theory of graphs, and mathematical morphology [132–134,146,147]. But, for many problems of texturology the simple PBUs shown in Figure 9.31a and Figure 9.31d are the most informative for analysis, which do not require a complicated mathematics. These PBUs include exactly one site and all its bonds (variant a) that correspond to a D-polyhedron, or PBU/C, with one void and windows, or one particle and all connections with neighbors (variant d) that correspond to V-polyhedron, or PBU/P. These two types of PBUs allow calculation of the coordination numbers Z_C and n_P, which are equal to the number of facets in corresponding polyhedrons, the interfacial surface area of a unit of volume A_V, mass A (the latter at known value of density ρ), porosity ε, maximum size of cavity d, which can be considered as a maximum diameter of an inscribed sphere, size of a window d_W, which is equal to the maximum diameter of a circle that can be inscribed in a window, etc.

Figure 9.32a and Figure 9.32b show two variants of the densest regular packing of monospheres: hexagonal closest packing (hcp; fragment a) and face-centered cubic (fcc; fragment b). These packings have similar porosity $\varepsilon = 1 - \pi/3\sqrt{2} \approx 0.2595$ and coordination number $n_p = 12$, but differ by the sequence of neighboring layers. This can be explained on the basis of the fact that the densest packings form under two conditions: (1) each level should have the most dense 2D structure (where each sphere must have six neighbors in a layer); and (2) all neighboring layers must be translated in a manner that allows the maximum possibility of the spheres form one layer to fill the voids between the spheres in another layer. Both conditions give six contacts each, resulting in $n_p = 12$. But, both of these conditions allow formation of packings that differ by an alternation of the layers. This possibility is demonstrated in Figure 9.33.

Let us call the arrangement of spheres shown in Figure 9.33 as A. The next layer may be arranged as B or C. Let us consider the case of B. The third layer will be A or C, etc. In the former case, alternation of the layers gives a scheme ABABAB…(hcp-packing) in the latter it gives AB-CABC… (fcc-packing). More complicated alternations are possible, for example, ABCBABCB…, ABCABABCAB…, etc.

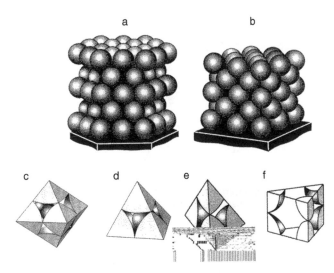

Figure 9.32 The densest ordered packings of monospheres: hcp (a) and fcc (b); octahedral (c) and tetrahedral (d) PBU/C closest packing ((a) or (b)), prismoidal PBU/C for packing with n_p=4.0 (e).

Such variations in layer alternation as a condition of formation of the densest packing do not affect the microtextural characteristics of a packing. However, layer alternation results in the possibility to allocate specific directions in the structure. For example, one can allocate only one direction perpendicular to the layers with the densest 2D packing in a hcp-packing; fcc-packing has four such directions. Such differences on a macrotextural level alter the physical properties that, for example, are characteristic to anisotropy of physical properties of metal crystals [125]. Moreover, it is noteworthy that hcp-packing is related to nonlateral packings [126].[*]

It is convenient to allocate PBU/C for hcp- and fcc-packings as a combination of N octahedral (Figure 9.30c) and $2N$ tetrahedral (Figure 9.30a) units (where N is the number of spheres in a packing). Such a combination allows filling of space without gaps and overlaps, which is a requirement for PBUs.

Alternation type AAAAA does not correspond to the condition of the densest packings since the centers of spheres in different layers are located one above (under) another (see Figure 9.33). This packing is called simple hexagonal packing (ph) with $n_p = 8$ and $\varepsilon = 1 - \pi/3\sqrt{3} \approx 0.3954$, and it is rare in crystallography. This is explained by a higher stability of a body-centered cubic packing (bcc) with n_p 8, but $\varepsilon = 1 - \sqrt{3}\pi/8 \approx 0.32$ [157].

A possible shift of neighboring hexagonally packed layers in the plane of Figure 9.34 allows a number of regular 3D packings to be obtained, whose porosity varies in the range $\pi/3\sqrt{3} < (1-\varepsilon) < \pi3\sqrt{2}$ and coordination number in the range $8 \le n_p < 12$. These packings are considered, for example, in Refs. [128,157,158,163]. In the cited literature the packings are also discussed, obtained for squarely packed layers using the alternations analogous to already considered (Figure 9.32). In this case, possible variations are $\pi/6 \le (1-\varepsilon) < \pi/3\sqrt{2}$ (where the left limit corresponds to Pc and right to fcc) and $4 \le n_p < 12$. According to Conway and Sloane [133], diamond-like packing is characterized by $n_p = 4$ and $\varepsilon = 1 - \sqrt{3}\pi/16 \approx 0.66$. More loose packings are possible if one moves the spheres apart in a layer or applies the chain-like structures [163–166]. Closed chains with the minimum value of $n_p \to 2.0$ can form packings with $\varepsilon \to 1$. The regular structures of this kind are described by Heesch and Laves [166] to model the texture of aerogels. At $n_p = 3$ porosity is $\varepsilon = 0.815$, and if one moves to $n_p = 2$ the porosity will reach the value of 1.0 quickly. For example, Smalley [167] discussed such issues in mathematics.

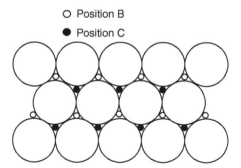

Figure 9.33 The variants of neighbor layers alternation in the densest regular packings: location of spheres in the plane of the figure corresponds to layer arrangement A, next layers may be located to arrangements B or C.

---

[*]Nonapplicable for hcc packing is a requirement for lateral packings, which by Ref. [136] can be formulated as if one allocates a straight line and the distance between two neighboring spheres on this line is a, then the distance between other neighboring spheres on this line is also a.

A number of correlative relations that bind n_P and ε were as proposed in the literature. These correlations are based on the use of stable regular packings of touching spheres [54,164–169]. For example, according to Meissner et al. [164],

$$n_{\mathrm{p}} = 2\exp[2.4(1-\varepsilon)] \tag{9.63}$$

Ridgway and Tarbuck proposed [169] that

$$\varepsilon = 1.072 - 0.1193n_{\mathrm{p}} + 0.0043n_{\mathrm{p}}^2 \tag{9.63a}$$

Lagemaat et al. [170], who modeled the random packings of monospheres, derived the following approximate relationship:

$$\varepsilon = \frac{3.08}{n_{\mathrm{p}} + 1.13} \tag{9.63b}$$

For many dense regular types of packing an accurate relation can be obtained:

$$(1-\varepsilon) = \frac{\pi}{6}\left(\frac{n_{\mathrm{p}}}{6}\right)^{1/2} \tag{9.64}$$

This relation originates from the above discussed exact values of ε for all packings with $n_{\mathrm{p}} = 12$ (including hcp- and fcc-packings), ph-packing with $n_{\mathrm{p}} = 8$, and Pc-packing with $n_{\mathrm{p}} = 6$. However, $(1-\varepsilon) = \pi\sqrt{3}/8 = (3\pi/16)n_{\mathrm{p}}/6)^{1/2}$ for a bcc-packing with $n_{\mathrm{p}} = 8$, and $(1-\varepsilon) = \pi\sqrt{3}/16 = (3\pi/16\sqrt{2})(n_{\mathrm{p}}/6)^{1/2}$ for a diamond-like packing with $n_{\mathrm{p}} = 4$, that is the relations differ from Equation 9.64 by a numerical factor.

Table 9.4 shows the sizes of cavities d/D, and windows d_{w}/D (D is the diameter of spheres) for five regular polyhedra (D-polyhedra) (see Figure 9.34).

The relative size of windows in the regular 2D packings is determined as [61]

$$\frac{d_{\mathrm{w}}}{D} = \frac{1 - \sin(\pi/N_{\mathrm{w}})}{\sin(\pi/N_{\mathrm{w}})} \tag{9.65}$$

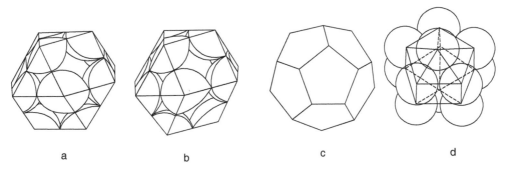

a   b   c   d

Figure 9.34 The typical coordinating polyhedra for a dense regular monospheres packing: (a) cuboctahedron, (b) anticuboctahedron, (c) dodecahedron, and (d) icosahedron.

where N_W is the number of spheres in contact that form a window. The 2D porosity ε_A for such packings is determined by the relation

$$(1-\varepsilon_A) = \frac{\pi(N_w-1)\mathrm{tg}(2\pi/N_w)}{2N_w}$$

(9.66)

which describes a 3D porosity of regular packings of parallel-packed rods as well, since a projection of these packings on the plane perpendicular to the axis is equivalent to the regular packing of circles in a plane.

The basic textural characteristics for the major types of regular packings of monospheres are summarized in Table 9.5.

As noted above, the maximum density in the regular packings of monospheres is achieved at $n_p=12$, where $(1-\varepsilon)=\pi/\sqrt{18}=0.74048\ldots$ However, the possibility of more dense packing of monospheres that is not lattice-like has been discussed for years. For example, the density of a tetrahedron element is 0.7796 (see Table 9.4), which is higher than the stated limit. However, it is impossible to fill a 3D space without gaps and overlaps with regular tetrahedrons. The ensembles of such PBUs can grow only in 1D dimensionality forming chains and rings. It seems that Bernal [127] observed first the pseudonuclei—the chains of five regular tetrahedrons. He demonstrated that these overdense ensembles are very plausible in random packings of monospheres. Their size is limited by several coordination spheres, and their porosity increases when one leaves the center sphere. Ultimately, the porosity reaches the values characteristic of the dense disordered packing (DDP)

Table 9.4 Sizes of Cavities d/D, and Windows d_W/D (for D-Polyhedra (PBU/C) in a Form of Regular Polyhedra

Type of PBU/C	Characteristics[a]	d/D[b]	d_W/D	ε
Tetrahedron	4 $\triangle$ facets, $v=4$	$\sqrt{6}/2 -1 \approx 0.2247$	0.155	0.2204
Octahedron	8 $\triangle$ facets, $v=6$	$\sqrt{2} - 1 \approx 0.414$	0.155	0.2795
Cube	6 $\square$ facets, $v=8$	$\sqrt{3} - 1 \approx 0.732$	0.414	0.3927
Dodecahedron	12 $\bullet$ facets, $v=20$	$(1+\sqrt{6})\sqrt{3}-1 \approx 1.802$	0.701	$1- \pi/\sqrt{18} \approx 0.2595$
Icosahedron	20 $\triangle$, $v=12$	$[2(5+\sqrt{5})^{1/2}]/2-1 \approx 1.902$	0.41	0.760

[a]Form of facets: $\triangle$, triangle; $\square$, square; $\bullet$, pentagonal; v is the number of vertices.
[b]$d/D=(d_{CR}-D)/D$, where d_{CR} is circumdiameter.

Table 9.5 Textural Characteristics of Ordered Packings of Monospheres of Size D

n_p	Packing Type	ε	PBU/C (shape)	Voids Size (d/D)	Z_c and Shape[a]	Windows Size (d_W/D)
12	fcc andhcp	0.2595	Tetrahedron +	0.2247	4$\triangle$	0.1547
			Octahedron	0.4142	8$\triangle$	0.1547
8	bcc	0.3198	Loose tetrahedron	0.291	4$\triangle$	0.225
8	Ph	0.3954	Trihedral prism	0.528	4$\triangle$	0.414
6	Pc	0.4764	Cube	0.732	6$\square$	0.4142
4	Diamond	0.6599	Diamond	1.00	–	−0.915

[a] $\triangle$, triangular; $\square$, square windows (see Figure 9.31 and Figure 9.32).
Source: By [54,157].

[171–173] (see Section 9.7.3). In practice, such ensembles may exist in the form of isolated clusters and inclusions in a disordered packing (such clusters were discovered, for example in Ref. [133]). Another direction for seeking the densest packings is an increase of coordination number $n_p>12$. Conway and Sloane [133] noted that this problem was a subject for discussion between Isaac Newton and David Gregory in 1694. Newton assumed that the value of $n_p=12$ for monosphere packings is the limit, but Gregory assumed the packings with $n_p=13$.

To understand the origin of this discussion we illustrate the typical coordination polyhedra, whose vertices allocate the centers of spheres surrounding a central sphere in Figure 9.34. The cuboctahedron is characteristic for fcc, but anticuboctahedron is characteristic for hcp-packing. Both of these polyhedra have six square and eight triangular facets, 36 edges and 24 vertices. A dodecahedron (12 pentagonal facets, 30 edges, 20 vertices) and icosahedron (20 triangle facets, 30 edges, and 12 vertices) are also shown.

The origin of the discussion between Newton and Gregory is based on the fact that the only regular polyhedron with 12 vertices that can be described around a sphere of diameter D is the icosahedron. But the length of the edge of the icosahedron is $D(4-\mathrm{cosec}^2(\pi/5))^{1/2} = D(2-2\sqrt{5})^{1/2}= 1.0515D$, which is slightly higher than D. This results in the formation of small gaps between the spheres. It seems to be very attractive to move the spheres slightly to insert a 13th sphere. Coxeter [174] has studied such icosahedral ensembles and demonstrated that the spheres that do not contact one another can be rolled around the central sphere at the expense of a void between them, but he did not find a possibility to admit a 13th sphere to contact the central one. However, it seems possible that a number of such 12-coordinated groups of spheres might be packed together, free space facing free space, with a sphere partially enclosed in the so formed hole in such a manner that the overall porosity might fall below the 0.2595 characteristic for the densest regular arrangement. Toth [145] showed that this configuration is possible if the centers of the 13 spheres are allocated at a distance of no less than $1.045\,D$ from the center of a central sphere, and do not touch it. By the Toth approach the density of such a system does not exceed $\pi/18$, but his estimations were made without taking into account the spheres of higher coordination spheres. Later, Rodgers [136] determined the limiting density of hard monospheres as $(1-\varepsilon)= \left[\mathrm{arccos}(1/3)\pi\right]/\sqrt{18}= 0.77963\ldots$ Lindsey [172] specified this limit as $(1-\varepsilon) \leqslant 0.7784$, and some smaller specifications were made by Muder [173]. These estimations show the possibility of formation of nonlateral packings of monospheres denser than $\pi/\sqrt{18}$ to about 10%. But the limiting value of $n_p = 12$ demonstrating that Newton was right in his discussion with Gregory, however only in the case of hard monospheres. Slight deformation of soft monospheres allows formation of packings with $n_p>12$. We will return to this discussion later.

The regular packings of the spheres of various composition and size are also considered in crystallography, including the filling of the voids between larger spheres with smaller ones. In the latter case, porosity decreases and the equations of type 9.39 become applicable. In a general case, the decrease of free volume stabilizes the packing because it decreases the potential energy of a system by means of increasing of interparticle (intermolecular) interactions. This tendency is clearly seen for nonspherical particles (molecules) [124,174–177], and can be described schematically by a packing shown in Figure 9.35.

9.7.3 Disordered Packings

Some quantitative characteristics of disordered packings have been discussed in Section 9.4 through Section 9.6. Let us begin this section with the specificity of determination of a coordination number for random packing (RP) of hard monospheres.

One can consider the total average number of contacts $\bar{n}_p$ in a DRP (see Figure 9.24) as a sum of full contacts $\bar{n}_{p,f}$ and incomplete contacts $\bar{n}_{p,inc}$ between the spheres divided by small gaps. These incomplete gaps exist in regular packings as well. For example, for bcc-packing with $\bar{n}_p=\bar{n}_{p,f}= 8$, one can allocate six incomplete contacts $\bar{n}_{p,inc}$ having a distance between the centers of spheres of

1.155 D; correspondingly, the total number of contacts is $\bar{n}_{p,\Sigma}=\bar{n}_{p,f}+\bar{n}_{p,inc}= 14$.[*] In this case, a PBU has the form of a truncated octahedron with 14 edges (see Figure 9.28c).

The distribution of the number of full contacts[†] in RP was studied experimentally in Ref. [106]. It was shown that the radial distribution function (RDF; see Figure 9.24) corresponds to a normal distribution. Table 9.6 illustrates the results of that study analyzed on the basis of Equation 9.63 and Equation 9.64 (see Section 9.7.2).

One can see from Table 9.6 that a good correlation between the values calculated with Equation 9.63b and Equation 9.63c and the experimental data, especially after taking into account a wall effect.

The mean value of $\bar{n}_p$ in DRP was experimentally measured in Refs. [127,158,178–182] by the residuals (stains) that retain at the contacts between the particles (the stains of paint, result of chemical interaction, etc.). These methods yield the values of $\bar{n}_{p,\Sigma}=\bar{n}_{p,f}+\bar{n}_{p,inc}$. By the thorough studies of Bernal et al. [127,180–182], and Refs. [158,178], the DRP with $\varepsilon \sim 0.36$ was found to have a first maximum at RDF, which corresponds to the distance between the centers of spheres of $\sim(1.05–1.1)D$; the value of $\bar{n}_{p,\Sigma}$ is 8 to 8.5, the second maximum is at $\sim 1.8D$ and $\bar{n}_{p,\Sigma} \sim 14$. Marvin [183] used another technique, the axial compression of a leaden shot, for counting the contacts by the results of deformation. The moderate compression gave a value of $\bar{n}_{p,\Sigma} = 8.46$, which correlates well with the total contact number of 8.5 by Bernal and Mason [180]. When compression increased, eliminating all voids, the average contact number was 14.16, indicating that the total number of the closest neighbors surrounding a sphere is about 14. This value of $\bar{n}_{p,\Sigma}$ corresponds to the V-polyhedron (PBU/P) with 14 facets — dodecahedron (see Figure 9.28c). Similar experiments by Bernal [182] gave the polyhedra with an average number of facets of 13.6. The shape of the facets varied from

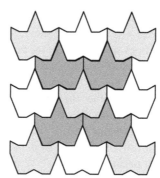

Figure 9.35 A scheme of the densest packing of particles of a complicated form.

Table 9.6 Interrelationship of ε and $\bar{n}_{p,f}$ by Experimental Data

Experiment by Ref. [106] $\bar{n}_{p,f}$	Calculation of Porosity for a Given Coordination Number $\bar{n}_{p,f}$ Porosity (ε)					
	ε^a	ε^b	Eq. 9.63	Eq.9.63a	Eq. 9.63b	Eq. 9.63c	Eq. 9.64
3.8	0.64	0.60.	0.732	0.681	0.625	0.637	0.416
4.7	0.54	0.52	0.644	0.606	0.528	0.515	0.463
6.0	0.49	0.48	0.542	0.511	0.432	0.403	0.523
5.9	0.45	0.45	0.549	0.517	0.438	0.410	0.519
7.1	0.35	0.345	0.472	0.442	0.374	0.341	0.569

[a] ε is the result of the experiment.

[b] ε is the result taking into account a wall effect.

Source: From [106].

---

[*]One can recollect a discussion between Newton and Gregory.
[†]The electrochemical deposition of Cu over the packing of steel spheres was studied in this work. The number of contacts was calculated by the dots not covered with Cu.

triangle to septangular with the maximum number of facets in the pentagonal shape. These experimental results correlate well with pure geometrical studies by Coxeter [174]. He has shown that the DRP are characterized by V-polyhedra (PBU/P) with an average number of facets of 13.56 and the average number of each facet's edges of 5.104. From Finney's [184] measurements, there are average values of 14.2 faces per polyhedron and 5.16 edges per facet. In Ref. [178] the number of such facets is 13.23; in Ref. [185] — 13.40. These close values were also obtained for polyhedra formed in metallography [50,52], for concentrated foams, emulsions, etc. [185–190]. The curious results were obtained by Oger et al. [191]. They studied the topological peculiarities and statistics of V-polyhedra in random (generated by Monte Carlo method) packings of monospheres and found a correlation between the average number of facets with the density of a packing. The obtained dependence is well described in the studied range of $\varepsilon > 0.38$ by the equation

$$f = 13.98 + 1.35\varepsilon \tag{9.67}$$

Table 9.7 shows the results of the calculations of average parameters of PBU/P for isotropic DRP, fulfilled by Serra [134] and Meijering [152]. Serra used VD-method while Meijering used the Johnson–Mehl's (JM) statistical model [150] of simultaneous growth of crystals until the total filling of the whole free space was accomplished. The parameter N_V in the table is the number of PBUs in a unit of system volume, thus N_V^{-1} is the mean volume of a single PBU, which is related to the relative density of the packing $(1-\varepsilon)$ with an interrelation

$$N_V^{-1} = \frac{\pi D^3}{6} \frac{1}{1-\varepsilon} \tag{9.68}$$

where D is the size of spherical particles. Table 9.7 additionally includes the values of statistical moments of distribution, obtained from 2D sections by image analysis technique [50,131,134]: $M(L_i)$, $M(A_i)$, and $M(V_i)$ are measure-weighted moments of distribution; L the linear size, A the surface area, and V the volume. Table 9.7 also includes the parameters of PBU/Ps in a form of regular tetrakaidecahedron (TKH)* and rhombododecahedron (or cubooctahedron; RDH) for comparison.

Table 9.7 Characteristics of PBU/Ps by V (Voronoi), JM (Johnson–Mehl), as well as TKH and RDH

Parameter		V	JM	TKH	RDH
Mean	Faces	15.54	>13.28	14	12
number	Edges	40.61	>33.84	36	24
	Vertices	27.07	22.56	24	14
Number of full neighbors $(\overline{n}_{p,f})$		8	7	14	12
Average size of PBU $(\overline{N}_L)$		$1.45\,N_V^{1/3}$	$1.28 N_V^{-1/3}$	$1.33 N_V^{-1/3}$	$N_V^{-1/3}$
Mean surface area of PBU, $\overline{A}_N$		$5.821 N_V^{-2/3}$	$5.143\,N_V^{-2/3}$	$5.315 N_V^{-2/3}$	$5.34 N_V^{-2/3}$
Moments of distribution	$M(L)$	$0.950 N_V^{-1/3}$	$1.10 N_V^{-1/3}$	$N_V^{-1/3}$	$N_V^{-1/3}$
	$M(A)$	$1.04 N_V^{-2/3}$	$1.50 N_V^{-2/3}$	$N_V^{-2/3}$	$N_V^{-2/3}$
	$M(V)$	$1.24 N_V^{-1}$	$2.20\,N_V^{-1}$	N_V^{-1}	N_V^{-1}

Source: By Serra [134] and Meijering [152].

*The term *tetrakaidecahedron* was introduced by Kelvin to designate a form of a polyhedron formed in the densest random packings (he particularly studied the cells in the foams) [189]. Kelvin assumed that the typical form of such tetrakaidecahedron is cubooctahedron (Figure 9.28c), that is, the truncated icosahedron with six square, and eight hexagonal facets. That is why the cubooctahedron is sometimes called a *Kelvin body*. But, the following studies have shown that although the total number of facets is close to 14, the majority of the facets are pentagonal.

It is noteworthy that the value of $n_{\mathrm{Pf}} = 14$ for a TKH may be realized only in a system of overlapping spheres. One can see from Table 9.7 that calculations by different models including the regular polyhedra TKH and RDH, result in close values of most of the parameters of PBU/P. For example, the values of the total number of the neighbors equal to $\bar{n}_{\mathrm{p}}$, and calculated assuming a distance between the centers of spheres up to $D\sqrt{2}$ are close. But the value of the mean number of full neighbors, $n_{\mathrm{p,f}}$, significantly depends on the algorithm of calculations and coincides with the experimentally measured [106] value of $n_{\mathrm{Pf}} = 7$ (at $\varepsilon = 0.35$) only for the JM model. As a result, one can assume that the value of the coordination number characteristic to the DRP of monospheres is $n_{\mathrm{Pf}} \approx 7$, and depends on ε (Equation 9.63b). Other typical characteristics correspond to values shown in Table 9.7 for V and JM models, but their values also depend on ε, and the values of N_{V} may be corrected by accounting for ε with Equation 9.68. Table 9.8 summarizes the quantitative characteristics of such PBU/Cs [127]; similar results were obtained elsewhere [184,192–195].

One can see from Table 9.8 that almost half of all D-polyhedra in a dense random packing have the form of a tetrahedron, and the total impact of the larger polyhedra types of prisms and dodecahedrons is relatively insignificant (Figure 9.36). According to Bernal [127,181] the transition to more friable packings can be achieved by (1) an increase in the fraction of larger polyhedra; and (2) formation of holes (removing a part of the spheres from the packing). Mechanism (1) gives a 17% increase in porosity even at total transition to PBU/C in the form of Archimedian antiprism; thus the main increase in ε is a result of the formation of holes — the unfilled vacancies in a lattice of particles with a characteristic size close to the size of particles D. An increase in porosity ε is accompanied with a decrease of the total average number of contacts $\bar{n}_{\mathrm{p}}$ at practically constant distances between the remaining neighboring spheres [127].

The approaches considered allow modeling of the primary texture of PS and the processes, limited by individual PBUs that mainly correspond to level III and partially to level IV in the hierarchical system of models (see Section 9.6.3). PBUs are identical in regular PSs, and simulation of numerous processes may be reduced to analysis of a process in a single PBU/C or PBU/P. An accurate modeling of the processes in irregular PSs requires the studies of the properties of structure and properties of the ensembles (clusters) of particles and pores (level IV of the system of models) and the lattices of such clusters (levels V to VII of the system of models). Let us consider the composition of clusters on the basis of fractal [127], and the lattices on the basis of percolation [8] theories.

9.8 MODELING THE ENSEMBLES (CLUSTERS) OF PARTICLES AND PORES ON THE BASIS OF A FRACTAL APPROACH

Fractal theory is a relatively new field of geometry, formulated by Mandelbrot [196] for irregular rough-surfaced objects. The major properties of such objects are the dependence of the measured length (perimeter), surface, or volume on the scale of measurement; and geometrical self-similarity

Table 9.8 Distribution of D-Polyhedrons in a Dense Random Packing of Monospheres

Form of PBU/C	Number		Relative Numerical Part (%)	Relative Volume (%)
	Facets *f*	Verciles *S*		
Tetrahedron	4	4	72.98	48.17
Half octahedron	5	5	20.26	26.75
Trigonal prism	14	9	3.22	7.82
Archimedean antiprism	14	10	0.46	2.47
Dodecahedron	10	8	3.07	14.78

Source: By Bernal [127,181].

in a certain range of scales.* Mandelbrot referred to these objects as fractals (i.e., the objects of non-integer dimension).

Fractals are widespread in nature. For example, fractals are the perimeter of coastline of islands and continents, cumulus clouds, crown and root of trees, bacterial colonies, bronchial tubes, branched polymers and proteins, liquid–vapor interface during impregnation of disordered PSs with liquids, assemblies of particles formed by a diffusion-limited aggregation (DLA) mechanism, and the clusters of joint particles or pores in a PS's bulk (see the clusters of pores in Figure 9.29, and Section 9.9). The fractal approach is successfully applied to model nucleation and propagation of fractures under breakdown of fragile porous and nonporous materials, formation and sedimentation of molecular aggregates in porous media, flows and transport through PS, which mainly occurs in the channels of interconnected large pores, a numerous processes of dynamic evolution and re-structuring of PS, etc. [196,200–203].

Let us demonstrate the geometrical features of rough-surfaced objects with an example of clusters from Figure 9.37. Such clusters are formed during aggregation of, for example, carbonaceous soot, Al_2O_3 or SiO_2 aerosols [203] according to DLA mechanism by Smoluchowski [204]. The simplest variant takes into account the Brownian diffusion of the particles that start from a random point of a sphere surrounding a cluster, and finish joining the cluster. The starting point of the cluster is the first particle, but all of the other particles aggregate around this first particle forming the cluster (a modern variant of DLA is considered, e.g., in Refs. [205–207]).

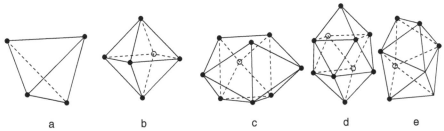

a   b   c   d   e

Figure 9.36   Five typical PBU/C(D-polyhedra), which Bernal [180–182] allocated DRP of monospheres (they are sometimes called the Bernal holes): (a) tetrahedron, (b) octahedron, (c) trigonal prism, (d) Archimedian's antiprism, and (e) dodecahrdron. It is northway that not all these D-polyhedrons are simple, but all are deltahedra, thus the relation 9.60a (see Section 9.7.1) is applicable for them.

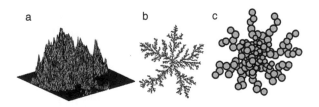

a   b   c

Figure 9.37   Computation of clusters formed by 3D (a), and 2D (b) Brownian motion; (c) a projection of a 3D cluster formed of monodisperse spherical patriclse.

---

*The properties characteristic to fractal objects were mentioned first by Leonardo da Vinci, but the term fractal dimension appeared in 1919 in a publication by Felix Hausdorff [197], a more poetic description of fractals was given by Lewis Richardson in 1922 [198] (cited by [199]), but the systematic study was performed by Benoit B. Mandelbrot [196]. Mandelbrot transformed pathological monsters by Hausdorff into the scientific instrument, which is widely used in materials science and engineering [200–202]. Geometrical self-similarity means, for example, that it is not possible to discriminate between two photographs of the same object taken with two very different scales.

The perimeter of a 2D cluster $L(R)$, measured by a bond radius of R (or probing yardstick) in the range of $L_{min} < L(R) < L_{max}$, is related to R (i.e., the scale of measuring) by

$$L(R) = L_0 R^{(1-D_1)} \tag{9.69a}$$

where L_0 and D_1 are constants, and $1 \leq D_1 \leq 2$. Similar measurements of the ulcerated surface area and volume of a 3D cluster gives

$$A(R) = A_0 R^{(2-D_2)} \tag{9.69b}$$

$$V(R) = V_0 R^{(3-D_3)} \tag{9.69c}$$

where A_0, D_2, V_0, and D_3 are constants, and $2 \leq D_2 \leq 3$, and $3 \leq D_3 \leq 4$, while $D_1 - 1 \approx D_2 - 2 \approx D_3 - 3$. The exponents D_i in the equations are limited in the ranges of $E \leq D_i \leq (E+1)$, where E is the ordinary Euclidean dimension of space (1 for a line, 2 for a surface, and 3 for a volume).[*] Mandelbrot designated D_i as object's fractal dimension, and L_0, A_0, and V_0 as fractal prefactors or lacunarity (Latin: lacune, meaning gap). In a majority of fractal applications the prefactors are assumed to be constant and are not considered, and fundamental equations of type 9.69 are given by the general form

$$X(R) \propto R^{(E-D_i)} \tag{9.69}$$

For the smooth objects, the morphology of which are described by ordinary Euclidean geometry, $D_i = E$ and linear sizes, surface area, and volume do not depend on the size of probe R. But for irregular morphology of rough surfaces it is essential that $E < D_i < E+1$. The larger the D_i, the higher the degree of irregularity and ruggedness. Thus, the geometrical characteristics of ulcerated systems are significantly dependent on the size of bond and scaling.

The range $L_{min} < L(R) < L_{max}$ determines the region of self-similarity of the object's geometrical features, or upper and lower cutoff scales in fractal behavior at D_i = Const. The constancy of D_i usually is observed in the range of several orders of magnitude regarding to L [208–216]. The values of the fractal dimension are calculated by Equation 9.69 via corresponding log–log plots. Most real systems are only fractal-like, that is, they approximately satisfy these equations, but do not possess the multiscale similarity. In these cases, the fractal dimension and fractal prefactors are defined by the least-squares straight line fit of the data. If the log–log plots consist of two linear segments or more, the objects are refereed to as bifractal or polyfractal, respectively, and formation of different fractal components is considered as evidence for different mechanisms of formation [199].

Application of the fractal approach is significant to many problems of texturology. The characteristic size of a cluster such as is shown in Figure 9.37 can be considered a radius of gyration, R_g, which is calculated by the relationship between R_g, the number of primary particles in the cluster, N, and the mean primary particle radius, r in a form [196,203]

$$N = N_0 \left(\frac{R_g}{r} \right)^{-D_2} \tag{9.70}$$

---

[*]In some publications (e.g., [208]), the E parameter is considered as a topological dimension and $E+1$ as the dimension of the corresponding Euclidean space.

where D_2 and N_0 are the fractal dimension and prefactor, respectively. This equation is applicable to an ensemble of aggregates on the average, that is, it is statistical in its nature. The porosity of the cluster, $\varepsilon(R)$, depends on the radius of the probe, R:

$$\varepsilon(R) = \varepsilon_0 R^{(3-D_2)} \tag{9.71}$$

that is, porosity increases with R. Equation 9.70 and Equation 9.71 were used in Refs. [100,101] in a complex with the VD-approach to the computer simulations of a fractal dimension of the Appolonium packing (Figure 9.18). From Equation 9.70, $D_2 = 2.40$, and from Equation 9.71, $D_2 = 2.50$.

Experimental methods of fractal dimension measurements are based on analysis by TEM (transmission electron microscopy) images or angular light scattering (ALS), small-angle x-ray or neutron scattering, etc. The intensity of scattering, $S(q)$, is related to scattering angle, θ, by a classical Rayleight–Debae–Gans equation [203,217]

$$S(q) = q^{-D_3} \tag{9.72}$$

where $q = (4\pi/\lambda)\sin(\theta/2)$ is a modulus of the scattering vector and λ the radiation length.

For example, studies of the morphology of a number of flame-generated aggregates (different carbonaceous soots, Al_2O_3 aerosols, etc.) [203] by both TEM and ALS procedures resulted in $D_2 = 1.7 \pm 0.15$, while analysis of silhouettes of computer models of seven proteins [211] gave a value of $D_2 = 1.120 \pm 0.025$, etc.

Another important field of the application of fractal approach to texturology is related to surface roughness. Anvir and Pfeifer [212,213] proposed characterization of surface irregularities by adsorption and established two methods, based on Mandelbrot's fundamental equations of type 9.69. According to the first method of D_i calculation, one uses the relations that interrelate a number of molecules in a complete monolayer during physisorption, n_m, or an accessible surface area, A, with a cross-sectional area, w, which correspond to one molecule in a monolayer:

$$n_m \propto w^{-D_2/2} \tag{9.73a}$$

or

$$A \propto w^{-(2-D_2)/2} \tag{9.73b}$$

In these cases, the values of w are used as a probing measure, and $w \propto R^2$ for the spherical molecules radius of R. As a result, $n_m \propto R^{-D_2}$. The second method by Pfeifer and Anvir is symmetric to the first one in the sense that instead of adsorbing a set of molecules on samples with a constant particle size distribution, one adsorbs a single adsorbate (e.g., N_2) on a set of samples with variable particles sizes, R_a. The corresponding equations for this method are

$$n_m \propto R_a^{(2-D_2)} \tag{9.74a}$$

or

$$A \propto R_a^{(2-D_2)} \tag{9.74b}$$

Both of these methods are commonly used in the literature for analysis of irregular surface aggregates and porous materials, and both physisorption and chemisorption have been applied

[200,208–216]. For example, for graphitized carbon blacks and aerosols [212], $D_2 \sim 2.0$; for silica gels, $D_2 = 2.94 \pm 0.04$; and active carbons with an increasing degree of activation show an increase in D_2 from 2.25 to 3.03 $\pm$ 0.25. Studies of O_2 chemisorption on Ag/SiO$_2$ catalysts with Ag particles of various size [213] resulted in an evaluation of fractal dimensions of the latter as $D_{Ag} = 1.82 \pm 0.07$.

Through such chemisorption studies, the values of D_i have been determined not only by geometric accessibility, but also by the chemical heterogeneity of the surface. This can result in abnormal values of D, and demonstrates the scale effect on the kinetics and selectivity of catalytic reactions. For such studies, Farin and Anvir [213] derived the equations that can be applied for characterization of supported catalysts:

$$a \propto R^{(D_R)} \tag{9.75a}$$

$$a_t \propto R^{(D_R - 2)} \tag{9.75b}$$

$$a_g \propto R^{(D_R - 3)} \tag{9.75c}$$

where Equation 9.75a binds the activity a (per catalyst particle) with the metal particle size R; Equation 9.75b determines the activity a_t (turnover units) per exposed surface atom, and Equation 9.75c interrelates the activity a_g and catalyst mass. According to Farin and Anvar [213], many catalytic reactions follow these equations, that is, the values of D_R contain information on the selectivity, activity, and distribution of the active sites on the surface of the catalyst. Also according to them, D_R may be very low (~0.2 for ethylene oxidation on Ag/SiO$_2$) or exceedingly high (Equation 9.55 for ammonia synthesis on Fe/MgO). D_R is affected not only by the type of molecule and metal interaction, but also by the morphology of the support, for example, a change in the support from a nonporous aerosol to flaky silica increased D_R of oxidation on the above-mentioned Ag catalyst, etc.

Energetic factors, including energetic heterogeneity of the surface may in general significantly influence D_i calculated by Equation 9.73 to Equation 9.75. These effects are possible even during physical adsorption on a flat surface, which possess specific interactions (formation of hydrogen or π-bonds, polarization, inductive polarization, etc. [53,218,219]). Unfortunately, reliable methods to distinguish morphological and energetic impacts to D_R are not yet available. Hence, the direct appropriation of monolayer formation with a mechanistic model of a packing of noninteracting probing measures may result in erroneous conclusions. Additional parameters should be considered for such cases, e.g., the density of packing of molecules in a monolayer, which depends on the form and orientation of adsorbate molecules, adsorbate/adsorbate and adsorbate/adsorbent interactions. Formation of linear plots in the log–log coordinates, accounting known smoothing effect of these coordinates, is only a necessary but not a sufficient condition for consideration of a material as a fractal. One may consider, for example, the attempts to determine the fractal dimension of coal using the second Pfeifer and Anvir [216] method. Adsorption of N$_2$ (77 K) and CO$_2$ (293 K) has been studied for the set of fractions prepared by splitting. A decrease in the fraction size from 3400 to 43 μm resulted in an increase in the surface area (from 1 to 15 m^2/g for N$_2$, and from 65 to 300 m^2/g for CO$_2$), while the surface area was constant in the range of fraction size 43 to 24 μm. The corresponding log–log plots gave $D_{f,N_2} = 2.38 \pm 0.14$ and $D_{f,CO_2} = 2.47 \pm 0.05$. At first it seems that these results confirm the fractal nature of the coal, but the observed increase in the surface area can be explained by a gradual opening of an internal void volume of closed pores without considering the fractal nature. This explains the constant values of the surface area for small fractions when all internal pores become opened, which cannot be explained by a fractal approach. Moreover, splitting usually decreases the porosity due to collapsing of large pores. For these situations, as shown by van Damme et al. [208], the specific surface area per unit apparent volume should be used in Equation 9.74b since the plot of the specific surface area per unit mass log A vs. log R yields the slope $D_2 - D_m$ instead of the expected $D_2 - 3$, where $D_m \leq 3$ is the fractal dimension of the dependence of the apparent density $\delta(R)$, defined by the equation similar to Equation 9.71:

$$\delta(R) \propto R^{(D_m - 3)} \tag{9.76}$$

Summarizing, one should note that a tree-like structure that corresponds to Bethe lattice or tree (Figure 9.38a) is characteristic of the truly fractal objects that follow the dependences type of Equation 9.69. The basic topological property of a Bethe lattice is the absence of intercrossing of neighboring branches to form closed cycles. The internal space of the majority of real PSs is formed by labyrinths of interconnected pores with an abundance of closed cycles (Figure 9.39b to Figure 9.39e). The path of a probe over such a lattice includes numerous widenings and

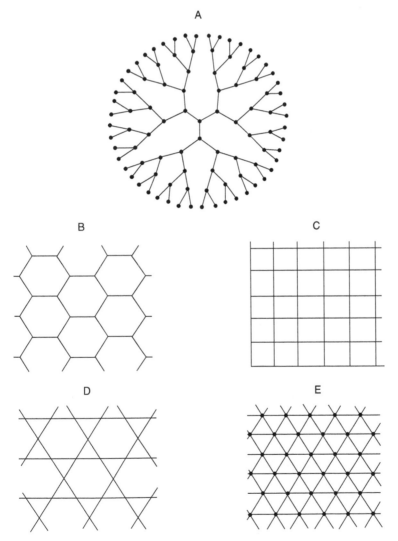

Figure 9.38 The simplest types of regular 2D lattices: (A) the Bethe lattice (Z=3); (B) the honeycomb lattice (Z=3), The simple square lattice (Z=4), (C) the simple square lattice (Z=4), (D) Kagomé lattice (Z=4), (E) the triangular lattice (Z=6).

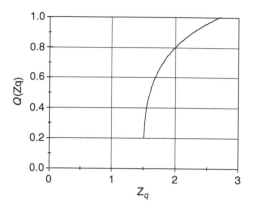

Figure 9.39   The plot of the dependence of percolation probability $Q(Zq)$ on Zq according to Equation 9.78.

narrowings that conflict with the gradual increase or decrease in sizes required for fractals. Therefore the fractal approach is applicable only for studies of irregular rough relief of PS surface without entering the bulk enclosed by this surface. The situation of probe movement through the bulk of PS is simulated by the lattices of pores and particles.

9.9   LATERAL AND STATISTICAL MODELS OF PORES AND PARTICLES ARRANGEMENT

For the analysis of the structure of a labyrinth of pores and particles, lateral models representing a PS as a lattice of sites and bonds are commonly used. The elementary examples of regular 2D lattices with different coordination numbers are shown in Figure 9.38. By analogy, one can construct the 3D lattices. The irregular lattices are usually constructed by randomizing, that is, casual removal of a part of bonds together with contiguous bonds.

The first model of porous space as a 2D lattice of interconnected pores with a variation of randomness and branchness was offered by Fatt [220]. He used a network of resistors as an analog PS. Further, similar approaches were applied in a number of publications (see, e.g., Refs. [221–223]). Later Ksenjheck [224] used a 3D variant of such a model (simple cubic lattice with coordination number 6, formed from crossed cylindrical capillaries of different radii) for modeling MP with randomized psd. The plausible results were obtained in these works, but the quantitative consent with the experiment has not been achieved.

9.9.1   Percolation Theory

A radically new stage of development of lateral models is related to the application of PT methodology. The term percolation was introduced by Broadbent and Hammersley [225] to describe the new class of mathematical problems connected to the analysis of infiltration of liquids or the path of an electrical current through a labyrinth of bonded and nonbonded elements. This theory has become very fashionable in various fields of physics, chemistry, and technical applications (e.g., a problem of displacement of oil from a porous medium [8,223,226,227]). The basics of this theory has been comprehensively discussed in several reviews [121,228–232] and monographs [8,233,234].

The percolation processes were first developed by Flory [235] and Stockmayer [236] to describe polymerization process, which result in gelation, that is, the formation of very large networks of molecules connected by chemical bonds. But, their theory was developed only for a special kind of network, namely, the Bethe lattice, an infinite branching structure without any closed loops. Broadbent and Hammersley have developed a more general theory and have introduced it into the mathematical literature. Essentially, the basic results of the PT are obtained by Monte Carlo simulation for regular or random networks or continua, the strict analytical solutions are received mainly for Bethe lattice [8,233,234].

The modern condition of PT allows three basic groups of problems: a problem of bonds, a problem of sites, and a mixed problem to be solved [8]. Briefly we shall consider these groups.

9.9.1.1 Problem of Bonds

Here, it is assumed that the permeability of a lattice is completely determined by the properties of bonds, for example, in a lattice of pores it is determined by the effective sizes of windows r_W only, whereas the sizes of all sites (cavities) more than r_W do not influence permeability. Let us illustrate the ideology of this approach using an example of the analysis of mercury intrusion in MP. For simplification, it is assumed that the whole volume of porous space is concentrated in the cavities (sites), and the windows (bonds) are only 2D cross sections with effective sizes of r_{Wi}.

To begin, let us consider the principle of MP measurements [55]. This method is based on the property of mercury not to wet the majority of solids, but with a possibility to be pressed through a window size of r_{Wi} under excess pressure:

$$P_{Hg,i} = \frac{2\sigma_{Hg}\cos(\theta)}{r_{Wi}} \quad (9.77)$$

where σ_{Hg} is mercury surface tension (~0.48 N/m), and θ a contact angle that is typically equal to 140°. The given numerical values simplify Equation 9.77

$$r_{Wi} = \frac{735}{P_{Hg,i}} \quad (9.77a)$$

where r_{Wi} is expressed in nanometers and $P_{Hg,i}$ in bars (1 bar = 0.101 MN/m²).

Relation 9.77 is usually called the Washburn equation [55,237]. One should consider it as a special case of the fundamental Young–Laplace equation [3,9–11]. Washburn was the first to propose the use of mercury for measurements of porosity. Now, it is a common method [3,8,53–55] of psd measurements for a range of sizes from several hundreds of microns to 3 to 6 nm. The lower limit is determined by the maximum pressure, which is applied in a mercury porosimeter; the limiting size of r_{Wi} = 3 nm is achieved under P_{Hg} = 4000 bar. The measurements are carried out after vacuum treatment of a sample and filling the gaps between pieces of solid with mercury. Further, the hydraulic system of a device performs the gradual increase of P_{Hg}, and the appropriate intrusion of mercury in pores of the decreasing size occurs.

If one models the porous space as a branch of cylindrical channels arranged parallel with various r_{Wi}, each channel is filled under $P_{Hg,i}$, which is determined by Equation 9.77 or Equation 9.77a. Under equilibrium conditions, all channels with size $r_W > r_{Wi}$ should be filled.

In the capillaries of variable cross section or the lattice of the interconnected sites and bonds of the various size the situation is essentially different. At the beginning, the mercury is at the external surface of particles of a sample, and only the pores that are directly contiguous to external surface can be filled according to the considered model of a bunch of capillaries. The cavities with windows of size $r_{Wi}(P_{Hg})$, which are adequate to an equilibrium condition but inside the bulk of a sample, can be filled only under a condition of their connection to an external surface through a circuit of cavities with windows of size $r_W > r_{Wi}$, already filled with mercury. Therefore, the condition for the filling of a cavity with a window of the size r_{Wi} can be expressed as the requirement of a direct contact of a considered cavity with mercury. Accordingly, under each pressure $P_{Hg,i}$ all windows of size r_{Wi} are only potentially permeable, and are not filled if they are connected to mercury through capillaries of the smaller size. The real filling begins only after the formation of a rather large (in the theory—infinite) coherent cluster of sites, which is connected to mercury by windows size of $r_W > r_{Wi}$.[*]

For relatively large lattices, the part of bonds (windows) $q = N/N_0$, where N_0 is the total number of bonds, is equal to the probability that the randomly chosen bond has a size of r_{Wi}, that is, it is potentially permeable. But, this bond is really permeable only with the percolation probability $Q(q) = M/M_0$, where M_0 is the total number of sites and M the number of sites that are bound by permeable bonds (the value of $Q(q)$ is numerically equal to the probability that the randomly selected site belongs to the interlinked cluster). At low values of q, formation of a relatively large cluster of interlinked sites is infinitely small, $Q(q) = 0$. This explains the low volume of pressed mercury, where the lattice is impermeable. If the lattice is large enough, the effect of filling of separate cavities near the external surface is insignificant. An increase in P_{Hg} decreases r_W, resulting in an increase in q. At some threshold value $q = q_{th}$, the probability of formation of an infinite cluster that permeates through the whole PS becomes nonzero.

One of the most significant results of the problem of bonds is in the establishment of invariant (i.e., independent from a type of lattice of the same dimension) dependence of $Q(q)$ from Zq, where $Z = 2N_0/M_0$ (factor 2 accounts belonging of each bond to two sites) is a coordination number of the lattice. Thus, $Zq = 2N/M_0 = Z_B(P)$ is the average number of potentially permeable bonds per site. For 2D lattices, $Q(Zq) = 0$ when $Zq \leq 2.0$, and $Q(Zq) \approx 1.0$ at $Zq > 4.0$. For 3D lattices, this dependence (shown in Figure 9.39) is expressed by an empirical relation [121]

$$Q(Zq) = \begin{cases} 0, & Zq < 1.5 \\ \dfrac{1.54\,(Zq - 1.5)^{0.4}}{\left[1 + 0.606\,(Zq - 1.5)^{0.4}\right]}, & 1.5 < Zq < 2.7 \\ 1, & Zq \geq 2.7 \end{cases} \tag{9.78}$$

Thus, the whole area of fillings for the large lattices considered is concentrated into a narrow range of threshold values of Z_B, and $Z_B < Z$. The windows, bonds of the largest sizes whose numerical part is lower than $q = 1.5/Z$, do not form an interconnected cluster and consequently are not measured. The bonds of the smallest size, whose numerical part exceeds $q = 2.7/Z$, essentially cannot be measured and do not participate in percolation processes, since the sites bound to them are filled through wider bonds. One should expect that nonpercolation bonds have a relatively small effect on diffusion mass transfer that extends from the external surface of a lattice to its bulk.

9.9.1.2 Problem of Sites

The subject of study in this case is permeability of regular or irregular 2D and 3D lattices that have some distinctive property. It can be, for example, the lattice of sites formed of different phases, A and B, and the problem is reduced to an establishment of interconnectivity of the system through phase A or B (in one of the phases there can be void). In other examples, there can be problems with the introduction of additional phases that regulate heat transfer or electrical conductivity of the catalyst, or additives, which are introduced into the volume of the catalyst, and further are dissolved or burned off to form a system of transport pores. In the latter case, the percolation approach allows estimations of a volumetric part of the additive that is necessary to form

---

[*]The similar mechanism determines the consequence of clearing of the cavities during the desorption of wetting liquid (DES) [3], and the main deviations are variations of wetting angle and desorption from a film in already cleaned cavities.

a coherent system of pores. In a 2D problem of sites, phase B may be a component that is deposited on a surface of a support (Figure 9.40). In this case, the interconnectivity and size of clusters of particles of phase B can determine the intensity of the sintering of the deposited component. All these problems can be drawn to a problem of sites, where the properties of a system are determined just in sites, instead of bonds.

Solving the problem of sites is similar to that of bonds. One can also calculate the percolation probability, $Q(q)$, of formation of an infinite cluster of sites of one type, for example, B, depending on their numerical part $q = B/(A+B)$, which is proportional to the part of the 3D volume occupied with phase B. The invariants of the problem of sites for 3D lattices of contiguous monospheres are [3,238]

$$Q(q) = \begin{cases} 0.0, & \alpha(B) \leq 0.16 \pm 0.02 \\ 1.0, & \alpha(B) \geq 0.32 \pm 0.2 \end{cases} \tag{9.79}$$

where $\alpha(B)=(1-\varepsilon_V)q$ is the volumetric part of phase B and ε_V an usual 3D porosity. For the 2D lattices, the corresponding thresholds of $(1-\varepsilon_A)q$ are ~0.45 and ~0.70 (ε_A is the 2D porosity).

Thus, phase B, which occupies the volumetric part $\alpha(B) < 0.16 \pm 0.02$ of a PS, is distributed in the form of separated clusters with no permeability through this phase, but with $\alpha(B) > 0.32 \pm 0.2$ the whole phase B is practically interlinked. A gap corresponds to coexistence of interlinked clusters of phase A and phase B. The thresholds in 2D lattices correspond to a degree of surface coverage. The whole phase B is interlinked into one cluster under $\alpha(B) > 0.7$, but coexistence of two infinite clusters of both phases is impossible because this contradicts one of the topology theorems [8].

In variants of the mixed problem of percolation, the situations are considered when availability of sites and bonds is correlated and the ranges of distribution of sites and bonds on the sizes overlap. This variant of a problem is considered, for example, in Refs. [120,121].

Generally, percolation media can be characterized not only by the percolation probability, but also by several important quantities [8,121,234]. Near the percolation threshold, q_{th}, a number of

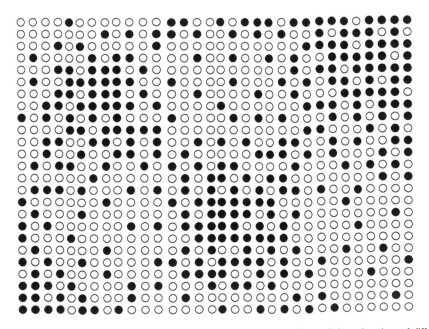

Figure 9.40 The primitive lattice of sites and bonds and bonds in the problem of sites; the sites of different types are as shown as black and white.

quantities X_i may be described by the power-law equations, which depend only on the dimensionality of the lattice rather than on the its structure:

$$X_i \propto (q - q_{th})^{Y_i} \tag{9.80}$$

Let us consider some of these quantities. Let $X_i \equiv Q(q)$, that is, the percolation probability. The values of $\beta_q = Y_i$ for various lattices are shown in Table 9.9.

Correlation length, $\xi(q)$, is the typical radius of the connected clusters at $q > q_{th}$ and the length scale over which the random network is macroscopically homogeneous, that is, the length scale over which the properties of the system are independent of its linear size L. At $L << \xi(q)$, the sample-spanning cluster is on average self-similar, that is, it looks the same at all length scales up to $\xi(q)$. Its mass M (total number of bonds or sites) that scales with $\xi(q)$ as $M \propto \xi(q)^{D_c}$, where $D_C = d - \beta_q/v$ is called the fractal dimension of the cluster (where d is the dimension of a lattice); for $L >> \xi(q)$ by Ref. [8] $D_C = d$. If $L < \xi(q)$, $M \propto L^{D_c}$. In addition, $\xi(q)$ is divergent at $q = q_{th}$, so that the sample-spanning cluster is a fractal object for any L. The values of critical exponents in Equation 9.79 are shown in Table 9.9.

A *backbone* cluster plays a fundamental role in all the transport properties in percolation systems. It is a cluster of occupied bonds in the infinite cluster, which actually carry flow or current, since some of the bonds in the cluster are dead ends and do not carry any flow. Stanley [239] divided the bonds of the backbone into two groups. In one group are those that are in the blobs, i.e., the multiply connected part of the backbone. In the second group are the red bonds, which if cut, would split the backbone. The reason for calling such bonds red is that in an electric percolation network they carry all the current between two blobs, and thus they are the hottest bonds in the backbone. The fraction of occupied bonds in the infinite cluster, which are the members of the backbone, is determined by Equation 9.79 with exponent $Y_i = \beta_B$. The backbone is a fractal object and its fractal dimension is $D_{BB} = d - \beta_B/v$, where d is the dimension of the lattice. For $L << \xi(q)$ the number of red bonds M_{red} scales with L as $M_{red} \propto L^{D_{red}}$, and thus D_{red} is the fractal dimension of the set of the red bonds with $D_{red} = 1/v$. The values of the exponents β_B and D_{BB} are shown in Table 9.9. Another important quantity is the minimal path, L_{min}, between two points of a percolation cluster, which is the shortest path between the two points; for $L << \xi(q)$ the length of the path scales with L as $L_{min} \propto L^{D_{min}}$ (and, therefore, D_{min} is the fractal dimension of the minimal path). All considered parameters correspond to infinitely large lattices. In the lattices of limited linear size L by Fisher [240] the variations of any property $X_{L,i}$ is written as $X_{L,i} \propto L^{-\beta q/v}G[(q-q_{th})L^{1/v}]$, where the scaling function $G(x)$ is a positive function of its argument and is analytical everywhere, even at the percolation threshold where $x = 0$ [121,234]. The finite size of the network also causes a shift in the q_{th} by Levinshtein et al. [241] $q_{th} - q_{th}(L) \propto L^{-1/v}$. Other scaling ratios may be found elsewhere [8,234].

9.9.2 The Stochastic and other Statistical Models of Long-Range Order

The rapid development of computer engineering initiated a simultaneous development and wide application of statistical methods of modeling of PS properties and processes occurring in them. It is

Table 9.9 Critical Exponents for d–Dimensional Systems

Exponent	Quantity	$d = 2$	$d = 3$	Bethe lattice
β_q	$Q(q)$, percolation probability	5/26	0.41	1.0
β_B	Characteristics of backbone	0.48	1.05	2.0
v	$\xi(q)$, correlation length	4/3	0.88	½
D_C	Fractal dimension of accessible cluster	91/48	2.52	4
D_{BB}	Fractal dimension of backbone	1.64	1.8	2.0
D_{min}	Fractal dimension of L_{min}	1.13	1.34	2.0

Note: Rational or integer values are exact results [8].

displayed in chemical engineering and many physicochemical studies of processes in heterogeneous catalysts and adsorbents, where the extensive assortment of mathematical models that describe the steady-state and dynamic behavior of PS in a wide range of scales is developed from a single separate pore up to a chemical reactor as a whole [242–245]. The methods of modeling of texture in these approaches can be conditionally divided into three groups: deterministics, stochastics, and heuristics [8,244]. The former assumes that the geometrical model is perfectly known (e.g., a lattice approach) as opposed to the stochastic techniques that only assume an approximate knowledge of this model and take a probabilistic view. The heuristic methods propose the rules for calculating plausible equivalent characteristics of PS. For each group of methods, different calculation techniques can be applied: analytical or numerical, exact or approximated. The algorithms of accounts are based on continuity, momentum, species, and energy balances and combining these with the various constitutive equations for the transport rate and other processes. The calculations can be based on a spectral or perturbation approach, method of moments or Monte Carlo method, field theory, dusty-gas model, population balance dynamics, etc. [244–251]. In particular, the stochastic models usually represent porous structure as a statistical ensemble of interconnected textural elements, properties of which and mutual arrangement are random and are susceptible to some probabilistic laws. In the calculations, the average values of the required parameters, dispersion, and moments of distribution are determined.

Some opportunities of such approximations are well illustrated by considering two characteristic examples. The first example will be a dusty-gas model, where porous media is considered as one of components of a gas mix of huge molecules (or particles of a dust), mobile or rigidly fixed in space [249,252,253]. Such a model allows a direct application of methods and results of kinetic theory of gases and is effectively applied to the description of mass transfer processes in PS. The history of such an approach, the origins of which can be found in the works by Thomas Graham (1830 to 1840) is considered in Ref. [249]. Actually, the model was first proposed by James Maxwell (1860), further it was independently reported by Deryagin and Bakanov (1957), and then also independently reported by Evans, Watson, and Mason (1961; see Refs. [249,252]).

For the second example, let us consider the random sphere model (RSM), which can be referred to as an intermediate deterministic–stochastic approach. This model and an appropriate mathematical apparatus were originally offered by Kolmogorov in 1937 for the description of metal crystallization [254]. Later, this model became widely applicable for the description of phase transformations and other processes in PS, and usually without references to the pioneer work by Kolmogorov [134,149–152,228,255,256].

The simplest form of RSM is constructed by placing N identical solid monospheres of radius R_{sp} in a single empty volume without any correlation between the positions of the spheres, so that some of the spheres remain isolated and others overlap. The porosity, ε, of such chaotic packing is determined by Kolmogorov as the probability of finding an arbitrarily chosen point outside of the space of the particles, and is equal to

$$\varepsilon = \exp(-V) = \exp\left[-\frac{4}{3}\pi R_{sp}^3 N\right] \tag{9.81}$$

where V is the total volume of all spheres. Multiplication of this probability with the total surface area of all spheres in a unit of volume gives the specific surface area of a unit volume

$$A_V = 4\pi R_{sp}^2 N\varepsilon = -\frac{3\varepsilon\ln\varepsilon}{R_{sp}} \tag{9.82}$$

Further, one can calculate the distribution of the number and size of spheres' intersections, but let us consider just an average number of crossings per sphere, which is $n = -8\ln(\varepsilon)$ and corresponds to an average coordination number for the packing of spheres n_p.

In Refs. [228,256,257], the RSM determination of the values of minimum density, $\eta_{min}=(1-\varepsilon_{max})$, and coordination number, $n_{p,min}$, appropriate to the moment of formation of a coherent cluster from spheres as well as from particles of other form are investigated by Monte Carlo. It was found that for the systems from convex particles (spheres, tetrahedrons, cubes) $\eta_{min} = 0.30 \pm 0.01$ and $n_{p,min} = 2.7$ to 3, while for the particles of the concave form (Maltese cross) $\eta_{min} = 0.29$ to 0.31 and $n_{p,min} = 2.5$ to 2.7. For 2D figures (circles and squares) the limiting values of 2D density $\eta_{S,min} = 0.67$ to 0.68, and $n_{S,p,min} = 4.4$ to 4.5 (though for the circles of various size $n_{S,p,min} = 4.0$ at the same value of $\eta_{S,min}$). These results specify a rather small dependence of critical values of η_{min} and $n_{p,min}$ on the form of particles. Simultaneously, these values are close to the results of the PT (see Equation 9.79 and below) and determine the lower limit of applicability of RSM that is appropriate to the minimal density of a formed coherent system. The upper limit for 3D space, $\eta_{max} = 0.70$ to 0.75, is determined by the condition that this model does not take into account threefold and more complex intersections, the contribution of which with a large number of crossed spheres N (i.e., with high density η) becomes significant.

Substitution of ε with η allows construction of a model of *random overlapping pores* (ROP). Equation 9.82 can be rewritten as:

$$A = \frac{A_V}{\delta} = \frac{A_V}{\rho(1-\varepsilon)} = -\frac{3}{R_{sp}\rho}\frac{\varepsilon\ln(\varepsilon)}{1-\varepsilon} \tag{9.83}$$

where A is the specific surface area of a unit mass of PS for RSM, and by analogy for ROP

$$A_{ROP} = -\frac{3}{R_{sp}\rho}\frac{\ln(\varepsilon)}{1-\varepsilon} \tag{9.84}$$

One can check that with a limit of $\varepsilon \to 0$ the surface area $A_{ROP} \to 0$, which is natural since the single spherical pores in a bulk solid do not form a coherent system, and their surface area is inaccessible.

The estimation of a number of other textural parameters, especially in the framework of a more complete model, which takes into account the statistics of the distribution of particles or cavities on the sizes [255] is possible. Application of this approach for modeling a number of processes related to textural changes is interesting. As an example, let us consider the changes of a surface with a uniform increase or decrease of the size of particles (or cavities). Consider, for example, a process of gradual deposition of an additional component on the surface of spherical particles, as a result of which their average radius increases from R_0 to $R=R_0+t$, where t is the average thickness of a film of the deposited component. The initial specific surface area, $A_{V,0}$, is transformed to A_V. It is convenient to express the quantity of the deposited component as a degree of filling pore volume, U, which appropriates the porosity of the modified product, ε, and initial porosity, ε_0, by the equation $\varepsilon=\varepsilon_0(1-U)$. According to Equation 9.81 $R/R_0=(\ln(\varepsilon)/\ln(\varepsilon_0))^{1/3}$, and taking into account Equation 9.82

$$\frac{A_V}{A_{V0}} = (1-U)\left[1+\frac{\ln(1-U)}{\ln(\varepsilon_0)}\right]^{2/3} \tag{9.85}$$

By analogy for ROP,

$$\frac{A_V}{A_{V0}} = \left(\frac{1-\varepsilon_0(1-U)}{1-\varepsilon_0}\right)\left\{\frac{\ln[1-\varepsilon_0(1-U)]}{\ln(1-\varepsilon_0)}\right\}^{2/3} \tag{9.86}$$

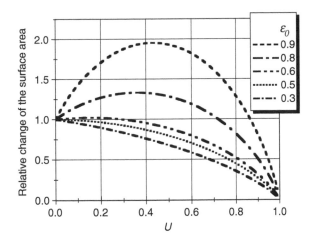

Figure 9.41   The relative changes of the surface area after gradual deposition of an additional component on the surface of monodisperse spherical particles depending on the degree of filling of pore volume, U, and initial porosity of a support ε_0.

The plots of A_V/A_{V0} vs. U under various values of ε_0 according to Equation 9.85 are shown in Figure 9.41. The calculations for ROP (Equation 9.86) give the plots that lay slightly higher under the same ε_0. The changes of the surface area of a corpuscular system with such modification are determined by two antibate effects: decrease of the surface area due to an increase of spheres overlapping in the places of their contacts, and increase in the surface area due to an increase of the radius of spheres outside the overlapping zones. At high initial porosity, ε_0, the number of contacts is low, and an increase in the surface area is observed at the first stage, which is further suppressed by the effect of overlapping. This explains the observable maximum. At ε_0, the number of contacts is high and the effect of overlapping dominates at already minimal values of U. For the ROP model, the situation is controversial — the effect of overlapping in the zones of contacts increases the surface area, and the decrease in the size of spheres reduces it. These models successfully explain and describe experimental data on the change of the surface area caused by the modification process during formation of carbon adsorbents of Sibunit type (corpuscular structure) and porous glasses (sponge structure), whose preparation technology includes similar operations [3].

9.10   CONCLUSIONS

Actual catalysis is a chemical phenomenon, since the intimate mechanisms of catalytic transformations are determined by chemical interaction of reagents with the catalyst, atomic structure, and the energy of formed intermediate active complexes [5]. It obviously is the world of molecular chemistry of the interface phenomena, kinetics, and mechanisms of catalytic transformations at a molecular level.

However, with a consistency of chemical (i.e., atomic) structure of a heterogeneous catalyst its ultimate efficiency in an industrial process essentially depends on its supramolecular structure at various hierarchical levels. Supramolecular structure determines the availability of the active sites, their stability, the degree of use of their activity, and also a complex of other major operational properties of the catalyst. The supramolecular world differs from the molecular one, since the rules of the game, methods of regulation, and control here are very specific. We have named this world as *texturology*, the laws (i.e., the constitution) of which are based on the laws of physical chemistry of superficial phenomena, supramolecular chemistry, topology, and morphology of PS. The main focus of this chapter is stimulation of the curious reader to recognize the structure of heterogeneous catalysts on various supramolecular levels. We have tried to allocate the major geometrical features of the supramolecular world, which essentially distinguish it from the world of molecular chemistry, and at the same time is general for the majority of PSs. Only a part of problems of texturology, which in our opinion should also include both methods of measurements of the major textural characteristics and laws of their formation during synthesis of PSs and evolution with their operation, are considered here.

REFERENCES

1. J.H. de Boer, in *Structure and Properties of Porous Materials*, D.H. Everett and F.S. Stone, Eds., Butterworth, London, 1958, p. 68.
2. J.C.P. Broekhoff and B.G. Linsen, in *Physical and Chemical Aspects of Adsorbents and Catalysts* (dedicated to J.H. de Boer), J.M.H. Fortuin, C. Okkerse, and J.J. Steggerda, Eds., Academic Press, London, 1970, p. 23.
3. V.B. Fenelonov, *Introduction to Physical Chemistry of Formation of Adsorbents and Catalysts Supramolecular Structure*, 2nd ed., Publ. House Sib. Branch Russian Academy of Science, Novosuibirsk 2004, p. 440 (in Russian).
4. J.-M. Lehn, *Supramolecular Chemistry. Concept and Perspectives*, New York, 1995.
5. G.K. Boreskov, *Heterogeneous Catalysis*, Nauka, Moscow, 1986.
6. A. Weller, *Adv. Catal.* 3 (1951) 249; *Catalysis. Vol. 5*, P.H. Everett, Ed., Van Nostrand, New York, 1955.
7. J. Rouquerol, D. Avnir, D.H. Everett, C. Fairbridge, M. Haynes, N. Pirnicone, J.D.F. Ramsay, K.S.W. Sing, and K.K. Unger, in *Characterization of Porous Solids III*, J. Rouquerol, F. Rodriguez-Reinoso, K.S.W. Sing, and K.K. Unger, Eds., Elsevier, Amsterdam, 1994, p. 1.
8. M. Sahimi, *Application of Percolation Theory*, Taylor & Francis, 1994.
9. A.W. Adamson and A.P. Gast, *Physical Chemistry of Surface*, 7th ed., Wiley, New York, 1997.
10. J.W. Gibbs, *Trans. Conn. Acad.* 3 (1878) 343; *The Scientific Papers of J. William Gibbs*, Vol. 1, Dover, New York, 1961.
11. M. Pasandideh-Farad, P. Chen, J. Mostaghimi, and A.W. Neumann, *Adv. Colloid Interface Sci.* 63 (1996) 151, 179.
12. G.A. Samorjai, *Introduction to Surface Chemistry and Catalysis*, Wiley, New York, 1994.
13. H. Knözenger and E. Taglauer, in *Handbook of Heterogeneous Catalysis. Vol. 1*, G. Ertl, H. Knözenger, and J. Weitkamp, Eds., VCH, Weinheim, 1997, p. 216.
14. C.F. Gauss, *Theorie der Gestalt von Flüssigkeiten*, Leipzig, 1903 (cited by J.M. Haynes, in *Problems in Thermodynamics and Statistical Physics*, P.T. Landsberg, Ed., PION, London, 1971, p. 267 (Russian edition)).
15. W. Ostwald, *J. Phys. Chem.* 34 (1900) 495.
16. J.J. Thomson, *Application of Dynamics*, London, 1888.
17. E. Rie, *Z. Phys. Chem.* 104 (1923) 354.
18. R.W. Batchelor and A.G. Foster, *Trans. Faraday Soc.* 40 (1944) 300.
19. C. Faivre, *Eur. Phys. J. B* 7 (1999) 19.
20. E. Ya. Aladko, Yu.A. Dyadin, V.B. Fenelonov, E.G. Larionov, M.S. Mel'gunov, A.Yu. Manakov, A.N. Nesterov, and F.V. Zhurko, *J. Phys. Chem. B.* 108 (2004) 16540.
21. S. Andersson, S.T. Hyde, K. Larsson, and S. Lidin, *Chem. Rev.* 88 (1988) 221.
22. J.S. Rowlinson and B. Widom, *Molecular Theory of Capillarity*, Clarenden Press, Oxford, 1982.
23. P.A. Kralchevsky, J.C. Eriksson, and S. Ljunggren, *Adv. Colloid Interface Sci.* 48 (1994) 19.
24. C.T. Kresge, M.E. Leonowicz, W.J. Roth, J.C. Vartuli, and J.S. Beck, *Nature* 359 (1992) 710; J.S. Beck, J.C. Vartuli, W.J. Roth, M.E. Leonowicz, C.T. Kresge et al. *J. Am. Chem. Soc.* 114 (1992) 10834.
25. V.B. Fenelonov, A.Yu. Derevyankin, S.D. Kirik, L.A. Solovyov, A.N. Shmakov, J.-L. Bonardet, A. Gedeon, and V.N. Romannikov, *Micropor. Mesopor. Mater.* 44–45 (2001) 33.
26. L.R. Fisher and J.N. Israelachvili, *Colloid. Surf.* 3 (1981) 303; *J. Colloid Interface Sci.* 80 (1981) 526.
27. H.K. Christensen, *J. Colloid Interface Sci.* 104 (1985) 234.
28. H. K. Christensen, *J. Phys. Condens. Mater.* 13 (2001) R95.
29. R. Schmidt, E.W. Hansen, M. Stöcker, D. Akporiaye, and O.G. Fllestad, *J. Am. Chem. Soc.* 117 (1995) 4049.
30. W.H. Qi, and M.P. Wang, *Mater. Chem. Phys.* 88 (2004) 280.

31. L.H. Liang, J.C. Li, and Q. Jiang, *Physica B* 334 (2003) 49.
32. Y. Xue, Q. Zhao, and Ch. Luan, *J. Colloid Interface Sci.* 243 (2001) 388.
33. V.B. Fenelonov, G.G. Kodenyov, and V.G. Kostrovsky, *J. Phys. Chem. B* 105 (2001) 1050.
34. V.G. Baidakov, S.P. Prostenko, G.G. Chernykh, and G.Sh. Boltachev, *Phys. Rev. E* 65 (2002) 041601; 70 (2004) 011603.
35. V.G. Baidakov and G.Sh. Boltachev, *J. Chem. Phys.* 121 (2004) 8594; V.G. Baidakov, A. Kaverin, and G.Sh. Boltachev, *J. Phys. Chem. B* 106 (2002) 167.
36. W.B. Haines, *J. Agric. Sci.* 20 (1930) 97.
37. A.V. Neimark, L.I. Kheifez, and V.B. Fenelonov, *Ind. Eng. Chem. Prod. Res. Dev.* 20 (1981) 439.
38. W. Yao, H.J. Maris, P. Pennington, and G.M. Seidel, *Phys. Rev. E* 71 (2005) 016309.
39. R. Ishiguro, F. Graner, E. Rolley, and S. Balibar, *Phys. Rev. Lett.* 93 (2004) 235301.
40. A. Baldan, *J. Mater. Sci.* 37 (2002) 2171.
41. G.W. Scherer, *J. Non-Crystal Solids*, 87 (1986) 199; 89 (1987) 217; 91 (1987) 83, 101; 92 (1987) 122; 99 (1988) 324; 107 (1989) 135; 109 (1989) 171; 225 (1998) 192.
42. R.K. Iler, *The Chemistry of Silica*, Wiley, New York, 1979.
43. V.B. Fenelonov, V.Yu. Gavrilov, and L.G. Simonova, in *Preparation of Catalysts III*, G. Poncelet, P. Grange, and P.A. Jacobs, Eds., Elsevier, Amsterdam, 1983, p. 665.
44. T. Heinrich, U. Klett, and J. Fricke, *J. Porous Mater.* 1 (1995) 7.
45. M.B. Pilling and R.E. Bedworth, *J. Inst. Met.* 1 (1923) 529.
46. V.A. Dzisko, *The Basis of the Methods of Catalysts Preparation*, Nauka, Novosibirsk, 1983 (in Russian).
47. O.P. Krivoruchko, B.P. Zolotovskii, L.M. Plyasova, R.A. Buyanov, and V.I. Zaikovskii, *React. Kinet. Catal. Lett.* 21 (1982) 103.
48. V.B. Fenelonov, *Poristyj Uglerod (Porous Carbon)*, Boreskov Institute of Catalysis, Novosibirsk, 1997 (in Russian).
49. H.F. Fishmeister, in *Pore Structure and Properties of Materials (Proc. I Inter. Sump. RILEM/IUPAC)*, Part IV, Academia, Prague, 1973, p. C-439.
50. R.T. DeHoff and F.N. Rhines, Eds., *Quantative Microscopy*, McGraw-Hill, New York, 1968.
51. A.G. Dixon, *Can. J. Chem. Eng.* 66 (1988) 705.
52. T. Allen, *Particle Size Measurement*, 3rd ed., Chapman & Hall, New York, 1981.
53. S.J. Gregg and K.S.W. Sing, *Adsorption, Surface Area and Porosity*, 2nd ed., Academic Press, London, 1982.
54. A.P. Karnaukhov, *Adsorption. Texture of Dispersed and Porous Materials*, Nauka, Novosibirsk, 1999 (in Russian); in *Pore Structure and Properties of Materials (Proc. I Inter. Sump. RILEM/IUPAC)*, Part I, Academia, Prague, 1973, p. A-3.
55. P.A. Webb and C. Orr, *Analytical Methods in Fine Particle Technology*, Micromeritics, Norcross, 1997.
56. I. Langmuir, *J. Am. Chem. Soc.* 38 (1916) 2219; 40 (1918) 1368.
57. S. Brunauer, P.H. Emmett, and E. Teller, *J. Am. Chem. Soc.* 60 (1938) 309.
58. IUPAC Manual of Symbols and Terminology, Appendix 2, Part 1, Colloid and Surface Chemistry, *Pure Appl. Chem.* 31 (1972) 578.
59. A.L. McClellan and H.F. Harnsburger, *J. Colloid Interface Sci.* 23 (1967) 577.
60. R.V. Zagrafskaya, A.P. Karnaukhov, and V.B. Fenelonov, *React. Kinet. Catal. Lett.* 16 (1981) 223.
61. V.B. Fenelonov, Physico-Chemical Basis, Doctoral dissertation, Institut Kataliza, Novosibirsk, 1987 (in Russian).
62. A.P. Karnaukhov, V.B. Fenelonov, and V.Yu. Gavrilov, *Pure Appl. Chem.* 61 (1989) 1913.
63. K.S.W. Sing, in *Proceedings of International Symposium on Surface Area Determination, 1969*, Butterworths, London, 1970, p. 25.
64. B.C. Lippens, B.G. Linsen, and J.H. de Boer, *J. Catal.* 3 (1964) 32; B.C. Lippens and J.H. de Boer, *J. Catal.* 4 (1965) 319.
65. S.J. Gregg, *J. Chem. Soc. Chem. Commun.* (1975) 699; S.J. Gregg and J.F. Langford, *Chem. Soc. Faraday Trans. I* 73 (1977) 747.
66. O. Kadlec, *Coll. Czech. Chem. Commun.* 36 (1971) 2415.
67. M.M. Dubinin, T.I. Izotova, and O. Kadlec, *Izvestia AN SSSR, Ser. Chim.* (1975) 1232 (in Russian).
68. R.Sh. Mikhail, S. Brunauer, and E.E. Bodor, *J. Colloid Interface Sci.* 26 (1968) 45.
69. C.G. Shull, *J. Am. Chem. Soc.* 70 (1948) 1405.
70. R.W. Cranston and F.A. Inkley, *Adv. Catal.* 9 (1957) 143.

71. A.V. Kiselev, in *Structure and Properties of Porous Materials*, D.H. Everett and F.S. Stone, Eds., Butterworths, London, 1958, p. 200.

72. A.P. Karnaukhov, A.V. Kiselev, and E.V. Khrapova, *DAN SSSR* 94 (1954) 915 (in Russian).

73. J.H. de Boer, B.G. Linsen, and Th.J. Osinga, *J. Catal.* 4 (1965) 643.

74. M.R. Harris and K.S.W. Sing, in *Proceedings of 3rd International Congress Surface Activity, II* (1960), 42; *Chem. Ind.* (1959) 487.

75. Pierce, *J. Phys. Chem.* 63 (1959) 1076.

76. V.Yu. Gavrilov, V.B. Fenelonov, V.I. Zheivot, and M.E. Shalaeva, *Kinetika i Kataliz* 28 (1987) 959 (in Russian).

77. V.B. Fenelonov, V.N. Romannikov, and A.Yu. Derevyankin, *Micropor. Mesopor. Mater.* 28 (1999) 57.

78. M. Kruk, M. Jaroniec, and A. Sayari, *Langmuir* 13 (1997) 6267; M. Kruk, V. Antochshuk, M. Jaroniec, A. Sayari, *J. Phys. Chem. B* 103 (1997) 10670.

79. A.M. Voloschuk, M.M. Dubinin, T.A. Moskovskaya et al., *Bull. AN SSSR, ser. chim.* (1988) 277 (in Russian).

80. P.J.M. Carrot, R.A. Roberts, and K.S.W. Sing, *Chem. Ind.* (1987) 855.

81. J. Gobet and Kovats, *Adsorption Sci. Technol.* 1 (1984) 255.

82. D.A. Payne and K.S.W. Sing, *Chem. Ind.* 27 (1969) 918.

83. V.Yu. Gavrilov, R.V. Zagrafskaya, A.P. Karnaukhov, and V.B. Fenelonov, *Kinetika i Kataliz* 22 (1981) 452 (in Russian).

84. N. Setoyama, T. Suzuki, and K. Kaneko, *Carbon* 35 (1998) 1459.

85. J. Fraissard and C.W. Conner, Eds., *Physical Adsorption: Experiment, Theory and Application*, Kluwer Academic Publishers, Dordrecht, 1997.

86. M. Jaroniec and E. Madey, *Physical Adsorption on Heterogeneous Solids*, Elsevier, Amsterdam, 1988.

87. W. Rudzinski and D.H. Everett, *The Adsorption of Gases on Heterogeneous Surfaces*, Academic Press, London, 1992.

88. N.E. Buyanova, A.P. Karnaukhov, N.G. Koroleva, N.T. Kulishkin, V.T. Rybak, and V.B. Fenelonov, *Kinetika i Kataliz* 16 (1975) 741 (in Russian).

89. M.S. Mel'gunov and V.B. Fenelonov, *React. Kinet. Catal. Lett.* 64 (1998) 153.

90. M.S. Mel'gunov, V.B. Fenelonov, R. Loboda, and B. Charmas, *Carbon* 39 (2001) 357.

91. T. On, D. Desplantier-Giscard, C. Danumah, and S. Kaliaguine *Appl. Catal. A: General* 253 (2003) 545.

92. A. Corma, *Chem. Rev.* 97 (1997) 2373.

93. B. Busham, Ed., *Handbook of Micro/Nano Tribology*, 2nd ed., CRS Press, New York, 1999.

94. B.C. Larson and B. Lenger, *MRS Bull.* 29 (2004), 152.

95. V.B. Fenelonov, M.S. Mel'gunov, and N.A. Baronskaya, *React. Kinet. Catal. Lett.* 63 (1998) 305.

96. R.B. Martin, in *Pore Structure and Properties of Materials (Proc. I Inter. Sump. RILEM/IUPAC)*, Part I, Academia, Prague, 1973, p. A-35.

97. L. Onzager, *Ann. NY Acad. Sci.* 51 (1949) 627.

98. M. Corti and T. Zemb, *Curr. Opin. Colloid Interface Sci.* 5 (2000) 1.

99. W.M. Gelbert and A. Ben-Shaul, *J. Phys. Chem. B* 100 (1996) 13169.

100. N.N. Medvedev, *The Voronoi–Delaunay Method for Investigation of Noncrystalline Systems Structure*, Izd. SO RAN, Novosibirsk, 2000, p. 214 (in Russian); S.V. Anishchik and N.N. Medvedev, *Phys. Rev. Lett.* 75 (1995) 4314.

101. E.D. Glandt, *J. Colloid Interface Sci.* 77 (1980) 512.

102. R.P. Zou and A.B. Yu, *Chem. Eng. Sci.*, 50 (1995) 1504; *Powder Technol.* 88 (1996) 71.

103. A.B. Yu and N. Standish, *Powder Technol.* 55 (1988) 171.

104. P.C. Carman, *Trans. Int. Chem. Eng.* 15 (1937) 150.

105. D.C.C. Lam and M. Nakagawa, *J. Ceram. Soc. Jpn.* 101(1993) 1187, 1234; 102 (1994) 133.

106. R.V. Zagrafskaya, A.P. Karnaukhov, and V.B. Fenelonov, *Kinetika i Kataliz* 16 (1975) 1583.

107. J.E. Ayer and F.E. Soppet, *J. Am. Ceram. Soc.* 48 (1965) 180.

108. M.S. Mel'gunov, E.A. Mel'gunova, V.I. Zaikovskii, V.B. Fenelonov, A.F. Bedilo, and K.J. Klabunde, *Langmuir* 19 (2003) 10426.

109. H. Wadell, *J. Geol.* 40 (1932), 243; 43 (1935) 250; *Physics* 5 (1934) 281.

110. H. Rumpf, *Particle Technology*, Carl Hanser Verlag, Munich, 1990.

111. A.M. Bouwman, J.C. Bosma, P. Vonk, J.A. Wesselingh, and H.W. Frijlink, *Powder Technol.* 164 (2004) 66.

112. H.J. Herrmann, *Physica A* 313 (2002) 188.
113. A.B. Stiles, *Catalyst Support and Supported Catalysts*, Butterworth, Boston, 1987.
114. M.M. Dubinin, *Zhur. Phys. Chem.* 34 (1960) 959; *Chem. Rev.* 60 (1960) 235.
115. Ch. Lastoskie, K.E. Gabbins, and N. Quirke, *J. Phys. Chem.* 97 (1993) 4786.
116. A.V. Kiselev, in *Methods of Study of Highly Dispersed and Porous Solids Structure*, Izd. AN. SSSR, Moscow, 1958, p. 47 (in Russian).
117. J.H. de Boer, in *The Structure and Properties of Porous Materials*, Butterworth, London, 1958, p. 68.
118. D. Everett, in *Characterisation of Porous Solids*, Soc. Chem. Ind., London, 1979, p. 229.
119. G.C. Wall and R.J. Brown, *J. Colloid Interface Sci.* 82 (1981) 141.
120. V.P. Zhdanov, V.B. Fenelonov, and D.K. Efremov, *J. Colloid Interface Sci.* 120 (1987) 218.
121. V.P. Zhdanov, *Adv. Catal.* 39 (1993) 1.
122. D.K. Efremov and V.B. Fenelonov, *React. Kinet. Catal. Lett.* 40 (1989) 177.
123. L.V. Radushkevich, in *Basic Problems of Physical Adsorption Theory*, M.M. Dubinin and V.V. Serpinskii, Eds., Nauka, Moscow, 1970, p. 270.
124. A.I. Kitaigorodskii, *Order and Disorder in the World of Atoms*, Nauka, Moscow, 1984.
125. B.K. Vainstein, *Modern Crystallography*, Vol. 1, Springer, Berlin, 1981.
126. J.V. Smith, *Geometrical and Structural Crystallography*, Wiley, New York, 1982.
127. J.D. Bernal and S.V. King, in *Physics of Simple Liquids*, H.N.V. Temperly, J.S. Rowlinson, and G.S. Rushbrooke, Eds., North-Holland, Amsterdam, 1968, p. 116.
128. G.D. Scott, *Nature*, 188 (1960) 908; 194 (1963) 956; 201 (1964) 382.
129. T. Astle and D. Weaire, *The Pursuit of Purfect Packing*, Inst. Phys. Publ., Bristol, 2000.
130. A. Menta, Ed., *Granular Matter: An Interdisciplinary Approach*, Springer, New York, 1994.
131. F.A.L. Dullien, *Porous Media, Fluid Transport and Porous Structure*, 2nd ed., Academic Press, New York, 1992.
132. G. Matheron, *Random Sets and Integral Geometry*, Wiley, New York, 1975.
133. J.H. Conway and N.J.A. Sloane, *Sphere Packing. Lattice and Groups*, Springer-Verlag, New York, 1988.
134. J. Serra, *Image Analysis and Mathematical Morphology*, Academic Press, London, 1982.
135. J.O. Hirschfelder, Ch.F. Curtiss, and R.B. Bird, *Molecular Theory of Gases and Liquids*, Wiley, New York, 1954.
136. C.A. Rodgers, *Packing and Covering*, Cambridge University Press, Cambridge, 1964; *Proc. Lond. Math. Soc.* 8 (1958) 609.
137. A.H. Thiessen, *Monthey eather, revier* 39 (1911) 1082.
138. E.P. Wigner and F. Seitz, *Phys. Rev.* 43 (1933) 804.
139. G.L. Dirichlet, *J. Reine Andew. Math* 40 (1850) 209.
140. Ch. Kittel, *Introduction to Solid State Physics*, 4th ed., Wiley, New York, 1976.
141. B. Boots, A. Okabe, and K. Sugihara, *Spatial Tessellation. Concepts and Applications of Voronoi Diagrams*, Wiley, New York, 1992.
142. G.F. Gauss, *J. Reine Andew. Math.* 20 (1840) 312.
143. G.F. Voronoi, *J. Reine Andew. Math.* 134 (1908) 198; 136 (1909) 67.
144. B.N. Delaunay, in *Proceedings of the International Math. Congress,* Toronto, August 1924, University of Toronto Press, 1928, p. 95; *Adv. Math. Sci.* 3 (1937) 16; 4 (1938) 102.
145. L.F. Toth, *Lagerungen in der ebene auf der Kugel und im Raum*, Springer-Verlag, Berlin, 1953.
146. R.J. Wilson, *Introduction to Graph Theory*, Oliver and Boyd, Edinburgh, 1972.
147. O. Ore, *Theory of Graphs*, American Math. Society, Providence, RI, 1962.
148. V.P. Voloshin, N.N. Medvedev, V.B. Fenelonov, and V.N. Parmon, *Doklady RAN*, 364 (1999) 337 (in Russian).
149. P. Barret, *Cinétique Hétérogéne*, Gauthier-Villars, Paris, 1973.
150. W.A. Johnson and R.F. Mehl, *Trans. Am. Inst. Min. Metall. Petrol. Eng.* 135 (1939) 416.
151. K.L. Mampel, *Z. Phys. Chem. A* 187 (1940) 43, 235.
152. J.L. Meijering, *Philips Res. Rep.* 8 (1953) 270.
153. N.N. Medvedev, in *Voronoi Impact of Modern Science*, P. Engel and H. Syta, Eds., Inst. Mathematic, Kiev State University, Kiev, 1998.
154. J. Brown and M. Sambridge, *Monthly Nat.* 3 (1995) 73.
155. H. Freundlich, *Kolloid Z.* 46 (1928) 51.
156. C.J. Plank and L.C. Drake, *J. Colloid Sci.* 2 (1947) 399.

157. H.E. White and S.F. Walton, *J. Am. Ceram. Soc.* 20 (1937) 155.

158. E. Manegold, R. Hoffman, and K. Soff, *Kolloid Z.* 56 (1931) 142.

159. K.A. Dawson, *Pure Appl. Chem.* 64 (1992) 1589.

160. S. Man, Ed., *Biomimetic Materials Chemistry*, VCH, New York, 1996.

161. S. Anderson, S.T. Hyde, K. Larsson, and S. Lidin, *Chem. Rev.* 88 (1988) 221.

162. D.H. Dollimore and G.R. Heal, in *Pore Structure and Properties of Materials (Proc. Int. Symp. Rilem/IUPAC)* Part 1, Academia, Prague, 1973, p. A73.

163. R.G. Avery and J.D.F. Ramsay, *J. Colloid Interface Sci.* 42 (1973) 597.

164. H.P. Meissner, A.S. Michaels, and R. Kaiser, *Ind. Eng. Chem. Proc. Des. Div.* 3 (1964) 202.

165. L.K. Frevel and L.J. Kressley, *Anal. Chem.* 35 (1963) 1492.

166. H. Heesch and F. Laves, *Z. Krist.* 85 (1933) 443.

167. I. Smalley, *Math. Mag.* 36 (1963) 295.

168. S. Kruer, *Trans. Faraday Soc.* 54 (1958) 1758.

169. K. Ridgway and K.J. Tarbuck, *Br. Chem. Eng.* 12 (1967) 384.

170. J. van de Lagemaat, K.D. Benkstein, and A.J. Frank, *J. Phys. Chem. B* 105 (2001) 2433.

171. F.F. Preparate and M.L. Shamos, *Computational Geometry, An Introduction*, Springer, Heidelberg, 1958.

172. J. Lindsey, *Math.* 33 (1986) 137.

173. D.H. Muder, *Proc. Lond. Math. Soc.* 56 (1988) 29.

174. H.S.M. Coxeter, *Introduction to Geometry*, Wiley, New York, 1961; *Proc. Symp. Pure Math. Am. Math. Soc.* 7 (1963) 53.

175. A.I. Kitaigorodskii, *Molecular Crystals*, Nauka, Moscow, 1971.

176. A.E. Larson and D.G. Grier, *Nature*, 385 (1997) 230.

177. C.A. Murray, *Nature*, 385 (1997) 203.

178. A. Gervois, M. Lichtenberg, L. Oger, and E. Guyon, *J. Phys. A Math. Gen.* 22 (1989) 2119.

179. W.O. Smith, P.D. Foote, and P.F. Busang, *Phys. Rev.* 34 (1929) 1271.

180. J.D. Bernal and J. Mason, *Nature* 188 (1960) 910.

181. J.D. Bernal, J. Mason, and K.R. Knight, *Nature* 194 (1962) 110.

182. J.D. Bernal, *Proc. Roy. Inst. Gr. Br.* 37 (1959) 355; *Nature* 183 (1959) 141; 185 (1960) 68; *Proc. Roy. Soc. A* 280 (1964) 299.

183. J.W. Marvin, *Am. J. Botany* 26 (1939) 280.

184. J.L. Finney, *J. Phys. (Paris)* 36 (1975) C2-1; *Roy. Soc. Lond.* 319 (1970) 479, 495.

185. J.A. Dodds, *J. Colloid Interface Sci.* 77 (1980) 317.

186. K.J. Mysels, K. Shinoda, and S. Frenkel, *Soap Films. Studies of Their Thinning and Bibliography*, Pergamon Press, New York, 1959.

187. Manegold, *Schaum. Strassenbau, Chemie und Technik*, Heidelberg, 1953.

188. J. Plateau, *Mem. Acad. Roy. Soc. Belgique*, 33 (1861) 6th series.

189. W. Thomson (Kelvin), *Philos. Mag.* 24 (1887) 503; *Proc. Roy. Soc.* 5 (1884) 333.

190. J.W. Gibbs, *Collected Works (Russian ed.)*, Gostechnisdat, Moscow, 1982; *Scientific Papers, v.1*, New York, 1961.

191. L. Oger, A. Gervois, J.P. Troadec, and N. Rivier, *Philos. Mag. B* 74 (1996) 177.

192. C.H. Bennett, *J. Appl. Phys.* 43 (1972) 2727.

193. D.J. Adams and A.J. Matheson, *J. Chem. Phys.* 56 (1972) 1989.

194. E.J.W. Whittaker, *J. Non-Crystal Solids* 43 (1978) 293.

195. H.J. Frost, *Acta Met.* 30 (1982) 889.

196. B.B. Mandelbrot, *Fractals: Form, Chance, and Dimension*, Freeman, San Francisco, 1977; *The Fractal Geometry of Nature*, Freeman, San Francisco, 1982.

197. F. Hausdorff, *Dimension und Ausseres Mass, Match. Ann.* 79 (1919) 157.

198. L.F. Richardson, *Weather Prediction by Numerical Process*, Cambridge University Press, 1922.

199. J.C. Graf, *Powder Technol.* 67 (1991) 83.

200. A. Bunde and H. Halvin, Eds., *Fractals and Disordered Systems*, Springer, Berlin, 1991.

201. P. Meakin, *Fractals, Scaling and Growth Far from Equilibrium*, Cambridge University Press, Cambridge, 1998.

202. M. Bransley, *Fractal Everywhere*, Academic Press, Boston, 1988.

203. J. Feder, *Fractals*, Plenum Press, New York, 1988.

204. M. Smoluchowski, *Graetz's Handb. D. Elektrizitai u. Magnetismus* 2 (1914) 366.
205. T.A. Witten and L.M. Sander, *Phys. Rev. Lett.* 47 (1980) 1400; *Phys. Rev. B* 27 (1983) 5696.
206. St.C. Pencea and M. Dumitrascu, in *Fractal Aspects of Materials*, F. Family, P. Meakin, B. Sapoval, and R. Wool, Eds., *Mat. Res. Soc. Proc.*, Pittsburg, PA, 1995, p. 373.
207. F. Family and D.P. Landau, Eds., *Kinetics of Aggregtion and Gelation*, North-Holland, Amsterdam, 1984.
208. H. van Damme, P. Levitz, L. Gatineau, J.F. Alcover, and J.J. Fripiat, *J. Colloid Interface Sci.* 122 (1988) 1.
209. K. Falconer, *Fractal Geometry: Mathematical Foundations and Application*, Wiley, New York, 1990.
210. R. Jullien and R. Botet, *Aggregation and Fractal Aggregates*, World Scientific, Singapore, 1987.
211. B.H. Kaye, *Powder Technol.* 21 (1978) 1; *Direct Characterization of Fine Particles*, Wiley, New York, 1981, pp. 367–378.
212. D. Anvir, D. Farin, and P. Pfeifer, *J. Chem. Phys.* 79 (1983) 3558, 3566; *Surf. Sci.* 126 (1983) 569; *J. Stat. Phys.* 36 (1984) 699; *J. Colloid Interface Sci.* 103 (1985) 112; *Langmuir* 1 (1985) 399.
213. D. Farin and D. Anvir, in *Characterization of Porous Solids*, K.K. Unger et al., Eds., Elsevier, Amsterdam, 1988, p. 421.
214. D. Anvir, Ed., *The Fractal Approach to Heterogeneous Chemistry*, Wiley, New York, 1990.
215. F. Brochard, *J. Physique* 46 (1985) 2117.
216. C. Fairbridge, A.D. Palmer, S.H. Ng, and E. Furimsky, *Fuel*, 65 (1986) 1789; 66 (1987) 688.
217. H. Bale and P. Schmidt, *Phys. Rev. Lett.* 53 (1984) 596.
218. A.V. Kiselev, *Disk. Farad. Sopc.* 40 (1965) 205.
219. A.Y. Meyer, D. Farin, and D. Anvir, *J. Am. Chem. Soc.* 108 (1986) 7897.
220. J. Fatt, *Trans. AIME* 207 (1956) 144, 160, 164.
221. G.G. Dodds and O.G. Kiel, *J. Phys. Chem.* 63 (1959) 1464.
222. A.C. Payatakis, Chi Tien and R.M. Turian, *AIChEJ* 19 (1973) 58.
223. F.A.L. Dullien, *AIChEJ* 21 (1975) 299.
224. O.S. Ksenjheck. *Z. Phys. Chem.* 37 (1963) 1297.
225. S.R. Broadbent and J.M. Hammersley, *Proc. Cambridge Philos. Soc.* 53 (1957) 629, 642.
226. R.G. Larson, L.E. Scriven, and H.T. Davis, *Chem. Eng. Sci.* 36 (1981) 57, 75.
227. Sh. Vossoughl and F.A. Seyer, *Ind. Eng. Chem. Fund.* 23 (1984) 64.
228. V.K.S. Shante and S. Kirpatric, *Adv. Phys.* 20 (1971) 315.
229. S. Kirpatric, *Rev. Mod. Phys.* 45 (1973) 574.
230. J.W. Essam, *Rep. Prog. Phys.* 43 (1980) 833.
231. M. Yanuka, *J. Colloid Interface Sci.* 127(1989) 35, 48; 134 (1990) 198.
232. H.-P. Wittmann, K. Kremer, and K. Binder, *J. Chem. Phys.* 96 (1992) 6291.
233. G. Grimmett, *Percolation*, Springer, Berlin, 1989.
234. D. Stauffer and A. Aharony, *Introduction to Percolation Theory*, Taylor & Francis, London, 1992.
235. P.J. Flory, *J. Am. Chem. Soc.* 63 (1941) 3083.
236. W.H. Stockmayer, *J. Chem. Phys.* 11 (1943) 45.
237. E.W. Washburn, *Proc. Natl. Acad. Sci.* (1921) 115.
238. H. Scher and R. Zallen, *J. Chem. Phys.* 53 (1970) 3759.
239. H.E. Stanley, *J. Phys. A* 11 (1977) L211.
240. M.E. Fisher, in *Critical Phenomena: Enrico Fermi Summer School*, M.S. Green, Ed., Academic Press, New York, 1971.
241. M. Levinshtein, M.S. Shur, and E.L. Efros, *Soviet Phys.-JETP* 42 (1976) 1120.
242. P.M. Adler, *Porous Media: Geometry and Transport*, Butterworth- Heinemann, Stoneham, MA, 1992.
243. M.A. Ioannidis and L. Chatzis, *J. Colloid Interface Sci.* 229 (2000) 323.
244. M. Sahimi, *Flow and Transport in Porous Media and Fractured Rock*, VCH, Weinheim, 1995.
245. J. Ghassemzadeh and M. Sahimi, *Chem. Eng. Sci.* 59 (2004) 2265.
246. G. Dagan, *Flow and Transport in Porous Formation*, Springer, New York, 1989.
247. L.W. Gelhar, *Stochastic Subsurface Hydrology*, Prentice-Hall, Englewood Cliff, NJ, 1993.
248. *Chem. Eng. Sci. (Special Issue)*, 57 (2002), 2123–322, 4255–4427; 58 (2003) 4757–4930.
249. E.A. Mason and A.P. Malinauskas, *Gas Transport in Porous Media: The Dusty-Gas Model*, Elsevier, Amsterdam, 1983.
250. S.P. Rigby and S. Daut, *Adv. Colloid Interface Sci.* 98 (2002) 87.

251. G.S. Armatas and P.J. Pomonis, *Chem. Eng. Sci.* 59 (2004) 5735.

252. R.B. Evans III, G.M. Watson and E.A. Mason, *J. Chem. Phys.* 35 (1961) 2076; 36 (1962) 1984.

253. A. Burghardt, *Chem. Eng. Process* 21 (1986) 229.

254. A. Kolmogoroff, *Bull. Acad. Sci. USSR, Reg. Math. Natur. Sci.* (1937) 355.

255. H.A.M. van Eekelen, *J. Catal.* 29 (1973) 75.

256. B.I. Shklovskii and A.L. Efros, *Electron Properties of Doped Semiconductors*, Nauka, Moscow, 1979 (in Russian).

257. N.V. Dalton, C. Domb, and M.F. Sykas, *Proc. Phys. Soc.* 83 (1964) 496.

CHAPTER 9 PROBLEMS

Problem 1

Catalytic activity of a unit of volume of catalytic reactor, W, is expressed as [5]

$$W = aAX\vartheta\Delta$$

where a is the specific catalytic activity and X the active part of the surface. All other textural parameters were introduced in the Introduction. How significant is the decrease of W if the specific catalytic activity is constant, but each of textural parameters in Equation 9.87 decreases by (a) 10%, (b) 30%?

Problem 2

This problem considers formation of a vapor bubble in water. The surface tension of water at 20°C is $\sigma = 72$ mN/m. Calculate the pressure in the bubble radius of (a) 1000 nm, (b) 10 nm, and (c) 1 nm.

Problem 3

The microporous active carbon was studied by nitrogen adsorption at 77 K. According to obtained data the multipoint BET method gave 771 m²/g; however, CM (t-method) method gave 123 m²/g. Explain the difference in the results of BET and comparative methods.

Problem 4

Consider a granule of a catalyst formed of nonporous particles (partition 1) and the pores (partition 2) between the particles (for illustration, consider all particles and all pores in a granule shown in Figure 9.17). The apparent density, δ, of the whole granule (according to Equation 9.25) is equal to the mass of a granule related to its volume. How is the porosity, ε, of the granule related to the apparent density of the granule, δ, and the true density, ρ, of a solid phase?

Problem 5

Consider a nonporous catalyst having an active component of a complex (unknown) composition, deposited over a known support (known parameters: the true density of a support and its partial mass, m_S). We have no information on the type of active component, only its partial mass, $m_C = 1 - m_S$. Also known is the density of solid phase of a catalyst as a whole. Allocate the partitions for this PS, and determine: (a) a partial volume of the active component; (b) the true density of the active component.

Problem 6

Let us complicate Problem 5 assuming that the supported catalyst is porous. Allocate the partitions for this PS, and determine (a) a partial volume of the active component and (b) the true density of the active component.

Problem 7

Let us consider a porous bed of biporous granules (Figure 9.17a) in a catalytic reactor (Figure 9.17b). Allocate the partitions, and determine the overall porosity of the bed (including pores between granules, pores between aggregates in granules, and pores inside the aggregates).

Hint: Use an approach of enclosed partitioning. At each level of partitioning divide a system into the particles (porous or nonporous) and pores between these particles. For a mathematical consideration it is convenient to substitute a real multilevel PS with a set of systems of particle–pore type in the sense of the previous sentence.

Problem 8

Let us consider a supported catalyst of two components: a support and an active component (see Problem 5). Adsorption of N_2 at 77 K gave a total surface area of the supported catalyst, A_{SC}. Adsorption of CO_2 at 273 K is different on support and active component, but one knows the reference values of $q_S(p)$ and $q_C(p)$, which were measured over pure support, and pure powder of active component with known surface areas, correspondingly. Propose a method of measuring of the surface area of the active component in the supported catalyst.

Hint: Use an approach of partitioning, and a principle of adsorption additivity (overall adsorption is the sum of adsorption over accessible support and active component surfaces).

Problem 9

The surface of a commercial catalyst ($A_0=100$ m²/g, $\rho_0= 3.2$ g/cm³) after exploitation in reaction covered with a layer of carbon ($\rho_C=1.7$ g/cm³) of thickness $l = 1$ nm. Assuming uniform coating with carbon of spherical particles of the catalyst calculate the resulting surface area of a carbonized catalyst.

Problem 10

Calculate the values of porosity $\varepsilon(n_P)$ depending on coordination number for a packing of particles with $3 \leq n_P \leq 12$ and compare these values with the correlation equation $\varepsilon=2.42/n_P$.

Problem 11

There is a routine procedure to form the transport pores in catalysts: catalyst is mixed with some additive, which can be burned off after preparation. Consider the catalyst with true density $\rho_C = 3.3$ g/cm³ and the additive (carbon black) with true density of $\rho_A= 1.2$ g/cm³. Calculate the amount of additive necessary to form the interlinked system of transport pores.

Understanding Catalytic Reaction Mechanisms: Surface Science Studies of Heterogeneous Catalysts

W.T. Wallace and D. Wayne Goodman

CONTENTS

10.1   Introduction 337
      10.1.1   Studies of CO Hydrogenation on Single-Crystal Surfaces 338
            10.1.1.1   Effects of Poisons and Promoters on CO Methanation 339
      10.1.2   CO Oxidation on Single-Crystal Surfaces 340
      10.1.3   Bimetallic Surfaces 340
      10.1.4   Experimental Techniques 344
10.2   Model Catalysts 344
      10.2.1   Thin-Film Growth 345
            10.2.1.1 TiO_2 Thin Films 345
            10.2.1.2   Amorphous SiO_2 Films 345
            10.2.1.3   Crystalline SiO_2 Films 348
            10.2.1.4   Highly Defective TiO_x Films 348
      10.2.2   Supported Metal Clusters (Au as an Example) 349
10.3   *In Situ* Studies 355
      10.3.1   Vibrational Spectroscopy 355
      10.3.2   *In Situ* STM 363
      10.3.3   Elevated Pressure XPS 363
10.4   Conclusions 368
Acknowledgments 368
References 368
Chapter 10 Questions 372

10.1   INTRODUCTION

The fields of heterogeneous catalysis and surface science have long been intertwined. Whether it is the studies of Faraday on oxidation reactions over platinum surfaces, Langmuir's studies of the surface properties of catalysts, or even work performed in the present day on the abilities of different surfaces to act as catalysts, advances in surface science often lead to the development of new heterogeneous catalysts, just as the discovery of new catalytic properties pushes surface scientists to discern the origin of these properties.[1,2] Surface science has played a tremendous role in determining the mechanisms of bond breaking and molecular rearrangements occurring on catalysts, as well as in determining the determination of the effects of surface structure on catalytic activity.

10.1.1 Studies of CO Hydrogenation on Single-Crystal Surfaces

Early surface-science studies related to catalysis used metal single-crystal surfaces to mimic the active sites of industrial catalysts.[3-7] One of the most widely studied reactions was the catalytic hydrogenation of CO on various metals.[7,8] Using an apparatus combining an ultrahigh-vacuum surface analysis chamber with a high-pressure reaction chamber, a group at the National Bureau of Standards (now NIST) studied this reaction on different planes of nickel. Figure 10.1 shows the rates of CH_4 production on Ni(100) and Ni(111) and compares with those determined from several different supported Ni catalysts by Vannice.[9] As is quite obvious, the rates compare very favorably within experimental error, showing that this reaction is structure-insensitive. To determine the reaction mechanism, Auger electron spectroscopy (AES) was used to study the surface following the reaction. A carbon species was observed on the surface, and analysis of its line shape showed it to be of the carbidic form. This carbidic species could also be produced by heating the sample in a CO background. By observing the rate of formation of the carbidic species and comparing it with the rate of CH_4 formation in a reaction mixture, it was determined that the rates were essentially the same. This fact suggests that the initial step in CH_4 synthesis requires the breaking of the CO bond and the formation of an active surface carbon species. Furthermore, it was found that the rates for carbide formation, carbide removal in H_2, and methane formation were very similar, indicating that a delicate balance of carbide formation and removal determines the reaction rate. The similarity in

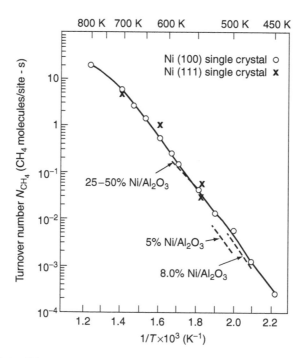

Figure 10.1 A comparison of the rate of methane synthesis over two different nickel single-crystal catalysts and supported Ni/alumina catalysts at 120 torr total reactant pressure. (Reprinted from Goodman, D.W., *J. Vac. Sci. Technol.*, 20, 522–526, 1982. With permission from the American Institute of Physics.)

the rate of carbide formation and methane formation from a CO–H_2 mixture was also observed on the Ru(110) and (001) surfaces,[10,11] as well as on polycrystalline Fe.[12] These striking similarities led to the proposal that a similar mechanism could be extended to all transition metals of interest in methanation and Fischer–Tropsch catalysis.

10.1.1.1 *Effects of Poisons and Promoters on CO Methanation*

The presence of impurities on a metal catalyst, whether intentionally present or not, can produce dramatic changes in the activity, selectivity, and resistance to poisoning of the catalyst. CO methanation studies similar to those described above were carried out on surfaces containing sulfur and phosphorous impurities.[13–15] As can be seen in Figure 10.2, the rate of methane production is affected dramatically by the introduction of these impurities. In the case of sulfur, low coverages result in a sharp decrease in the catalytic activity, while increasing sulfur coverages show few further effects. The initial attenuation of the catalytic activity of the nickel suggested that ten or more nickel sites were being affected by a single sulfur atom. This could be interpreted as an electronic effect or an ensemble effect, in which a certain number of surface atoms are necessary for the reaction to occur. If electronic effects are dominating the poisoning process, then a change in the electronegativity of the poison should result in a change in the reaction rate. The substitution of phosphorus for sulfur as a poison results in a noticeable change in the poisoning effect at low coverages. As phosphorus is less electronegative than sulfur, it affects fewer nickel sites surrounding it. This suggests that electronic effects, rather than simple site blocking, are dominant in the catalytic deactivation of nickel by sulfur, similar to effects seen on rhodium and ruthenium.[4]

As electronegative elements were shown to decrease the rate of methane synthesis on metal surfaces, it may be expected that electropositive elements would increase the rate of the catalytic reaction. Studies of CO methanation on nickel surfaces containing potassium show that this is not the case.[15,16] Evidence of this can be seen in Figure 10.3. As is apparent, increasing potassium coverage results in lower reaction rates. The kinetic studies also indicated that the presence of potassium did not affect the apparent activation energy of the reaction. However, the addition of potassium did cause a noticeable increase in the amount of active carbon on the nickel surface. Presumably, the donation of electron density from the potassium to the surface aided in CO bond scission. This assumption is confirmed by the fact that the dissociation probability increases in the order Na, K, and Cs.[5] As the electronegativity of the impurity decreases, the amount of electron donation increases along with CO dissociation.

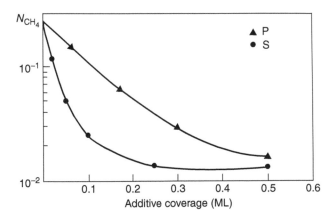

Figure 10.2 Methanation rate as a function of phosphorus and sulfur coverage on a Ni(100) catalyst. Pressure = 120 torr, H_2/CO = 4. Reaction temperature = 600 K. (From Goodman, D.W., *Appl. Surf. Sci.* 19, 1–13, 1984. Used with permission from Elsevier Scientific Publishers.)

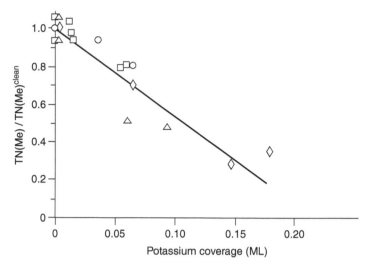

Figure 10.3   Relative methanation reaction rate as a function of potassium coverage at various reaction conditions: ($\diamond$) P_{CO}= 1.0 torr, P_{H2} = 99.9 torr, T = 600 K; ($\bigcirc$) P_{CO} = 24 torr, P_{H2} = 97.6 torr, T = 538 K; ($\triangle$) P_{CO} = 24 torr, P_{H2} = 96 torr, T = 600 K; ($\square$) P_{CO} = 24 torr, P_{H2} = 96 torr, T=594 K. (From Campbell, C.T. and Goodman, D.W., *Surf. Sci.* 123, 413–426, 1982. Used with permission from Elsevier Scientific Publishers.)

10.1.2 CO Oxidation on Single-Crystal Surfaces

Carbon monoxide oxidation is a relatively simple reaction, and generally its structurally insensitive nature makes it an ideal model of heterogeneous catalytic reactions. Each of the important mechanistic steps of this reaction, such as reactant adsorption and desorption, surface reaction, and desorption of products, has been studied extensively using modern surface-science techniques.[17] The structure insensitivity of this reaction is illustrated in Figure 10.4. Here, carbon dioxide turnover frequencies over Rh(111) and Rh(100) surfaces are compared with supported Rh catalysts.[3] As with CO hydrogenation on nickel, it is readily apparent that, not only does the choice of surface plane matters, but also the size of the active species.[18–21] Studies of this system also indicated that, under the reaction conditions of Figure 10.4, the rhodium surface was covered with CO. This means that the reaction is limited by the desorption of carbon monoxide and the adsorption of oxygen.

Like Rh, CO oxidation over Ru surfaces is structure insensitive, as seen in Figure 10.5. However, postreaction studies on the Ru(0001) surface indicated that the highest rates of CO oxidation were achieved on those surfaces with the largest amount of surface oxygen remaining after reaction, in contrast to Rh(111) and Pd(110). Kinetics measured under less than optimal conditions for CO oxidation, i.e., the surface oxygen concentration was low, showed significant deviations from those obtained under optimal conditions, indicating the possible importance of a chemisorbed oxygen intermediate. Therefore, even for a simple reaction such as CO oxidation, it is sometimes dangerous to assume that similar metals will act similarly under reaction conditions.

10.1.3 Bimetallic Surfaces

The ability of bimetallic systems to enhance various reactions, by increasing the activity, selectivity, or both, has produced a great deal of interest in understanding the different roles and relative importance of ensemble and electronic effects. Deposition of one metal onto the single-crystal face of another provides an advantage by which the electronic and chemical properties of a well-defined bimetallic surface can be correlated with the atomic structure.[5,22,23] Besenbacher et al.[24] used this method to study steam reforming (the reverse of the CO methanation process) on Ni(111) surfaces

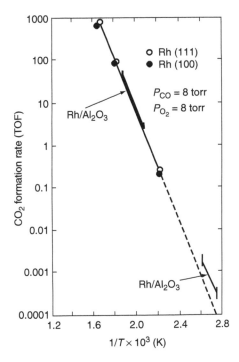

Figure 10.4 A comparison of the rates of CO oxidation by O_2 over two different rhodium single-crystal catalysts and supported Rh/alumina catalysts. (Reprinted from Goodman, D.W., *Chem. Rev.*, 95, 523–536, 1995. Copyright 1995. With permission from American Chemical Society.)

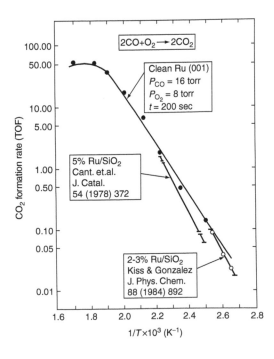

Figure 10.5 A comparison of the rates of CO_2 formation for a Ru(001) single crystal with two different supported Ru/silica catalysts. (Reprinted from Peden, C.H.F. and Goodman, D.W., *J. Phys. Chem.*, 90, 1360–1365, 1986. Copyright 1986. With permission from American Chemical Society.)

containing various amounts of Au. They found that the addition of a small amount of Au lowered the rate of CH$_4$ dissociation and, therefore, the overall rate of the steam reformation process. However, the formation of graphite (by which the surface is poisoned) occurred much less quickly on these alloy surfaces. It was found that carbon adsorption was favored at threefold Ni sites on the clean surface. At threefold sites near Au, though, the carbon was much less stable and more likely to react with oxygen to produce CO. This work showed that the addition of Au produced a less active, but more stable, nickel catalyst for steam reforming.

Rodriguez and Goodman[23] studied Pd, Ni, and Cu films deposited on different transition metal substrates to understand the nature of the bonding between surface metal species. A 100 K increase in the Pd desorption temperature was seen when changing from a Ru to a Ta substrate. This behavior can be seen in Figure 10.6A. This temperature-programmed desorption (TPD) data seemed to indicate that Pd was more strongly bound to substrates with less-occupied valence d bands. As the valence band of Pd is almost fully occupied, it was suggested that electron donor–acceptor interactions could explain the TPD results. Figure 10.6B presents x-ray photoelectron spectroscopy (XPS) results that seem to confirm this belief. Changes in the Pd(3d$_{5/2}$) core-level binding energy from the films with respect to the Pd(100) surface are presented in the figure. An increase in binding energy is a result of charge transfer between the Pd and the substrate, and the amount of charge transfer from Pd increases as the number of occupied states in the valence band of the substrate decreases. Similar results were also

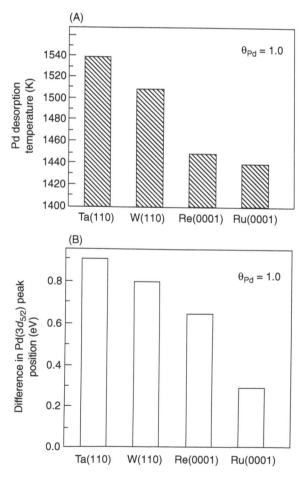

Figure 10.6 (A) Desorption temperature of a 1 ML Pd film from different metal substrates. (B) Change in Pd(3d$_{5/2}$) binding energy as a function of the metal substrate. The changes are referenced to the Pd(3d$_{5/2}$) binding energy of Pd(100). (Reprinted from Rodriguez, J.A. and Goodman, D.W., *Science*, 257, 897–903, 1992. Copyright 1992. With permission from AAAS.)

seen for Ni films grown on the same substrates. However, Cu shows quite different behavior when deposited on different substrates, as shown in Figure 10.7A. In this case, the strongest bonds formed between Cu and the substrates occur at the extremes of the transition series. Even though the behavior of Cu is different from Pd and Ni, the same explanation can be used to describe the interactions of Cu with the substrates. With a half-full 4s valence band, Cu interacts more strongly with metals in which the valence band is either almost empty or full. How the interaction actually occurs depends on the particular system. The core-level shifts shown in Figure 10.7B indicate that Cu donates charge to transition metals with almost empty valence bands and withdraws charge from those with almost full valence bands. Therefore, it appears that the formation of a heteronuclear metal–metal bond results in the flow of electron density toward the metal with a greater number of initial empty valence band states. This charge transfer can have dramatic effects of the chemical behavior of these systems and is also contrary to the behavior observed in bulk alloys. Thus, the effects of structure (surface vs. bulk) on the formation of a bimetallic bond cannot be underestimated.

Reports such as those described above show how surface structure and energetics can affect catalytic reactions, and more of these studies have been performed on a multitude of different surfaces possessing every manner of crystal face for almost any catalytic reaction, as described in detail by Somorjai[1] and Zaera.[2] This experimental work has led, in turn, to many theoretical treatments of the reactions in question. Mavrikakis and coworkers[25] recently presented a review of the theoretical techniques used in these studies, as well as a review of the latest activity in this field.

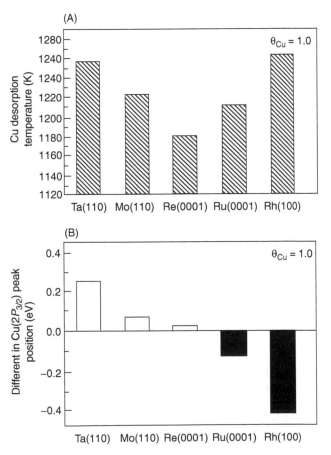

Figure 10.7 (A) Desorption temperature of a 1 ML Cu film from different metal substrates. (B) Change in $Cu(2p_{3/2})$ binding energy as a function of the metal substrate. The changes are referenced to the $Pd(2p_{3/2})$ binding energy of $Cu(100)$. (Reprinted with permission from Rodriguez, J.A. and Goodman, D.W., *Science*, 257, 897–903, 1992. Copyright 1992. With permission from AAAS.)

10.1.4 Experimental Techniques

There are numerous surface-sensitive techniques that can be applied to the study of catalyst surfaces; in fact, a complete treatment of these is beyond the scope of this discussion. Therefore, the reader is directed toward some excellent resources for a more complete discussion.[1,26–29] Here, we aim only to introduce some of the more popular techniques as well as to familiarize the reader with the alphabet soup of surface-science acronyms that will be used below.

Some of the most widely used techniques for determining surface cleanliness and structure (as well as to observe possible surface contaminations or reconstructions following reactions) are AES and low-energy electron diffraction (LEED). In AES, core-hole excitations are created by a beam of monoenergetic electrons (<10 keV). As excited atoms relax, they emit electrons with characteristic energies. These electrons can be analyzed to provide information on the near-surface chemical composition. Surface structure can be determined by LEED. In this technique, electrons of <500 eV are elastically backscattered from a surface and detected as a function of energy and angle. Changes in the surface structure owing to the presence of adsorbates or reconstructions lead to changes in the observed diffraction pattern.

Another technique that provides very surface-specific information is ISS. As noble gas ions (often He) with kinetic energies of only a few keV strike a surface, they lose energy through inelastic collisions with surface atoms. The energies of the outgoing ions are measured and compared with reference samples, providing information on the surface composition. As He ions are almost completely neutralized after passing through a single substrate layer, most of the ions that reach the detector are scattered from the outermost surface layer.

X-ray photoelectron spectroscopy also provides information on the chemical composition of a surface. An incoming photon causes electrons to be emitted from atomic core levels, which are then analyzed as a function of kinetic energy. The shifts of these core-level energies provide information about the chemical environment surrounding the excited atom. This information also includes changes in the oxidation state of the sample.

Information on the bonding of adsorbates to surfaces can be obtained by a variety of techniques. Two of the most popular vibrational techniques are infrared reflection absorption spectroscopy (IRAS) and electron energy loss spectroscopy (EELS). In IRAS, the absorption or emission of infrared (IR) radiation by the vibrational modes of adsorbates is monitored as a function of incoming photon energy. In EELS, monoenergetic electrons are scattered off of a surface, and the energy losses arising from excitation of vibrational modes are monitored as a function of angle and energy. In addition to providing information on the electronic structure of the surface, this technique can also provide details of the surface structure.

More detailed studies of surface structure can be carried out using scanning tunneling microscopy (STM). For the use of this technique, a metal tip is scanned over the surface, and the distance from the tip to the surface is determined by the tunneling current between them. Images of surfaces with subnanometer resolution are often obtained using this method.

10.2 MODEL CATALYSTS

As described earlier, metal single crystals have found wide use as model catalysts. The use of surface analytical techniques to study reactions on these surfaces has provided a wealth of information that has helped in the understanding of working catalysts. However, single crystals preclude the assessment of important catalyst variables such as cluster size and support-cluster interactions. Also, while single crystals can possess a variety of defects, such as steps, kinks, adatoms, and vacancies, it is impossible to mimic the tremendous amount of defects present on real catalysts using a single crystal. These drawbacks led to the development of model catalysts. In these systems, metal clusters are supported on metal oxide thin films. These thin oxide films are structurally and electronically similar to the corresponding bulk oxides, yet are thin enough to permit the use of electron spectroscopic techniques.[30] By preparing samples in this manner, traditional surface-science techniques can be used and, additionally, many more of the complexities of real catalysts can be introduced.

10.2.1 Thin-Film Growth

There are several ways to prepare thin films for use as model catalyst supports.[30,31] For the purposes of this review, we will point the reader toward other sources that discuss two of these methods: direct oxidation of a parent metal and selective oxidation of one component of a binary alloy.[32–34] The remaining method consists of the deposition and oxidation of a metal on a refractory metal substrate. This method has been used extensively in our group[30,31,35–51] and by others[33,52–68] and will be the focus of the discussion here. The choice of the metal substrate is important, as lattice mismatch between the film and the substrate will determine the level of crystallinity achieved during film growth.

10.2.1.1 TiO$_2$ Thin Films

Owing to its wide use as a catalyst support, titanium dioxide has long been a standard support for clusters used in surface-science studies.[37,69–82] However, because of its insulating nature, when TiO$_2$(110) single crystals are used, they must be made suitably conducting by sputtering the surface and annealing in vacuum, thereby making an n-type semiconductor. A way to circumvent this problem is through the growth of thin TiO$_2$ films. Oh et al.[47] prepared thin titania film (<100 Å) by evaporating Ti onto an Mo(100) surface in an oxygen background (5×10^{-7} torr). During deposition, the Mo substrate was kept between 500 and 700 K and the films were annealed between 900 and 1200 K in vacuum after depostion. This particular substrate was chosen because the lattice mismatch between Mo(100) surface and TiO$_2$(001) is only 4%, leading to the expectation of epitaxial growth. As can be seen in Figure 10.8A, XPS shows the presence of a peak corresponding to Ti^{4+} after deposition of Ti in an O$_2$ atmosphere. This indicates the presence of a stoichiometric thin film. However, further annealing of the film to 1200 K leads to the introduction of defects in the film, as can be seen by the appearance of shoulder peaks in Figure 10.8B. Subsequent ion scattering studies of these films before and after annealing confirm that the annealing step leads to the reappearance of the Mo substrate, which could arise because of clustering or faceting of the films or diffusion of Ti into the Mo substrate. STM images taken after annealing a film of the same thickness show that the Mo surface is no longer covered (Figure 10.9). Thicker titania films maintain their stoichiometric character after annealing, and ISS and STM indicate that the Mo substrate is still covered, though STM shows that the titania surface is made up of small crystallites of several nanometers in diameter.

10.2.1.2 Amorphous SiO$_2$ Films

Silica is also a standard industrial support, and, therefore, its use for model catalyst studies is quite desirable. Goodman and coworkers[83–85] formed amorphous silica films on Mo(110) and Mo(100) substrates by depositing Si on the surface in an oxygen background. Figure 10.10 presents the AES spectra obtained of silica thin films deposited at different oxygen pressures. After depositing silicon in a background containing no oxygen, only the Si peak at 91 eV can be seen, indicating elemental silicon. By increasing the oxygen pressure during deposition to 1×10^{-6} torr, a new peak at 76 eV appears, corresponding to silicon dioxide. A further increase to 4×10^{-6} torr during deposition results in the presence of only the silicon dioxide peak at 76 eV, with an Si/O AES ratio suggesting that the films are stoichiometric, i.e., SiO$_2$. An XPS investigation of the growth of these films suggested that, at a dosing rate of ~0.12 nm/min, deposition in an oxygen background of 2×10^{-5} torr, followed by annealing at 1300 K, would lead to a stoichiometric SiO$_2$ film. The x-ray photoelectron spectra can be seen in Figure 10.11, showing the effects of different annealing temperatures. The loss of the peak near 99.5 eV after annealing at 1300 K indicates a stoichiometric film containing no elemental silicon. Further studies, including vibrational and LEED, suggested that the

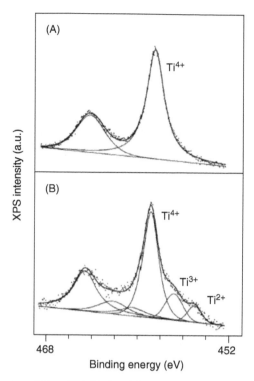

Figure 10.8 Ti 2p XP spectra for a 1.5 nm TiO$_2$/Mo(100) thin film collected at 300 K: (A) after preparation at 600 K; (B) after annealing to 1200 K. (Reprinted from Oh, W.S. et al., *J. Vac. Sci. Technol. A*, 15, 1710–1716, 1997. With permission from the American Institute of Physics.)

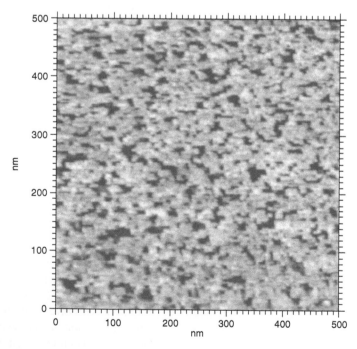

Figure 10.9 Room-temperature scanning tunneling microscopy (STM) image (500 × 500 nm) of a 1.5 nm TiO$_2$/Mo(100) thin film after annealing to 1200 K. After the annealing step, the Mo substrate is no longer completely covered. $V = +2$ V and $I = 0.5$ nA. (Reprinted from Oh, W.S. et al., *J. Vac. Sci. Technol. A*, 15, 1710–1716, 1997. With permission from the American Institute of Physics.)

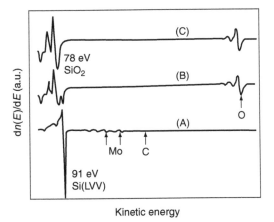

Kinetic energy

Figure 10.10 Auger electron spectra of SiO_2 thin films showing the effects of varying oxygen pressure: (A) no oxygen present during Si deposition, (B) 1×10^{-6} torr O_2, and (C) 4×10^{-6} torr O_2. (Adapted from Xu, X. and Goodman, D.W., *Appl. Phys. Lett.*, 61, 774–776, 1992.)

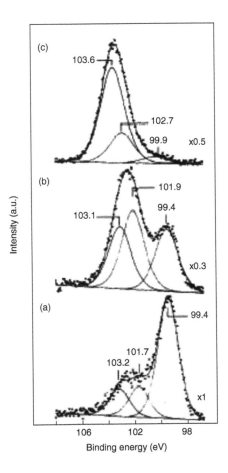

Figure 10.11 Si 2p XP spectra of 3.6 nm SiO_2 thin films grown on Mo(100). The background oxygen pressures during Si deposition were (a) 1×10^{-8} torr, (b) 5×10^{-8} torr, and (c) 2×10^{-5} torr. (From He, J.-W. et al., *Surf. Sci.*, 279, 119–126, 1992. Used with permission from Elsevier Scientific Publishers.)

structure of the films as deposited at low temperatures are disordered and consist of [SiO$_4$] units. Long-range [SiO$_4$] networks form after annealing to 1300 K, leading to a more ordered film.

10.2.1.3 Crystalline SiO$_2$ Films

While the amorphous silica films described above have found wide use for supported catalyst studies, their noncrystalline nature presents difficulties when attempting to control the structural aspects of model catalysts. To circumvent this problem, Freund and coworkers[54,56,57] developed a method by which they could grow crystalline ultrathin SiO$_2$ films. Using a modified version of their method, we have also produced these films.[35,43–46,49] Briefly, a Mo (112) single crystal was first oxidized in 1×10^{-7} torr O$_2$ at ~800 K for 5 min to produce a $p(2 \times 3)$-O structure. Approximately 1 monolayer (ML) of Si was then deposited on the oxygen-covered Mo surface. The Si-covered surface was then oxidized and annealed in O$_2$ (1×10^{-7} torr) at ~1150 K for 30 min. This step forms a highly ordered SiO$_2$ thin film that exhibits an atomically resolved STM image and a sharp LEED pattern with $c(2 \times 2)$ periodicity (Figure 10.12). The formation of a stoichiometric film was confirmed by the absence of Si^0 or Si^{2+} AES features. The types of defects on these films and their relative stabilities have been studied using Au cluster nucleation by Min et al.[43] They found that Au clusters preferentially nucleate at line defects upon deposition at room temperature, while nucleation occurred at step edges when deposition was carried out at higher temperatures.

10.2.1.4 Highly Defective TiO$_x$ Films

Recently, a new type of titania film, (8×2)-TiO$_x$, grown on the Mo(112) surface was found to be wetted by gold, leading to monolayer and bilayer Au films that showed unprecedented activity for CO oxidation.[86–88] This TiO$_x$ film could be prepared by two different methods. The first method requires the growth of the $c(2 \times 2)$-SiO$_2$ film described above. Approximately 1 to 1.5 ML of Ti was then deposited on the silica surface, with the titanium amount being estimated from the AES break point after direct deposition and oxidation on the Mo(112) surface. After depositing 1 to 1.5 ML of Ti onto the SiO$_2$ surface, the sample was oxidized at 800 K in 5×10^{-8} torr O$_2$ for 10 min. Figure 10.13 presents typical HREEL and AES spectra of the TiO$_x$–SiO$_2$ surface at different stages. Spectra obtained after oxidizing at 600 K (Figure 10.13Ab and Figure 10.13 Bb) are shown to illustrate the attenuation of Si and Mo AES lines and the Si–O phonon by the TiO$_x$ overlayer. The details of the TiO$_x$–SiO$_2$ interface interactions were discussed previously, where it was found that the annealing temperature plays a critical role in determining the properties of the mixed oxides.[89] The sample was then annealed at 1200 K in 5×10^{-8} torr O$_2$ for 10 min. The AES spectrum following this step (Figure 10.13Bd) showed that most of the

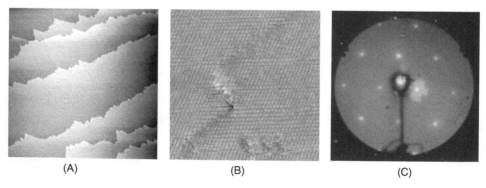

(A)   (B)   (C)

Figure 10.12   (A) STM image (100 × 100 nm) of a clean SiO$_2$ thin film grown on Mo(112) ($V = -1.7$ V and $I = 0.18$ nA). (B) STM image (25 × 25 nm) of the SiO$_2$ thin film. (C) Low-energy electron diffraction pattern of the SiO$_2$ thin film showing $c(2 \times 2)$ periodicity ($V = 56$ eV).

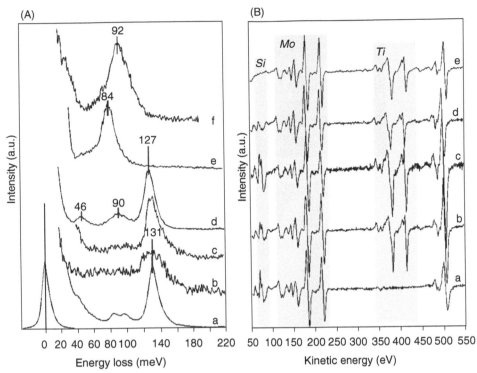

Figure 10.13 (A) HREEL and (B) Auger electron spectroscopy (AES) spectra for 1.5 monolayer (ML) TiO$_x$ on the SiO$_2$(ML)/Mo(112) surface following annealing at various temperatures: (a) SiO$_2$(ML)/Mo(112), (b) oxidized at 600 K, (c) oxidized at 800 K, (d) annealed at 1200 K, (e) annealed at 1400 K, and (f) 1 ML Ti oxidized at 800 K. (From Chen, M.-S. et al., *Surf. Sci.*, 581, L115–L121, 2005. Used with permission from Elsevier Scientific Publishers.)

SiO$_2$ was decomposed and desorbed, while an intense Si–O phonon in HREELS (Figure 10.13Ad) was still present. This is because of the much stronger Si–O phonon intensity compared to the Ti–O phonon. Subsequent annealing at 1400 K in 1×10^{-8} O$_2$ for an additional 5 min completely removes the SiO$_2$ film and any residual Si, as reflected by the lack of an Si AES line (see Figure 10.13Be) and no Si–O phonon (Figure 10.13Ae). The TiO$_x$ film formed exhibits a very sharp (8×2) LEED pattern (Figure 10.14A), and a very smooth and well-ordered surface as displayed from the large terraces in STM images (Figure 10.14B). The possible structure of this surface is also shown in Figure 10.14C.

The (8×2)-TiO$_x$ film can also be synthesized by the stepwise direct deposition of Ti onto an oxygen-covered Mo(112) surface followed by subsequent oxidation–annealing cycles. However, the quality and reproducibility are not comparable with growth on the SiO$_2$ film. In either case, analysis of the HREELS and XPS results indicate that the oxidation state of the Ti is probably +3. This reduced Ti state is apparently responsible for the ability of Au and other metals to wet the surface.

10.2.2 Supported Metal Clusters (Au as an Example)

Owing to the sheer number of studies that have been carried out to understand the properties of supported metal clusters, it is impossible to discuss all of them here. Several excellent reviews can provide more information in this area.[36,53,64] For the purposes of this discussion, we simply focus on some of the work from our laboratory regarding supported Au clusters.

Following the pioneering work of Haruta and coworkers,[90–92] numerous studies of Au as a catalyst have been reported. In their studies, Haruta and coworkers found that small Au clusters (<5 nm), highly dispersed on reducible metal-oxide supports, are quite active for a variety of catalytic

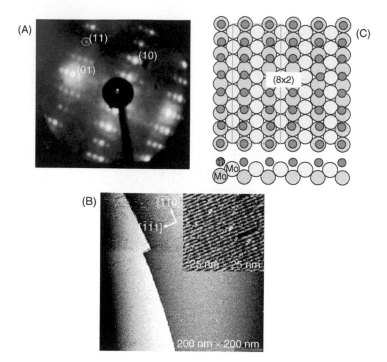

Figure 10.14 (A) Low-energy electron diffraction (LEED) pattern for (8×2) TiO$_x$/Mo(112). (B) STM images (200 × 200 nm) of the (8×2) TiO$_x$ Mo(112) surface. I = 0.18 nA, V = −1.7 V (inset: 25 × 25 nm, I = 0.18 nA, V = +1.2 V). (C) Structural model, top and side view of the (8×2) TiO$_x$ Mo(112) surface. The oxygen atoms are omitted for clarity. (Reprinted from Chen, M.S. and Goodman, D.W., *Science*, 306, 525–555, 2004; Chen, M.-S. et al., *Surf. Sci.*, 581, L115–L121, 2005. Copyright 2004. With permission from AAAS.)

reactions at or below room temperature, including the oxidation of various hydrocarbons and carbon monoxide. They also found that the catalytic properties depended critically on the choice of oxide support, the preparation method, and, most importantly, the size of the Au clusters. The need to understand the microscopic underpinnings of these surprising results have led to a number of studies on Au model catalysts, both theoretically and experimentally.[31,37,43,44,48,53,67,71–75,77,78,80,93–101] Goodman and coworkers[96,102] studied Au clusters on TiO$_2$(110) using STM and tunneling spectroscopy, which allows the measurement of the surface electronic states. They found that in small Au clusters, the size of the bandgap depended critically on the size of the Au clusters (Figure 10.15). For example, a cluster with a diameter of 4 nm and a height of 1.5 nm, corresponding to approximately five atomic layers, is almost completely metallic in nature. However, for a cluster with dimensions of 15 × 0.6 nm (~2 atomic layers), the bandgap is >1.5 eV, even more than the bandgap of the TiO$_2$ substrate.

As mentioned above, Haruta and coworkers found that the size of the Au clusters was critical in determining the catalytic activity. The work of Valden et al.[96] helped to explain this fact, as shown in Figure 10.16. Here, it can be seen that the peak of catalytic activity for CO oxidation corresponds to an increase in the bandgap of the clusters.

Goodman and coworkers[80] also noted that, despite the high initial activity of the Au-based catalysts for CO oxidation, the activity decreased essentially to zero after exposing the sample to a reaction mixture for 120 min: An answer to the question of what can be seen in the STM images of Figure 10.17. After exposure of the sample to reactant gases, the average size of the Au clusters increases to ~3.6 × 1.4 nm. From the STS data presented above, this would indicate that the Au clusters are in the size range possessing metallic properties and would, therefore, no longer be expected to maintain their catalytic activity.

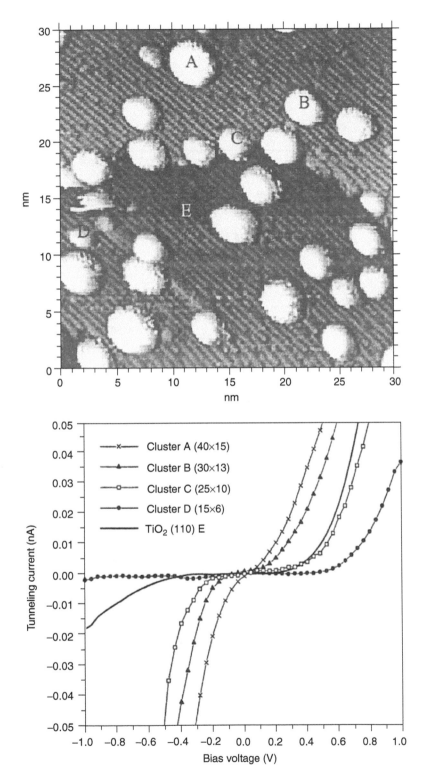

Figure 10.15 CCT STM image (I = 2.0 nA, V = +2.0 V) and the corresponding STS data acquired for Au clusters of varying sizes on the $TiO_2(110)$-(1×1) surface. The STS curve of the bare titania surface is also shown. (Reprinted from Meier, D.C. et al., in *Surface Chemistry and Catalysis*, A.F. Carley et al., Eds., Kluwer, New York, 2002, pp. 147–189. With permission from Springer Science and Business Media.)

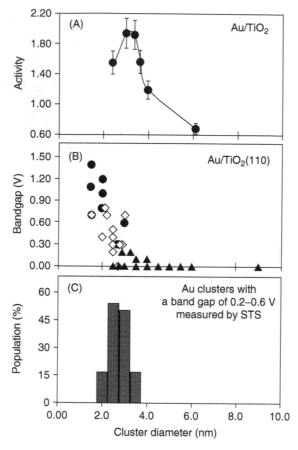

Figure 10.16 (A) Activity for CO oxidation at 350 K as a function of the Au cluster size supported on TiO$_2$(110)-(1×1), assuming total dispersion of the Au. (B) Cluster bandgap measured by STS as a function of the Au cluster size supported on TiO$_2$(110)-(1×1). The bandgaps were obtained while the corresponding topographic scan was acquired on various Au coverages ranging from 0.2 to 4.0 ML. (•) Two-dimensional (2D) clusters; (□) 3D clusters, two atom layers in height;(▲) 3D clusters with three atom layers or greater in height. (C) Relative population of the Au clusters (two atom layers in height) that exhibited a bandgap of 0.2 to 0.6 V as measured by STS from Au/TiO$_2$(110). (Reprinted from Valden, M. et al., *Science*, 281, 1647–1650, 2004. Copyright 1998. With permission from AAAS.)

As long as Au catalysts could not maintain their activity owing to sintering, their wide application was prevented, regardless of their high initial activity. Therefore, an attempt was made to alleviate the problem of cluster sintering under reaction conditions.[44] Owing to its wide use in industrial catalysts, silica was chosen as a starting support. The crystalline SiO$_2$ film described above was grown and ~0.4 ML Au was deposited on it. Compared to reducible oxide supports, SiO$_2$ is known to have a relatively weak metal-support interaction with noble metals.[103] In this case, sintering, owing to increased temperatures or pressures, was anticipated to be quite pronounced. For Au clusters supported on SiO$_2$ this was indeed the case, where thermally induced sintering was observed. Room-temperature deposition of Au generally leads to cluster nucleation at line defects, whereas for annealing to 700 to 850 K, decoration of line defects no longer occurs. A dramatic decrease in cluster density and an increase in cluster size accompany the diffusion of clusters to the step edges. STM images of this sintering process can be seen in Figure 10.18A.

Gold clusters supported on a mixed TiO$_2$–SiO$_2$ surface show activity for propylene epoxidation comparable to TiO$_2$-supported Au catalysts but with enhanced stability.[104] Therefore, it was decided

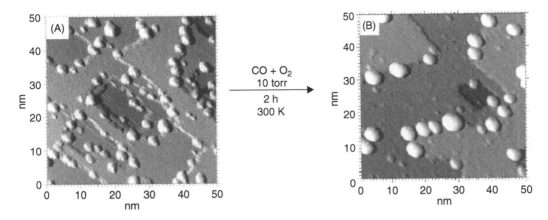

Figure 10.17 CCT STM images (50 × 50 nm) of Au/TiO₂(110)-(1×1) (A) before and (B) after 120 min of CO/O₂ (2:1) exposure at 10 torr. (Reprinted from Valden, M. et al., *Science*, 281, 1647–1650, 2004. Copyright 1998. With permission from AAAS.)

that this type of surface presented a good starting point in an attempt to make sinter-resistant supports. Various amounts of titanium were added to the silica thin film, oxidized, and annealed. More information on the effects of these different treatments can be found elsewhere.[44] After the addition of ~17% Ti to the surface, elongated TiO_x islands were seen on the surface and a maximum in the Au cluster number density was achieved upon Au deposition. Even more exciting was the lack of movement of the Au clusters after annealing the surface to 850 K. Figure 10.18B shows the effects of this annealing treatment on the Au/mixed oxide sample. In contrast to the bare silica surface (and surfaces containing less Ti), there is no apparent change in this surface upon annealing. This behavior is even more apparent in Figure 10.19, in which the number of clusters seen after annealing is normalized to the number seen before annealing. The number density of the Au clusters on the bare SiO_2 surface decreases by 60% after annealing, while a surface containing ~8% Ti shows a decrease of almost 50% in the number of Au clusters. In contrast, a surface containing 17% Ti shows little to no change in the number of clusters after annealing. Also interesting is the information presented to the far right of the figure. Exposure of the Au/TiO$_x$(17%)–SiO$_2$/Mo(112) sample to a 60 torr reaction mixture for 2 h at elevated temperatures also leads to no change in the number of clusters present. While the exact mechanism leading to the sinter-resistant nature of this support is unknown, it is nevertheless exciting and points toward a method of using highly active Au-based catalysts in industrial situations.

Spectroscopic studies have also been carried out to understand adsorption on supported Au clusters.[105–107] Lemire et al.[106,107] used TPD and IRAS to study CO adsorption on Au/FeO(111). They found, using TPD, that the desorption of CO is almost completely independent of particle thickness. Figure 10.20A presents their TPD results in which it can be seen that all samples have a low-temperature feature at 130 K and a high-temperature feature at 200 K; only the relative intensities change with increasing particle size. Comparing these results with desorption from stepped and flat Au surfaces, the authors suggest that the low-temperature feature is associated with terrace sites on the Au particles while the high-temperature feature is associated with areas of lower coordination, such as the particle edges. IRAS results showing the effects of Au coverage on CO are shown in Figure 10.20B. Here, only a single peak can be seen at 2108 cm^{-1}, in agreement with studies of single-crystal Au surfaces. By flashing the sample to 150 K (thereby removing the CO corresponding to the low-temperature state), it was found that there was little to no change in the intensity of the 2108 cm^{-1} IRAS feature, allowing the authors to assign it to the high-temperature desorption state. The low-temperature desorption state was attributed to CO molecules lying parallel to the surface (or highly tilted), as the selection rules of IRAS state that only those vibrations having a dipole moment normal to the surface will be excited.

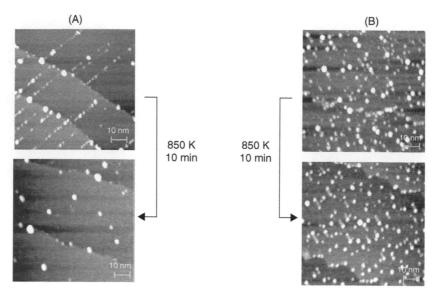

Figure 10.18 STM images (100 × 100 nm, $B = -1.7$ V and $I = 0.18$ nA) of Au (0.4 ML) clusters: (A) nucleated on a Ti-free SiO_2 thin at room temperature and (B) nucleated on a TiO_x(17%)–SiO_2 thin film at room temperature. Thin film frames of (A) and (B) show the effects of an 850 K anneal on the respective samples.

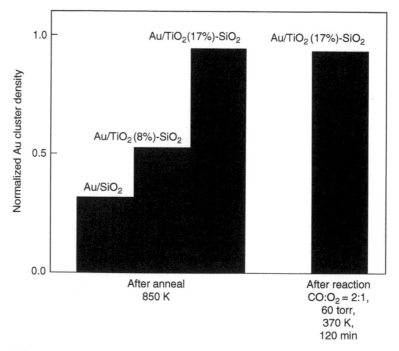

Figure 10.19 A histogram of Au cluster density after the indicated treatment normalized to the cluster density after nucleation at room temperature. The Au coverage in each of the experiments was 0.4 ML Au. The first three columns compare the cluster density on Ti-free SiO_2, TiO_x(8%)–SiO_2, and TiO_x(17%)–SiO_2 thin films, respectively, after a 850 K anneal. The fourth column shows the normalized Au cluster density of a TiO_x(17%)–SiO_2 thin film after a CO oxidation reaction (CO/O_2 = 2:1, 60 torr, 370 K, and 120 min). (Reprinted from Min, B.K. et al., *J. Phys. Chem. B*, 108, 14609-14615, 2004. Copyright 2004. With permission from American Chemical Society.)

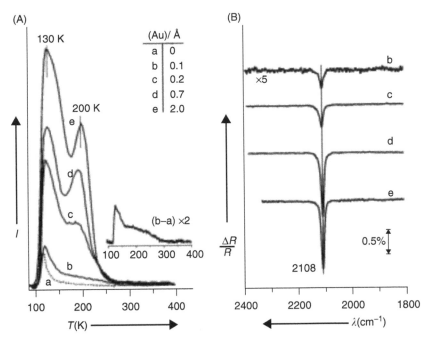

Figure 10.20 (A) Temperature-programmed desorption (TPD) and (B) Infrared reflection absorption spectroscopy (IRAS) spectra of CO adsorbed on Au/FeO(111) as a function of Au coverage. (Reprinted from Lemire, C. et al., *Angew. Chem. Int. Ed.*, 43, 118–121, 2004. With permission from Wiley-VCH.)

Meier and Goodman[105] also studied CO adsorption on Au/TiO$_2$. In particles having a size that would be expected to show metallic properties, they observed a vibrational feature at 2122 cm^{-1}. As this frequency is somewhat blue-shifted from that seen on bulk gold, it was suggested that the substrate might have an effect on CO binding to the Au clusters. With decreasing cluster size, IRAS continues to show only a single peak that shifts only by a small amount. However, there are dramatic changes in the heats of adsorption. The effects of changing cluster size on the isosteric heats of adsorption are presented in Figure 10.21. While the heat of adsorption for an Au coverage of 0.5 ML is only slightly higher than that of bulk gold (10.9 kcal/mol) at the zero coverage limit, lowering the gold coverage significantly increases the heat of adsorption. For the smaller gold clusters, the heat of CO adsorption at the highest coverage even remains higher than the zero coverage limit for bulk Au and the 0.5 ML Au coverage.

Meier and Goodman[108] noted that the trends in the heat of adsorption found using IRAS compare favorably with those found using approximation methods to analyze CO TPD of Au on a variety of supports.[67] Each of the studies using different surface-science models indicated that the Au clusters of ~3 nm in diameter have higher CO heats of adsorption than bulk gold. Therefore, the highest heat of CO adsorption, the peak in CO oxidation activity, and the onset of the metal-to-nonmetal transition all occur at approximately the same cluster size, once again showing the particular importance of cluster size in this highly interesting system.

10.3   *IN SITU* STUDIES

10.3.1 Vibrational Spectroscopy

While the development of planar model catalysts has largely led to the closing of the material gap, the pressure gap still remains. Rupprechter et al.[109] nicely cover some of the concerns of the pressure

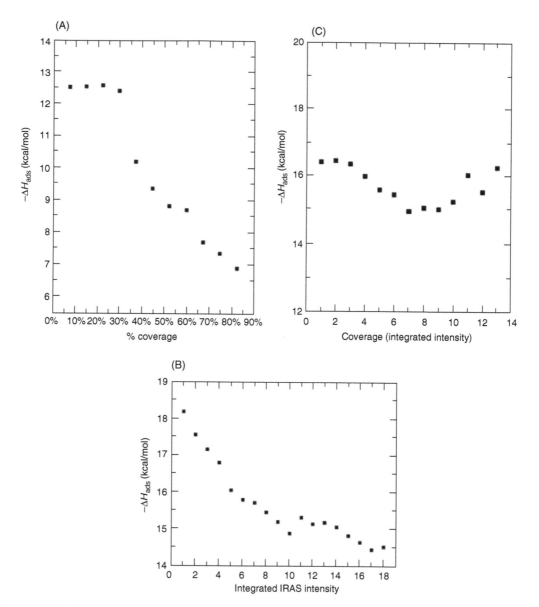

Figure 10.21   Heats of adsorption for CO on (A) 0.5 ML Au/TiO$_2$, (B) 0.25 ML Au/TiO$_2$, and (C) 0.125 ML Au/TiO$_2$. For comparison, the heat of adsorption of CO on bulk gold at the zero coverage limit is 10.9 kcal/mol. (Reprinted from Meier, D.C. and Goodman, D.W., *J. Am. Chem. Soc.*, 126, 1892–1899, 2004. Copyright 2004. With permission from American Chemical Society.)

gap. They state that some species that are not present under UHV conditions may play a role as active species under pressures ≥1 torr. This could be owing to the presence of surface coverages unattainable under UHV, leading to different adsorbate structures. While these higher coverages could possibly be obtained under UHV at low temperatures, the low temperatures could have a pronounced effect on the ability of catalytic reactions to occur, and, if they did, they could be forced to proceed through a different mechanism. Another possibility is that the surface may restructure under high temperatures and pressures. On lowering the temperature and pressure for study using surface analytical techniques, it is possible that one would not be studying the same surface postreaction as during reaction!

Over the past 15 to 20 years, new spectroscopic techniques have been developed, which present an opportunity to study reaction intermediates and products under working conditions. The first of these, pioneered for use in catalysis by the group of Somorjai,[110] is sum frequency generation (SFG). In this nonlinear optical technique, visible light of a fixed frequency is focused onto the catalysts surface while IR light is scanned over a wide frequency range with the sum of the two beams being detected. When the frequency of the IR beam hits a vibrational resonance of a surface species, enhancement of the output signal is seen, thereby providing a vibrational spectrum of the surface.[109,111,112] Owing to the fact that this is a highly surface-sensitive technique, it can be used to study surfaces under reaction conditions with very little background signal. Somorjai and coworkers[110] used this technique to study the adsorption of CO on Pt(111) over a pressure range of 10^{-7} to 700 torr. At low pressures, it was found that CO was bound to bridge (1845 cm^{-1}) and atop (2095 cm^{-1}) sites, as had been seen in previous HREELS and IR studies.[113,114] While the bridge-bonded CO was no longer observable after raising the pressure to 1 torr, the atop feature surprisingly was also lost above ~12 torr along with the growth of a feature at ~2045 cm^{-1}. This feature was attributed to multiple bonded platinum–carbon monoxide cluster analogs on the surface arising from reconstruction of the platinum surface, in which individual Pt atoms could be raised out of the lattice. High-pressure STM studies (to be discussed in more detail later) confirmed the roughening of the Pt surface under high CO pressures.

The group of Freund[109,115–122] has also been an active participant in the study of catalytic surfaces using SFG. Their work has focused extensively on the interaction of CO with the Pd(111) surface and also on its interaction with small metal-oxide-supported Pd clusters. Their initial studies on the CO/Pd(111) system were carried out to discover whether the well-known ordered CO structures seen under UHV conditions would be joined by any high-pressure species.[109,120] Additionally, they wished to discover if CO would dissociate on the Pd(111) surface under higher pressures. Freund and coworkers found that, in general, the CO structures formed under elevated pressures were very similar to those seen under UHV conditions. Under high pressures and low temperatures, some less-ordered structures, however, could be seen. Combining SFG with high-pressure XPS (discussed below), it was found that CO would not dissociate on a perfect Pd(111) surface at 400 K and pressures up to 1 torr (for XPS)/760 torr (for SFG). As previous studies had presented evidence of partial CO dissociation on supported Pd clusters,[100,123,124] the authors suggested that low-coordinated (defect) sites present only on nanoparticles or rough surfaces were necessary for decomposition to be seen. Even by subjecting the Pd(111) surface to Ar$^+$ sputtering, though, there was no evidence of CO dissociation, in contrast to methanol, in which C–O bond cleavage was seen.[119] On the basis of these facts, the perpendicular adsorption geometry of CO on Pd(111) was blamed for the lack of dissociation.

The group of Freund[109,115] has also used SFG to study the interaction of CO with alumina-supported Pd clusters, thereby bridging both the material and pressure gaps. By measuring the SFG intensity on 3 nm Pd particles under 10^{-7} torr CO, the authors found the presence of both atop and bridge-bonded CO, as could be seen using IRAS under CO saturation coverages. By noting the frequency shift of the bridge-bonded CO, it was suggested that the CO was bound to defects or the edges of the particles. On larger (6 nm) Pd clusters, even less atop CO was seen, a fact attributed to the larger size and more ordered surface of the larger particles, which help in the formation of bridge bonds. It should be noted, however, that an increase in the CO pressure to ~200 torr led to an increase in the intensity of the atop peak to a point that was almost the same as that of the smaller particles. These studies demonstrated that the distribution of CO on adsorption sites depends critically on particle size, structure, and temperature, but only at low pressure. At higher pressures, the preferential adsorption occupancies were seen to be relatively independent of the physical characteristics of the Pd particles.

In situ studies of catalytic reactions have also been a prime focus of our group. The high-pressure spectroscopic technique used in our research is polarization modulation IR reflection absorption spectroscopy (PM-IRAS). Like SFG, PM-IRAS is a highly surface-sensitive technique that yields vibrational information about adsorbed surface species. Unlike SFG, however, PM-IRAS does not require the use of pulsed lasers, making it a much more versatile technique. In PM-IRAS, modulation is applied to a linearly polarized IR beam.[125–131] The surface selection rules of reflected IR light state that species on a metal surface can only absorb light polarized perpendicularly to the surface (p-polarized), while gas- (or liquid-) phase species can absorb both s- and p-polarized light. Therefore, by subtracting the s-polarized signal from the p-polarized signal and normalizing to the total signal intensity, a normalized surface-specific signal can be obtained. In contrast to SFG, loss of beam intensity because of gas-phase CO only becomes a problem when the CO pressure is increased sufficiently to become opaque to IR light within the desired frequency range.

An early study using PM-IRAS by Ozensoy et al.[130] was aimed at understanding the pressure-dependent phase transitions of CO on the Pd(111) surface. Kuhn et al.[132] and Szanyi et al.[133,134] had previously studied this system using IRAS in the pressure regime of 1×10^{-6} to 10 torr and noted the phase transitions in CO overlayer structure that occurred as a function of pressure. Using their high-pressure data (up to 600 torr), Ozensoy et al. extended the phase diagram previously reported for the transition from bridging CO to a threefold hollow/atop structure. The phase diagram including the high-pressure data is shown in Figure 10.22. Linearity in the Arrhenius plot can be easily seen, and an activation energy of 44.35 ± 1.63 kJ/mol was obtained from the slope. These types of studies, covering over nine orders of magnitude, show that there are no high-pressure species or adsorbate-induced surface restructuring. Additionally, this study shows that the pressure gap indeed can be bridged; the surface structure of an adsorbate overlayer can be known over a wide pressure range.

Most of the *in situ* studies using PM-IRAS have focused on the CO + NO system on Pd(111), as palladium has come to be increasingly used in the automotive industry because of its advantages over previous three-way catalysts for NO reduction and CO oxidation.[128] An initial study of this system showed the necessity of using *in situ* techniques to study reactions.[131] When in the presence of a 180 torr CO + NO mixture at 350 K, features appeared in the PM-IRA spectra at 1922, 1876, and 1745 cm^{-1}, corresponding to CO (bridge/threefold), CO (threefold), and NO(atop) binding, respectively. However, under the same pressures, when the temperature is increased to 600 K, the spectrum was dominated by a new feature at 2255 cm^{-1}, which was assigned to the asymmetric stretching mode of an isocyanate species, as seen in Figure 10.23. This species was found to be stable even when the temperature was lowered and the chamber pressure was lowered to 10^{-7} torr range. This result was interesting in that, while an isocyanate species had been seen for Pd/Al$_2$O$_3$, it had been described as substrate bound isocyanate. In further studies to determine the onset of isocyanate growth, the sample temperature was set at 600 K, and it was exposed to a constant CO/NO mixture at different total pressures. Regardless of the amount of time that the sample was held at these lower pressures, it was eventually found that a threshold of ~0.1 torr must overcome for isocyanate to be seen. Combined with isotope experiments, these results implied NO dissociation on Pd(111), which is a crucial step for both isocyanate formation and the CO + NO reaction. Note the reaction pathways for both reactions in Figure 10.24.

Subsequent experiments on the same system aimed to determine the stability of the isocyanate species and to measure the reactivity of the Pd(111) model catalyst for the CO + NO reaction.[125] When exposing the sample to different CO/NO ratios (2 and 1.5) at room temperature, peaks were obtained which corresponded to threefold NO, atop NO, and threefold CO, with the higher CO/NO ratio leading to a greater amount of CO binding. When the samples were flashed to 650 K and cooled back to 300 K in the presence of the reaction mixtures, isocyanate was formed. However, as is apparent from Figure 10.25, an increase in the CO/NO ratio strongly favored isocyanate formation.

The kinetic studies of this reaction on Pd(111) at total pressure of 116 torr (CO/NO = 1.67) showed that the rate of the reaction increased above 550 K. This behavior was similar to that found at lower pressures, where it was attributed to the presence of energy barriers to NO dissociation and N$_2$ formation. With regard to the high-pressure work, the presence of the isocyanate species did not lead to measurable changes in the reactivity compared to the studies carried out at low pressure. It, therefore, seems that the isocyanate species acts as little more than a spectator in the reaction.

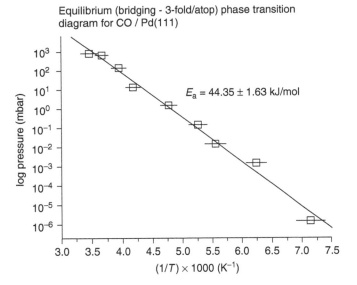

Equilibrium (bridging - 3-fold/atop) phase transition diagram for CO / Pd(111)

$E_a = 44.35 \pm 1.63$ kJ/mol

Figure 10.22   Isoseric plot showing the phase transition from bridging CO to threefold hollow/atop CO on Pd(111) over approximately nine orders of magnitude. (Reprinted from Ozensoy, E. et al., *J. Phys. Chem. B,* 106, 9367–9371, 2002. Copyright 2002. With permission from American Chemical Society.)

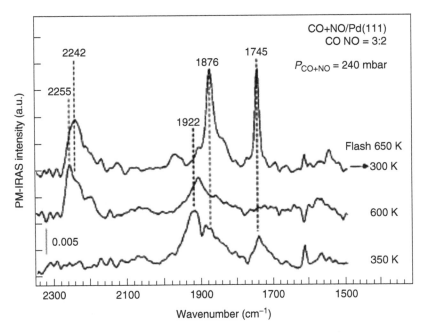

Figure 10.23   *In situ* PM-IRA spectra of Pd(111) in the presence of a CO + NO mixture at 182 torr (CO/NO = 3:2). The initial dosage was at 300 K. (Reprinted from Ozensoy, E. et al., *J. Am. Chem. Soc.,* 124, 8524–8525, 2002. Copyright 2002. With permission from American Chemical Society.)

Ozensoy et al.[127] also used PM-IRAS to study the CO adsorption behavior on SiO_2-supported Pd clusters. As mentioned above, these crystalline, ultrathin silica films possess the structural and electronic properties of the bulk analogues, but are thin enough to permit the use of vibrational and electronic spectroscopic techniques (and tunneling microscopy) without charging.[39,40] As with the

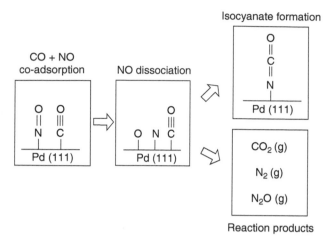

Figure 10.24 Reaction pathways of the CO + NO reaction on Pd(111). (Reprinted from Ozensoy, E. et al., *J. Am. Chem. Soc.*, 124, 8524–8525, 2002. Copyright 2002. With permission from American Chemical Society.)

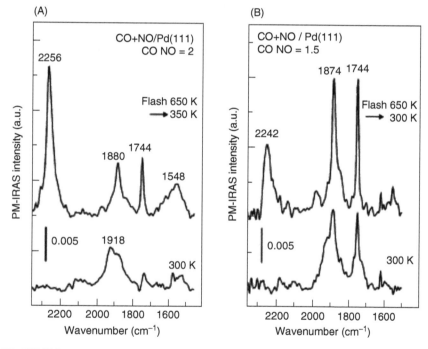

Figure 10.25 PM-IRA spectra of the CO + NO reaction on Pd(111) at different CO/NO pressure ratios. (A) CO/NO = 2 with total pressure of 137 torr. (B) CO/NO = 1.5 with total pressure of 182 torr. (Reprinted from Hess, C. et al., *J. Phys. Chem. B*, 107, 2759–2764, 2003. Copyright 2003. With permission from American Chemical Society.)

work of Freund and coworkers,[115] both atop- and bridge-bonded CO vibrations were observed on the supported particles using PM-IRAS at higher pressures, though, as can be seen in Figure 10.26, the resolution of the spectra for the supported particles is much lower than that for the Pd(111) surface. This is due in large part to the roughness of the surface containing the particles. Figure 10.26 also shows that the changes in CO adsorption behavior as a function of temperature on both the single crystal and model catalyst surface are similar.

Annealing and cooling cycles were used to study the reversibility of CO adsorption on the supported Pd particles. As can be seen in the top frame of Figure 10.27a, at 300 K in the presence of 141 torr CO, two major peaks are present in the PM-IRA spectrum, corresponding to atop binding at 2089 cm^{-1} and bridge/threefold binding at 1957 cm^{-1}. When the sample was heated to 600 K and cooled to 300 K, a dramatic decrease in the amount of atop CO binding is seen, demonstrating the irreversibility of CO binding on the supported particles. By annealing again (to 680 K), the atop intensity is decreased by 70%, while the bridge/threefold intensity is decreased by 60%. In contrast to the supported particles, similar annealing treatments of the Pd(111) surface in high-pressure CO do not lead to significant changes in the PM-IRA spectra (Figure 10.27b). Clearly, some kind of reaction is taking place on the Pd/SiO$_2$ surface. The use of AES provides information on the nature of that reaction. Figure 10.28 shows Auger spectra for the silica-supported Pd particles before and after exposure to 141 torr CO. By focusing on the carbon region of the spectra, it is apparent that the exposure of the sample to high-pressure CO changes the spectrum, as shown more clearly in the difference spectrum presented in the inset. While it could be suggested that the annealing procedure could lead to a change in the morphologies of the Pd clusters and thereby change the CO binding preferences, STM images of the clusters before and after annealing to 650 K (without CO) show no noticeable changes in the cluster morphologies. Therefore, a possible explanation is that CO dissociates on the Pd clusters and the dissociation products poison the adsorption sites. The high number of defect sites present on clusters likely bind CO more strongly and help in C–O bond scission. As studies of CO adsorption at lower pressures showed no evidence of CO dissociation, this study presents another example of why it is important to carry out spectroscopic surveys under realistic conditions when determining the possible reaction mechanisms for catalysts.

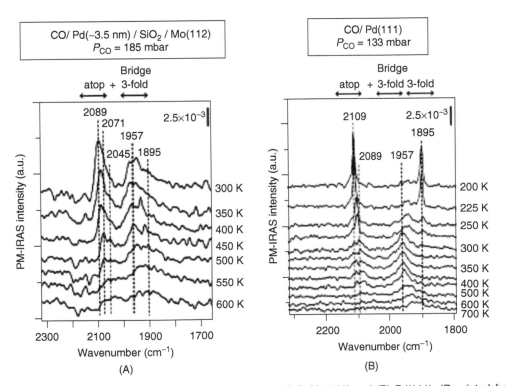

Figure 10.26 PM-IRA spectra of CO adsorption on (A) Pd/SiO$_2$/Mo(112) and (B) Pd(111). (Reprinted from Ozensoy, E. et al., *J. Phys. Chem. B*, 108, 4351–4357, 2004. Copyright 2004. With permission from American Chemical Society.)

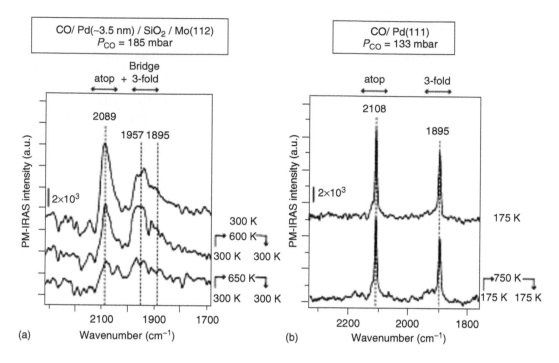

Figure 10.27 PM-IRA spectra showing the effects of annealing–cooling cycles on the CO adsorption characteristics of (a) Pd/SiO$_2$/Mo(112) and (b) Pd(111). (Reprinted from Ozensoy, E. et al., *J. Phys. Chem. B*, 108, 4351–4357, 2004. Copyright 2004. With permission from American Chemical Society.)

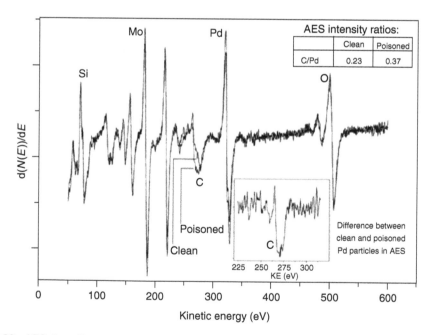

Figure 10.28 AES for silica-supported Pd nanoparticles before and after 141 torr CO exposure. Inset: Difference spectrum between the AES for clean and poisoned Pd particles. (Reprinted from Ozensoy, E. et al., *J. Phys. Chem. B*, 108, 4351–4357, 2004. Copyright 2004. With permission from American Chemical Society.)

10.3.2 *In Situ* STM

While *in situ* vibrational spectroscopy can provide important information concerning adsorbate binding on surfaces, determinations of the actual structures of supported metal catalysts using these techniques can be only made indirectly, namely by comparing vibrational frequencies obtained from the catalysts with those from metal single crystals. The use of STM under UHV onditions has provided a great deal of information about the structures of planar model talysts.[31,37,43–46,48,49,53,75,78,79,87,99,102,127,135–138] However, the ability to study the catalysts under reaction conditions and to follow their evolution helps to provide information concerning the problems of deactivation and regeneration of industrial catalysts.[71] As with the early *in situ* SFG experiments, some of the earliest work with high-pressure STM related to catalysis was performed in the group of Somorjai.[110] This group studied the reconstruction of the Pt(110) surface under an atmosphere of various reactant gases. In the presence of hydrogen, the surface was observed to adopt a "missing-row" reconstruction. When oxygen was substituted for hydrogen, however, the surface was found to develop (111) microfacets. Finally, a switch to carbon monoxide resulted in an unreconstructed (110) surface. Images of these different surfaces are shown in Figure 10.29. These results show how important it is to observe the surface directly under high pressures to find possible active sites.

As described earlier, the use of Au-based catalysts for industrial processes is inhibited by their tendency to deactivate over time, a process blamed on the agglomeration of the clusters owing to the interaction of oxygen with the clusters. Recently, we have followed this agglomeration using a variable-temperature STM, especially designed for *in situ* studies.[37,48,72–74,78] In Figure 10.30, the effects of reaction conditions on individual clusters (and the metal-oxide surface) are shown. On the left side of the figure, an image of Au/TiO$_2$(110) is shown under UHV conditions. The change in the surface during the introduction of a 5.4 torr CO/O$_2$ mixture (1:5) is shown in the image on the right side of the figure. As can be easily seen, the presence of the reactant gases cause quite noticeable changes in the surface. Some clusters seem to grow larger, while others decrease in size. Additionally, the titania surface seems to roughen in several areas and even seems to grow around the clusters.

Similar experiments were carried out on Ag, Au, and bimetallic Ag–Au clusters on TiO$_2$.[73] These studies compared the relative effects of water vapor and air on the clusters. As can be seen in Figure 10.31, the exposure of the sample to ~7 torr of water vapor does not lead to dramatic changes in the cluster distributions or size, demonstrating the stability of the interface toward this elevated pressure treatment. It should be noted that a small number of the smallest clusters (of all types) disappear on exposure to water. These are marked with circles in Figure 10.31. Exposure of the sample to the room-temperature saturation pressure of water or to ambient conditions results in dramatic changes in the surface morphology. In Figure 10.32, the sample is exposed to ~200 torr of ambient air containing 70% humidity. As in the case of studies with water vapor, the Au and bimetallic clusters remain intact under elevated pressure conditions. However, extreme sintering occurs for the Ag clusters. Because there are many possible chemical and physical effects that could be taking place under these pressures, an interpretation of the sintering effect is not straightforward.

10.3.3 Elevated Pressure XPS

An *in situ* technique that has been available for quite some time but has not found much use is *in situ* XPS. Because of the high probability of inelastic scattering of electrons emitted from surfaces, the overall gas pressure must be kept lower than ~10^{-5} torr, and it is generally kept much lower. However, in the late 1970s, Roberts and coworkers[139] developed a method by which photoelectron spectra could be acquired at pressures of up to 1 torr. This was accomplished by using differential pumping and a specially designed sample/gas cell. These authors determined that, if the path length through the gas were short enough, then a sufficient number of electrons would pass elastically through the gas and into the spectrometer, even at a pressure of 1 torr. This instrument was used to measure the adsorption of oxygen on a silver foil at elevated pressures.[140] At the time,

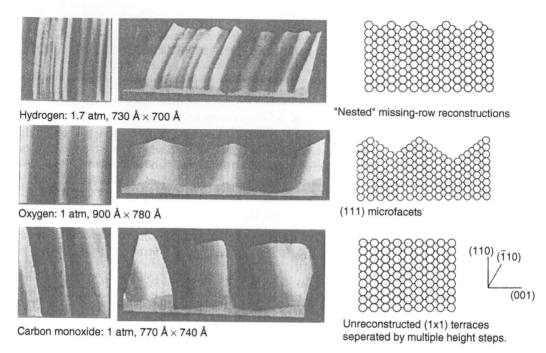

Hydrogen: 1.7 atm, 730 Å × 700 Å

"Nested" missing-row reconstructions

Oxygen: 1 atm, 900 Å × 780 Å

(111) microfacets

Carbon monoxide: 1 atm, 770 Å × 740 Å

(110) $(\bar{1}10)$

(001)

Unreconstructed (1x1) terraces seperated by multiple height steps.

Figure 10.29 STM images of the Pt(110) surface after heating to 425 K for several hours in each of the noted background gases. The presence of different gases leads to different reconstructions of the Pt(110) surface. (From Somorjai, G.A., *Appl. Surf. Sci.*, 121/122, 1–19, 1997. Used with permission from Elsevier Scientific Publishers.)

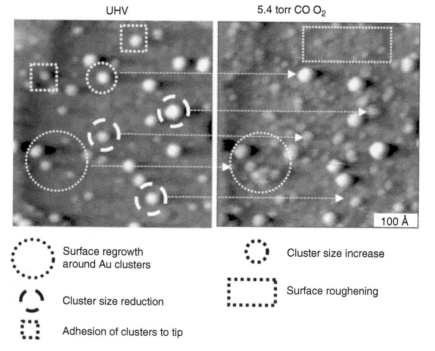

UHV   5.4 torr CO O_2

100 Å

Surface regrowth around Au clusters

Cluster size increase

Cluster size reduction

Surface roughening

Adhesion of clusters to tip

Figure 10.30 STM images (50 × 50 nm) of the same area of a Au/TiO$_2$(110) sample acquired at 450 K: (left) under ultrahigh vacuum conditions and (right) during exposure of the sample to a CO/O$_2$ mixture. (From Goodman, D.W., *J. Catal.*, 216, 213–222, 2003. Used with permission from Elsevier Scientific Publishers.)

UHV 10^3 Pa, 30 min

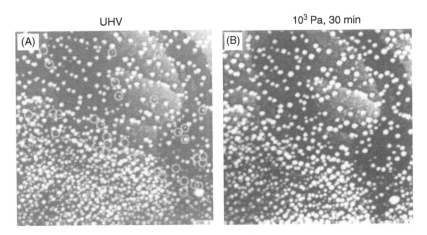

Figure 10.31 STM images of Ag, Ag–Au, and Au clusters deposited on TiO$_2$. (A) Imaged under UHV conditions. (B) Imaged in the presence of ~10 torr of water vapor. The small circles indicate clusters that disappear during exposure to the gas. (Reprinted from Kolmakov, A. and Goodman, D.W., *Rev. Sci. Instrum.*, 74, 2444–2450, 2003. With permission from American Institute of Physics.)

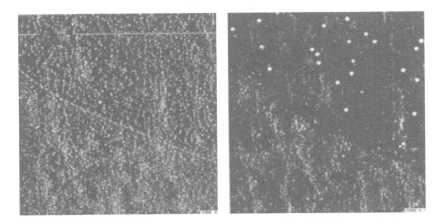

Figure 10.32 STM image (500 × 500 nm) of the same area imaged in Figure 10.31 containing Ag, Ag–Au, and Au clusters on TiO$_2$. Left: imaged under UHV conditions. Right: imaged in the presence of ~200 torr of air. (Reprinted from Kolmakov, A. and Goodman, D.W., *Rev. Sci. Instrum.*, 74, 2444–2450, 2003. With permission from the American Institute of Physics.)

there was a consensus that oxygen could exist as chemisorbed adatoms, dissolved lattice oxygen, and molecularly adsorbed oxygen on silver. However, there was no experimental evidence for all three forms of oxygen in any study. Figure 10.33 presents spectra from this study showing the development of the O(1s) spectrum with increasing pressure. As can be seen, the initial exposure of the silver to oxygen results in the appearance of a peak at 528.3 eV. With further exposure and increasing pressure, the main peak shifts to 530.3 eV. This peak also possesses a tail at high binding energy (532.5 eV). By varying the electron take-off angle from almost normal to 75° from normal, it was hoped that a differentiation could be made between surface and subsurface oxygen species. When the exit angle was 75° from normal (grazing), the ratio of the 532.5/530.3 peaks increased by a factor of 2. The authors, therefore, suggested that the peak at 530.3 eV reflected the presence of subsurface oxygen species while the 532.5 eV peak obviously denoted a surface species. Because peak at 528.3 eV had not been seen in previous XPS studies using molecular oxygen, this feature was assigned to atomic oxygen chemisorbed on the surface. Finally, the authors assigned the peak

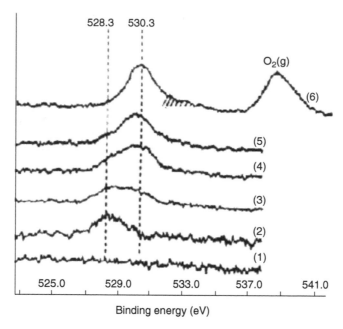

Figure 10.33 O(1s) XP spectra showing the interaction of a silver surface with oxygen at 473 K: (1) clean surface; (2) O_2 pressure of 5×10^{-3} torr; (3) after 30 min of exposure of O_2 at 5×10^{-3} torr; (4) after 10 min at 5×10^{-2} torr; (5) after 10 more min at 5×10^{-2} torr; (6) after 10 min at 0.5 torr O_2. At this pressure, the gas-phase spectrum can also be seen. (Adapted from Joyner, R.W. and Roberts, M.W., *Chem. Phys. Lett.*, 60, 459–462, 1979.)

at 532.5 eV to a molecularly chemisorbed species on the surface, showing, for the first time, direct experimental evidence of the presence of three different oxygen species on the silver surface.

More recently, *in situ* XPS has been used to study the selective oxidation of methanol to formaldehyde using copper and used in conjunction with the SFG studies described above to determine the forms of CO binding on Pd(111).[118–121,141,142] In the case of copper-catalyzed methanol oxidation, it is necessary to obtain information on the system under reaction conditions because the form of adsorbed oxygen that is active in the reaction is only present under reaction conditions and the active sites of the system are only apparent in the presence of the reaction medium. Figure 10.34 shows x-ray photoelectron spectra obtained under conditions of oxygen and methanol coadsorption (O_2/CH_3OH = 1:3, 0.1 torr total pressure). The middle spectra are almost identical to the bottom spectra (obtained after the preparation of methoxy and formate groups on the surface). This indicates that, at 420 K, the surface is mostly covered with methoxy groups. The top spectra show the absence of any carbon-related features, indicating that the oxygen species remaining are not because of formate groups. The authors assign the oxygen peak with lower binding energy to atomic oxygen adsorbed on the surface. The higher binding energy feature (531.2 eV) is assigned to suboxide oxygen based on a previous *in situ* XANES study of the system. They state that this assignment makes sense as the formation of this state only occurs under reaction conditions and only on the metallic copper surface (copper remains in its metallic state throughout the reaction). A plot of the temperature dependence of formaldehyde formation is compared with the oxygen intensities in Figure 10.35. The quick decrease of the peak at 530 eV is attributed to the decomposition of methoxy species on the surface. On the other hand, the suboxide coverage increases to ~520 K, at which point it begins to decrease. This decrease coincides with an increase in formaldehyde production, indicating that the suboxide species is the active species in selective methanol oxidation. The remaining adsorbed atomic oxygen increases the ability of incoming methoxy groups to form. This example shows the

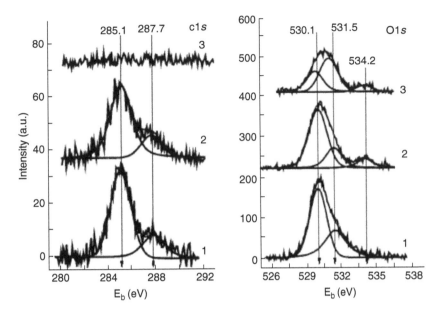

Figure 10.34 C(1s) and O(1s) photoelectron spectra recorded at (2) 420 K and (3) 670 K during an *in situ* experiment in which the total pressure of O_2 and CH_3OH was 4×10^{-3} torr. Spectrum (1) was obtained after the model preparation of methoxy and formate groups on the surface. (Reprinted from Prosvirin, I.P. et al., *Kinetics and Catal.*, 44, 724–730, 2003. With permission from Springer Science and Business Media.)

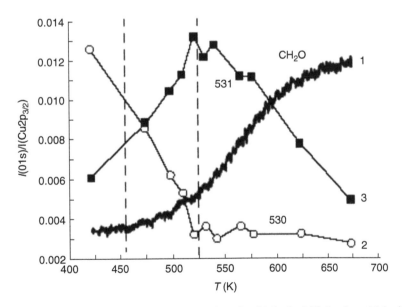

Figure 10.35 The temperature dependence (during methanol oxidation) of (1) the formaldehyde mass spectrometer intensity and the intensities of the O1s peaks at (2) 530 eV and (3) 531 eV normalized to the $Cu2p_{3/2}$ intensity. (Reprinted from Prosvirin, I.P. et al., *Kinetics and Catal.*, 44, 724–730, 2003. With permission from Springer Science and Business Media.)

important role that *in situ* XPS can play in understanding reaction mechanisms, as it is unlikely that the other *in situ* methods described above would be able to differentiate these different species on the copper surface under reaction conditions.

10.4   CONCLUSIONS

The past several decades have seen a great deal of progress from the surface-science community in the understanding of catalytic reaction mechanisms. Initial studies using metal single crystals and high-pressure reactors provided a great deal of information on reaction mechanisms but were often criticized because of the relative lack of complexity in comparison to actual catalysts. To bridge this materials gap, model catalysts have been prepared, which consist of metal clusters on planar oxide surfaces. These model catalysts more accurately mimic the complexities of the working catalysts, yet are still amenable to a wide range of surface-science techniques. The current goal is to study these catalysts under reaction conditions, an objective made possible by the development of techniques such as SFG, PM-IRAS, and even high-pressure STM. The use of these techniques to study metal-oxide-supported clusters makes the goal of a complete understanding of catalytic reaction mechanisms much more realistic.

ACKNOWLEDGMENTS

We acknowledge the support of this work by the Department of Energy, Office of Basic Energy Sciences, Division of Chemical Sciences, and the Robert A. Welch Foundation.

REFERENCES

1. G.A. Somorjai, *Introduction to Surface Chemistry and Catalysis.* 1st ed., Wiley, New York, 1994, p. 667.
2. F. Zaera, *Prog. Surf. Sci.* **69** (2001) 1–98.
3. D.W. Goodman, *Chem. Rev.* **95** (1995) 523–536.
4. D.W. Goodman, *Surf. Sci.* **299/300** (1994) 837–848.
5. D.W. Goodman, *Ann. Rev. Phys. Chem.* **37** (1986) 425–457.
6. D.W. Goodman, *Acc. Chem. Res.* **17** (1984) 194–200.
7. D.W. Goodman, *J. Vac. Sci. Technol.* **20** (1982) 522–526.
8. T.E. Madey, D.W. Goodman, and R.D. Kelley, *J. Vac. Sci. Technol.* **16** (1979) 433–434.
9. M.A. Vannice, *J. Catal.* **44** (1976) 152–162.
10. D.W. Goodman and J.M. White, *Surf. Sci.* **90** (1979) 201–203.
11. R.D. Kelley and D.W. Goodman, *Surf. Sci.* **123** (1982) L743–L749.
12. H.J. Krebs, H.P. Bonzel, and G. Gafner, *Surf. Sci.* **88** (1979) 269–283.
13. M. Kiskinova and D.W. Goodman, *Surf. Sci.* **108** (1981) 64–76.
14. D.W. Goodman and M. Kiskinova, *Surf. Sci.* **105** (1981) L265–L270.
15. D.W. Goodman, *Appl. Surf. Sci.* **19** (1984) 1–13.
16. C.T. Campbell and D.W. Goodman, *Surf. Sci.* **123** (1982) 413–426.
17. T. Engel and G. Ertl, *The Chemical Physics of Solid Surfaces and Herogeneous Catalysis*, eds. D.A. King and D.P. Woodruff, Vol. 4, Elsevier, Amsterdam, 1982.
18. S.H. Oh, G.B. Fisher, J.E. Carpenter, and D.W. Goodman, *J. Catal.* **100** (1986) 360–376.
19. C.H.F. Peden, D.W. Goodman, D.S. Blair, P.J. Berlowitz, G.B. Fisher, and S.H. Oh, *J. Phys. Chem.* **92** (1988) 1563–1567.
20. C.H.F. Peden and D.W. Goodman, *J. Phys. Chem.* **90** (1986) 1360–1365.
21. D.W. Goodman and C.H.F. Peden, *J. Phys. Chem.* **90** (1986) 4839–4843.
22. C.T. Campbell, *Annu. Rev. Phys. Chem.* **41** (1990) 775–837.
23. J.A. Rodriguez and D.W. Goodman, *Science* **257** (1992) 897–903.

24. F. Besenbacher, I. Chorkendorff, B.S. Clausen, B. Hammer, A.M. Molenbroek, J.K. Nørskov, and I. Stensgaard, *Science* **279** (1998) 1913–1915.

25. J. Greeley, J.K. Norskov, and M. Mavrikakis, *Annu. Rev. Phys. Chem.* **53** (2002) 319–348.

26. I. Chorkendorff and J.W. Niemantserdriet, *Concepts of Modern Catalysis and Kinetics*. Wiley-VCH, Weinheim, 2003.

27. K.W. Kolasinski, *Surface Science: Foundations of Catalysis and Nanoscience*. Wiley, West Sussex, 2002.

28. J.W. Niemantsverdriet, *Spectroscopy in Catalysis*. VCH, New York, 1993.

29. D.P. Woodruff and T.A. Delchar, *Modern Techniques of Surface Science-Second Edition*, Cambridge Solid State Science Series. Cambridge University Press, Cambridge, 1994.

30. S.C. Street, C. Xu, and D.W. Goodman, *Annu. Rev. Phys. Chem.* **48** (1997) 43–68.

31. A.K. Santra and D.W. Goodman, Size-dependent electronic, structural, and catalytic properties of metal clusters supported on ultrathin oxide films, in *Catalysis and Electrocatalysis at Nanoparticle Surfaces*, eds. A. Wieckowski, E.R. Savinova, and C.G. Vayenas. Marcel Dekker, New York, 2003, pp. 281–309.

32. S.A. Chambers, *Surf. Sci. Reports* **39** (2000) 105–180.

33. H.-J. Freund, H. Kuhlenbeck, and V. Staemmler, *Rep. Prog. Phys.* **59** (1996) 283–347.

34. R. Franchy, *Surf. Sci. Reports* **38** (2000) 199–294.

35. M.S. Chen, A.K. Santra, and D.W. Goodman, *Phys. Rev. B* **69** (2004) 155404.

36. D.W. Goodman, *J. Phys. Chem.* **100** (1996) 13090–13102.

37. D.W. Goodman, *J. Catal.* **216** (2003) 213–222.

38. Q. Guo, W.S. Oh, and D.W. Goodman, *Surf. Sci.* **437** (1999) 49–60.

39. Y.D. Kim, T. Wei, S. Wendt, and D.W. Goodman, *Langmuir* **19** (2003) 7929–7932.

40. Y.D. Kim, T. Wei, and D.W. Goodman, *Langmuir* **19** (2003) 354–357.

41. A. Kolmakov, J. Stultz, and D.W. Goodman, *J. Chem. Phys.* **113** (2000) 7564–7570.

42. X. Lai, Q. Guo, B.K. Min, and D.W. Goodman, *Surf. Sci.* **487** (2001) 1–8.

43. B.K. Min, W.T. Wallace, A.K. Santra, and D.W. Goodman, *J. Phys. Chem. B* **108** (2004) 16339–16343.

44. B.K. Min, W.T. Wallace, and D.W. Goodman, *J. Phys. Chem. B* **108** (2004) 14609–14615.

45. B.K. Min, A.K. Santra, and D.W. Goodman, *Catal. Today* **85** (2003) 113–124.

46. B.K. Min, A.K. Santra, and D.W. Goodman, *J. Vac. Sci. Technol. B* **21** (2003) 2319–2323.

47. W.S. Oh, C. Xu, D.Y. Kim, and D.W. Goodman, *J. Vac. Sci. Technol. A* **15** (1997) 1710–1716.

48. A.K. Santra and D.W. Goodman, *J. Phys.: Condens. Matter* **14** (2002) R31–R62.

49. A.K. Santra, B.K. Min, and D.W. Goodman, *Surf. Sci.* **515** (2002) L475–L479.

50. X. Xu, J. Szanyi, Q. Xu, and D.W. Goodman, *Catal. Today* **21** (1994) 57–69.

51. C. Xu and D.W. Goodman, *Chem. Phys. Lett.* **263** (1996) 13–18.

52. H.-J. Freund, *Farad. Discuss.* **114** (1999) 1–31.

53. H.-J. Freund, M. Baumer, and H. Kuhlenbeck, Catalysis and surface science: what do we learn from studies of oxide-supported cluster model systems?, in *Advances in Catalysis,* eds. W.O. Haag, B.C. Gates, and H. Knozinger. Elsevier, Oxford, 2000, pp. 333–384.

54. T. Schroeder, M. Adelt, B. Richter, M. Naschitzki, M. Baumer, and H.-J. Freund, *Microelect. Reliab.* **40** (2000) 841–844.

55. H. Hakkinen, S. Abbet, A. Sanchez, U. Heiz, and U. Landman, **42** (2003) 1297–1300.

56. T. Schroeder, M. Adelt, B. Richter, M. Naschitzki, M. Baumer, and H.-J. Freund, *Surf. Rev. Lett.* **7** (2000) 7–14.

57. T. Schroeder, J.B. Giorgi, M. Bäumer, and H.-J. Freund, *Phys. Rev. B* **66** (2002) 165422.

58. U. Heiz, F. Vanolli, L. Trento, and W.-D. Shneider, *Rev. Sci. Instrum.* **68** (1997) 1986–1994.

59. U. Heiz, *Appl. Phys. A* **67** (1998) 621–626.

60. U. Heiz, F. Vanolli, A. Sanchez, and W.-D. Schneider, *J. Am. Chem. Soc.* **120** (1998) 9668–9671.

61. U. Heiz, A. Sanchez, S. Abbet, and W.-D. Schneider, *Eur. Phys. J. D* **9** (1999) 35–39.

62. U. Heiz, A. Sanchez, S. Abbet, and W.-D. Schneider, *Chem. Phys.* **262** (2000) 189–200.

63. U. Heiz and W.-D. Schneider, Physical chemistry of supported clusters, in *Metal Clusters at Surfaces*, ed. K.-H. Meiwes-Broer. Springer, Berlin, 2000, pp. 237–273.

64. C.R. Henry, *Surf. Sci. Rep.* **31** (1998) 235–325.

65. S.K. Purnell, X. Xu, D.W. Goodman, and B.C. Gates, *J. Phys. Chem.* **98** (1994) 4076–4082.

66. M.-H. Schaffner, F. Patthey, and W.-D. Schneider, *Eur. Phys. J. D* **9** (1999) 609–612.

67. S.K. Shaikhutdinov, R. Meyer, M. Naschitzki, M. Baumer, and H.-J. Freund, *Catal. Lett.* **86** (2003) 211–219.

68. A. Sanchez, S. Abbet, U. Heiz, W.-D. Schneider, H. Hakkinen, R.N. Barnett, and U. Landman, *J. Phys. Chem. A* **103** (1999) 9573–9578.

69. M. Bowker, P. Stone, R. Bennett, and N. Perkins, *Surf. Sci.* **497** (2002) 155–165.

70. D.A. Chen, M.C. Bartelt, S.M. Seutter, and K.F. McCarty, *Surf. Sci.* **464** (2000) L708–L714.

71. A. Kolmakov and D.W. Goodman, *Catal. Lett.* **70** (2000) 93–97.

72. A. Kolmakov and D.W. Goodman, *Surf. Sci.* **490** (2001) L597–L601.

73. A. Kolmakov and D.W. Goodman, *Rev. Sci. Instrum.* **74** (2003) 2444–2450.

74. A. Kolmakov and D.W. Goodman, *Chem. Rec.* **2** (2002) 446–457.

75. X. Lai, T.P.S. Clair, M. Valden, and D.W. Goodman, *Prog. Surf. Sci.* **59** (1998) 25–52.

76. X. Lai, T.P.S. Clair, and D.W. Goodman, *Farad. Discuss.* **114** (1999) 279–284.

77. C.E.J. Mitchell, A. Howard, M. Carney, and R.G. Egdell, *Surf. Sci.* **490** (2001) 196–210.

78. A.K. Santra, A. Kolmakov, F. Yang, and D.W. Goodman, *Jpn. J. Appl. Phys.* **42** (2003) 4795–4798.

79. A.K. Santra, F. Yang, and D.W. Goodman, *Surf. Sci.* **548** (2004) 324–332.

80. M. Valden, S. Pak, X. Lai, and D.W. Goodman, *Catal. Lett.* **56** (1998) 7–10.

81. J. Zhou, Y.C. Kang, and D.A. Chen, *J. Phys. Chem. B* **107** (2003) 6664–6667.

82. J. Zhou, Y.C. Kang, S. Ma, and D.A. Chen, *Surf. Sci.* **562** (2004) 113–127.

83. X. Xu and D.W. Goodman, *Surf. Sci.* **282** (1993) 323–332.

84. X. Xu and D.W. Goodman, *Appl. Phys. Lett.* **61** (1992) 774–776.

85. J.-W. He, X. Xu, J.S. Corneille, and D.W. Goodman, *Surf. Sci.* **279** (1992) 119–126.

86. C.T. Campbell, *Science* **306** (2004) 234–235.

87. M.-S. Chen and D.W. Goodman, *Science* **306** (2004) 252–255.

88. M.-S. Chen, W.T. Wallace, D. Kumar, Z. Yan, K.K. Gath, Y. Cai, Y. Kuroda, and D.W. Goodman, *Surf. Sci.* **581** (2005) L115–L121.

89. M.-S. Chen and D.W. Goodman, *Surf. Sci.* **574** (2005) 259–268.

90. M. Haruta, N. Yamada, T. Kobayashi, and S. Iijima, *J. Catal.* **115** (1989) 301.

91. M. Haruta, T. Kobayashi, H. Sano, and N. Yamada, *Chem. Lett.* **2** (1987) 405–408.

92. M. Haruta, S. Tsubota, and T. Kobayashi, *J. Catal.* **144** (1993) 175.

93. T.S. Kim, J.D. Stiehl, C.T. Reeves, R.J. Meyer, and C.B. Mullins, *J. Am. Chem. Soc.* **125** (2003) 2018–2019.

94. J.A. Rodriguez, G. Liu, T. Jirsak, J. Hrbek, Z. Chang, J. Dvorak, and A. Maiti, *J. Am. Chem. Soc.* **124** (2002) 5242–5250.

95. T.G. Schaaff and D.A. Blom, *Nano Lett.* **2** (2002) 507–511.

96. M. Valden, X. Lai, and D.W. Goodman, *Science* **281** (1998) 1647–1650.

97. E. Wahlstrom, N. Lopez, R. Schaub, P. Thostrup, A. Ronnau, C. Africh, E. Laegsgaard, J.K. Norskov, and F. Besenbacher, *Phys. Rev. Lett.* **90** (2003) 026101.

98. M. Bäumer and H.-J. Freund, *Prog. Surf. Sci.* **61** (1999) 127–198.

99. X. Lai and D.W. Goodman, *J. Mol. Catal. A* **162** (2000) 33–50.

100. D.R. Rainer, C. Xu, P.M. Holmblad, and D.W. Goodman, *J. Vac. Sci. Technol. A* **15** (1997) 1653–1662.

101. C. Xu, W.S. Oh, G. Liu, D.Y. Kim, and D.W. Goodman, *J. Vac. Sci. Technol. A* **15** (1997) 1261–1268.

102. D.C. Meier, X. Lai, and D.W. Goodman, Surface chemistry of model oxide-supported metal catalysts: An overview of gold on Titania, in *Surface Chemistry and Catalysis*, eds. A.F. Carley et al. Kluwer, New York, 2002, pp. 147–189.

103. S.J. Tauster, S.C. Fung, R.T.K. Baker, and J.A. Horsley, *Science* **211** (1981) 1121–1125.

104. T.A. Nijhuis, B.J. Huizinga, M. Makkee, and J.A. Moulijn, *Ind. Eng. Chem. Res.* **38** (1999) 884–891.

105. D.C. Meier and D.W. Goodman, *J. Am. Chem. Soc.* **126** (2004) 1892–1899.

106. C. Lemire, R. Meyer, S.K. Shaikhutdinov, and H.-J. Freund, *Surf. Sci.* **552** (2004) 27–34.

107. C. Lemire, R. Meyer, S. Shaikhutdinov, and H.-J. Freund, *Angew. Chem. Int. Ed.* **43** (2004) 118–121.

108. P.A. Redhead, *Vacuum* **12** (1962) 203–211.

109. G. Rupprechter, H. Unterhalt, M. Morkel, P. Galletto, L. Hu, and H.-J. Freund, *Surf. Sci.* **502–503** (2002) 109–122.

110. G.A. Somorjai, *Appl. Surf. Sci.* **121/122** (1997) 1–19.

111. Y.R. Shen, *Nature* **337** (1989) 519–525.

112. J.H. Hunt, P. Guyot-Sionnest, and Y.R. Shen, *Chem. Phys. Lett.* **133** (1989) 189–192.

113. B.E. Hayden and A.M. Bradshaw, *Surf. Sci.* **125** (1983) 787–802.

114. N.R. Avery, *J. Chem. Phys.* **74** (1981) 4202–4203.

115. T. Dellwig, G. Rupprechter, H. Unterhalt, and H.-J. Freund, *Phys. Rev. Lett.* **85** (2000) 776–779.

116. M. Morkel, G. Rupprechter, and H.-J. Freund, *J. Chem. Phys.* **119** (2003) 10853–10866.

117. G. Rupprechter, M. Morkel, H.-J. Freund, and R. Hirschl, *Surf. Sci.* **554** (2004) 43–59.

118. G. Rupprechter, V.V. Kaichev, H. Unterhalt, M. Morkel, and V.I. Bukhtiyarov, *Appl. Surf. Sci.* **235** (2004) 26–31.

119. V.V. Kaichev, M. Morkel, H. Unterhalt, I.P. Prosvirin, V.I. Bukhtiyarov, G. Rupprechter, and H.-J. Freund, *Surf. Sci.* **566–568** (2004) 1024–1029.

120. V.V. Kaichev, I.P. Prosvirin, V.I. Bukhtiyarov, H. Unterhalt, G. Rupprechter, and H.-J. Freund, *J. Phys. Chem. B* **107** (2003) 3522–3527.

121. M. Morkel, V.V. Kaichev, G. Rupprechter, H.-J. Freund, I.P. Prosvirin, and V.I. Bukhtiyarov, *J. Phys. Chem. B* **108** (2004) 12955–12961.

122. M. Morkel, H. Unterhalt, M. Salmeron, G. Rupprechter, and H.-J. Freund, *Surf. Sci.* **532–535** (2003) 103–107.

123. N. Tsud, V. Johanek, I. Stara, K. Veltruska, and V. Matolin, *Surf. Sci.* **467** (2000) 169–176.

124. D.L. Doering, H. Poppa, and J.T. Dickinson, *J. Catal.* **73** (1982) 104–119.

125. C. Hess, E. Ozensoy, and D.W. Goodman, *J. Phys. Chem. B* **107** (2003) 2759–2764.

126. E. Ozensoy, C. Hess, D. Loffreda, P. Sautet, and D.W. Goodman, *J. Phys. Chem. B* **109** (2005) 5414–5417.

127. E. Ozensoy, B.K. Min, A.K. Santra, and D.W. Goodman, *J. Phys. Chem. B* **108** (2004) 4351–4357.

128. E. Ozensoy, C. Hess, and D.W. Goodman, *Top. Catal.* **28** (2004) 13–23.

129. E. Ozensoy and D.W. Goodman, *Phys. Chem. Chem. Phys.* **6** (2004) 3765–3778.

130. E. Ozensoy, D.C. Meier, and D.W. Goodman, *J. Phys. Chem. B* **106** (2002) 9367–9371.

131. E. Ozensoy, C. Hess, and D.W. Goodman, *J. Am. Chem. Soc.* **124** (2002) 8524–8525.

132. W.K. Kuhn, J. Szanyi, and D.W. Goodman, *Surf. Sci.* **274** (1992) L611–L618.

133. J. Szanyi, W.K. Kuhn, and D.W. Goodman, *J. Vac. Sci. Technol. A* **11** (1993) 1969–1974.

134. J. Szanyi, W.K. Kuhn, and D.W. Goodman, *J. Phys. Chem.* **98** (1994) 2978–2981.

135. E. Wahlstrom, N. Lopez, R. Schaub, P. Thostrup, A. Ronnau, C. Africh, E. Laegsgaard, J.K. Norskov, and F. Besenbacher, *Phys. Rev. Lett.* **90** (2003) 026101.

136. J. Wintterlin, Scanning tunneling microscopy studies of catalytic reactions, in *Advances in Catalysis*, eds. W.O. Haag, B.C. Gates, and H. Knozinger. Elsevier, Oxford, 2000, pp. 131–206.

137. F.B.D. Mongeot, M. Scherer, B. Gleich, E. Kopatzki, and R.J. Behm, *Surf. Sci.* **411** (1998) 249–262.

138. A.F. Carlsson, M. Naschitzki, M. Baumer, and H.-J. Freund, *J. Phys. Chem. B* **107** (2003) 778–785.

139. R.W. Joyner, M.W. Roberts, and K. Yates, *Surf. Sci.* **87** (1979) 501–509.

140. R.W. Joyner and M.W. Roberts, *Chem. Phys. Lett.* **60** (1979) 459–462.

141. V.I. Bukhtiyarov, I.P. Prosvirin, E.P. Tikhomirov, V.V. Kaichev, A.M. Sorokin, and V.V. Evstigneev, *React. Kinet. Catal. Lett.* **79** (2003) 181–188.

142. I.P. Prosvirin, E.P. Tikhomirov, A.M. Sorokin, V.V. Kaichev, and V.I. Bukhtiyarov, *Kinet. Catal.* **44** (2003) 724–730.

CHAPTER 10 QUESTIONS

Question 1

Compare the effects of the addition of Li and Se with that of a clean Ni catalyst in the methanation reaction.

Question 2

Describe the differences between ensemble effects and ligand effects (in relation to bimetallic surfaces), and how these may play a role in the adsorption activity of a surface.

Question 3

What is a possible problem that one might see with regard to spectroscopies using electrons (or ions) as the species striking a surface?

Question 4

Why might it be important to understand cluster nucleation on an oxide surface?

Question 5

What is a possible reason for the small amount of studies carried out using *in situ* XPS?

Question 6

In addition to the fact that SFG requires the use of pulsed lasers, what are some other possible reasons that cause it to be less versatile than PM-IRAS?

Question 7

What is the molecular gas density for an ideal gas at 300 K when the pressure is 10^{-6} torr? (in molecules/m^3)

Question 8

With the answer to Question 7 in hand, what would be the gas density at a pressure of 10^{-9} torr?

Question 9

CO is often used as a probe molecule to aid in the determination of surface structures. Why would the CO stretching frequency change depending on where it was bound on the surface?

Question 10

When the coverage of CO (and many other gases) on surfaces reach a critical point, it has been found that the heat of adsorption decreases rapidly. What could be the explanation for this behavior?

High-Throughput Experimentation and Combinatorial Approaches in Catalysis

Stephan Andreas Schunk, Oliver Busch, Dirk G. Demuth, Olga Gerlach, Alfred Haas, Jens Klein, and Torsten Zech

CONTENTS

11.1 High-Throughput Experimentation and Combinatorial Catalysis —
Definition and Scope 374
11.2 Experimental Planning and Data Handling 376
    11.2.1 Descriptor-Driven Approaches 376
    11.2.2 Approaches Based on Classical Statistical Designs 377
    11.2.3 Other Techniques for Finding Local Optima 378
11.3 Stage I and Stage II Screening 380
11.4 Analytical Techniques for Screening 382
11.5 Synthetic Approaches for High-Throughput Experimentation and
Combinatorial Chemistry 385
    11.5.1 Synthetic Approaches for Molecular Catalysts 386
    11.5.2 Synthetic Approaches for Solid-State Inorganic Catalysts 387
    11.5.3 Combinatorial Synthetic Approaches for Solid-State Inorganic and
        Molecular Catalysts 389
11.6 Testing of Catalysts in Gas-Phase Reactions 390
    11.6.1 Stage I Testing of Catalysts for Gas-Phase Reactions 396
        11.6.1.1 General Considerations 396
        11.6.1.2 Reactant Distribution for Stage I Screening Systems 396
        11.6.1.3 Single-Bead Reactors 398
        11.6.1.4 Optimal Use of Stage I in Screening Programs 401
    11.6.2 Stage II Testing in Gas-Phase Applications 402
        11.6.2.1 The Epoxidation of 1,3-Butadiene with Ag-Based Catalysts 405
        11.6.2.2 Dynamic Experiments in Stage II Screening for Automotive
                Applications 407
        11.6.2.3 Refinery Catalysis Applications in High-Throughput
                Experimentation 409
11.7 Testing of Catalysts in Liquid–Liquid, Gas–Liquid, and Gas–Liquid–Solid Reactions 411
    11.7.1 Stage I Screening for Liquid-Phase Catalysis 413
        11.7.1.1 Alternative Stage I Screening Concepts 417

      11.7.2   Stage II Screening for Liquid-Phase Catalysis 418
11.8 Summary and Outlook 420
References 421
Chapter 11 Questions 425

11.1 HIGH-THROUGHPUT EXPERIMENTATION AND COMBINATORIAL CATALYSIS — DEFINITION AND SCOPE

High-throughput experimentation (HTE) and combinatorial chemistry (CombiChem) have been the most powerful trends in research related to catalysis and materials science in the late 1990s and the early millennium. The wealth that these technologies have brought to the scene is obvious: catalysis in chemical production and R&D is one of the most laborious and know-how intensive fields. For certain application processes or materials developments, it may take decades before an economic return on investment can be expected. It is more than evident that the promise to be able to accelerate the time to market from an industrial perspective and the systematic mapping of large arrays of compositional and structural parameters from a scientific perspective are more than welcome in the world of catalysis. The fast development of processes goes back to the last century, and it is more than fair to nominate BASF chemist Alwin Mittasch as one of the first scientists who was active in the field of HTE and catalysis [1]. Although Mittasch was less renowned than Haber and Bosch, to whom the honor of the invention of the process for the synthesis of ammonia from hydrogen and nitrogen is usually granted, it was he who discovered the useful catalyst formulation for the industrial process. The search for a useful elemental combination that would meet the economical target for a process realization was realized by massive assignment of workforce. Although it was already known from Nernst and Haber's research that among osmium, uranium, and tantalum the much cheaper and readily available iron was a lead element, yet a total of 20,000 catalyst formulations consisting of iron-containing catalyst candidates were tested before the final composition was identified, which could then be introduced into the industrial process. All of the work that Mittasch and coworkers carried out was done manually, as laboratory automation was not very advanced in the early 1910s. Today manual labor throughout the chemical industry is much more cost intensive, yet 100 years later automation is far more advanced and can increase the efficiency of R&D processes, leading to a fast identification of the crucial parameters that make a catalytically driven chemical process an economical and ecological success story. Combinatorial chemistry and HTE have turned out to be the key elements for acceleration of the automation process, and this chapter is devoted to transmit the essence and principles that govern the field to the reader.

Many scientists, and maybe even experts in the field of HTE and CombiChem, may find it hard to give the appropriate definitions for both the terms. The definition of the term CombiChem as given by IUPAC is use of a combinatorial process for the preparation of sets of compounds from sets of building blocks [2]. The term combinatorial means relating to combinations or the arrangement of, selection from and operation on discrete elements belonging to finite sets. Conclusively, we define HTE as the rapid completion of two or more experimental stages in a concerted and integrated fashion. Combinatorial chemistry can, on occasion, contribute to certain stages of HTE. The scientist working in the field of catalysis has to be aware that HTE and CombiChem are additional toolkits that are an aid in the total development and research effort in catalysis. Both the methods enable the researcher to identify prospective candidates, also called leads, which can possibly be developed into industrial catalysts at a later stage. Still the power of numbers does not discharge the scientist from detailed studies of the inorganic catalyst compound materials, the ideal reactor design, and the in-depth search for the optimum reaction conditions in order to find a performance optimum for the process.

Combinatorial chemistry was developed in the early 1990s with pharmaceutical industry being the main driver [3]. The rapid screening of potential drug molecules against well-defined assays had quickly become one of the major tools for the identification of novel lead candidates.

With the potential of the tools enabling the rapid testing of molecular compounds, efficient tools that allowed access to large molecular libraries were also requested. Highly automated parallel synthesis, synthesis on solid phase, and split and pool methodologies for the synthesis of small molecular libraries were the outcomes of this revolution (for further reading see [4]). If scientists talk about CombiChem today, in many cases screening and synthetic methodologies are summed up under this term. The substantial differences existing between inorganic materials and molecular entities will be addressed later. The facts essential for the development of the first approaches for inorganic catalysis were:

- Most of the screening technology for CombiChem was developed for the use under conditions of 20 to 40°C and ambient pressure.
- The wealth of analytical techniques that was applied for CombiChem was devoted to biochemical challenges.
- The synthetic equipment was hardly compatible for inorganic materials synthesis.

On this basis, in the mid- and late 1990s the researchers started to develop the first concepts for integrated synthesis and screening approaches, which were adaptable with the conditions usually applied in inorganic catalysis of several hundred degree centigrade and pressures up to 250 bar [5]. But earlier roots for approaches in materials chemistry are yet to be found: following the history of HTE for inorganic materials, the importance of the concepts of Hanak [6,7] is more than evident. To our knowledge, Hanak was the first scientist who developed the concept of substrate-bound material libraries, which he produced by cosputtering of different elements on substrates. His goal was not to synthesize materials for the purpose of monitoring their catalytic performance, but rather for their physical properties. As Hanak worked with gradients of the elements he deposited on substrates, he did not produce single defined compositions on his substrates but rather an endless number of materials with the relative space on the substrate defining the difference in composition. Further development of this approach by Schultz and coworkers [8] led to the sputtering of defined spots on substrates for the discovery of materials with outstanding magneto-resistive properties . With this approach they introduced synthesizing materials libraries of defined candidate materials on substrates and their characterization with regard to a certain property. It is worthwhile to mentioning that even though a small number of industrial and academic groups had already started to employ the concept of parallel reactors for gas-phase reactions, still the usual degree of parallelization employed was not larger than 4- to 6-fold, and in some cases the reactors were not operated in parallel but sequentially (for a review of the work of Moulijn see [9], earlier work on similar technologies is mainly reflected in the patent literature and is beyond the scope of this chapter). It is important to mention these early approaches, as they nicely illustrate that the search for efficient catalyst screening tools has always been an issue for scientists working in the field of catalysis. Still, the integrated approach that aroused a broader interest in the scientific community was due to the efforts of Mallouk and Moates [10,11]. These were the first illustrative examples that proved that the screening of inorganic catalysts on substrates via parallel analytical techniques is a viable way for the identification and differentiation of catalyst candidates. In both cases, (IR) infrared-thermography (a quasioptical technique which will be described in detail later) was used as an integral analytical tool. These approaches are important as pioneering technologies but still lack essential technical features that render them as powerful tools with regard to a later technical application. Yet together these pioneering studies, with the work of other groups in the field, led to a paradigm shift and a broader acceptance for the principal thoughts of CombiChem and HTE in the community. It took further efforts to create a workflow, which would allow the synthesis and testing of catalysts under industrially relevant conditions; Schüth and coworkers [12] were the first to demonstrate that HTE in inorganic catalysis can be performed under close-to-conventional conditions with at least the same experimental quality that is obtained in state-of-the-art bench laboratory tests. Their work was performed using titration robots for catalyst synthesis and testing in a 16-fold reactor system with nondispersive IR as an analytical tool. From the point of integration, the test unit is clearly ahead of the six-fold reactor developments of the 1970s and 1980s, but massively behind the mass screening techniques described in [8,10,11]. An important milestone that was achieved with this technology was the validation that close-to-conventional synthesis and testing can be performed in an HTE mode. Since these developments, massive efforts have led to a technological platform that is applicable to a wealth of challenges which inorganic catalysis brings along. In this chapter, the reader will get acquainted with some basic definitions and will then be introduced by case studies to the potential of HTE (see Figure 11.1).

11.2   EXPERIMENTAL PLANNING AND DATA HANDLING

Combinatorial chemistry and HTE are powerful tools in the hands of a scientist, as they are a source for meaningful consistent records of data that would be hard to obtain via conventional methods within a decent timeframe. This blessing of fast data acquisition can turn into a curse if the experimentalist does not take precautions to carefully plan the experiments ahead and the means of handling the data and analyzing them afterwards. The two essential elements that ensure a successful execution of ambitious projects on a rational and efficient basis are, therefore, tools that enable the scientist to carefully plan experiments and get the most out of the minimum number of experiments in combination with the possibility of fast and reliable data retrieval from databases. Therefore, experimental planning and data management are complementary skillsets for the pre- and post experimental stages.

In-depth coverage of data handling and databases is beyond the scope of this chapter. A number of programmable databases are available on the market and can be custom manufactured to the individual project needs. The original roots of these databases go back to the 1980s when analytical laboratories started to digitalize results with the aid of databases for the purpose of higher efficiency. The databases and visualization tools that are offered by the leading companies of the market today in HTE [13,14] fulfill the requirements to master the challenges that the scientist is confronted with. It should be mentioned here that visualization tools are niche software products, all by themselves, with a high degree of specialization [15]. Generally, these tools, which are a great aid for the visualization of complex data packages, interface nicely with most data formats extractable from databases.

In the stage of experimental design, a number of different methods to carry out a sensible structuring of experiments are available. The choice of the different approaches will largely depend on the nature of the problem to be solved, the knowledge and the data already available, the data quality that can be obtained, the accuracy at which an answer is expected, and the time available for the experimental campaign/the single experiment and evaluation of the data. In the following sections, we will focus on the different approaches for experimental planning and discuss their pros and cons.

11.2.1   Descriptor-Driven Approaches

Descriptors are vectors or scalars that describe certain properties of a molecular or solid-state compound (the descriptors for molecular compounds are also described as SMILES or broad

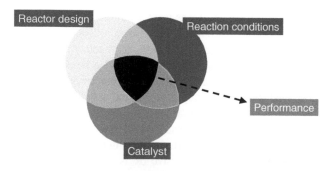

Figure 11.1   Illustration of the interplay of reactor design, reaction conditions, and catalyst.

SMILES [16] or as QSAR approach [17]; a similar approach was developed by Klanner et al. [18,19] for the description of solid-state materials). They can be used for experimental planning as the display of the molecular or solid entity as vectorial or scalar units enables the scientist to rationalize the potentially different behavior of molecular units and solid-state compounds with regard to a certain chemical test. These descriptors can therefore be used very efficiently for library planning as they give the scientist the possibility to oversee the quality of a library of molecular or solid-state catalysts with respect to its quality regarding diversity or clustering of catalyst species. It should be mentioned here that, it is of course evident that the same or similar descriptors can be applied for molecular catalysts and molecular compounds that are potential candidates for a screening vs. their bioactivity. The descriptors needed for solid-state compounds differentiate substantially from the latter and go far beyond them with regard to the numerical description of physicochemical properties. One has to keep in mind that the descriptor approach is novel for solid-state compounds and that it will take more time to get to the state which the descriptor approach for molecular compounds has already reached. Figure 11.2 shows a display of descriptors that can be employed together with molecular and solid-state compounds.

What is the most useful way to make use of the descriptor approach? The descriptor approach is extremely useful when it is applied together with algorithms seeking to cover the experimental space equidistant from points. An excellent example is the optimal coverage algorithm [20]. Through the vectorial and the scalar display of the properties of the catalyst candidates, the experimentalist can get an overview of the diversity of the library and the coverage of the multidimensional space being spanned by the properties of the catalyst candidates. It is clear that this approach is especially useful if a large number of samples are intended to be screened. The drawback of this method is connected to the generation of the descriptors; the descriptors that need to be developed usually require an additional pre-experimental evaluation stage.

11.2.2 Approaches Based on Classical Statistical Designs

The application of classical statistical designs has a long tradition in the evaluation of properties, including catalysis, of inorganic solid-state compounds. Generally, when statistical approaches such as factorial designs, Box & Hunter, Hunter, Plackatt–Burmann, Box–Behnken, Taguchi, or even NK (full factorial) model designs are employed, it is very important to be aware of the nature of the variables

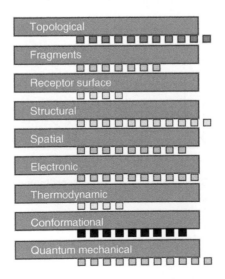

Figure 11.2 List of descriptors that can be employed for describing molecular and solid-state compounds.

and the expected results/answers from the experiment (see classic textbooks for elemental statistics and basic design of experiments, for example [21]). First of all, the scientist has to be aware what the control and response variables are. During the experimental planning, he also has to evaluate whether the design is suitable for the questions that he wants to answer with the experimental approach. In addition, chemical knowledge is essential for an appropriate application of a statistical design. When the experimentalist has an overview of all the input variables, the first control step is to make sure that the design is orthogonal. Second, variations that do not make sense from a chemical point of view have to be excluded to avoid failure of the design. Additional points, especially center points in the design may prove useful, particularly where it is obvious that the density of points may not be sufficient. The success of the design strategy is ensured if one is careful with these control steps. Figure 11.3 shows a graphical representation of common factorial designs.

Figure 11.4 illustrates with three examples why chemical knowledge plays such an important role in the planning of an experiment. The highest risks are associated with an inadequate choice of the density of experimental points. The only way to overcome this problem is by carefully considering the relevant chemical processes in order to carefully plan the experiment and also carefully evaluate the data obtained during the experiment. It is obvious that not every behavior of a newly explored system can be foreseen and a risk of missing singularities or sharp global optima in the behavior will always remain.

11.2.3 Other Techniques for Finding Local Optima

The search for optima within a given experimental space can also be realized by methodologies different from those that we have discussed before. We want to highlight two of them in this context, namely genetic algorithms and neural networks.

The principle of genetic algorithms is largely connected to the idea of evolutionary improvement of the properties of a given specimen within a defined space of variables. A genetic algorithm is, in other words, an algorithm that permutes and converts variables in search of a global maximum or minimum of a multidimensional parameter space. Within this context, genetic algorithms have to be considered under the same precautions that apply for statistical designs as listed above. Also, here the parameterization and the chemical know-how play a large role in the setting up of an adequate parameter space and the choice of the right variables. The rules according to which the algorithm operates are the selection of "winners" from a given set of samples and their potential "improvement" via "mutation" and "crossover". The so-called selection of winners is an algorithm that tests whether the system has

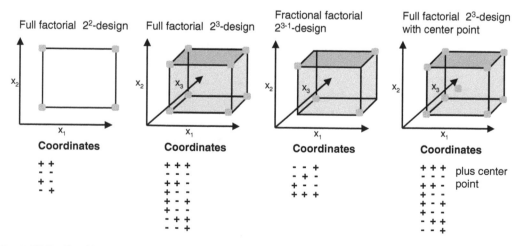

Figure 11.3 Graphic representation of factorial designs.

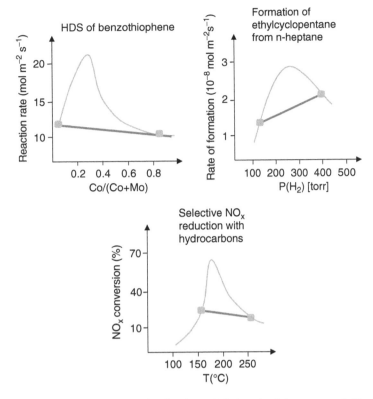

Figure 11.4 Three illustrative examples why the density and choice of points are essential for the experimental planning stage. It can be seen in all three cases that the wrong choice of experimental points can result in missing the opportunity of finding the local optimum.

reached a local extremum and suggests new experiments to confirm and to reach the predicted minimum or maximum. Both "mutation" and "crossover" are algorithms by which new experiments are created by choosing from a given set of experimental parameters. Figure 11.5 illustrates this approach graphically.

Genetic algorithms have been successfully tested for applications in heterogeneous catalysis [22] and, although the results of this example and the general approach look very promising one should keep in mind that generally several thousand experimental data points are required to reach a local maximum or minimum. The original applications that were the target applications for the application of this algorithm were improvements in production chains or other logistical tasks where a single 24 h run delivers generally several thousand data points. If sufficient data points can be generated in the catalytic testing within a suitable variation of chemical parameters, then genetic algorithms provide an interesting approach for optimization.

Even more sophisticated than genetic algorithms are neural networks: these are mathematical models which target the correlation between input and output data by application of an underlying system of neurons and connections (as depicted in Figure 11.6). The neurons and connections are also named "hidden layers," generally the experimentalist has no influence on the structure of these layers. Neural networks were designed to imitate the functionality of biological neural networks; therefore, similar to these systems they also have to undergo a learning or training process prior to predicting optimum compositions to obtain local or global optima.

Additionally, neural networks are an interesting approach for system optimization; still one has to take into account that (1) the training phase requires a certain amount of time and experience (both over- and under-trained networks will tend to give false readouts) and (2) generally the data

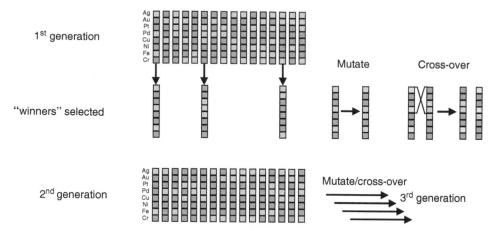

Figure 11.5 Graphical illustration of the evolutionary approach for optimization.

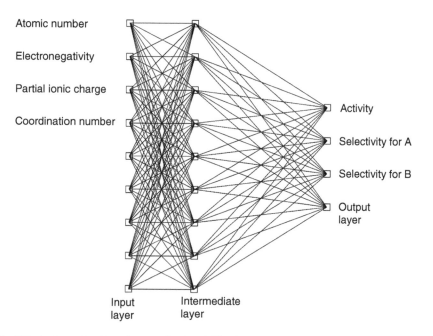

Figure 11.6 Principal construction of a neural network. The input and output data are examples of potential data sources of interest.

correlation will be logical but nondeterministic, which means without insight into the underlying physicochemical correlation. From the point of view of the author, further developmental work is required to make neural networks a powerful tool for HTE.

11.3   STAGE I AND STAGE II SCREENING

The complexity of an HTE screening workflow has led to two different approaches that are employed for the identification of prospective catalyst candidates and useful reaction conditions. Both approaches are defined as experimental stages and are conclusively called Stage I and Stage II. Stage I screening is also known as mass screening and is mainly devoted to the rapid screening for new, potentially useful compounds for a given conversion. The emphasis is clearly on the large number of compounds that is tested for useful properties. Therefore, it is also accepted that the reaction conditions at which the compounds are tested may sometimes be different from the conditions of a potential technical process.

With regard to the ranking of the large amounts of material, a low level of analytical information is accepted to accelerate the screening process. In many cases, conversion and selectivity that are calculated from Stage I analysis data will have quite a substantial relative error attributed to them, so the thresholds for classification of the catalyst samples have to be chosen correctly. The reduction in analytical information often extends to include systems for which the presence or absence of a target component is a key classification criterion. Stage I screening is generally considered extremely useful for the exploration of catalytic conversions where little is known about useful catalytic element combinations or potentially active inorganic structures. The libraries employed in a Stage I screening approach will, in most cases, be very diverse and the same strategy of diversity has a greater value for the reaction conditions: even "esoteric" conditions may lead to a success. The goal of the Stage I approach is the systematic mapping of elements and element combinations within the given framework of reaction conditions. As a final result, the scientist obtains a panoramic overview and hopefully ends up with a few lead compounds for further investigation.

Complementary to Stage I screening is the Stage II approach: here the emphasis is mainly on the technical proximity of the screening conditions with regard to the target conditions of the relevant technical process. In line with this focus of relevant reaction conditions is the focus of industrially relevant catalyst preparation and preparation techniques that are imaged in focused libraries of identified catalytic lead structures or compounds. The unit operations employed for the preparation of the active compounds should be as realistic as possible and issues such as purity of sources, prices of materials, or compounds employed in the synthetic effort should be carefully considered. Expensive or wasteful synthetic efforts for ligand synthesis, potential virulence of the inorganic catalyst, and other factors also govern the search.

The analytical effort in Stage II screening is generally a very detailed one. Here, a greater emphasis is put on the particular analysis of the compounds. The error of the analysis should be as small as possible and in many cases even traces of side products play a role. The systematic approach in Stage II screening also entails the introduction of all knowledge available from, e.g., scientific and patent literature. In many cases, development projects in Stage II will be connected to existing catalyst systems and their improvements; therefore, a knowledge-based design is essential for the success of the screening approach. Table 11.1 sums up the most important features of Stage I and Stage II screening.

Table 11.1 again illustrates the complementary orientation of both the stages. In most research programs, the simultaneous use of both the stages has a tremendous value for the progress of finding a solution for the catalytic challenge. Stage I will usually be a source of catalytic candidates that can be evaluated in detail in Stage II. Vice versa, refined hits of Stage II screening can be checked on a broader basis with regard to composition, dopants, or physical treatments in Stage I. In the latter case of course, only effects that have significance over the analytical threshold of the Stage I process will be visible.

Table 11.1 Distinction of Stage I and Stage II Screening

Stage I	Stage II
Maximum sample throughput	Approaching real conditions
Reduced information $(+/0/-)$	Existing system knowledge
Analysis of target components	Detailed analysis of products
Used for new discoveries	Used for continuous improvements and key advances

The overall strategy of the screening process is depicted in Figure 11.7. The concerted approach is devoted to a funneling strategy, where the number of samples that are relevant is systematically reduced and the final goal of a single catalyst that is superior for the industrial application is never lost as the project focus. Owing to the above-described funneling strategy, it is also desirable to keep the number of lead compounds limited. An overflow of hits does not contribute to the fast identification of the best catalyst and, in many cases, the screening approach will have to be thought over.

11.4 ANALYTICAL TECHNIQUES FOR SCREENING

Analytical precision and time required for a certain analysis are, generally speaking, tradeoffs that have to be made in a lot of cases. As the goal within the framework of Stage I and Stage II screening is the acceleration of conventional testing procedures, rapid analytical methods are required in order to keep up with the desired screening speed. Here, the philosophy can be to push conventional analytical tools to the limits with regard to analysis times, while trying to still be able to have an acceptable error of the analysis. The other option is to come up with new analytical solutions either on the side of new analytical techniques or the "intelligent" use of known techniques. It is fair to make a division of the analytical tool sets that can be employed for Stage I and Stage II screening, as the needs for the two screening stages differ quite substantially with regard to speed and accuracy.

Table 11.2 enlists analytical techniques employed in the two screening stages. From the table it is evident that there are a lot of parallel or quasiparallel analysis techniques but fewer techniques that are employed in a fast sequential mode. Evidently, rapid GC analysis is in the focus of Stage II screening as the most useful analytical technique.

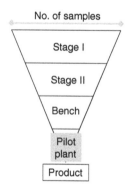

Figure 11.7 Illustration of the funneling technique in HTE.

Table 11.2 Analytical Techniques Employed in Stage I and Stage II

Stage I: Parallel and quasi-parallel analytical techniques	IR thermography Photoacoustic analysis Photothermal deflection REMPI Adsorption technique
Stage I: Sequential analytical techniques	Mass spectrometry
Stage II: Parallel analytical techniques	—
Stage II: Sequential analytical techniques	Gas chromatography Gas chromatography coupled with mass spectrometry, multidimensional gas chromatography

We focus the following discussion on analysis techniques for small molecular entities. The reader should be aware that there are also a number of solutions for analytical tools that are useful for fast HPLC, GPC, or even DSC. The classification principles that apply for these techniques are the same as for the analytical tools described in this chapter.

For IR thermography and mass spectrometry, it is rather obvious that their main application lies within Stage I screening. Especially, IR thermography still receives substantial criticism as a screening tool because the technique is usually only indicative of the activity of a given sample, regardless of the product composition [23–25]. This is of course a major drawback as product selectivities remain obscure with this technique. In combination with spectrally resolving detectors, the thermographic technique has certainly more potential in terms of accurate product identification and quantification than initially perceived by many groups [26–28]. For clarification, it should be emphasized that IR thermography is a true parallel analysis technique with simultaneous analysis of the samples. The following techniques can only be subsumed under the term quasiparallel (photoacoustic technique, thermodeflection technique, resonance enhanced multiphoton ionization [REMPI]) or fast sequential (mass spectrometric analysis, gas chromatographic analysis).

Mass spectrometric analysis systems are advantageous when it comes to product analysis with regard to selectivity of desired or undesired reaction products and can thus overcome the disadvantages connected to IR thermography [29–32]. A number of different ionization and isolation techniques contribute to the resolution of the analytical tool. Typically, these techniques are used for Stage I screening where the amount of information is reduced and the primary goal is rapid analysis. It should also be mentioned here that IR thermography can be employed as a preselection tool to identify suitable candidates for the more time consuming MS analysis on candidates in the right activity window. Klein et al. [33] introduced an integrated and fully automated reactor system that allows the combination of both analytical techniques in an efficient manner.

Senkan et al. [34] introduced REMPI analysis technique as a Stage I tool and exemplified its applicability with the example of a dehydrogenation reaction. The principle of this analysis method is based on sample ionization via laser light and subsequent detection of the ionized reactor effluent at dedicated electrodes at the reactor exit. Owing to a number of limitations connected with the analysis technique, it has to be considered of restricted applicability.

Photothermal deflection spectroscopy [35], a technique used by Symyx for the rapid optical identification of product effluents by IR laser light, can also be regarded as a Stage I analysis technique. Also taking advantage of IR adsorption of product molecules is the photoacoustic analysis of reactor effluents with the aid of laser light [36]. The principle of this technique is based on the acoustic analysis of light absorption by vibrational or stretching modes of a molecule in the IR. The absorption of light will lead to excitation of the molecule and the molecular relaxation to a thermal signal which is detectable as sound via a microphone setup. This technique also has limitations with regard to the spectral absorptions in the IR of the entities, which shall be detected and available laser wavelengths of sufficient laser power. In some cases one may find that certain absorptions of a product and side product molecules may overlap (examples are, for instance, ethylene and propylene, or acrylonitrile, acetonitrile, and HCN) and it may be hard to find alternative absorption bands, especially in the fingerprint region. This limits the capabilities of the analysis systems to reaction product mixtures with limited ambiguity. It should be pointed out here that with regard to the accuracy of analysis method, the photoacoustic analysis certainly exceeds the capabilities of commonly applied Stage I analysis techniques. The sensitivity range with regard to product molecules lies within the ppm range and shows a high linearity with regard to the concentration range over several orders of magnitude [36].

Product detection via product adsorption and subsequent analysis has also become a standard technique for Stage I screening [28,37–39]. In general, techniques, which are capable of indicating a physical change that can easily be detected like a change in color of the adsorbate, are used in accord with this method. The analysis of the adsorbate can also take place via optical methods in a parallel fashion.

The list of analysis techniques that is available for Stage I screening is quite notable; nevertheless, there is still a large demand for new analysis methods for Stage I screening which bring forward the field of HTE. It should especially be noted that the need for truly parallel techniques is high, but these should preferably be optical to empower very short analysis times.

Analysis in Stage II has to essentially fulfill different tasks than analysis methods for Stage I screening. Of course, for Stage II analysis techniques time of analysis is as essential as for the Stage I, but the accuracy of the analysis underlies the highest quality criteria. As Stage II screening is dedicated to obtaining process relevant data, it is essential that no compromises concerning any errors caused by the analysis method are made. This follows the same ratio as of the engineering of Stage II reactor unit and makes Stage II screening essential in modern industrial catalyst development.

In most cases, GC analysis methods will be the methods of choice for Stage II screening, especially when online analytical measurements are required. Gas chromatography is a well-established methodology for a broad range of analytical applications and the workhorse in many analytical laboratories. The history of the use of online process GCs ranges back decades and since then major developments have been achieved. High-throughput experimentation in this case can take advantage of those years of development in online GC and process GC analyses [40,41].

Gas chromatography is especially useful for gas-phase analysis of partial oxidation, hydrogenation, or hydroconversion products as in many cases a full carbon balance (educts, products, and all side products), in order to evaluate sample performance. As the detection and quantification of permanent gases such as N_2, O_2, CO, and CO_2 and also of higher boiling compounds are standard separation problems for gas chromatography, it is wise to employ the method regarding this problem.

The key to attractive solutions to high-throughput compatible GC-analytical technologies will require the development of fast separations. There are again several ways to shorten the analysis time required for a GC separation. The choice of the GC column is essential but will not be discussed in this context as this would go far beyond the framework of this chapter.

An alternative option that is attractive is the use of multidimensional GC analysis. Figure 11.8 a shows a simplified column setup, which is an example for multidimensional analytical purposes.

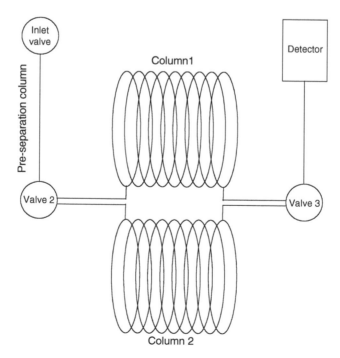

Figure 11.8 Potential column setup for multidimensional GC analysis.

The two different columns are connected via valves, which can be actuated at the corresponding residence time to give access to the sample fraction of interest. The sample fraction of interest is then separated on the second column. This setup is especially attractive with regard to high-throughput screening for the analysis of high- and low-boiling compounds on different column types (Figure 11.9). It should be noted that in many cases back-flushing of the column or the use of parking volumes is mandatory. The reader is again advised to refer to [40,41] for further details.

11.5 SYNTHETIC APPROACHES FOR HIGH-THROUGHPUT EXPERIMENTATION AND COMBINATORIAL CHEMISTRY

As vitally important as the capabilities for experimental planning, screening, and data analysis are the procedures for preparation of inorganic catalysts. In contrast to the procedures usually applied in conventional catalyst synthesis, the synthetic techniques have to be adapted to the number of catalysts required in the screening process. Catalyst "production" can become a "bottleneck" and it is therefore necessary to ensure that HTE- and CombiChem-capable synthesis technologies are applied to ensure a seamless workflow.

In this context, from a chemical point of view it is very important to differentiate between molecular compounds and solid-state materials. Table 11.3 enlists and summarizes the differences of inorganic solid-state materials and organic entities. Both classes of inorganic catalysts are essentially different in their nature and therefore the synthetic approaches differ substantially. For molecular entities in general, a number of analytical methods can be applied in order to ensure the successful completion of a synthesis effort. The synthetic pathways that are known today in organic and metal-organic chemistry, offer a rich portfolio of methodologies that enable the scientist to carefully plan the synthetic effort and choose the most efficient synthetic pathway in order to obtain the desired molecule with appropriate purity. The challenge for the scientist active in the field of HTE and CombiChem is now to translate the recipes to HTE- and CombiChem-capable methodologies to achieve adequate throughput and to avoid bottlenecks.

Especially for inorganic functional materials, it is evident that for the majority of cases the situation is completely different. Even though today a full range of analytical methods is at hand, it is impossible to fully characterize a solid-state material. This does not even take into account that catalysis (with the help of solid-state catalysts) is an interface phenomenon and traces of impurities very

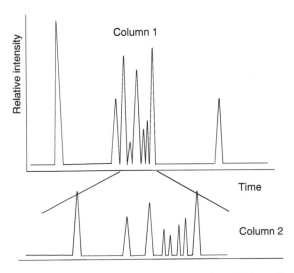

Figure 11.9 Example for separation of a complex product mixture via multidimensional GC analysis.

Table 11.3   Differences in the High-Throughput Experimentation for Molecular Entities and Inorganic Solids Applied for Catalysis

Organic Molecular entities	Inorganic Functional Materials
Discrete molecular entities	Three-dimensional structures, often multielement composites, potentially metastable materials
Finite number of active centers, often well characterized, can in many cases be modeled, good understanding of chemistry at a molecular level	Only for few materials basic understanding of type and function of active centers
Purities of above 85% achievable via combinatorial synthesis	Pure substances often without catalytic effect, especially multicomponent systems with defect structures show largest wealth in catalysis
Highly developed screening procedures for biological activity available	Characterization of numerous, often independent parameters, challenging
Descriptors for library diversity well developed	Basic development work for descriptors still required
Good availability of synthetic building blocks and methods	Adaptation of complete unit operation for automated synthesis required

often "make the day." In numerous cases, solid-state materials that serve as catalysts are far better characterized by the synthetic procedure that is employed to obtain the desired material [47]. The challenges to translate recipes from conventional synthetic methods to methods that are adequate in an HTE- or CombiChem-screening campaign becomes a totally different challenge than that for molecular entities. The quintessence for success is to break down the conventional synthetic procedure into discrete steps, the so-called unit operations, which mimic the procedures conventionally applied on a laboratory or technical scale in order to control the properties of the final solid-state material.

11.5.1   Synthetic Approaches for Molecular Catalysts

In general, for organic entities two different approaches for the synthetic generation can be distinguished. The first approach is the synthesis on solid phase and the second is the parallel synthesis not using polymer bound or unsupported reaction pathways. For both approaches a wealth of automation technology for parallel and pooled synthesis is available today. Most technologies work with solution-based molecular chemistry and the techniques are at hand for automated and semi-automated solution-based synthesis of organic and metal-organic compounds; typical technology platforms supplied by a number of specialized companies can be used for the purposes of liquid handling and reacting liquids in appropriate vessels [42–45]. Preferentially, for purposes of automated handling with non-solid-bound compounds, compounds and products should only occur as fluids. As soon as solid formation takes place usually by precipitation of the product or the need of solid-reactant addition, most of the technology suppliers lack sufficient solutions that allow the automatic handling of the resulting precipitates and powders, and additional manual workforce is therefore required. Fortunately, organic crystalline solids can, in many cases, be obtained in the form of crystals in the micrometer range, which can be filtered off and washed with solvents easily. Technical solutions for the implementation of crystallization reactions as separation steps are available with many manufacturers [43]. These solutions usually contain filtration devices operated over glass frits, Teflon frits, or membranes. It is understandable that precipitation steps with subsequent filtration can be avoided, typically if solvent separation is performed under evaporative conditions, either thermally or via application of vacuum conditions, sometimes even by vacuum centrifugation [43,46]. The technologies on hand are far advanced and appropriate chemistries can easily be performed under inert conditions.

Many of the metal–organic catalysts that are utilized today in catalytic processes can be considered as metal ligand compounds. The general synthetic approach is to identify scaffolds, which lead to an easy access of a library of considerable diversity; Figure 11.10 illustrates this approach. Usually ligands that contain heteroatoms (such as imines or phosphines) are the preferred compounds that give

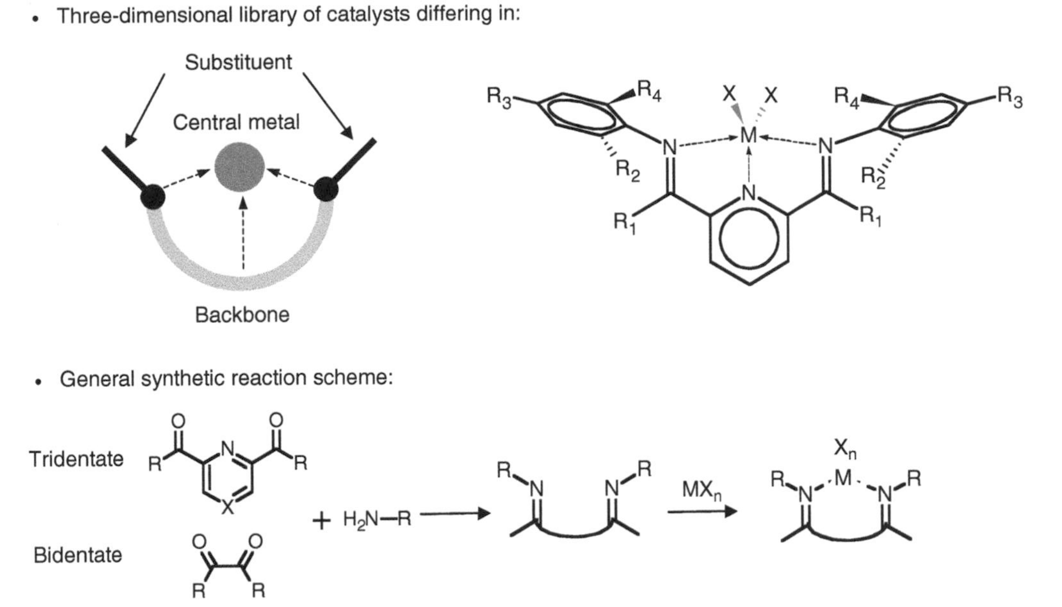

- Three-dimensional library of catalysts differing in:

- General synthetic reaction scheme:

Figure 11.10   The setup of a library based on imine ligands. The general synthetic route is established by reacting a dicarbonyl backbone with an amine substituent. Libraries of different amines and dicarbonyls offer the perspective of obtaining sets of different ligands.

access to libraries with regard to the synthetic strategy; more generally speaking, also other chemical transformations in organic chemistry offer similar perspectives for the scaffolding technique.

The scaffolding technique still requires intense validation studies to develop and optimize the different synthesis conditions required for the different backbones and ligand systems. Different electronic and steric properties may lead to synthetic conditions and in some cases even to alternative synthetic routes to arrive at a given ligand structure.

Furthermore, an important factor concerning the synthesis of molecular catalysts is the purity of the resulting molecular compounds. In the case of impurities beyond a certain threshold, it is questionable whether testing makes sense at all. Impurities may act as catalyst poisons and may lead to a completely false indication of the catalyst performance. Still, the analysis of the compounds obtained during a synthetic step is essentially important. Even if the synthetic step leads to partially impure products, a purification step with highly advanced methods such as HPLC in a high-throughput mode might be necessary and can lead to acceptable results with regard to the quality of the library. (Figure 11.11)

11.5.2 Synthetic Approaches for Solid-State Inorganic Catalysts

The synthesis of inorganic solids from solution is a very complex process aiming at more than the pure separation of a compound from solution and further purification as for organic or metal–organic catalysts [47,48]. Chemical conditions play a larger role in terms of performance properties of the obtained inorganic solids, than for organic materials, as the resulting solid-state materials can, in nearly all of the cases, not be purified after the formation step and the synthesis history will generally be the reason for the performance-limiting behavior. For inorganic precipitates, the work-up of the resulting precipitate, factors such as aging conditions or washing steps are crucial synthetic steps and have to be closely monitored. Similar essential synthetic features can be identified for impregnation catalysts; the usual factors are the choice of the salt/precursor, pH of the solution vs. the isoelectric point of the support material, potential additives to the impregnation solution, amount of

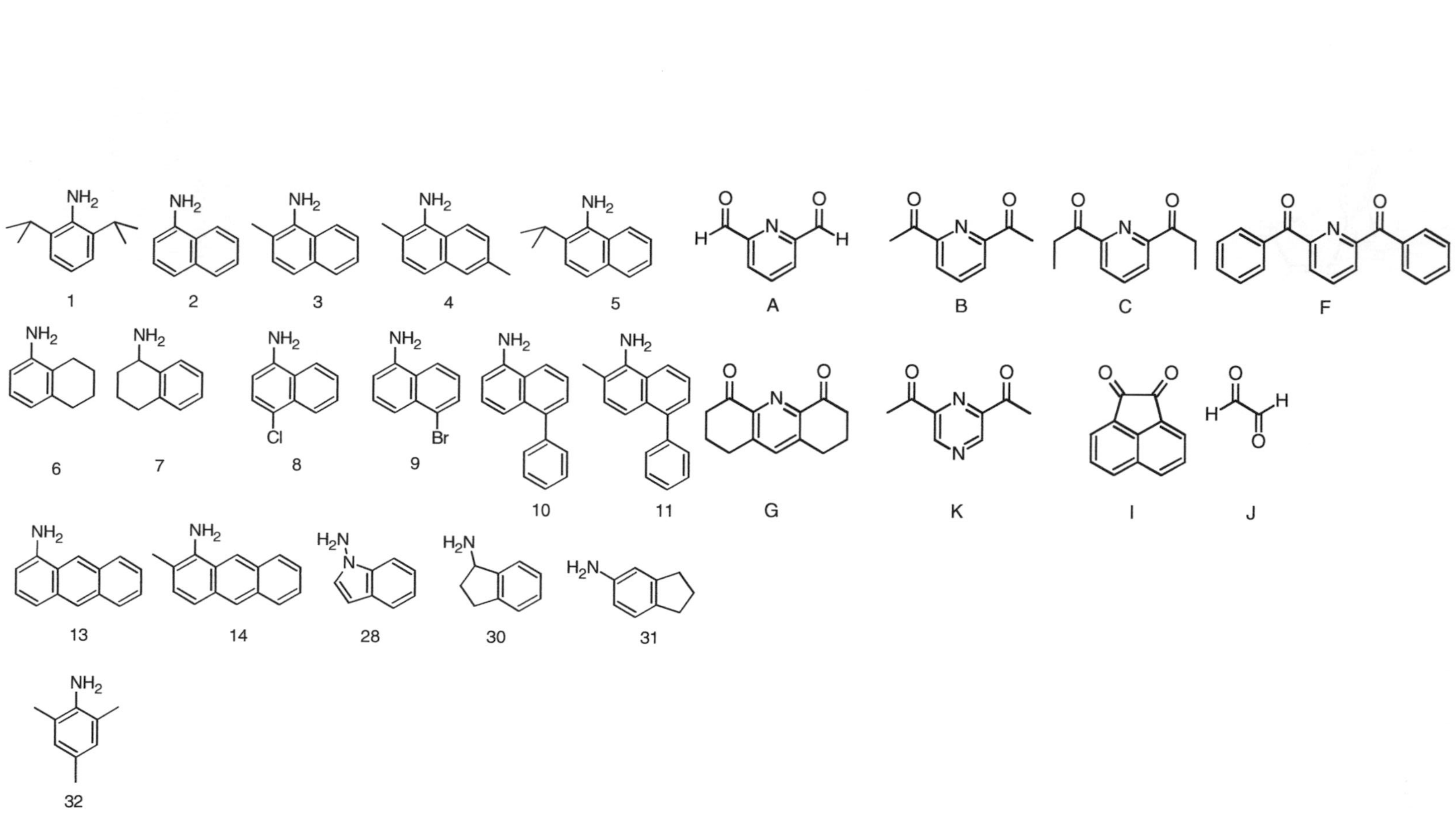

Figure 11.11 Library of amines and diketo compounds, which was employed in a polymerization catalyst study described in Section 11.7 [87]. For simplicity, amines have a number and diketones a letter as library name.

liquid used for the impregnation, and other sensitive factors that influence the resulting catalyst. Typical for a large range of functional inorganic materials is the thermal post treatment, often under specific gas mixtures. These treatments are essential as the oxidation state (usually of oxides) of the active metal centers can be influenced by these procedures. Treatment under reactive gas atmospheres can also be used for thermal alloying of metals or formation of mixed oxides, or the transformation of oxides into metals and vice versa. Heat ramps and gas compositions of the thermal treatment will also influence the resulting particle sizes and the phase composition of the solid-state sample. An additional preparative aspect involved with inorganic solids are shaping procedures, even simple operations like sieving and tabletting are far beyond the scope of the automatic synthesis platforms used for organic synthesis. To conclude all of these preparation procedures that tune the properties of the inorganic material, detailed control over every single synthetic operation (the "unit operations") is essential. In Section 11.6, Stage II testing in gas-phase catalysis, we will discuss some of the issues related to the performance as a function of different synthetic procedures.

There is still a strong debate with regard to the issue as to whether synthetic operations should be divided into Stage I and Stage II operations. The reason for this debate is the fact that synthetic operations, at whatever scale they are performed, should in any case be scalable and translate into the later technical application. The following becomes clear from the discussion above: the guideline for Stage II synthetic operations will in general be such that the unit operations employed for the synthesis of the material will directly lead to scalable synthetic operations that close the link to laboratory procedures and ultimately the technical production. In extreme cases one may even employ catalyst libraries which stem completely from commercially produced materials to ensure straightforward production of a material in case it proves to be a hit.

For these reasons it is doubtful whether synthetic operations that do not only fulfill requirements of Stage I screening with regard to the synthesis of large numbers of samples also do make sense in an integrated screening approach. There are numerous arguments for the use of these synthetic methods, as they can be regarded as an integral piece of the Stage I screening process. Still as soon as the issue of scalability occurs and the questions of relevance with regard to the later technical material production are addressed, a missing piece occurs. As long as a translational function in between the Stage I synthetic process is known, which can be directly linked to larger scale synthesis, the integrity of the screening process is assured and Stage I screening has a proven value in the screening process.

All issues about the preparation of inorganic solid catalysts that we have discussed above apply for the rapid sequential synthesis of inorganic solids by automated methodologies. As for HTE - and combinatorial approaches in organic chemistry, technical solutions for compound synthesis can be obtained commercially for a number of synthetic problems [42–45] and it is beyond the scope of this chapter to discuss all the technical details.

It should also be mentioned here that a number of publications deal with the true parallel synthesis of inorganic solids by sputtering and chemical vapor deposition; however, these approaches are of major use for other fields in materials science than for catalysis. For a broad overview of synthetic and screening efforts refer to [49].

11.5.3 Combinatorial Synthetic Approaches for Solid-State Inorganic and Molecular Catalysts

A clear transition from a synthetic point of view in comparison to parallel or fast sequential approaches is the application of the split and pool concept for the synthesis of molecular compounds (for an overview of synthetic technologies see [50] or for inorganic materials see [51–53]. The split and pool concept goes beyond the pure application of parallel reaction sequences, but relates to the concept of single bodies as entities of catalytic materials themselves (in the case of inorganic catalysts) or carriers binding molecular entities. Figure 11.12 illustrates the schematic steps of the split and pool synthesis. After the first synthetic step of binding the single components (A, B, and C) to the bodies a second step is performed, where all bodies are united (the so-called pool step). This step

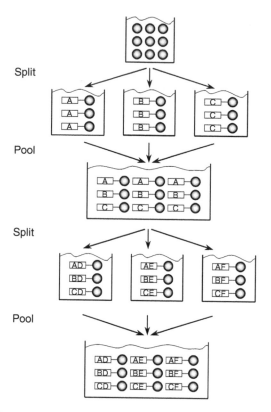

Figure 11.12   Illustration of a sequence of split and pool steps for the generation of single bodies containing either different organic molecules or inorganic materials.

is followed by a split step, which is used for a statistic separation of the pooled bodies followed by further binding (of the entities D, E, and F), pool and split steps. The beauty of this synthesis technology lies in the fact that with a minimum amount of synthetic operations all possible permutations of single components that can be attached to the single bodies can be obtained utilizing the synthetic steps. The only disadvantage is that the identity of the samples is lost. Tracking of the identity can usually just take place by the so-called tagging technologies: for each synthetic step a specific tag is added to the single body that enables a posthistoric deconvolution of the synthetic history.

The split and pool technology has a proven track record in organic chemistry; however, the translation into inorganic materials synthesis is far from trivial. The considerations for inorganic materials synthesis go beyond purely synthetic operations but address questions as to whether a different sequence for the addition of components also results in a different material. This issue can easily be exemplified; for example, if 3 different amino acids are synthetically attached to a support, 27 different combinations of oligomers can be distinguished. In the case of inorganic materials the question now occurs is whether or not the synthetic sequence has an influence on the resulting product (good examples for materials where little effect of the sequence is expected are alloys of metals)— if this is the case the number of materials obtained is reduced to 10. The second question related to the fact of catalyst testing is how single bodies can be efficiently tested.

11.6   TESTING OF CATALYSTS IN GAS-PHASE REACTIONS

The performance testing of inorganic catalysts for a target reaction on a technical scale close to the process conditions is the most critical test of the research effort of a given scientific project.

One has to keep in mind that even today with the level of sophistication that is available for characterizing inorganic molecular compounds and inorganic materials, there are practically no examples of industrial catalysts that have been developed without the repeated testing and cross-checking of all potential improvements via the direct test of the inorganic compound in the target reaction. Among academic and industrial chemists, it is very often questioned and debated whether or not the procedure of performance testing for catalyst development is too simplistic with regard to the wealth of analytical techniques available today. The answer from an industrial viewpoint will usually be that "the truth comes out of the reactor," meaning that no performance test is as reliable as a test that is as close as possible to the conditions of the target industrial process.

Owing to the reasons discussed above, it is essential for a scientist working in the field of catalysis that the basic principles of performance testing of inorganic catalysts are well understood and applied. Analyzing developments concerning "good laboratory practice" and their success, especially throughout the pharmaceutical industry, the authors are urged to ask the reader to adapt the general rules denoted below as a suggestion for "good catalyst testing practice."

The development of inorganic catalysts does not only concern the development of inorganic compounds and materials but also the optimization of mass and heat transfer. The essence of performance testing is to test candidate catalysts under conditions in which their performance (1) can be compared so that the catalysts can be ranked and (2) the test result has a significant relevance with regard to the target application. Most ideally, catalytic test conditions are employed, which provide true microkinetic data that enable the scientist to distinguish between the behavior of the "active centers" of the different candidate compounds. In many cases, this estimation of the microkinetic properties will not be possible unless model systems are used as catalyst candidates. In other cases it may be unwise to look for the microkinetic behavior as the intrinsic properties of the inorganic catalyst demand testing under different conditions or process-related issues leave no options for variations in the modes of testing. In both the mentioned cases, the comparison of the catalytic compound materials will take place on the basis of macrokinetic data representing the combination of effects caused by diffusion and reaction. It is important that the scientist always has an awareness if the performance test is taking place under limitations with regard to mass and heat transfer and has to interpret the obtained results regarding the limiting technical background of the test procedure. The following pieces of literature are recommended for further reading on practices of catalyst testing that dwell deeper into the issues related to mass and heat transfer, and the construction and the pros and cons of laboratory equipment for catalyst testing [54–56].

The main focus of the following considerations is on catalysis using inorganic materials. Similar considerations come into play for catalysis with molecular compounds as catalytic components; of course, issues related to diffusion in porous systems are not applicable there as molecular catalysts, unless bound or attached to a solid material or contained in a polymeric entity, lack a porous system which could restrict mass transport to the active center. It is evident that the basic considerations for mass transport-related phenomena are also valid for liquid and liquid–gas-phase catalysis with inorganic materials.

Catalyst testing can be performed in a number of different reactor configurations all having pros and cons: the scope of this chapter is to present an outline for integral catalyst testing in tubular reactors, as this has the greatest importance in HTE. For further reading on other useful reactor types and configurations, and general procedures for catalyst testing we recommend [56–58].

Any given catalytic material can be abstracted based on the same underlying similar architecture — for ease of comparison, we describe the catalytic material as a porous network with the active centers responsible for the conversion of educts to products distributed on the internal surface of the pores and the external surface area. Generally, the conversion of any given educt by the aid of the catalytic material is divided into a number of consecutive steps. Figure 11.13 illustrates these different steps. The governing transport phenomenon outside the catalyst responsible for mass transport is the convective fluid flow. This changes dramatically close to the catalyst surface: from a certain boundary onwards, named the hydrodynamic boundary layer, mass transport toward and from the catalyst surface only takes place

1. Convection towards
   the catalyst particle

2. Diffusion of reactants through
   hydrodynamic boundary layer

3. Pore diffusion

4. Physisorption or chemisorption

5. Surface reaction

6. Desorption of products

7. Pore diffusion of products

8. Diffusion of reactants through
   hydrodynamic boundary layer

9. Convection away from the
   catalyst particle

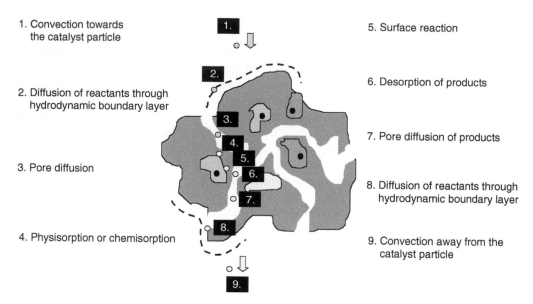

Figure 11.13 Elementary steps of a catalytic reaction.

by diffusion. This is also the case for mass transport in the pore system and the diffusion taking place inside the pore is called pore diffusion. It is different compared with the diffusion in the boundary layer and the free gas, due to frequent collisions of the molecule with the pore walls and thus the free pathway of the gas molecule is shortened substantially. The diffusion coefficient for pore diffusion usually changes due to frequent collisions of the molecules with the pore wall as a function of the pore size and the relative pore volume (still as a rule of thumb it can be approximated with one-tenth of the diffusion coefficient of the same molecule in the free gas; for an exact calculation see [54], Chapter 3.4.2). The next step on the reaction coordinate is the adsorption of the educt by chemi- or physisorption on the catalyst surface. Typical adsorption energies for physisorption are in the range of <40 kJ/mol, energies for chemisorption are higher (typically >80 kJ/mol, still one should keep in mind that the original definition for the distinction between physi- and chemisorption is an energy of 40 kJ/mol [59]). The next step is the conversion of the adsorbed educts (e.g., the catalytic reaction, transfer of electrons, regrouping of atoms), followed by desorption of the formed product. Again this process is followed by diffusion out of the pore system and the hydrodynamic boundary layer.

In the case that the chemical reaction proceeds much faster than the diffusion of educts to the surface and into the pore system a starvation with regard to the mass transport of the educt is the result, diffusion through the surface layer and the pore system then become the rate limiting steps for the catalytic conversion. They generally lead to a different result in the activity compared to the catalytic materials measured under non-diffusion-limited conditions. Before solutions for overcoming this phenomenon are presented, two more additional terms shall be introduced: the Thiele modulus and the effectiveness factor.

For reasons of simplicity, the Thiele modulus will be defined and calculated for a catalyst plate with pore access at both ends of the plate and not at the bottom or top. Note that for most cases in real-life applications the assumptions have to be modified using polar coordinates for the calculations. The Thiele modulus φ is therefore defined as the product of the length of the catalyst pore, l, and the square root of the quotient of the constant of the speed of the reaction, k, divided by the effective diffusion coefficient D_{eff}:

$$\varphi = l\sqrt{k/D_{\text{eff}}}$$

(11.1)

The Thiele modulus is related to the concentration dependence in a catalyst body by the following equations representing the ratios of the hyperbolic cosines:

$$\Gamma = \frac{\cosh \varphi (1 - \xi)}{\cosh \varphi} \tag{11.2}$$

$$\frac{c}{c_0} = \frac{\cosh \varphi \left(1 - \frac{z}{l}\right)}{\cosh \varphi} \tag{11.3}$$

where Γ represents the quotient of the educt concentrations, $\Gamma = c/c_0$, and $\zeta = z/l$ is the quotient of the coordinate along the catalyst pore z and the length of the pore l (both again related to a catalyst plate).

The concentration dependence of z/l vs. c/c_0 is plotted in Figure 11.14a. It can be seen that from a Thiele modulus $\varphi \geq 3$ the educt does not reach the internal part of the pore. The inner part of the pore system is useless for catalysis. This is especially relevant if expensive metals serve as active components on a porous carrier, which are then wasted. There are chances to master this diffusion limitation, which will be discussed later in detail. Another important variable is the efficiency factor η. The efficiency factor η is defined as the quotient of the speed of reaction r_s to the maximal possible speed of reaction r_{smax}. η is related to φ as the quotient of the hyperbolic tangent of the Thiele modulus φ:

$$\eta = \frac{\tanh \varphi}{\varphi} \tag{11.4}$$

Figure 11.14b illustrates the dependence of the Thiele modulus φ on the efficiency factor η. Two characteristic regimes can be seen: for small values of the Thiele modulus, $\varphi \leq 0.18$, the efficiency factor will be close to 1, $\eta \approx 1$; for values of the Thiele modulus larger than 3, $\varphi \geq 3$, the efficiency factor will be antiproportional to the Thiele modulus, $\eta \approx 1/\varphi$. Practically this means that from a value of the efficiency factor of 3 onwards no educt reaches the interior of the catalyst particle (which in our case is approximated as a plate).

Qualitatively, one can now deduct the apparent activation energy E_{As} for the case of a diffusion-limited reaction. If we plot the logarithm of the product of the efficiency factor and the constant for the speed of reaction $\ln[k\eta]$ against $1/T$, a typical curve with three regimes can be seen (see Figure 11.15).

- *Kinetic regime*: This is the regime of small Thiele modules. The pore system of the catalyst with its interior surface is completely accessible for the educt. $\eta \approx 1$ and $k\eta \approx k$; as $\ln k \approx \ln k_0 - E_{As}/RT$, that the apparent activation energy is practically identical with the true activation energy.

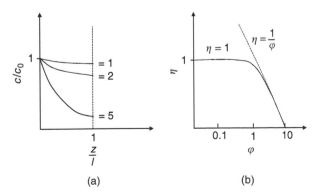

Figure 11.14 (a) Plot of c/c_0 vs. z/l displaying three different concentration gradients for $\phi = 1$, 2, and 5; (b) Plot of η vs. ϕ displaying the regimes of $\phi = 1$ and $\phi = 1/r$.

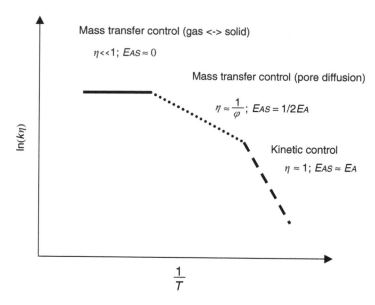

Figure 11.15 Plot of ln[kn] vs. 1/T displaying the three different reaction regimes.

- *Regime of pore diffusion*: In this regime $\eta = 1/\varphi$ and $\phi = \sqrt{k / D_{eff}}$; this leads to the fact that: ln*[kη]* $= \frac{1}{2} \ln k_0 - E_A/2RT - \ln l + \frac{1}{2} \ln D_{eff}$. Therefore, E_{As} in the pore diffusion regime equals half of the value of the true activation energy of the reaction, $E_A : E_{As} = E_A$.
- *Regime of transport limitation*: here $\varphi \gg 1$ and $\eta \ll 1$. In this regime the reaction is clearly dominated by the diffusion through the hydrodynamic boundary layer. The apparent activation energy under these conditions gets close to zero. Every educt molecule reacts instantaneously on the outer catalyst surface, no educt diffusion inside the catalyst particle takes place.

It can be seen in the plot in Figure 11 that E_{As} shows a clear temperature dependence. For rising temperatures the mass transport limitation can be observed, which leads to a lowering of E_{As} by a factor of ½ in the pore diffusion regime down to 0, owing to the shift of the reaction from the interior of the pore system of the catalytic particle to the outer surface. In the final state, the diffusion through the boundary layer becomes the rate-limiting step of the reaction.

There are a number of generally applicable ways of overcoming diffusion limitations. First is the variation of the particle size (and thereby the pore length l). To maximize the effectiveness factor and avoid intraparticle diffusion limitations, smaller particles can be employed as test fractions. The usual procedure is to grind the catalytic material and use a certain sieve fraction as test granulate. Generally, there is a natural limit concerning the particle size for the catalytic material in the reactor, which will be determined by the increasing pressure drop resulting from the smaller particle fractions. Beyond a certain pressure threshold, either the limitations of the testing equipment or a possible change in the reaction behavior are the usual limitations that account for the boundaries of the approach described above. A good rule of thumb is that the pressure drop should not be greater than 20% of the operating pressure. Another approach for overcoming intraparticle diffusion limitations is the use of catalytic materials that have a thin layer of the active component on or slightly below the catalyst surface (so-called egg-shell catalysts).

An approach for checking and overcoming extraparticle diffusion limitations is the variation of the gas hourly space velocity (GHSV) and the amount of catalyst used to maintain the ratio catalyst weight per molar flow of the educt constant. Figure 11.16 illustrates the typical behavior that can be observed upon checking for intraparticle and extraparticle diffusion limitations by varying catalyst particle size and GHSV at constant ratio of catalyst weight to molar flow. It is highly recommended

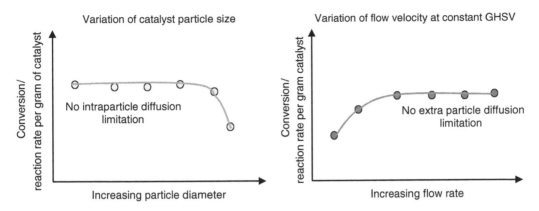

Figure 11.16 Illustration of the effects observed upon variation of increasing catalyst particle size and increasing flow rate at constant ratio of catalyst weight to molar flow ratio.

to study this behavior with select reference samples, which are representative of the libraries to be tested prior to entering the screening mode (see also Figure 11.16).

Another important aspect of testing is the avoidance of limitations concerning heat transfer. This is usually a problem that occurs with intrinsically exothermic reactions where excessive heat evolution leads to thermal runaway of the reactor. If the heat inside the reactor is not monitored via a temperature measurement, the phenomenon of thermal runaway sometimes becomes hard to diagnose. Consequently, procedures for avoiding thermal runaway are extremely important. Excellent heat conductivity of the reactor material and a dilution of the catalytic bed with inert materials are, in general, the precautions necessary to ensure that the heat transfer does not generate falsified catalytic results. Again, a good rule of thumb is that the catalyst should not be diluted over 5 to 10 times by volume with an inert material. The dilutant particles should be smaller than the catalyst particles (factor of 2 to 10) for good fluid dispersion and transport properties within the catalyst bed and should be inert with regard to the catalytic reaction. Apart from the positive effect on the avoidance of hot spots, diluting a catalyst sample may also have the positive side effect that plug-flow behavior in the reactor is improved.

Concerning the hydrodynamics and the dimensioning of the test reactor, some rules of thumb are a valuable aid for the experimentalist. It is important that the reactor is operated under plug-flow conditions in order to avoid axial dispersion and diffusion limitation phenomena. Again, it has to be made clear that in many cases testing of monolithic bodies such as metal gauzes, foam ceramics, or monoliths used for environmental catalysis, often needs to be performed in the laminar flow regime.

The dimensioning of the reactor is generally done according to the rule that the minimum reactor diameter for gas–solid reactions should be ten times as large as the catalyst particle diameter employed for testing, in order to be able to neglect the wall effects. Concerning the catalyst bed length, it is highly recommended to keep the catalyst bed at a length of 3 to 15 times the diameter of the catalyst bed. This is done to avoid axial dispersion due to inhomogeneities of the differences in the packing of the catalyst particles in the center of the bed and close to the reactor wall. It is important to mention here that the hydrodynamics for liquid fluids and gas/liquid fluid mixtures over a solid catalyst bed often demand a thorough testing to determine whether a good dispersion over the reactor has been achieved, channeling avoided, and testing performed under non-diffusion-limited conditions.

All the rules mentioned above, of course, also apply for catalyst testing in the HTE mode. It should be mentioned here that the trained individual will perceive that especially the testing configurations that a number of groups have developed for Stage I testing are in a critical range when it comes to testing under conditions without transport limitations concerning heat and mass transfer. To prevent falsified results of numerous false hits or missed development candidates, it is of central importance that the testing conditions are illuminated by critical consideration.

Generally, two technical features are the most critical ones when it comes to parallel reactor units: the fluid management and the thermal management over the reactor unit play an important role when it comes to the performance of the HTE reactor. In the section about Stage I testing in the gas phase, we will devote some paragraphs to the basic rules that apply to fluid management. Thermal management is also a critical issue. It is surely an art to ensure isothermal conditions over a larger piece of HTE equipment, like a 48-fold Stage II testing reactor. Here, the main focus is on adequate control and feedback control systems which match the needs of a specific technology. Beyond these rather basic features one has to ensure that thermal runaway does not become a critical issue in multitube reactors for HTE applications. The most critical case is the so-called thermal cross-contamination. Different catalytic substances in a common reaction vessel may be affected by the thermal behavior of neighboring active substances if neighboring catalysts produce large amounts of heat under reaction conditions which cannot be effectively removed. Here, a smart reactor construction and the dilution of the catalyst bed by inert materials are the only ways to avoid falsification of the results.

11.6.1 Stage I Testing of Catalysts for Gas-Phase Reactions

11.6.1.1 General Considerations

The main target in primary screening is the identification of hits out of large catalyst libraries. The target for the number of catalysts tested in parallel should be a few hundred on a small reactor format. Therefore, specific library preparation methods resulting in an efficient way for preparing the single materials are required. The information content of the analytical methods employed is often reduced in favor of a high analytical speed. Accordingly, only the most promising materials from a library are characterized in order to gain additional insights into the properties of these materials. Primary screening approaches are usually employed in ambitious catalyst discovery programs with little or no previous knowledge about the target reaction and the class of materials to be studied. Primary screening approaches further require an integration of catalyst preparation, reactor, and analysis methods. This may result in unconventional preparation methods or reaction conditions that can complicate the transfer of results from primary to secondary screening. Generally, some challenging requirements that have to be met by primary screening reactors are: the reactor design has to accommodate catalysts made by a simple, fast, affordable, and scalable catalyst preparation method. Many catalysts should be tested simultaneously under steady-state conditions and continuous flow. The analysis method should have a high spatial resolution and allow a good estimate of conversion degree and selectivity, while still being very fast and reproducible. Microstructured reactors with cavities in the 100 μm range present a number of interesting properties for applications in primary screening [60,61]. These reactors provide excellent heat and mass transfer properties and can be operated at isothermal conditions. The response times to changes of the reaction conditions are very low and a number of additional static or active elements (mixers, valves, etc.) can be integrated into the device besides the reactor itself. Owing to their potential mass fabrication, miniaturized reactors may prove cost efficient. However, a lot of developmental work is necessary to set up a microstructured reactor system, as many microfabrication technologies are still being considered "experimental" and as the understanding of the behavior of miniaturized reactors is presently evolving .

11.6.1.2 Reactant Distribution for Stage I Screening Systems

In order to carry out comparable catalytic experiments in a high-throughput parallel mode, equal reaction conditions for each library member have to be ensured. This is especially true for reactant distribution, as variances in residence time and space velocity can have a dramatic impact on the performance. For conventional laboratory reactors, reactant flow is usually controlled by individual mass flow controllers for each reactor that use valves to adjust the pressure drop for the fluid flow through the controller, thus regulating mass flow. For reactors having several hundred parallel reaction chambers, such an active control of fluid flow for each reaction chamber would probably not be possible or at least not be economic as long as we do not see dramatic developments and improvements in microstructured devices. Therefore, those fluidic networks have to be realized where a limited number of feed sources is connected to the parallel reactors and further downstream to the analytics. To achieve equal mass flows through the parallel reactors in such a complex network, the basic principles of fluid dynamics should be taken into consideration. The Bernoulli equation, one of the basic equations of fluid dynamics, gives the balance of the potential energy along one flow path. This equation becomes especially simple for the flow of incompressible liquids (with p denoting pressure, ρ the density of the fluid, v the velocity of the fluid, g the acceleration of the fluid, and h the height difference along the fluid path):

$$p + \tfrac{1}{2} \cdot \rho.v + \rho.g.h = \text{const} = p_{\text{total}} \tag{11.5}$$

This means that the pressure drops, i.e., the flow resistances along the separate flow paths, determine the flow distribution in a fluidic network. Furthermore, such fluidic networks can be calculated analogous to electrical networks with Kirchhoff's law for parallel and series connection of electrical resistances:

in series:

$$\Delta p_{\text{total}} = \sum_{i=1}^{n} \Delta p_i \tag{11.6}$$

in parallel:

$$\frac{1}{\Delta p_{\text{total}}} = \sum_{i=1}^{n} \frac{1}{\Delta p_i} \tag{11.7}$$

Therefore, adjusting the pressure drop along the fluid flow path is the key for controlling the fluid flow rate [62]. This means that an equal flow distribution can be achieved by equal flow resistances along the parallel flow paths. If so, a high-throughput system should be designed in such a way that the materials to be tested do not affect the fluid distribution, as the materials are intended to be different, which means that they may generate different pressure drops in a continuous flow situation. Hence, the flow resistance through or around the material should be small compared to total flow resistance along this particular material. The common solution to this problem is to include a defined flow resistance element into each flow path in order to generate a particular pressure drop. In Stage I systems, miniaturized elements inside the flow path are usually used to form passive flow restrictors. An approach of using capillary manifolds or microchannel manifolds, the so-called "binary trees", was, among others, described by Bergh and Guan [63] (see Figure 11.17). A channel network is built onto a flat wafer substrate which splits a common feed channel recursively into two resulting new feed channels. As a result, a tree of splitting channels is generated in order to connect 256 reaction chambers with a common feed channel. The channel length and channel geometry of each of the chambers is the same, resulting in an equal pressure drop and therefore equal reactant distribution. However, the generated channel network is very complex, and all reaction chambers are connected by the network. These interconnections may present a considerable limitation. It may, for instance, occur that one single particle in the feed stream may clog and block a whole subtree of the channel network. Furthermore, the tree becomes increasingly complex with increasing number of reaction chambers, therefore limiting scalability. Finally, the channel network consumes a considerable amount of space on the wafer, thereby limiting the density of the

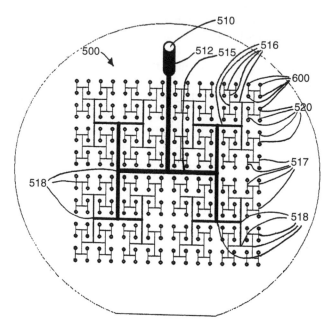

Figure 11.17 Fluid reactant distribution in a microstructured reactor by a binary tree channel network. (From Berg, S.H. and Guan, S., W000/51720 to Symyx Technologies, Inc., March 1999.)

reaction chambers. As presented, there may be a number of drawbacks resulting from the single fact that the channels, i.e., the flow restrictors causing pressure drop, are connected to each other. However, regarding only the required pressure drop, there is no need for connecting all the channels in a tree format. An obvious solution of the aforementioned problem is the use of single, independent flow restrictors for each reaction chamber. For an early design study, the pressure drop of fluid flow through narrow and short channels or pores was calculated using the law of Hagen–Poiseuille:

$$\Delta p = \frac{8}{\pi} \cdot l_\text{P} \cdot \dot{V}_\text{B} \cdot \eta \cdot \frac{1}{r_\text{P}^4 \cdot N} \tag{11.8}$$

with l_P being the pore length, $\dot{V}_\text{B}$ the flow rate, η the viscosity, r_P the pore radius, N the pore number. As can be seen from Figure 11.18, a considerable pressure drop can be generated depending on the pore number and the pore radius. Especially, reduced pore radii result in a strong increase in the pressure drop. Considering this rough calculation, it becomes clear that such arrangement of narrow and short pores can be sufficient to generate an equal flow distribution. Therefore, the fluid access ports of the microreaction chambers have been designed as flow restrictors in HTE's "single-bead approach." These flow restrictors are microstructured membranes, which are a collection of a defined number of short straight pores with a very small diameter, as described further.

11.6.1.3 Single-Bead Reactors

The basis of the single-bead concept is the use of single-shaped bodies as the catalytic material of interest. These particles may in principle be of any shape, but usually spherical particles are applied. In accordance with approaches known from CombiChem, such spherical particles are called "beads," although they fulfill very different functions in comparison to their application in combinatorial setups

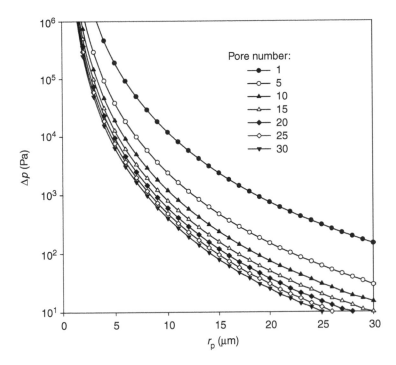

Figure 11.18 Pressure drop for fluid flow through narrow and short pores as a function of pore radius and pore number; $l_P = 100$ μm, $V_B = 1$ mL/min, $\eta = 28$ Pa s.

in organic- and biochemistry. Each bead represents one catalyst as a member of a library of solid catalysts. It consists of nonporous material such as α-Al$_2$O$_3$ or steatite (a magnesium silicate) or of typical porous support materials such as γ-Al$_2$O$_3$, SiO$_2$, or TiO$_2$. These beads can be subjected to different synthesis procedures and sequences such as impregnation, and coating, etc. In addition, fully mixed metal oxide catalysts can also be formed as spherical particles. Using single beads for HTE has a number of advantages. First, such beads are comparable to well-known fixed-bed catalysts and the synthesis pathways can be the same as for conventional materials, which may facilitate scale-up procedures. A number of common preparation procedures for these beads are available and can be carried out in standard laboratory environments. The best example for efficient library preparation for single-bead reactors is the split and pool synthesis (see Section 11.5.3).

The second advantage is that each bead represents a single entity, which can be handled independently from other beads or the final reactor configuration. Starting from masterbatches, a large diversity of materials can easily be prepared. Furthermore, different beads may be treated individually, for instance, subjected to different preparation steps or pretreatments such as calcination or steaming, rather than handling the complete library, as is necessary for substrate-bound thin- or thick-film catalysts. This characteristic allows the use of synthesis procedures different from the parallel approach. The potential use of *ex situ* synthesis procedures may furthermore present a significant advantage especially in microchemical systems, as an *in situ* preparation of the catalysts in the reactor could cause problems with contamination or thermal and chemical stability of the reactor material. The reactor suitable for testing single-bead catalysts in a highly parallel fashion, the so-called "single-bead reactor", is designed as two-dimensional arrangement of microreaction chambers that can hold one catalytic bead each (see Figure 11.19). The reactor can be divided into two parts, a base part and a top part. Having filled the base part with beads, the top part is pressed, sealed, or bonded against the base part to encapsulate each bead in a single, independent microreaction chamber. The beads sit loosely in the reaction chambers, allowing continuous fluid flow from fluid openings in the top part of the reactor, around the bead, and through fluid openings in the base part. To avoid mass transfer problems while reacting the fluid on the bead, egg-shell-type catalytic beads should be used. Using the top–down direction through a wafer-like material or steel plate as fluid flow path, the small layout of a single microreaction chamber allows the extensions of the reactor in two dimensions. In contrast, this is not possible using microchannel-based approaches, as these microchannels extend as straight, long channels on a substrate and therefore block one dimension. Parallel microchannel reactors on a substrate are in most cases one-dimensional arrangements of channels, having a much larger scale than a two-dimensional arrangement of reaction chambers as the single-bead reactor. The resulting array density is much higher for the latter, reaching up to 60 or more catalytic reaction chambers per cm^2 [64]. The partial cross-section of a single-bead reactor is shown in Figure 11.20. The complete setup consists of four wafers, where wafers 1 to 3 belong to the base part and wafer 4 to the top part of the reactor. The microreaction chambers contain means for fixing the position of the bead in the reaction chambers. For this purpose, a number of possible designs can be visualized, however, a simple frustum of pyramid was chosen for compatibility reasons with the well-known wet etching processes. Another advantage of the pyramid is the tolerance to variations of size and form of the beads. A spherical bead cannot block the quadratic

Figure 11.19   384-parallel single-bead reactor with independent microreaction chambers and integrated flow restrictors. Typical bead diameter:1 mm.

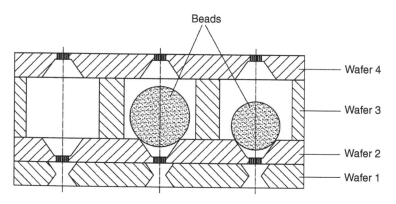

Figure 11.20   Typical configuration of a single-bead reactor shown as partial cross-section with and without beads in the reaction chambers.

cross-section of the frustum. Furthermore, different bead sizes can be employed; however, the bead size is usually kept constant for one experiment.

For validation purposes, a partial oxidation reaction was chosen and the reactor was filled with inactive and active catalysts according to the pattern shown in Figure 11.21a. A continuous flow screening experiment is carried out at 400°C and a reactant flow of $V = 1$ mL/min per bead, using spatially resolved mass spectrometry for sequential product analysis. The normalized results for the conversion degree are shown in Figure 11.21b. According to the mass spectrometric intensities, white color represents low conversions, while dark color represents high conversions of the hydrocarbon. The maximum conversion in this experiment is approximately 60%. As can be seen, the results for the conversion degree correspond very well with the filling pattern of the reactor. Cross talking between adjacent reaction chambers can be observed to some minor extent, although not limiting the applicability of the single-bead reactor for primary screening purposes.

The results of a typical screening experiment for a partial oxidation reaction are shown in Figure 11.22. The patterns of the different components in the mixture correspond very well to the expected behavior. Catalysts with interesting properties can easily be identified.

11.6.1.4　*Optimal Use of Stage I in Screening Programs*

The technologies used for Stage I catalyst screening for heterogeneously catalyzed gas-phase reactions differ quite substantially from traditional catalyst-screening approaches for such reactions. Therefore, a number of (technical and structural) guidelines should be followed to achieve optimal efficiency in the use of these highly sophisticated screening systems:

- To avoid bottlenecks in the overall screening workflow, the different elements of a high-throughput research program need to be tightly integrated. In general, the handling of your materials and your libraries/reactors is of critical importance. This is especially true for a close adaptation of preparation and screening techniques, as the overall cost per screened material has to be substantially lower than the cost for a Stage II screen.
- Reduce complexity and design modular systems: As far as possible, use proven techniques and combine them in new ways, but let them be independent from each other, e.g., allow different synthesis methods that are independent from your reactor configuration. Furthermore, allow different sequential or parallel analysis methods. This is not a contradiction to the aforementioned "integration" because workflow integration does not necessarily mean technology integration.
- Allow quick turnaround: Usually, the first technology generation will not be perfect. Thus, design your systems to allow easy improvements and new generations.

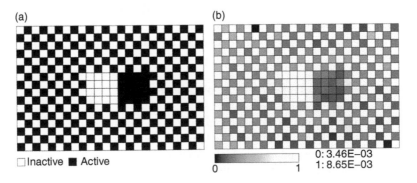

Figure 11.21　Results of high-throughput screening of catalysts in a 384-parallel single-bead reactor in a partial oxidation reaction. (a) Arrangement of inactive and total oxidation catalysts in the reactor, (b) screening results for the conversion of a hydrocarbon at 400°C, 1 mL/min per bead.

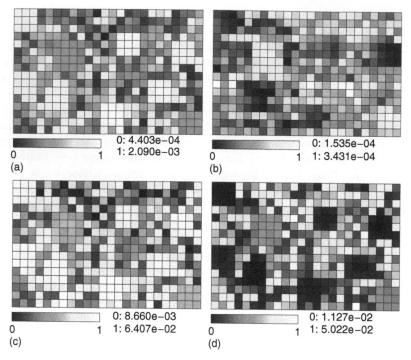

Figure 11.22 Color-coded normalized MS-signals recorded during a typical partial oxidation experiment. (a) Educt, (b) desired partial oxidation product, (c) O_2, (d) CO_2 (dark: low amount; light: high amount).

- Be aware of the qualitative or semiquantitative nature of a Stage I screen. Do not raise the expectations to your data quality too high. Instead, always use reference materials and give relative performances to these references.
- Screen under realistic conditions. Screen the materials under continuous flow and steady-state conditions and apply different sets of process conditions (reactant concentrations, residence time, temperature, etc.).
- Transfer results as often as you can. If possible, transfer materials from your Stage I screen to a Stage II screen on a regular basis. This will give you invaluable feedback on the quality of your Stage I system.

If so, a good comparison between Stage I and Stage II screening can be achieved, which is essential for the success of the overall screening program. A typical example of the comparison of a Stage I screen (384-parallel single-bead reactor) and a Stage II screen (48-parallel fixed-bed reactor) is given in Figure 11.23. As can be seen, a good comparison can be achieved and the same catalytic trends are observed in both reactor systems, although the catalyst mass differs by three orders of magnitude.

11.6.2 Stage II Testing in Gas-Phase Applications

Partial oxidation plays a central, if not crucial, role in the chemical industry for the functionalization of especially alkanes and olefins [65–67]. In particular, gas-phase oxidation is attractive with regard to feedstock conversion and usually high space time yields can be achieved and also product separation is, in many cases, far easier than for product mixtures produced by liquid-phase oxidation. Gas-phase oxidation, however, faces major challenges which need to be considered when tackling the complex technological challenges. Starting from alkanes, olefins, or aromatics, the formation of CO and CO_2 are favored thermodynamically, but are undesirable with regard to the

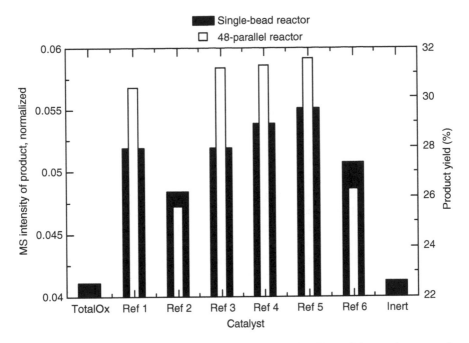

Figure 11.23　Comparison of a 384-parallel single-bead reactor and a 48-parallel secondary screening reactor for a set of reference catalysts in a partial oxidation reaction.

products that usually represent partial oxidation products of the relevant feedstock. Starting from relatively low boiling feedstock compounds, increasing boiling points (or decreasing partial pressures) can be observed for hydrocarbon oxidation products with increasing oxidation state of the functionalized hydrocarbon molecule. This will lead to changed adsorption behavior and larger "sticking coefficients" of the molecule with functional groups to the catalytic surface and increase the chance of further oxo-functionalization and finally of total combustion. Partial oxidation and total oxidation are exothermic and therefore sources of heat and processes on an industrial basis can rarely be run isothermally over the full catalytic bed. In most cases explicit hot spots are formed, which in extreme scenarios can display temperatures exceeding the average temperature of the catalyst bed by 5 to 15°C, sometimes temperatures of even 50°C are reported [68]. Usually, if a hot spot is displayed in a catalyst bed, the degree of conversion for single-pass engineered processes is far above 60% with regard to the hydrocarbon feedstock — for multipass processes the degree of conversion can be much lower. Still, this fact leads to a bizarre scenario: the catalyst is expected to deliver high product selectivity over a temperature range of approximately ±10°C and a large range of varying partial pressures of educt, oxygen, desired products, and side products of the reaction. Keeping these facts in mind, it is even more surprising that there are many examples of highly selective gas-phase oxidation catalysts and processes. Therefore, it is not astonishing that screening efforts with regard to changes in material properties, new prospective catalyst candidates, and variations in engineering conditions for partial oxidation reactions are a highly attractive topic for the chemical industry.

As discussed above, the level of sensitivity of most gas-phase oxidation catalysts is high with regard to the reaction temperature. Therefore, any Stage II-screening tool should operate under isothermal conditions for all active materials, and it is a given prerequisite that thermal equilibrium of the reactor and a homogeneous temperature distribution are essential.

Having in mind the parallelization of multiple reaction tubes into a single, e.g., 48-fold-Stage II-screening reactor system, the control of heat distribution and thermal crosstalk is much more difficult than in a single fixed-bed plug-flow reactor. Computational fluid dynamics (CFD) modeling as shown in Figure 11.24 is a very important and helpful tool for the construction and geometric dimensioning of the multifold reactor system under reaction, i.e., realistic industrial conditions. The dimensions for a single-reactor tube in the calculated example are 11 cm length and 7 mm inner diameter.

Although the reaction is highly exothermic, no thermal cross talk of the neighboring reactor sites is experienced, indicated through the blue color (ca. 523 K) in the vicinity of the reactor tube in Figure 11.13. This close to ideal isothermic behavior can be reached by the use of inserted, stainless-steel cartridges as reaction tubes (see also the photograph of the reactor in Figure 11.14).

For demonstration purposes, the distribution of the speed of gas streamlines is presented for an industrial laboratory-scale reactor (length 1.2 m, diameter 21 mm) and an HTE reactor type (length 11 cm, diameter 7 mm) in Figure 11.25. Here, also a good agreement among the reactor types of different scale can be achieved by the right choice of reactor dimensions, reaction conditions, and particle sizes for the construction of HTE Stage II-screening equipment. By comparing the color-coded distribution of the streamline speed for the different reactor dimensions, one can clearly observe the strong similarity between both reactor "behaviors. Owing to the difference in reactor diameters of the industrial vs. the HTE reactor by a factor of 3, the density of streamlines is higher in the case of the industrial tubular reactor.

Nevertheless, these modeling efforts are of little value when the practical implementation does not corroborate the above-calculated results. Ensuring the constancy of any parameter in the catalytic testing workflow, the reactor performance with regard to temperature distribution, gas distribution, constant feed

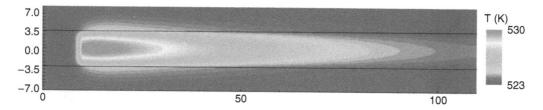

Figure 11.24 Simulation of the thermal behavior of a catalyst in a multitube test reactor. ($\Delta H > 100$ kJ/mol, T_{inlet} :523 K, $T_{reactor}$: 523 K, porosity: 80%, λ_{bed}: 3.0 W/mK, GHSV: 10000 h^{-1}, reactor tube geometry: 0.11 × 0.007 m).

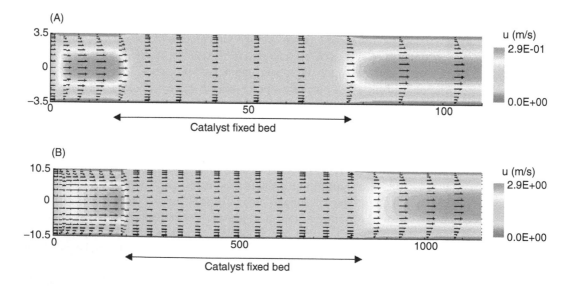

Figure 11.25 Calculated speed of streamlines of (A) an HTE-multifold reactor (11 cm length and 7 mm diameter) with 1.5 mm pellets and (B) an industrial reactor (1.2 m length and 21 mm diameter) with 5 mm pellets.

concentration, and constant gas (space) velocities can be critically evaluated. This is the crucial, decision-making step for the application of HTE equipment to the screening of gas-phase partial oxidation.

As a test reaction in this case, the oxidation of propylene to acrolein was chosen, a standard bismuth–molybdate catalyst was employed [69,70]. Then 48 identical catalysts were introduced into the reactor system and exposed to reaction conditions (reaction conditions were 2% hydrocarbon in air at GHSV of 3000 h^{-1}). The goal of this experiment was to prove that (1) practically identical values for activity and selectivity are obtained for all catalysts introduced into the reactor system and (2) the typical behavior as known from conventional testing in laboratory reactors and pilot plant could be reproduced in the Stage II-reactor system. Finding (2) also illustrates that good thermal and fluid distribution at nearly identical temperatures at all positions in the reactor as well as a good distribution of the reaction gases over the reactor system are achieved.

As Figure 11.26 undoubtedly demonstrates, the deviation between the same catalytic material under practically identical reaction conditions is in the range of ±2% conversion (if appropriate measures are taken this error can be reduced to ±0.5%). These experimental data points lead to the important verification of the above-discussed CFD modeling results and confirm the assumption of realizing identical reaction conditions over the whole reactor system independent from the position of a catalyst to be tested. By testing inert carrier material in reactor column number 8, the inertness and catalytic inactivity of the reactor steel can be proven.

In the following, the screening power and the scientific potential applying HTE Stage II technologies in gas-phase oxidation are demonstrated for two illustrative case studies: (1) the epoxidation of 1,3-butadiene with Ag-based catalysts and (2) dynamic experiments for automotive applications (DeNO$_x$).

11.6.2.1 The Epoxidation of 1,3-Butadiene with Ag-Based Catalysts

Since the development of Stage II-screening reactor systems in 1998 [71], the 48-fold reactor technology is, in the meantime, a state-of-the-art methodology in the context of robustness, cost, and efficiency, and operates 24 h a day, 7 days a week, which is eased by the use of a smart control

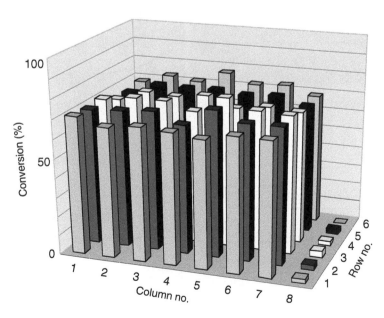

Figure 11.26 Plot of the position sensitivity of the degree of conversion for a set of 48 bismuth–molybdate catalysts (same batch) in propylene to acrolein conversion in a Stage II 48-fold-screening reactor (reaction conditions: 2% hydrocarbon in air at GHSV of 3000 h^{-1}, column no. 8 contains only inert carrier material).

and data evaluation software environment, and overall workflow. Many industrial requirements such as high pressures, high temperatures or, e.g. multiphase reactions are, regarding only the screening technology, already implemented and realized in HTE reactor technology. It seems astonishing that a major part of the synthesis of the optimization processes connected to Stage II screening is only covered in a few scientific publications dealing with highly sophisticated synthesis operations, which are realized in an automated and reproducible fashion (see [72]).

All the results concerning the gas-phase epoxidation were obtained using HTE equipment, for all procedures involved, namely, the synthesis and testing of the Ag-based materials (Figure 11.27).

A catalytic reaction system involving great synthetic challenges on the side can be seen in the epoxidation of small olefins. While the conversion of ethylene in the gas phase to the corresponding epoxide (ethylene oxide, EO) with supported silver catalysts is efficiently realized in industry [73,74], the reaction behavior of the corresponding (di-) olefins with or without allylic hydrogen (propylene or 1,3-butadiene) changes drastically, resulting in low selectivities to the C_3 or C_4 epoxides [75]. Conversions under similar reaction conditions are high, but the main reaction pathway is the total oxidation to CO_x. For a long period of time, the scientific community accepted the opinion that Ag-based catalyst is only feasible for olefins without allylic hydrogen [76,77]. Next to propylene oxide (PO), vinyloxirane (3,4-epoxy-1-butene, VO), the product of the epoxidation of 1,3-butadiene, is a very important C_4 intermediate due to its intrinsic double functionality (double bond and epoxy ring). Main products with VO as reaction intermediate are 2,5-dihydrofurane, THF and 1,4-butane diol. The reaction scheme for VO is shown in Figure 11.28.

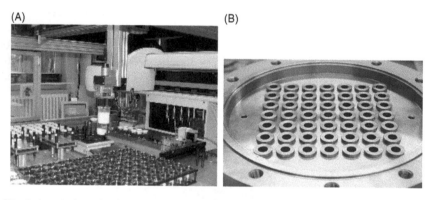

Figure 11.27   Automated synthesis station (A) and top view of a 49-fold parallel reactor (B).

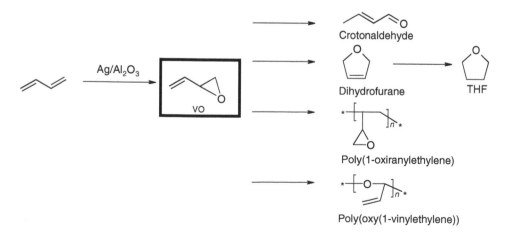

Figure 11.28   Reaction scheme of vinyloxirane (VO) via 1,3-butadiene.

The showcase is also attractive, as the epoxidation of 1,3-butadiene to vinyloxirane in air is as challenging as the sophisticated synthesis of the Ag/α-Al$_2$O$_3$ catalysts.

For more than 30 years, the synthetic procedure to obtain a very active and selective, Ag-based epoxidation catalyst has been an established and well-understood technology [78]. The classical impregnation techniques via silver nitrate or silver oxide precursors lead to Ag particle sizes after calcination, in the range of 1 to 2 mm, while the use of a mixture of solubilizing/complexing agents (ethanol amine, ethylene diamine) in combination with a reducing agent (oxalic acid) induces Ag particles in the desired range of 100 to 300 nm. Low-surface-area carriers such as α-alumina or steatite are the carrier materials of choice for deposition of the active component. In this study, we optimized the corresponding molar ratios of silver and organic amines and introduced two additional synthesis parameters, significantly influencing the Ag particle growth: crystallization temperature and — dedicated to the Ag chemistry — light. By varying these two parameters, the catalytic performance of identically synthesized materials changes drastically, as shown in the temperature-dependent conversion vs. selectivity diagrams in Figure 11.28A through Figure 11.28D . For detailed information about the exact synthesis procedures see [79] and references therein. Starting with the organic amine-based synthesis at room temperature and daylight, one can reach modest yields at around 20% VO (A). Cooling down to 0°C under daylight improves the yield to approximately. 40% (B). The maximum VO yield while performing the synthesis in darkness at room temperature is around 60% (C). Nearly 90% yield of VO is found by additional, literature-known doping with Cs at 0°C and darkness (D)!

From Figure 11.29A to Figure 11.28D it is obvious that minor changes in preparation have a major effect on catalyst performance due to its influence on particle nucleation and particle growth. The temperature window for maximum yield can be fine-tuned via the type of dopant and the dopant level, a fact that is usually exploited in an industrial process. Apart from adding amines to the impregnation solution, an impregnation temperature below 5°C and the preparation of the solution under the exclusion of light deliver materials with the best performance data in the desired oxidation reaction. Table 11.4 gives—a summary of Figure 11.29A through Figure 11.28D—the maximum yields obtained via the different preparation procedures. From the differences it becomes evident that a high degree of sophistication for (1) the automated preparation sequences and (2) the accuracy of the analysis of the Stage II-screening reactor is required.

11.6.2.2 *Dynamic Experiments in Stage II Screening for Automotive Applications*

A particularly interesting case of catalyst screening for gas-phase reactions under dynamic conditions is related to the catalyst development in the field of automotive catalysis. Several technologies for exhaust treatment systems such as three-way catalysts for Otto engines, oxidation catalysts for diesel and lean-burn engines, and particulate filters for diesel engines have been developed and implemented into vehicles over the last several years. One of the biggest challenges that emission control technology developments still face is the development of advanced DeNO$_x$ technology for lean-burn applications (e.g., diesel and gasoline direct injection engine [GDI]), since three-way catalysts exert DeNO$_x$ activity only at near stoichiometric air-to-fuel ratios.

Table 11.4   Comparison of Influence of the Synthesis Procedure on the Catalytic Performance of Ag-Containing Catalysts in the Epoxidation of 1,3–Butadiene

Ag precursor	Ethylene–diamine complex	Ethylene–diamine complex	Ethylene–diamine complex	Ethylene–diamine complex
Dopant	—	—	Cs or Rb	Cs or Rb
Synthesis conditions	Room temperature / day light	Ice bath cooling / darkness	Room temperature / day light	Ice bath cooling / darkness
Optimal temperature (°C)	260–280	240–270	230–240	220–240
Maximum yield (%)	24	38	84	87

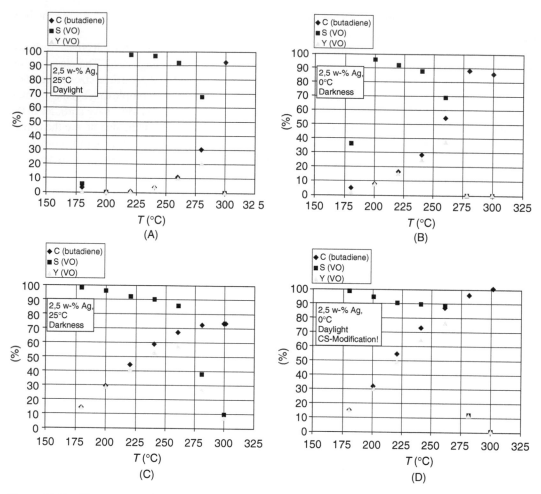

Figure 11.29 Temperature dependent conversion/selectivity diagrams A–D, varying synthesis conditions as indicated in the upper left corner (1% 1,3-butadiene in synthetic air, GHSV 3,000 to 12,000 h^{-1}, ambient pressure, 150 to 350°C, catalyst volume, 2 mL, diluted with corundum (1/1 mL)).

NO$_x$ adsorption catalysts are based on the concept of "NO$_x$ storage and reduction," which convert NO$_x$ in the course of precisely controlled lean–rich cycles. This technology has been known since the mid-1990s and entails a great potential as DeNO$_x$ technology for purification of diesel exhaust gases. State-of-the-art NO$_x$ adsorber catalysts contain a patinum group component for the oxidation of NO to NO$_2$ and BaCO$_3$, as NO$_x$ is the adsorbing medium. During the lean periods, NO$_x$ is temporarily adsorbed (stored) on the catalyst and then reduced during intermittently rich periods. The problems to be solved for application in diesel passenger cars are related to DeNO$_x$ activity at low temperatures, improvement of thermal stability, and particularly sulfur tolerance.

For the development of a new, advanced generation of NO$_x$ adsorption catalysts, high-throughput Stage II screening systems capable of mimicking the dynamic operation that occurs during lean–rich cycling in the real engine were developed.

The Stage II screening reactor used in this study is a fully automated 48-fold Stage II HTE and entails a gas–liquid mixing or dosing, 48 reaction channels, a process control unit and on line gas analysis system with time-resolved monitoring of NO, NO$_2$, N$_2$O, O$_2$, CO, CO$_2$, H$_2$, and HC. An important feature of this reactor system is its capability of rapid switching of feed gases, thus realizing dynamic spikes of fuel-rich feed concentrations in the range of milliseconds. The lean phases are generally run from 50 to 300 sec. Typical feed gas which is passed over the catalyst consists of 100 to 500 vppm NO_x, 500 to 10,000 vppm CO, up to 100 vppm propylene, 5 to 15% O_2 and 5 to 10% H_2O; with a balance of N_2. Gas hourly space velocities of 50,000 to 150,000 h^{-1} are applied. Upon switching to rich conditions, for instance, for 1 to 10 sec, the oxygen level is sharply decreased to 0–0.3%, simultaneously a mixture of $CO/H_2=3:1$ in the vol% range is added to the feed gas, which causes the regeneration of the NO_x storage. By returning the lean exhaust condition, NO_x adsorption occurs and the lean–rich cycle starts again. Tests are typically performed at several temperatures between 150 and 450°C.

Figure 11.30 shows the NO_x and O_2 responses monitored in the cyclic lean–rich operation over two different catalysts. At the beginning of the lean phase, a significant decrease in NO_x concentration at the reactor outlet is observed for both catalysts. The best candidate (catalyst B) shows, however, a better activity within time-on-stream in the lean period and enables a lower overall level within lean–rich cycles.

Considerable progress in evaluation of novel $DeNO_x$ adsorber catalysts can be achieved by applying a 48-fold reactor system for catalyst development with dynamic operation enabling testing of up to 1000 catalysts in just 1 month.

11.6.2.3 Refinery Catalysis Applications in High-Throughput Experimentation

In the previous sections we have discussed two different applications of gas-phase catalysis. Although different in nature, both case studies had the common feature that (1) all employed educts and the obtained products were gaseous and (2) the change in volume of the gas phase resulting from the educt conversion is negligible. For a number of reactions, especially for reactions in the petrochemical industry both of the above-mentioned features cannot be neglected, we will discuss classic examples and present technical approaches to overcome the obstacles related to the chemical transformations.

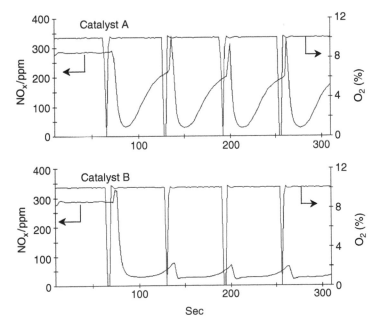

Figure 11.30 Time-resolved NO_x and O_2 monitoring in the rich and lean cycling operation over two different catalysts.

A typical example for a reaction with substantial contraction of volume is the synthesis of methanol from syngas. Formally, 1 mol CO and 2 mol of H_2 react to form 1 mol of methanol. This means that at high degrees of conversion, the contraction in volume can be a factor of three. This has dramatic implications for the pressure: as the gas volume drops, the total pressure also drops. As a surplus, the analytical evaluation of the reaction is also complicated owing to the change in volume as a function of the degree of conversion.

The second prototypic example for a reaction with volume contraction and as a surplus the evolution of liquid products is the synthesis of synthetic fuels, lubricants, and waxes by the Fischer–Tropsch (FT) synthesis. From the viewpoint of process engineering and catalyst testing, the FT synthesis is a demanding reaction, owing to the complex product spectrum, comprising liquefied waxes and gases, elevated reaction pressures, inconsistency of volume (products vs. educts, drastic volume contraction), and high reaction exothermicity. The engineering of appropriate laboratory equipment for FT synthesis is challenging with particular difficulties arising from tailoring of parallelized high-throughput reactor systems.

Since its invention in the early 1920s the FT synthesis (for further reading we recommend [80]), named according to its inventors Franz Fischer and Hans Tropsch, only found some industrial niche applications in countries such as South Africa (Sasol) and Malaysia (Shell), with limited access to crude oil but is currently being reinvigorated owing to several challenges in the gas and oil sector. The increasing demand for better industrial use of natural gas and low-value refinery products, rising crude oil prices, cleaner fuels, higher diesel cetane numbers, and tightening emission standards have shifted the attention of the oil industry toward gas-to-liquid (GTL) technologies. Major oil companies are performing programs in the GTL area and the volume of FT products may increase by approximately a factor of 10 within the next years, then contributing 2 to 3% of the total refinery output. Besides the FT-synthesis route, the conversion of methanol to hydrocarbons has been proven commercially to make C_2+C_3 olefins (methanol to olefins or MTO), gasoline (methanol to gasoline or MTG), or diesel fuels.

The spectrum of FT primary products depends on the process and the catalyst used but normally encompasses unbranched paraffins, α-olefins, and, to a lesser extent, oxygenated products. The carbon numbers may range from C_1 to C_{40} or even higher; the carbon number distribution often follows a certain chain-growth pattern leading to the so-called Schulz–Flory or Anderson–Schulz–Flory distribution. Recent developments in the FT area are targeting catalysts with a higher productivity, better attrition resistance, or new catalyst chemistries, which are supposed to shift the product spectrum into a more desirable region, for instance, to increase the yield in α-olefins within a certain range of carbon numbers for subsequent use in polymerization or hydroformylation reactions. For fuel application, the FT-synthesis products require further processing such as hydrocracking and hydroisomerization.

The main technical challenges to "translate" the requirements of GTL chemistry into a viable HTE workflow are mainly: (1) control of the pressure/pressure-flow behavior and (2) gas–liquid separation in a parallelized fashion.

11.6.2.3.1 Pressure Flow and Pressure Control

The basic components for achieving pressure flow and pressure control are briefly listed as follows:

- Closed-loop pressure control of the reactor
- Pressure reduction from the reactor to the sampling lines and on-line analytics
- Hot flow gas measurement and analysis of the liquid products
- Compensation with inert gas to offset volume contraction

These components guarantee consistency of pressures and flows at any reaction condition independent of the level of conversion.

11.6.2.3.2 *Liquid–Gas Separation*

The waxy products, which are normally generated in FT reactions, have often been reported as the cause for plugging of the downstream line. The problem can be minimized by an arrangement of proprietary hot-gas condensers fitted downstream to the reactors. The condensers should be regularly depleted and analyzed at certain intervals. The unconverted amounts of syngas and the inert gases added to the products guarantee a thorough separation of the lighter hydrocarbons from the liquid products, so that these products can be analyzed in the gas phase.

An additional challenge is that in many cases the catalysts will deactivate rapidly. Provisions have to be taken to start the single reactors of a multitube reactor sequentially to be able to compare time-on-stream behavior. If regeneration procedures are advisable or necessary, it is even recommended to have independent regeneration of the single reactors.

The technology for gas–liquid separation is also valuable for product separation under trickle-bed operation conditions. The commencement of a major refinery catalysis program in 2000, hte Aktiengesellschaft, already has developed automated multifold high-pressure fixed-bed (gas-solid) and trickle-bed (gas–liquid–solid) systems and has proven their reliability and process stability in continuous operation even in trials lasting up to several months. During the course of development of the units, significant efforts were devoted to the upstream and downstream flow management at reactor pressures and temperatures approaching 120 bar and 500°C, respectively. The existing units have been hardened through day-to-day use in several refinery catalysis development programs.

11.7 TESTING OF CATALYSTS IN LIQUID–LIQUID, GAS–LIQUID, AND GAS–LIQUID–SOLID REACTIONS

Generally speaking, most of the catalytic reactions that are performed in the liquid phase are devoted to chemical products with smaller yearly tonnage produced by the chemical industry. Most of these chemicals are, in general, summarized under the term "fine chemicals." It is not surprising that together with the development of HTE technology for the investigation of heterogeneously catalyzed gas-phase reactions, developments to study catalysis in liquid phase in a parallel way were also a key incentive. Nevertheless, it should be mentioned here that the parallel investigation of catalyzed reactions in the liquid phase is a topic that was already technologically feasible in the 1980s. The tools developed at that time did not have the degree of automation of Stage II tools available today and the workflow management was not as far developed, however, Figure 11.31 illustrates that the capabilities were already at hand and could be purchased off the "shelf".

The HTE characteristics that apply for gas-phase reactions (i.e., measurement under non-diffusion-limited conditions, equal distribution of gas flows and temperature, avoidance of cross-contamination, etc.) also apply for catalytic reactions in the liquid-phase. In addition, in liquid phase reactions mass-transport phenomena of the reactants are a vital point, especially if one of the reactants is a gas. It is worth spending some time to reflect on the topic of mass transfer related to liquid–gas-phase reactions. As we discussed before, for gas-phase catalysis, a crucial point is the measurement of catalysts under conditions where mass transport is not limiting the reaction and yields "true" microkinetic data. As an additional factor for mass transport in liquid–gas-phase reactions, the rate of reaction gas saturation of the liquid can also determine the kinetics of the reaction [81]. In order to avoid mass-transport limitations with regard to gas/liquid mass transport, the transfer rate of the gas into the liquid (saturation of the liquid with gas) must be higher than the consumption of the reactant gas by the reaction. Otherwise, it is not possible to obtain true kinetic data of the catalytic reaction, which allow a comparison of the different catalyst candidates on a microkinetic basis, as only the gas uptake of the liquid will govern the result of the experiment (see Figure 11.32a). In three-phase reactions (gas–liquid–solid), the transport of the reactants to the surface of the solid (and the transport from the resulting products from this surface) will also

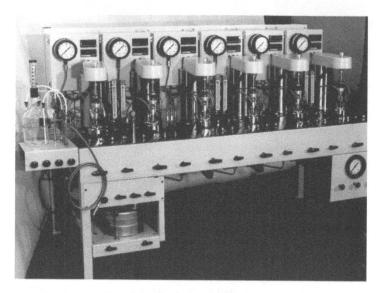

Figure 11.31   Photograph of a sixfold autoclave bench developed by Parr Instrument Company (www.par-rinst.com). The apparatus was built in 1988, similar technologies that Parr commercialized in the early 1990s were already fully computer controlled.

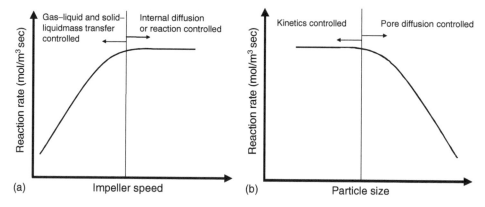

Figure 11.32   Influence of essential reaction parameters exerted on the reaction rate of gas–liquid–solid reactions (a, impeller speed; b, particle size).

influence the kinetics of the chemical reaction. As discussed for gas-phase reactions, in three-phase reactions with porous catalyst bodies the reaction rate will also depend on the size and the tortuosity of the porous solid particles. Figure 11.32b illustrates that the particle size has to be adjusted in order to avoid that a reaction is carried out under pore diffusion control.

An optimal mass transfer of a gas into a liquid can be achieved by adjusting the mechanical factors which control the shear force and therefore the size of the interface between gas and liquid which is responsible for gas introduction into the liquid. The adjustment of the shear force can be achieved by adaptation of the stirring rate for a given stirrer and baffle system. Alternatively, the utilization of different stirrer and baffle constructions, sparger geometry, and the geometry and number of the baffles can be a way of facilitating gas saturation of the liquid. Figure 11.33 illustrates the effect of the stirrer speed on the gas dispersion (0.5 NL of gas is introduced by a sparger below the stirrer). It can be seen that a better dispersion of the gas can be achieved with higher stirrer speeds. The pictures in the figure also illustrate how, in many cases, so-called cold flow experiments can be utilized to visually inspect

Figure 11.33 Photographs of the gas dispersion in a 300 mL reactor. To be able to follow the dispersion visually the reactor was made of acrylic glass. The figure illustrates the influence of the stirrer speed on the distribution of gas in a liquid (rounds per minute indicated in the pictures). At higher stirrer speed the gas bubbles are smaller and well distributed over the whole reactor volume.

the operation mode of a reactor. In the special case of stirred-tank reactors for gas/liquid reactions, to visually inspect the reactor performance under cold flow conditions prior to running experiments is highly recommended. Usually the price for transparent reactor systems is comparatively low with regard to the increased knowledge that is obtained with these experiments.

11.7.1 Stage I Screening for Liquid-Phase Catalysis

There are numerous publications in the scientific literature about the different technical approaches that have been realized for a fast Stage I screening of catalysts in liquid media. As stated previously, the aim of Stage I screening is to get information about the relative performance of a number of tested potential catalysts in order to accelerate the time to find suitable compositions for conventional or Stage II testing. In order to survey, we listed the most relevant approaches in Table 11.5 and attributed the advantages and disadvantages of the techniques.

In Table 11.6 illustrative examples from the literature are enlisted which give the reader an impression about the different chemistries that can be mastered with these screening techniques. As can be seen, in some cases the individual advantages of a technique might overcome its general drawbacks, or at least render them less important. This strongly depends on the investigated chemistry, e.g., mass-transfer limitations are in a number of cases not the main problem of liquid/liquid catalytic reactions.

It should be mentioned here that apart from the possibility of performing Stage I screening as batch reactions, the screening in a continuous fashion has also been explored. In particular, the utilization of microreaction devices for liquid-phase reactions in a tubular reactor format was the focus of the research and the developments of Abdallah et al. [82]. In their setup, different homogeneous catalysts were sequentially added in the reactant stream and the resulting products of the reaction were sampled again sequentially at the end of the reactor. The separation of the reaction segments was achieved by inserting bubbles of inert gas into the reactor. The risk associated with this technique is coupled to the potential cross-contamination of different catalyst species that are insufficiently removed from the reactor walls.

In the following some of the aforementioned techniques for Stage I screening in liquid-phase catalysis are presented in further depth.

A parallel reactor system for liquid–liquid phase reactions such as oxidation reactions with H_2O_2 at ambient pressure was reported from hte Aktiengesellschaft . If compared with other chemistries, rather mild-reaction conditions (ambient pressure, moderate temperature) are often applied in liquid-phase oxidation for fine chemical production with terminal oxidants that can be dosed as liquids (e.g., aqueous H_2O_2 or organic peroxides). The reaction that was investigated was the partial oxidation of

Table 11.5 Most Popular Technical Approaches for Stage I-Screening in Liquid Phase Catalysis

	Detection of activity	Detection of selectivity	Information level	Low risk of mass transfer limitations	Reaction under pressure	Integral analysis	No common head space	Individual heating	Individual stirring	Easy to parallelize
(Micro) titer plate and indicator, (dye, fluorescent, etc.)	⊕	⊖	⊖	⊖	⊖	⊕	⊖	⊖	⊖	⊕
(Micro) titer plate in a pressure vessel	⊕	⊖	⊖	⊖	⊕	⊕	⊖	⊖	⊖	⊕
Vortex-agitated vials	⊕	⊕	⊕	⊖	⊕	⊖	⊕	⊕	⊖	⊕
Stirbar-stirred vials	⊕	⊕	⊕	⊖	⊕	⊖	⊕	⊕	⊕	⊕
Bubble columns	⊕	⊕	⊕	⊕	⊕	⊖	⊕	⊕	⊖	⊕
Overhead-stirred vials	⊕	⊕	⊕	⊕	⊕	⊖	⊖	⊕	⊕	⊖

⊕ Stands for a positive feature with regard to the technology, ⊖ for a negative one

Table 11.6 Examples of Applications of Some of the Most Popular Technical Approaches for Stage I-Screening in Liquid Phase Catalysis

Examples for the application of Stage I-screening techniques:	
(Micro)titer plate and indicator (dye, fluorescent, etc.)	1. Catalytic electrooxidation of methanol + pH indicator[a] 2. Catalytic hydrosilylation of alkenes and imines + dye[b]
(Micro)titer plate in a pressure vessel	Selective catalytic oxidation of alcohols to aldehydes and ketones[c]
Vortex agitated vials	Biphasic phase transfer catalysis: carben addition to C-C double bonds[d]
Stirbar stirred vials	Catalytic oxidation with hydrogen peroxide: selective oxidation of methylcyclohexene[e]
Bubble columns	Heterogeneously catalyzed hydrogenation of p-nitrotoluene[f]
Overhead stirred vials	Heterogeneously catalyzed hydrogenation of crotonaldehyde[g]

Note: Review the citations for a more detailed insight in the chemistry performed.

[a] Reddington, E., Sapienza, A., Guraou, B., Viswanathan, R., Sarangapani, S., Smotkin, E.S., Mallouk, T.E., *Science* 1998, 280, 1735.

[b] Cooper, A.C., McAlexander, L.H., Lee, D.-H., Torres, M.T., Crabtree, R.H., *J. Am. Chem. Soc.* 1998, 120, 9971.

[c] Desrosiers, P. Guram, A., Hagemeyer, A., Jandeleit, B., Poojary, D.M., Turner, H., Weinberg, H., Catal. *Today* 2001, 67, 397.

[d] Wessjohann, L., Schmidt, J., Ostermann, L., Brändli, C., www.chemspeed.ch/applications.html, application note no. 011.

[e] Schüth, F., Busch, O., Hoffmann, C., Johann, T., Kiener, C., Demuth, D., Klein, J., Schunk, S., Strehlau, W., Zech, T., *Top. Catal.* 2002, 21, 55.

[f] Zech, T., Bohner, G., Li, Q., Kaiser, H., Haas, A., Schunk, S.A., Proceedings of XXXVII. *Jahrestreffen Deutscher Katalytiker*, Weimar, March, 17-19, 2004, p. 145.

[g] Thomson, S., Hoffmann, C., Ruthe, S., Schmidt, H.-W., Schüth, F., *Appl. Catal. A: Gen.* 2001, 220, 253.

methylcyclohexene as a prototypical model reaction for functionalization of sterically hindered olefins such as steroids or terpenes. In Figure 11.34, the substrate methylcyclohexene and for comparison a terpene and a steroid, both also displaying double bonds in sterically hindered positions, are illustrated. The potential reaction products that stem from the oxidative conversion of methylcyclohexene with H_2O_2 are the epoxide, the diole, the ketone, and the three isomers of the allyl alcohols. The oxidation reaction was performed in a Stage I screening system designed for performing parallel reactions with on-line analysis. The testing system consists of a 48-fold reaction block with 5 to 15 mL disposable glass vials as reaction units. Each reactor is stirred with a magnetic stirbar and can be run within a temperature range 20 to 150°C. The reaction block is accommodated in a fully automated liquid handling system that is capable of charging the reactors with fluids and to take samples for the quasi online HPLC analysis and offline GC analysis. With this system it was possible to analyze the correlation of catalyst concentration, reaction pH, and concentration of the substrate (for a picture of the reaction platform see Figure 11.35).

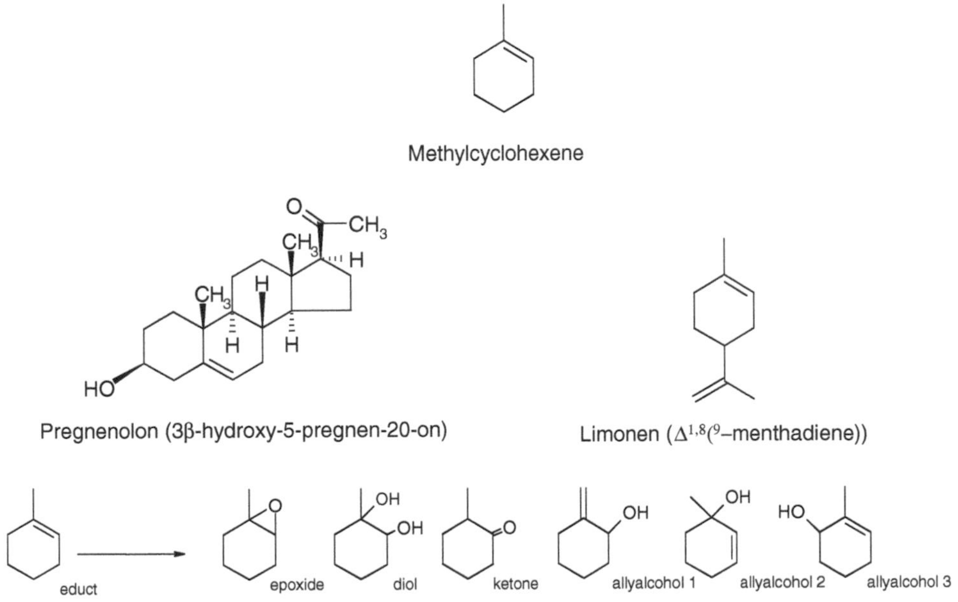

Methylcyclohexene

Pregnenolon (3β-hydroxy-5-pregnen-20-on)

Limonen ($\Delta^{1,8}(^9$–menthadiene))

educt   epoxide   diol   ketone   allyalcohol 1   allyalcohol 2   allyalcohol 3

Figure 11.34 Illustration of the similarity of the double bond of methylcyclohexene with typical steroids or terpenoids and potential reaction products for methylcyclohexene conversion with H_2O_2.

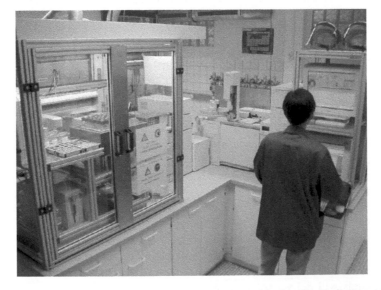

Figure 11.35 Image of the integrated Stage I screening system employed for liquid-phase catalysis.

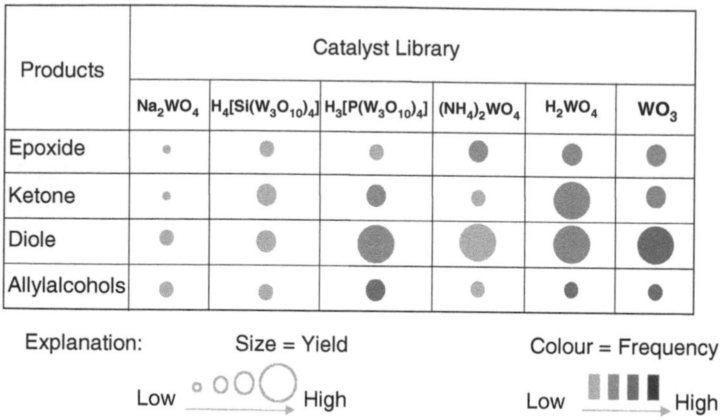

Products	Catalyst Library					
	Na_2WO_4	$H_4[Si(W_3O_{10})_4]$	$H_3[P(W_3O_{10})_4]$	$(NH_4)_2WO_4$	H_2WO_4	WO_3
Epoxide	·	◉	◉	●	●	●
Ketone	·	◉	●	◦	●	●
Diole	◉	◉	●	●	●	●
Allylalcohols	◉	◉	●	◦	●	●

Explanation: Size = Yield   Colour = Frequency

Low ○ ○ ○ ○ High   Low ▮▮▮▮ High

Figure 11.36   Frequency count for the oxidative conversion of methylcyclohexene of a tungsten-based library. The variation of the reaction conditions included a pH variation of the H_2O_2 solution, variation of the ratios of catalyst, oxidant and substrate and the reaction temperature. A total of 800 experiments were performed.

In Figure 11.36 a display of the performance of a tungsten-based catalyst library is shown: a total of more than 800 experiments were performed to obtain this set of data. The variation included pH variation of the H_2O_2 solution, variation of the ratios of catalyst, oxidant, and substrate, and the reaction temperature. The data were visualized by the so-called frequency count where the color encoding visualizes the number of experiments where the product was found and the size of the circle visualizes the yield that was found most frequently. The basis of this frequency count is in this case the total number of experiments that have been performed. The frequency count enables the experimentalist to obtain a qualitative overview of a large set of experiments in a Stage I screening. The smaller the number of experiments visualized by the plot, the more exactly the reaction conditions can be filtered out at which the prospective experiments to reach the target performance can be performed. For the case depicted in Figure 11.36, one can conclude that Na_2WO_4 and $H_4Si(W_3O_{10})_4$ are rather poor catalysts for converting cyclomethylhexene into oxygenates. High yields of dioles can be obtained with the catalysts $H_3P(W_3O_{10})_4$, $(NH_3)_2WO_4$, H_2WO_4, and WO_3. With H_2WO_4 also high yields of the corresponding ketone can be obtained. Of course, the conditions under which these products were obtained cannot be extracted out of the current graph, still one can conclude from the frequency count that with the WO_3 catalyst (more often than for the other catalyst species) high yields of the diole were obtained very often. In the following study $(NH_3)_2WO_4$ was converted to WO_3 by calcination. The graph in Figure 11.37 illustrates the catalytic results obtained from this set of approximately 200 experiments, again a variation was made including a pH variation of the H_2O_2 solution, variation of the ratios of catalyst, oxidant, and substrate, and the reaction temperature. It can clearly be seen that the conversion of $(NH_3)_2WO_4$ into WO_3 results in a better catalyst for producing higher yields of the diole. On the outer left, the reference catalyst WO_3 delivers even better results, as samples of WO_3 stemming from $(NH_3)_2WO_4$ calcined at 500°C under N_2 or air. The inset in Figure 11.38 shows the color of the different samples after calcination and the resulting diffractograms indicating the structural conversion of $(NH_3)_2WO_4$ into WO_3.

The interpretation of this result is not straightforward. From experience one would speculate that the more soluble $(NH_4)_2WO_4$ would also be the more active catalyst in the reaction. The finding is contrary: WO_3 calcined at high temperatures is also more active than samples calcined at lower temperatures. Additional experiments where the heterogeneous catalyst WO_3 was filtered off after starting the reaction revealed that the same results were obtained as in the presence of the heterogeneous catalyst. The stringent interpretation goes along with the basic chemistry of soluble oxoanions of tungsten [83] and findings of Venturello et al. [84] and Aubry et al. [85]. Better soluble tungsten compounds will deliver solutions with a rich polyanion chemistry including polyanions

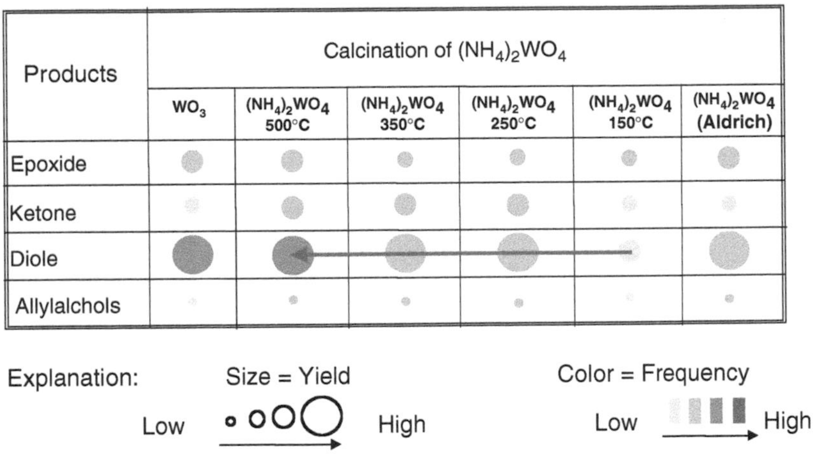

Products	Calcination of $(NH_4)_2WO_4$					
	WO_3	$(NH_4)_2WO_4$ 500°C	$(NH_4)_2WO_4$ 350°C	$(NH_4)_2WO_4$ 250°C	$(NH_4)_2WO_4$ 150°C	$(NH_4)_2WO_4$ (Aldrich)
Epoxide	●	●	●	●	●	●
Ketone	●	●	●	●	●	●
Diole	●	●	●	●	●	●
Allylalchols	●	●	●	●	●	●

Explanation: Size = Yield   Color = Frequency

Low ° ○ ○ ◯ High   Low ▮▮▮▮ High

Figure 11.37   Frequency count for the oxidative conversion of methylcyclohexene of a library based on of WO_3 stemming from the calcination of $(NH_3)_2WO_4$. The variation of the reaction conditions included a pH variation of the H_2O_2 solution, variation of the ratios of catalyst, oxidant and substrate and the reaction temperature. A total of 200 experiments were performed.

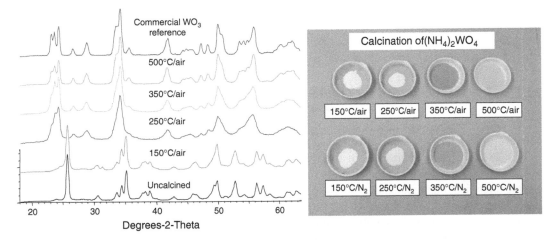

Figure 11.38   Diffractograms and appearance of color of different samples of $(NH_4)_2WO_4$ after thermal treatment in air. The decomposition of $(NH_4)_2WO_4$ and subsequent formation of WO_3 can be followed in the diffractograms.

with higher degrees of polymerization. The contrary is true for precursors, which are not especially soluble (like WO_3 especially if calcined at high temperatures): the solution chemistry of the tungsten-oxo-anion species will be dominated by smaller units such as monomers, dimers, and trimers. These small dimeric phosphor-bridged units are suggested to be the most active and selective species for the oxo-functionalization of olefins by H_2O_2 in the presence of tungsten polyoxoanions as catalytic species.

11.7.1.1 Alternative Stage I Screening Concepts

A new alternative approach for Stage I screening in liquid phase is the use of bubble column-type reactors. These parallel bubble columns can operate in batch and fed-batch mode regarding the reaction mixture, while a continuous stream of gas is used as reactant (H_2, O_2, or others) as well as for the intense agitation of the reaction mixture (Figure 11.39).

Figure 11.39   Photographs of cold-flow experiments studying the flow regimes and catalyst suspension in laboratory bubble columns. Left: low gas flow; middle: high gas flow; right: high gas flow with catalyst suspension.

The advantages of this kind of reactors are that no mechanical stirring device is required, a higher degree of parallelization can easily be achieved, the time-consuming step of cleaning is shortened noticeably compared to stirred reactors, and the technology is comparably cheap as the only parts needed are tubes and fittings. An experimental setup with 24-parallel pressurized bubble column system was realized at hte Aktiengesellschaft. The reactor consists of stainless-steel tubes containing 2 µm frits, which do not only serve as gas-dispersion devices but can also be used as filtration devices for heterogeneous catalysts. The current system can be applied up to temperatures of 200°C and pressures of 20 bar. The system can be used to investigate hydrogenation or oxidation reactions, and a good correlation with results gained in Stage II screening is generally obtained.

11.7.2 Stage II Screening for Liquid-Phase Catalysis

In order to achieve results with close-to-conventional testing conditions, the parallel reactor setup for liquid-phase reaction must mimic the real process conditions of the later process as nicely as possible. The main efforts to be realized lie in the miniaturization and integrated construction of the parallel testing setup and the automation of process control combined with suitable online and offline analytical methodologies.

We have illustrated above that the principal idea of running parallel test equipment for liquid-phase catalysis is not very new. The suppliers of conventional autoclave equipment have offered basic systems of parallel pressure vessels for several decades. Since the late 1990s several companies have offered commercial solutions for parallel autoclave rigs for different purposes [86]. The quality of the resulting analytical data is as good as that for the data obtained from conventional single autoclave units. For the daily work in the laboratory and the total workflow, one should keep in mind that, in fact, reaction and data acquisition can be done in parallel and often fully automated, but manual cleaning of parallel setups can become a severe bottleneck, especially depending on the chemistries performed.

We have discussed the structure and synthesis of the library of molecular catalysts for polymerization in Section 11.5.1. In the present section we want to take a closer look at the performance of the catalyst library and discuss the results obtained [87]. The entire catalyst library was screened in a parallel autoclave bench with exchangeable autoclave cups and stirrers so as to remove the bottleneck of the entire workflow. Ethylene was the polymerizable monomer that was introduced as a gas, the molecular catalyst was dissolved in toluene and activated by methylalumoxane (MAO), the metal to MAO ratio was 5000. All reactions were carried out at 50°C at a total pressure of 10 bar. The activity of the catalysts was determined by measuring the gas uptake during the reaction and the weight of the obtained polymer. Figure 11.40 gives an overview of the catalytic performance of the entire library of catalysts prepared. It can clearly be seen that different metals display different activities. The following order can be observed for the activity of the different metals: $Fe(III) > Fe(II) > Cr(II) > Co(II) > Ni(II) > Cr(III)$. Apparently iron catalysts are far more active than any of the other central metal

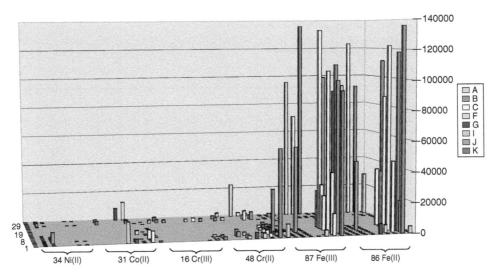

Figure 11.40 Overview of the entire library of catalysts based on the metals Fe(II) and (III), Ni(II), Co(II) and Cr(II) and (III). The activity on the z axis is displayed as grams of polyethylene per millimole of catalyst per hour. The dicarbonyl backbones are visualized as color encoding; the amine substituents are encoded as number on the y axis.

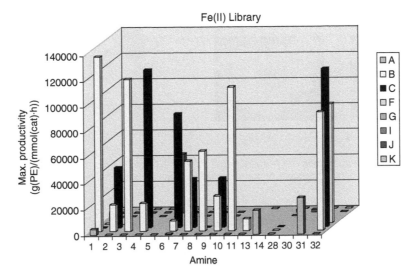

Figure 11.41 Overview of the Fe(II) library. The activity on the z axis is displayed as grams of polyethylene per millimole of catalyst per hour. The dicarbonyle backbones are visualized as color encoding; the amine substituents are encoded as number on the y axis.

atoms tested. Iron catalysts also produced the polymers with the highest molecular weight (110 kDa), for nickel- and cobalt-based catalysts only very low molecular weights were obtained (below 20 kDa), and chromium-based catalysts delivered intermediate molecular weights.

Apart from the analysis of the results with regard to the most active metal, the structure–property relationships with regard to the ligand structure were also a focus of the project. The results for the Fe (II) complexes are depicted in Figure 11.41 and Figure 11.42. From these results, two hypotheses were concluded (see also Figure 11.43):

- The sterics of α-substituent of the dicarbonyl-backbone is crucial for the activity.
- An α-substituent next to the amino group is beneficial for the activity of the complex.

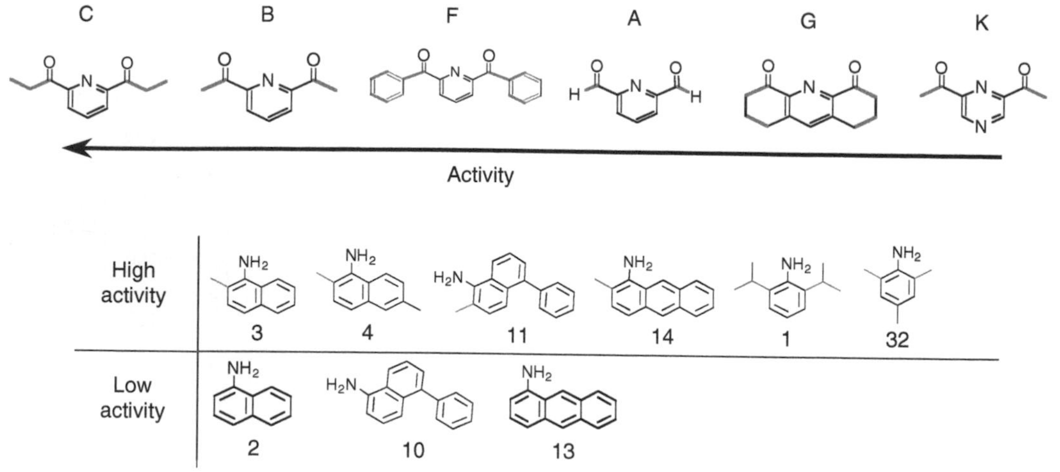

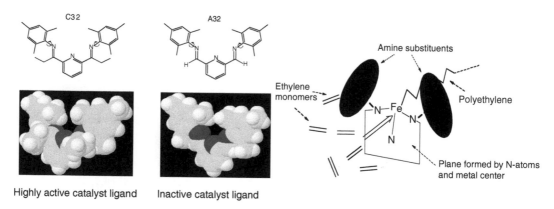

Figure 11.42 Classification of the amines and dicarbonyl backbones of the Fe(II) library.

Highly active catalyst ligand   Inactive catalyst ligand

Figure 11.43 Space-filling models of the ligands C32 (high activity of Fe(II) complex) and A32 (low activity of Fe(II) complex). Structure of the metal ligand complex under polymerization conditions.

In Figure 11.43 a space-filling model illustrates these hypotheses for two different complexes of different activities. The N–C bond between the imino group and the carbon atom in the aromatic ring has a rotational degree of freedom. The two groups that can hinder this rotation in combination are the α-substituent next to the amino group and the α-substituent of the dicarbonyle backbone. They can act as a "lock" for the aromatic rings attached to the dicarbonyl backbone and keep the access to the active metal center free for monomers. The structure in Figure 11.43 illustrates this finding.

The results obtained in the Stage II screening for over 420 active compounds in 10 weeks would hardly have been possible with conventional nonparallized equipment. Furthermore, the results prove that through intelligent library design and stringent interpretation of the results the user can quickly gain insight into chemical details on a molecular basis and deepen understanding of the system explored.

11.8 SUMMARY AND OUTLOOK

In this chapter we tried to outline the key aspects and challenges of the application of HTE to catalysis, especially heterogeneous catalysis. As an outcome it can be summarized that the HTE technologies are now established and available for broad application throughout the scientific world.

Since the main focus of application of the technologies described in this chapter is on heterogeneous catalysis, we also need to include the chemical industry as the potential key user. Nevertheless, this depends to a large degree on the R&D future of the chemical industry — if the chemical industry will convert to a commodity- or innovation-driven business. In this chapter we just want to highlight these economical issues and would like to refer the interested reader to "The Power of Innovation — How Innovation Can Energize the Chemical Industry, Festel Capital, 2003." Based on a summary of chemical industry strategies and related success factors such as increased competitivness, need for new "high-chem" products, increased R&D productivity, decreased R&D costs, and faster result realization, the industry needs an improved ratio of R&D spendings vs. resulting return on investment. On the basis of the discussion in this chapter it can be clearly concluded that HTE offers a new strategic approach: an affordable, systematic research tool fulfilling the efficiency requirements of the innovation-driven chemical industry. Therefore, it is not surprising that this strategic tool is considered nowadays, and from the first major players used, to improve their research approach and to increase the success probability of R&D.

From a technological and scientific point of view it can be summarized that HTE is not magic but just a new technology available as additional R&D tool within the full spectrum of R&D technology. High-throughput experimentation for catalysis research is a robust methodology, yields reproducible results, and allows affordable, broad-range systematic R&D projects. The applicability of the technology covers all major application fields, including continuous gas–phase, liquid–phase as well as the highly sophisticated triple–phase applications. Temperature and pressure range are — as well as the chemical application base, reactor design, and materials synthesis — close or even identical to the large–scale application parameters and ensure an appropriate scale-up to the next application stage. Currently, it is feasible to achieve a 10- to 100-fold gain in discovery and optimization speed, and it contributes already to an early phase scientific understanding of catalysis.

From a future perspective, we assume that HTE for catalysis will be a broadly accepted, well–established, and widely used methodology. The new HTE tools will be integrated in the overall available research technology platform, and routine use will considerably contribute to the development and discovery of improved and new chemical processes and applications thereof. The overall screening efficiency will increase with future technology developments, e.g. new and faster analytical methods, sampling technologies, and higher degrees of parallelization, and will yield in total a higher throughput of optimizations and discovery. Owing to the necessary standardization and the broad systematic approach of possible research projects, HTE will in the long term facilitate the collection of a consistent database, the generation of valuable knowledge toward the rational understanding and design of catalytic processes, and therefore we would like to conclude this chapter with a key final statement: The HTE tool base for catalysis will significantly increase R&D productivity and substantially contribute to scientific system understanding.

REFERENCES

1. Mittasch, A., *Adv. Catal.* 1950, II, 81.
2. Maclean, D., Baldwin, J.J., Ivanov, V.T., Kato, Y., Shaw, A., Schneider, P., Gordon, E.M., *Pure Appl. Chem.* 1999, 71, 2349.
3. Balkenhohl, F., Busche-Hünnefeld, C., Lansky, A., Zechel, C., *Angew. Chem.* 1996, 108, 2436.
4. *Chem. Rev.* 1997, 2, total volume.
5. Weinberg, H.W., Jandeleit, B., Self, K., Turner, H., *Curr. Opin. Solid State Mater. Sci.* 1998, 3, 104.
6. Hanak, J.J., *J. Mater. Sci.* 1970, 5, 964.
7. Hanak, J.J., *J. Vac. Sci. Technol.* 1971, 8, 172.
8. Briceno, G., Chang, H., Sun, X., Schultz, P.G., Xiang, X.D., *Science* 1995, 270, 273.
9. Perez-Ramirez, J., Berger, R.J., Mul, G., Kapteijn, F., Moulijn, J.A., *Catal. Today* 2000, 60, 93.
10. Mallouk, T.E., Reddington, E., Pu, C., Ley, K.L., Extended abstracts, *Fuel Cell Seminar*, Orlando, FL, 1996, 686.

11. Moates, F.C., Somani, M., Annamalai, J., Richardson, J.T., Luss, D., Wilson, R.C., *Ind. Eng. Chem. Res.* 1996, 35, 4801.

12. Hoffmann, C., Wolf, A., Schüth, F., *Angew. Chem.* 1999, 38, 2800.

13. www.hte-company.de

14. www.symyx.com

15. www.spotfire.com

16. Hinze, J., Welz, U., Broad smiles, in *Software Development in Chemistry 10*, ed., Gasteiger, J., Gesellschaft Deutscher Chemiker (GDCh), Frankfurt/Main, 1996, p. 59.

17. Hansch, C., Leo, A., Hoekman, D., *Exploring QSAR — Hydrophobic, Electronic, and Steric Constants*, American Chemical Society, Washington, DC, 1995.

18. Klanner, C., Farrusseng, D., Baumes, L., Mirodatos, C., Schüth, F., *QSAR Combin. Sci.* 2003, 22, 729.

19. Klanner, C., Farrusseng, D., Baumes, L., Lengliz, M., Mirodatos, C., Schüth, F., *Angew. Chem.* 2004, 40, 5347.

20. Bem, D., Gillespie, D., Erlandson, E.J., Harmon, L.A., Schlosser, S.G., Vayada, A.Y., in *Combinatorial Experimental Design Using the Optimal-Coverage Algorithm in Experimental Design for Combinatorial and High Throughput Materials Development*, ed., J.N. Cawse, Wiley-Interscience, Hoboken, 2003, p. 89.

21. Mason, R.L., Gunst, R.F., Hess, J.L., *Statistical Design and Analysis of Experiments: with Applications to Engineering and Science*, 2nd ed., Wiley-Interscience, Hoboken, 2003.

22. Wolf, D., Buyevskaya, O., Baerns, M., *Appl. Catal.* 2000, 200, 63.

23. Taylor, S.J., Morken, J.P., *Science* 1998, 280, 267.

24. Moates, F.C., Somani, M., Annamalai, J., Richardson, J.T., Luss, D., Willson, R.C., *Ind. Eng. Chem. Res.* 1996, 35, 4801.

25. Holzwarth, A., Schmidt, H.-W., Maier, W.F., *Angew. Chem.* 1998, 110, 2788.

26. Snively, C.M., Okarsdottir, G., Lauterbach, J., *Angew. Chem.* 2001, 113, 3117.

27. Snively, C.M., Okarsdottir, G., Lauterbach, J., *Catal. Today* 2001, 67, 357.

28. Schüth, F., Busch, O., Hoffmann, C., Johann, T., Kiener, C., Demuth, D., Klein, J., Schunk, S., Strehlau, W., Zech, T., *Topics Catal.* 2002, 21, 55.

29. Cong, P., Doolen, R.D., Fan, Q., Giaquints, D.M., Guan, S., McFarland, E.W., Pooraj, D.M., Self, K., Turner, H.W., Weinberg, W.H., *Angew. Chem. Int. Ed. Engl.* 1999, 39, 484.

30. Senkan, S., Ozturk, S., *Angew. Chem. Int. Ed. Engl.* 1999, 38, 791.

31. Orschel, M., Klein, J., Schmidt, H.W., Maier, W.F., *Angew. Chem. Int. Ed. Engl.* 1999, 38, 2791.

32. Claus, P., Hönicke, P.D., Zech, T., *Catal. Today* 2001, 67, 319.

33. Klein, J., Stichert, W., Strehlau, W., Brenner, A., Demuth, D., Schunk, S.A., Hibst, H., Stork, S., *Catal. Today* 2003, 81, 329.

34. Senkan, S., *Nature* 1998, 394, 350.

35. Cong, P., Dehestani, A., Doolen, R., Giaquinta, D.M., Guan, S., Markov, V., Poojary, D., Self, K., Turner, H., Weinberg, W.H., *Proc. Natl. Acad. Sci. USA* 1999, 96, 11077.

36. Johann, T., Brenner, A., Schwickardi, M., Busch, O., Marlow, F., Schunk, S., Brenner, A., *Catal. Today* 2003, 81, 449.

37. Brenner, A. et al., DE 19830607, 1998, EP 9911311 hte Aktiengesellschaft, 1999.

38. Bergh, S., Guan, S., WO 00/51720, Symyx Technologies, 1999.

39. Bergh, S., Engstrom, J.R., Hagemeyer, A., Lugmair, C., Self, K., Cong, P., Guan, S., Liu, Y., Markov, V., Turner, H., Weinberg, W.H., High-throughput screening of combinatorial heterogeneous catalyst libraries, IMRET 4, *4th International Conference on Microreaction Technology, Topical Conference Proceedings*, Atlanta, GA, AICHE National Spring Meeting 2000, oral presentation.

40. Grob, K., *Split and Splitless Injection for Quantitative Gas Chromatography*, Wiley-VCH, Weinheim, 2001.

41. Huebschmann, H.J., *Handbook of GC/MS*, Wiley-VCH, Weinheim, 2001.

42. http://www.gilson.com

43. http://www.chemspeed.com

44. http://www.zinsser-analytic.com

45. http://www.accelab.de

46. Brümmer, H., Markert, R., Schwemler, C., *GIT Laborfachzeitschrift*, 1999, 43, 598.

47. Stiles, A.B., *Catalyst Manufacture—Laboratory and Commercial Preparations*, Marcel Dekker, New York, 1983.
48. Matijevic, E., *Acc. Chem. Res.* 1981, 14, 22.
49 Archibald, B., Brümmer, O., Devenney, M., Giaquinta, D.M., Jandeleit, B., Weinberg, W.H., Weskamp, T., in *Handbook of Combinatorial Chemistry, Drugs, Catalysts, Materials*, eds. Nicolaou, K.C., Hanko, R., Hartwig, W., Vol. 2, Wiley-VCH, Weinheim, 2002, p. 1017.
50. Nicolaou, K.C., Hanko, R., Hartwig, W., eds., *Handbook of Combinatorial Chemistry, Drugs, Catalysts, Materials*, Vol. 1, Wiley-VCH, Weinheim, 2002.
51. Sun, Y., Chan, B.C., Ramnarayanan, R., Leventry, W.N., Mallouk, T.E.J., *Comb. Chem.* 2002, 4, 569.
52. Klein, J., Laus, O., Newsam, J.M., Strasser, A., Sundermann, A., Vietze, U., Zech, T., Schunk, S.A., *Petroleum Chemistry Division Preprints* 2002, 47, 254.
53. Klein, J., Zech, T., Newsam, J.M., Schunk, S.A., *Appl. Catal.* 2003, 254, 121.
54. Wijngaarden, R.J., Kronberg, A., Westerterp, K.R., *Industrial Catalysis — Optimizing Catalysts and Processes*, Wiley VCH, Weinheim, New York, 1998.
55. Levenspiel, O., *Chemical Reaction Engineering*, Wiley, New York, 1999.
56. Levenspiel, O., *The Chemical Reactor Omnibook*, OSU Book Stores, Inc., Corvalis, OR, 1996.
57. Kapteijn, F., Moulijn, J.A., in *Handbook of Heterogeneous Catalysis*, eds. Ertl, G., Knözinger, H., Weitkamp, J., Vol. 3, VCH, Weinheim, 1996, Chapter 9.1.
58. Weitkamp, J., in *Handbook of Heterogeneous Catalysis*, eds. Ertl, G., Knözinger, H., Weitkamp, J., Vol. 3, VCH, Weinheim, 1996, Chapter 9.2.
59. Wedler, G., *Adsorption, Chemische Taschenbücher*, Vol. 9, VCH, Weinheim, 1970.
60. Zech, T., Miniaturisierte Screening-Systeme für die kombinatorische heterogene Katalyse, Dissertation, Technische Universität Chemnitz, VDI-Verlag Düsseldorf, Reihe 3, No. 732, ISBN 3-18-373203-3, 2002.
61. Claus, P., Hönicke, D., Zech, T., Miniaturization of screening devices for the combinatorial development of heterogeneous catalysts, *Catal. Today* 2001, 67, 319–339.
62. Idelchik, I.E., *Handbook of Hydraulic Resistance*, 3rd ed., CRC Press, New York, 1994.
63. Bergh, S.H., Guan, S., Chemical processing microsystems, diffusion-mixed microreactors and methods for preparing and using the same, WO 00/51720, Symyx Technologies, Inc., March 1999.
64. Zech, T., Hönicke, D., Klein, J., Schunk, S.A., Demuth, D., A novel system architecture for o high-throughput primary screening of heterogeneous catalysts, IMRET 5, *Proceedings of the 5th International Conference on Microreaction Technology*, Strasbourg, 2001, poster presentation.
65. Weissermehl, K., Arpe, H.-J., *Industrielle Organische Chemie*, Wiley-VCH, Weinheim, 1998.
66. Ertl, G., Knözinger, H., Weitkamp, J., *Handbook of Heterogenous Catalysis*, Vol. 5, Wiley-VCH, Weinheim, 1997.
67. Sheldon, R.A., van Bekkum, H., *Fine Chemicals through Heterogeneous Catalysis*, Wiley-VCH, Weinheim, 2001.
68. Hodnett, B.K., *Heterogeneous Catalytic Oxidation, Fundamental and Technological Aspects of the Selective and Total Oxidation of Organic Compounds*, Wiley-VCH, Weinheim, 1999.
69. Tenten, A., Proll, T., Schildberg, H., DE 19622331, BASF AG, 1996.
70. Tenten, A., Hibst, H., Machhammer, O., Hechler, C., Müller-Engel, K., Unverricht, S., DE 19955176, BASF AG, 1999.
71. Brenner, A., DE Oliveira, A., Lange, de, Schüth, F., Schunk, S., Stichert, W., Unger, K., DE 19830607, 1998.
72. Hoffmann, C., Wolf, A., Schüth, F., *Angew. Chem.* 1999, 111, 2971.
73. Matusz, M., Mesters, A., Buffum, J., WO 9507139 (Shell).
74. Nielsen, R.P., La Rochelle, J.H., US 3962136 (Shell).
75. Monnier, J.R., Mühlbauer, P.J., US 5081096 (Eastman Kodak).
76. Monnier, J.R., Mühlbauer, P.J., WO 8907101 (Eastman Kodak).
77. Monnier, J.R., Mühlbauer, P.J., US 4897498 (Eastman Kodak).
78. Nielsen, R.P., US 3702259, 1970, Shell.
79. Stichert, W. et al., DE 10300526 (hte Aktiengesellschaft).
80. Schulz, H., Claeys M., eds., Recent advances in Fischer–Tropsch synthesis. *Appl. Catal.* 1999, A 186 [Special Issue].
81. Patwardhan, A.W., Joshi, J.B., *Ind. Eng. Chem. Res.* 1999, 38, 49.

82. Abdallah, R., Caravieilhes, S., Grenouillet, P., de Bellefon, C., *Proceedings of 4th International Symposium on Catalysis in Multiphase Reactors*, Lausanne, September 22–25, 2002, p. 43.

83. Brinker, C.J., Scherrer, G.W., *Sole-Gel Science, The Physics and Chemistry of Sol-Gel Processing*, Academic Press, San Diego, 1990.

84. Venturello, C., DÁloisio, R., Bart, J.C.J., Ricci, M., *J. Mol. Catal.* 1985, 32, 107.

85. Aubry, C., Chottard, G., Platzer, N., Bregault, J.M., Thouvenot, R., Cauveau, F., Huet, C., Ledon, H., *Inorg. Chem.* 1991, 30, 4409.

86. www.parrinst.com

87. Kolb, P., Demuth, D., Newsam, J.M., Smith, M.A., Sundermann, A., Schunk, S.A., Bettonville, S., Breulet, J., Francois, P., *Macomol. Rapid Commun.* 2004, 25, 280.

CHAPTER 11 QUESTIONS

Question 1

- Define HTE and CombiChem.
- Explain the goals of HTE and CombiChem.
- Summarize in short catchwords the history of HTE.

Question 2

- What are the important facts that make DOE and databases within HTE irreplaceable?
- Try and classify different approaches of DOE according to their usefulness for different types of work in HTE and CombiChem. Try to rank the approaches and name the pros and cons.

Question 3

- Define Stage I and Stage II screening?
- Why is an overflow of hits not desirable? How can it be avoided?

Question 4

- Make a matrix of some parallel and sequential analytical techniques for screening and enlist their advantages and limitations.
- What are the essential differences for a Stage I and a Stage II analytical method?

Question 5

- Define the term "unit operation" and discuss its usefulness within synthetic operations.
- Search the web for automated labware for the synthesis of organic molecular compounds. Try to classify the technologies according to automation degree, flexibility of synthetic operations, and price.
- Make a plan for a split and pool library consisting of four different amino acids: (1) try to synthesize as many different combinations as possible with a sequence of two split and pool steps. (2) Applying the procedure of Figure 11.12 will deliver dimers for question (1), how can you produce higher oligomers using the same number of split and pool steps in the sequence?
- Take care: dimers like AB and BA are different chemical entities.

Question 6

- Why is it important to control mass and heat transfer?
- Define the effectiveness factor and describe the relation with the Thiele modulus.
- Name some diagnostic criteria to diagnose mass transfer limitations.
- Name some technical measures to avoid thermal runaway and hotspot formation.

Question 7

- How can you insure scalability from Stage I to Stage II testing?
- How can you avoid bottlenecks in Stage I screening? What is the role of the synthetic approach?
- What are the basic laws for fluid distribution? How critical is even distribution in a reactor module? Discuss.

Question 8

- Why are partial gas-phase oxidations so important for the chemical industry?
- How can you check the temperature and gas distribution over a multitube reactor?
- What are the challenges of dynamic catalyst testing? When is it applied?

Question 9

- Enlist some of the means of overcoming mass transport limitations for gas–liquid-phase catalysis.

Question 10

- Name some of the limitations of certain Stage I screening tools in liquid-phase catalysis.
- Can certain technical challenges become a bottleneck during screening in liquid-phase catalysis? Discuss.

Question 11

- Develop a theoretical project for liquid-phase catalysis (Stage I and II screening) and describe the technologies you want to employ. What are the potential challenges you are facing? How can you overcome or work around bottlenecks?

Heterogeneous Photocatalysis

Vasile I. Pârvulescu and Victor Marcu

CONTENTS

12.1 Introduction 428
12.2 Photocatalysis 429
    12.2.1 General Principles 430
    12.2.2 TiO$_2$ Photocatalysts 433
        12.2.2.1 Preparation Procedures 435
    12.2.3 Modified Titania Photocatalysts 438
    12.2.4 Mixed Oxide and Composites Containing Titania 441
    12.2.5 Noble Metal Deposited on Titania Surfaces 441
    12.2.6 Monoliths Containing Titania 443
    12.2.7 ZnO 443
    12.2.8 Other Oxide Semiconductors 444
    12.2.9 Calcium Hydroxyapatite Modified with Ti(IV) 444
    12.2.10 Polyoxometalates 444
    12.2.11 Heterogeneous Fenton-Type Catalysts 445
12.3 Kinetic Studies 445
12.4 Combinatorial Approaches in Preparation and Testing Photocatalysts 446
12.5 Example of Reactions under Photocatalytic Conditions 448
    12.5.1 Photoinduced Deposition of Various Metals onto Semiconductor 449
    12.5.2 Energy Storage 450
12.6 Solar Photocatalysis 450
12.7 Sonophotocatalysis 450
12.8 Photocatalysis Associated to Microwave Radiation 452
12.9 Photocatalyst Deactivation 452
12.10 Conclusions 452
References 453
Chapter 12 Questions 461

12.1 INTRODUCTION

Photochemical reactions occur when a gas, a solution, or a solid mixture of chemicals absorbs light to produce an excited state, which further reacts generating different reaction products. Part of the excited-state particles might not convert into new species, but rather revert to the ground-state species. The research field of heterogeneous photochemistry is focused toward the investigation of each of the above-mentioned steps and elucidation of their mechanism and kinetics.

The absorption of monochromatic light [1a] is governed by the Beer–Lambert law that connects the decrease in intensity of the light with the optical path and the concentration of the absorbing species by an exponential equation

$$I = I_0 \exp alc \tag{12.1}$$

where I is the intensity of the monochromatic light after it passed through a cell of length l, containing a solution of concentration c of the absorbing species. I_0 is the incident light intensity and a is a constant characteristic of the absorbing material.

The Einstein law of photochemical equivalence [1b] gives the ideal yield of a photochemical reaction (when no secondary interaction occurs); for each absorbed quantum of light, one particle of the reactant undergoes a reaction. The primary quantum yield gives the number of particles that disappear directly as a result of the absorption of one quantum of light and it can vary between 0 and 1. The quantum yield may be different depending on the nature of the excited state. For instance, the quantum yield of biacetyl formation in a polymer matrix [2a] changes from 0.045 for the first excited triplet to 0.7 for a higher triplet.

An obvious condition for a photochemical reaction to occur is that the particle absorbs light and that the excited state be stable enough to further react, but not so stable as to remain indefinitely unchanged. For instance, the normal triplet lifetimes of aromatic compounds without chemical quenching are in the range of msec to sec [2b]. In practice, either more or less particles than the absorbed number of quanta of light react due to secondary reactions. The ratio between the practical and ideal number of particles that reacted for each absorbed quantum of light is called the *quantum yield*. Often, the reaction proceeds through the formation of radicals and chain reactions and very high quantum yields are attained. The quantum yield can be very low and still high enough for practical application of the reaction.

The absorption process is governed by the laws of quantum chemistry that give a description of the energy levels that are available for electronic transitions due to the absorption of light. The levels have different energies due to contributions of electronic, vibrational, or rotational structure of the particle. The energy necessary for electronic transitions is higher than that due to a change in the vibrational state of the atoms in the molecule, which in turn is smaller than that due to a rotational change of the molecule. For a given energy level the distance between the atoms can change from a minimal value given by the left branch of the energy curve and a maximal value given by the intersection of a parallel to the O*x* axis and the right-hand branch of the enegy curve. As long as the latter point exists, the molecule remains stable. When that point moves to infinity, the molecule dissociates. In a transition between energy states of the molecule, depending on the relative position of the two curves, a stable molecule will be maintained or else dissociation will occur.

The Franck–Condon principle [1c] states that owing to the fact that light absorption is a fast process, there is no change in the relative atomic positions during the first step of a photoreaction and there is no need to take into account the change in energy level owing to the movement of the atoms. The electronic states of the molecules are characterized by the number of unpaired electrons. If all the electrons are coupled the state is called a singlet, while one unpaired electron gives a triplet state. When the ground state is a singlet and the excited state a triplet, manifold intersystem crossing is said to take place. The lifetime of the reaction intermediates is usually shorter than that of either the reactants or the reaction products. The degradation process of the reaction intermediates involves, on top of the already mentioned possible return to the ground state, a chemical reaction or an energy transfer to another species.

Photochemical reactions, like any chemical reaction, can be classified into various groups, depending on the reactants and products, for example, elimination, isomerization, dimerization, reduction, oxidation, or chain reaction. One important practical field of photochemistry is organic photochemistry. In solution photochemical reactions, the nature of the solvent can markedly influence the reaction. The absorbtion of the solvent and of the reaction products is an important parameter for the choice of the reaction conditions. It is useful to have a solvent with a relatively low absorption in the desired wavelength. Sometimes photosensitizers are used: these are substances that absorb light to further activate another substance, which decomposes.

Several techniques are used to follow the photoreactions: their intermediates and the reaction products. Fluorescence, phosphorescence, UV or visible spectra, or chromatography can be used to follow the lifetime of the excited state. Recently, the thermal grating method was used to study various processes involved in photoreactions [2]. Stabilizers of a certain type of intermediate (singlet or triplet) or the opposite (quencher) were used to determine which kind of intermediate is active.

Semiconductors play a special role in photoreactions [3]. They are used either as photosensitisers or photocatalysts (see below). In the semiconductors, similar to large organic molecules, two different groups of energy states are present. The lower energy group is called a valence band and the higher energy group is called the conduction band. The reactants that undergo a reaction are usually adsorbed on the semiconductor surface and then are desorbed as they form products, provided it does not decompose due to light absorption. The semiconductor is ready for a further reaction, thus functioning much like a classical catalyst and is called photocatalyst.

In classical kinetic theory the activity of a catalyst is explained by the reduction in the energy barrier of the intermediate, formed on the surface of the catalyst. The rate constant of the formation of that complex is written as $k = k_0 \exp(-\Delta G/RT)$. Photocatalysts can also be used in order to selectively promote one of many possible parallel reactions. One example of photocatalysis is the photochemical synthesis in which a semiconductor surface mediates the photoinduced electron transfer. The surface of the semiconductor is restored to the initial state, provided it resists decomposition. Nanoparticles have been successfully used as photocatalysts, and the selectivity of these reactions can be further influenced by the applied electrical potential. Absorption chemistry and the current flow play an important role as well. The kinetics of photocatalysis are dominated by the Langmuir–Hinshelwood adsorption curve [4], where the surface coverage $PHY = KC/(1 + KC)$ (K is the adsorption coefficient and C the initial reactant concentration). Diffusion and mass transfer to and from the photocatalyst are important and are influenced by the substrate surface preparation.

12.2 PHOTOCATALYSIS

Since the Honda–Fujishima effect [5] was reported in the early 1970s, extensive studies of photocatalysis on semiconductors, in particular on illuminated surfaces of titanium dioxide (TiO_2), have been carried out [6–19]. Through the 1970s to the 1980s, the main interest was focused on hydrogen photoevolution from water or organic wastes. At this stage, the properties of semiconductors were thoroughly investigated and described, including semiconductor modifications and sensitization, and improvement of hydrogen evolution ability. Although water splitting is not in practical use yet, some progress has been accomplished in the basic science. In the 1990s, the topic shifted to applications of environmental remediation using TiO_2 photocatalysts and significant progress has been realized. Thus, photocatalytic chemistry involving semiconductor materials has grown from a subject of esoteric specialty interest to one of central importance in both academic and technological research. In this context, environmental pollution and its control through nontoxic treatment and easy recovery processes is a matter of serious concern. The number of publications concerning mineralization of dyes, pesticides, accaricides, fungicides, etc., increased enormously during the last decade. Many products utilizing TiO_2 photocatalyst with antibacterial, antifogging, and self-cleaning functions are already on the market. As a new trend, fossil fuels and the ever-increasing problem of carbon dioxide concentrations represent a particular issue of interest to researchers in this field.

12.2.1 General Principles

Photocatalysis is the segment of catalysis, which covers the range of the reactions proceeding under the action of light. Among them, we can distinguish phenomena such as catalysis of photochemical reactions, photo-activation of catalysts, and photochemical activation of catalytic processes. This term is defined by the IUPAC as follows "Photocatalysis is the catalytic reaction involving light absorption by a catalyst or a substrate." A more detailed definition may be the following "Photocatalysis is a change in the rate of chemical reactions or their generating under the action of light in the presence of the substances (photocatalysts) that absorb light quanta and are involved in the chemical transformations of the reaction participants, repeatedly coming with them into intermediate interactions and regenerating their chemical composition after each cycle of such interactions" [20].

Usually, the most typical processes that are covered by "'photocatalysis" are the photocatalytic oxidation (PCO) and the photocatalytic decomposition (PCD) of substrates, which most often belong to the organic class of compounds. The former process employs the use of gas-phase oxygen as direct participant to the reaction, while the latter takes place in the absence of O_2.

Photocatalysis uses semiconductor materials as catalysts. The photoexcitation of semiconductor particles generates electron–hole pairs due to the adsorption of 390 nm or UV light of low wavelength (for TiO_2). If the exciting energy employed comes from solar radiation, the process is called solar photocatalysis [21].

As previously mentioned, to exhibit photocatalytic properties, inorganic solid compounds must have electronic properties described by a semiconducting behavior [22]. A relatively wide range of metal oxides and sulfides has been successfully tested as photocatalysts, for example, TiO_2, WO_3, WS_2, ZnO, Fe_2O_3, V_2O_5, CeO_2, CdS, and ZnS [23–46]. However, even in the early stage of development of this new branch of chemistry, the interest was focused on the reactions which occurred inside illuminated suspensions of titania. For example, the formation of •OH radicals was observed in aqueous TiO_2 suspensions and their role was explained by Bard and co-workers [47,48].

The structure of the energy levels in semiconductor materials is well described by the quantum theory of solids [49]. Electron–hole pairs are generated upon interband electronic transition, as an effect of interaction with photons having an energy comparable (equal or higher) to the bandgap that is separating the conduction and the valence bands (Figure 12.1). For titanium dioxide, this energy can be supplied by photons with energy in the near ultraviolet range. This property promoted

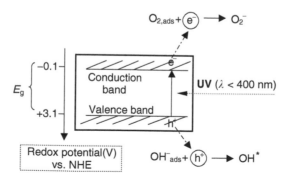

Figure 12.1 Generation of photocatalytic active species at the surface of titania particles (NHE = normal hydrogen electrode).

titanium dioxide as a good and promising candidate for use as a photocatalyst using solar light as energy source. However, solar daylight provides only 3 to 5% of its radiation intensity within the absorption spectral range of titanium dioxide.

The generation of electron–hole pairs under the action of the light is the initiating step in photocatalysis:

$$TiO_2 \overset{h\nu}{\leftrightarrow} e_{cb}^-(TiO_2) + h_{vb}^+(TiO_2) \tag{12.2}$$

In the case of anatase allotropic form of titania, the redox potential for the photogenerated holes vs. the standard hydrogen electrode is of +3.1 V, and that for the conduction band electrons of +0.5 V (Figure 12.2). These values show that holes created by light excitation have a strong oxidizing potential.

The mobile charge carrier species may either recombine or reach the semiconductor surface, where they can be trapped by the surface adsorbates or other sites. The lifetime of electron–hole (e^-/h^+) pairs that are generated is important in determining the reaction yield. The holes are mainly trapped by water molecules or hydroxyl ions, giving rise to very reactive hydroxyl radicals:

$$Ti^{IV} - OH^- + h^+ \leftrightarrow Ti^{IV}:OH^\bullet \tag{12.3}$$

$$Ti^{IV} - H_2O + h^+ \leftrightarrow Ti^{IV}:OH^\bullet + H^+ \tag{12.4}$$

which are the principal oxidative species in the photocatalytic systems. These species may easily attack organic substrates leading to their oxidation in a more or less degradative way (Equation 12.5 to Equation 12.8). It is already known that the $OH^\bullet$ radical species are nonselective nucleophylic oxidants [50,51]:

$$Ti^{IV}:OH^\bullet + R_{1,ads} \rightarrow Ti^{IV} + R_{2,ads} \tag{12.5}$$

$$OH^\bullet + R_{1,ads} \rightarrow R_{2,ads} \tag{12.6}$$

$$Ti^{IV}:OH^\bullet + R_1 \rightarrow Ti^{IV} + R_2 \tag{12.7}$$

$$OH^\bullet + R_1 \rightarrow R_2 \tag{12.8}$$

The sign ($:$) means that the $OH^\bullet$ radical is associated with the Ti^{IV} site. The chemical bond between the two species is altered or even completely destroyed due to the formation of the oxidized organic species.

The formation of hydroxyl radicals is the rate-limiting reaction step in the photocatalyzed oxidation using TiO_2 and ZnO slurries. The rate of the reaction in D_2O is approximately three times slower than in H_2O, due to the fact that the O–D bond has a lower energy level than the O–H bond and, consequently, it needs a higher energy to generate radicals. But substitution of the hydrogen atoms with deuterium in the reactant molecule has no effect on the initial reaction rate [48]. Experiments carried out in water-free aerated organic solvents resulted only in partial degradation of the organic compounds. Complete mineralization to CO_2 that typically occurs in aqueous solutions has not been realized under these conditions. Although the oxidation potential for many organic compounds is above the valence band energy of anatase, which from a thermodynamic point of view would allow a direct interaction with the holes at the photocatalyst surface, the presence of hydroxyl radicals is very important for the complete photocatalytic destruction of the organic substrates. The dynamics of the hydroxyl radical ($OH^\bullet$), which is one of the active species generated, has been studied by ESR spectroscopy [52], while the recombination dynamics of the generated charge carrier has been revealed by laser spectroscopic studies [53–58].

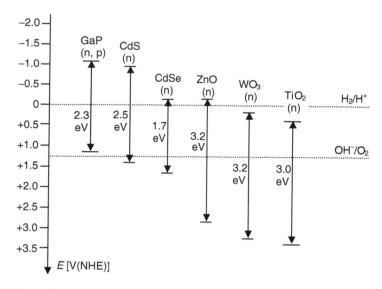

Figure 12.2 Position and width of energy bands of several illuminated semiconductors, with respect to the electrochemical scale (NHE = normal hydrogen electrode).

The contribution of the holes in the photocatalytic oxidation is decisive for the overall process efficiency. Ishibashi et al. [51] estimated the quantum yields of hydroxyl generation for an aqueous TiO_2 suspension by monitoring the formation of 2-hydroxyterephtalic acid, which is a product of the photocatalytic oxidation of the terephtalic acid. The quantum yield for OH$^\bullet$ production was found to be ~7×10^{-5} per absorbed photon. The photocatalytic oxidation of I$^-$ ions to I_2, preformed under the same conditions as terephtalic acid photooxidation, led to an iodide formation yield that clearly showed the direct implication of photogenerated holes in the oxidation process. The calculated value for the photogeneration of holes was ~5.7×10^{-2} per absorbed photon, which is much higher than the corresponding value for the hydroxyl formation. Guillard [59] suggested that one of the photodegradation paths of butanoic acid on TiO_2 involves a direct reaction between a hole and one butanoate ion, resulting in CO_2 and propyl radical production. They suggested that the propyl radical further reacts with oxygen molecules, leading to propanoic acid and propylic alcohol.

The photoexcited electrons are trapped at the surface by Ti^{IV} sites, subsequent to this process resulting Ti^{III} sites, which may further transfer charge carriers to molecular oxygen resulting in the formation of a superoxide radical:

$$Ti^{IV} + e^- \leftrightarrow Ti^{III} \tag{12.9}$$

$$Ti^{III} + O_2 \leftrightarrow Ti^{IV} - O_2^{\bullet-} \tag{12.10}$$

The superoxide radical may interact with the protons (resulting, for example, from the reaction described in Equation 12.4), with or without implying other electrons from the conduction band (the species between parentheses may be adsorbed or in the aqueous phase):

$$Ti^{IV} - O_2^{\bullet-} + 2(H^+) + e^- \leftrightarrow Ti^{IV}(H_2O_2) \tag{12.11}$$

$$Ti^{IV} - O_2^{\bullet-} + (H^+) \leftrightarrow Ti^{IV}(HO_2^\bullet) \tag{12.12}$$

Studies carried out by Cermenati et al. [60] also gave evidence about the contribution of radical species in the photocatalytic oxidation of organic compounds. The addition of superoxide dismutase, which is an enzyme that catalyzes the dismutation of the superoxide radical, to an aqueous solution of quinoline (pH 6), in the presence of TiO_2 particles, as photocatalysts, led under UV irradiation to a considerable decrease in the photocatalytic disappearance of the organic compound.

Another method to control photoassisted transformations is to use molecules that may be excited under visible light (Equation 12.13), such as dye molecules. These molecules (denoted by the letter X in the following equations) are able to inject electrons into the conduction band of a catalyst (Equation 12.14). The cation radical formed under these conditions may undergo chemical decomposition, unless it regenerates in a period of time smaller than the period needed for its decomposition (Equation 12.15) [29]

$$X + h\nu \rightarrow X^*$$ (12.13)

$$X^*_{ads} + TiO_2 \rightarrow X^+_{ads} + TiO_2(e)$$ (12.14)

$$X^+_{ads} \rightarrow product$$ (12.15)

TiO_2 photoactivity depends on the efficiency of electron–hole generation, charge transfer, and separation [57]. An elegant way to investigate the interfacial electron transfer between photocatalyst particles and charge carriers is to use a photoelectrochemical experiment which measures the working electrode potential in the same working condition as photocatalysis in the bulk organic molecule sensitizer/TiO_2 reaction system. These measurements are mostly applicable to dyes. The effect is appreciated from the analysis of photocurrent–time profiles.

The sensitization process involves the excitation of the organic molecules by absorbing visible light photons and subsequent electron injection from excited states to the TiO_2 conduction band [61]. Then electrons undergo transfer processes from the conduction band to a conductive film on the working electrode and finally lead to the out-circuit. In this process, TiO_2 particles function as a bridge connecting the dye and working electrode. The difference in photocurrent under light-on conditions resulted from different electron transfer efficiencies in different reaction systems. The photocurrent correlates perfectly with the adsorption properties of the photocatalyst. These adsorptive properties can be improved by introducing ions bound to the TiO_2 that can act as good scavengers to trap electrons. As a result, only a small amount of photogenerated electrons can transfer to the working electrode to form an out-circuit. Figure 12.3 presents such curves for a dye X-3B/TiO_2 suspension system.

12.2.2 TiO_2 Photocatalysts

In the area of advanced oxidation technology, semiconductor TiO_2 photocatalysis has been widely studied because of its potential application in air cleanup and water purification [62,63]. TiO_2 is largely used as a photocatalyst owing to its beneficial characteristics: high photocatalytic efficiency, physical and chemical stability, low cost and low toxicity. In addition to the wide bandgap, titania exhibits many other interesting properties, such as transparency to visible light, high refractive index, and a low absorption coefficient. The two principal polymorphs of TiO_2 are anatase and rutile which are associated with bandgap energies of 3.2 and 3.1 eV, respectively. Figure 12.4 shows the XRD patterns of anatase and rutile. It has been pointed out that the photodegradation reaction rate is much more rapid over anatase than in the rutile [28,64,65], and it is mainly affected by the crystalline state and textural properties, particularly, surface area and particle size of the TiO_2 powder. However, these factors often vary in-opposite ways, since a high degree of crystallinity is generally achieved through a high-temperature thermal treatment leading to a reduction in the surface area. Thus, optimal conditions for synthesis have been sought to obtain materials of high photoreactivity [31,66]. In addition, since photocatalytic reactions are generally studied in aqueous suspensions,

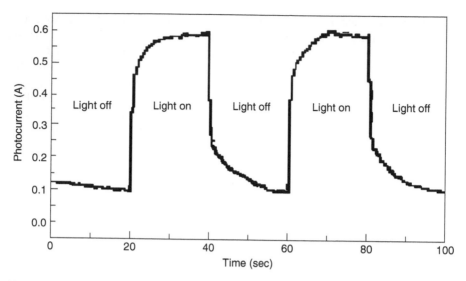

Figure 12.3 Photocurrent of (a) X-3B/TiO2 suspension system in terms of irradiation time at visible light. (From Xie, Y. and Yuan, C., *Mater. Res. Bull.*, 39, 533, 2004.)

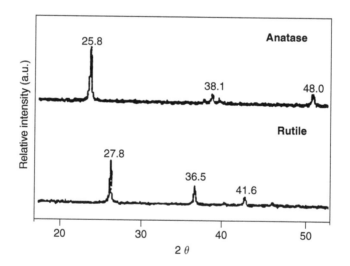

Figure 12.4 XRD patterns of anatase and rutile.

problems arising from the formation of hard agglomerates such as diffusion of reactants and products and light absorption have to be considered [66].

The phase structure of TiO_2 greatly affects the photoactivity. In addition, amorphous TiO_2 seldom displays photocatalytic activity due to the presence of nonbridging oxygen atoms in bulk TiO_2, whose Ti–O atomic arrangement defects could act as recombination centers of photogenerated electron–hole pairs.

The photocatalytic performance of TiO_2 depends not only on its bulk energy band structure but, to a large extent, on surface properties. The larger the surface area, the higher the photocatalytic activity. Decomposition of methylene blue solution over TiO_2 thin film photocatalysts indicated that TiO_2 thin films on a glass substrate exhibited better photocatalytic decomposition than those on polycarbonate and polymethyl methacrylate because of surface morphology. When titania was deposed on aluminum, inferior decomposition reaction occurred, which was explained by the fact that it captures the holes generated from the bulk of TiO_2 and thereby decreases the photocatalytic efficiency [67].

The type and density of surface states are affected, among others, by the synthesis process. A soft mechanical treatment of TiO_2 powder, for instance, was found to reduce significantly its photocatalytic activity for Cr(VI) reduction [68], while treatment in either H_2 or N_2 plasma was found to enhance the activity within the visible-light range for certain reactions [69,70]. However, the interplay between processing conditions and photocatalytic activity remains largely a state-of-art and are beyond prediction at this point. Crystalline TiO_2 has typically been calcined or crystallized in oxidizing atmospheres, such as air and oxygen. The effect of the so-called inert atmospheres, such as N_2, Ar, and vacuum ($\sim 5 \times 10^3$ torr), has mostly been overlooked. The effect of other calcination atmospheres, including air and H_2 on the photocatalytic properties of TiO_2, was investigated as well [71]. Thus, the calcination atmosphere has been found to have significant effects on the photocatalytic activity of TiO_2 in hydrogen production from a methanol/water solution [71]. Calcination in either hydrogen or in vacuum results in a high density of defects and low surface hydroxyl coverage, thus yielding low activity. Calcination in Ar, in contrast, enhances visible-light excitation and high hydroxyl coverage, leading to high activity.

12.2.2.1 Preparation Procedures

To extend the applicability of titania, the ease of production and reproducibility has to be secured in terms of production cost and product quality. The literature is abundant in methodologies to prepare TiO2 photocatalysts, and as a consequence a very large number of materials have been prepared and tested from colloidal to large surface area mesoporous materials.

Colloidal systems were reported to be much more effective than immobilized or supported catalysts for photodegradation of any hazardous molecule [72,73]. Powdered materials exhibit an important inconvenience. The main difficulty in employing an insoluble, powdered semiconductor in aqueous dispersion is the need to remove the solids after treatment and subsequent redispersion in a second aqueous solution to be purified. However, in the case of colloids it was speculated that they may either be dispersed in the irradiated aqueous solution as a colloidal suspension, or attached to a suitable support as a fixed or mobile fluidized bed [72,73].

Anatase was demonstrated to be the most effective phase of titania. Different methods for the preparation of anatase powders were proposed. TiO_2 materials need to be fabricated with high specific surface areas and highly porous structures (preferred in nanoscale) for effective contact with reactant substances to acquire high photocatalysis in the decomposition of various harmful substances in gas or liquid. Among the reported procedures several proved to provide such photocatalysts: hydrolysis under hydrothermal condition of titanium tetraethoxide above 250°C gave particles of 20 to 30 nm size [74], vapor hydrolysis of titanium tetraisopropoxide at 260°C results in nanosized particles [75], destabilization of aqueous titanium lactate below 100°C gave thin films on various substrates [76], decomposition of titania-hydrate coated on hollow glass spheres [77], pulsed laser deposition [78], electrodeposition [79,80], etc. Through application of these procedures or modified variants, extensive studies have been performed to achieve various TiO_2 nanomaterials with large surface areas (powders) [81–83], nanotubules [84–88], nanofibers [78], and thin films comprised of nanocrystals [79,80]. Among these, nanotubules, nanofibers, and thin films have attracted much interest in this decade because of the large surface areas and various potential applications of these materials. Nanotubules and nanofibers were fabricated through a template synthesis method that was pioneered by Martin's group [84]. In the template-synthesis method, commercial porous alumina membranes (with pore diameters of 22 or 200 nm, several centimeters in diameter and 50 to 100 μm thick) were used as templates in a sol–gel process. After removing the alumina templates, nanotube-shaped TiO_2 powder [78] or bush-like TiO_2 nanotubules or nanofibers arrays [84] were obtained. Several potential applications of the TiO_2 nanostructures were investigated, including photocatalysis. Nevertheless, from the viewpoint of practical applications, exploring nanostructured TiO_2 materials with high photocatalytic performance and improved mechanical strength is vital and still remains a challenge for scientists.

TiO_2 films have been prepared by various vacuum techniques such as chemical vapor deposition [89], flame pyrolysis [90], DC sputtering [91], and precipitation [92]. The alkoxide sol–gel process [93] has emerged as one of the most promising techniques for growing TiO_2 thin films. The TiO_2 thin films prepared by an alkoxide sol–gel process can be of high purity and low cost due to the availability of high-purity chemicals and simplicity of the process. Low processing temperatures may be desired to minimize the interfacial interaction at the film–substrate interface. Unfortunately, the TiO_2 thin films prepared using alkoxides or other titanium precursors, require processing temperatures higher than 300°C for the crystallization of TiO_2 particles and removal of organic materials [94,95]. However, appropriate choice of alkoxide and its solvent may reduce processing temperatures. Experiments carried out by deposition of TiO_2 on glass, polycarbonate, polymethyl methacrylate, and aluminum via the sol–gel process using different alkoxide precursors evidenced the role of the carrier [96]. Decomposition of methylene blue solution over these TiO_2 thin film photocatalysts indicated that the TiO_2 thin films on glass substrate exhibited better photocatalytic decomposition than those on polycarbonate and polymethyl methacrylate because of the surface morphology. The lower rate of decomposition reaction of films deposited on aluminum was explained by the fact that it captures the holes generated from the bulk of TiO_2 and thereby decreases the photocatalytic efficiency [96].

The deposition of nanostructured TiO_2-based material directly on glass is expected to achieve enhanced photocatalysis and mechanical strength. Sputtering and a combination of sputtering and sol–gel techniques seem to be the most adequate for such a purpose [91,97,98]. In the latter case, porous alumina films on glass formed by anodizing sputter-deposited Al layers were used as templates in the successive sol–gel process.

High-surface-area inorganic materials with ordered mesoporous structures have also been of major interest for numerous applications including photocatalysis [99–102]. The ultra-high-surface-area of mesoporous materials is appealing in applications of heterogeneous photocatalysis where it is desirable to minimize the distance between the site of photon absorption and electron–hole redox reactions to improve efficiency [103–105].

Degussa TiO_2 P-25 represents a commercially available photocatalyst that was intensively investigated both at the laboratory- and pilot-scale plant. However, to enhance the applicability, it was thought that its immobilization on preformed inert support surfaces was beneficial. Coating a granular silica support with 0.1% TiO_2 (Degussa P-25) suspension and air-drying overnight followed by pretreating under UV radiation in clean air led to granular photocatalysts in the range of 2 to 3 mm in size and possessing an effective surface area of about 300 m^2/g [106]. The deposition of the same photocatalysts on glass Rasching rings [107–110] was found to remove the color and increase the biodegradability of chlorinated cellulose effluent [109] or Orange II photocatalysis [110]. Other attempts have been made to prepare titania-supported catalysts, using glass beads [111], fiber glass [112], silica [113], stainless steel [114], fiber textile [115], honeycomb [116], and zeolites [117]. Besides a large adsorption capacity, high-silica zeolites (HSZ) notably exhibit a very low dependence on effluent humidity and on compound polarity. This may be of particular interest since nonpolar molecules hardly adsorb on pure TiO_2. To develop enhanced photoactivity of TiO_2, investigators elaborated very different methods to support active titanium on zeolites, from simple amalgam of powders through mechanical mixing [118], or with papermaking techniques [89] to chemical vapor deposition [89], cation exchange [119], or direct synthesis [120]. Zeolites in pellet form (faujasite Y and ZSM-5) were also used as supports for a titanium sol–gel impregnation, and physical characterization of materials showed that a great intimacy of mixing was obtained without significant modification of the adsorbent. Zeolites are inert and do not interfere with the photocatalytic mechanism in the degradation of volatile organic compounds (VOC) from industrial waste gas streams [121]. However, most impregnation methods are based on acid-catalyzed sol–gel formation [122], which may alter the framework structure of zeolites and more studies are necessary to define to what extent the support undergoes physical or chemical modifications.

All of these reports indicate that the deposition of titania upon available supports seems to be a good strategy to prepare active photocatalysts.

TiO$_2$-pillared clay has also attracted significant attention as a photocatalyst, since its superior adsorption properties were reported to accelerate photocatalytic reactions [123–125]. TiO$_2$-pillared clay has a mesoporous structure owing to its small TiO$_2$ particles that are located between silicate layers as pillars and shows high adsorption ability owing to its large specific surface area [126,127]. The interlayer surfaces of the pillared clay are generally hydrophobic [128], which is an advantage needed to adsorb and enrich diluted hydrophobic organic compounds in water [123]. In addition, it also demonstrates photocatalytic activity owing to the nanosized TiO$_2$ pillars. Photocatalytic degradation on TiO$_2$-pillared clay for gaseous substances in air has also been found to be influenced by surface hydrophobicity of the pillared clay, because the air contains humidity [129]. The surface hydrophobicity of TiO$_2$-pillared clay can be controlled by selection of the clay, as it is known to increase in the sequence: Saponite-Ti < Montmorilonite-Ti < Mica-Ti [129]. This increase enhances the adsorption of highly hydrophobic substances such as toluene and trichloroethylene, but does not influence the adsorption of less hydrophobic substances such as ethylene and ethanol. Enrichment of toluene and trichloroethylene by adsorption on TiO$_2$-pillared clay enhances photocatalytic degradation and the sequence of events is in agreement with the surface hydrophobicity of the pillared clay.

As is described in detail by Fenelonov and Melgunov elsewhere in this book, texture is a very important parameter for heterogeneous catalysts including photocatalysts. The effect of the size of the primary particles (crystal size) and secondary particles (aggregates of primary particles) have been shown to be very important. For the decomposition of trichloroethylene gas, the maximum decomposition rate was observed on primary particles with a size of about 8 nm and the rate was observed to decrease gradually with increasing particle size [130]. Amongst three anatase powders synthesized under hydrothermal conditions with different crystallite sizes (6, 11, and 21 nm) anatase of the intermediate size (i.e., 11 nm) was shown to have the highest photoactivity for the decomposition of gaseous CHCl$_3$ [131], while anatase particles with average size of 30 nm were reported to have the highest decomposition rate for methylene blue in water [132]. For C$_2$HCl$_3$ in aqueous solution a crystallite size of about 15 nm was the most effective [133]. Other reports also pointed out that crystallinity of anatase phase is one of the important factors governing photocatalytic performances and the maximum rate for decomposition might be different between pollutants in water and those in gaseous state, and also might depend on the nature of the pollutant [134]. Maximum methylene blue decomposition in water was obtained on anatase powders with an apparent crystallite size of about 36 nm [134].

Another problem to overcome is the growth of TiO$_2$ nanocrystallites, prepared during sintering processes at high temperatures. Serious aggregation of the ultrafine powder occurs when it is dispersed in aqueous solutions, and should therefore be taken into account for the development of a photocatalyst that can yield high reactivity under visible light.

The effect of the surface area is far from being a simple one. It was shown for titania that when the surface area changes from 110 to 12 m^2/g, the average time required for a complete mineralization of organic substrates increased from 40 to 75 and 50 to 75 min for salicylic acid and phenol, respectively [135]. These results clearly show that textural properties, particularly the surface area, strongly affect the photoreactivity, although a high-temperature treatment improved their crystallinity [18]. Therefore, this phenomenon may be explained only in connection with the catalyst surface dehydroxylation.

The number of surface OH groups can be determined by the method described by Van Veen et al. [136] according to the reaction

$$3(TiS-OH) + Fe(acac)_3 \rightarrow (TiS-O)_{3-y}Fe(acac)_y + y(TiS-OH) + (3-y)Hacac \quad (12.16)$$

The texture of the catalysts also controls the half-lifetime of charge carriers. This can be easily obtained from microwave absorption experiments using a time-resolved microwave conductivity

(TRMC) method [83,137]. The decrease in half-lifetime of photogenerated charge carriers has two major causes, which contribute to the photocatalytic reactivity in opposite ways. The first one is associated with fast charge recombination rates in the semiconductor particles, which may result from impurities or small crystal size and leads to a drastic reduction in the photoreactivity. The second one originates in the charge carrier being trapped by defect sites or surface hydroxyl groups, thus enhancing the probability of reaction with the present organic substrates.

12.2.3 Modified Titania Photocatalysts

Titanium dioxide has attracted much attention in the field of photocatalytic applications for environmental purification, decomposition of harmful substances, and utilization of solar energy. Solar photocatalysis for wastewater treatment has already proven to be a highly effective technology and TiO_2 the most commonly applied catalyst system [16,18,22,138]. TiO_2, as a wide bandgap semiconductor (3.2 eV), is only responsive to radiation in the UV and near-UV range ($\lambda < 388$ nm). This bandgap is however not ideal for solar applications, and limits application to the visible range. Therefore, the development of photocatalysts that can be excited by visible light ($\lambda > 400$ nm) is of considerable interest [70,139,140].

To perform decontaminations utilizing solar light, many attempts to develop second-generation photocatalysts that can drive the photodegradation reaction with visible light were undertaken [141]. These efforts concentrated either on organic molecule sensitization or transition metals doping of crystal TiO_2-based photocatalysts [142–145]. The chemical stability of organic photosensitizers represents an important issue in photocatalytic systems [141,146]. As one of the major environmental pollutants, dyes and their derivatives possess good absorption in the visible-light range. As a consequence, a dye/TiO_2/visible-light reaction system should be a reasonable solution for dyestuff wastewater treatment.

On the other hand, recombination of generated e^-/h^+ pairs affects the amount of active oxygen species produced by charge trapping. To suppress annihilation of the photogenerated e^-/h^+ pairs, studies of selective metal ion doping of the crystalline TiO_2 matrix have been carried out. This technique has proven to be the most popular for modification of TiO_2 surfaces.

Studies have demonstrated that rare-earth ion modification can greatly improve the photocatalytic activity of TiO_2 [147], although recent data disclosed uncertain doping effects (even controversial results) about transition metal ion dopants [140].

The addition of a second species can cause a decrease in charge recombination and an increase in the TiO_2 photocatalytic efficiency. Such behavior was examined by loading a series of species on the surface or into the crystal lattice of photocatalysts: inorganic ions [148–152], noble metals [153,154], and other semiconductor metal oxides [155]. It was thus proven that modifications produced by these species can change semiconductor surface properties by altering interfacial electron-transfer events and thus the photocatalytic efficiency.

Rare-earth elements have been largely used in these studies. Among them, europium (III) was considered owing to its unusual electronic structure. Eu^{3+}-ion-modified TiO_2 samples were prepared by a chemical coprecipitation–peptization method [156], which consists of prehydrolysis of $TiCl_4$ by frozen distilled water. The Eu_2O_3 powder was added to the above solution to produce a transparent aqueous solution according to the required Eu^{3+} modifying content (Eu^{3+} ion equivalent to 3.0 at% of Ti in bulk solution).

The local structure around the lanthanide ions with differing redox potentials: Eu(III)/(II) (-0.35 V vs. NHE), Yb(III)/(II) (-1.05 V), and Sm(III)/(II) (-1.55 V) in TiO_2 particles was investigated by EXAFS. The photocatalytic reaction and EXAFS studies were also carried out for a calcined Yb(III) ion adsorbed-TiO_2 catalyst [157]. The photocatalytic activity of these lanthanides toward methyl blue photodecomposition was very similar, suggesting that adsorbed lanthanide ions on TiO_2 particles scarcely assist the high-photocatalytic activity of TiO_2 catalyst. However, the way in which the catalysts were activated was of high importance. The photocatalytic activity of the calcined Yb/TiO_2

catalyst was greater than that of a noncalcined catalyst. This is in contrast with the photocatalytic activity exhibited by Fe/TiO$_2$, for example, for which the calcined catalyst was much less active than the noncalcined one. These observations clearly show that the role of adsorbed metal ions on TiO$_2$ particles changes before and after the calcinations of the system, and that the change involves the local structure and charge of the metal ions adsorbed-TiO$_2$ particles.

In the case of 2-propanol, since there is an imbalance of charge between OH and CH, the Yb(III) ion attracts the OH side. Consequently, the photocatalytic activity of Yb/TiO$_2$ catalyst was slightly higher than that of pure TiO$_2$ catalyst, since the reaction of 2-propanol with OH* radicals produces acetone via the CH$_3$C*(OH) CH$_3$ radical. However, the Fe/TiO$_2$ catalysts again exhibited better photoactivity.

Finally, for adenosine 50-triphosphate (ATP) (an anionic substance), lanthanide-ion-modified TiO$_2$ catalysts were found to be more selective. This suggests that the selectivity of photocatalytic decomposition for target substances can be controlled by a combination of the properties of the adsorbed metal ions (e.g., ionic radius, d- or f-electron, redox potential) and the coordination structure of their ions on the surface of TiO$_2$. These data indicate that Sm/TiO$_2$, Eu/TiO$_2$, and Yb/TiO$_2$ are more selective toward anionic substances than Fe/TiO$_2$, which is excellent for decomposition of cationic and neutral substances.

Neodymium has also been used as an ion modifier for a TiO$_2$ sol photocatalyst made of nanoparticles with an anatase semicrystalline structure. Under visible light irradiation this system was able to decompose phenol and many phenol derivatives, which confirms other data obtained for lanthanides, revealing the selective photocatalysis and weak photomineralization power of Nd^{3+}–TiO$_2$ sol. The fact that visible light-induced photocatalysis suggests that introducing lanthanoid neodymium ion acceptor impurities or clusters until reaching the necessary amount for the splitting of discrete energy levels into an impurity subband and overlapping of this sub-bandgap with the conduction band of the TiO$_2$ semiconductor is possible (Figure 12.5) [158].

Choi et al. have made a comprehensive investigation of the relative efficiency of metal ion-doped TiO$_2$ by laser spectroscopy, in which the metal ions have an ionic radius similar to Ti(IV) (0.75). They concluded that the efficiency depends on whether the metal ion acts as a mediator of interfacial charge or as a recombination center [159], while the existence of the positive charge of ion species bound on the surface of TiO$_2$ nanoparticles may enhance the chemisorption of anionic reactant species. This conclusion also emerges from the studies of Ranjit et al. [160], who investigated the photocatalytic degradation of organic pollutants and salicylic acid in the presence of lanthanide oxide-doped TiO$_2$ catalyst, in which the lanthanide ions were expected to interact with organic functional groups via the f-orbitals rather than as an interfacial charge mediator, since lanthanide ions have much larger ionic radii (ca. 1.1) than the Ti(IV) ion and are known for their ability to form complexes with various Lewis bases.

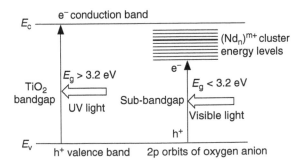

Figure 12.5 Photocatalytic mechanism of Nd^{3+}–TiO$_2$ sol/visible light system (From Xie, Y. and Yuan, C., *Appl. Surf. Sci.*, 221, 7, 2004.)

Transition metal ions have also been incorporated into TiO_2-based matrices. A bandgap shift to the visible range was observed when they are impregnated on large bandgap semiconductors [161–163]. Intimately mixed aerogels of TiO_2–SiO_2 with transition metal ions like V, Cr, Mn, Fe, and Ni were investigated for the decomposition of acetaldehyde under UV irradiation [164]. The reaction rates for V and Mn were nearly five and ten times, respectively, greater than their corresponding rate constants under visible light irradiation. In addition, these authors found that the activity of these catalysts under UV light irradiation is double than that of the standard Degussa P-25 TiO_2 catalyst. These results indicate that the extent of dispersion and local structure of the metal oxides play a significant role in the decomposition reaction of acetaldehyde under visible as well under UV light irradition. This report in conjunction with other data shows that, indeed, visible light photocatalysts for gas-phase oxidation reactions are available. Another example is the M–TiO_2–SiO_2 catalysts, in which M = Co or Ni. These visible light photocatalysts exhibit comparable activity to analogous chromium systems in gas-phase photoxidation of acetaldehyde to CO_2. However, under the conditions employed [164], only the M–TiO_2–SiO_2 formulations were active, and not M–TiO_2. The presence of silica seemed to be essential for the photooxidation of acetaldehyde.

Cr-ion-doped titania loaded MCM-41 (TiO_2/Cr-MCM-41) has exhibited reasonable photocatalytic activity with visible light for aqueous formic acid conversion [165]. Also, Cr ion doping by ion beam methods has been used with some success [166–168].

The way in which the activation of transition metal ions incorporated photocatalysts is carried out is also very important for the photocatalytic behavior. Simply using adsorbed Fe(III) ions on TiO_2 without calcination, Ohno et al. [169] have reported a remarkable effect on the oxidation of water. This result suggests the possibility that contact of TiO_2 with metal ions in water affects photocatalytic activity. Conversely, a binary mixed oxide of Fe_2O_3/TiO_2 (1:1, by weight) calcined at 700°C showed reduced activity in the photocatalytic degradation of o-cresol, compared with the oxide calcined at 500°C [170], owing to an increased proportion of the inactive pseudobrookite phase [171]. EXAFS studies and x-ray absorption near-edge structure (XANES) studies [172–174] have been employed to find answers to the nature of interatomic bonding and electronic states, and to provide information on the structural relationship between a metal ion and an oxide surface or a reactant. However, a complete explanation has yet to be attained, although some interpretation has been reported for several oxides [175–177].

As was mentioned above, photocatalytic dye reactions represent an important part of photocatalysis. The combination of organic molecule sensitization and transition metal doping on TiO_2 crystals may enhance photocatalyst performance. The sensitization process involves the excitation of dye molecules with visible light and the subsequent electron injection and electron transfer. The electron–cation radical (e^- – dye^+) generation capacity depends mainly on the incident light intensity. The charge transfer and separation efficiency depend on the electron acceptor, transferring route, and adsorption effect between the dye molecule and TiO_2 particles. The addition of Eu^{3+} may result in a higher scavenging capability than that of the oxygen molecule for the TiO_2 conduction band electrons, thus resulting in a better transfer and separation efficiency of charge pairs. Considering the standard redox potentials of Eu^{3+}/Eu^{2+} and O_2/O_2^- ($E^0(Eu^{33+} = Eu^{2+}) = -0.36$ V, $E^0(O_2/O_2^-) = 0.338$ V and $E_{vb}(TiO_2) = -0.5$ V vs. NHE), it was found that the presence of Eu^{3+} ions on nanoparticle surfaces may result in the following [178]:

First step: $dye^* + CB(TiO_2) \Rightarrow dye^+ + e\text{-}CB(TiO_2)$ electron injection

Second step: $Eu^{3+} + e^- \Rightarrow Eu^{2+}$ electron trapping

Third step: $Eu^{2+} + O_2 \Rightarrow Eu^{3+} + O_2^-$ electron transferring

The above electron trapping process promotes the electron transfer effects, which improves the photocatalytic activity of Eu–TiO_2. At the same time, the positively charged vacancies (h^+) remaining on the dye molecule can extract electrons from hydroxyl species in solution to produce hydroxyl radicals

(*OH). These radicals are strong electrophiles capable of oxidizing and destroying organic molecules in an aqueous medium. Less stable dye cation radicals (dye*$^+$) are also very susceptible to photooxidative degradation by molecular oxygen and superoxide anion radicals (O_2^-). Throughout the process, the Eu^{3+} species act as an electron scavenger to trap the TiO_2 CB electrons injected from excited dye molecules. The electron trapping capability of Eu^{3+} ion is apparently superior to that of the oxygen molecule (O_2) [179]. The electrons trapped by Eu^{3+} ions are subsequently transferred to the surrounding O_2 adsorbed on TiO_2 particles. Meanwhile, the destruction of dye molecules occurs by the interaction with holes (h^+) and hydroxyl radicals (*OH).

12.2.4 Mixed Oxide and Composites Containing Titania

Titania–silica aerogels possess very high surface areas (600 to 1000 m^2/g) and large pore volumes (1 to 4 cm^3/g), and thereby have attracted considerable interest for photocatalysis. A number of studies have shown that titania–silica intimate mixtures exhibit enhanced UV photocatalytic activity compared with pure titania [180–183].

More complicated mixed oxide photocatalysts have been reported as well. Chu et al. [184] have fabricated highly porous TiO_2–SiO_2–TeO_2/Al_2O_3/TiO_2 composite nanostructures on Ti/glass substrates. In these preparations, porous alumina films were used as the carrier material acquiring the necessary mechanical strength. The porosity and pore distribution of the anodic alumina nanostructures were controlled by adjusting the anodizing conditions (solution, voltage, or current) and the successive pore-expanding time. These photocatalysts were investigated by decomposing acetaldehyde gas under UV illumination.

Moon et al. [185–187] prepared titanium–boron binary oxides by the sol–gel method. They reported that Pt-loaded Ti/B photocatalysts could decompose water into O_2 and H_2, stoichiometrically. More recently, Jung et al. [188] investigated the local structure and photoactivity of B_2O_3–SiO_2/TiO_2 ternary mixed oxides (the SiO_2 content was fixed as 30 at% with respect to TiO_2). Boron was incorporated into the framework of a titania matrix by replacing Ti–O–Si with Si–O–B or Ti–O–B bonds. During this process, paramagnetic species such as O^- and Ti^{3+} defects were formed by boron incorporation. All B_2O_3–SiO_2/TiO_2 samples had pure anatase phase structure even though the calcination temperature was over 900°C. Incorporating more than 10% of boron oxides enlarges the grain size of the anatase phase and causes a red shift of the light absorption spectrum. As a result, the photoactivity of B_2O_3–SiO_2/TiO_2 ternary mixed oxides was greatly influenced by the content of boron.

Other composite photocatalysts were prepared by mounting immobilized anatase particles on mesoporous silica and silica beads [189–191]. The behavior of anatase-mounted activated carbons was also studied in detail [192–194]. It was even suggested that carbon-coated anatase exhibits better performance in photocatalysis than anatase itself, demonstrating high adsorptivity, inhibition of interaction with organic binders, etc. [195,196].

Homogeneous, nanosized, copper-loaded anatase titania was synthesized by an improved sol–gel method [197]. These titania composite photocatalysts were applied to the photoreduction of carbon dioxide to evaluate their photocatalytic performance. Methanol was found to be the primary hydrocarbon product [198]. Under calcination conditions, small copper particles are well dispersed on the surface of anatase titania. According to XAS and XPS analysis, the oxidation state of Cu(I) was suggested to be the active species for CO_2 photoreduction [199]. Higher copper dispersion and smaller copper particles on the titania surface are responsible for a great improvement in the performance of CO_2 photoreduction.

12.2.5 Noble Metal Deposited on Titania Surfaces

One of the major limitations in semiconductor photocatalysis is the relatively low value of the overall quantum efficiency mainly due to the high rate of recombination of photoinduced electron–hole pairs at or near the surface. Some success in enhancing the efficiency of photocatalysts has been achieved by using nanosized semiconductor crystallites, instead of bulk materials and modifying photocatalysts by depositing noble metals like Ag and Pd on their surface [200,201]. These approaches demonstrate once again that semiconductor particle size and surface properties are two important factors influencing the performance of photocatalysts in that they can influence the separation efficiency of photoinduced electron–hole pairs, besides the photocatalytic oxidation reactions take place at or near the surface.

To understand the role of the noble metal in modifying the photocatalysts we have to consider that the interaction between two different materials with different work functions can occur because of their different chemical potentials (see [200] and references therein). The electrons can transfer from a material with a high Fermi level to another with a lower Fermi level when they contact each other. The Fermi level of an n-type semiconductor is higher than that of the metal. Hence, the electrons can transfer from the semiconductor to the metal until thermodynamic equilibrium is established between the two when they contact each other, that is, the Fermi level of the semiconductor and metal at the interface is the same, which results in the formation of an electron-depletion region and surface upward-bent band in the semiconductor. On the contrary, the Fermi level of a p-type semiconductor is lower than that of the metal. Thus, the electrons can transfer from the metal to the semiconductor until thermodynamic equilibrium is established between the two when they contact each other, which results in the formation of a hole depletion region and surface downward-bent band in the semiconductor. Figure 12.6 shows the formation of semiconductor surface band bending when a semiconductor contacts a metal.

The characterization of semiconductors, which relies on analyzing illumination-induced changes in the surface voltage, can be easily made using surface photovoltage (SPS) spectroscopy [200]. It was thus found that the SPS response of semiconductor nanoparticles (TiO_2 or ZnO) becomes much weaker after a noble metal (Pd or Ag) is deposited on their surfaces, which may result from the noble metal clusters with an appropriate size effectively trapping photoinduced electrons. However, when the loading or size of the noble metal clusters becomes too large, the advantages of metal deposition are lost and these sites begin to function as recombination centers. The concentration of the noble metal at which these changes occur is a function of the nature of the metal. For

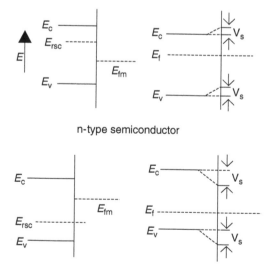

Figure 12.6 Plot showing the formation of semiconductor surface band bending when a semiconductor contacts a metal (E_C, the bottom of conduction band; E_V, the top of valence band; E_F, the fermi energy level; SC, semiconductor; M, metal; V_S, the surface barrier). (From Liqiang, J. et al., *Solar Energy Mater. Solar Cells*, 79, 133, 2003.)

example, the SPS response of Pd/semiconductor composite nanoparticles with a Pd content of 0.5 wt% was weaker than that of 0.75 wt%, while the SPS response of Ag/composite nanoparticles with the Ag content of 0.5 wt% was stronger than that of 0.75 wt%, which may demonstrate that the Ag content of 0.75 wt% is also appropriate. Concerning the role played by the noble metal clusters, it was suggested that they help fill the holes of the TiO_2 semiconductor surface with anions [201]. These holes on Ag–TiO_2 catalyst are much more common, which may explain the increased amount of Ag which can accommodate with the semiconductor support.

In addition, the rate of O_2 reduction, forming O_2^- by electron, is of importance in preventing carrier recombination during photocatalytic processes utilizing semiconductor particles. O_2^- formation may be the slowest step in the reaction sequence for the oxidation of organic molecules by OH radicals or directly by positive holes. Cluster deposition of noble metals such as Pt, Pd, and Ag on semiconductor surfaces has been demonstrated to accelerate their formation because the noble metal clusters of appropriate loading or size can effectively trap the photoinduced electrons [200]. Therefore, the addition of a noble metal to a semiconductor is considered as an effective method of semiconductor surface modification to improve the separation efficiency of photoinduced electron and hole pairs.

12.2.6 Monoliths Containing Titania

To be used as photocatalysts, especially in the so-called clean technologies, active materials must fulfill the following requirements: (1) very low toxicity, (2) resistance to photo-corrosion, (3) high availability, (4) high catalytic efficiency, and (5) low cost. From all the materials cited above, titanium dioxide and its derivatives seem to offer the best answer to these requirements, being by far the most commonly utilized photocatalysts. To be used in gas-phase photocatalysis, in addition to the above requirements, two other conditions are still necessary, that is, a very small pressure drop and an easy recovery.

Inclusion of photocatalysts in monolithic structures, namely solid structures with bored parallel channels, enables the pressure drop caused by the passage of the gas through the catalyst to be reduced by several orders of magnitude and improves both chemical and photon contact surfaces [202]. Studies on the gas-phase photocatalytic destruction of VOCs have shown the efficiency of such catalysts in the destruction of chlorinated compounds [203], even using relatively small amounts of TiO_2 on the wall of the monolith [204].

Avila et al. [205] extruded a titania monolith using a mixture of TiO_2 and fibrous magnesium silicate as the binder and calcined the extruded green bodies at 500°C. For comparison, the same group prepared a titania monolith photocatalyst starting from a ceramic monolith made of $MgSiO_4$, which was coated by immersion in a basic aqueous suspension of hydroxylated TiO_2 gel, prepared by hydrolysis of titanium isopropoxide in abundant water acidulated with nitric acid. This procedure yields particles of 5 to 10 nm diameter, aggregated into 100 nm clusters. In the latter case, the TiO_2 film should be stabilized by an appropriate thermal treatment that allows the particles to be sintered with the closest neighboring particles and with the magnesium silicate substrate. Firing conditions are critical to avoid sudden evaporation of water that may crack (and peel off) the TiO_2 layer. The stabilization of the layers of titanium is also important to avoid loss of material by erosion.

From the results obtained in this study [205], it was concluded that the catalyst prepared by extrusion of a paste containing the TiO_2 precursor together with natural silicates, is more active in the photocatalytic mineralization of organic compounds than the coated film.

12.2.7 ZnO

As a well-known photocatalyst, ZnO has received much attention in the degradation and complete mineralization of environmental pollutants [109,206–208]. Since ZnO has almost the same bandgap energy (3.2 eV) as TiO_2, its photocatalytic capacity is anticipated to be similar to that of

TiO$_2$. However, in the case of ZnO, photocorrosion frequently occurs with the illumination of UV lights. This phenomenon is considered as one of the main reasons for the decreased photocatalytic activity of ZnO in aqueous solutions [209,210]. But for gas-phase applications, this disadvantage should not exist [211]. Additionally, some studies have confirmed that ZnO exhibits a better efficiency than TiO$_2$ in the photocatalytic degradation of some dyes, even in aqueous solutions [212,213].

ZnO particle morphologies are very complex and diversiform in comparison with TiO$_2$. Thus, monodispersed ZnO particles with well-defined morphological characteristics, such as spherical, ellipsoidal, needle, prismatic, and rod-like shapes, have been obtained. Aggregates composed of these basic shape particles have also been achieved. The methods used for synthesis of these ZnO powders include alkali precipitation [214–216], thermal decomposition [217], hydrothermal synthesis [218], organo-zinc hydrolysis [219], spray pyrolysis [220], and other routes.

In contrast to TiO$_2$, the effects of ZnO particle morphologies on photocatalysis have not been completely elucidated. Although crystallinity is regarded as an important factor in photocatalysis [208,221], the meaning of this terminology is ambiguous. It is often estimated qualitatively by the peak intensity in x-ray diffractograms, while its quantitative evaluation is less commonly reported. Recently, Li and Haneda [222] synthesized ZnO powders with different morphologies and examined the photocatalysic properties of these powders toward acetaldehyde decomposition. The crystallinity of ZnO powders was calculated from lattice strain data. The authors found a dependence of the photocatalytic activity on crystallinity, surface area, and particle morphology. The morphology of ZnO particles is very sensitive to the preparation conditions and methods.

As with TiO$_2$, ZnO has also been fixed on glass Raschig rings to remove the color and to increase the biodegradability of the chlorinated cellulose effluent [109] or Orange II photocatalysis [110].

12.2.8   Other Oxide Semiconductors

CeO$_2$, WO$_3$, and Fe$_2$O$_3$ have also been investigated as photocatalysts in the oxidation of water to oxygen [46,223]. Their performance is, however, inferior to those exhibited by other photocatalysts.

12.2.9   Calcium Hydroxyapatite Modified with Ti(IV)

Calcium hydroxyapatite (HAP) modified with Ti(IV) is a photocatalyst with very specific behavior. The modification of HAP particles was performed by coprecipitation and ion-exchange methods [235]. A UV radiation was absorbed by Ti(IV)-modified HAP particles but not by the unmodified particles. Similarly, the decomposition of acetaldehyde and albumin by Ti(IV)-doped particles was observed under UV irradiation, while the unmodified HAP particles were inactive for the decomposition of both materials. Further, unlike TiO$_2$, Ti(IV)-doped HAP showed a bactericidal function in the dark, although TiO$_2$ is active only under UV irradiation. HAP has a feature that Ca(II) ions in the crystal are exchanged with Ti (IV) up to about 0.1 Ti in atomic ratio. The added Ti(IV) exists as a divalent cation such as [Ti(OH)$_2$]$^{2+}$ and [TiHPO$_4$]$^{2+}$ in the HAP crystals. Ti(IV)-substituted HAP has surface OH groups originating from HPO$_4^{2-}$ and OH$^-$ coordinating to Ti(IV). The HAP particles penetrated by Ti(IV) absorb UV radiation at a wavelength <380 nm. This character of Ti(IV)-modified HAP resembles the case of the oxidative decomposition of organisms by TiO$_2$. However, HAP has a high affinity to organisms such as proteins whereas TiO$_2$ shows no affinity to organisms. Therefore, unlike TiO$_2$ photocatalyts, Ti(IV)-modified HAP has both adsorption affinity and photocatalytic activity for organisms.

12.2.10   Polyoxometalates

Polyoxometalates (POMs) have been the object of growing interest as photocatalysts for the oxidation or reduction of organic or inorganic compounds [225–230]. Many of these POM systems share the same general photochemical characteristics as semiconductor photocatalysts. For example, they can act as electron relays, and undergo stepwise multielectron redox reactions while their structure remains intact. They have also been combined with photoactive and inactive support materials to assist in POM recovery and to improve catalytic activity [231–234].

POMs have also been employed as electron scavengers in connection with photoexcited TiO_2 to decrease charge recombination, resulting in significant enhancement of the rate of photoreaction. It was shown [235] that addition of POM anions to TiO_2 dispersions resulted in significant rate enhancement in the oxidation of dichlorobenzene. Other studies reported enhanced photocatalytic reduction rates for methyl orange in a composite $TiO_2/PW_{12}{}^{3-}$ system [236] or in the photocatalytic degradation of 2,4-dichlorophenol (DCP) in aqueous media on α-12-tungsto-phosphatic acid ($PW_{12}{}^{3-}$, POM), loaded on the surface of TiO_2 particles as a cocatalyst [237].

Loading the $PW_{12}{}^{3-}$ species on the surface of TiO_2 enhances charge separation in UV-illuminated TiO_2, thereby accelerating the hydroxylation of the initial DCP substrate, but not the mineralization of DCP, which is somewhat suppressed in the presence of POM. An increase in the loading of the POM increases the concentration of aromatic intermediates, and more toxic intermediates, such as 2,6-dichlorodibenzo-p-dioxin and 2,4,6-trichlorophenol, which were detected in the $PW_{12}{}^{3-}/TiO_2$ system.

12.2.11 Heterogeneous Fenton-Type Catalysts

FeZSM-5 was reported as a heterogeneous Fenton-type catalyst for photocatalytic oxidation of organic substances using H_2O_2 [238]. FeZSM-5 was prepared by hydrothermal crystallization in the presence of an iron salt. In contrast to the homogeneous Fenton system, the catalyst prepared demonstrated minimal leaching of iron ions, being stable through 30 catalytic runs and not losing activity in the presence of complexing agents, e.g., $P_2O_7{}^{4-}$. The catalyst was active in the oxidation of organic substances at pH from 1.5 to 8, with maximum activity being observed at pH 3. The FeZSM-5 effectively oxidized a simulant of the warfare agent, diethylnitrophenil phosphate, which is hardly detoxified by other methods. The rate of oxidation of formic acid, ethanol, and benzene over FeZSM-5 increased under the action of visible light ($\lambda > 436$ nm), the quantum efficiency being in the range of 0.06 to 0.14.

12.3 KINETIC STUDIES

The affinity between reactants and photocatalysts plays an important role on the overall reaction rate [239]. To understand the mechanistic details of the photocatalytic processes, numerous kinetic studies have been investigated [28,50,240]. From these studies, there is general agreement that the principal oxidizing agent involved in photodegradation reactions of organic compounds is the hydroxyl radical, $OH^{\bullet}$. This species can be generated by reactions of a hole with surface OH groups and electrons with adsorbed O_2, followed by an attack on water. Furthermore, it is indicated that the presence of a large amount of OH groups on the surface of the photocatalyst enhances O_2 adsorption and consequently, its photoreactivity. Thus, the knowledge of this quantity is interesting in comparative studies as mentioned by several recent works [83,241,242].

In the presence of molecular oxygen as a surface electron acceptor, the mechanistic pathway, generally given, is

$$O_{2(ads)} + e^- \rightarrow O_2{}^* + H^+ \rightarrow HO_2{}^* + HO_2{}^* \rightarrow O_2 + H_2O_2 + e - O_2 + OH^* + OH \qquad (12.17)$$

The above mechanism suggests that the presence of adsorbed oxygen $O_{2(ads)}$ is essential for photocatalysis. It allows an increase of hole lifetime by reaction with an electron and the formation of oxidizing OH* radicals.

Kinetics of the photooxidation of organic water impurities on illuminated titania surfaces has been generally regarded to be based on the Langmuir–Hinshelwood equation with first-order reaction kinetics vs. initial substrate concentration was established univocally by many authors

[50,243–246]. Full mineralization of organic compounds, including phenols and other aromatics, on the surface of illuminated titania proceeds via many steps, made of one-electron oxidation or reduction reactions. A full set of the reactions which take place (or are likely to) on illuminated semiconductors has been given by Gerischer and Heller [247]. Under illumination, organic molecules react with photogenerated holes or, more probably, with photoinduced •OH radicals, resulting in a number of hydroxylated reaction intermediates. Determination of these compounds has been the subject of many investigations [248–251]. However, until now a full mechanism of photooxidation of organic derivatives has not been established.

Okamoto et al. [243] proposed a kinetic expression, which included the initial substrate concentration, amount of semiconductor powder, oxygen partial pressure, and intensity of incident light. In this study, phenol concentration was the only variable. Therefore, the reaction shows first-order behavior with kinetic constants, k and K, which can be easily calculated from the afore mentioned Langmuir–Hinshelwood equation. When kinetics and intermediates of catechol, resorcinol, and hydroquinone decomposition on illuminated TiO_2 were investigated, three factors remained constant in all experiments, i.e., amount of semiconductor powder, oxygen partial pressure, and intensity of incident light. Therefore, assuming the Langmuir–Hinshelwood model for the photocatalytic reaction, the rate equation is as follows:

$$r = -dc/dt = kK_{cl} + K_c \qquad (12.18)$$

where c is the substrate concentration in water, k a rate constant, and K a constant of adsorption equilibrium. For first-order reaction kinetics, the dependence of $1/r$ on $1/c$ should be linear.

For the case of large content of various organic intermediates, their competitive reactions with TiO_2-photogenerated •OH radicals require the use of initial photoreaction rates for kinetic calculations of photocatalytic oxidation [252]. During Diuron decomposition on TiO_2 it was emphasized that the concentration of decomposition products (DP) varies throughout the reaction up to their mineralization and, thus, the following equation (also based on Langmuir–Hishelwwod kinetic model) can describe the kinetics [36,253], where k_r is the reaction rate constant, K the reactant (diuron) adsorption constant, C diuron concentration at any time, K_i the DP adsorption constant, and C_i DP concentration at any time.

$$r = \frac{k_r KC}{1 + KC + \sum_{i=1}^{n} K_i C_i (i = 1, n)} \qquad (12.19)$$

When C is sufficiently small, Equation (12.19) can be simplified to a first-order reaction rate equation.

After the induction period, when DPs are mineralized, total organic compound (TOC) mineralization is the sum of the degradation of different compounds. Under these conditions, the evolution of the reaction can be described as apparent zero-order kinetics:

$$r_{TOC} = \sum r_i = \frac{\sum_{i=1}^{n} kr_i K_i C_i \quad (i = 1, n)}{1 + \sum_{i=1}^{n} K_i C_i \quad (i = 1, n)} \approx \frac{\sum_{i=1}^{n} kr_i K_i C_i \quad (i = 1, n)}{\sum_{i=1}^{n} K_i C_i \quad (i = 1, n)} = \sum_{i=1}^{n} kr_i \qquad (12.20)$$

12.4 COMBINATORIAL APPROACHES IN PREPARATION AND TESTING PHOTOCATALYSTS

The high-throughput, combinatorial methodology has been widely applied in heterogeneous catalysis discovery, and there have been preliminary reports of combinatorial investigations into micro- and macroporous structures [254–256]. McFarland and co-workers [257,258] used high-throughput combinatorial methods to identify cost-effective metal oxide photocatalyst systems for hydrogen photosynthesis, investigating mesoporous materials considering electrochemical methods of combinatorial library synthesis and screening of thin film photocatalysts.

In these studies, diversity can be achieved by (1) variations in composition (by variable doping, electrochemical synthesis conditions, and surface redox catalysts), and (2) variations in structure (by diverse surfactant [SDA] and block copolymer templating agents, synthesis conditions, and doping). The capabilities of automated electrochemical synthesis and screening systems to investigate the relationship between mesoporous structure, SDA concentration, and performance in water-splitting photoelectrocatalysis were examined in the preparation of ZnO films [257,258]. Thus, mesoporous ZnO were synthesized by electrodeposition in a library format with poly(ethylene oxide)-block-poly(propylene oxide)-block-poly(ethylene oxide) — $EO_{20}PO_{70}EO_{20}$ — as a SDA. Mesoporous ZnO synthesized with concentrations of 1.0 to 5.0 wt% $EO_{20}PO_{70}EO_{20}$ exhibited greater photoactivity than control polycrystalline ZnO samples, with 3.0 wt% of the SDA demonstrating an improvement of 30% in unbiased photocurrent. However, a further increase IN SDA concentration led to a substantially decreased photocurrent, presumably due to poor electronic properties. TEM and XRD confirmed the presence of a variety of mesoporous structures including lamellar planes and disordered mesopores (Figure 12.7).

As demonstrated, the fact that added chemicals alter the photocatalytic activity of TiO_2 opens new paths toward combinatorial investigation of photocatalysis. Not only the TiO_2 layer itself may be locally modified by adding dopants or impurities for systematic improvement of photoactivity but different chemicals could also be adsorbed or immobilized on the surface layer of a single sample to study the relative performance in photocatalysis. With homogeneous samples, this would allow comparison to the catalytic activity for different reagents directly, because layer properties, morphology, and other experimental conditions are kept constant [259].

Electrodeposition of mesoporous materials has also been reported, thus taking advantage of tunable charges at the surface–liquid interface control assembly patterns while depositing electrically active films, even at reduced concentrations of the SDA [260–264].

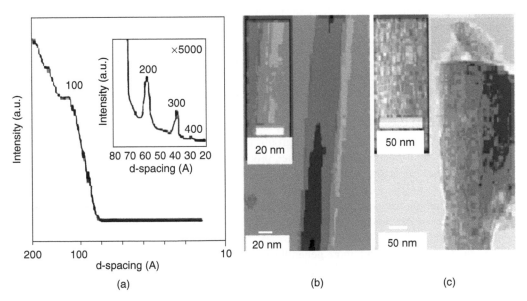

Figure 12.7 (a) Small-angle XRD of ZnO electrodeposited with 15 wt% $EO_{20}PO_{70}EO_{20}$, indicating a well-ordered lamellar phase. TEM images confirming (b) lamellar structure and (c) disordered mesopores. (From Jaramillo, T.F., Ivanovskaya, A., and McFarland, E.W., *J. Comb. Chem.*, 4, 17, 2002.)

12.5 EXAMPLE OF REACTIONS UNDER PHOTOCATALYTIC CONDITIONS

Several articles have reviewed the ongoing work in the photocatalytic degradation of pollutants that involve oxidation or reduction processes (depending on the experimental conditions) [16,18,187,265–273]. The addition of external oxidants such as ozone or hydrogen peroxide during the photocatalytic process can improve the degradation of the organic material when they are added in suitable doses [274–275].

Titanium dioxide is a well-known powerful photocatalyst for decomposing a wide range of substances, including inorganic compounds [276–279], organic compounds including 2-propanol and methylene blue [280–292], and microorganisms such as bacteria and plankton [293–296] and decomposition of NO_x in ambient air [297].

However, the decomposition of organic compounds does not always lead to nontoxic solutions. Mineralization of melamine (2,4,6-triamino-1,3,5-triazine) and other chemicals was accomplished on TiO_2 Degussa P-25. The formation of cyanuric acid prevents the complete mineralization of melamine as observed for atrazine and other s-triazines on TiO_2 photocatalysts, and the toxicity of the photocatalyzed solutions was higher than initially found for melamine [199].

Various types of dyes can be photomineralized effectively in aqueous TiO_2 dispersions under visible light irradiation. In these photochemical processes, the first step following the photoexcitation of dyes by visible light is the injection of electrons from the excited dyes to the conduction band of the TiO_2, with subsequent formation of radical dye cations. The succeeding reaction(s) lead to the mineralization of the dyes [298–301]. Although the mechanistic details of the degradation of dyes under visible irradiation greatly differ from those in the photodegradation of organic compounds under UV irradiation, there should be very little, if any, difference(s) in the scavenging of photogenerated electrons by O_2 and formation of $O_2\bullet^-$ radical anions between the two cases.

The roles played by O_2 in the photoassisted degradations of organic compounds are twofold: (1) suppress charge recombination and regenerate the photocatalysts by acting as an electron acceptor and (2) participate directly in the chemical oxidative process of substrates through active oxygen species produced or through ground-state molecular O_2. Most of the reports on photocatalytic processes have primarily emphasized the former function for O_2. The formation of superoxide radical anions is only a side-reaction that regenerates the photocatalysts, and the degradation of substrates results primarily from (direct/indirect) hole oxidation. Several studies have reported the requisite of superoxide species for the TiO_2-assisted photodegradation of aromatic compounds under UV irradiation [60], suggesting that the oxidative steps in the photodegradation of quinoline involve not only •OH radicals but also substrate activation by direct hole oxidation, followed by addition of the superoxide radical anion to the resulting radical cation. It has also been noted that organic radicals can react with molecular O_2 to yield the corresponding peroxides, with the subsequent reactions accounting for the degradation of the original organic substrates.

The strongest evidence for the OH• radical attack mechanism is the formation of hydroxylated intermediates through addition of hydroxyl groups to aromatic rings. However, other pathways can also account for the formation of hydroxylated products. No doubt some reactions can also be initiated by direct hole oxidation, especially when adsorption of the substrates is rather extensive and concentrations of substrates are relatively high. For instance, direct hole-organic reactions generate cation radicals, which may be subsequently hydrated to generate hydroxylated products [237,302]. The participation of the radicals in the reactions of photocatalytic oxidation of organic compounds has been studied [250,302–305].

The selective oxidation and, more generally, the activation of the C–H bond in alkanes is a topic of continuous interest. Most methods are based on the use of strong electrophiles, but photocatalytic methods offer an interesting alternative in view of the mild conditions, which may increase selectivity. These include electron or hydrogen transfer to excited organic sensitizers, such as aryl nitriles or ketones, to metal complexes or POMs. The use of a solid photocatalyst, such as the suspension of a metal oxide, is an attractive possibility in view of the simplified work up. Oxidation of the hydrocarbon occurs via hydrogen abstraction by the highly reactive hydroxyl radicals and is followed by trapping dioxygen. The products of intermediate oxidation, such as alcohols, aldehydes, or ketones are more reactive than the starting alkanes and are not accumulated to a significant extent on the way to full mineralization to carbon dioxide. However, it has been recently shown that carrying out the photooxidation in an organic medium may lead to useful yields of alcohols or ketones, as is shown in the case of the TiO_2 photocatalytic oxidation of cyclohexane (neat or admixed with organic solvents) [306].

The same authors [306] found that the functionalization of the C–H bond in alkanes can be achieved by TiO_2 photocatalysis under mild conditions, both through oxygen incorporation and through C–C bond-forming reactions. Both oxygen and an inorganic or organic oxidant can be used as the electron scavenger. Although conversions are incomplete, yields are reasonable.

Incorporation of CO_2 in valuable products is a hot topic. Methanol has been reported to be favorably produced on Cu/TiO_2 catalysts. A positive zeta potential at pH 7 was found to promote the photoactivity of CO_2 photoreduction [199].

The photoreactivity of the involved catalyst depends on many experimental factors such as the intensity of the absorbed light, electron–hole pair formation and recombination rates, charge transfer rate to chemical species, diffusion rate, adsorption and desorption rates of reagents and products, pH of the solution, photocatalyst and reactant concentrations, and partial pressure of oxygen [19,302,307]. Most of these factors are strongly affected by the nature and structure of the catalyst, which is dependent on the preparation method. The presence of the impurities may also affect the photoreactivity. The presence of chloride was found to reduce the rate of oxidation by scavenging of oxidizing radicals [151,308]:

$$^*ox + Cl^- \rightarrow Cl^{*-} \tag{12.21}$$

Others impurities such as carbon present in the sample may play a major role in the reduction of the photocatalytic efficiency so far as they behave as recombination sites of electron-hole pairs [309].

12.5.1 Photoinduced Deposition of Various Metals onto Semiconductor

Photoinduced deposition of various noble metals onto semiconductor particles has been extensively reported [310–315]. Several factors are controlling this reaction. To control the morphology of metal clusters with desired size and distribution pattern on a given surface area of titania, the most relevant factors are the surfactant, pH, local concentration of cations, and the source of cation [316]. In the case of the Ag clusters, the reaction steps proposed include the creation of electron (e^-)–hole (p^+) pairs, the reaction of holes with OH^- surface species, and the reaction of electrons with adsorbed Ag^+ ions:

$$TiO_2 + h\nu \rightarrow e^- + p + (h\nu > E_{bg}) \tag{12.22}$$

$$H_2O \leftrightarrow H^+ + OH^- \tag{12.23}$$

$$OH^- + p^+ \rightarrow OH^0 \rightarrow 1/2H_2O + 1/4O_{2(g)} \tag{12.24}$$

$$Ag^+ + e^- \rightarrow Ag^0 \tag{12.25}$$

$$mAg^0 \rightarrow Ag \tag{12.26}$$

Nanosized selenium particles were deposited onto TiO_2 by the photocatalytic reduction of selenate (Se(VI)) and selenite (Se(IV) ions) [317]. The deposition of Se particles on TiO_2 was only observed in the presence of formic acid, which acted as the organic hole scavenger. Se particles of different sizes could be deposited onto the TiO_2 particles by manipulating experimental parameters such as pH and the Se precursor used. When Se(VI) ions were used as the precursor, the Se particles formed on TiO_2 were found to be spherical in shape, up to six times larger than the TiO_2 particles (up to 145 nm) and discretely formed on the TiO_2 particles. The growth and shape of the Se particles were explained in terms of electron transfer across the p–n junctions formed by the p-type Se and n-type TiO_2 semiconductors under illumination and the adsorption of the Se(VI) ions. The size of the Se particles was found to be dependent on the amount of Se(VI) photoreduced. When Se(IV) ions were used as the precursor for Se particles formation, the particles formed were much smaller than that of TiO_2 crystals (<25 nm) and also more evenly dispersed on the TiO_2 particles.

12.5.2 Energy Storage

Nanosized functional materials made of semiconductors promise the prospect of application in many high-technique fields such as photovoltaic conversion and energy storage because of their characteristics of surface structure and high surface area [318–322].

12.6   SOLAR PHOTOCATALYSIS

Photocatalytic degradation of environmental pollutants by solar energy is a very attractive technology for the remediation of contaminated water [253,323]. In some variants of this process, solar UV radiation is absorbed by semiconductor catalyst particles suspended in water. TiO_2 photocatalytic particles are the most widely used for these applications.

Photocatalytic degradation of the pesticide Aldrin dissolved in water was carried out with concentrated solar radiation [324]. These experiments demonstrated that the effects of catalyst concentration, oxidant agent concentration, and solar irradiation are important. In the case of the nonconcentrated solar exposure, the achieved results indicated that the use of H_2O_2 increased the degradation rate. The addition of H_2O_2 reduces the energy requirements, and thus permits the utilization of such a system under low solar radiative levels. The presence of ozone and certain ions such as phosphates on the photocatalytic degradation of organic matter was also studied. These results are comparable to literature reports for photodegradation of other organochloride pesticides using cylindrical lamp reactors.

Another important photocatalytic application is the synthesis of hydrogen from water. However, low solar efficiency and photocorrosion have proven to be hindrances limiting the process economics of photocatalysis [325]. The most efficient systems to date consist of compound semiconductor heterostructures that operate with efficiencies of approximately 16%, however, cost and stability are still problematic [325].

Presently, many technologies are verified at the pilot scale [253,323,326], which is a very important key for a successful application. Therefore, the selection of the reactor and its optimization are of significant interest. The mathematical modeling required for the study and design of solar photocatalytic reactors includes radiation transport [323]. This is a complicated task owing to the large amount of radiation scattering produced by the catalyst particles. These methodologies require large amounts of computer calculations and do not provide analytical formulas. The investigated reactors include layers divided into cells, assuming that there is only one particle (agglomerate) of catalyst in each cell [327], flat plate configurations [328], etc.

12.7   SONOPHOTOCATALYSIS

The simultaneous use of ultrasound and photocatalysis on semiconductors, i.e., the so-called sonophotocatalysis, has recently attracted the attention of several research groups [329–334].

Although the results are not always comparable, it is clear that sonophotocatalysis led to an enhancement in the degradation reaction rate, when compared to sonolysis [329]. A synergy between photocatalysis and sonolysis was observed when operating at a relatively low ultrasound frequency (30 kHz) [330], i.e., under conditions which are less effective than a higher frequency for the degradation of organic species [335], or when employing small particle size semiconductors [332]. Moreover, the simultaneous use of sonolysis and photocatalysis, though leading only to additive effects, was found to be more effective than their sequential combination in the degradation of dyes [332]. The degradation of 2-chlorophenol, chosen as a model water pollutant, and of an azo dye (constituting the largest class of dyes used in the textile and paper industries [336]) also demonstrated evidence for synergy between photocatalysis on TiO_2 and sonolysis [334] when employing a relatively low ultrasound frequency and a low ratio between ultrasound power and reaction volume.

The effects of substrate concentration and photocatalyst type and amount are indicative of a synergistic effect of ultrasound involving the aqueous phase, and not the photocatalyst-water interface, and of a key role played by H_2O_2. Sonication of water suspensions produces some H_2O_2, and a reasonable hypothesis of the origin of the observed synergistic effect is the formation of additional $*OH^-$ radicals by photocatalysis from the extra H_2O_2 produced by ultrasound [337]. Selli [337] has shown that the H_2O_2 concentration under sonophotocatalysis is not the sum of that produced under separate sonolysis and photocatalysis processes. However, it is even lower than that measured under photocatalysis without ultrasound, while in the case of the above hypothesis one would have expected a somewhat higher H_2O_2 concentration (or at least equal) to that measured under photocatalysis. Moreover, the synergistic effect is increased when employing ZnO as photocatalyst, i.e. when a greater amount of hydrogen peroxide accumulates during the run. These observations led to the conclusion that the main effect of ultrasound is to contribute, together with photocatalysis, to the scission of H_2O_2, produced by both photocatalysis and sonolysis, with an increase in the reactive radical species inducing degradation of the substrate.

Much interest has recently been shown in artificial photosynthesis. Photosynthesis is a system for conversion or accumulation of energy. It is also interesting that some reactions occur simultaneously and continuously. Fujishima et al. [338] pointed out that a photocatalytic system resembles the process of photosynthesis in green plants. They described that there are three important parts of the overall process of photosynthesis: (1) oxygen generation by the photolysis of water, (2) photophosphorylation, which accumulates energy, and (3) the Calvin cycle, which takes in and reduces carbon dioxide. The two reactions, reduction of CO_2 and generation of O_2 from water, can occur simultaneously and continuously by a sonophotocatalytic reaction.

By the combined effects of sonolysis and photocatalysis, water has been decomposed into O_2 and H_2 stoichiometrically and continuously [339]. Overall water splitting was achieved using a sonophotocatalytic technique in the presence of a TiO_2 photocatalyst under the white light illumination of a xenon lamp. Effects of ultrasonic irradiation on heterogeneous photocatalytic reaction systems have been reported by Mason [340], Kado et al. [341], Suzuki et al. [342,343], and Ragaini et al. [344]. This combined system consisted of sonochemical and photocatalytic reactions. In the case of water splitting, sonolysis and photocatalysis proceed simultaneously as shown by the following reaction equations:

$$\text{Sonochemical reaction}: \quad 4H_2O \rightarrow 2H_2 + 2H_2O_2 \tag{12.27}$$

$$\text{Photocatalytic reaction}: \quad 2H_2O_2 \rightarrow O_2 + 2H_2O \tag{12.28}$$

$$\text{Sonolysis of } CO_2: \quad 2CO_2 + 2H_2 \rightarrow 2CO + 2H_2O \tag{12.29}$$

$$\text{Total reaction}: \quad 2CO_2 \rightarrow 2CO + O_2 \tag{12.30}$$

A combination of photocatalysis and sonolysis can also perform the overall CO_2 deoxygenation into CO and O_2 [345].

12.8 PHOTOCATALYSIS ASSOCIATED TO MICROWAVE RADIATION

Recent work has shown that the photooxidative degradation of waste-water substances could be achieved with reasonable efficiency in titania dispersions under simultaneous irradiation by UV light and microwave radiation [346–348]. In these studies, the UV radiation was provided by a microwave-powered plasma light source [348,349]. The effect of microwave radiation in these photooxidations was attributed to thermal and nonthermal interactions of microwaves with the TiO_2 particle surface. In particular, it was deduced that the latter interactions were responsible for the increased efficiency through the likely formation of additional oxidizing species (e.g., $*OH^-$ radicals or equivalent reactive oxygen species) typically produced when such semiconductor photocatalysts as TiO_2 are photoactivated with UV light of energy equal to or greater than their bandgap energy [350].

12.9 PHOTOCATALYST DEACTIVATION

The gas-phase photocatalytic oxidation of organic contaminants is a subject of current interest due to its potential for removal of low concentration pollutants in slightly contaminated enclosed atmospheres (closed intelligent buildings, factory buildings, especially enclosed atmospheres, etc.). Furthermore, comparing the photocatalytic processes in gas and aqueous phase indicates some advantages and few drawbacks [351]. In particular, (1) the reaction mechanisms in the gas phase seem to be different from those that prevail in aqueous systems, and (2) relatively low levels of UV light energy are needed for the occurrence of gas-phase processes, in contrast to reactions in water. However, in the gas phase, extremely high decomposition rates have been found, making the technique very useful for practical applications [351]. Additional advantages are: (1) the diffusion of reagents and products is favored; (2) HO* scavengers present in aqueous phase such as chlorides or alkalines do not interfere; (3) electron scavengers such as O_2 are rarely limiting; and (4) photons absorbed by air are rare. Despite these advantages gas-phase photocatalysis presents an important drawback. While in aqueous phase water helps to remove reaction intermediates and reaction products from the catalyst surface (specially those compounds of polar character), in the gas phase these species tend to accumulate on the surface causing the deactivation of the catalyst.

The deactivation of TiO_2 Degussa P-25 has been studied during the gas-phase photocatalytic oxidation of ethanol [351]. Water vapor plays a clear competitive role for surface site adsorption, thus hampering the ethanol photooxidation. Dark adsorption of ethanol on a fresh catalyst shows a Langmuirian behavior with the formation of a monolayer of adsorbate. Dark adsorption in a TiO_2 surface that has been used in consecutive photocatalytic experiments of ethanol degradation gives non-Langmuirian isotherms, indicating the existence of noticeable changes of the catalyst surface structure. After several irradiations the catalyst activity decreases.

12.10 CONCLUSIONS

Photocatalysis has become a very important tool both in organic syntheses and environmental chemistry. It represents not only a very selective instrument to achieve synthesis or decomposition of complex structures but also a *green chemistry* approach since the electron is a very clean reagent.

Practical application of the results of photocatalysis will depend much on the capability of the catalysts to work under visible light conditions. Therefore, the design of new photocatalysts either by modification of the existent TiO_2 or by discovery of new materials will be a key step in developing this domain. On the other hand, the synthesis of new photosensitizers, which will be associate to heterogeneous photocatalysts would represent a second important way in extending the practical applications of photocatalysis.

REFERENCES

1. D.C. Neckers, *Mechanistic Organic Photochemistry*, Reinhold, New York, 1967, a p. 3, b p. 4, c p. 19.
2. M. Terazima, in D.C. Neckers, D.H. Volman, G.v. Bunau, Eds., *Advances in Photochemistry*, Vol. 24, Wiley, New York, 1998, a. p. 272, b. 284, c p. 255.
3. H. Kisch, in D.C. Neckers, G.v. Bunau, W.S. Jenks, Eds., *Advances in Photochemistry*, Vol. 26, Wiley, New York, 1998, p. 93.
4. M.A. Fox, in V. Balzani, Ed., *Electron Transfer in Chemistry*, Vol. 1, Wiley-VCH, Weinheim, 2001, p. 271.
5. A. Fujishima, K. Honda, *Nature* 238 (1972) 37.
6. J. Bard, *J. Photochem.* 10 (1979) 59.
7. H. Van Damme, W.K. Hall, *J. Am. Chem. Soc.* 101 (1979) 4373.
8. S. Sato, J.M. White, *J. Am. Chem. Soc.* 102 (1980) 7206.
9. R. Memming, *Electrochim. Acta* 25 (1980) 77.
10. E. Borgarello, J. Kiwi, M. Graetzel, E. Pelizzetti, M. Visca, *J. Am. Chem. Soc.* 104 (1982) 2996.
11. H. Courbon, J.M. Herrmann, P. Pichat, *Int. J. Hydrogen Energy* 9 (1984) 397.
12. A. Sobczynski, J.M. White, *J. Mol. Catal.* 29 (1985) 379.
13. A. Sobczynski, A.J. Bard, A. Campion, M.A. Fox, T. Mallouk, S.E. Webber, J.M. White, *J. Phys. Chem.* 91 (1987) 3316.
14. A.J. Bard, *Ber. Bunsenges. Phys. Chem.* 92 (1988) 1187.
15. R. Venkatadri, R.W. Peters, *Hazard. Waste Hazard. Mater.* 10 (1993) 107.
16. O. Legrini, E. Oliveros, A.M. Braun, *Chem. Rev.* 93 (1993) 671.
17. P.V. Kamat, *Chem. Rev.* 93 (1993) 267.
18. M.R. Hoffmann, S.T. Martin, W. Choi, D.W. Bahnemann, *Chem. Rev.* 95 (1995) 69.
19. A.L. Linsebigler, L. Guangquan Jr., J.T. Yates, *Chem. Rev.* 95 (1995) 735.
20. V. Parmon, *Catal. Today* 39 (1997) 137.
21. S. Malato, J. Blanco, J. M. Herrmann, *Catal. Today* 54 (1999) 89.
22. A. Mills, S. Le Hunte, *J. Photochem. Photobiol. A* 108 (1997) 1.
23. T. Ibusuki, K. Takeuchi, *J. Mol. Catal.* 88 (1994) 93.
24. I. Sopyan, S. Murasawa, K. Hashimoto, A. Fujishima, *Chem. Lett.* (1994) 723.
25. S.A. Larson, J.A. Widegren, J.L. Falconer, *J. Catal.* 157 (1995) 611.
26. C. Bouquet-Somrani, A. Finiels, P. Graffin, J.-L. Olive, *Appl. Catal. B Environ.* 8 (1996) 101.
27. K.T. Ranjit, B. Viswanathan, *Indian J. Chem.* 35A (1996) 351.
28. O.M. Alfano, M.I. Cabrera, *J. Catal.* 172 (1997) 370.
29. F. Zhang, J. Zhao, L. Zang, T. Shen, H. Hidaka, E. Pelizzetti, N. Serpone, *J. Mol. Catal. A* 120 (1997) 173.
30. S.A. Larson, J.L. Falconer, *Catal. Lett.* 44 (1997) 57.
31. S.-J. Tsai, S. Cheng, *Catal. Today* 33 (1997) 227.
32. R. Mendez-Roman, N. Cardona-Martinez, *Catal. Today* 40 (1998) 353.
33. A.D. Modestov, O. Lev, *J. Photochem. Photobiol. A* 112 (1998) 261.
34. D.S. Muggli, K.H. Lowery, J.L. Falconer, *J. Catal.* 180 (1998) 111.
35. S.G. Zhang, S. Higashimoto, H. Yamashita, M. Anpo, *J. Phys. Chem. B* 102 (1998) 5590.
36. J.-M. Herrmann, *Catal. Today* 53 (1999) 115.
37. K.-H. Wang, Y.-H. Hsieh, M.-Y. Chou, C.-Y. Chang, *Appl. Catal. B Environ.* 21 (1999) 1.
38. K.J. Buechler, R.D. Noble, C.A. Koval, W.A. Jacobi, *Ind. Eng. Chem. Res.* 38 (1999) 892.
39. D.D. Dionysiou, A.P. Khodadoust, A.M. Kern, M.T. Suidan, I. Baudin, J.-M. Laine, *Appl. Catal. B Environ.* 24 (2000) 139.
40. K.Y. Jung, S.B. Park, *Appl. Catal. B Environ.* 25 (2000) 249–256.
41. M.R. Dhananjeyan, J. Kiwi, K. Ravindranathan Thampi, *Chem. Commun.* (2000) 1443.
42. E. Piera, J.C. Calpe, E. Brillas, X. Domenech, J. Peral, *Appl. Catal. B Environ.* 27 (2000) 169.

43. A. Di Paola, L. Palmisano, V. Augugliaro, *Catal. Today* 58 (2000) 141.

44. I. Arslan, I.A. Balcioglu, D.W. Bahnemann, *Appl. Catal. B Environ.* 26 (2000) 193.

45. A. Assabane, Y.A. Ichou, H. Tahiri, C. Guillard, J.-M. Herrmann, *Appl. Catal. B* 24 (2000) 71.

46. G.R. Bamwenda, T. Uesigi, Y. Abe, K. Sayama, H. Arakawa, *Appl. Catal. A* 205 (2001) 117.

47. B. Kraeutler, C.D. Jaeger, A.J. Bard, *J. Am. Chem. Soc.* 100 (1978) 4903.

48. C.D. Jaeger, A.J. Bard, *J. Phys. Chem.* 83 (1979) 3146.

49. A.J. Nozik, R. Memming, *J.Phys. Chem.* 100 (1996) 13061.

50. C.S. Turchi, D.F. Ollis, *J. Catal.* 122 (1990) 178.

51. K. Ishibashi, A. Fujishima, T. Watanabe, K. Hashimoto, *J. Photochem. Photobiol. A* 134 (2000) 139.

52. M.A. Grela, M.E.J. Coronel, A.J. Colussi, *J. Phys. Chem.* 100 (1996) 16940.

53. D.E. Skinner, D.P. Colombo Jr., J.J. Cavaleri, R.M. Bowman, *J. Phys. Chem.* 99 (1995) 7853.

54. N. Serpone, D. Lawless, R. Khairutdinov, *J. Phys. Chem.* 99 (1995) 16655.

55. D.P. Colombo Jr., R.M. Bowman, *J. Phys. Chem.* 100 (1996) 18445.

56. B. Ohtani, R.M. Bowman, D.P. Colombo Jr., H. Kominami, H. Noguchi, K. Uosaki, *Chem. Lett.* 1998 (1998) 579.

57. J. Rabani, K. Yamashita, K. Ushida, J. Stark, A. Kira, *J. Phys. Chem. B* 102 (1998) 1685.

58. K. Fujihara, S. Izumi, T. Ohno, M. Matsumura, *J. Photochem. Photobiol. A: Chem.* 132 (2000) 99.

59. C. Guillard, *J. Photochem. Photobiol. A* 135 (2000) 65.

60. L. Cermenati, P. Pichat, C. Guillard, A. Albini, *J. Phys. Chem. B* 101 (1997) 2650.

61. H. Hidaka, K. Ajisaka, S. Horikoshi, *J. Photochem. Photobiol. A* 138 (2001) 185.

62. A. Fujishima, T.N. Rao, D.A. Tryk, *J. Photochem. Photobiol. C* 1 (2000) 1.

63. R. Andreozzi, V. Caprio, A. Insola, *Catal. Today* 53 (1999) 51.

64. K. Tanaka, T. Hisanaga, A.P. Rivera, Eds., *Photocatalytic Purification and Treatment of Water and Air*, Elsevier, Amsterdam, 1993, p. 169.

65. A. Sclafani, L. Palmisano, M. Schiavello, *J. Phys. Chem.* 94 (1990) 829.

66. H.Y. Chen, O. Zahraa, M. Bouchy, F. Thomas, J.Y. Bottero, *J. Photochem. Photobiol. A: Chem.* 85 (1995) 179.

67. C.H. Kwon, H. Shin, J.H. Kim, W.S. Choi, K.H. Yoon, *Mat. Chem. Phys.* 86 (2004) 78.

68. M.C. Hidalgo, G. Colon, J.A. Navio, *J. Photochem. Photobiol. A: Chem.* 148 (2002) 341.

69. M. Nakamura, N. Negishi, S. Kutsuna, T. Ihara, S. Sugihara, K. Takeuchi, *J. Mol. Catal. A: Chem.* 161 (2000) 205.

70. R. Asahi, T. Morikawa, T. Ohwaki, K. Aoki, Y. Taga, *Science* 293 (2001) 269.

71. N.-L. Wu, M.-S. Lee, Z.-J. Pon, J.-Z. Hsu, *J. Photochem. Photobiol. A: Chem.* 163 (2004) 277.

72. D.W. Bahnemann, J. Cunningham, M.A. Fox, E. Pelizzetti, N. Serpone, in G.R. Helz, R.G. Zepp, D.G. Crosby, Eds., *Aquatic and Surface Photochemistry*, Lewis Publishers, Boca Raton, FL, 1994. p. 261.

73. R.L. Pozzo, M.A. Baltanas, A.E. Cassano, *Catal. Today* 39 (1998) 219.

73a. Y. Xie, C. Yuan, *Mater. Res. Bull.* 39 (2004) 533.

74. K. Yanagisawa, Y. Yamamoto, Q. Feng, N. Yamasaki, *J. Mater. Res.* 13 (1998) 825.

75. C.K. Chan, J.F. Porter, Y.-G. Li, W. Guo, C.-M. Chan, *J. Am. Ceram. Soc.* 83 (1999) 566.

76. S. Baskaran, L. Song, J. Liu, Y.L. Chen, G.L. Graff, *J. Am. Ceram. Soc.* 81 (1998) 401.

77. D.-Y. Shin, K. Kimura, *J. Ceram. Soc. Jpn.* 107 (1999) 775.

78. M. Terashima, N. Inoue, S. Kashiwabara, R. Fujumoto, *Appl. Surf. Sci.* 167 (2001) 535.

79. P. Hoyer, H. Masuda, *J. Mater. Sci. Lett.* 15 (1996) 1228.

80. Y. Ishikawa, Y. Matsumoto, *Electrochim. Acta* 46 (2001) 2819.

81. S.D. Burnside, V. Shklover, C. Barbe, P. Comete, F. Arendse, K. Brooks, M. Gratzel, *Chem. Mater.* 10 (1998) 2419.

82. B.L. Bischoff, M.A. Anderson, *Chem. Mater.* 7 (1995) 1772.

83. L. Znaidi, R. Seraphimova, J.F. Bocquet, C. Colbeau-Justin, C. Pommier, *Mater. Res. Bull.* 36 (2001) 811.

84. C.R. Martin, *Science* 266 (1994) 1961.

85. J.C. Hulteen, C.R. Martin, *J. Mater. Chem.* 7 (1997) 1075.

86. B.B. Lakshmi, P.K. Dorhout, C.R. Martin, *Chem. Mater.* 9 (1997) 857.

87. M. Zhang, Y. Bando, K. Wada, *J. Mater. Sci. Lett.* 20 (2001) 167.

88. D.-S. Seo, J.-K. Lee, H. Kim, *J. Crystal Growth* 229 (2001) 428.

89. Z. Ding, X. Hu, P.L. Yue, G.Q. Lu, P.F. Greenfield, *Catal. Today* 68 (2001) 173.

90. S.E. Pratsinis, *J. Aerosol Sci.* 27 (1996) s153.
91. D. Dumitriu, A.R. Bally, C. Ballif, V.I. Pârvulescu, P.E. Schmid, R. Sanjines, F. Levy, *Stud. Surf. Sci. Catal.* 118 (1998) 485.
91a. D. Dumitriu, A.R. Bally, C. Ballif, P. Hones, P.E. Schmid, R. Sanjines, F. Levy, V.I. Pârvulescu, *Appl. Catal. B: Environ.* 25 (2000) 83.
92. H.D. Nam, B.H. Lee, S. Kim, C. Jung, J. Lee, S. Park, *J. Appl. Phys.* 37 (1998) 4063.
93. K.Y. Jung, S.B. Park, *J. Photochem. Photobiol. A* 127 (1999) 117.
94. K.A. Vorotilov, E.V. Orlova, V.I. Petrovsky, *Thin Solid Films* 207 (1992) 180.
95. M. Mosaddeq-ur-Rahman, G. Yu, T. Soga, T. Jimbo, H. Ebisu, M. Umeno, *J. Appl. Phys.* 88 (2000) 4634.
96. C.H. Kwon, H. Shin, J.H. Kim, W.S. Choi, K.H. Yoon, *Mat. Chem. Phys.* 86 (2004) 78.
97. S.Z. Chu, K. Wada, S. Inoue, S. Todoroki, *Chem. Mater.* 14 (2002) 266.
98. S.Z. Chu, K. Wada, S. Inoue, *Adv. Mater.* 14 (2002) 1752.
99. G.J. de A.A. Soler-Illia, E.L. Crepaldi, D. Grosso, C. Sanchez, *Curr. Opin. Colloid Interface Sci.* 8 (2003) 109.
100. X. He, D. Antonelli, *Angew. Chem. Int. Ed.* 41 (2001) 214.
101. D. Trong On, D. Desplantier-Giscard, C. Danumah, S. Kaliaguine, *Appl. Catal. A Chem.* 222 (2001) 299.
102. V.F. Stone, R.J. Davis, *Chem. Mater.* 10 (1998) 1468.
103. Y. Takahara, J.N. Kondo, T. Takata, D. Lu, K. Domen, *Chem. Mater.* 13 (2001) 1194.
104. Y. Takahara, J.N. Kondo, D. Lu, K. Domen, *Solid State Ionics* 151 (2002) 305.
105. M. Uchida, J.N. Kondo, D. Lu, K. Domen, *Chem. Lett.* 5 (2002) 498.
106. M. Mohseni, A. David, *Appl. Catal. B: Environ.* 46 (2003) 219.
107. H.M. Coleman, B.R. Eggins, J.A. Byrne, F.L. Palmer, E. King, *Appl. Catal. B: Environ.* 24 (2000) L1.
108. H.T. Chang, N. Wu, F. Zhu, *Water Res.* 34 (2000) 407.
109. M.C. Yeber, J. Rodrýguez, J. Freer, N. Durán, H.D. Mansilla, *Chemosphere* 41 (2000) 1193.
110. J. Fernández, J. Kiwi, J. Baeza, J. Freer, C. Lizama, H.D. Mansilla, *Appl. Catal. B: Environ.* 48 (2004) 205.
111. Y. Xu, X. Chen, *Chem. Ind. (London)* 6 (1990) 497.
112. D. Robert, A. Piscoro, O. Heintz, J.V. Weber, *Catal. Today* 54 (1999) 291.
113. S. Sato, *Langmuir* 4 (1988) 1156.
114. Y. Zhu, L. Zhang, L. Wang, L. Fu, L. Cao, *J. Mater. Chem.* 11 (2001) 1864.
115. Y. Ku, C.M. Ma, Y.S. Shen, *Appl. Catal. B: Environ.* 34 (2001) 181.
116. A. Fernandez, G. Lassaletta, V.M. Jimenez, A. Justo, A.R. Gonzalez-Elipe, J.M. Herrmann, H. Tahiri, Y. Ait-Ichou, *Appl. Catal. B: Environ.* 7 (1995) 49.
117. K.J. Green, R.J. Rudham, *J. Chem. Soc., Faraday Trans.* 89 (1993) 1867.
118. K. Hashimoto, K. Wasada, M. Osaki, E. Shono, K. Adachi, N. Toukai, H. Kominami, Y. Kera, *Appl. Catal. B: Environ.* 30 (2001) 429.
119. Y. Kim, M. Yoon, *J. Mol. Catal. A: Chem.* 168 (2001) 257.
120. M.G. Kang, H.S. Park, K.-J. Kim, *J. Photochem. Photobiol. A* 148 (2002) 546.
121. P. Monneyron, M.-H. Manero, J.-N. Foussard, F. Benoit-Marquie, M.-T. Maurette, *Chem. Eng. Sci.* 58 (2003) 971.
122. M.A. Anderson, M.J. Gieselman, Q. Xu, *J. Membr. Sci,* 39 (1988) 243.
123. H. Murayama, K. Shimizu, N. Tsukada, A. Shimada, T. Kodama, Y. Kitayama, *Chem. Commun.* (2002) 2678.
124. H. Yoneyama, S. Haga, S. Yamanaka, *J. Phys. Chem.* 93 (1989) 4833.
125. C. Ooka, S. Akita, Y. Ohashi, T. Horiuchi, K. Suzuki, S. Komai, H. Yoshida, T. Hattori, *J. Mater. Chem.* 9 (1999) 2943.
126. S. Yamanaka, T. Nishihara, M. Hattori, Y. Suzuki, *Mater. Chem. Phys.* 17 (1987) 87.
127. A. Bernier, L.F. Admaiai, P. Grange, *Appl. Catal.* 77 (1991) 269.
128. S. Yamanaka, P.B. Malla, S. Komarneni, *J. Colloid Interface Sci.* 134 (1990) 51.
129. C. Ooka, H. Yoshida, K. Suzuki, T. Hattori, *Appl. Catal. A: Gen.* 260 (2004) 47.
130. A.J. Maira, K.L. Leung, C.Y. Lee, P.L. Yue, C.K. Chan, *J. Catal.* 192 (2000) 185.
131. C.C. Wang, Z. Zhang, J.Y. Ying, *Nano-Str. Mater.* 9 (1997) 583.
132. N. Xu, Z. Shi, Y. Fan, J. Dong, J. Shi, M.Z.-C. Hu, *Ind. Eng. Chem. Res.* 38 (1999) 373.

133. A.P. Rivera, K. Tanaka, T. Hisanaga, *Appl. Catal. B: Environ.* 3 (1993) 37.
134. M. Inagaki, Y. Nakazawa, M. Hirano, Y. Kobayashi, M. Toyoda, *J. Inorg. Mater.* 3 (2001) 809.
135. K. Chhor, J.F. Bocquet, C. Colbeau-Justin, *Mater. Chem. Phys.* 86 (2004) 123.
136. J.A.R. Van Veen, F.T.G. Velmaat, G. Jonkers, *J. Chem. Soc. Chem. Commun.* (1985) 1656.
137. M. Kunst, G. Beck, *J. Appl. Phys.* 60 (1986) 3558.
138. D.F. Ollis, E. Pelizzetti, N. Serpone, *Environ. Sci. Technol.* 25 (1991) 1523.
139. K. Yogo, M. Ishikawa, *Catal. Surv. Jpn.* 4 (2000) 83.
140. L. Zang, W. Macyk, C. Lange, *Chem. Eur. J.* 6 (2000) 379.
141. G.A. Epling, C. Lin, *Chemosphere* 46 (2002) 561.
142. Z.X. Zhou, S.P. Qian, S.D. Yao, *Dyes Pigments* 51 (2001) 137.
143. K. Wilke, H.D. Breuer, *J. Photochem. Photobiol. A Chem.* 121 (1999) 49.
144. M. Anpo, *Stud. Surf. Sci. Catal.* 130 (2000) 157.
145. A. Fuerte, M.D. Hernandez-Alonso, A.J. Maira, A. Martinez-Arias, M. Fernandez-Garcia, J.C. Conesa, J. Soria, *Chem. Commun.* (2001) 2718.
146. T. Ohno, F. Tanigawa, K. Fujihara, *J. Photochem. Photobiol. A* 127 (1999) 107.
147. Y.B. Xie, C.W. Yuan, *J. Mol. Catal. A Chem* 206 (2003) 419.
148. K.E. Karakitsou, X.F. Verykios, *J. Phys. Chem.* 97 (1993) 1184.
149. C. Minero, G. Marirlla, V. Maurino, E. Pelizzetti, *Langmuir* 16 (2000) 2632.
150. C. Anderson, A. Bard, *J. Phys. Chem.* 99 (1995) 9882.
151. M. Abfullah, G.K.-C. Low, R.W. Matthews, *J. Phys. Chem.* 94 (1990) 6820.
152. C. Wang, D.F. Bahnemann, J.K. Dohrmann, *Chem. Commun.* (2000) 1539.
153. A. Sclafani, M.N. Mozzanega, P. Pichat, *J. Photochem. Photobiol. A Chem.* 59 (1991) 181.
154. M. Sadeghi, W. Liu, T.G. Zhang, P. Stavropoulos, B. Levy, *J. Phys. Chem.* 100 (1996) 19466.
155. B. O'Regan, D.T. Schwartz, *J. Appl. Phys.* 80 (1996) 4749.
156. L.N. Wang, M. Muhammed, *J. Mater. Chem.* 9 (1999) 2871.
156a. Y. Wang, H. Cheng, L. Zhang, Y. Hao, J. Ma, B. Xu, W. Li, *J. Mol. Catal. A: Chem.* 151 (2000) 205.
157. S. Matsuo, N. Sakaguchi, K. Yamada, T. Matsuo, H. Wakita, *Appl. Surf. Sci.* 228 (2004) 233.
158. Y. Xie, C. Yuan, *Appl. Surf. Sci.* 221 (2004) 17.
159. W. Choi, A. Termin, M.R. Hoffmann, *J. Phys. Chem.* 98 (1994) 13669.
160. K.T. Ranjit, I. Willner, S.H. Bossmann, A.M. Braun, *J. Catal.* 204 (2001) 305.
161. H. Yamashita, N. Harada, J. Misaka, M. Takeuchi, Y. Ichihashi, F. Goto, M. Ishida, T. Sasaki, M. Anpo, *J. Synchrotron Rad.* 8 (2001) 569.
162. P.H. Maruska, A.K. Ghosh, *Sol. Energy Mater.* 1 (1979) 237.
163. A.D. Paola, G. Marc, L. Palminsano, K. Uosaki, S. Ikeda, B. Ohtani, *J. Phys. Chem. B* 106 (2002) 637.
164. J. Wang, S. Uma, K.J. Klabunde, *Appl. Catal. B: Environ.* 48 (2004) 151.
165. L. Davydov, P. Reddy, P. France, P.G. Smirniortis, *J. Catal.* 203 (2001) 157.
166. V.Y. Gusev, X. Feng, Z. Bu, G.L. Haller, J.A. O'Brien, *J. Phys. Chem.* 100 (1996) 1989.
167. T. Ishikawa, M. Matsuda, A. Yasukawa, K. Kandori, S. Inagaki, T. Fukushima, S. Kondo, *J. Chem. Soc., Faraday Trans.* 92 (1996) 1985.
168. K.A. Koyano, T. Tatsumi, Y. Tanaka, S. Nakata, *J. Phys. Chem. B* 101 (1997) 9436.
169. T. Ohno, D. Haga, K. Fujihara, K. Kaizaki, M. Matsumura, *J. Phys. Chem. B* 101 (1997) 6415.
170. F.C. Gennari, D.M. Pasquevich, *J. Mater. Sci.* 33 (1998) 1571.
171. B. Pal, M. Sharon, G. Nogami, *Mater. Chem. Phys.* 59 (1999) 254.
172. L.X. Chen, T. Rajh, Z.Wang, M.C. Thurnauer, *J. Phys. Chem. B* 101 (1997) 10688.
173. S. Matsuo, N. Sakaguchi, E. Obuchi, K. Nakano, R.C.C. Perera, T. Watanabe, T. Matsuo, H. Wakita, *Anal. Sci.* 17 (2001) 149.
174. Y. Izumi, F. Kiyotaki, H. Yoshitake, K. Aika, T. Sugihara, T. Tatsumi, Y. Tanizawa, T. Shido, Y. Iwasawa, *Chem. Commun.* (2002) 2402.
175. T. Yoshida, T. Tanaka, S. Yoshida, S. Hikita, T. Baba, T. Hinode, Y. Ono, *Appl. Surf. Sci.* 156 (2000) 65.
176. T. Yoshida, T. Tanaka, S. Yoshida, S. Hikita, T. Baba, Y. Ono, *Solid State Commun.* 114 (2000) 255.
177. T. Yamamoto, T. Matsuyama, T. Tanaka, T. Funabiki, S. Yoshida, *J. Mol. Catal. A: Chem.* 155 (2000) 43.
178. P. Yang, C. Lu, N.P. Hua, *Mater. Lett.* 57 (2002) 794.
179. R.B. Gratian, U. Takashi, A. Yoshimoto, *Appl. Catal. A: Gen.* 205 (2001) 157.

180. X. Fu, L.A. Clark, Q. Yang, M.A. Anderson, *Environ. Sci. Technol.* 30 (1996) 647.
181. M. Machida, K. Norimoto, T. Watanabe, *J. Mater. Sci.* 34 (1999) 2569.
182. Y. Xu, W. Zheng, W. Liu, *J. Photochem. Photobiol. A. Chem.* 122 (1999) 57.
183. Z. Ding, G.Q. Lu, P.F. Greenfield, *J. Colloids Interface Sci.* 232 (2000) 1.
184. S.-Z. Chu, S. Inoue, K. Wada, D. Li, H. Haneda, S. Awatsu, *J. Phys. Chem. B* 107 (2003) 4201.
185. S.C. Moon, H. Mametsuka, E. Suzuki, M. Anpo, *Chem. Lett.* (1998) 117.
186. S.C. Moon, H. Mametsuka, E. Suzuki, Y. Nakahara, *Catal. Today* 45 (1998) 79.
187. S.C. Moon, H. Mametsuka, S. Tabata, E. Suzuki, *Catal. Today* 58 (2000) 125.
188. K.Y. Jung, S.B. Park, S.-K. Ihm, *Appl. Catal. B: Environ.* 51 (2004) 239–245.
189. N. Takeda, T. Torimoto, S. Sampath, S. Kuwabara, H. Yoneyama, *J. Phys. Chem.* 99 (1995) 9986.
190. H. Nishikawa, Y. Takahara, *J. Mol. Catal. A Chem.* 172 (2001) 247.
191. H. Chun, W. Yizhong, T. Hongxiao, *Appl. Catal. B: Environ.* 35 (2001) 95.
192. T. Torimoto, Y. Okada, N. Takeda, H. Yoneyama, *J. Photochem. Photobiol. A: Chem.* 103 (1997) 153.
193. J.-M. Herrmann, H. Tahiri, Y. Ait-Icho, G. Lassaletta, A.R. Gonzalez-Elipe, A. Fernandez, *Appl. Catal. B: Environ.* 13 (1997) 219.
194. B. Tryba, A.W. Morawski, M. Inagaki, *Appl. Catal. B: Environ.* 41 (2003) 427.
195. T. Tsumura, N. Kojitani, I. Izumi, N. Iwashita, M. Toyoda, M. Inagaki, *J. Mater. Chem.* 12 (2002) 1391.
196. T. Tsumura, N. Kojitani, H. Umemura, M. Toyoda, M. Inagaki, *Appl. Surf. Sci.* 196 (2002) 386.
197. J.C.S. Wu, I-H. Tseng, W.-C. Chang, *J. Nanoparticle Res.* 3 (2001) 113.
198. I-H. Tseng, W.-C. Chang, J.C.S. Wu, *Appl. Catal. B: Environ.* 37 (2002) 37.
199. I.-H.Tseng, J. C.S. Wu, H.-Y. Chou, *J. Catal.* 221 (2004) 432.
200. J. Liqiang, S. Xiaojun, S. Jing, C. Weimin, X. Zili, D. Yaoguo, F. Honggang, *Solar Energy Mater. Solar Cells* 79 (2003) 133.
201. A. Özkan, M.H. Özkan, R. Gürkan, M. Akçay, M. Sökmen, *J. Photochem. Photobiol. A: Chem.* 163 (2004) 29.
202. P. Ávila, B. Sánchez, A.I. Cardona, M. Rebollar, R. Candal, *Catal. Today* 76 (2002) 271.
203. B. Sánchez, M. Romero, A. Cardona, B. Fabrellas, E. Garcia, J. Blanco, P. Avila, A. Bahamonde, *J. Phys.* V 9 (1999) 271.
204. C. Knapp, F.J. Gil-Llambýas, M. Gulppi-Cabra, P. Avila, J. Blanco, *J. Mater. Chem.* 7 (1997) 1641.
205. P. Avila, A. Bahamonde, J. Blanco, B. Sánchez, A. Cardona, M. Romero, *Appl. Catal. B: Environ.* 17 (1998) 75.
206. C. Richard, F. Bosquet, J.F. Pilichowski, *J. Photochem. Photobiol. A: Chem.* 108 (1997) 45.
207. M.D. Driessen, T.M. Miller, V.H. Grassian, *J. Mol. Catal. A: Chem.* 131 (1998)
208. J. Villaseooor, P. Reyes, G. Pecchi, *J. Chem. Technol. Biotechnol.* 72 (1998) 105.
209. A.V. Dijken, A.H. Janssen, M.H.P. Smitsmans, D. Vanmaekelbergh, A. Meijerink, *Chem. Mater.* 10 (1998) 3513.
210. B. Neppolian, S. Sakthivel, B. Arabindoo, M. Palanichamy, V. Murugesan, *J. Environ. Sci. Health, Part A: Toxic/Hazardous Subst. Environ. Eng.* 34 (1999) 1829.
211. Y. Yamaguchi, M. Yamazaki, S.Yoshihara, T. Shirakashi, *J. Electroanal. Chem.* 442 (1998) 1.
212. C.A.K. Gouvea, F. Wypych, S.G. Moraes, N. Duran, N. Nagata, P. Peralta-Zamora, *Chemosphere* 40 (2000) 433.
213. B. Dindar, S. Icli, *J. Photochem. Photobiol. A: Chem.* 140 (2001) 263.
214. A. Chittofrati, E. Matijevic, *Colloids Surf.* 48 (1990) 65.
215. M.A. Verges, A. Mifsud, C.J. Serna, *J. Chem. Soc. Faraday Trans.* 86 (1990) 959.
216. T. Trindade, J.D.P. de Jesus, P.O. Brien, *J. Mater. Chem.* 4 (1994) 1611.
217. J. Auffredic, A. Boultif, J.I. Langford, D. Louer, *J. Am. Ceram. Soc.* 78 (1995) 323.
218. C.H. Lu, C.H. Yeh, *Ceram. Inter.* 26 (2000) 351.
219. K. Kamata, H. Hosono, Y. Maeda, K. Miyokawa, *Chem. Lett.* (1984) 2021.
220. O. Milosevic, B. Jordovic, D. Uskokovic, *Mater. Lett.* 19 (1994) 165.
221. L. Jing, Z. Xu, X. Sun, J. Shang, W. Cai, *Appl. Surf. Sci.* 180 (2001) 308.
222. D. Li, H. Haneda, *Chemosphere* 51 (2003) 129.
223. G.R. Bamwenda, H. Arakawa, *Appl. Catal. A: Gen.* 210 (2001) 181.
224. M. Wakamura, K. Hashimoto, T. Watanabe, *Langmuir* 19 (2003) 3428.
225. M.K. Awad, A.B. Anderson, *J. Am. Chem. Soc.* 112 (1990) 1603.

226. D. Sattari, C.L. Hill, *J. Am. Chem. Soc.* 115 (1993) 4649.
227. A. Maldotti, R. Amadelli, G. Varani, S. Tollari, F. Porta, *Inorg. Chem.* 33 (1994) 2968.
228. J.F. Kirby, L.C.W. Baker, *J. Am. Chem. Soc.* 117 (1995) 10010.
229. A. Hiskia, A. Mylonas, E. Papaconstantinou, *Chem. Soc. Rev.* 30 (2001) 62.
230. A. Troupis, A. Hiskia, E. Papaconstantinou, *Angew. Chem. Int. Ed.* 41 (2002) 1911.
231. A. Molinari, R. Amadelli, L. Andreotti, A. Maldotti, *J. Chem. Soc. Dalton. Trans.* (1999) 1203.
232. Y. Guo, Y.Wang, C. Hu, Y. Wang, E. Wang, *Chem. Mater.* 12 (2000) 3501.
233. R.C. Schroden, C.F. Blanndford, B.J. Melde, B.J.S. Johnson, A. Stein, *Chem. Mater.* 13 (2001) 1074.
234. B.J.S. Johnson, A. Stein, *Inorg. Chem.* 40 (2001) 801.
235. R.R. Ozer, J.L. Ferry, *Environ. Sci. Technol.* 35 (2001) 3242.
236. M. Yoon, J.A. Chang, Y. Kim, J.R. Choi, *J. Phys. Chem. B* 104 (2000) 2539.
237. C.C. Chen, P. X. Glei, H.W. Ji, W. Ma, J.C. Zhao, H. Hidaka, N. Serpone, *Environ. Sci. Technol.* 38 (2004) 329.
238. E.V. Kuznetsova, E.N. Savinov, L.A. Vostrikova, G.V. Echevskii, *Water Sci. Technol.* 49 (2004) 109.
239. M.S. Chiou, H.Y. Li, *Chemosphere* 50 (2003) 1095.
240. C. Minero, E. Pelizzetti, S. Malato, J. Blanco, *Solar Energy* 56 (1996) 421.
241. V. Augugliaro, S. Coluccia, V. Loddo, L. Marchese, G. Marta, L. Palmisano, M. Schiavello, *Appl. Catal. B: Environ.* 20 (1999) 15.
242. Y. Nosaka, M. Kishimoto, J. Nishino, *J. Phys. Chem. B* 102 (1998) 10279.
243. K. Okamoto, Y. Yamamoto, H. Tanaka, A. Itaya, *Bull. Chem. Soc. Jpn.* 58 (1985) 2023.
244. R.W. Matthew, *J. Catal.* 111 (1988) 264.
245. W.-Y. Wei, C. Wan, *J. Photochem. Photobiol. A: Chem.* 69 (1992) 241.
246. E. Pelizzetti, C. Minero, *Electrochim. Acta* 38 (1993) 47.
247. H. Gerischer, A. Heller, *J. Phys. Chem.* 95 (1991) 5261.
248. J.-C. D'Oliveira, C. Minero, E. Pelizzetti, P. Pichat, *J. Photochem. Photobiol. A: Chem.* 72 (1993) 261.
249. A. Mills, S. Morris, R. Davies, *J. Photochem. Photobiol. A: Chem.* 70 (1993) 183.
250. U. Stafford, K.A. Gray, P.V. Kamat, *J. Phys. Chem.* 98 (1994) 6343.
251. R. Zona, S. Schmid, S. Solar, *Water Res.* 33 (1999) 1314.
252. A. Sobczynski, L. Duczmal, W. Zmudzinski, *J. Mol. Catal. A: Chem.* 213 (2004) 225.
253. S. Malato, J. Caceres, A.R. Fernandez-Alba, L. Piedra, M.D. Hernando, A. Aguera, J. Vial, *Environ. Sci. Technol.* 37 (2003) 2516.
254. P. Cong, R.D. Doolen, Q. Fan, D.M. Giaquinta, S. Guan, E.W. McFarland, D.M. Poojary, K. Self, H.W. Turner, W.H. Weinberg, *Angew. Chem. Int. Ed.* 38 (1999) 484.
255. J.M. Newsam, T. Bein, J. Klein, W.F. Maier, W. Stichert, *Micropor. Mesopor. Mater.* 48 (2001) 355.
256. C. Brändli, T.F. Jaramillo, A. Ivanovskaya, E.W. McFarland, *Electrochim. Acta* 47 (2001) 553.
257. S.-H. Baeck, T.F. Jaramillo, C. Brändli, E.W. McFarland, *J. Comb. Chem.* 4 (2000) 563.
258. T.F. Jaramillo, A. Ivanovskaya, E.W. McFarland, *J. Comb. Chem.* 4 (2002) 17.
259. A. Hagen, A. Barkschat, J.K. Dohrmann, H. Tributsch, *Solar Energy Mater. Solar Cells* 77 (2003) 1.
260. G.S. Attard, P.N. Bartlett, N.R.B. Coleman, J.M. Elliott, J.R. Owen, J.H. Wang, *Science* 278 (1997) 838.
261. K.-S. Choi, H.C. Lichtenegger, G.D. Stucky, E.W. McFarland, *J. Am. Chem. Soc.* 124 (2002) 12402.
262. S.-H. Baeck, K.-S. Choi, T.F. Jaramillo, G.D. Stucky, E.W. McFarland, *Adv. Mater.* 15 (2003) 1269.
263. P. Liu, S.-H. Lee, C.E. Tracy, Y. Yan, J.A. Turner, *Adv. Mater.* 14 (2002) 27.
264. T.F. Jaramillo, S.-H. Baeck, A. Kleiman-Shwarsctein, E.W. McFarland, *Macromol. Rapid Commun.* 25 (2004) 297.
265. J.M. Hermann, in F. Jansen, R.A. van Santen, Eds., *Water Treatment by Heterogeneous Photocatalysis in Environmental Catalysis*, Catalytic Science Series, Vol. 1, Imperial College Press, London, 1999, chap. 9, pp. 171–194.
266. G. Colón, M.C. Hidalgo, J.A. Navýo, *J. Photochem. Photobiol. A: Chem.* 138 (2001)79.
266a. D. Chatterjee, *J. Mol. Catal. A: Chem.* 154 (2000) 1.
267. L. Cermenati, A. Albini, P. Pichat, C. Guillard, *Res. Chem. Intermed.* 26 (2000) 221.
268. D.F. Ollis, H.Al. Ekabi, Eds., *Photocatalytic Purification and Treatment of Water and Air*, Elsevier, Amsterdam, 1993.
269. J.-M. Herrmann, M.-N. Mozzanega, P. Pichat, *J. Photochem.* 22 (1983) 333.
270. A.L. Pruden, D.F. Ollis, *J. Catal.* 82 (1983) 404.

271. C.-Y. Hsiao, C.-L. Lee, D.F. Ollis, *J. Catal.* 82 (1983) 418.
272. M. Barbeni, E. Pramauro, E. Pelizzetti, E. Borgarello, M. Graetzel, N. Serpone, *Nouv. J. Chim.* 8 (1984) 547.
273. R.W. Matthew, *J. Catal.* 97 (1986) 565.
274. F.J. Beltrán, F.J. Rivas, R. Montero-de-Espinosa, *Appl. Catal. B: Environ.* 39 (2002) 221.
275. S.G. De Moraes, R.S. Freire, N. Duran, *Chemosphere* 40 (2000) 369.
276. K.T. Ranjit, B. Viswanathan, *J. Photochem. Photobiol. A: Chem.* 108 (1997) 79.
277. H. Yamashita, Y. Ichihashi, S.G. Zhang, Y. Matsumura, Y. Souma, T. Tatsumi, M. Anpo, *Appl. Surf. Sci.* 121–122 (1997) 305.
278. C.N. Rusu, J.T. Yates Jr., *J. Phys. Chem. B* 104 (2000) 1729.
279. M. Takeuchi, H. Yamashita, M. Matsuoka, M. Anpo, T. Hirao, N. Itoh, N. Iwamoto, *Catal. Lett.* 66 (2000) 185.
280. J. Lin, J.C. Yu, *J. Photochem. Photobiol. A: Chem.* 116 (1998) 63.
281. Y. Ohko, D.A. Tryk, K. Hashimoto, A. Fujishima, *J. Phys. Chem. B* 102 (1998) 2699.
282. T. Wu, G. Liu, J. Zhao, H. Hidaka, N. Serpone, *J. Phys. Chem. B* 102 (1998) 5845.
283. A. Topalov, D. Molna´r-Ga´bor, J. Csana´di, *Wat. Res.* 33 (1999) 1371.
284. K.T. Ranjit, H. Cohen, I. Willner, S. Bossmann, A.M. Braun, *J. Mater. Sci.* 34 (1999) 5273.
285. Y. Ohko, K. Hashimoto, A. Fujishima, *J. Phys. Chem. A* 101 (1997) 8057.
286. H. Yamashita, M. Honda, M. Harada, Y. Ichihashi, M. Anpo, T. Hirao, N. Itoh, N. Iwamoto, *J. Phys. Chem. B* 102 (1998) 10707.
287. S. Hager, R. Bauer, *Chemosphere* 38 (1999) 1549.
288. T. Ohno, K. Tokieda, S. Higashida, M. Matumura, *Appl. Catal. A: Gen.* 244 (2003) 383.
289. S. Naskar, S.A. Pillay, M. Chanda, *J. Photochem. Photobiol. A: Chem.* 113 (1998) 257.
290. T. Tatsuma, S. Tachibana, T. Miwa, D.A. Tryk, A. Fujishima, *J. Phys. Chem. B* 103 (1999) 8033.
291. B.-N. Lee, W.-D. Liaw, J.-C. Lou, *Environ. Eng. Sci.* 16 (1999) 165.
292. M.E. Fabiyi, R.L. Skelton, *J. Photochem. Photobiol. A: Chem.* 132 (2000) 121.
293. J.C. Ireland, P. Klostermann, E.W. Rice, R.M. Clark, *Appl. Environ. Microbiol.* 59 (1993) 1668.
294. S. Lee, M. Nakamura, S. Ohgaki, *J. Environ. Sci. Health A* 33 (1998) 1643.
295. W.S. Kuo, Y.T. Lin, *J. Environ. Sci. Health A* 35 (2000) 671.
296. S. Matsuo, Y. Anraku, S. Yamada, T. Honjo, T. Matsuo, H. Wakita, *J. Environ. Sci. Health A* 36 (2001) 1419.
297. Y. Ohko, Y. Utsumi, C. Niwa, T. Tatsuma, K. Kobayakawa, Y. Satoh, Y. Kubota, A. Fuhishima, *J. Biomed. Mater. Res.* 58 (2001) 97.
298. F. Zhang, J. Zhao, L. Zang, T. Shen, H. Hidaka, E. Pelizzetti, N. Serpone, *J. Mol. Catal.* 120 (1997) 235.
299. J. Zhao, T. Wu, K. Wu, K. Oikawa, H. Hidaka, N. Serpone, *Environ. Sci. Technol.* 32 (1998) 2394.
300. T. Wu, T. Lin, J. Zhao, H. Hidaka, N. Serpone, *Environ. Sci. Technol.* 33 (1999) 1379.
301. G. Liu, X. Li, J. Zhao, H. Hidaka, N. Serpone, *Environ. Sci. Technol.* 34 (2000) 3839.
302. M. Fox, M. Dulay, *Chem. Rev.* 93 (1993) 341.
303. R.W. Matthews, *J. Chem. Soc., Faraday Trans.* 1 (1984) 457.
304. V. Brezova, A. Staško, L. Lapcik, *J. Photochem. Photobiol. A: Chem.* 59 (1991) 115.
305. D. Lawless, N. Serpone, D. Meisel, *J. Phys. Chem.* 95 (1991) 5166.
306. L. Cermenati, D. Dondi, M. Fagnoni and A. Albini, *Tetrahedron* 59 (2003) 6409.
307. N. Serpone, E. Pelizzetti, Eds., *Photocatalysis, Fundamentals and Applications*, Wiley Interscience, New York, 1989.
308. F. Fernandez-Ibanez, J. Blanco, S. Malato, F.J. de las Nieves, *Water Res.* 37 (2003) 3180.
309. K. Chhor, J.F. Bocquet, C. Colbeau-Justin, *Mater. Chem. Phys.* 86 (2004) 123.
310. J.M. Herrmann, D. Jean, P. Pierre, *J. Catal.* 113 (1988) 72.
311. W.W. Dunn, A. Bard, J. *Nouv. J. Chim.* 5 (1981) 651.
312. S. Nishimoto, B. Ohtani, H. Kajiwara, T. Kagiya, *J. Chem. Soc., Faraday Trans. 1* 79 (1983) 2685.
313. P.D. Fleischauer, H.K. Alan Kan, J.R. Shepherd, *J. Am. Chem. Soc.* 94 (1972) 283.
314. B. Krauetler, A. Bard, *J. Am. Chem. Soc.* 100 (1978) 4317.
315. T. Ohno, K. Sarukawa, M. Matsumura, *New J. Chem.* (2002) 1167.
316. F. Zhang, N. Guan, Y. Li, X. Zhang, J. Chen, H. Zeng, *Langmuir* 19 (2003) 8230.
317. C.A. Arancibia-Bulnes, E.R. Bandala, C.A. Estrada, *Catal. Today* 76 (2002) 149.

318. B. O'Regan, M. Graetzel, *Nature (London)* 353 (1991) 737.

319. S.Y. Hung, G. Schlichthorl, A.J. Nozik, *J. Phys. Chem.* 101 (1997) 2576.

320. Y. Hao, M. Yang, W. Li, *Sol. Energy Mater. Sol. Cells* 60 (2000) 349.

321. Y. Murata, S. Fukuta, S. Ishikawa, S. Yokoyama, *Sol. Energy Mater. Sol. Cells* 62 (2000) 157.

322. D.A. Tryk, A. Fujishima, K. Honda, *Electrochim. Acta* 45 (2000) 2363.

323. T.T.Y. Tan, M. Zaw, D. Beydoun, R. Amall, *J. Nanoparticle Res.* 4 (2002) 541.

324. E.R. Bandala, S. Gelover, M.T. Leal, C. Arancibia-Bulnes, A. Jimenez, C.A. Estrada, *Catal. Today* 76 (2002) 189.

325. C.C. Elam, C.E. Gergoire Padro, G. Sandrock, A. Luzzi, P. Lindblad, E.F. Hagen, *Int. J. Hydrogen Energy* 28 (2003) 601.

326. J. Araña, J.A. Herrera Melián, J.M. Doña Rodrýguez, O. González Dýaz, A. Viera, J. Pérez Peña, P.M. Marrero Sosa, V. Espino Jiménez, *Catal. Today* 76 (2002) 279.

327. D. Curcó, J. Giménez, A. Addardak, S. Cervera-March, S. Esplugas, *Catal. Today* 76 (2002) 177.

328. R.J. Brandi, G. Rintoul, O.M. Alfano, A. E. Cassano, *Catal. Today* 76 (2002) 161.

329. I.Z. Shirgaonkar, A.B. Pandit, *Ultrason. Sonochem.* 5 (1998) 53.

330. P. Theron, P. Pichat, C. Guillard, C. Petrier, T. Chopin, *Phys. Chem. Chem. Phys.* 1 (1999) 4663.

331. N.L. Stock, J. Peller, K. Vinodgopal, P.V. Kamat, *Environ. Sci. Technol.* 34 (2000) 1747.

332. L. Davydov, E.P. Reddy, P. France, P.G. Smirniotis, *Appl. Catal. B: Environ.* 32 (2001) 95.

333. P. Theron, P. Pichat, C. Petrier, C. Guillard, *Water Sci. Technol.* 44 (2001) 263.

334. V. Ragaini, E. Selli, C.L. Bianchi, C. Pirola, *Ultrason. Sonochem.* 8 (2001) 251.

335. C. Petrier, M.-F. Lamy, A. Francony, A. Benahcene, B. David, V. Renaudin, N. Gondrexon, *J. Phys. Chem.* 98 (1994) 10514.

336. H. Zollinger, *Color Chemistry — Synthesis, Properties and Application of Organic Dyes and Pigments*, VCH, New York, 1987.

337. E. Selli, *Phys. Chem. Chem. Phys.* 4 (2002) 6123.

338. A. Fujishima, K. Hashimoto, T. Watanabe, TiO_2 *Photocatalysis — Fundamentals and Applications*, BKC Inc., Tokyo, 1999, p. 146.

339. H. Harada, *Ultrason. Sonochem.* 8 (2001) 55.

340. T.J. Mason, in G.J. Price, Ed., *Current Trends and Future Prospects in Current Trends in Sonochemistry*, The Royal Society of Chemistry, Cambridge, 1992, p. 171.

341. Y. Kado, M. Atobe, T. Nonaka, *Ultrason. Sonochem.* 8 (2001) 69.

342. Y. Suzuki, Warsito, H. Arakawa, A. Maezawa, S. Uchida, *Int. J. Photoenergy* 1 (1999) 1.

343. Y. Suzuki, A. Maezawa, S. Uchida, *Jpn. J. Appl. Phys.* 39 (1998) 2958.

344. V. Ragaini, E. Selli, C.L. Bianchi, C. Pirola, *Ultrason. Sonochem.* 8 (2001) 247.

345. H. Harada, C. Hosoki, M. Ishikane, *J. Photochem. Photobiol. A: Chem.* 160 (2003) 11.

346. S. Horikoshi, H. Hidaka, N. Serpone, *Environ. Sci. Technol.* 36 (2002) 1357.

347. S. Horikoshi, H. Hidaka, *Chem. Ind.* 53 (2002) 740.

348. S. Horikoshi, N. Serpone, H. Hidaka, *J. Photochem. Photobiol. A: Chem.* 159 (2003) 289.

349. S. Horikoshi, H. Hidaka, N. Serpone, *J. Photochem. Photobiol. A: Chem.* 153 (2002) 185.

350. S. Horikoshi, H. Hidaka, N. Serpone, *Chem. Phys. Lett.* 376 (2003) 475.

351. E. Piera, J.A. Ayllón, X. Doménech, J. Peral, *Catal. Today* 76 (2002) 259.

CHAPTER 12 QUESTIONS

Question 1

What does quantum yield mean and how can it be calculated?

Question 2

Which is the role of semiconductors in photocatalytic reactions?

Question 3

How are electron–hole pairs generated? Exemplify for the case of titania.

Question 4

What is the role of OH radical species in photocatalysis?

Question 5

Which is the rate-limiting step in photocatalyzed oxidations on oxide-based photocatalysts?

Question 6

What does sensitization process mean?

Question 7

What are the factors controlling TiO_2 photoactivity?

Question 8

Which are the most investigated photocatalytic reactions?

Question 9

What are the reasons for the modification of TiO_2 photocatalysts?

Question 10

What is sonophotocatalysis and what are its advantages?

Liquid-Phase Oxidations Catalyzed by Polyoxometalates

Noritaka Mizuno, Keigo Kamata, and Kazuya Yamaguchi

CONTENTS

13.1   Introduction 463
13.2   Homogeneous Catalysts with Polyoxometalate-Based Compounds 465
    13.2.1   Mixed-Addenda Polyoxometalates 465
    13.2.2   Transition-Metal-Substituted Polyoxometalates 465
    13.2.3   Peroxometalates 472
    13.2.4   Lacunary Polyoxometalates 474
13.3   Heterogeneous Catalysts with Polyoxometalate-Based Compounds 474
    13.3.1   Dispersion onto Inert Supports 475
        13.3.1.1   Active Carbon 475
        13.3.1.2   Silica and MCM-41 477
        13.3.1.3   Others..478
    13.3.2   Formation of Insoluble Solid Ionic Materials 478
        13.3.2.1   Metal Ions 478
        13.3.2.2   Alkylammonium Ions 478
        13.3.2.3   Crosslinking of Copolymer with POM 479
    13.3.3   Intercalation into Anion-Exchange Materials 480
    13.3.4   Immobilization on Surface-Modified Supports 482
        13.3.4.1   Anion–Cation Pairing 482
        13.3.4.2   Covalent Bond Formation 484
        13.3.4.3   Others 484
13.4   Conclusions and Future Opportunities 485
Acknowledgments 487
References 487
Chapter 13 Questions 492

13.1   INTRODUCTION

The enormous range of chemical products make much contribution to the quality of our lives. However, the manufacturing processes of these products also lead to vast amounts of waste. The reduction or elimination of these wastes is now our central issue. To minimize waste in the manufacture of chemical products, the catalytic method is a reliable solution, which replaces synthetic processes of low atom efficiency, using hazardous stoichiometric reagents [1–3].

Stoichiometric oxidants such as permanganate and dichromate are traditionally used in the selective oxidation of hydrocarbons. In the manufacture of large-scale petrochemicals as well as the laboratory-scale syntheses, the environment-unfriendly processes should be replaced with greener catalytic oxidations. Clean, safe, and inexpensive oxidants such as H_2O_2 and O_2 in combination with catalysts can provide useful and safer catalytic oxidation methods [1–10]. In the preceding two decades, many efficient catalytic oxidations in combination with the above-mentioned greener oxidants have been developed [11–24]. However, most of them are homogeneous systems and there are a few excellent heterogeneous catalysts in spite of offering significant advantages from environmental and economical standpoints [1–10]. Oxidations by Ti/SiO_2 and TS-1 (titanium silicalite with an MFI structure) are the successful methods [11–14], leading to the development of a variety of heterogeneous oxidation systems with organic hydroperoxide and 30% aqueous H_2O_2 and much research work on the synthesis of related heterogeneous catalysts for liquid-phase oxidations.

The versatility and accessibility of polyoxometalates (POMs) have led to the various applications in the fields of structural chemistry, analytical chemistry, surface science, medicine, electrochemistry, photochemistry, and catalysis. Especially, POMs have received much attention in the area of oxidation and acid catalysis [25–31]. Several categories of POMs are formed by proper selection of the starting components and by the adjustment of pH and temperature. Typical examples are shown in Figure 13.1: (i) isopolyoxometalates of the general formula, $[M_xO_y]^{q-}$, produced by condensation

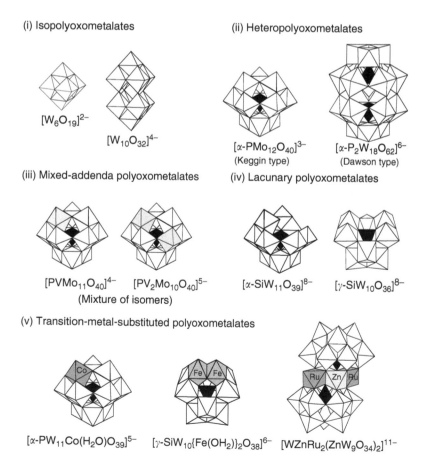

Figure 13.1 Structure of polyoxometalates.

of MO_4^{2-} (M = Mo^{6+}, W^{6+}, $\cdots$); (ii) heteropolyoxometalates of general formula, $[X_aM_xO_y]^{q-}$, produced by condensation of MO_4^{2-} and the heteroatom X (A = P^{5+}, Si^{4+}, Zn^{2+}, Co^{2+}, $\cdots$), for example, Keggin type $[PMo_{12}O_{40}]^{3-}$ and Wells–Dawson type $[P_2W_{18}O_{62}]^{6-}$; (iii) mixed-addenda POMs with various ratios of elements such as Mo, W, and V, for example, $[PMo_{12-n}V_nO_{40}]^{(3+n)-}$; (iv) transition–metal-substituted POMs, for example, $[PW_{11}O_{39}M'(OH_2)]^{n-}$ (M' = Mn^{2+}, Co^{2+}); and (v) lacunary POMs, for example, $[\alpha\text{-}SiW_{11}O_{39}]^{8-}$ and $[\gamma\text{-}SiW_{10}O_{36}]^{8-}$. Some review articles and books describe structures and properties of POMs in more detail [25–33].

Various types of POMs are effective catalysts for the H_2O_2- and O_2-based environment-friendly oxidations. Most of these oxidations are carried out in homogeneous systems and share common drawbacks, that is, catalyst/product separation and catalyst recycling are very difficult. The heterogenization of POMs can improve the catalyst recovery and recycling. This chapter focuses on the development of (1) homogeneous catalysts with POMs and (2) the heterogenization for liquid phase-oxidations.

13.2 HOMOGENEOUS CATALYSTS WITH POLYOXOMETALATE-BASED COMPOUNDS

Various catalytic systems for H_2O_2- and O_2-based oxidations catalyzed by POMs have been developed. Typical examples are listed in Table 13.1. The systems can be classified into four groups according to the structures of POMs: (1) mixed-addenda POMs, (2) transition-metal-substituted POMs, (3) POMs, and (4) lacunary POMs. In this chapter, liquid-phase homogeneous oxidations by POMs with H_2O_2 and O_2 are described according to the above classification.

13.2.1 Mixed-Addenda Polyoxometalates

A wide variety of hydrocarbons including alkanes, alcohols, amines, and arenes could be oxidized with O_2 in the presence of mixed-addenda POMs $[PMo_{12-n}V_nO_{40}]^{(3+n)-}$ [27,28,30,31,34–50]. The oxidation mechanism included one (or more) electron oxidation of the substrate by $[PMo_{12-n}V_nO_{40}]^{(3+n)-}$ to give the product and the reduced form of $[PMo_{12-n}V_nO_{40}]^{(3+n)-}$, followed by reoxidation of the reduced form by O_2. The $[PMo_{12-n}V_nO_{40}]^{(3+n)-}/O_2$/hydrocarbon redox system can be used to oxidize a wide variety of hydrocarbons because the redox potential of $[PMo_{12-n}V_nO_{40}]^{(3+n)-}$ (0.7 V vs. SHE) is lower than that of O_2 and is higher than that of a wide variety of hydrocarbons [30,31,53]. It is noted that the system consisting of $[PMo_{12-n}V_nO_{40}]^{(3+n)-}$ and Pd^{2+} salts could catalyze the Wacker oxidation and Moritani–Fujiwara reaction with the use of O_2 [30,31,34,51,52,54].

Thermodynamically controlled self-assembly of an equilibrated ensemble of POMs with $[AlVW_{11}O_{40}]^{6-}$ as the main component could act as a catalyst for the selective delignification of wood (lignocellulose) fibers (Figure 13.2) [55]. Equilibration reactions typical of POMs kept the pH of the system near 7 during the catalysis that avoided acid or base degradation of cellulose.

13.2.2 Transition-Metal-Substituted Polyoxometalates

Hill and Brown for the first time reported, in 1986, that transition-metal-substituted POMs $[PW_{11}O_{39}M'(OH_2)]^{5-}$ (M' = Mn^{2+}, Co^{2+}) catalyzed the oxidation of alkenes with PhIO [56]. The M' atom easily becomes coordinatively unsaturated by elimination of the aquo ligand, and the resulting polyanion is regarded as an inorganic metalloporphyrin analog [57]. Transition-metal-substituted POMs are oxidatively and hydrolytically stable compared with organometallic complexes, and their active sites can be controlled. These advantages have been applied to the development of biomimetic catalysis relating to the heme enzyme, cytochrome P-450, and the non-heme enzyme, methane monooxygenase analogs (Figure 13.3). Until now, numerous catalytic oxidations by transition-metal-substituted POMs have been developed [58–77] and some of them can activate O_2 as described below.

Table 13.1　Examples of H$_2$O$_2$– and O$_2$–Based Homogeneous Oxidations Catalyzed by POMs

Catalyst	Reaction	Oxidant	Solvent	Temperature (K)	Reference
$[PV_2Mo_{10}O_{40}]^{5-}$	anthracene → (98)	O_2	Tetraglyme/ClCH$_2$CH$_2$Cl	343	[44]
$[PV_2Mo_{10}O_{40}]^{5-}$	cyclohexanone → COOH/COOH (40)	O_2	AcOH/H$_2$O	343	[45]
$[PV_6Mo_6O_{40}]^{9-}$	alkene → epoxide (77)	O_2/aldehyde	ClCH$_2$CH$_2$Cl	298	[46]
$[PV_2Mo_{10}O_{40}]^{5-}$	2 dimethylphenol → quinone (80)	O_2	n-Hexanol	363	[47]
$[PV_2Mo_{10}O_{40}]^{5-}$	benzyl alcohol (OH) → benzaldehyde (81)	O_2	Dacalin/H$_2$O	363	[48]
PdSO$_4$/CuSO$_4$/$[PV_6Mo_6O_{40}]^{19-}$/β-CD	alkene → ketone (98)	O_2	H$_2$O	363	[49]
Pd(OAc)$_2$/$[PMo_{12-n}V_nO_{40}]^{(3+n)-}$	benzene → phenol OH (10)	O_2	AcOH/H$_2$O	403	[50]
Pd(OAc)$_2$/$[PMo_8V_4O_{40}]^{7-}$	benzene + OEt acrylate → Ph–cinnamate OEt (72)	O_2	AcOH	363	[51]
$[PV_2Mo_{10}O_{40}]^{5-}$	anthracene → anthraquinone (96)	O_2/quinone	CH$_3$CN	333	[38]

Catalyst	Substrate → Product (yield)	Oxidant	Solvent	T	Ref.
$[AlV^VW_{11}O_{40}]^{6-}$	(90)	O_2	H_2O	403	[55]
$[WZnRu^{III}_2(ZnW_9O_{34})_2]^{11-}$	(57)	O_2	$ClCH_2CH_2Cl$	353	[58,59]
$[WZnMn^{II}_2(ZnW_9O_{34})_2]^{12-}$	(55)*	H_2O_2	$ClCH_2CH_2Cl$	275	[63]
$[WZnMn^{II}_2(ZnW_9O_{34})_2]^{12-}$	(88)	H_2O_2	$ClCH_2CH_2Cl$	RT	[64]
$[WZn_3(ZnW_9O_{34})_2]^{12-}$	(94)	H_2O_2	H_2O	358	[65]
$[(Mn^{II}(H_2O))_3(SbW_9O_{33})_2]^{12-}$	(90)	H_2O_2	$ClCH_2CH_2Cl$	RT	[66]
$[Fe^{III}_4(H_2O)_2(PW_9O_{34})_2]^{6-}$	(7) (selectivity)	H_2O_2	CH_3CN	298	[67]
$[Fe^{III}_4(H_2O)_2(P_2W_{15}O_{56})_2]^{12-}$	(11) (selectivity)	H_2O_2	CH_3CN	298	[68]
$\alpha\beta\beta\alpha\text{-}(Mn^{II}OH_2)_2(Mn^{II})_2(AsW_{15}O_{62})_2]^{16-}$	(46)*	H_2O_2	CH_3CN	298	[69]
$[Fe^{III}_2(NaOH_2)_2(P_2W_{15}O_{56})_2]^{12-}$	(32)*	H_2O_2	CH_3CN	298	[70]

(Continued)

Table 13.1 (Contiuned)

Catalyst	Reaction	Oxidant	Solvent	Temperature (K)	Reference
$[M(OH_2)_2Fe^{III}_2(P_2W_{15}O_{56})_2(P_2M_2(OH_2)_4W_{13}O_{52})]^{16-}$ (M=Cu^{II} or Co^{II})	(22)*	H_2O_2	CH_3CN	298	[71]
$[((Mn^{II}(OH_2))Mn^{II}_2PW_9O_{34})_2(PW_6O_{26})]^{17-}$	(55)*	H_2O_2	CH_3CN	353	[72]
$[PW_9O_{37}\{Fe_2Ni(OAc)_3\}]^{10-}$	(22) (total)	O_2	Benzene	356	[73]
$[\gamma\text{-}SiW_{10}\{Fe^{III}(OH_2)\}_2O_{38}]^{6-}$	(80)	O_2	$ClCH_2CH_2Cl/CH_3CN$	356	[61]
$[\gamma\text{-}SiW_{10}\{Mn^{III}(OH_2)\}_2O_{38}]^{6-}$	(6) (total)	O_2	$ClCH_2CH_2Cl/CH_3CN$	356	[74]
$[\gamma\text{-}SiW_{10}\{Fe^{III}(OH_2)\}_2O_{38}]^{6-}$	(66)* (total)	H_2O_2	CH_3CN	305	[60]
$[\gamma\text{-}SiW_{10}\{Fe^{III}(OH_2)\}_2O_{38}]^{6-}$	$CH_4 \longrightarrow HCOOCH_3$ (7)*	H_2O_2	CH_3CN	353	[75]
$[Ni(H_2O)H_2F_6NaW_{17}O_{35}]^{9-}$	(33)*	H_2O_2	CH_3CN	333	[76]
$[H_2F_6NaVW_{17}O_{56}]^{8-}$	(38)	O_2	$ClCH_2CH_2Cl$	363	[77]

Catalyst	Reaction	Oxidant	Solvent	Temperature (K)	Ref.
$[(CH_3CN)_xFeSiW_9V_3O_{40}]^{5-}$	2,4-di-t-Bu phenol → di-t-Bu tropolone type product (45)	O_2	$ClCH_2CH_2Cl$	338	[62]
$H_3PW_{12}O_{40}$/CPC (CPC=cetyl pyridinium chloride)	octene → epoxide (80)	H_2O_2	$CHCl_3$	333	[18]
$H_3PW_{12}O_{40}$/CPC	geraniol → epoxy alcohol (96)	H_2O_2	CH_3Cl_3	RT	[18,78]
$H_3PW_{12}O_{40}$/CPC	crotonic acid → epoxy acid (90)	H_2O_2	H_2O	333–338	[80]
$H_3PW_{12}O_{40}$/CPC	2-octanol → 2-octanone (99)	H_2O_2	t-BuOH	Reflux	[18,78]
$H_3PW_{12}O_{40}$/CPC	tetrahydroisoquinoline → nitrone (90)	H_2O_2	$CHCl_3$	RT	[81]
$H_3PW_{12}O_{40}$/CPC	diol → hydroxy ketone (93)	H_2O_2	$CHCl_3$	RT	[18]
$WO_4^{2-}/PO_4^{3-}/Q^+X^-$	octene → epoxide (82)*	H_2O_2	$ClCH_2CH_2Cl$ or C_6H_6	343	[17,83]

(Continued)

Table 13.1 (Continued)

Catalyst	Reaction	Oxidant	Solvent	Temperature (K)	Reference
$WO_4^{2-}/NH_2CH_2PO_3H_2/[(CH_3(C_8H_{17})_3N]HSO_4$	(86)	H_2O_2	Without	363	[89,90]
$WO_4^{2-}/[(CH_3(C_8H_{17})_3N]HSO_4$	(97)	H_2O_2	Without	363	[91,92]
$WO_4^{2-}/[(CH_3(C_8H_{17})_3N]HSO_4$	(93)	H_2O_2	Without	348–363	[88]
$[\{W(=O)(O_2)_2(H_2O)\}_2(\mu\text{-}O)]^{2-}$	(95)	H_2O_2	H_2O	305	[93,94]
$[HPO_4\{WO(O_2)_2\}_2]^{2-}$	(72)	H_2O_2	$CHCl_3$	274	[86]
$[\beta_3\text{-}Co^{III}W_{11}O_{35}(O_2)_4]^{10-}$	(91) (total)	H_2O_2	$CHCl_3$/buffered H_2O	RT	[96]
$[\gamma\text{-}SiW_{10}O_{34}(H_2O)_2]^{4-}$	(90)*	H_2O_2	CH_3CN	305	[99]
$[\gamma\text{-}SiW_{10}O_{34}(H_2O)_2]^{4-}$	(83)*	H_2O_2	CH_3CN	305	[100]

Values in the parentheses are product yiedls based on substrates. RT Room Temperature.

*Yields based on H_2O_2.

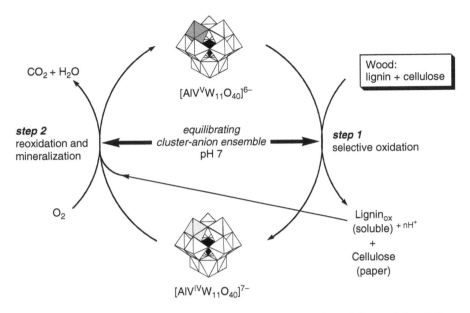

Figure 13.2 Proposed catalytic cycle for the selective delignification of wood (lignocellulose) fibres with an equilibrated ensemble of POMs. (From Weinstock, I. A. et al., *Nature*, *414*, 191, 2001.)

Active sites of cytochrome P-450 and methane monooxygenase

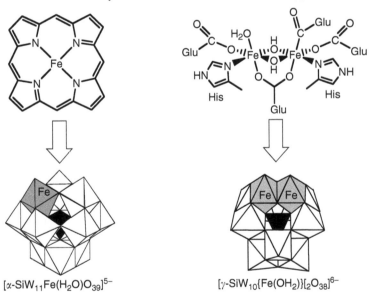

Figure 13.3 Application of transition-metal-substituted polyxometalates as biomimetic catalysts.

The ruthenium-substituted sandwich-type POM $[WZnRu_2(OH)(H_2O)(XW_9O_{34})_2]^{11-}$ ($X = Zn^{2+}$ or Co^{2+}, Figure 13.1) catalyzed the selective hydroxylation of adamantane with O_2 as an oxidant[58,59]:

$$\text{adamantane} \xrightarrow[\text{1,2-C}_2\text{H}_4\text{Cl}_2, \ 353 \text{ K}, \ 72 \text{ h}]{[WZnRu^{3+}{}_2(ZnW_9O_{34})_2]^{11-}, \ O_2 \ (1 \text{ atm})} \text{1-adamantanol} \quad \mathbf{57\%} \qquad (13.1)$$

The catalyst oxygenated selectively tertiary C—H bond of adamantane, giving only 1-adamantanol. The oxidation featured an adamantane/O_2 stoichiometry of 2:1. On the basis of the stoichiometry, spectroscopic evidence, and kinetic studies, a dioxygenase-type mechanism was proposed: activation of O_2 via complexation to a Ru^{2+} species, followed by the formation of a $Ru^{4+}=O$ species via a $Ru^{3+}-O-O-Ru^{3+}$ intermediate (Figure 13.4). Di-iron-substituted silicotungstate, $[\gamma\text{-SiW}_{10}\{Fe(OH_2)\}_2O_{38}]^{6-}$ (Figure 13.1), was synthesized by the reaction of $[\gamma\text{-SiW}_{10}O_{36}]^{8-}$ with $Fe(NO_3)_3$ in acidic aqueous solution [60,61]. This POM catalyzed the selective oxidation of alkenes and alkanes with highly efficient utilization of H_2O_2[60]:

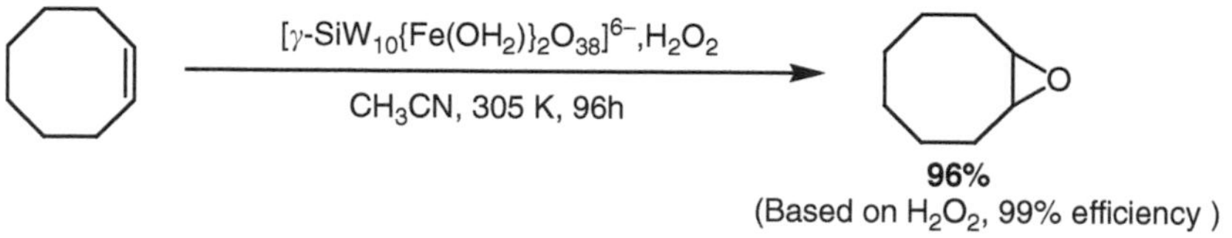

$$(13.2)$$

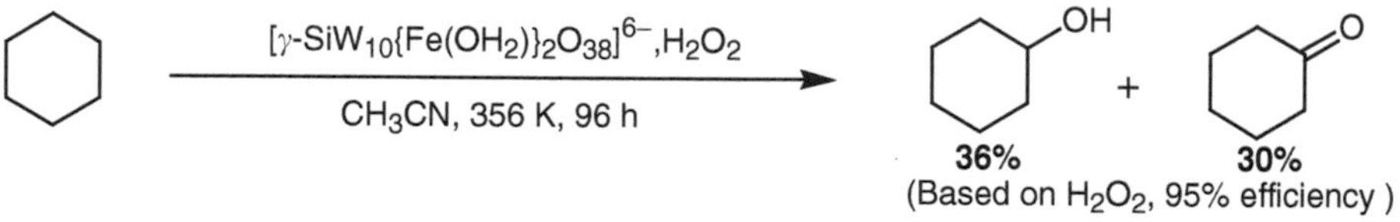

96%
(Based on H_2O_2, 99% efficiency)

$$(13.3)$$

36%   **30%**
(Based on H_2O_2, 95% efficiency)

The POM also catalyzed the selective epoxidation of alkenes with only 1 atm of O_2 as an oxidant [61]:

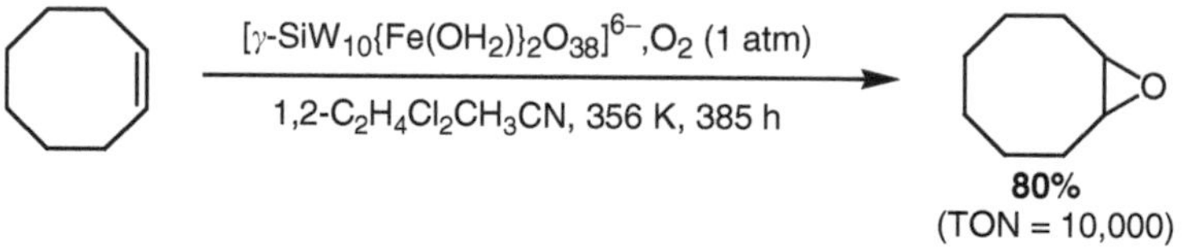

$$(13.4)$$

80%
(TON = 10,000)

The reaction was also proposed to proceed via a dioxygenase-type mechanism. Finke and Weiner [62] have reported that vanadium- and iron-containing polyanion-based precatalysts show high catalytic activity for catechol dioxygenation with O_2 [62]. For example, 3,5-di-*tert*-butylcatechol was oxidized in the presence of $(n\text{-Bu}_4N)_5[(CH_3CN)_xFe\cdot SiW_9V_3O_{40}]$ with extremely high turnover number (100000) and the value was far superior to those for man-made and natural enzymes:

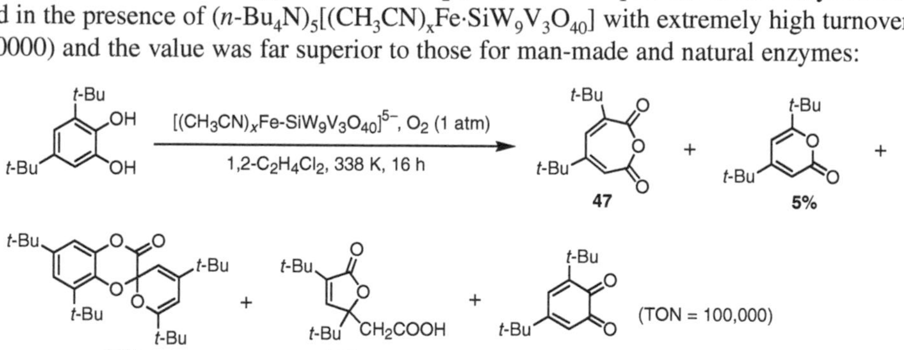

$$(13.5)$$

13.2.3 Peroxometalates

Tungsten-based catalysts including POMs show high efficiency of H_2O_2 utilization [17,18,78–100]. Ishii and coworkers [18] have reported effective H_2O_2-based epoxidation of alkenes catalyzed by $H_3PW_{12}O_{40}$ combined with cetyl pyridinium chloride (CPC) as a phase-transfer agent:

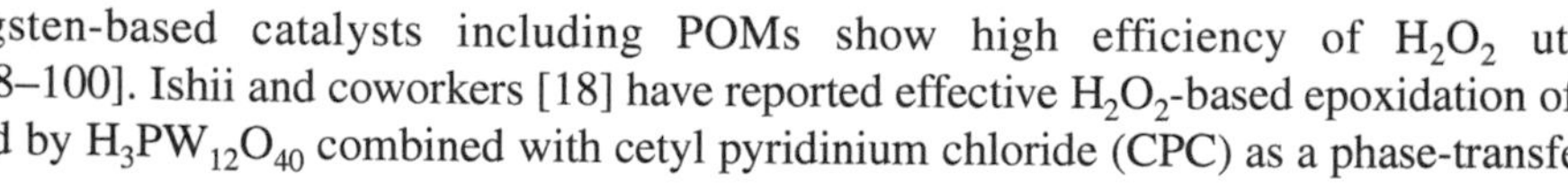

$$(13.6)$$

80%

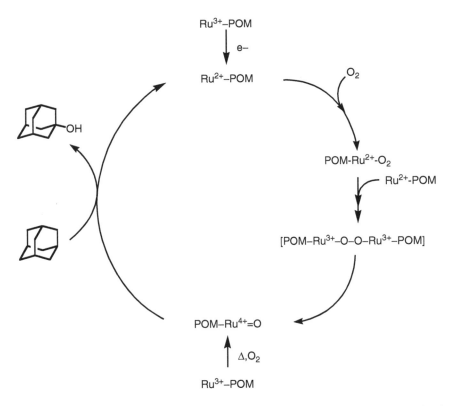

Figure 13.4 Proposed catalytic cycle for activation of molecular oxygen and substrate by the ruthenium-substituted sandwich-type [WZnRu$_2$$^{3+}$(XW$_9O_{34}$)$_2$]$^{11-}$ (X=Zn^{2+} or Co^{2+}). (From Neumann, R., and Dahan, M., *Nature, 388*, 353, 1997 and Neumann, R., and Dahan, M., *J. Am. Chem. Soc., 120*, 11969, 1998.)

The use of a biphasic system prevented the hydrolytic cleavage of the oxirane rings, resulting in high selectivity to epoxides (Figure 13.5). This system could also be applied to the epoxidation of allylic alcohols [18,78], monoterpenes [79], and α,β-unsaturated carboxylic acids [80]; oxidation of alcohols [18,78], amines [81], and alkynes [82]; and oxidative transformation of diols [18]. The [PO$_4${WO(O$_2$)$_2$}$_4$]$^{3-}$ peroxotungstate was isolated and crystallographically characterized by Venturello and coworkers in 1982. The POM comprises PO$_4$$^{3-}$ and two [W$_2$O$_2$(O$_2$)$_4$]. The peroxo anion has been postulated to be catalytically the most active species in the H$_3$PW$_{12}$O$_{40}$/H$_2$O$_2$ system because [PO$_4${WO(O$_2$)$_2$}$_4$]$^{3-}$ exhibited catalytic reactivity similar to that of Ishii system [17,83–87]. Noyori and coworkers [19,88–92] reported an H$_2$O$_2$-based oxidation system of tungstate and methyltrioctylammonium hydrogensulfate without halides and organic solvents:

$$\text{WO}_4{}^{2-}/\text{NH}_2\text{CH}_2\text{PO}_3\text{H}_2$$
$$\xrightarrow[\text{solvent free, 383 K, 2 h}]{[\text{CH}_3(n\text{-C}_8\text{H}_{17})_3\text{N})]\text{HSO}_4,\ \text{H}_2\text{O}_2}$$

86% (13.7)

$$\xrightarrow[\text{solvent free, 383 K, 2 h}]{\text{WO}_4{}^{2-}/[\text{CH}_3(n\text{-C}_8\text{H}_{17})_3\text{N})]\text{HSO}_4,\ \text{H}_2\text{O}_2}$$

45%
(TON = 179,000) (13.8)

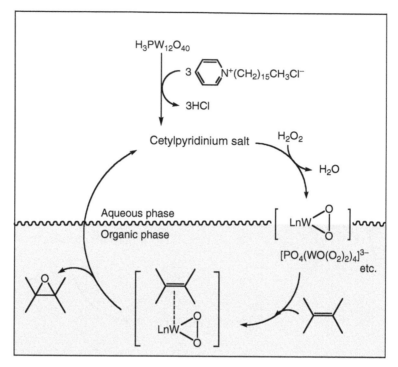

Figure 13.5 Proposed catalytic cycle for the epoxidation of alkenes with H_2O_2 by $H_3PW_{12}O_{40}$. (From Ishii, Y. et al., *J. Org. Chem.*, *53*, 3587, 1988.)

The highly chemo-, regio-, and diastereoselective and stereospecific epoxidation of various allylic alcohols with only 1 equivalent of hydrogen peroxide in water solvent could be efficiently catalyzed by an isolated dinuclear peroxotungstate $[W_2O_3(O_2)_4(H_2O)_2]^{2-}$ [93,94]:

$$\text{(13.9)}$$

$$\text{OH} \xrightarrow[\text{water, 305 K, 24 h}]{[\{W(=O)(O_2)_2(H_2O)\}_2(\mu\text{-}O)]^{2-}, H_2O_2} \underset{\substack{\textbf{84\%} \\ (\text{TON} = 4200)}}{\text{O} \diagup\diagdown \text{OH}}$$

13.2.4 Lacunary Polyoxometalates

The vacant sites of lacunary POMs have the possibility to activate hydrogen peroxide catalytically [96]. While a lacunary phosphotungstate, $Na_7[PW_{11}O_{39}]$, reacts with hydrogen peroxide to form $[PO_4\{WO(O_2)_2\}_4]^{3-}$ [87,97], silicotungstates are rather stable in water compared with phosphotungstates, and the chemistry has been well established by Tézé and Hervé [98]. A silicotungstate compound, $[\gamma\text{-}SiW_{10}O_{34}(H_2O)_2]^{4-}$ (Figure 13.6), synthesized by protonation of the well-known dilacunary Keggin-type POM of $[\gamma\text{-}SiW_{10}O_{36}]^{8-}$, exhibited high catalytic performance for the epoxidation of various alkenes including propylene with an H_2O_2 oxidant at 305 K [99,100]. The effectiveness of this catalyst was evidenced by $\geqslant 99\%$ selectivity to epoxide, $\geqslant 99\%$ efficiency of H_2O_2 utilization, high stereospecificity, and the easy recovery of the catalyst from the homogeneous reaction mixture.

13.3 HETEROGENEOUS CATALYSTS WITH POLYOXOMETALATE-BASED COMPOUNDS

As mentioned earlier, POMs can act as very effective catalysts for environment-friendly oxidations with H_2O_2 and O_2, and these oxidations are homogeneous in many cases. Therefore, the development

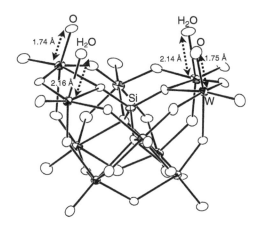

Figure 13.6 Molecular structure of $[\gamma\text{-}SiW_{10}O_{34}(H_2O)_2]^{4-}$. (From Kamata, K. et al., *Science, 300*, 964, 2003.)

of heterogeneous catalysts based on POMs is very interesting. The development of efficient heterogeneous catalysts with POMs and the related compounds for the liquid-phase oxidations has been attempted and the strategies can be classified into the following four groups (Figure 13.7): (1) dispersion onto inert supports, (2) formation of insoluble solid ionic materials, (3) intercalation into anion-exchange materials, and (4) immobilization on surface-modified supports. In this chapter, the developments of heterogeneous catalysts with POMs are described according to this classification.

13.3.1 Dispersion onto Inert Supports

Dispersion of POMs onto inert solid supports with high surface areas is very important for catalytic application because the surface areas of unsupported POMs are usually very low (~10 m^2g). Another advantage of dispersion of POMs onto inert supports is improvement of the stability. Therefore, immobilization of POMs on a number of supports has been extensively studied. Silica and active carbon are the representative supports [25]. Basic supports such as MgO tend to decompose POMs [101–104]. Certain kinds of active carbons firmly entrap POMs [105,106]. The maximum loading level of POMs on active carbons is ~14 wt% [107]. Dispersion of POMs onto other supports such as zeolites, mesoporous molecular sieves, and apatites, is of considerable interest because of their high surface areas, unique pore systems, and possibility to modify their compositions, morphologies, and sorption properties. However, a simple impregnation of POM compounds on inert supports often results in leaching of POMs.

13.3.1.1 Active Carbon

Izumi and Urabe [105] found first that POM compounds could be entrapped strongly on active carbons. The supported POMs catalyzed etherization of *tert*-butanol and *n*-butanol, esterification of acetic acid with ethanol, alkylation of benzene, and dehydration of 2-propanol [105]. In 1991, Neumann and Levin [108] reported the oxidation of benzylic alcohols and amines catalyzed by the neutral salt of $Na_5[PV_2Mo_{10}O_{40}]$ impregnated on active carbon. Benzyl alcohols were oxidized efficiently to the corresponding benzaldehydes without overoxidation:

$$\text{PhCH}_2\text{OH} \xrightarrow[\text{PhCH}_3,\ 373\ K,\ 22\ h]{Na_5PV_2Mo_{10}O_{40}/C,\ \text{air (1 atm)}} \text{PhCHO} \quad \mathbf{97\%} \tag{13.10}$$

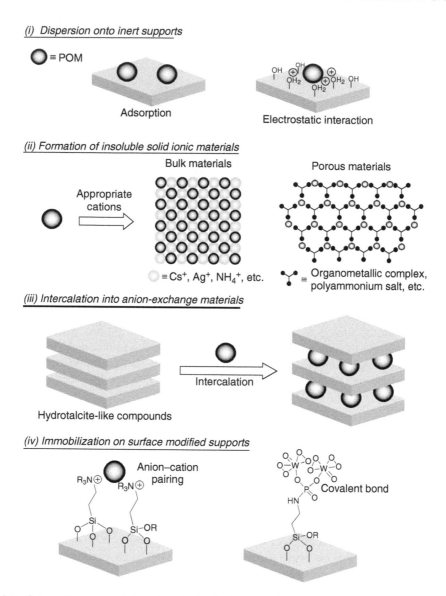

Figure 13.7 Schematic representation of strategies for heterogenization of POMs.

Benzyl amines gave the corresponding imines followed by hydrolysis to afford benzaldehydes. Then, the benzaldehydes formed were reacted with the starting benzyl amines to give the corresponding Schiff bases:

$$\text{(13.11)}$$

Ishii and co-workers [109] reported the aerobic oxidation of various organic compounds catalyzed by $(NH_4)_5H_6[PV_8Mo_4O_{40}]$ supported on active carbon. The catalyst showed high activity for oxidative dehydrogenation of various benzylic and allylic alcohols to give the corresponding carbonyl compounds in moderate to high yields. The catalyst can be recycled without loss of activity for the alcohol oxidation. Under similar reaction conditions, 2,3,6-trimethylphenol was selectively oxidized to trimethyl-p-benzoquinone, which is a precursor of vitamin E:

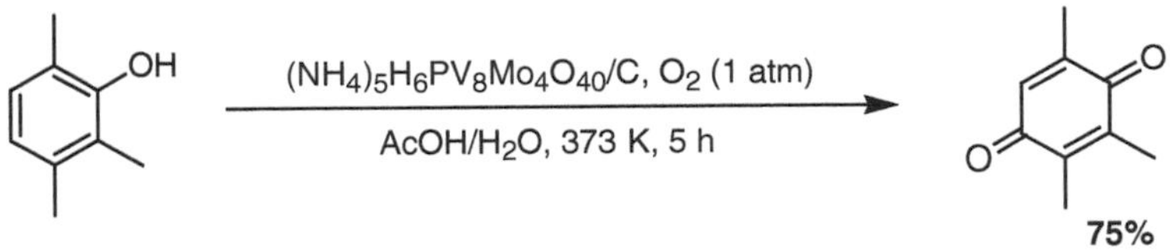

$$(13.12)$$

75%

Alkyl-substituted phenols and alkanes were also oxidized to give the corresponding oxygenated products. In the case of isophorene, 3-formyl-5,5-dimethyl-2-cyclohexen-1-one was obtained selectively [110]. The regioselectivity of the oxidation was remarkably different from that observed with the corresponding homogeneous analog, which produced 1,4-diketone as a major product [110]:

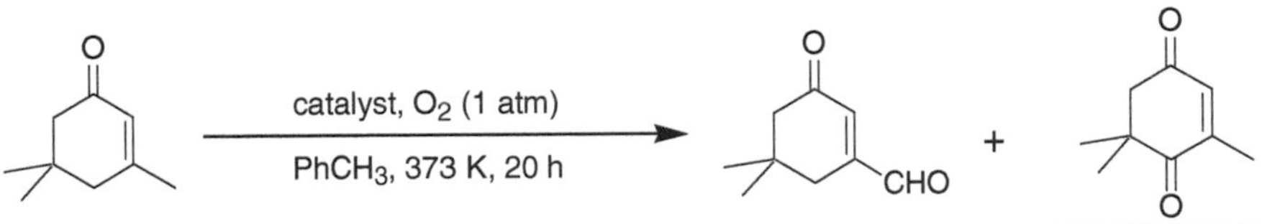

	select.	
$(NH_4)_5H_6PV_8Mo_4O_{40}$/C	65%	12%
$(NH_4)_5H_6PV_8Mo_4O_{40}$	<1%	50%

$$(13.13)$$

This unusual selectivity was explained by the pore size effect of the active carbon.

The acid form of $H_5[PV_2Mo_{10}O_{40}]$ was supported on porous carbons and carbon fiber materials for oxidation of the highly toxic mustard gas analog bis(2-chloroethyl)sulfide using *tert*-butyl hydroperoxide (TBHP) [111,112]. High selectivity to the corresponding sulfoxide is imperative because the sulfoxide has low toxicity whereas the sulfone is highly toxic. Over 99% selectivity to the sulfoxide, even at 75% conversion of the substrate was obtained for the catalyst of 10 wt% $H_5[PV_2Mo_{10}O_{40}]$ on porous carbon:

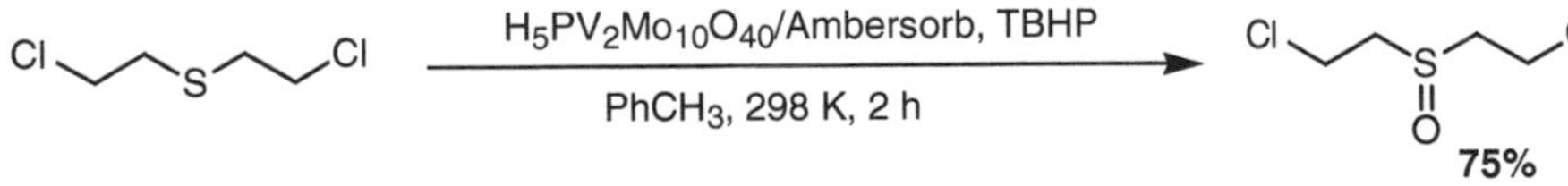

$$(13.14)$$

75%

Washing for over 100 h in a solvent such as toluene before the reaction resulted in no significant loss of the catalytic activity, and recovery and reusability studies at high turnover number also indicated the catalyst stability. Same catalysts, $H_5[PV_2Mo_{10}O_{40}]$ supported on fiber and fabric carbon materials, catalyzed O_2-based oxidations of acetaldehyde and 1-propanethiols [113]. This aerobic heterogeneous oxidation proceeded under mild reaction conditions.

A mixed-valent polymolybdate on active carbon was prepared from molybdenum metal and H_2O_2, followed by the addition of active carbon to the aqueous solution [114,115]. This catalyzed the epoxidation of several alkenes in 2-propanol using H_2O_2 as an oxidant, while the efficiency of H_2O_2 utilization was very low ($\leqslant 25\%$). The epoxidation likely proceeded mainly on the surface of the catalyst because the recovered catalyst showed almost similar catalytic activity.

13.3.1.2 Silica and MCM-41

The supported Co^{2+}-substituted Wells–Dawson POM, $Cs_6H_2[P_2W_{17}O_{61}Co(OH_2)]$, on silica was stable up to 773 K and catalyzed the heterogeneous oxidation of various aldehydes to the corresponding carboxylic acids with O_2 as a sole oxidant [116]. The $H_5PV_2Mo_{10}O_{40}$ POM, impregnated onto mesoporous MCM-41, catalyzed the aerobic oxidation of alkanes and alkenes using isobutyraldehyde as a co-oxidant [117]. The isostructural phosphotungstate, $H_3PW_{12}O_{40}$, supported on MCM-41 demonstrated high catalytic activity for acid-catalyzed reactions [118]. The oxidations of alkanes and alkenes gave similar product selectivities to those observed in the corresponding homogeneous reactions while reaction rates (conversions) were somewhat reduced. With low aldehyde concentration ($\leqslant 1.1$ M), there was no leaching of $H_5[PV_2Mo_{10}O_{40}]$ and the solid catalyst could be recovered and reused without significant loss of catalytic activity.

13.3.1.3 Others

Ichihara and coworkers [119–122] developed new catalytic reaction systems using calcium apatite as a support. Calcium apatite, $Ca_{10}(PO_4)_6X_2$ ($X = OH^-$ [HAP], F^- [FAP]), which forms the mineral component of bone and teeth, is handled as a harmless solid to environment, and has strong affinity to various organic compounds [123]. Various kinds of POMs such as $(CP)_{10}[H_2W_{12}O_{42}]$ (CP = cetylpyridinium) and $(NH_4)_3[PMo_{12}O_{40}]$ could be supported on FAP. Oxidation of alkenes and sulfides using urea–H_2O_2 (UHP) as an oxidant proceeded efficiently with the catalyst under solvent-free conditions. The recovered solid phase was reused several times with the maintenance of the catalytic activity. Ichihara suggested that *in situ* solid–phase activation of the catalyst with UHP proceeds to form microcrystals of the active species and the FAP phase plays an important role in dispersing and stabilizing the active species.

POM/SiO_2 composites were synthesized by a sol–gel technique using tetraethyl orthosilicate (TEOS) and POMs such as $H_3[PW_{12}O_{40}]$, $H_4[SiW_{12}O_{40}]$, and $Na_4[W_{10}O_{32}]$ [124]. During the sol–gel process, the POM was entrapped into the silica network, resulting in the formation of POM/SiO_2 composites with microporous structures. In the POM/SiO_2 composites, surface silanols ($\equiv Si-OH$) were protonated under the acidic sol–gel conditions to form cationic $\equiv Si-OH_2^+$ groups, which may act as counter cations for POM anions. These materials were used for heterogeneous photocatalytic oxidations [124]. The sandwich-type Fe-containing POM, $[(Fe(OH_2)_2)_3(A-\alpha-PW_9O_{34})_2]^{9-}$, was electrostatically immobilized on the surface of cationic silica/alumina nanoparticles ((Si/AlO_2)Cl) [125–127]. In this case, there were ~58 POM molecules per nanoparticle bound to the surface $\equiv Al-OH_2^+$ groups. This material efficiently catalyzed the aerobic oxygenation of sulfides and autooxidation of aldehydes. Interestingly, the activity per POM was higher than that of the parent POM.

13.3.2 Formation of Insoluble Solid Ionic Materials

Selection of appropriate counter cations can control the solubility of POMs. Usually alkylammonium ions are used as counter cations of POM anions to dissolve the compounds in organic solvents (homogeneous system). The introduction of metal counter cations can reduce the solubility in organic solvents [128–132].

13.3.2.1 Metal Ions

Hill and coworkers [133] reported that selective aerobic oxidation of 2-chloroethyl ethyl sulfide to the corresponding sulfoxide is catalyzed by $Ag_5[PV_2Mo_{10}O_{40}]$, which is synthesized by a simple cation-exchange reaction between $Na_5[PV_2Mo_{10}O_{40}]$ and $Ag(NO)_3$ in water. In this system, 2,2,2-trifluoroethanol was used as a solvent because the substrate was soluble but $Ag_5[PV_2Mo_{10}O_{40}]$ was insoluble in the solvent. The activity of the recovered catalyst remained unchanged after several turnovers and the filtrate showed no activity, indicating that the $Ag_5[PV_2Mo_{10}O_{40}]$ powder is an actual catalyst.

13.3.2.2 Alkylammonium Ions

The mono-ruthenium-substituted silicotungstate, $[SiW_{11}O_{39}Ru^{3+}(H_2O)]^{5-}$, synthesized by reaction of the lacunary POM $[SiW_{11}O_{39}]^{8-}$ with Ru^{3+} in acetone, was an efficient catalyst for the oxidation of various alkanes and alcohols using 1 atm of O_2 with high turnover numbers [134]. The tetrabutylammonium salt of $[SiW_{11}O_{39}Ru^{3+}(H_2O)]^{5-}$ was insoluble in isobutyl acetate, *tert*-butyl acetate, and trifluorotoluene, and could therefore be used as a heterogeneous catalyst in these solvents. After several turnovers of the substrate, the catalyst could be removed easily by filtration. Ru and W were not detected in the filtrate and then the reaction was repeated with the filtrate. No additional conversion of the substrate was observed. When the oxygenation of adamantane was repeated with the recovered catalyst by simple filtration under the same conditions, the reaction proceeded at almost the same rate and selectivity as those for the first run. These results indicate that any Ru and W species that leached into the reaction solution are not active homogeneous catalysts and that the observed catalysis is truly heterogeneous.

Xi and coworkers [135–137] reported on the epoxidation of alkenes performed with $(CP)_3[PO_4(WO_3)_4]$ catalyst. This insoluble catalyst formed soluble active species, $(CP)_3[PO_4\{WO_2(O_2)\}_4]$, by the reaction with H_2O_2. When H_2O_2 was consumed completely, the catalyst became insoluble again. Therefore, the catalyst recovery was simple. When coupled with the 2-ethylanthraquinone/2-ethylanthrahydroquinone redox process for H_2O_2 production, O_2 could be used as an oxidant for the epoxidation of propylene in 85% yield based on 2-ethylanthrahydroquinone which was obtained without any by-products (Figure 13.8).

13.3.2.3 Crosslinking of Copolymer with POM

Crosslinking of the copolymer was achieved by the presence of POMs, and insoluble materials with micro- or mesopores were formed. Ikegami and coworkers [138,139] prepared a novel insoluble catalyst of a crosslinked complex consisting of $PW_{12}O_{40}^{3-}$ and *N*-isopropylacrylamide with ammonium cations (PWAA) (Figure 13.9). This PWAA catalyst heterogeneously catalyzed the oxidations of various substrates including allylic alcohols, amines, and sulfide with 30% aqueous H_2O_2 under mild reaction conditions without organic solvents. The turnover number reached up to 35,000 and was very high for the epoxidation of phytol:

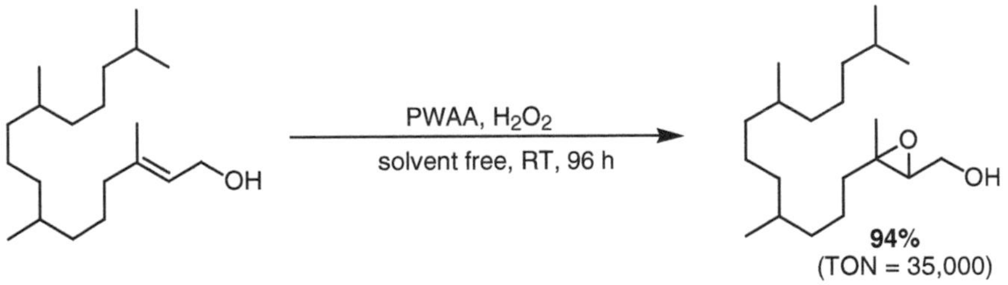

$$(13.15)$$

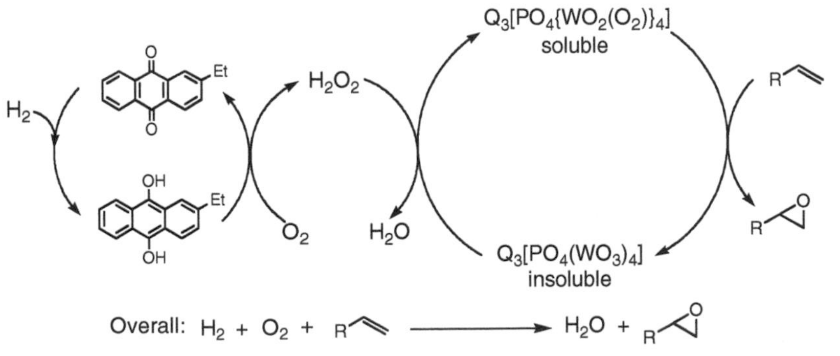

Figure 13.8 Proposed catalytic cycle for the epoxidation of alkenes by $[\pi\text{-}C_5H_5NC_{16}H_{33}]_3[PO_4(WO)_4]$ catalyst coupled with the 2-ethylanthraquinone/2-ethylanthrahydroquinone redox process. (From Xi, Z. et al., *Science*, 292, 1139, 2001.)

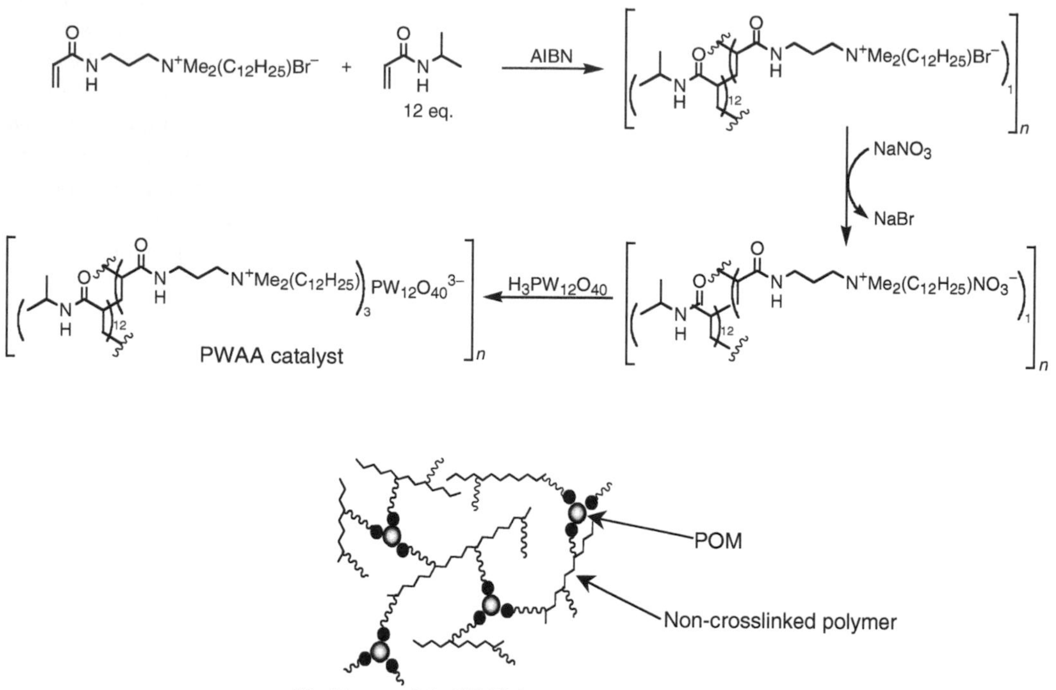

Working model of PWAA catalyst

Figure 13.9 Preparation of PWAA catalyst. (From Yamada, Y. M. A. et al., *Org. Lett., 3*, 1837, 2001 and Yamada, Y. M. A. et al., *Tetrahedron, 60*, 4087, 2004.)

The oxidation of methyl phenyl sulfide using fresh PWAA gave methyl phenyl sulfone in 97% yield. In the repeated use of recovered catalyst, the yields of sulfone in runs 2 to 5 ranged from 82 to 88%. The activities of recovered catalysts were somewhat reduced. The deactivation may be caused by catalyst pulverization or degradation of the $PW_{12}O_{40}^{3-}$ species. While the reusability and stability of the catalyst should be improved, this concept would be useful for the creation of sophisticated solid catalysts. After Ikegami's reports, Neumann and coworkers [140] applied this strategy to the development of alkylated polyethyleneimine/POM synzymes.

Neumann and coworkers [141] prepared inorganic–organic mesoporous materials by complexation of the sandwich-type $[ZnWZn_2(H_2O)_2(ZnW_9O_{34})_2]^{12-}$ with branched tripodal organic polyammonium salts such as tris[2-(trimethylammonium)ethyl]-1,3,5-benzenetricarboxylic acid and 1,3,5-tris [4-(N,N,N-trimethylammoniumethylcarboxy)-phenyl]benzene triciations. However, SEM and TEM analyses showed the formation of three-dimensional (3D), perforated, coral-shaped, amorphous materials with organic cations surrounding the polyoxoanions. The hybrid materials had a BET surface area in the range 30 to 50 m²/g and an average pore diameter of 36 ± 6 Å. The materials were tested as heterogeneous catalysts for the epoxidation of allylic alcohols and secondary alcohols with H_2O_2. The authors reported that the catalytic activity and chemo-, regio-, and diastereoselectivity of the hybrid materials in acetonitrile were very similar to those of the corresponding homogeneous analogs (alkylammonium salt of $[ZnWZn_2(H_2O)_2(ZnW_9O_{34})_2]^{12-}$ dissolved in 1,2-dichloroethane) [64]. The catalyst recovery–recycling experiments and characterization of recovered catalysts show that the above heterogeneous catalysis takes place on the surface of the hybrid materials.

13.3.3 Intercalation into Anion-Exchange Materials

Hydrotalcitelike compounds (HTs) are layered double hydroxides with the general formula $[M_n^{2+}M_m^{3+}(OH)_{2(n+m)}]^{m+}A_{m/x}^{x-}\cdot yH_2O$, where M^{2+} and M^{3+} are divalent and trivalent metal cations, respectively [142]. The A^{x-} anions required to compensate the net positive charge of Brucite-like layers are located in the interlayer space [142]. The HTs have been used as catalysts and catalyst supports.

Recently, various kinds of HTs have been synthesized by substitution of the cations in the Brucite-like layer. These compounds have anion-exchange ability and various kinds of anions including POMs can be intercalated. For example, $Zn_2Al(OH)_6NO_3 \cdot 4H_2O$ (ZnAl-HT) underwent facile and complete intercalation with POMs such as $[SiW_{11}O_{39}]^{8-}$ and $[SiV_3W_9O_{40}]^{7-}$ by an anion–exchange reaction [143,144]. The XRD patterns for ZnAl-HT-$(SiW_{11}O_{39})$ and ZnAl-$(SiV_3W_9O_{40})$ showed several (001) harmonics corresponding to a basal spacing of 14.5 Å. If the thickness of the Brucite-like layer is taken to be 4.7 Å, the interlayer space is 9.8 Å, in accordance with the POM size. The ZnAl-HT-$(SiW_{11}O_{39})$ catalyzed epoxidation of cyclohexene with H_2O_2. Tatsumi and coworkers [145] applied MgAl-HT intercalated with POMs such as $[Mo_7O_{24}]^{6-}$ or $[H_2W_{12}O_{42}]^{10-}$ to the shape-selective epoxidation of 2-hexene and cyclohexene with H_2O_2.

Monomeric oxoanions such as WO_4^{2-} and MoO_4^{2-} were also intercalated into the interlayer of HTs. These materials could act as efficient oxidation catalysts with H_2O_2. MgAl- or NiAl-HT intercalated with WO_4^{2-} (HT-WO_4^{2-}) showed excellent activities for oxidative bromination of unsaturated hydrocarbons using bromide-H_2O_2 as a bromine source under very mild reaction conditions (Figure 13.10) [146,147]. This material catalyzed the oxidation of bromide with H_2O_2 to give the bromonium species, which reacts rapidly with the substrate to provide the corresponding brominated product. The bromination of monochlorodimedone by HT-WO_4^{2-} showed a turnover frequency of 71 h^{-1}, which was two orders of magnitude higher than that of the corresponding homogeneous analog using Na_2WO_4. The recovered catalyst by filtration after the reaction showed the same catalytic performance, suggesting that the observed catalysis is truly heterogeneous. In some alkenes, such as 1-methyl-1-cyclohexene and linalool, bromide-assisted selective formation of the corresponding epoxides in water proceeded via ring closure of intermediate bromohydrines:

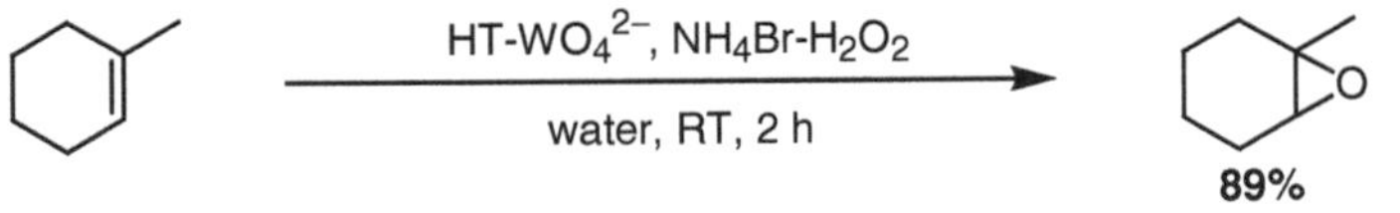

$$\text{(13.16)}$$

89%

The HT–WO_4^{2-} was also active for the bromide-assisted oxidation of substituted phenols to p-quinol derivatives with NH_4Br–H_2O_2 [148], and the epoxidation of allylic alcohols with H_2O_2 as the sole oxidant [149]. Hydrotalcitelike compounds intercalated with MoO_4^{2-} (HT-MoO_4^{2-}) catalyzed the decomposition of H_2O_2 to singlet oxygen and water [150–152]:

$$MoO_4^{2-} + nH_2O_2 \rightleftharpoons MoO_{4-n}(O_2)_n^{2-} + nH_2O \qquad (13.17)$$

$$MoO_{4-n}(O_2)_n^{2-} \longrightarrow {}^1O_2 + MoO_{6-n}(O_2)_{n-2}^{2-} \qquad (13.18)$$

This system has been successfully applied to the peroxidation of various unsaturated hydrocarbons [150–152].

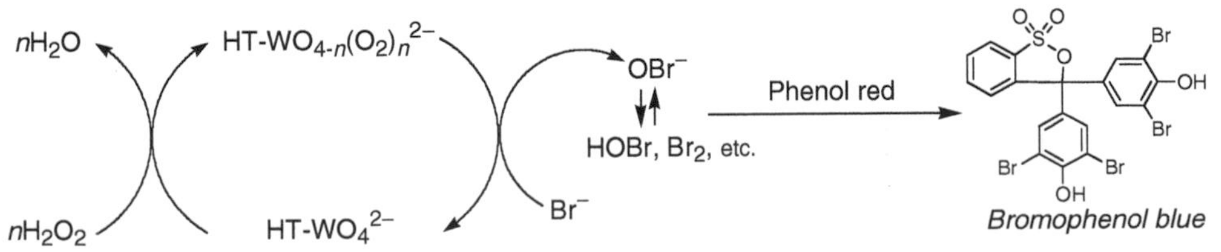

Figure 13.10 Catalytic cycle of bromination with HT-WO_4^{2-} catalyst. (From Sels, B. F. et al., *Nature*, 400, 855, 1999.)

13.3.4 Immobilization on Surface-Modified Supports

Polyoxometalates can be immobilized on surface-modified supports via appropriate spacer ligands. There are two main strategies for the immobilization: (1) formation of anion–cation pairing and (2) formulation of a covalent bond. Surface modification of polymers and silica can be carried out using various methods [153–157]. According to the review by Davis et al. [156], organic groups can be easily grafted onto a silanol-containing surface using a trialkoxy- or trichloro-organosilane, for example. The grafted organic groups are covalently attached, and are stable. The surface properties such as hydrophobicity and hydrophilicity can be controlled by changing the organic groups. For tethering POMs, quaternary ammonium cation-functionalized supports have been most widely used.

13.3.4.1 Anion–Cation Pairing

Srinivasan and Ford [158] reported that epoxidation of cyclooctene using excess H_2O_2 was catalyzed by polymolybdate such as $[Mo_7O_{24}]^{6-}$ tethered on the colloidal polymer with alkylammonium cations. Cyclooctene was oxidized to give 1,2-epoxycyclooctane with >99% selectivity at 90% conversion for 24 h at 313 k:

$$\text{polym.--CH}_2\text{N}^+(\text{CH}_3)_3(1/6)[\text{Mo}_7\text{O}_{24}]^{6-},\text{H}_2\text{O}_2$$
$$\text{solvent free, 313 K, 24 h}$$

90%

(13.19)

However, styrene and cyclohexene gave complex product mixtures, and 1-octene did not react under the same reaction conditions. Thus, the activity of this catalyst is intrinsically low. Jacobs and co-workers [159,160] applied Veturello's catalyst $[PO_4(WO(O_2)_2)_4]^{3-}$ (tethered on a commercial nitrate-form resin with alkylammonium cations) to the epoxidation of allylic alcohols and terpenes. The regio- and diastereoselectivity of the parent homogeneous catalysts were preserved in the supported catalyst. For bulky alkenes, the reactivity of the POM catalyst was superior to that of Ti-based catalysts with large pore sizes such as Ti-β and Ti-MCM-48. The catalytic activity of the recycled catalyst was completely maintained after several cycles and the filtrate was catalytically inactive, indicating that the observed catalysis is truly heterogeneous in nature.

Degradation (or pulverization) of the organic polymer support is disadvantageous for the polymer-supported POM catalysts. To overcome this disadvantage, inorganic-based hybrid supports have been used. Neumann and Miller [161] grafted POMs such as $[PO_4(WO(O_2)_2)_4]^{3-}$ and $[ZnWZn_2(H_2O)_2(ZnW_9O_{34})_2]^{12-}$ on modified SiO_2 particles with various quaternary ammonium cation moieties, prepared by copolymerization of TEOS (tetraethyl orthosilicate) and trialkoxy organosilanes using the sol–gel technique for the heterogeneous epoxidation using H_2O_2. The catalytic activities were greatly influenced by the type of quaternary ammonium cation moieties introduced into the modified SiO_2. Octyldimethyl benzyl ammonium cations showed the best result. With these catalysts, however, the epoxidation was not selective and efficient except for a few common alkenes. Catalytic epoxidation with analogous catalysts based on surface-modified MCM-41 has also been reported [162].

The peroxotungstate $[W_2O_3(O_2)_4(H_2O)_2]^{2-}$ (W2), immobilized on dihydroimidazolium-based ionic liquid-modifed SiO_2 (W2/1-SiO_2), has been synthesized [163]. The synthetic procedure is shown in Figure 13.11. For the epoxidation of cis- and trans-alkenes catalyzed by W2/1-SiO_2, the configuration around the C = C double bonds was completely retained in the corresponding epoxides, suggesting that free-radical intermediates were not involved in the epoxidation. Theegioselective epoxidation of geraniol took place at the electron-deficient 2,3-allylic double bond to afford 2,3-epoxy alcohol in high yield. Epoxidation of secondary β,β-disubstituted allylic

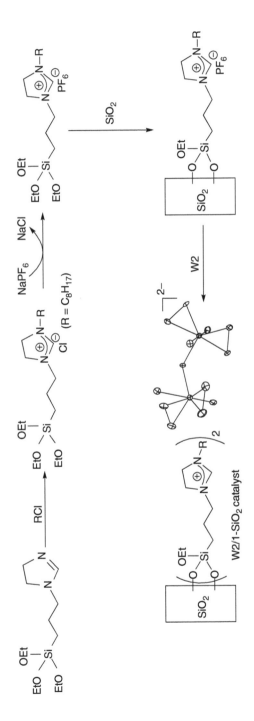

Figure 13.11 Procedures for the preparation of W2/1-SiO₂ catalyst. (From Yamaguchi, K. et al., *J. Am. Chem. Soc.*, *127*, 530, 2005.)

alcohols of 4-methyl-3-penten-2-ol proceeded diastereoselectively to form the *threo*-epoxy alcohol (*threo/erythro* = 90/10). In addition, the epoxidation of (Z)-3-methyl-3-penten-2-ol, in which 1,3- and 1,2-allylic strains compete with each other, was more *erythro*-selective (*threo/erythro* = 40/60). A similar regio- and diastereoselectivity for allylic alcohols was also observed for homogeneous W2-catalyzed epoxidations [93,94]. Epoxidation was terminated completely by catalyst removal and no tungsten species could be detected in the filtrate. Furthermore, W2/1-SiO$_2$ could be easily recovered by filtration and reused at least three more times for the epoxidation of cyclooctene under the same conditions without loss of catalytic activity and selectivity. The above results rule out any contribution to the observed catalysis from tungsten species that might have leached into the reaction solution and therefore the observed catalysis is truly heterogeneous in nature. The catalytic activity of W2/1-SiO$_2$ was compared with that of the corresponding homogeneous analog of [n-C$_{12}$H$_{25}$N(CH$_3$)$_3$]$_2$[W$_2$O$_3$(O$_2$)$_4$(H$_2$O)$_2$] (DTMA-W2) under the same conditions. The reaction rates of W2/1-SiO$_2$ were comparable to those of DTMA-W2. This fact indicates that homogeneous catalysis can be heterogenized with retention of the W2 catalyst performance by using ionic liquid-modified SiO$_2$ as a support.

13.3.4.2 Covalent Bond Formation

Various kinds of supports with phosphorylated spacers were synthesized to form covalent bonds with peroxotungstate species (Figure 13.12) [162,164,165]. The resulting materials were shown to act as effective catalysts in the epoxidation with H$_2$O$_2$ [162,164,165]. Very recently, the peroxo compound [HPO$_4${W(O)(O$_2$)$_2$}$_2$]$^{2-}$ (PW2) was synthesized on the surface of mesoporous HMS by reacting HMS–CH$_2$CH$_2$CH$_2$NH(PO$_3$H$_2$) with W2, and then palladium ions were exchanged into the channels of HMS to form a hybrid catalyst [166]. The catalyst was active in the oxidation of propylene to propylene oxide in methanol using O$_2$ as an oxidant. For example, oxidation of propylene at 373 K for 6 h gave propylene oxide in 83.2% selectivity at 34.1% conversion. In this system, the actual epoxidation catalyst may be the PW2 species. H$_2$O$_2$ might be formed by the reaction of methanol and O$_2$ over the palladium catalyst and regenerates peroxo species. The active components such as palladium and tungsten did not leach into the reaction medium and this catalyst could be reused.

13.3.4.3 Others

Solvent-anchored, supported catalysts have also been reported [167]. As shown in Figure 13.13, polyethers such as polyethylene oxide (PEO) and polypropylene oxide (PPO) were covalently attached to the surface of SiO$_2$, acting as a solvent or complexing agent for POMs such as H$_5$PV$_2$Mo$_{10}$O$_{40}$ and [(C$_6$H$_{13}$)$_4$N]$_3$[PO$_4${W(O$_2$)$_2$}$_4$]. Oxidative dehydrogenation of 9,10-dihydroanthracene with TBHP and epoxidation of cyclooctene with isobutyraldehyde/O$_2$ proceeded by H$_5$PV$_2$Mo$_{10}$O$_{40}$-based catalyst:

$$\text{H}_5\text{PV}_2\text{Mo}_{10}\text{O}_{40}\text{–PEO–SiO}_2,\ \text{TBHP}$$
$$\text{PhCH}_3,\ \text{RT},\ 24\ \text{h}$$

96%

(13.20)

In the case of supported Venturello catalyst of [(C$_6$H$_{13}$)$_4$N]$_3$[PO$_4${W(O$_2$)$_2$}$_4$], epoxidation of cyclooctene was performed by using 30% aqueous H$_2$O$_2$ without organic solvents:

$$Q_3[\text{PO}_4(\text{WO(O}_2)_2)_4]\text{–PEO/PPO–SiO}_2,\ \text{H}_2\text{O}_2$$
$$\text{solvent free},\ 296\ \text{K},\ 24\ \text{h}$$

>99%

(13.21)

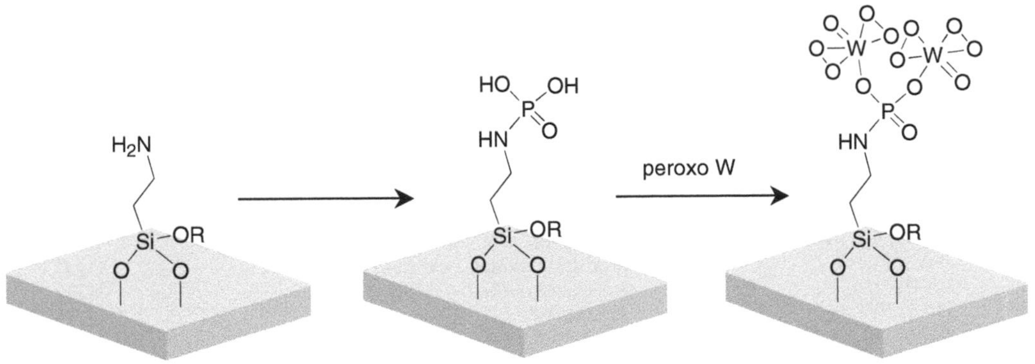

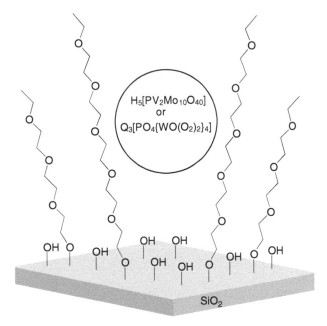

Figure 13.12   Covalent binding of peroxotungstate species with phosphorylated spacers.

Figure 13.13   Illustration of the solvent-anchored liquid-phase catalysts. (From Neumann, R., and Cohen, M., *Angew. Chem. Int. Ed. Engl.*, 36, 1738, 1997.)

Similarly, heterogeneous catalyst prepared by immobilization of POMs on chemically modified hydrophobic SiO_2 has been applied to the selective epoxidation of various alkenes with 15% aqueous H_2O_2 without organic solvents [168].

13.4   CONCLUSIONS AND FUTURE OPPORTUNITIES

In liquid-phase oxidations, POMs can be used as efficient homogeneous catalysts in combination with the environment-friendly oxidants H_2O_2 and O_2. For H_2O_2-based oxidations, tungsten-based compounds are effective. Some transition-metal-substituted POMs (e.g., $[WZnRu_2(OH)(H_2O)(XW_9O_{34})_2]^{11-}$ (X = Zn^{2+} or Co^{2+}) [58,59], $[\gamma\text{-}SiW_{10}\{Fe(OH_2)\}_2O_{38}]^{6-}$ [61], and $[(CH_3CN)_xFe\cdot SiW_9V_3O_{40}]^{5-}$ 62]) can activate O_2. Practical applications of POMs to oxidation catalysis will require methods for perfect catalyst recovery and recycling. One promising approach

Table 13.2   Porous Polyoxometalates

Group	Compound	Comment	Reference
(i)	$Cs_{2.5}H_{0.5}PW_{12}O_{40}$	Adsorption of N_2, hydrocarbons; hydrolysis of esters; hydration of olefins	[169–173]
(i)	$Cs_{2.1}H_{0.9}PW_{12}O_{40}$	Adsorption of N_2, n-butane	[170, 174]
(i)	$Pt\text{-}Cs_{2.5}H_{0.9}PW_{12}O_{40}$ (Pt: 0.5 wt%)	Adsorption of N_2, hydrogenation of olefins	[174, 175]
(i)	$Pt\text{-}Cs_{2.1}H_{0.9}PW_{12}O_{40}$ (Pt: 0.5–1.5 wt%)	Adsorption of N_2, methane, n-butane; oxidation of CO, alkanes (shape-selective)	[174–176]
(i)	$Pt\text{-}Rb_{2.1}H_{0.9}PW_{12}O_{40}$ (Pt: 0.5 wt%)	Adsorption of N_2, benzene; hydrogenation of aromatics (shape-selective)	[174, 176]
(i)	$Cs_3PW_{12}O_{40}$	Adsorption of N_2	[177]
(i)	$(NH_4)_3PW_{12}O_{40}$	Adsorption of N_2	[177–180]
(i)	$A_3PM_{12}O_{40}$, $A_3SiW_{12}O_{40}$ (A = Ag; M = W, Mo)	Adsorption of N_2	[181]
(i)	$(C_{52}H_{60}NO_{12})_{12}[\{Mn(H_2O)\}_3 (SbW_9O_{33})_2]$		[182]
(i)	$Ag_4SiW_{12}O_{40}$	Adsorption of N_2	[183]
(i)	Organic polyammonium salts of $[ZnWZn_2(H_2O)_2(ZnW_9O_{34})_2]^{12-}$	Adsorption of N_2, oxidation of allylic alcohols, secondary alcohols	[141]
(ii)	$[M_3V_{18}O_{42}(H_2O)_{12}(XO_4)]\cdot24H_2O$ (M = Fe, Co; X = V, S)	Sorption of water	[184, 185]
(ii)	$[Cu(tpypor)Cu_2Mo_3O_{11}]\cdot nH_2O$, $[\{Fe(tpypor)\}_3Fe(Mo_3O_{19})_2]$ $\cdot nH_2O$	Sorption of water, alcohols	[186]
(ii)	$[AlO_4Al_{12}(OH)_{24}(H_2O)_{12}][Al(OH)_6 Mo_6O_{18}]_2(OH)\cdot29.5H_2O$	Sorption of water	[187]
(ii)	$(pipzH_2)(H_3O)[Al_{15}(\mu_3\text{-}O)_4 (\mu_3\text{-}OH)_6(\mu\text{-}OH)_{14}(hpdta)_4]$	Sorption of water	[188]
(ii)	$K_3[Cr_3O(OOCH)_6(H_2O)_3]$ $[SiW_{12}O_{40}]\cdot16H_2O$; $Cs_3[Cr_3O (OOCH)_6(H_2O)_3]$ $[CoW_{12}O_{40}]\cdot7.5H_2O$	Sorption of water, alcohols, nitriles; noxidation of allylic alcohols (shape selective)	[189–191]
(ii)	$(calix[4]arene\text{-}Na^+)_3[PW_{12}O_{40}]$	Sorption of water, alcohols (amphiphilic)	[192]
(iii)	$Na_{16}[Mo^{VI}_{124}Mo^{V}_{28}O_{429}(\mu_3\text{–}O)_{28} H_{14}(H_2O)_{66.5}]\cdot300H_2O$		[193]
(iii)	$Na_{16}[\{Mo^{VI}_2O_5(H_2O)_2\}_{16} \{Mo^{VI/V}_8O_{26}(\mu_3\text{-}O)_2H(H_2O)_3 Mo^{VI/V}\}_{16} \{Mo^{VI/V}_{36}O_{96}(H_2O)_{24}\}_2]$		[194]
(iii)	$(NH_4)_{72-n}[(H_2O)_{81-n}(NH_4)_n \{(Mo^{VI}Mo^V_5O_{21}(H_2O)_6)_{12} \{Mo^V_2O_4(SO_4)\}_{30}\}$ $\cdot200H_2O$		[195]
(iii)	$(NH_4)_{26}[(H_4Cu^{II}_5)Mo^V_{28}Mo^{VI}_{114}O_{432}(H_2O)_{58}]\cdot300H_2O$		[196]
(iii)	$Na_{15}[Mo^{VI}_{126}Mo^V_{28}O_{462}H_{14} (H_2O)_{70}]_{0.5}[Mo^{VI}_{124}Mo^V_{28}O_{457} H_{14}(H_2O)_{68}]_{0.5}\cdot400H_2O$		[197]
(iii)	$(DODA)_{40}(NH_4)_2[(H_2O)_{50} Mo_{132}O_{372}(CH_3COO)_{30}(H_2O)_{72}]$		[198]
(iii)	$Cs_2[Mo_{12}S_{12}O_{12}(OH)_{12} (H_{10}C_7O_4)]\cdot20H_2O$		[199]
(iii)	$A_4[W_{16}S_{16}O_{16}(OH)_{16}(H_2O)_4 (C_5H_6O_4)_2]$ (A = Cs, Li)		[200]
(iii)	$K_{28}Li_5H_7P_8W_{48}O_{184}\cdot92H_2O$		[201]
(iii)	$K_{24}[\{\beta\text{-}Ti_2SiW_{10}O_{39}\}_4]\,50H_2O$		[202]

Notes: (1)Tpypor = tetrapyridlporphyrin; pipz = piperazine; H_5hpdta = 2-hydroxypropane-1,3-di-amine-N,N,N', N''-tetraaceticacid; organic polyammonium cation = tris[2-(trimethylammonium)ethyl]-1,3,5-benzenetricarboxylate; 1,3,5-tris[4-(N,N,N-trimethylammoniumethylcarboxyl)phenyl]benzene; DODA = dimethyl dioctadecylammonium.

(2) Group (i), compounds with pores between particles (crystallites); Group (ii), compounds with pores in their crystal lattices; Group (iii), compounds with pores in the molecules.

is to heterogenize the active homogeneous POM catalysts and many strategies have been proposed in which the observed catalysis is truly heterogeneous in nature, i.e., recyclable catalysts and no leaching of the active species. However, in most cases the catalytic activities of the parent homogeneous POMs are much decreased by the heterogenization. The results of Hill [125], Jacobs [146], Neumann [161,167], and Mizuno [163] represent the few successful examples in which the catalytic activities of heterogenized catalysts are comparable or even higher than those of the corresponding homogeneous analogs. Future desirable goals in this area will be to overcome the decrease in activities by the heterogenization.

Recently, various kinds of porous POMs have been synthesized (Table 13.2). They have controlled pores between particles (crystallites) [169–183], in their crystal lattices [184–192], or in the molecules [193–201], and show unique sorption and catalytic properties. Worthy targets of future research are shape- and stereo-selective oxidations using porous POMs.

ACKNOWLEDGMENTS

This work was supported by the Core Research for Evolutional Science and Technology (CREST) program of the Japan Science and Technology Corporation (JST) and a Grant-in-Aid for Scientific Research from the Ministry of Education, Culture, Science, Sports and Technology of Japan.

REFERENCES

1. Anastas, P.T., and Warner, J.C., *Green Chemistry: Theory and Practice*, Oxford University Press, London, 1998.
2. Sheldon, R. A., *Green Chem.*, 2, G1 (2000).
3. Anastas, P. T., Bartlett, L. B., Kirchhoff, M. M., and Williamson, T. C., *Catal. Today, 55*, 11 (2000).
4. Sheldon, R. A., and van Bekkum, H., *Fine Chemical through Heterogeneous Catalysis*, Wiley, Weinheim, 2001.
5. Sheldon, R. A., and Kochi, J. K., *Metal-Catalyzed Oxidations of Organic Compounds*, Academic Press, New York, 1981.
6. Hill, C. L., in *Advances in Oxygenated Processes* (Baumstark, A. L., Ed.), Vol. 1, p. 1, JAI Press, London, 1988.
7. Hudlicky, M., *Oxidations in Organic Chemistry*, ACS Monograph Series, American Chemical Society, Washington, DC, 1990.
8. Sheldon, R. A., in *New Development in Selective Oxidation* (Centi, G., and Trifiró, F., Eds.), p. 1, Elsevier, Amsterdam, 1990.
9. de Vos D. E., Sels, B. F., and Jacobs, P. A., *Adv. Catal., 46*, 1 (2001).
10. Grigoropoulou, G., Clark, J. H., and Elings, J. A., *Green. Chem., 5*, 1 (2003).
11. Notari, B., *Stud. Surf. Sci. Catal., 37*, 413 (1988).
12. Perego, G., Bellussi, G., Corno, C., Taramasso, M., Buonomo, F., and Esposito, A., *Stud. Surf. Sci. Catal., 28*, 129 (1986).
13. Notari, B., *Adv. Catal., 41*, 253 (1996).
14. Sheldon, R. A., and van Doorrn, J. A., *J. Catal., 31*, 427 (1973).
15. Battioni, P., Renaud, J. P., Bartoli, J. F., Reina-Artiles, M., Fort, M., and Mansuy, D., *J. Am. Chem. Soc., 110*, 8462 (1988).
16. Romão, C. C., Kühn, F. E., and Herrmann, W. A., *Chem. Rev., 97*, 3197 (1997).
17. Venturello, C., Alneri, E., and Ricci, M., *J. Org. Chem., 48*, 3831 (1983).
18. Ishii, Y., Yamawaki, K., Ura, T., Yamada, H., Yoshida, T., and Ogawa, M., *J. Org. Chem., 53*, 3587 (1988).
19. Noyori, R., Aoki, M., and Sato, K., *Chem. Commun.*, 1977 (2003).

20. de Vos, D. E., Meinershagen, J. L., and Bein, T., *Angew. Chem. Int. Ed. Engl.*, *35*, 2211 (1996).

21. Lane, B. S., Vogt, M., DeRose, V. J., and Burgess, K., *J. Am. Chem. Soc.*, *124*, 11946 (2002).

22. White, M. C. G., Doyle, A., and Jacobsen, E. N., *J. Am. Chem. Soc.*, *123*, 7194 (2001).

23. Chen, K., Costas, M., and Que Jr. L., *J. Chem. Soc., Dalton Trans.*, 672 (2002).

24. Montanari, F., and Casella, L. (Eds.), *Metalloporphyrin Catalyzed Oxidations*, Kluwer, Dordrecht, 1994.

25. Okuhara, T., Mizuno, N., and Misono, M., *Adv. Catal.*, *41*, 113 (1996).

26. Mizuno, N., and Misono, M., *Chem. Rev.*, *98*, 199 (1998).

27. Neumann, R., *Prog. Inorg. Chem.*, *47*, 317 (1998).

28. Hill, C. L., Chrisina, C., and Prosser-McCartha, M., *Coord. Chem. Rev.*, *143*, 407 (1995).

29. Hill, C. L., in *Comprehensive Coordination Chemistry II* (McCleverty, J. A., and Meyer, T. J., Eds.), Vol. 4, p. 679, Elsevier, Amsterdam, 2004.

30. Kozhevnikov, I. V., *Chem. Rev.*, *98*, 171 (1998).

31. Kozhevnikov, I. V., *Catalysis by Polyoxometalates*, Wiley, Chichester, 2002.

32. Pope, M. T., in *Comprehensive Coordination Chemistry II* (McCleverty, J. A., and Meyer, T. J., Eds.), Vol. 4, p. 635, Elsevier, Amsterdam, 2004.

33. Pope, M. T., *Heteropoly and Isopoly Oxometalates*, Springer, Berlin, 1983.

34. Matveev, K. I., *Kinet. Katal.*, *18*, 862 (1977).

35. Kozhevnikov, I. V., and Matveev, K. I., *J. Mol. Catal.*, *5, 135* (*1983*).

36. Kozhevnikov, I. V., *J. Mol. Catal. A*, *117*, 151 (1997).

37. Khenkin, A. M., Rosenberger, A., and Neumann, R., *J. Catal.*, *182*, 82 (1999).

38. Khenkin, A. M., Weiner, L., Wang, Y., and Neumann, R., *J. Am. Chem. Soc.*, *123*, 8531 (2001).

39. Neumann, R., and Assael, I., *J. Chem. Soc. Chem. Commun.*, 1285 (1988).

40. Ben-Daniel, R., and Neumann, R., *Angew. Chem. Int. Ed.*, *42*, 92 (2003).

41. Shinachi, S., Yahiro, H., Yamaguchi, K., and Mizuno, N., *Chem. Eur. J.*, *10*, 6489 (2004).

42. Shinachi, S., Matsushita, M., Yamaguchi, K., and Mizuno, N., *J. Catal.*, *233*, 81 (2005).

43. Weinstock, I. A., Atalla, R. H., Reiner, R. S., Moen, M. A.,Hammel, K. F., Houtman, C. J., Hill, C. L., and Harrup, M. K., *J. Mol. Catal.*, *116*, 59 (1997).

44. Neumann, R., and Lissel, M., *J. Org. Chem.*, *54*, 4607 (1989).

45. Atlamsani, A., Brégeault, J.-M., and Ziyad, M., *J. Org. Chem.*, *58*, 5663 (1993).

46. Hamamoto, M., Nakayama, K., Nishiyama, Y., and Ishii, Y., *J. Org. Chem.*, *58*, 6421 (1993).

47. Lissel, M., in de Wal, H. J., and Neumann, R., *Tetrahedron Lett.*, *33*, 1795 (1992).

48. Neumann, R., Khenkin, A. M., and Vigdergauz, I., *Chem. Eur. J.*, *6*, 875 (2000).

49. Monflier, E., Blouet, E., Barbaux, Y., and Mortreux, A., *Angew. Chem. Int. Ed. Engl.*, *33*, 2100 (1994).

50. Passoni, L. C., Cruz, A. T., Buffon, R., and Shuchardt, U., *J. Mol. Catal. A*, *120*, 117 (1997).

51. Yokota, T., Tani, M., Sakaguchi, S., and Ishii, Y., *J. Am. Chem. Soc.*, *125*, 1476 (2003).

52. Yokota, T., Sakaguchi, S., and Ishii, Y., *Adv. Synth. Catal.*, *344*, 849 (2002).

53. Kozhevnikov, I. V., and Matveev, K. I., *Appl. Catal.*, *5*, 135 (1983).

54. Grate, J. H., *J. Mol. Catal. A*, *106*, 57 (1996).

55. Weinstock, I. A., Barbuzzi, E. M. G., Wemple, M. W., Cowan, J. J., Reiner, R. S., Sonnen, D. M., Heintz, R. A., Bond, J. S., and Hill, C. L., *Nature*, *414*, 191 (2001).

56. Hill, C. L., and Brown Jr., R. B., *J. Am. Chem. Soc.*, *108*, 536 (1986).

57. Hill, C. L., in *Activation and Functionalization of Alkanes*, (Hill, C. L., Ed.), p. 243, Wiley, New York, 1989.

58. Neumann, R., and Dahan, M., *Nature*, *388*, 353 (1997).

59. Neumann, R., and Dahan, M., *J. Am. Chem. Soc.*, *120*, 11969 (1998).

60. Mizuno, N., Nozaki, C., Kiyoto, I., and Misono, M., *J. Am. Chem. Soc.*, *120*, 9267 (1998).

61. Nishiyama, Y., Nakagawa, Y., and Mizuno, N., *Angew. Chem. Int. Ed.*, *40*, 3639 (2001).

62. Weiner, H., and Finke, R. G., *J. Am. Chem. Soc.*, *121*, 9831 (1999).

63. Neumann, R., and Gara, M., *J. Am. Chem. Soc.*, *117*, 5066 (1995).

64. Adam, W., Alsters, P. L., Neumann, R., Saha-Möller, C. R., Sloboda-Rozner, D., and Zhang, R., *J. Org. Chem.*, *68*, 1721 (2003).

65. Sloboda-Rozner, D., Saha-Möller, C. R., and Neumann, R., *J. Am. Chem. Soc.*, *125*, 5280 (2003).

66. Bösing, M., Nöh, A., Loose, I., and Krebs, B., *J. Am. Chem. Soc.*, *120*, 7252 (1998).

67. Zhang, X., Chen, Q., Duncan, D. C., Lachicotte, R. J., and Hill, C. L., *Inorg. Chem.*, *36*, 4381 (1997).

68. Zhang, X., Chen, Q., Duncan, D. C., Campana, C. F., and Hill, C. L., *Inorg. Chem.*, *36*, 4208 (1997).

69. Mbomekalle, I. M., Keita, B., Nadjo, L., Berthet, P., Neiwert, W. A., Hill, C. L., Ritorto, M. D., and Anderson, T. M., *Dalton Trans.*, 2646 (2003).
70. Zhang, X., Anderson, T. M., Chen, Q., and Hill, C. L., *Inorg. Chem.*, *40*, 418 (2001).
71. Anderson, T. M., Hardcastle, K. I., Okun, N., and Hill, C. L., *Inorg. Chem.*, *41*, 6418 (2002).
72. Ritorto, M. D., Anderson, T. M., Neiwert, W. A., and Hill, C. L., *Inorg. Chem.*, *43*, 44 (2004).
73. Mizuno, N., Nozaki, C., Hirose, T., Tateishi, M., and Iwamoto, M., *J. Mol. Catal. A: Chem.*, *117*, 159 (1997).
74. Hayashi, T., Kishida, A., and Mizuno, N., *Chem. Commun.*, 381 (2000).
75. Mizuno, N., Seki, Y., Nishiyama, Y., Kiyoto, I., and Misono, M., *J. Catal.*, *184*, 550 (1999).
76. Ben-Daniel, R., Khenkin, A. M., and Neumann, R., *Chem. Eur. J.*, *6*, 3722 (2000).
77. Khenkin, A. M., and Neumann, R., *Inorg. Chem.*, *39*, 3455 (2000).
78. Ishii, Y., Yamawaki, K., Yoshida, T., Ura, T., and Ogawa, M., *J. Org. Chem.*, *52*, 1868 (1987).
79. Sakaguchi, S., Nishiyama, Y., and Ishii, Y., *J. Org. Chem.*, *61*, 5307 (1996).
80. Oguchi, T., Sakata, Y., Takeuchi, N., Kaneda, K., Ishii, Y., and Ogawa, M., *Chem. Lett.*, 2053 (1989).
81. Sakaue, S., Sakata, Y., Nishiyama, Y., and Ishii, Y., *Chem. Lett.*, 289 (1992).
82. Ishii, Y., and Sakata, Y., *J. Org. Chem.*, *55*, 5545 (1990).
83. Venturello, C., D'Aloisio, R., Bart, J. C. J., and Ricci, M., *J. Mol. Catal.*, *32*, 107 (1985).
84. Aubry, C., Chottard, G., Platzer, N., Brégeault, J.-M., Thouvenot, R., Chauveau, F., Huet, C., and Ledon, H., *Inorg. Chem.*, *30*, 4409 (1991).
85. Dengel, A. C., Griffith, W. P., and Parkin, B. C., *J. Chem. Soc., Dalton Trans.*, 2683 (1993).
86. Salles, L., Aubry, C., Thouvenot, R., Robert, F., Dorémieux-Morin, C., Chottard, G., Ledon, H., Jeannin, Y., and Brégeault, J.-M., *Inorg. Chem.*, *33*, 871 (1994).
87. Duncan, D. C., Chambers, R. C., Hecht, E., and Hill, C. L., *J. Am. Chem. Soc.*, *117*, 681 (1995).
88. Sato, K., Aoki, M., and Noyori, R., *Science*, *281*, 1646 (1998).
89. Sato, K., Aoki, M., Ogawa, M., Hashimoto, T., and Noyori, R., *J. Org. Chem.*, *61*, 8310 (1996).
90. Sato, K., Aoki, M., Ogawa, M., Hashimoto, T., Panyella, D., and Noyori, R., *Bull. Chem. Soc. Jpn.*, *70*, 905 (1997).
91. Sato, K., Aoki, M., Takagi, J., and Noyori, R., *J. Am. Chem. Soc.*, *119*, 12386 (1997).
92. Sato, K., Aoki, M., Takagi, J., Zimmermann, K., and Noyori, R., *Bull. Chem. Soc. Jpn.*, *72*, 2287 (1999).
93. Kamata, K., Yamaguchi, K., Hikichi, S., and Mizuno, N., *Adv. Synth. Catal.*, *345*, 1193 (2003).
94. Kamata, K., Yamaguchi, K., and Mizuno, N., *Chem. Eur. J.*, *10*, 4728 (2004).
95. Gelbard, G., Raison, F., Lachter, E. R., Thouvenot, R., Ouahab, L., and Grandjean, D., *J, Mol. Catal. A*, *114*, 77 (1996).
96. Server-Carrió, J., Bas-Serra, J., González-Núnez, M. E., García-Gastaldi, A., Jameson, G. B., Baker, L. C. W., and Acerete, R., *J. Am. Chem. Soc.*, *121*, 977 (1999).
97. Schwegler, M., Floor, M., van Bekkum, H., *Tetrahedron Lett.*, *29*, 823 (1988).
98. Tézé, A., and Hervé, G., *Inorg. Synth.*, *27*, 85 (1990).
99. Kamata, K., Yonehara, K., Sumida, Y., Yamaguchi, K., Hikichi, S., and Mizuno, N., *Science*, *300*, 964 (2003).
100. Kamata, K., Nakagawa, Y., Yamaguchi, K., and Mizuno, N., *J. Catal.*, *224*, 224 (2004).
101. Cheng, W.-C., and Luthra, N. P., *J. Catal.*, *109*, 163 (1988).
102. Nowinska, K., Fiedorow, R., and Adamiec, J., *J. Chem. Soc. Faraday Trans.*, *87*, 749 (1991).
103. Rao, K. M., Gobetto, R., Lannibello, A., and Zecchina, A., *J. Catal.*, *119*, 512 (1989).
104. van Veer, J. A. R., Hendriks, P. A. J. M., Andrea, R. R., Romers, E. J. M., and Wilson, A. E., *J. Phys. Chem.*, *94*, 1831 (1990).
105. Izumi, Y., and Urabe, K., *Chem. Lett.*, 663 (1981).
106. Schwegler, M. A., van Bekkum, and H. de Munok, N. A., *Appl. Catal.*, *92*, 80 (1992).
107. Izumi, Y., Urabe, K., and Onaka, M., *Zeolite, Clay, and Heteropolyacids in Organic Reactions*, Kodansha, Tokyo, VCH, Weinheim, 1992.
108. Neumann, R., and Levin, M., *J. Org. Chem.*, *56*, 5707 (1991).
109. Fujibayashi, S., Nakayama, K., Hamamoto, M., Sakaguchi, S., Nishiyama, Y., and Ishii, Y., *J. Mol. Catal. A*, *110*, 105 (1996).
110. Hanyu, A., Sakurai, Y., Fujibayashi, S., Sakaguchi, S., and Ishii, Y., *Tetrahedron Lett.*, *38*, 5659 (1997).
111. Gall, R. D., Hill, C. L., and Walker, J. E., *J. Catal.*, *159*, 473 (1996).

112. Gall, R. D., Hill, C. L., and Walker, J. E., *Chem. Mater.*, *8*, 2523 (1996).
113. Xu, L., Boring, E., and Hill, C. L., *J. Catal.*, *195*, 394 (2000).
114. Kurusu, Y., *Bull. Chem. Soc. Jpn.*, *54*, 293 (1981).
115. Inoue, M., Itoi, Y., Enomoto, S., and Watanabe, Y., *Chem. Lett.*, 1375 (1982).
116. Kharat, A. N., Pendleton, P., Badalyan, A., Abedini, M., and Amini, M. M., *J. Mol. Catal. A*, *175*, 277 (2001).
117. Khenkin, A. M., Neumann, R., Sorokin, A. B., and Tuel, A., *Catal. Lett.*, *63*, 189 (1999).
118. Kozhevnikov, I. V., Sinnema, A., Jansen, R. J. J., Pamin, K., and van Bekkum, H., *Catal. Lett.*, *30*, 241 (1995).
119. Ichihara, J., *Tetrahedron Lett.*, *42*, 695 (2001).
120. Ichihara, J., Yamaguchi, S., Nomoto, T., Nakayama, H., Iteya, K., Naitoh, N., and Sasaki, Y., *Tetrahedron Lett.*, *43*, 8231 (2002).
121. Ichihara, J., Iteya, K., Kambara, A., and Sasaki, Y., *Catal. Today*, *87*, 163 (2003).
122. Sasaki, Y., Ushimaru, K., Iteya, K., Nakayama, H., Yamaguchi, S., and Ichihara, J., *Tetrahedron Lett.*, *45*, 9513 (2004).
123. Elliott, J. C., *Structure and Chemistry of the Apatites and Other Calcium Orthophosphates*, Elsevier, Amsterdam, 1994.
124. Guo, Y.-H., and Hu, C.-W., *J. Cluster Sci.*, *14*, 505 (2003).
125. Okun, N. M., Anderson, T. M., and Hill, C. L., *J. Am. Chem. Soc.*, *125*, 3194 (2003).
126. Okun, N. M., Anderson, T. M., and Hill, C. L., *J. Mol. Catal. A*, *197*, 283 (2003).
127. Okun, N. M., Ritorto, M. D., Anderson, T. M., Apkarian, R. P., and Hill, C. L., *Chem. Mater.*, *16*, 2551 (2004).
128. Misono, M., *Catal. Rev. Sci. Eng.*, *29*, 269 (1987).
129. Okuhara, T., Nishimura, T., and Misono, M., *Chem. Lett.*, 155 (1995).
130. Yoshinaga, Y., Seki, K., Nakato, N., and Okuhara, T., *Angew. Chem. Int. Ed. Engl.*, *36*, 2833 (1997).
131. Yamada, T., Yoshinaga, Y., and Okuhara, T., *Bull. Chem. Soc. Jpn.*, *71*, 2727 (1998).
132. Inumaru, K., Nakajima, M., Ito, T., and Misono, M., *Chem. Lett.*, 559 (1996).
133. Phule, J. T., Neiwert, W. A., Hardcastle, K. I., Do, B. T., and Hill, C. L., *J. Am. Chem. Soc.*, *123*, 12101 (2001).
134. Yamaguchi, K., and Mizuno, N., *New J. Chem.*, *26*, 972 (2002).
135. Xi, Z., Zhou, N., Sun, Y., and Li, K., *Science*, *292*, 1139 (2001).
136. Sun, Y., Xi, Z., and Cao, G., *J. Mol. Catal. A*, *166*, 219 (2001).
137. Xi, Z., Wang, H., Sun, Y., Zhou, N., Cao, G., and Li, M., *J. Mol. Catal. A*, *168*, 299 (2001).
138. Yamada, Y. M. A., Ichinohe, M., Takahashi, H., and Ikegami, S., *Org. Lett.*, *3*, 1837 (2001).
139. Yamada, Y. M. A., Tabata, H., Ichinohe, M., Takahashi, H., and Ikegami, S., *Tetrahedron*, *60*, 4087 (2004).
140. Haimov, A., Cohen, H., and Neumann, R., *J. Am. Chem. Soc.*, *126*, 11762 (2004).
141. Vasylyev, M. V., and Neumann, R., *J. Am. Chem. Soc.*, *126*, 884 (2004).
142. Cavani, F., Trifiró, F., and Vaccari, A., *Catal. Today*, *11*, 173 (1991).
143. Narita, E., Kaviratna, P., and Pinnavaia, *Chem. Lett.*, 805 (1991).
144. Kwon, T., Tsigdinos, G. A., and Pinnavaia, *J. Am. Chem. Soc.*, *110*, 3653 (1988).
145. Tatsumi, T., Yamamoto, K., Tajima, H., and Tominaga, H., *Chem. Lett.*, 815 (1992).
146. Sels, B. F., de Vos, D. E., Buntinx, M., Pierard, F., Mesmaeker, A. K.-D., and Jacobs, P. A., *Nature*, *400*, 855 (1999).
147. Sels, B. F., de Vos, D. E., and Jacobs, P. A., *J. Am. Chem. Soc.*, *123*, 8350 (2001).
148. Sels, B. F., de Vos, D. E., and Jacobs, P. A., *Angew. Chem. Int. Ed.*, *44*, 310 (2005).
149. Sels, B. F., de Vos, D. E., and Jacobs, P. A., *Tetrahedron Lett.*, *37*, 8557 (1996).
150. van Laar, F., de Vos, D. E., Vanoppen, D. L., Sels, B. F., Jacobs, P. A., Pierard, A., and Kirsch-De Mesmaeker, A., *Chem. Commun.*, 267 (1998).
151. Sels, B. F., de Vos, D. E., Grobet, P. T., Pierard, F., Kirsch-De Mesmaeker, A., and Jacobs, P. A., *J. Phys. Chem. B*, *103*, 11114 (1999).
152. de Vos D. E., Sels, B. F., van Rhijn, W. M., and Jacobs, P. A., *Stud. Surf. Sci. Catal.*, *130*, 137 (2000).
153. Leadbeater, N. E., and Marco, M., *Chem. Rev.*, *102*, 3217 (2002).
154. McNamara, C. A., Dixon, M. J., and Bradley, M., *Chem. Rev.*, *102*, 3275 (2002).
155. Bergbreiter, D. E., *Chem. Rev.*, *102*, 3345 (2002).
156. Wight, A. P., and Davis, M. E., *Chem. Rev.*, *102*, 3589 (2002).

157. de Vos, D. E., Dams, M., Sels, B. F., and Jacobs, P. A., *Chem. Rev.*, *102*, 3615 (2002).
158. Strinvasan, S., and Ford, W. T., *New J. Chem.*, *15*, 693 (1991).
159. Villa, A. L., Sels, B. F., de Vos, D. E., and Jacobs, P. A., *J. Org. Chem.*, *64*, 7267 (1999).
160. Sels, B. F., Villa, A. L., Hoegaerts, D., de Vos, D. E., and Jacobs, P. A., *Top. Catal.*, *13*, 223 (2000).
161. Neumann, R., and Miller, H., *J. Chem. Soc. Chem. Commun.*, 2277 (1995).
162. Hoegaerts, D., Sels, B. F., de Vos, D. E., Verpoort, F., and Jacobs, P. A., *Catal. Today*, *60*, 209 (2000).
163. Yamaguchi, K., Yoshida, C., Uchida, S., and Mizuno, N., *J. Am. Chem. Soc.*, *127*, 530 (2005).
164. Gelbard, G., Breton, F., Quenard, M., and Sherrington, D. C., *J. Mol. Catal. A*, *153*, 7 (2000).
165. Duprey, E., Maquet, J., Man, P. P., Manoli, J. M., Delamar, M., and Brégeault, J.-M., *Appl. Catal. A*, *128*, 89 (1995).
166. Liu, Y., Murata, K., and Inaba, M., *Green Chem.*, *6*, 510 (2004).
167. Neumann, R., and Cohen, M., *Angew. Chem. Int. Ed. Engl.*, *36*, 1738 (1997).
168. Sakamoto, T., and Pac, C., *Tetrahedron Lett.*, *41*, 10009 (2000).
169. Okuhara, T., *Chem. Rev.*, *102*, 3641 (2002).
170. Yamada, T., Yoshinaga, Y., and Okuhara, T., *Bull. Chem. Soc. Jpn.*, *71*, 2727 (1998).
171. Nakato, T., Kimura, M., Nakata, S., and Okuhara, T., *Langmuir*, *14*, 319 (1998).
172. Okuhara, T., Watanabe, H., Nishimura, T., Inumaru, K., and Misono, M., *Chem. Mater.*, *12*, 2230 (2000).
173. Okuhara, T., Mizuno, N., and Misono, M., *Appl. Catal. A*, *222*, 63 (2001).
174. Okuhara, T., *Appl. Catal. A*, *256*, 213 (2003).
175. Yoshinaga, Y., and Okuhara, T., *J. Chem. Soc. Faraday Trans.*, *95*, 2235 (1998).
176. Yoshinaga, Y., Seki, K., Nakato, T., and Okuhara, T., *Angew. Chem. Int. Ed.*, *36*, 2833 (1997).
177. Inumaru, K., Ito, T., and Misono, M., *Micropor. Mesopor. Mater.*, *21*, 629 (1998).
178. Ito, T., Inumaru, K., and Misono, M., *Chem. Mater.*, *13*, 824 (2001).
179. Ito, T., Inumaru, K., and Misono, M., *J. Phys. Chem. B*, *101*, 9958 (1997).
180. Ito, T., Hashimoto, M., Uchida, S., and Mizuno, N., *Chem. Lett.*, 1272 (2001).
181. Parent, M. A., and Moffat, J. B., *Langmuir*, *12*, 3733 (1996).
182. Volkmer, D., Bredenkötter, B., Tellenbröker, J., Kögerler, P., Kurth, D. G., Lehmann, P., Schnablegger, H., Schwahn, D., Pipenbrink, M., and Krebs, B., *J. Am. Chem. Soc.*, *124*, 10489 (2002).
183. Kang, Z., Wang, E., Jiang, M., Lian, S., Li, Y., and Hu, C., *Eur. J. Inorg. Chem.*, 370 (2003).
184. Khan, M. I., Yohannes, E., and Powell, D., *Inorg. Chem.*, *38*, 212 (1999).
185. Khan, M. I., Yohannes, E., and Doedens, R. J., *Angew. Chem. Int. Ed.*, *38*, 1292 (1999).
186. Hagrman, D., Hagrman, P. J., and Zubieta, J., *Angew. Chem. Int. Ed.*, *38*, 3165 (1999).
187. Son, J. H., Choi, H., and Kwon, Y. U., *J. Am. Chem. Soc.*, *122*, 7432 (2000).
188. Schmitt, W., Baissa, E., Mandel, A., Anson, C. E., and Powell, A. K., *Angew. Chem. Int. Ed.*, *40*, 3577 (2001).
189. Uchida, S., Hashimoto, M., and Mizuno, N., *Angew. Chem. Int. Ed.*, *41*, 2814 (2002).
190. Uchida, S., and Mizuno, N., *Chem. Eur. J.*, *9*, 5850 (2003).
191. Uchida, S., and Mizuno, N., *J. Am. Chem. Soc.*, *126*, 1602 (2004).
192. Ishii, Y., Takenaka, Y., and Konishi, K., *Angew. Chem. Int. Ed.*, *43*, 2702 (2004).
193. Müller, A., Krickemeyer, E., Bögge, H., Schmidtmann, M., Beugholt, C., Das, S. K., and Peters, F., *Chem. Eur. J.*, *5*, 1496 (1999).
194. Müller, A., Shah, S. Q. N., Bögge, H., and Schmidtmann, M., *Nature*, *397*, 48 (1999).
195. Müller, A., Krickemeyer, E., Bögge, H., Schmidtmann, M., Botar, B., and Talismanova, M. O., *Angew. Chem. Int. Ed.*, *42*, 2085 (2003).
196. Müller, A., Krickemeyer, E., Bögge, H., Schmidtmann, M., Körgerler, P., Rosu, C., and Bockmann, E., *Angew. Chem. Int. Ed.*, *40*, 4034 (2001).
197. Tianbo, L., Diemann, E., Li, H., Dress, A. W. M., and Müller, A., *Nature*, *426*, 59 (2003).
198. Volkmer, D., Chesne, A. D., Kurth, D. G., Schnablegger, H., Lehmann, P., Koop, M. J., and Müller, A., *J. Am. Chem. Soc.*, *122*, 1995 (2000).
199. Salignac, B., Riedel, S., Dolbecq, A., Sécheresse, F., and Cadot, E., *J. Am. Chem. Soc.*, *122*, 10381 (2000).
200. Cadot, E., Marrot, J., and Sécheresse, F., *Angew. Chem. Int. Ed.*, *40*, 774 (2001).
201. Contant, R., and Tézé, A., *Inorg. Chem.*, *24*, 4610 (1985).
202. Hussain, F., Bassil, B. S., Bi, L.-H., Reiche, M., and Kortz, U., *Angew. Chem. Int. Ed.*, *43*, 3485 (2004).

CHAPTER 13 QUESTIONS

Question 1

Provide the 12 principles of "green chemistry."

Question 2

There are some ways to evaluate environmental burden in chemical synthetic processes. The atom economy is the fraction of the atoms in the product to the fraction of all atoms in the reactants. The E factor is the ratio of the mass of all waste to the mass of product. For the following reaction concerning the production of propylene oxide, calculate the theoretical atom economy and E factor.

Question 3

H_2O_2 and O_2 are desirable as oxidants for oxidation reactions? Give reasons.

Question 4

Various types of polyoxometalates are effective catalysts for the H_2O_2- and O_2-based oxidations. Give some examples.

Question 5

Polyoxometalate-catalyzed oxidations are carried out in homogeneous or heterogeneous systems. Write advantages and disadvantages of homogeneous and heterogeneous systems, respectively.

Question 6

The system consisting of $[PMo_{12-n}V_nO_{40}]^{(3+n)-}$ and Pd^{2+} salts can catalyze the Wacker-type oxidation. What is the role of $[PMo_{12-n}V_nO_{40}]^{(3+n)-}$?

Asymmetric Catalysis by Heterogeneous Catalysts

Simona M. Coman, Georges Poncelet, and Vasile I. Pârvulescu

CONTENTS

14.1 Introduction 493
14.2 Principles of Stereodifferentiation and Asymmetric Catalysis 496
14.3 Historical Developments 499
14.4 Tartaric Acid-Modified Me/Support Hydrogenation Catalysts and Related Systems 502
    14.4.1 Catalyst Preparation Process 502
    14.4.2 The Modification Process..503
    14.4.3 Substrate and Hydrogenation Parameters..503
    14.4.4 Mechanistic Investigations and Hypotheses for Enantioselection 504
    14.4.5 Proposed Mechanisms 507
14.5 Chiral-Modified Platinum Hydrogenation Catalysts and Related Systems.........510
    14.5.1 Kinetic Models 512
    14.5.2 Mechanistic Investigations 512
14.6 Heterogeneized Homogeneous Catalysts...517
14.7 Organic Polymers 519
14.8 Diastereoselective Catalysis 519
14.9 Conclusions 523
References 523
Chapter 14 Questions 531

14.1 INTRODUCTION

Enantioselective heterogeneous catalytic reactions are of growing interest, as optically pure chiral compounds are of great importance in different areas like fine chemical, pharmaceutical, agrochemical, flavor, and fragrance industries.

It is now well recognized that different enantiomers can possess different physiological properties, sometimes the wrong enantiomer being a ballast, sometimes a pollutant, and sometimes a structure with an opposite biological activity. From the application point of view, the last case is evidently the most dangerous. Unfortunately, there are many examples of drugs manufactured and sealed as racemic mixture without a previously detailed characterization of both isomers. Such was the case in the tragic example of thalidomide. In the late 1950s, thalidomide was prescribed for controlling morning sickness during pregnancy, but by 1952 the drug was withdrawn from the market after thousands of deformed children were born to mothers who had taken the drug. Later, it was discovered that D-thalidomide exhibited safe, therapeutic effects but L-thalidomide was a potent mutagen. This represented perhaps the first application example showing the importance of obtaining pure optical isomers. However, literature reports many other examples in which the enantiomers exhibit opposite biological activity. The S-isomer of penicillamine is a useful antiarthritic drug, whereas the R-isomer is highly toxic. The S-isomer of propranolol is a beta-blocker drug useful in the treatment of heart diseases, whereas the R-isomer is an active contraceptive. But the pharmaceutical industry is not the only one to know such examples. In the flavor industry, for example, it is very well known that the S-isomer of limonene exhibits the odor of lemon, whereas the R-isomer has the odor of orange. These only represent a few examples which show how important are the asymmetric syntheses producing only a single enantiomer of a chiral compound. In accordance, the Food and Drug Administrations were obliged to move to a regulatory philosophy enforcing the delivery of only the therapeutically active optical isomer for a given drug. This philosophy determined minimization or even elimination of any trace of other isomers, even in instances where these are toxicologically harmless [1].

Among the various methods proposed to produce selectively isolated enantiomers, the use of enantioselective catalysis is by far the most attractive one. The control of stereochemistry by use of a minute amount of an asymmetric catalyst offers clear advantages. Therefore, the design and development of catalytic enantioselective organic reactions is considered as one of the most attractive and challenging frontiers in synthetic organic chemistry.

A large number of reactions have been successfully subjected to asymmetric catalysis in homogeneous phase and many of these are already applied in industry. An important example in this sense is the development of an asymmetric hydrogenation catalysts based on rhodium complexes with chiral phosphine ligands (DIPAMP) and their application in the manufacture of L-DOPA [2], the process occurring with 95% enantiomeric excess (e.e.). However, from the viewpoint of scope of applications and industrial potential, BINAP, another chiral phosphine ligand, is the most important [3], being used in many asymmetric catalytic processes. It is, for example, highly efficient (97% e.e.) for the synthesis of (S)-naproxen based on the asymmetric hydrogenation of the unsaturated precursors and for the synthesis of L-menthol based on the asymmetric isomerization of diethylgeranylamine to citronellaldiethylamine (99% e.e.) [4].

It is now clear that asymmetric catalytic hydrogenation is rather successful. However, the initial research work of Sharpless [5] in the asymmetric epoxidation, followed by the results of Jacobsen et al. [6] opened large opportunities for liquid-phase asymmetric oxidation. Sharpless epoxidation has been widely applied in bench-scale organic synthesis, and more recently, salene derivatives emerged among the most effective catalysts in this reaction [7,8].

These are only few examples of successful asymmetric processes in homogeneous catalysis but, in the last four decades, this chemistry has rapidly grown from the level of an academic curiosity to an essential research and, in some cases, even to a giant-scale industrial production [9,10]. In spite of the fact that asymmetric catalysis is undoubtedly one of the most promising methods for these complicate syntheses [11], industrial applications are surprisingly still underdeveloped. The main reason is not associated with qualitative aspects of the transformations where high e.e.'s and conversions are obtained for a wide variety of reactions but to rather high costs of the ligand or metal. These drawbacks could be avoided if the catalyst is separated and reused after reaction. An evident way for achieving it is the use of heterogeneous catalysis, allowing easy work-up procedures and possible recycling.

However, the use of the heterogeneous catalysts in applicative enantioselective syntheses has a limited success. Several factors contribute to this situation: (1) a long time is required to achieve an effective heterogeneous enantioselective catalyst compared with the homogeneous ones, (2) a more complex structure of the heterogeneous catalyst surface on which centers coexist with different catalytic activity and selectivity, which can lead to undesired secondary reactions, and (3) an increased difficulty to create an effective asymmetric environment and to accommodate it with the multitude of reactions that are interesting to be carried out under enantioselective restrictions.

In fact, there are only two heterogeneous catalysts that reliably give high enantioselectivities (e.s.'s) (90% e.e. or above). These are Raney nickel (or Ni/SiO$_2$) system modified with tartaric acid (TA) or alanine for hydrogenation of β-ketoesters [12–30], and platinum-on-charcoal or platinum-on-alumina modified with cinchona alkaloids for the hydrogenation of α-ketoesters [31–73].

To overcome the mentioned disadvantages, several approaches have been envisaged. Among these, the heterogenization of a homogeneous catalyst is trying to combine the advantages of heterogeneous and homogeneous catalysis. Using this approach, a homogeneous catalyst is bonded to an inert surface support such as silica or a resin by different techniques such as: (1) formation of a covalent bond with a ligand; (2) ion-pair formation; and (3) entrapment. In such heterogenized systems the catalytically active species resembles its homogeneous counterpart in solution, whereas the catalyst can be separated from the product by filtration, similar to a truly heterogeneous system. Although in principle such an idea is extremely attractive, gave many examples in which heterogenization of homogeneous catalysts resulted in a significant loss of activity and asymmetric induction capability with respect to that of homogeneous phases is found in the literature [74]. Binding of a ligand to a solid via a covalent bond has become the most frequently employed method of heterogenization of an enantioselective catalyst. The simplest method available is the grafting of a suitably functionalized monomeric ligand onto a support. This method benefits from being able to use a well-characterized monomer in the grafting with a well-characterized, often commercially available polymer, leading to an immobilized ligand of known origin. Furthermore, the same functionalized monomer may be used to graft onto a variety of supports, and therefore, this method is widely applicable. An alternative method of immobilization is to perform a copolymerization reaction with a monomer functionalized with, for example, a styrene unit. However, this method has the drawback that a particular monomer may only be suited to a certain type of polymer. A third, less desirable method, is to prepare the ligand on the polymer support [75].

One of the problem that is frequently questionable for heterogenized asymmetric catalysts on inorganic support is the leaching of the complex due to instability of the silicon-complex bond. Understanding this aspect, a considerable progress was achieved in the last years leading to highly stable enantioselective catalysts. In this context, Corma et al. [76,77] anchored chiral transition metal complexes on USY zeolites, starting from β-aminoalcohols such as prolinol. The zeolite-supported Ni complexes catalyzes the conjugate addition of ZnEt$_2$ to enones, with similar yields of the β-ethylated ketones as in homogeneous conditions [78,79].

Metal oxides were also chirally modified and few of them showed a significant or at least useful e.s. Thus, while Al$_2$O$_3$/alkaloid [80] showed no enantiodifferentiation, Zn, Cu, and Cd tartrate salts were quite selective for a carbene addition (45% e.e.) [81] and for the nucleophilic ring opening of epoxides (up to 85% e.e.) [82]. Recently, it was claimed that β-zeolite, partially enriched in the chiral polymorph A, catalyzed the ring opening of an epoxide with low but significant e.s. (5% e.e.) [83]. All these catalysts are not yet practically important but rather demonstrate that amorphous metal oxides can be modified successfully.

Alternatively, one might perform reactions with a nonenantioselective catalyst, and with a substrate that already contains at least one stereogenic center besides the prochiral group. Molecules with a prochiral group and one or more stereocenters are frequently encountered in complex organic synthesis, and these molecules can be used as such in diastereoselective transformations. Alternatively, relatively simple molecules with a prochiral group can be derivatized with a chiral auxiliary (i.e., an optically active molecule that is used in stoichiometric amounts to orchestrate asymmetric induction at a newly formed stereogenic center without being incorporated into the product). Therefore, chiral auxiliaries provide another alternative to asymmetric catalysis. In principle, chiral auxiliaries suffer from disadvantages that additional steps are required for attaching and cleaving the chiral auxiliary, and that a stoichiometric by-product is formed that needs to be separated after the auxiliary cleavage [84]. Even so, chiral auxiliaries can be useful for stereoselective synthesis of pharmaceutical intermediates on a large scale. Diastereoselective hydrogenations have been surveyed rather recently in two excellent reviews by Besson and Pinel [85] and De Vos and coworkers [86].

14.2   PRINCIPLES OF STEREODIFFERENTIATION AND ASYMMETRIC CATALYSIS

The enantiomers (i.e., stereoisomers that are not superimposable on their mirror images) of a racemate have identical chemical properties but they can exhibit different chemical reactivities when placed in a chiral environment. This means that a chiral substance (i.e., a molecule of its image and mirror image are not superimposable), such as an enzyme, is able to differentiate between two enantiomers on the basis of their different stereochemical configuration (i.e., the actual three-dimensional spatial arrangement of the atoms in a molecule). This phenomenon is referred to as chiral discrimination or chiral recognition and can also be observed in interactions between certain achiral molecules (i.e., a molecule of its image and mirror image are superimposable) and a chiral reagent or catalyst.

There are two possible approaches for the preparation of optically active products by chemical transformation of optically inactive starting materials: kinetic resolution and asymmetric synthesis [44,87]. For both types of reactions there is one principle: in order to make an optically active compound we need another optically active compound. A kinetic resolution depends on the fact that two enantiomers of a racemate react at different rates with a chiral reagent or catalyst. Accordingly, an asymmetric synthesis involves the creation of an asymmetric center that occurs by chiral discrimination of equivalent groups in an achiral starting material. This can be done either by enantioselective (which involves the reaction of a prochiral molecule with a chiral substance) or diastereoselective (which involves the preferential formation of a single diastereomer by the creation of a new asymmetric center in a chiral molecule) synthesis.

The e.e., also called optical yield (OY), is defined as the selectivity of an enantioselective reaction and is expressed as e.e. It can be calculated from the formula

$$\text{e.e.} = OY(\%) = ([R] - ([S])([R] + [S])$$

where [R] is the concentration of R-isomer and [S] is the concentration of S-isomer.

As mentioned above, the formation of an e.e. is a kinetic phenomenon. This assumes that in a nonchiral environment, the activation energy for the formation of two enantiomers is same while in a chiral environment, there is a difference in the activation energy between two diastereomeric transition states leading to preferential formation of one enantiomer. In other words, the difference in activation energy can be rationalized by interaction of a chiral agent with the substrates in the product-determining step. At 25°C, an energy difference of 2.66 kJ/mol leads to about 98% (99%:1%) e.e. (Figure 14.1).

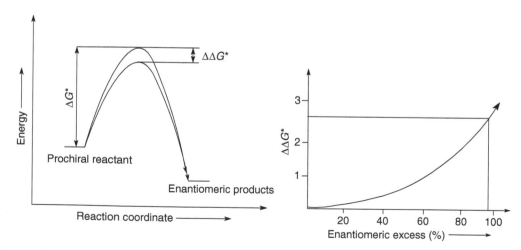

Figure 14.1   Reaction coordinate of an enantioselective synthesis.

The kinetic resolution and enantioselective syntheses are illustrated in Scheme 14.1.

In the enantioselective synthesis, the asymmetry (i.e., the stereoselectivity) is induced by the external chiral catalyst, while the diastereoselective synthesis does not require a chiral catalyst. The stereogenic center already present in the molecule is able to induce stereoselectivity, assuming that the synthesis starts with a single enantiomer. For instance, imagine that an α,β-substituted product is formed, and that the reactant already contains a stereogenic carbon at α. If the reaction of (αS) leads, e.g., largely to (αS, βR) and hardly to the (αS, βS) diastereomer (i.e., stereoisomers that are not mirror-images of each other), the reaction is diastereoselective (Scheme 14.2).

Starting from a racemic mixture of (αS) and (αR), the same diastereoselective reaction will mainly lead to the racemic mixture of the enantiomers (αS, βR) and (αR, βS). The minor products are then (αS, βS) and (αR, βR). Diastereoselectivity only leads to enantiomerically pure products if the starting product is enantiomerically pure. With a nonenantioselective catalyst, the absolute value of the diastereomeric excess (d.e.) (i.e., the mole fraction expressing the ratio of two diastereomers in a mixture, analogous to e.e. for enantiomers) is the same whether starting from (αS), (αR), or a 50:50 mixture of (αS) and (αR). In contrast, if the catalyst is itself enantioselective, the d.e. values may well be different depending on the configuration in the starting product. Scheme 14.2 gives an example of enantio- vs. diastereoselective process.

The energy profiles for an enantioselective and a diasteroselective synthesis are compared in Figure 14.2. An interesting feature of the asymmetric catalytic synthesis is the nonlinear correlation between the optical purity of the chiral catalyst or auxiliary and that of the reaction product, reported

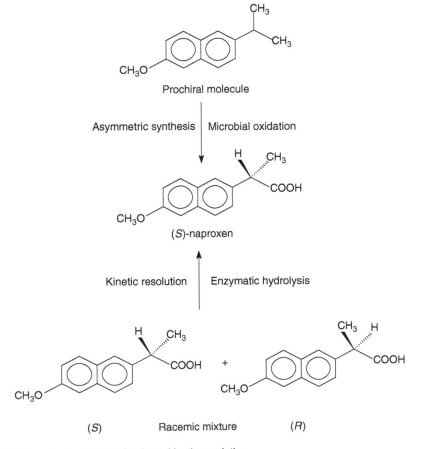

Scheme 14.1 Enantioselective synthesis vs. kinetic resolution.

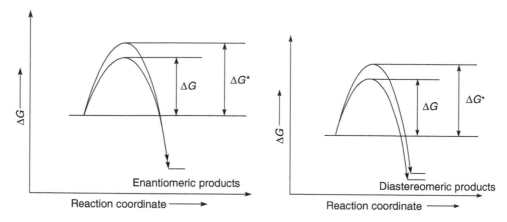

Figure 14.2 Energy profiles for enantioselective and diastereoselective syntheses.

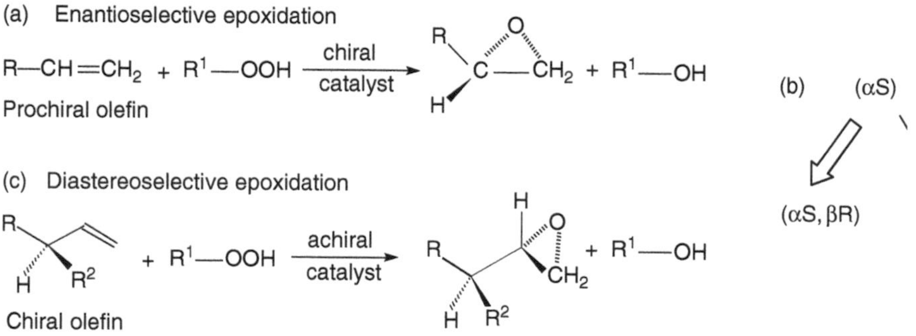

(a) Enantioselective epoxidation

R—CH=CH₂ + R¹—OOH $\xrightarrow[\text{catalyst}]{\text{chiral}}$ (epoxide product) + R¹—OH

Prochiral olefin

(b) (αS)

(c) Diastereoselective epoxidation

(chiral olefin) + R¹—OOH $\xrightarrow[\text{catalyst}]{\text{achiral}}$ (product) + R¹—OH

Chiral olefin

(αS, βR)

Scheme 14.2 (a) Enantio- vs. (b) Concept of the diastereoselective synthesis, (c) diastereoselective epoxidation.

for the first time by Puchot and coworkers [88] and observed by different other research groups [89–91]. Such an effect may be exemplified by the fact that in some cases the products of an asymmetric catalytic reaction exhibit an e.e. higher than the enantiopurity of the chiral catalyst, and this departure from proportionality is termed as a *positive asymmetric amplification* (Figure 14.3).

Kagan's group developed mathematical models to describe this behavior for systems in which an organometallic catalyst contains two or more different chiral ligands, pointing out that an understanding of such inexplicable behavior may lead to the description of the reaction mechanism and the structure of active catalytic species. Recently, it has been shown that the theoretical models developed by Kagan may also predict the reaction rate [92]. Therefore, a strong amplification in product chirality may come at the expense of a severely suppressed rate of product formation; in comparison, a system exhibiting a negative nonlinear effect in product e.e. can provide a significantly amplified rate of formation of the desired product.

An enantioselective catalyst must fulfill two functions: (1) activate the different reactants (activation) and (2) control the stereochemical outcome of the reaction (controlling function). As an accepted general model, it is postulated that this control is achieved by specific interactions between the active centers of the catalyst, the adsorbed substrates, and the adsorbed chiral auxiliary (Figure 14.4). Experience has shown that most substrates that can be transformed in useful enantiomers have an additional functional group that can interact with the chiral active center.

As a result of the catalytic center–chiral entity interaction the reaction rate accelerates substantially. This phenomenon was described for the first time by Sharpless [6], who coined the term *ligand accelerated catalysis*. Unfortunately, the reasons for this phenomenon are still not well

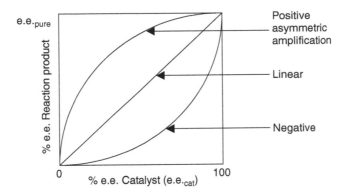

Figure 14.3 Relationship between the optical purity of the catalyst and the enantioselectivity obtained in the products of the reaction it catalyses (the straight line shows the linear relationship expected in most catalytic reactions while the negative deviation from linearity means that the optical yield of the re-action products is lower than would be expected from a simple algebric sum of the enantioselec-tivities obtained separately from the individual catalysts in the mixture).

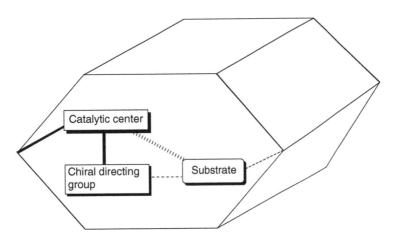

Figure 14.4 Catalytic center–chiral entity–substrate interaction in the confinement concept [93].

understood, since very little is known about the interactions occurring between substrate, chiral modifier, and catalyst. The same phenomenon was claimed to explain the effect of low modifier concentrations on the rate and e.e. of the ethyl pyruvate hydrogenation catalyzed by Pt/Al_2O_3 mod-ified with 10,11-dihydrocinchonidine [31].

14.3 HISTORICAL DEVELOPMENTS

The first synthesis of an heterogeneous enantioselective catalyst was reported in 1956 by Izumi and coworkers [100]. The heterogeneous catalyst was prepared by reducing palladium chloride that had been previously adsorbed on a silk fibroin support. It was suggested that the enantioselective reduction of precursors of amino acids was the result of an optically active configuration of the silk fibroin fiber, which is a natural protein. The study of the silk fibroin system was extended by Izumi [94–96], who investigated the hydrogenation of various C=C, C=O, and C–NO$_2$ bonds on palla-dium, platinum, and rhodium-based catalysts. Each metal had a particularly strong selectivity for the hydrogenation of a certain bond. However, no optically active product was obtained, and therefore Akabori's findings were not confirmed. Rather surprising results were reported for the epoxidation of chalcones using H_2O_2 in a two-phase reaction system in the presence of polyamino acid derivative as chiral catalyst [97]. Several parameters were shown to be of importance for the catalytic performance: type of amino acid, degree of polymerization, substituent at the terminal amino group, and organic solvent. From these and other observations, it was inferred that supramolecular interactions inside the polymer aggregate are important to catalysis and stereocontrol. Changes in the polymers properties led to remarkable improvements. Therefore, 10 years later, Itsuno and coworkers [98] described the preparation of polyamino acids grafted onto 2% crosslinked polystyrene with a high selectivity in the epoxidation of several substituted chalcones (OY up to 99%). Recently, the use of polypeptides as catalysts in the same epoxidation of chalcones led, also, to e.e. of 99% [99]. Synthetic chiral polymers were also evaluated as supports, but the early work experienced a range of problems including lack of reproducibility and low e.s.

One of the important conclusions of the early attempts was that it is fruitful to place the functionality near an optically active support. Already in 1958, Isoda and coworkers reported for the first time the enantioselective hydrogenation with a Raney nickel catalyst modified with optically pure amino acids. Optical yields reported at that time were from low (2.5%) to moderate (36%) values (for references see [12]). Subsequently, in 1963, Izumi and coworkers [100] initiated an extended study of the modified Raney nickel system with TA. As a result of their initial researches, this system was the first heterogeneous chiral catalyst to give high enantioselectivities in the hydrogenation of β-ketoesters (95%) [101,102].

Hydrogenation of enantioselective α-ketoesters over platinum catalysts, modified by treatment with cinchona alkaloids and first reported in 1979 by Orito and coworkers [103,104], represents another example which pushed the development of heterogeneous asymmetric catalysis. In the 1980s, several European groups began to use the *in situ* modification technique for cinchona-modified Pt and Pd systems. Since its discovery, a large amount of research has been done on this catalyst system, in an attempt to develop a sufficiently detailed understanding of the processes taking place on the catalyst surface so that this information can be used to design other catalyst systems for other reactions. Therefore, a variety of catalyst–chiral modifier–reactant systems have been tested resulting in a wide range of OYs [34,37,42,57,105–133]. From this body of work some specific features of the catalyst system have come to light. First, platinum is the only viable metal for the hydrogenation of α-ketoesters and acids. No other metal will work in this reaction. However, palladium/cinchonidine system also led to reasonable high enantioselectivity in hydrogenation of α-phenylcinnamic acid [134]. Secondly, it appears that the modifier must have at least a binuclear aromatic ring system attached to a chiral β-oxo tertiary amine for optimum selectivity. This subject has been much reviewed [43, 44, 65, 102, 135–138].

In 1990, Choudary [139] reported that titanium-pillared montmorillonites modified with tartrates are very selective solid catalysts for the Sharpless epoxidation, as well as for the oxidation of aromatic sulfides [140]. Unfortunately, this research has not been reproduced by other authors. Therefore, a more classical strategy to modify different metal oxides with histidine was used by Moriguchi et al. [141]. The catalyst showed a modest e.s. for the solvolysis of activated amino acid esters. Starting from these discoveries, Morihara et al. [142] created in 1993 the so-called molecular footprints on the surface of an Al-doped silica gel using an amino acid derivative as chiral template molecule. After removal of the template, the catalyst showed low but significant e.s. for the hydrolysis of a structurally related anhydride. On the same lines, Cativiela and coworkers [143] treated silica or alumina with diethylaluminum chloride and menthol. The resulting modified material catalyzed Diels–Alder reaction between cyclopentadiene and methacrolein with modest e.s. (30% e.e.). As mentioned in the Introduction, all these catalysts are not yet practically important but rather they demonstrate that amorphous metal oxides can be modified successfully.

Starting from the Pt-cinchona modified system, more recently an interesting concept has been developed by Feast and coworkers [144]. A chiral acidic zeolite was created by loading one molecule of R-1,3-dithiane-1-oxide per supercage of zeolite Y, either during or after the zeolite synthesis. Other chiral zeolites were formed by adsorbing ephedrine as a modifier on zeolites X and Y for the Norrish–Yang reaction [145].

Ni/SiO$_2$–TA modified and Pt/Al$_2$O$_3$–cinchona modified systems are the most performing heterogeneous enantioselective systems but they can transform only a limited number of substrates. Therefore, the most important problem was to develop a similar performing enantioselective system for the transformation of other prochiral substrates. This is why in the past two decades, starting in 1974 [146], many attempts have been made to heterogenize the most versatile enantioselective homogeneous catalysts, the primary aim to maintain the reaction activity and e.s. of the homogeneous species while at the same time to increase significantly the ease of separation from the reaction medium [93,147–154]. One approach to achieve heterogenization involved reacting a metal complex or salt with a solid support as a polymer or a metal oxide that had been previously modified by the addition of phosphine or amine ligands to the surface of the support. From a practical approach, these catalysts are not widely used since their activities are frequently lower than those of the corresponding homogeneous analogues. In addition, problems associated with polymer swelling and related mass transport difficulties can be encountered, as well as the finding that activity is frequently lost or attempted reuse. Some success has been reported in preparing polymer-supported chiral complexes, but the selectivity observed with the use of such heterogenized species has generally been lower than obtained using the homogeneous catalyst itself [155–158]. In 1994, Wan and Davis [159] reported a heterogeneous catalyst for asymmetric hydrogenation reactions, in which an organometallic catalytic complex is held in a film of water on a porous hydrophilic support, while the reactants and products remain within a hydrophobic organic phase. This system gave selectivity of only about 70% e.e., whereas values of at least 95% are needed for practical utility. Only few months later, the same authors [147a] described a modified process in which an e.e. of 96% was obtained for the asymmetric reduction of 2-(6′-methoxy-2′-naphtyl) acrylic acid to the commercially important anti-inflammatory agent naproxen. It seems that one reason for their success is the method of immobilization. Rather than immobilizing the active organometallic complex through a covalent bond to a supporting surface (which usually leads to a significant loss in activity), the phase containing the organometallic complex and ethylene glycol was directly immobilized onto the support. Maybe, this is the best demonstration of the catalyst design.

In parallel to these studies, cinchona alkaloid derivatives have been immobilized on insoluble supports [160–165] and used in the asymmetric dihydroxylation of alkenes. The use of an insoluble polymer as support allowed one to attain high e.s. levels, in many cases comparable to those obtained in the homogeneous phase with soluble ligands. This approach might be useful for the development of technologies for small- and large-scale production of optically active diols. On the contrary, the use of inorganic matrices as insoluble supports for the chiral ligands has not met with the same success [166].

From the above, the next logical extension to the idea of confinement was the use of confined spaces large enough to accommodate complicated organometallic catalytically active molecules as well as relevant substrate and product species. Therefore, the research groups turned their attention to the recently discovered families of mesoporous oxides, particularly the mesoporous siliceous oxides with channel apertures ranging from 25 to 100 Å [167]. They unlock new possibilities in solid-state catalysis as they provide unique supports for the development of new chiral catalysts. It opens routes to the preparation of novel catalysts consisting of large concentrations of accessible, well-spaced, and structurally well-defined active sites [93]. It might be expected that the spatial constraints induced by a carrier such as MCM-41, will significantly increase the influence of the chiral ligands. There are some examples showing that enantioselective reduction catalyzed by a Pd complex immobilized inside the pores of MCM-41 can give a large increase in enantioselectivity (e.e.) compared to the homogeneous Pd complex (e.e.'s increased from 43 to 96%) [10].

As it can be seen, the investigations of heterogeneous chiral catalysts started in the late 1950s in Japan and has known a worldwide renaissance in the last few years. Because of a multitude of catalysts discovered and developed in the recent years, combinatorial methods have become an important focus of research in asymmetric catalysis [168,169]. In the last few years, efficient techniques have been developed for the high throughput parallel screening of chiral catalysts [170–172]. However, parallel screening based on product analysis has potential pitfalls, since the e.s. of a reaction is often lower than the inherent selectivity of the catalyst because of an unselective background reaction, catalytically active impurities, or partial dissociation of a chiral ligand from a metal catalyst. Problems of this nature could be avoided if the catalyst's ability for enantiodiscrimination could be determined directly from the examination of catalyst–reactant complexes rather than from product analysis.

14.4 TARTARIC ACID-MODIFIED ME/SUPPORT HYDROGENATION CATALYSTS AND RELATED SYSTEMS

Enantioselective metal surfaces may be prepared from the theoretical point of view by three basic methods: (1) deposition of the metal onto a chiral support; (2) rearrangement of the metal surface atoms into a chiral arrangement; and (3) deposition of a chiral moiety onto the metal surface. However, up to now, no enantioselective metal catalyst has been obtained using method (2), while several catalysts have been prepared using procedures (1) and (3). The tartrate-modified nickel catalysts following procedure (3) has been developed over many years. At present, this is the best studied family of catalysts. Various substrates were hydrogenated with low to very high e.s. The hydrogenation of β-ketoesters [12–30,100], β-diketones [173], and prochiral ketones [23,174,175] on these catalysts led to the highest OYs. The selective preparation of secondary or tertiary asymmetric fatty amines from fatty acids, fatty esters or nitriles, hydrogen and ammonia, in the presence of solid copper or nickel-based catalysts was also studied [176,177].

To develop effective enantioselective catalysts for asymmetric hydrogenation, the research groups in this area have fully investigated the factors controlling the OY (Figure 14.5).

As Figure 14.5 shows, the enantio-differentiating (e.d.) hydrogenation consists of three processes: (1) catalyst preparation, (2) chiral modification, and (3) hydrogenation reaction. These processes imply preparation variables for activated nickel, as a base catalyst for modified Ni, modification variables for the activated catalyst, and reaction variables of the hydrogenation processes, respectively. All these factors should be optimized for each type of substrate.

14.4.1 Catalyst Preparation Process

Literature reports hundreds of procedures for preparation of metallic nickel catalysts for hydrogenation. Among them, only limited types of nickel can be utilized as a base catalyst of modified nickel: Raney nickel [178–180], silica-supported nickel [19,181,182], commercial nickel powder [16], fine nickel powder obtained by condensation of nickel vapor in vacuum [183], and nickel black obtained by the hydrogenolysis of pure nickel oxide or nickel carbonate [184,185]. A comparison between the research results obtained by different groups shows that freshly prepared Raney nickel gives the best results. Bimetallic and noble metal catalysts have been also studied, but with the exception of some $NiPd/SiO_2$ catalysts these give lower e.e. values than pure nickel [186].

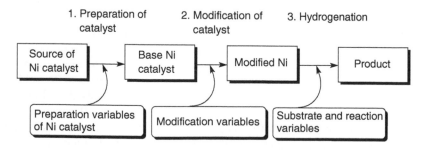

Figure 14.5 Factors controlling optical yield of e.d. hydrogenation [26].

The characterization and control of the structural features of Raney-type catalysts is not an easy task. Anyway, Raney-type leaching at high temperature results in a higher e.d. ability than leaching at lower temperature [179]. The catalysts with a larger mean crystallite size of nickel tend to give higher e.d. ability [185]. For supported nickel, the carbonate precipitation method leads to Ni/SiO$_2$ with a relatively large mean crystal size and a narrow crystal size distribution [187,188]. The correlation with catalytic performances of these catalysts suggest that the relationship between mean crystal size and e.d. ability is similar to that for unsupported Ni. However, for materials with micro- and mesopores, the mesopores affect negatively the e.d. ability, while the micropores have no effect [188].

Representative results of TA-modified Ni black and supported Ni catalysts are compiled by Tai and Sugimura [26]. Summarizing the hydrogenation activity, and the e.d. ability widely varies with the precursor preparation method. The supported catalysts display the highest activity, while Raney-modified catalysts give the best e.d. abilities.

14.4.2 The Modification Process

The modification of nickel catalysts is carried out in a separate step in aqueous medium. During this process the e.d. site is generated on the surface, and the conditioning of the catalyst surface occurs, leading to a suitable surface for e.d. hydrogenation. The nature of modification reagent (hydroxy acids and their derivatives, amino acids and their derivatives, peptides, amino alcohols) and the conditions in which the modification is carried out (modifier concentration, pH, temperature, time, and sometimes procedures) are definite factors in controlling the e.d. ability of modified Ni catalyst [101,189]. Among the different modifiers, the screening of different water-soluble chiral compounds indicated that TA was by far superior. TA was found to be the best modification reagent in the hydrogenation of prochiral alkanones and β-ketoesters (see, e.g., methylacetoacetate, MAA) (Scheme 14.3).

Heterogeneous asymmetric catalysis is still an art and, therefore, the experimental conditions must be optimized for each type of catalyst. Inorganic salts in the modification solution are known to enhance the e.d. ability of unsupported [180] and supported [190–191] modified nickel catalysts in the MAA hydrogenation. NaBr is the most important comodifier as it enhances OY by 10 to 30%. Osawa et al. [23] proposed that NaBr deactivates the non-e.d. site to reduce the production of racemic products. The effect of NaBr was also investigated in hydrogenation over supported Ni catalyst. However, this explanation is not unique and some other explanations have also been supplied in the literature. Bostelaar and Sachtler [192], although admitting the e.d. ability enhancement effect of NaBr, interpreted the effects of NaBr as a change in the intrinsic e.s. of the product-determining surface complex. Webb and Wells [102] demonstrated that the addition of NaBr increased the e.d. ability even when a thiophene-poisoned catalyst was used and, in consequence, they suggested that NaBr could not be simply regarded as a site-blocking agent.

Anyway, all results have a common point. They suggest that there are two types of active sites on the catalyst surface: (1) one site exhibiting affinity for TA where e.d. hydrogenation (e.d. site) is supposed to take place and (2) a second site without affinity for TA where racemic products are produced (non-e.d. site).

14.4.3 Substrate and Hydrogenation Parameters

Izumi [189] examined more than 50 prochiral substrates and found that only hydrogenation of β-ketoesters and β-diketones is accompanied by good e.e. values. At the same time, the effects of the hydrogenation temperature on OYs were found to depend on the types of substrate. In the hydrogenation of MAA, changing the reaction temperature (60 to 120°C) in high-pressure liquid-phase hydrogenation did not affect the optical yield. On the contrary, in the hydrogenation of 2-alkanones, it was observed that OY strongly depended on the hydrogenation temperature. The optimal temperature was about 50 to 60°C and OY reached 80% [193]. However, no simple correlation between e.e.

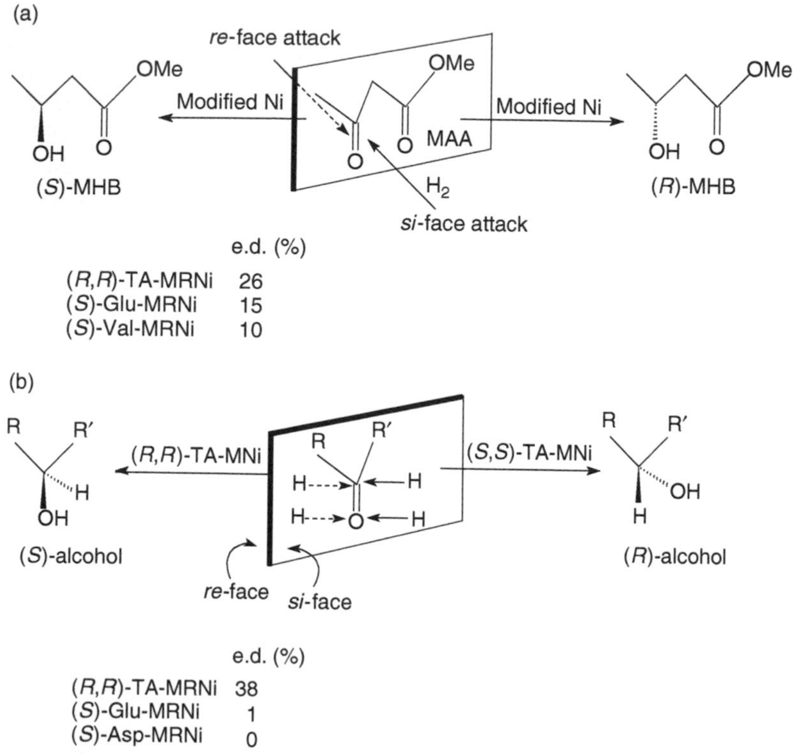

Scheme 14.3 (a) Stereochemistry and the e.d. for some representative chiral modifiers in the hydrogenation of MAA to MHB. (b) Stereochemistry and the e.d. for some representative chiral modifiers in hydrogenation of prochiral 2-alkanones to chiral alcohols.

and temperature or pressure was reported [101]. In the hydrogenation of MAA, Webb [194] found that only pressures >90 bar afford OYs above 80%. The reaction could also be carried out in the gas phase, but the OYs are lower [195]. Aprotic semipolar solvents, in articular methyl propionate and THF, give the highest e.e. values [101]. At the same time, Webb [194] gave the following sequence: *n*-alkanols > methyl propionate = ethyl acetate ≫ THF ≫ toluene = acetonitrile. The addition of water significantly decreases the e.d. ability [196], while small amounts of acids, such as acetic acid, increase the e.d. ability [197]. In particular, the addition of pivalic acid in the hydrogenation of 2-alkanones was found to be really necessary [198,199].

14.4.4 Mechanistic Investigations and Hypotheses for Enantioselection

The most conventional investigations on the adsorption of both modifier and substrate looked for the effect of pH on the amount of adsorbed tartrate and MAA [200]. The combined use of different techniques such as IR, UV, x-ray photoelectron spectroscopy (XPS), electron microscopy (EM), and electron diffraction allowed an in-depth study of adsorbed tartrate in the case of Ni catalysts [101]. Using these techniques, the general consensus was that under optimized conditions a corrosive modification of the nickel surface occurs and that the tartrate molecule is chemically bonded to Ni via the two carbonyl groups. There were two suggestions as to the exact nature of the modified catalyst: Sachtler [195] proposed adsorbed nickel tartrate as chiral active site, whereas Japanese [101] and Russian [201] groups preferred a direct adsorption of the tartrate on modified sites of the Ni surface.

The emphasis in surface science research is increasingly focused on the understanding of surface functionality at the nanoscale. Recently, Lorenzo and coworkers [202] created model stereodirecting surfaces by adsorption of R, R-TA on a Cu(1 1 0) single-crystal surface. Their work clearly showed that a variety of surface phases are created upon adsorption of a single molecular layer of (R, R)-TA on the Cu(1 1 0) metal surface, each possessing different local chemical and bonding characteristics and different two-dimensional organizational structures, depending on adsorption temperature, coverage, and holding time. On these various phases, the ordered bitartrate phase is believed to be catalytically relevant and provides important insight into the creation of the enantioselective site, since it is the only phase capable of coadsorbing the reactant, MAA, within its structure.

The same authors utilized a multidisciplinary approach to identify the adsorption of (R, R)-TA on Ni surface [28]. For the investigation of the local ordering at the nanoscale, they used a scanning tunneling microscope (STM), while the long-range ordering of the adsorbed molecules was established by low-energy electron diffraction (LEED). The chemical nature and orientation of the adsorbed species were analyzed by reflection absorption infrared spectroscopy (RAIRS), whereas to have a better understanding of the adsorption energetics, they performed extensive *ab initio* calculations within the density functional theory (DFT) approach. Such a combination of experimental and theoretical methods has provided a detailed insight into the mode of chiral induction by adsorption of the modifier on the defined Ni(1 1 0) surface. The authors demonstrated in a very elegant way that the adsorption of (R, R)-TA at low coverage and room temperature takes place in its bitartrate form with two-point bonding to the surface via both carboxylate groups. The molecule is preferentially located above the fourfold hollow site with each carboxylate functionality adsorbed at the short bridge site via O atoms placed above the adjacent Ni atoms. The most stable adsorption structure was achieved by a chiral relaxation of atoms in the bulk-truncation Ni(1 1 0) surface so that a large footprint with a long distance of 7.47 Å between pairs of Ni atoms could be accommodated at the surface. In this way, all the local mirror planes associated with the clean surface are destroyed locally by the adsorbed complex (Figure 14.6). Calculations show only one chiral footprint to be favored by the (R, R)-TA (Figure 14.6a), with the mirror adsorption site being less favored energetically by about 6 kJ/mol (Figure 14.6b). This means that at room temperature the same local chiral motif is expected to be repeated over 90% of the metal surface, leading to an overall chiral and very enantiospecific system.

Interestingly, it is found that chiral exhibition is not simply limited to systems in which chiral molecules are adsorbed at achiral surfaces (i.e., adsorption of (R, R)-TA on the Ni or Cu surface) but it can also be displayed in systems where no initial chirality is present, i.e., from the adsorption of achiral molecules at achiral surfaces [203–211]. Raval and coworkers [203] reported on the adsorption of succinic acid on Cu(1 1 0) and compared the results with those found for (R, R)-TA. Structurally, succinic acid is very similar to TA, with the only difference being that the two hydroxyl groups present in TA are replaced by hydrogen atoms, leading to a consequent loss of both chiral centers (Figure 14.7).

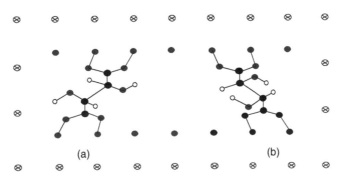

(a)   (b)

Figure 14.6 Depiction of the relaxed bitartrate–Ni₄ species adsorbed in twin mirror chiral footprints at the Ni(1 1 0) surface [28].

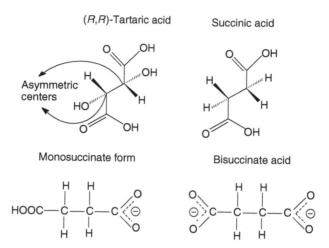

Figure 14.7 Diagram showing the (R,R)-tartaric acid and succinic acid molecules and the different chemical forms that can be adopted by the latter upon adsorption [203].

To compare these two systems at both the local adsorption motif level and at the extended self-organization level, they have applied the same ensemble of surface science techniques. As in the case of Ni(1 1 0), the chiral influence of (R,R)-TA on Cu(1 1 0) can be discerned at two levels: first, at the local level where adsorption events conserve the chiral centers and, thus, give rise to point chirality and, second, at the organizational level where self-assembled structures form that are chiral in arrangement and, thus, destroy the mirror symmetry elements possessed by the underlying surface. From the adsorption models presented in Figure 14.8 [203], it is evident that a similar manifestation of organizational surface chirality exists for the chiral bitartrate molecule and the achiral bisuccinate adsorbate. However, a critical difference between the two systems also emerges; whereas for TA only one chiral handedness exists, which is sustained over the entire surface, for succinic acid both the chiral (9 0, −2 2) phase and its mirror image (2 2, −9 0) phase coexist at the surface, so that by integrating over the entire surface one obtains an overall zracemic conglomerate. This is a crucial difference in the expression of chirality between the two systems in which the bitartrate possesses global chirality while the bisuccinate is locally chiral but globally achiral.

For the (R, R)-bitartrate system, the major factors governing the creation of the chiral super-structure must be very similar to that for succinic acid, i.e., a similar molecular distortion or recon-struction must be induced by the bicarboxylate–Cu interaction to create a similar superstructure. However, (R,R)-bitartrate yields a single domain of one-handedness only suggesting that the presence of the OH groups at the chiral centers crucially restricts the distortion/reconstruction to one-handedness only. DFT calculations on (R, R)-bitartarte on Ni(1 1 0) [28] show an energy differ-ence of 6 kJ/mol between the two mirror image distorted/reconstructed adsorption motifs, sufficient to ensure that over 90% of the nucleation points at 300 K would be of the lower energy form. The role of the OH groups as chiral directors of the supramolecular assembly is illustrated when adsorp-tion of the (S,S)-bitartrate unit is examined (Figure 14.8b). The rigid adsorption structure of the bitartarte unit forces the OH groups to lie in a uniquely defined direction, which is reflected in space compared to the (R, R)-bitartrate unit.

As a result, the energy preference of the local adsorption unit is switched to the opposite dis-tortion/reconstruction and, thus, chiral lateral interactions are switched in the direction of the in-duction and propagation of the chiral assembly occurring in the mirror image construct, leading to a mirror chiral surface. Therefore, from this work, one may conclude that the overall global or local chirality is determined principally at the nucleation stage.

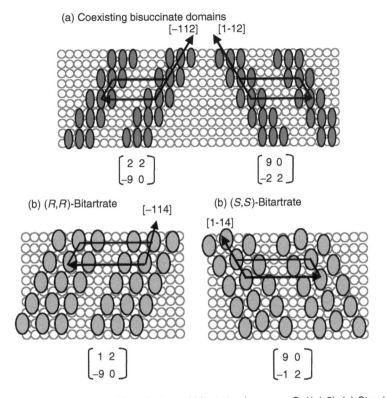

Figure 14.8 Adsorption models of the bisuccinate and bitartrate phases on Cu(1 1 0). (a) Structural models for the two coexisting chiral domains for bisuccinate on Cu(1 1 0). The (2 2, −9 0) and (9 0, −2 2) unit cells of the overall structure are shown as are the (2 2, −2 0) and (2 0, −2 2) unit cells representing the packing within each chain. (b) Structural models of the bitartrate phases of the two tartaric acid enantiomers on Cu(1 1 0): (S,S)-bitartrate (9 0, −1 2) and (R,R)-bitartrate (1 2, −9 0). The (3 1, −2 1) unit cell is also shown for the (R,R)-bitartrate phase showing the packing within the chain [203].

In conclusion, the extensive research of Raval research group [203] on the succinate system suggests that when the chiral centers are destroyed by replacing the OH groups with H, chirality at the local level may still exist in the system, but within a globally racemic system. This puts a different perspective on chiral enantioselective strategies for heterogeneous systems, where successful routes may not be restricted to designing a chiral modifier, but now also include the possibility of spontaneous chiral creation with simpler achiral modifiers where, subsequently, domains of the unrequired handedness can be neutralized by coadsorbing specific blocking molecules, as is often used in homogeneous enantioselective catalysis.

14.4.5 Proposed Mechanisms

One of the oldest mechanisms of interaction between adsorbed reactant and adsorbed TA has been proposed by Klabunovskii and Petrov [212]. They suggested that the reactant adsorbs stereoselectively onto the modified catalyst surface. The subsequent surface reaction is itself nonstereospecific. Therefore, the optically active product is a result of the initial stereoselective adsorption of the reactant, which in turn, is a consequence of the interactions between reactant, modifier, and catalyst. The entities form an intermediate chelate complex where reactant and modifier are bound to the same surface atom (Scheme 14.4). The orientation of the reactant in such a complex is determined by the most stable configuration of the overall complex intermediate. The mechanism predicts that OY only depends on the relative concentrations of keto and enol forms of the reactant, and not on the modifier surface concentration. Unfortunately, the prediction of the independence of OY with respect to modifier coverage cannot accommodate the well-documented dependence of OY on modifying conditions.

Sachtler [195] proposed a dual-site mechanism in which the hydrogen is dissociated on the Ni surface and then migrates to the substrate that is coordinated to the adsorbed dimeric nickel tartrate species. In their model, adsorption of modifier and reactants takes place on different surface atoms in contrast to Klabunovskii's proposal. Adsorbed modifier and reactant are presumed to interact through hydrogen bonding (Scheme 14.5). The unique orientation of adsorbed modifier molecules leads to a sterically favored adsorbed reactant configuration to achieve this bonding.

These models, however, are less instructive because the authors incorporate insufficient information from organic stereochemistry.

The work of Tai et al. [213] in this area allows the isolation of optically pure compounds in high yields, which is the eventual goal of synthetic organic chemists. His model, the so-called 2P model, is based on the formation of two hydrogen bonds between the two hydroxyl groups of (R, R)-TA and the two carbonyl groups of MAA. In Scheme 14.6a, the carbonyl group of MAA to be hydrogenated is fixed on site 1, and comes within 0.1 nm from the surface with its si-face toward the catalyst. When MAA is adsorbed in this way, it will be hydrogenated to (R)-MHB (hydrogen attack from si-face; Scheme 14.6b). When (S,S)-TA is used in the modification process, (S)-MHB is obtained (hydrogen attack from re-face; Scheme 14.6c).

This stereochemical model explains the stereochemistry of the hydrogenation of MAA over TA–Ni system. It also predicts that the TA–Ni catalyst can be effective for the enantioselective hydrogenation of some prochiral ketones with excellent e.e. values (70%).

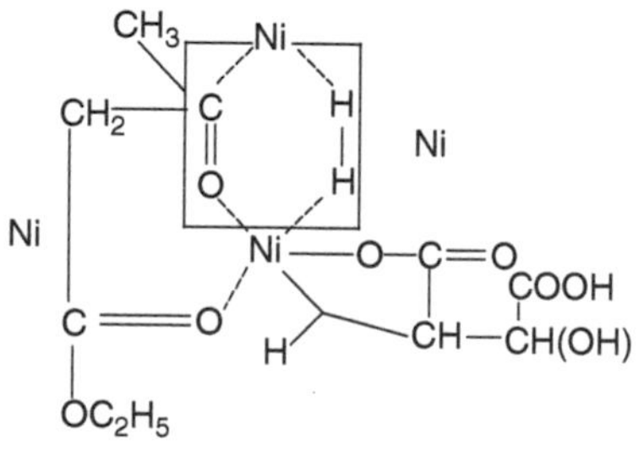

Scheme 14.4 Klabunovskii model.

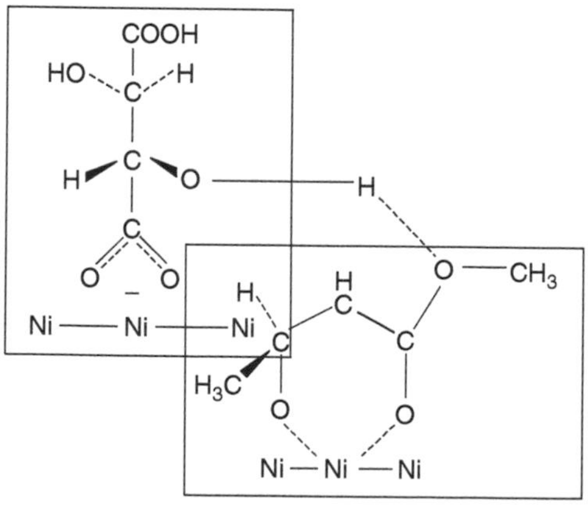

Scheme 14.5 Sachtler model.

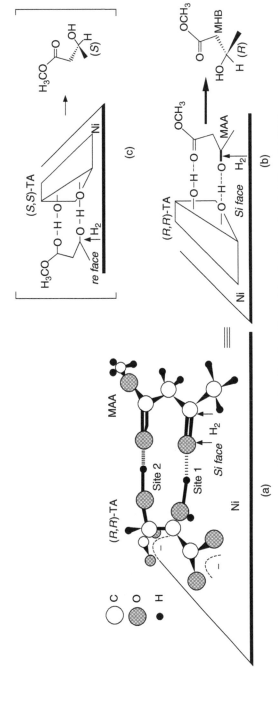

Scheme 14.6 Interaction between MAA and tartaric acid adsorbed on the Ni surface through two hydrogen bonds 2P model [213].

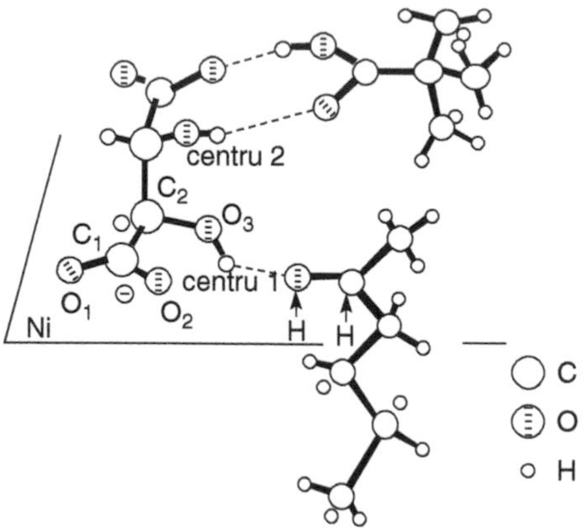

Scheme 14.7 Interaction model between 2-hexanone and TA adsorbed on the Ni surface through a hydrogen bond and a steric repulsion (1P model) [213].

As mentioned above, the TA–Ni system can catalyze the hydrogenation of not only MAA but also of 2-alkanones. However, hydrogenation of 2-alkanones following the optimum conditions for MAA hydrogenation led to poor results. But the addition of small amount of acetic acid to the reaction mixture, as an additive, turned out to be essential for the enantioselective hydrogenation of these substrates. Further experiments indicated that acetic acid is not the only organic acid leading to such e.e. and the best results were obtained with pivalic acid (PA) in a concentration more than twice that of the substrate [214].

In the case of 2-hexanone, the adsorption can only be governed by the relative size of the methyl and butyl groups connected to the carbonyl group. The carbonyl group of 2-hexanone may interact with one of the hydroxy groups of TA in the same manner as MAA. The other hydroxy group of TA will repel the large hydrophobic butyl group to the other side (the hydroxy groups of TA are expected to form a highly polar region on the catalyst). This interaction model, compressing one hydrogen bond and a steric repulsion is called the 1P model (Scheme 14.7).

The models predicted by Tai and coworkers [213] allow one to simulate the functions of modified Ni catalysts and to propose ways to improve the e.d. ability of the catalyst, without claiming to imply a real mechanism. Full understanding of modified Ni will require further significant advances in the physicochemical approach.

14.5 CHIRAL-MODIFIED PLATINUM HYDROGENATION CATALYSTS AND RELATED SYSTEMS

Cinchona-modified platinum catalysts received special interest mainly because of to the results obtained for the hydrogenation of methyl pyruvate (MP) or ethyl pyruvate. E.e.'s up to 80% or even higher (98% for ethyl pyruvate) made this system one of the most interesting ones from the application point of view. At present, 5% Pt/alumina with low dispersion (metal particles > 2 nm) and a rather large pore volume constitutes one of the best catalysts commercially available.

Other substrates were also chiral hydrogenated but with lower OYs: cyclohexane-1,2-dione (80% e.e.) [132], 4-hydroxy-6-methyl-2-pyrone (50 to 80% e.e.) [215], 1-phenyl-1,2-propanedione (64% e.e.) [130], butane-2,3-dione (85 to 90% e.e.) [59,117,216], isopropyl 4,4,4-trifluoroacetoac-etate (90% e.e.) [133] or ethyl 4,4,4-trifluoroacetoacetate (90% e.e.) [58,68], 2,2,2-trifluoroace-tophenone (61 to 89.5% e.e.) [110,114], hydroxymethylpyrone (85% e.e.) [127], pyruvic acid oxime (26% e.e.) [107], isophorone (55% e.e.) [46,122] and phenylcinnamic acid (28% e.e.) [122], 2,4-diketo acid derivatives (23 to 87% e.e.) [128], dialkyl 2-oxoglutarates (96% e.e.) [57],

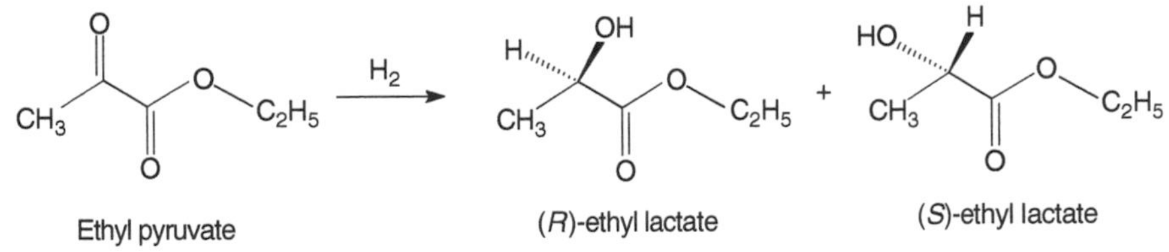

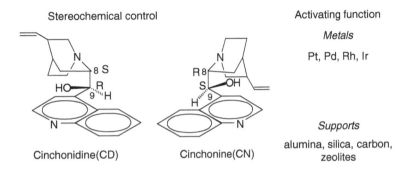

Scheme 14.8 The enantioselective hydrogenation of the ethyl pyruvate.

Stereochemical control

Activating function

Metals

Pt, Pd, Rh, Ir

Cinchonidine(CD)

Cinchonine(CN)

Supports

alumina, silica, carbon, zeolites

Scheme 14.9 Main features of cinchona-modified metal systems [35].

ketopantolactone (83.4 to 92% e.e.) [115,123], α-ketoamides (60% e.e.) [109], and 1-ethyl-4,4-dimethylpyrrolidine-2,3,5-trione (91% e.e.) [119].

Extended studies demonstrated that Pt [31–73] and, to a lesser degree, Pd [107,121,127,217–225], Rh [131,226], and Ir [227] are the most suitable active metals. With one exception (colloidal Pt [111,125,228]) all these metals have been used in supported form. The most used supports are alumina, carbon, silica, and titania [56,226], and more recently zeolites [106,229–230].

The most successful modifier is cinchonidine and its enantiomer cinchonine, but some work in expanding the repertoire of substrate/modifier/catalyst combinations has been reported: (S)-(-)-1-(1-naphthyl)ethylamine or (R)-1-(1-naphthyl)ethylamine for Pt/alumina [108,231], derivatives of cinchona alkaloid such as 10,11-dihydrocinchonidine [36,71], 2-phenyl-9-deoxy-10,11-dihydrocinchonidine [55], and O-methyl-cinchonidine for Pt/alumina [133], ephedrine for Pd/alumina [107], (-)-dihydroapovincaminic acid ethyl ester (-)-DHVIN for Pd/TiO₂ [122], (-)-dihydrovinpocetine for Pt/alumina [42], chiral amines such as 1-(1-naphthyl)-2-(1-pyrrolidinyl) ethanol, 1-(9-anthracenyl)-2-(1-pyrrolidinyl)ethanol, 1-(9-triptycenyl)-2-(1-pyrrolidinyl)ethanol, (R)-2-(1-pyrrolidinyl)-1-(1-naphthyl)ethanol for Pt/alumina [37,116], D- and L-histidine and methyl esters of D- and L-tryptophan for Pt/alumina [35], morphine alkaloids [113].

Interesting catalysts were described by Bhaduri and coworkers [105], who anchored anionic Pt and Ru carbonyl clusters on a 20% crosslinked polystyrene functionalized with cinchonidine and ephedrine. On the same lines, Huang et al. [131] studied the hydrogenation of ethyl pyruvate catalyzed by polyvinylpyrrolidone-stabilized Rh nanoclusters modified with cinchonidine and quinine leading to 42.2% e.e. of (R)-ethyl lactate. These results are surprising because all known supported small Pt crystallites are unselective and Rh/alumina catalysts give very small e.e. This means that these cluster catalysts must have a different mode of action compared with the classical systems. Blaser and Müller [232] grafted a functionalized cinchona derivative onto the surface of a Pt/SiO₂ catalyst. Although the grafted modifier exhibits rates and e.e. values comparable to that of the normally modified system, the catalyst was not reusable.

As mentioned, the most studied reaction using these modified catalysts is the enantioselective hydrogenation of MP or ethyl pyruvate to the corresponding lactates using cinchona alkaloids

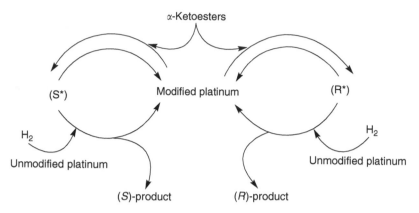

Figure 14.9 Two cycle mechanisms proposed for enantioselective hydrogenation of α-ketoesters on chirally modified platinum [66].

modifiers (Scheme 14.8). The reaction rate of the enantioselective reaction is significantly higher than the rate of the racemic hydrogenation; the e.s. and the reaction rate show a maximum as a function of the modifier concentration.

The main features of the cinchona alkaloid-modified metal system are illustrated in Scheme 14.9.

The alkaloid-modified catalyst can be easily prepared either by stirring the metal catalyst with a solution of the alkaloid in air and subsequent separation by decantation, as described by Orito and coworkers [103], or by *in situ* addition of alkaloid to the reactant mixture [226]. Good OYs are achieved with both methods. Reactions are generally carried out at room temperature, or slightly above, and at hydrogen pressures in the range 1 to 10 MPa. The best solvent is AcOH. Under optimal reaction conditions the decrease in e.e. can be ascribed to the hydrogenation of the modifier.

14.5.1 Kinetic Models

Kinetic models for quantifying the remarkable increase in the reaction rate occurring upon modification are based on the two-cycle mechanism as predicted for a ligand-accelerated reaction [6]. The chiral modification of the platinum surface is supposed to lead to two enantiofacially distinct types of sites. The α-ketoesters from the fluid phase will adsorb reversibly on these sites in its two enantiofacial forms, (R*), leading to (R)-product and (S*) to the (S)-product on hydrogenation (Figure 14.9). The modified sites have been suggested to interact with the adsorbed α-ketoesters via the hydrogen bonding between the quiniclidine N- and O-atom of the α-carbonyl moiety. The rate of acceleration and the e.d. are considered to originate from the preferential stabilization of one of the two diastereomeric intermediates ([R*] and [S*]) formed via the interaction of the α-ketoester and the adsorbed cinchona modifier. However, the question of whether the e.s. is thermodynamically (stability of adsorbed R* vs. adsorbed S*) or kinetically controlled (different activation energies of pro-R- and pro-S-routes) has not yet been definitively answered.

14.5.2 Mechanistic Investigations

Trying to understand the mode of action of a catalyst in enantioselective processes is fascinating but also a very difficult endeavor. For the hydrogenation of α-ketoesters using Pt catalysts modified with cinchonidine derivatives, investigations are focused on explaining the very good enantiodiscrimination and on finding an explanation for the remarkable rate enhancement. The early systematic studies [233] of the influence of structure variation of cinchonidine indicated three structural elements that are crucial for the functioning of cinchona alkaloids as chiral modifiers: (1) an *anchoring part*, represented by the flat aromatic ring system (quinoline) which is assumed to be adsorbed on the platinum by multicenter π-bonding; (2) a stereogenic center embracing C-9 and C-8; the latter is decisive for the sense of e.d.; and (3) a basic nitrogen (quinuclidine) that is directly involved in the interaction with the reactant α-ketoester, resulting in a 1:1 interaction.

One of the first attempts to explain e.s. was made by Wells and coworkers [234], who proposed that the L-shaped modifier could generate a chiral surface, by adsorption on Pt in ordered nonclose-packed arrays, allowing preferential adsorption on the metal surface of one of the faces of the prochiral substrate (*template model*).

This mechanistic model is based on the planar adsorption of MP on the solid surface in the same configuration as in the liquid phase, (that is, the most stable configuration, from the energetical point of view, is the one in which the C=O groups are in a parallel *trans* position.

A striking feature of the template model is the restriction of the role of the modifier to that of a template, which does not take into account direct binding interactions of the reactant with the modifier. Furthermore, there exists no experimental evidences for the formation of ordered arrays of cinchona molecules on a platinum surface. In 1995, Margitfalvi and Hegedus [235] criticized this model showing that the model is too idealistic and oversimplified.

Later, the first *1:1 interaction model* was proposed by Baiker and coworkers [236,237], suggesting that modifier and reactant interact via hydrogen bonding. The interaction models differ greatly in the way the modifier is adsorbed on the platinum surface and in the structure of the adduct formed upon interaction of the pyruvate with the modifier. Augustine and coworkers [34] proposed a model that relies on N_1 being adjacent to the hydroxyl group to enable six-membered ring interaction between the quinuclidine N_1 and the keto carbon atom as well as the C_9 oxygen and the ester carbon atoms. The nucleophilic character of the quinuclidine nitrogen (QN) of the alkaloid would be much stronger than that of the oxygen atom in water or alcohol. Therefore, the more nucleophilic QN should be attracted to the more electron-deficient ketone carbon rather than the ester carbon. The C_9 oxygen could also be involved in a similar attraction. There are two possibilities of the modifier adsorption on the catalyst providing a chiral environment to the corner atoms and adatoms (Scheme 14.10). In mode A, the adsorption takes place between the face of the quinoline ring of the alkaloid (Q) and an ensemble of face atoms, which are adjacent to an active corner atom site. The C_8 and C_9 chiral portion of the alkaloid sites over the corner atom provides the chiral environment. The QN may or may not be adsorbed on the corner atom. The e.s. is produced by hydrogen bonding between the C_9 OH and the ester oxygen. Transfer of hydrogen to the pyruvate takes place to

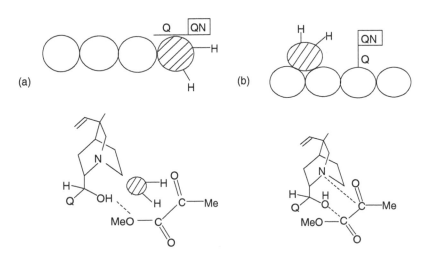

Scheme 14.10 The 1:1 interaction model [34].

the side of the carbonyl group facing the alkaloid resulting in (S)-lactate formation. In mode B, the adsorption occurs through the nitrogen of the quinoline ring so this ring is adsorbed upright on the metal surface. Edge-on adsorption of the quinoline ring of the alkaloid on face atoms adjacent to an adatom places the quinuclidine ring with the C_8 and C_9 chiral centers somewhat above and sufficiently away from the site so that the pyruvate can be adsorbed between the chiral centers of the alkaloid and the active site of the metal. In this configuration hydrogen is transferred to the keto group from the side away from the alkaloid to give the (R)-lactate.

However, the assumption of a rigid six-membered ring is not plausible and attaches the same importance to N_1 and to oxygen at C_9 in the interaction between pyruvate and cinchonidine; this is against experimental observation, since substitution of OH by OCH_3 at C_9 afforded the best values of e.e. [233].

A different model (the *shielding effect model*) was proposed by Margitfalvi and coworkers [60] according to which e.s. originates via a complex between the cinchona alkaloid and the substrate in the liquid phase, which, subsequently stereoselectively reduced, should lead to the observed e.d. The above complex formation can also be considered as host–guest-type supramolecular interaction. This model was based on the closed conformations of the alkaloid, and cannot explain e.s. achieved with the modifiers with rigid open conformations as studied by Bartók and coworkers [238]. Moreover, Margitfalvi and Tálas [53] investigated the enantioselective hydrogenation of ethyl pyruvate over a Pt/alumina catalyst containing silicium organic moieties anchored to the support and cinchonidine, the chiral template molecule. The introduction of $-Si(CH_2)_nR$ moieties onto the alumina significantly decreased both rate and e.s. of the hydrogenation reaction. The results show that the e.d. step is more complex than it is predicted by the existing models and that the e.d. step cannot be exclusively attributed to the interaction between the half-hydrogenated substrate and cinchonidine taking place at the platinum surface.

Recently, Vayner and coworkers [239] have revisited the model proposed by Augustine et al. [34] which is based on the assumption that the QN can make a nucleophilic attack to an activated carbonyl. According to this model the two possible zwitterionic intermediates that can thus be formed have different energies, which leads to the selective formation of one of the two intermediates, and, therefore, to e.s. after hydrogenolysis by surface hydrogen. This model nevertheless does not explain the e.d. of nonbasic modifiers, such as the one reported by Marinas and coworkers [240], which have no quinuclidine moiety and no nitrogen atom, and thus no possibility to form zwitterionic intermediates. Furthermore, *in situ* spectroscopic evidence for hydrogen bond formation between the quinuclidine moiety of cinchonidine and the ketopantolactone has been provided recently [241], which supports the hypothesis of the role of weak bond formation rather than the formation of intermediates such as those proposed by Vayner and coworkers.

A striking deficiency of all model calculations performed so far is the lack of an explicit treatment of the role of metal surface, although in some cases the geometrical constraints were inserted [67,242–243].

Very recently, Baiker and coworkers [244] filled this gap by explicitly treating the adsorption of the chiral modifier and the reactant on the platinum surface. Based on the *in situ* spectroscopic knowledge collected in the past years [245–247], and on the modeling of the adsorption of activated ketones on platinum, they proposed a possible scenario for e.d. By analyzing the minimum energy structures that resulted from the rotation of C_8–C_9 and C_4–C_9 bonds, four surface conformations were found for cinchonidine, two of which, named tilted surface open(5) (TSO5) and parallel surface open(5) (PSO5), had the quinuclidine moiety sufficiently close to the surface to admit interactions with a chemisorbed ketone, while the other two, named TSO3 and PSO3, had the quinuclidine moiety more distant from the surface (Figure 14.10).

Furthermore, for the two surface open(5) conformations the interactions with the reactant take place far from the anchoring group and thus the interaction between the latter and the reactant is unimportant and independent of the rotation around the C_4–C_9 bond.

The alkaloid adsorbed on the platinum surface could form a tridimensional space within which the hydrogenation reaction can preferentially occur, due to a close interaction with QN. This space was called *chiral pocket* in analogy to biological systems that show high differentiation ability due to shape discrimination (Figure 14.11).

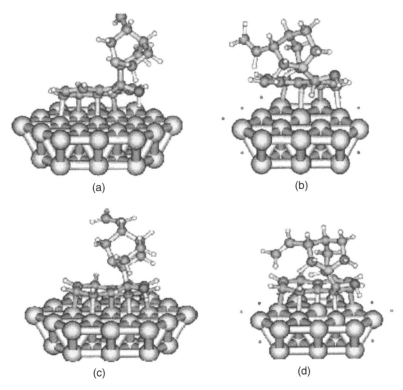

Figure 14.10   Surface conformations of cinchonidine: (a) TSO(3); (b) TSO(5); (c) PSO(3); and (d) PSO(5) [244].

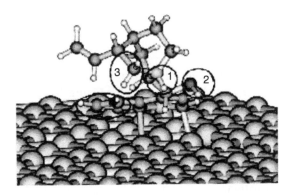

Figure 14.11   Example of the chiral pocket for the tilted surface open(5) conformation. Functional groups 1 and 2 are able to give bonding interactions, while the hydrocarbon skeleton of the alkaloid gives repulsive interactions, allowing only some molecular shapes to adjust themselves in the proximity of the quinuclidine nitrogen [244].

The four main elements of the *chiral pocket* are the following: (1) The QN: whether or not protonated, it is able to interact with a surface species, either by promoting proton transfer or by stabilizing a semihydrogenated surface ketone. It has been shown that alkylation of this nitrogen leads to a complete loss of selectivity [233]. (2) The *hydroxylic moiety*: in the surface-open(5) and TSO5 conformations, the hydroxylic proton does not point toward the surface, but toward the space where the reaction takes place. The O–H can not only take part in hydrogen-bonding interactions with an adsorbed substrate, but can also regulate the equilibrium between surface conformations. Any O-alkylation would induce the alkyl group to occupy the space where the reaction takes place, at a short distance from QN, and may have the effect of altering the

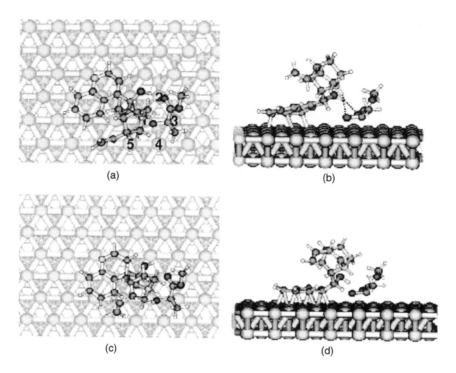

(a)   (b)

(c)   (d)

Figure 14.12 Interaction of the TSO(5) (a, b) and PSO(5) (c, d) conformations of cinchonidine with an adsorbed *cis* pro-(*R*) methyl pyruvate [244].

selectivity or, if the space becomes too crowded, to reduce it by hindering access to the QN interaction, thus favoring racemic hydrogenation. (3) The C_4–C_9 and C_8–C_9 bonds: these two bonds account for the flexibility of the quinuclidine moiety. It has already been shown that the conformational complexity of cinchonidine in solution is due to rotational freedom around these two bonds. When cinchonidine is bound to the surface, rotation around their axes allows the flexibility of the quinuclidine moiety, and a dynamic variation of size and position of the chiral pocket. (4) The *skeleton of the alkaloid* (numbers 3 and 4 in the Figure 14.12): the hydrocarbon structure of cinchonidine forms a space for repulsive interaction, either with the anchoring group or with the quinuclidine moiety. Repulsive interactions are more pronounced for bulky substituents, and, therefore, cinchonidine mostly produces R-alcohols. The repulsive interactions are all on the side where the bulkier group of a pro-(S)-adsorbed species should be accommodated, due to the anchoring moiety for the surface open(3) conformations, and due to the quinuclidine skeleton for the surface open(5) conformations.

For the surface open(5) conformations the surface was graphically extended and MP accommodated in a position that allowed for interactions with QN. Figure 14.12 shows the interaction geometries proposed by Baiker and coworkers [244]. They found that for the TSO(5) the distance between protonated quinuclidine and keto-carbonyl oxygen was 2.5 Å, while for PSO(5) the same distance was only 2.2 Å. For both tilted and parallel conformations the interaction between the hydroxylic proton and ester carbonyl oxygen is within reach. This double interaction is possible also for the pro-(S) adsorption mode of MP, but previous studies of the same group [236] reported on the interaction of an ammonium ion with MP and showed that an ammonium ion has a stronger interaction energy with the keto-carbonyl than with the ester carbonyl, which would lead to an energy difference between pro-(R) and pro-(S) complexes that favors the pro-(R).

Cinchonidine promotes the enantioselective hydrogenation of ethyl pyruvate even when the hydroxy group is O-methylated. This behavior supports the shape discrimination rather than the participation of a guiding group that is able to produce a second interaction. Nevertheless, it is also possible that the double interaction takes place when the hydroxy group is present, and when it becomes O-methylated, the sole shape discrimination could act to differentiate the binding of a pro-(R)- and a pro-(S)-adsorbed species.

The modulation of the chiral pocket, possible by O-alkylation of the hydroxylic moiety, was found to be in line with the observed experimental variations of enantio-discrimination when O-alkyl-modified cinchonidine was used as surface modifier.

14.6 HETEROGENEIZED HOMOGENEOUS CATALYSTS

One promising strategy to combine the best properties of the homogeneous and heterogeneous catalysts is the heterogenization or immobilization of active metal complexes on supports, which may be separated by filtration or precipitation [248]. Polymers (linear, noncrosslinked polymers; swellable, slightly crosslinked polymers; and highly crosslinked polymers) or inorganic materials (amorphous oxides such as silica, alumina, zirconia, and ZnO; clay minerals, pillared clays, and LDHs; zeolites such as β-zeolite and Y-zeolite; regular mesoporous structures such as MCM-41 and MCM-48) are usually used as supports.

The most important approaches to immobilize or heterogenize soluble catalysts that have been described in the literature are summarized in Table 14.1 [249].

Anchoring of a highly enantioselective complex on a solid surface, even if it is pure, may reduce its asymmetric induction capability, owing to its interaction with the solid surface, but it is possible to increase e.s. of the anchored complex to the values obtained in solution. To achieve this goal, the tether linking the complex and the solid should be long enough to permit the complex to have a large conformational freedom, and the solid surface has to be modified to reduce the presence of residual silanol groups. An interesting example of such a catalyst is given by Corma and coworkers [153]: chiral vanadyl salen complexes covalently anchored to silica, ITQ-2, or MCM-41 surface. Basically, the methodology consists in anchoring on the surface the 3-mercaptopropylsilyl groups followed by silylation of the rest of the silanol groups. The catalysts were tested in the enantioselective reaction of aldehydes with trimethylsilyl cyanide. Unfortunately, the catalysts are rapidly deactivated and from 85% e.e. the value dropped fast to 63%. The reason for the deactivation of the vanadium complex could be poisoning of the solid, loss of the vanadium metal, degradation of the ligand, or a combination of these possibilities.

However, many of heterogenized chiral catalysts suffer from the leaching of the active metal or the chiral auxiliary into the solvent and from the decrease of e.s. Another problem associated with catalysts made from metal complexes that are attached to either a modified polymer or metal oxide surface is that the techniques used for their preparation are rather specific and are driven by the nature

Table 14.1 Schematic View and Important Properties of Immobilized Complexes

Immobilization method	Covalent binding	Adsorption	Ion pair formation	Entrapment or "ship-in-a-bottle"
Applicability	Broad	Restricted	Restricted	Restricted
Problems	Preparation	Competition with solvents, substrates	Competition with ionic substrates, salts	Size of substrate, diffusion

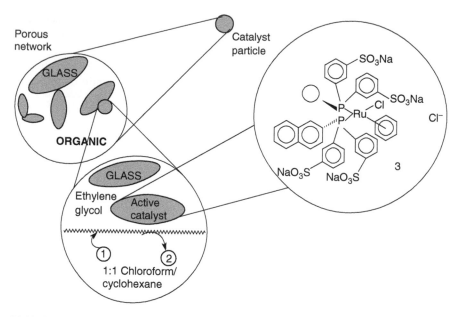

Scheme 14.11 Schematic diagram of the designed heterogeneous catalyst [147a].

of the ligand to be attached. To develop a method to prepare a stable catalyst to avoid these disadvantages is a challenge to the academic community. There are already several successful approaches in this sense and the reason for the catalysts success seems to be the method of immobilization. Therefore, Davis and coworkers [147a] used ethylene glycol as the hydrophilic liquid phase containing the chiral catalyst, which effectively prevents leaching of the complex into the organic phase during the reaction. Used in the enantioselective synthesis of anti-inflammatory agent naproxen, the e.e. values reach 96%. These authors previously created and investigated [159] a supported aqueous-phase catalyst comprising a Ru(II)-sulphonated–BINAP complex, but the loss of a chloro ligand from the ruthenium through hydrolysis was detrimental to the e.s. Therefore, the authors replaced water with a nonvolatile hydrophobic liquid (ethylene glycol) that can still immobilize the water-soluble complex without cleavage of Ru–Cl. The Ru-leaching, observed due to significant solubility of ethylene glycol in the organic phase, was eliminated by using a hydrophobic organic solvent (1:1 mixture of cyclohexane and chloroform) immiscible with the components of the supported film, but miscible withreactants and products (Scheme 14.11). At the molecular level, this method of immobilization yields a heterogeneous catalyst that is basically the same as its homogeneous analog, thus allowing for high e.s. and activity.

Another interesting example refers to the development of a new type of asymmetric heterogeneous catalysts, prepared by using techniques derived from the surface organometallic chemistry on metals (SOMC/M) [250]. This well-proven methodology comprises the reaction between a supported and a reduced transition metal catalyst with an organometallic compound [251,252]. Previously published papers demonstrated that it was possible to anchor $Sn(C_4H_9)_4$ onto a monometallic catalyst and to generate an organometallic phase (retaining butyl groups on the surface) with very interesting properties in the selective hydrogenation of carbonyl compounds [253,254]. Starting from these informations, Ferretti and coworkers [250] synthesized a new kind of asymmetric heterogeneous catalysts, based on silica-supported Ni, Rh, and Pt, chemically modified with chiral organotin compounds. The systems were tested in the enantioselective hydrogenation of acetophenone to (S)-phenylethanol. The highest e.e. values obtained in this reaction were 20% and selectivity over 97%. Reuse of the catalysts is possible, keeping the selectivity and the e.e. values.

A somewhat similar approach has been developed by Smith and coworkers [224], who published a method to deposit a chiral silyl ether moiety onto a Pd surface through Si–metal bonds.

The so-modified catalysts were tested in the enantioselective hydrogenation of prochiral molecules containing C=C bonds.

To improve the chances for the application of immobilized homogeneous catalysts the following obstacles have to be addressed: (1) preparation of inexpensive chiral ligands with a suitable anchoring group; (2) development of efficient and versatile immobilization methods; and (3) better understanding of the interactions between supports and catalysts.

Sometimes, the activities, the chemo- and enantioselectivities of the heterogeneized systems can be superior to the ones of the homogeneous catalysts. Thomas and coworkers [93] state that an improved performance of a heterogeneous catalyst compared with the homogeneous one can sometimes be explained by the *confinement concept* (Figure 14.4), in which the key point is the favorable substrate-pore wall-chiral directing group interaction. Compared to the homogeneous system, the confinement of the substrate in small pores leads to a larger influence of the chiral directing group on the orientation of the substrate relative to the reactive catalytic center. Owing to their very regular structure and tunable pore diameter, the MCM-type materials seem to be very interesting in this respect. However, in these cases there are two possible drawbacks that lead to lower e.s.'s and activities. The lower activities can be owing to reduced or even blocked mass transport in the small pores, imposing limits to the effective range of substrates that can be utilized, while the reduced e.s.'s can be owing to either a too strong sorption of the complex on the walls of the support, or a restricted environment that prevents the chiral complex to adopt the spacial configuration that is required to induce chiral recognition.

Another solution to the problem of catalyst/product separation is the *biphasic catalysis*. The liquid biphasic catalysis became an attractive technology for potential commercial application of enantioselective homogeneous catalysis. The most important features of such systems are related to the fact that both reaction rate and e.s. may be influenced by the number of ionic groups in water-soluble ligand or by addition of surfactants. Descriptions of water-soluble ligands and the recent results in the rapidly progressing area of biphasic enantioselective catalysis are available in recent reviews [255,256].

14.7 ORGANIC POLYMERS

Chiral polymers represent another group of heterogeneous enantioselective catalysts that have already been applied to the synthesis of active compounds. However, their utilization is still very narrow and only few transformations are catalyzed with useful enantioselectivities.

A more versatile method to use organic polymers in enantioselective catalysis is to employ these as catalytic supports for chiral ligands. This approach has been primarily applied in reactions as asymmetric hydrogenation of prochiral alkenes, asymmetric reduction of ketone and 1,2-additions to carbonyl groups. Later work has included additional studies dealing with Lewis acid-catalyzed Diels–Alder reactions, asymmetric epoxidation, and asymmetric dihydroxylation reactions. Enantioselective catalysis using polymer-supported catalysts is covered rather recently in a review by Bergbreiter [257].

14.8 DIASTEREOSELECTIVE CATALYSIS

Many of the molecules with biological activity contain more than two asymmetric centers and the synthesis of these occurs via a *diastereoselective reaction*.

Almost 50 years ago, Cram outlined a rule (Cram's rule), which proved to be fruitful in understanding, predicting, and controlling diastereoselectivity induced by a remote stereocenter [258,259]. Numerous examples of 1,2 induction have confirmed over the time the predictive character of this rule [260]. Afterwards, other important contributions of Felkin and coworkers and Anh enlarged the concept [261,262]. This rule concerns several aspects: (1) Cram-chelate rule refers to the reactions facilitated by a metal cation when the chelation between the carbonyl group (as example) and one of the substituents is locked in a conformation which allows the attack of a nucleophile species from the side of the small substituent (S), and (2) Cram's rule refers to the cases where chelation cannot occur. The reactive conformation is the consequence of steric reasons. Therefore, the substituents of the vicinal asymmetric atom plays an important role. The attack of the nucleophile will be again preferred from the side of the small substituent. The diastereoselectivity is in this case very sensitive to polarity of the substituents and of the solvents. Even the first rule seems to be more specific to heterogeneous catalysis where both situations are found.

The principal classes of reported heterogeneously catalyzed reactions and the synthesis of biologically active molecules by heterogeneous diastereoselective catalysis are covered in a recent review by De Vos and coworkers [86].

Temperature plays an important contribution in kinetically controlled reactions. High temperature supplies enough energy to the system as the barrier leading to both diastereoisomers can be surmounted, whereas a low temperature makes more probable only the surmount of the barrier corresponding to one of the diastereoisomers.

Another possibility to increase the diastereoselectivity in an asymmetric synthesis can arise from different thermodynamic stabilities of the diastereoisomeric products. If the thermodynamic stabilities of these are different enough, then, under conditions of equilibrium, a complete conversion of the less stable into the more stable can be achieved. For example, the diastereoselective hydrogenation of naphthalene derivates over Pd/C catalyst leads to a mixture of dihydronaphtalenes in which the *cis*-isomer predominates. The conversion of this isomer into the *trans* occurs by changing the properties of the reaction medium, namely by equilibration with a base. For such a purpose, NaOMe in THF can be used [263]. Generally, such an increase in stability in the six-membered rings can result from a rearrangement of the substituents from an axial to an equatorial position.

Catalytic reduction of bridgehead enol lactone over Pd/C indicates that, indeed, the *syn* addition from the *exo* face of the bridgehead double bound establishes the relative configuration of all substituents [264]. Equilibration studies performed in EtONa/EtOH also established that the ratio of the epimers corresponds to an equilibrium mixture. Under mild basic conditions (Na$_2$CO$_3$/ EtOH), the product isomerization occurs to a very small extent. The product distribution is best understood by rapid conformational relaxation to one of the two low-energy half-chair conformations. The stereochemistry is established at the subsequent protonation step. This takes place with a strong preference for axial protonation from the β face at carbon 2 to produce the most stable chair conformation (Scheme 14.12).

The diastereoselectivity in heterogeneous hydrogenation can be directed by coordination between the metal catalyst and a neighbor group that could be OH, NHR, or COOR [265]. In this way, it is possible to facilitate the delivery of hydrogen to only one diastereoface.

The solvent could also exhibit a strong influence in these reactions. Thompson and coworkers [266] have shown that the hydrogenation of a polycyclic alkene on Pt/C is greatly dependent on the

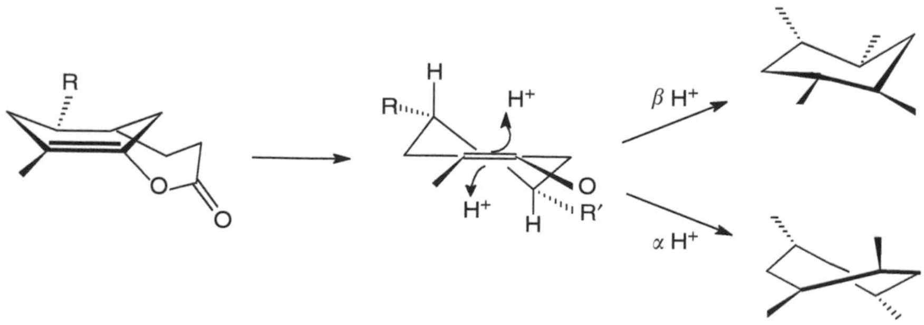

Scheme 14.12 Diastereoselective synthesis of ketoester from an enol lactone.

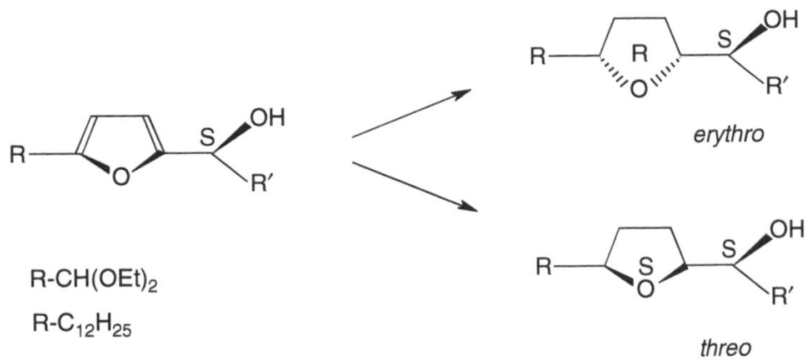

Scheme 14.13 The diastereoselective hydrogenation of 2- and 2,5-substituted furons.

solvent polarity. This effect is due to a competition between the solvent and the catalyst for coordination to the hydroxyl group. This behavior is interpreted, in the opinion of the same authors, as an intramolecular delivery. A solvent dependence of the diastereoselective hydrogenation was also observed for Raney nickel [267]. The hydrogenation of 2- and 2,5-substituted furans showed that, owing to the alcohol used as solvent, it is possible to control the direction of the stereoselection process (*erythro* vs. *threo*).

In isopropyl alcohol, the *erythro*-isomer is formed predominantly. An increase in the polarity of the solvents results in a shift of the diastereoselectivity to the *threo*-isomer and in methanol this becomes the main product. This example offers another possible influence of the solvent. An explanation of the determined diastereoselectivies can be different conformations preferred in the various solvents. In isopropyl alcohol the intramolecular hydrogen bonds can stabilize the conformation and the attack of hydrogen can occur only with the *erythro*-isomer formation (Scheme 14.13). In a polar solvent, this possibility is much diminished and therefore the *threo*-isomer is preferred.

The presence of more chiral centers in the reactant molecule exhibits a positive influence in diastereoselective synthesis. The *conformation of the reactants* mediates these syntheses. In addition, part of the molecules could exhibit steric or electronic effects which can amplify or diminish the diastereoselectivity.

The *chiral auxiliaries* anchored to the substrate, which is subjected to diastereoselective catalysis, is another factor that can control these reactions. These *chiral auxiliaries* should be easily removed after reduction without damaging the hydrogenated substrate. A representative example in this sense is given by Gallezot and coworkers [268]. They used (-)mentoxyacetic acid and various (S)-proline derivates as *chiral auxiliaries* for the reduction of o-cresol and o-toluic acid on Rh/C. A successful use of proline derivates in asymmetric catalysis has also been reported by Harada and coworkers [269,270]. The nature of the solvent only has a slight influence on the d.e. [271].

The pretreatment of the catalysts could also influence their catalytic behavior. Besson and coworkers [271a] reported a change in the diastereoselectivity in the synthesis of N-(2-methylbenzoyl)-(S)-proline methyl esther as a function of the conditions in which the catalysts (Rh/C, Rh/graphite, and Rh/Al$_2$O$_3$) were treated before the reaction. They interpreted these modifications considering (1) changes in the oxidation states of the metal, stressing on the idea that the reduction state can govern the selective interaction between the substrate and the metal; (2) changes in the morphology of the metal, and (3) changes in the support surface.

An important example of heterogeneous diastereoselective synthesis by catalytic way is the synthesis of prostaglandines (a family of compounds having the 20-carbon skeleton of the prostanoic acid) (Scheme 14.14). Naturally, these molecules are biosynthesized via a cyclooxygenase enzyme system that is widely distributed in mammalian tissues. Many of the synthetic routes [272] involve the diastereoselective hydrogenation of a carbonylic bond having a C=C double bond

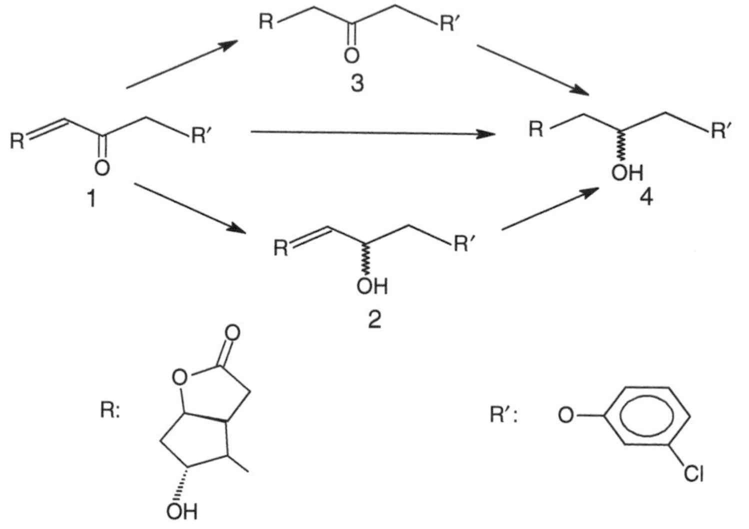

Scheme 14.14 Possible routes in hydrogenation of a prostaglandin intermediate.

in a α-position. From the thermodynamic point of view, it is possible to obtain a mixture of diastereomers but from a practical point of view, only the allylic alcohol is important and, to exhibit biological activity, this should be in the form (11S, 15S).

The hydrogenation of this molecule using heterogeneous catalyst as the classical Pd/C leads to hydrogenation of both the C=C and C=O double bonds. Therefore, the identification of new selective solid catalysts is very important. As a first change, it was found that replacing Pd by Ru had a beneficial effect because of the preference of this metal to hydrogenate the C=O double bond. But the support was also found to exhibit a strong influence in this reaction [273].

These results should be interpreted in terms of the different accessibility of the organic molecule to the active sites and of differences in the metal size. However, it is worth noting that using Ru supported on these molecular sieves, it is possible to reach a 100% diastereomeric excess in the (11S, 15R) form for about 80% selectivity to allyl alcohol. These results also indicated the fact that the reaction is very sensible to the solvent nature. Methanol was found to enhance the reaction rate compared with THF or ethanol.

Another interesting observation using molecular sieves supported by Ru catalysts is the one related to the effect of the modifier. The use of an alkaloid-type modifier, like cinchonidine, enhances the d.e. to the same (11S, 15R) form. But the use of TA leads to a decrease in the d.e., and the addition of pivalic acid in mixture with TA leads to a spectacular inversion in diastereoselectivity, resulting in the isomer with natural-like configuration, namely (11S, 15S).

The effect of the modifiers appears to be more evident in the case of Ru-MCM-41 catalysts. The size-controlled mesoporous channels of MCM-41 affords a better interaction of the prostaglandin intermediates with modifiers [274,275].

One of the most active and selective catalysts in this kind of reaction is undoubtedly Ir/support, as recently demonstrated by Jacobs and coworkers [276]. Therefore, by combining the carbonyl affinity of metallic iridium with the promotion effect of the H-β zeolite, which is a strong Brönsted acid, one can reduce a large variety of unsaturated ketones and aldehydes to allylic alcohols with high conversions, selectivities, and diastereoselectivities.

From the chemical point of view, in diastereoselective syntheses, several kinds of reactions like hydrogenation [273,277–286], hydrogenolysis [287–293], isomerization [294], and epoxidation [295–300] are involved. Hydrogenation is the most important application of heterogeneous catalysis because of its potential to produce a wide variety of chiral functional groups.

Up to date, the scientists involved in these investigations used only few commercial catalysts, predominantly Pd/C, Pt/C, PtO_2, or Raney nickel. But except for a few cases, these catalysts are not selective enough and only a mixture of diastereomers results from the reactions. The separation of the pure diastereomers is expensive and requires a lot of energy and reagents. Even more complicate is the case of diastereoselective oxidation.

14.9 CONCLUSIONS

After a period of stagnation, the area of heterogeneous asymmetric catalysis has again become a fashionable topic. More and more research groups, both from academic and industry, are involved in this interesting and important field. From the catalytic preparation point of view, three general strategies have been successful: (1) the chiral modification of metallic heterogeneous catalysts; (2) the heterogenization of homogeneous chiral complexes or chiral ligands; and (3) the application of organic polymers as catalysts or as supports for chiral complexes. However, there are differences among these categories. Modified catalysts and chiral polymers have already been applied for the synthesis of active compounds or have been developed for technical applications. But their application is still narrow and only few transformations are catalyzed with useful e.e.'s. However, the use of polymer supports continues to expand because of potential practical applications.

On the same lines, the large interest concerning the synthesis of biological active molecules requires more and more effective catalysts and many of these syntheses involve diastereoselective reactions. Unfortunately, at this moment, there is a very large gap between the homogeneous and the heterogeneous catalysis in this field. Therefore, it is very important in the future, that as scientists working in the field of heterogeneous catalysis have to pay more attention not only to enantioselective catalysis but also to heterogeneous diastereoselective catalysis for natural total synthesis.

REFERENCES

1. J.A. Cusumano, in *Perspectives in Catalysis*, J.M. Thomas, K.J. Zamaraev (Eds.), International Union of Pure & Applied Chemistry, London, Oxford Blackwell, Scientific Publications, 1992.
2. W.S. Knowles, M.J. Sabacky, B.D. Vineyard, D.J. Weinkauf, *J. Am. Chem. Soc.* 97 (1975) 2567.
3. S. Otsuka, K. Tani, *Synthesis* (1991) 665.
4. T. Ohta, H. Takaya, M. Kitamura, K. Nagai, R. Noyori, *J. Org. Chem.* 52 (1987) 3176.
5. K.B. Sharpless, *Chemtech.* 15 (1985) 692.
6. E.N. Jacobsen, I. Markó, W.S. Mungall, G. Scröder, K.B. Sharpless, *J. Am. Chem. Soc.* 110 (1988) 1968.
7. E.N. Jacobsen, in *Comprehensive Organometallic Chemistry II*, E.W. Abel, F.G.A. Stone, E. Wilkinson (Eds.), Pergamon, New York, 12, 1995, p. 1097.
8. T. Katsuki, *J. Mol. Catal. A* 113 (1996) 87.
9. H. Brunner, W. Zettlmeier, in *Handbook of Enantioselective Catalysis with Transition Metal Compounds*, VCH, Weinheim, 1993.
10. E.N. Jacobsen, A. Pfaltz, H. Yamamoto, in *Comprehensive Asymmetric Catalysis*, Springer, Berlin, 1999.
11. R. Noyori, *Asymmetric Catalysis in Organic Synthesis*, Wiley Interscience, New York, 1994.
12. M.J. Fish, D.F. Ollis, *Catal. Rev., Sci. Eng.* 18 (1978) 259.
13. Y. Nitta, T. Imanaka, S. Teranishi, *J. Catal.* 80 (1983) 31.
14. D.R. Richards, H.H. Kung, W.M.H. Sachtler, *J. Mol. Catal.* 36 (1986) 329.
15. L. Fu, H.H. Kung, W.M.H. Sachtler, *J. Mol. Catal.* 42 (1987) 29.
16. H. Brunner, M. Muschiol, T. Wischert, *Tetrahedron Asymm.* 1 (1990) 159.
17. M.A. Keane, G. Webb, *J. Chem. Soc. Chem. Commun.* (1991) 1619.
18. A. Tai, T. Kikukawa, T. Sugimura, Y. Inone, T. Osawa, S. Fujii, *J. Chem. Soc. Chem. Commun.* (1991) 795.
19. M.A. Keane, G. Webb, *J. Catal.* 136 (1992) 1.

20. M.A. Keane, G. Webb, *J. Mol. Catal.* 73 (1992) 91.
21. M.A. Keane, *Langmuir* 10 (1994) 4560.
22. T. Harada, T. Kawamura, S. Haikawa, T. Osawa, *J. Mol. Catal.* 93 (1994) 211.
23. T. Osawa, T. Harada, A. Tai, *Catal. Taday* 37 (1997) 465.
24. T. Osawa, A. Ozawa, T. Harada, O. Takayasu, *J. Mol. Catal. A* 154 (2000) 271.
25. T. Osawa, S. Mita, A. Iwai, O. Takayasu, H. Hashiba, S. Hashimoto, T. Harada, I. Matsuura, *J. Mol. Catal. A* 157 (2000) 207.
26. A. Tai, T. Sugimura, in *Chiral Catalyst Immobilization and Recycling*, D.E. De Vos, I.F.J. Vankelecom, P.A. Jacobs (Eds.), Wiley-VCH, Weinheim, 2000, p. 173.
27. T. Osawa, Y. Hayashi, A. Ozawa, T. Harada, O. Takayasu, *J. Mol. Catal. A* 169 (2001) 289.
28. V. Humblot, S. Haq, C. Muryn, W.A. Hofer, R. Raval, *J. Am. Chem. Soc.* 124 (2002) 503.
29. M.O. Lorenzo, V. Humblot, P. Murray, C.J. Baddeley, S. Haq, R. Raval, *J. Catal.* 205 (2002) 123.
30. T. Sugimura, S. Nakagawa, A. Tai, *Bull. Chem. Soc. Jpn.* 75 (2002) 355.
31. M. Garland, H.-U. Blaser, *J. Am. Chem. Soc.* 112 (1990) 7049.
32. J.L. Margitfalvi, B. Minder, E. Tálas, L. Botz, A. Baiker, in *New Frontiers in Catalysis*, L. Guczi, F. Solymosi, P. Tétényi (Eds.), Elsevier Science Publishers, Amsterdam, The Netherlands and Akadémiai Kiadó, Budapest, Hungary, 1993, p. 2471.
33. H.-U. Blaser, M. Garland, H.P. Jallet, *J. Catal.* 144 (1993) 569.
34. R.L. Augustine, S.K. Tanielyan, L.K. Doyle, *Tetrahedron Asymm.* 4 (1993) 1803.
35. K.E. Simons, P.A. Meheux, S.P. Griffits, I.M. Sutherland, P. Johnston, P.B. Wells, A.F. Carley, M.K. Rajumon, M.W. Roberts, A. Ibbotson, *Recl. Trav. Chim. Pays-Bas* 113 (1994) 465.
36. U.K. Singh, R.N. Landau, Y. Sun, C. LeBlond, D.G. Blackmond, S.K. Tanielyan, R.L. Augustine, *J. Catal.* 154 (1995) 91.
37. B. Minder, M. Schürch, T. Mallat, A. Baiker, *Catal. Lett.* 31 (1995) 143.
38. J.L. Margitfalvi, in *Catalysis of Organic Reactions*, M.G. Scases, M.L. Rouniev (Eds.), Marcel Dekker, Inc., New York, 1995, p. 189.
39. J.L. Margitfalvi, M. Hegedüs, *J. Mol. Catal. A* 107 (1995) 281.
40. Y. Sun, R.N. Landau, J. Wang, C. LeBlond, D.G. Blackmond, *J. Am. Chem. Soc.* 118 (1996) 1348.
41. J.L. Margitfalvi, M. Hegedüs, E. Tfirst, *Tetrahedron Asymm.* 7 (1996) 571.
42. A. Tungler, T. Máthé, K. Fodor, R.A. Sheldon, P. Gallezot, *J. Mol. Catal. A* 108 (1996) 145.
43. H.-U. Blaser, H.-P. Jallet, M. Müller, M. Studer, *Catal. Lett.* 37 (1997) 441.
44. A. Baiker, H.-U. Blaser, in *Handbook of Heterogeneous Catalysis*, G. Earl, H. Knözinger, J. Weitkamp (Eds.) Wiley-VCH, Weinheim, Vol. 5, 1997, p. 2422.
45. R.L. Augustine, S.K. Tanielyan, *J. Mol. Catal. A* 118 (1997) 79.
46. K. Fodor, A. Tungler, T. Máthé, S. Szabó, R.A. Sheldon, in *Catalysis of Organic Reactions*, F.E. Herkes (Ed.), Marcel Dekker, Inc., New York, 1998, p. 115.
47. B. Török, K. Felföldi, G. Szakonyi, K. Balázsik, M. Bartók, *Catal. Lett.* 52 (1998) 81.
48. F.M. Bohnen, A. Gamez, D.G. Blackmond, *J. Catal.* 179 (1998) 335.
49. D.G. Blackmond, *J. Catal.* 176 (1998) 271.
50. H.-U. Blaser, H.-P. Jalett, M. Garland, M. Studer, H. Thies, A. Wirth-Tijani, *J. Catal.* 173 (1998) 282.
51. J.L. Margitfalvi, E. Tfirst, M. Hegedüs, E. Tálas, in *Catalysis of Organic Reactions*, F.E. Herkes (Ed.), Marcel Dekker, Inc., New York, 1998, p. 531.
52. G.J. Hutchings, *Chem. Commun.* (1999) 301.
53. J.L. Margitfalvi, E. Tálas, *Appl. Catal. A* 182 (1999) 65.
54. J.L. Margitfalvi, E. Tálas, M. Hegedûs, *Chem. Commun.* (1999) 645.
55. T. Bürgi, Z. Zhou, N. Künzle, T. Mallat, A. Baiker, *J. Catal.* 183 (1999) 405.
56. P.B. Wells, K.E. Simons, J.A. Slipszenko, S.P. Griffiths, D.F. Ewing, *J. Mol. Catal. A* 146 (1999) 159.
57. K. Balázsik, K. Szöri, K. Felföldi, B. Töröc, M. Bartók, *Chem. Commun.* (2000) 555
58. M. von Arx, T. Mallat, A. Baiker, *J. Catal.* 193 (2000) 161.
59. S.P. Griffiths, P. Johnston, P.B. Wells, *Appl. Catal. A* 191 (2000) 193.
60. J.L. Margitfalvi, E. Tálas, E. Tfirst, C.V. Kumar, A. Gergely, *Appl. Catal. A* 191 (2000) 177.
61. A. Baiker, *J. Mol. Catal. A* 163 (2000) 205.
62. D. Ferri, T. Bürgi, K. Borszeky, T. Mallat, A. Baiker, *J. Catal.* 193 (2000) 139.
63. J.M. Bonello, F.J. Williams, A.K. Santra, R.M. Lambert, *J. Phys. Chem. B* 104 (2000) 9696.
64. J.M. Bonello, R.M. Lambert, K. Künzle, A. Baiker, *J. Am. Chem. Soc.* 122 (2000) 9864.

65. P.B. Wells, R.P.K. Wells, in *Chiral Catalyst Immobilization and Recycling*, D.E. De Vos, I.F.J. Vankelecom, P.A. Jacobs (Eds.), Wiley-VCH, Weinheim, 2000, p. 123.
66. A. Baiker, in *Chiral Catalyst Immobilization and Recycling*, D.E. De Vos, I.F.J. Vankelecom, P.A. Jacobs (Eds.), Wiley-VCH, Weinheim, 2000, p. 155.
67. A. Vargas, T. Bürgi, A. Baiker, *J. Catal.* 197 (2001) 378.
68. M. von Arx, T. Mallat, A. Baiker, *Angew. Chem. Int. Ed.* 40 (2001) 2302.
69. J.W. de M. Carneiro, C. da S.B. de Oliveira, F.B. Passos, D.A.G. Aranda, P.R.N. de Souza, O.A.C. Antunes, *J. Mol. Catal. A* 170 (2001) 235.
70. K. Szöri, B. Török, K. Felföldi, M. Bartók, in *Catalysis of Organic Reactions*, M.E. Ford (Ed.), Marcel Dekker, Inc., New York, 2001, p. 489.
71. M. Bartók, K. Balázsik, G. Szöllösi, T. Bartók, *J. Catal.* 205 (2002) 168.
72. K. Szöri, M. Sutyinszki, K. Felföldi, M. Bartók, *Appl. Catal. A* 237 (2002) 275.
73. M. Bartók, G. Szöllösi, K. Balázsik, T. Bartók, *J. Mol. Catal. A* 177 (2002) 299.
74. H. Brunner, in *Applied Homogeneous Catalysis with Organometallic Compounds*, B. Cornils, W.A. Herrmann (Eds.), Wiley-VCH, Weinheim, 1996.
75. D.J. Bayston, M.E.C. Polywka, in *Chiral Catalyst Immobilization and Recycling*, D.E. De Vos, I.F.J. Vankelecom, P.A. Jacobs (Eds.), Wiley-VCH, Weinheim, 2000, p. 211.
76. A. Corma, M. Iglesias, C. del Pino, F. Sanchez, *J. Chem. Soc. Chem. Commun.* (1991) 1253.
77. A. Corma, M. Iglesias, C. del Pino, F. Sanchez, *J. Organometal. Chem.* 431 (1992) 233.
78. A. Corma, M. Iglesias, M.V. Martin, J. Rubio, F. Sanchez, *Tetrahedron Asymm.* 3 (1992) 845.
79. A. Corma, H. Garcia, A. Moussaif, M.J. Sabater, R. Zniber, A. Redouane, *Chem. Commun.* (2002) 1058.
80. T.L. Jacobs, D. Dankner, *J. Org. Chem.* 22 (1957) 1424.
81. A.R. Daniewski, T. Kowalczyk-Przewloka, *J. Org. Chem.* 50 (1985) 2976.
82. H. Yamashita, *Bull. Chem. Soc. Jpn.* 61 (1988) 1213.
83. M.E. Davis, *Acc. Chem. Res.* 26 (1993) 111.
84. A. Choudhury, J.R. Moore, M.E. Pierce, J.M. Fortunak, I. Valvis, P.N. Confalone, *Org. Process Res. Dev.* 7 (2003) 324.
85. M. Beson, C. Pinel, *Top. Catal.* 5 (1998) 25.
86. D.E. De Vos, M. De bruyn, V.I. Pârvulescu, F.G. Cocu, P.A. Jacobs, in *Chiral Catalyst Immobilization and Recycling*, D.E. De Vos, I.F.J. Vankelecom, P.A. Jacobs (Eds.), Wiley-VCH, Weinheim, 2000, p. 284.
87. R.A. Sheldon, in *Chirotechnology*, Marcel Dekker, Inc, New York, 1993.
88. C. Puchot, O. Samuel, E. Dunach, S. Zhao, C. Agami, H.B. Kagan, *J. Am. Chem. Soc.* 108 (1986) 2353.
89. N. Oguni, Y. Matsuda, T. Kaneko, *J. Am. Chem. Soc.* 110 (1988) 7877.
90. R. Noyori, M. Kitamura, *Angew. Chem., Int. Ed. Engl.* 30 (1991) 49.
91. K. Mikami, Y. Motoyama, M. Terada, *J. Am. Chem. Soc.* 116 (1994) 2812.
92. D.G. Blackmond, *J. Am. Chem. Soc.* 119 (1997) 12934.
93. J.M. Thomas, T. Maschmeyer, B.F.G. Johnson, D.S. Shephard, *J. Mol. Catal. A* 141 (1999) 139.
94. Y. Izumi, *Bull. Chem. Soc. Jpn.* 32 (1959) 932.
95. Y. Izumi, *Bull. Chem. Soc. Jpn.* 32 (1959) 936.
96. Y. Izumi, *Bull. Chem. Soc. Jpn.* 32 (1959) 942.
97. S. Julia, J. Masana, J.C. Vega, *Angew. Chem. Int. Ed. Engl.* 19 (1980) 929.
98. S. Itsuno, M. Sakakura, K. Ito, *J. Org. Chem.* 55 (1990) 6047.
99. W. Kroutil, P. Mayon, M.E. Lasterra-Sanchez, S.J. Maddrell, S.M. Roberts, S.R. Thornton, C.J. Todd, M. Tuter, *J. Chem. Soc. Chem. Commun.* (1996) 845.
100. Y. Izumi, M. Imaida, H. Fukawa, S. Akabori, *Bull. Chem. Soc. Jpn.* 36 (1963) 21.
101. Y. Izumi, *Adv. Catal.* 32 (1983) 215.
102. G. Webb, P.B. Wells, *Catal. Today* 12 (1992) 319.
103. Y. Orito, S. Imai, S. Niwa, N.G. Hung, *J. Synth. Org. Chem. Jpn.* 37 (1979) 173.
104. S. Niwa, S. Imai, Y. Orito, *J. Chem. Soc. Jpn.* (1982) 137.
105. S. Bhaduri, V.S. Darshane, K. Sharma, D. Mukesh, *J. Chem. Soc. Chem. Commun.* (1992) 1738.
106. W. Reschetilowski, U. Böhmer, J. Wiehl, in *Zeolites and Related Microporous Materials: State of the Art*, J. Weitkamp, H.G. Karge, H. Pfeifer, W. Hölderich (Eds.), *Stud. Surf. Sci. Catal.* Elsevier Science BV, Amsterdam, The Netherlands. 84 (1994) 2021.
107. K. Borszeky, T. Mallat, R. Aeschiman, W.B. Schweizer, A. Baiker, *J. Catal.* 161 (1996) 451.
108. B. Minder, M. Schürch, T. Mallat, A. Baiker, T. Heinz, A. Pfaltz, *J. Catal.* 160 (1996) 261.

109. G.-Z. Wang, T. Mallat, A. Baiker, *Tetrahedron Asymm.* 8 (1997) 2133.
110. T. Mallat, M. Bodmer, A. Baiker, *Catal. Lett.* 44 (1997) 95.
111. J. U. Köhler, J.S. Bradley, *Catal. Lett.* 45 (1997) 203.
112. T. Mallat, C. Brönnimann, A. Baiker, *Appl. Catal. A* 149 (1997) 103.
113. S.P. Griffiths, P.B. Wells, K.G. Griffin, P. Johnston, in *Catalysis of Organic Reactions*, F.E. Herkes (Ed.), Marcel Dekker, Inc., New York, 1998, p. 89.
114. M. Bodmer, T. Mallat, A. Baiker, in *Catalysis of Organic Reactions*, F.E. Herkes (Ed.), Marcel Dekker, Inc., New York, 1998, p. 75.
115. M. Schürch, N. Künzle, T. Mallat, A. Baiker, *J. Catal.* 176 (1998) 569.
116. M. Schürch, T. Heinz, R. Aeschimann, T. Mallat, A. Pfaltz, A. Baiker, *J. Catal.* 173 (1998) 187.
117. M. Studer, V. Okafor, H.-U. Blaser, *Chem. Commun.* (1998) 1053.
118. J.A. Splipszenko, S.P. Griffiths, P. Johnston, K.E. Simons, W.A.H. Vermeer, P.B. Wells, *J. Catal.* 179 (1998) 267.
119. N. Künzle, A. Szabo, M. Schürch, G. Wang, T. Mallat, A. Baiker, *Chem. Commun.* (1998) 1377.
120. K. Balázsik, B. Török, G. Szakonyi, M. Bartók, *Appl. Catal. A* 182 (1999) 53.
121. K. Borszeky, T. Bürgi, Z. Zhaohui, T. Mallat, A. Baiker, *J. Catal.* 187 (1999) 160.
122. A. Tungler, Y. Nitta, K. Fodor, G. Farkas, T. Máthé, *J. Mol. Catal. A* 149 (1999) 135.
123. N. Künzle, R. Hess, T. Mallat, A. Baiker, *J. Catal.* 186 (1999) 239.
124. M. Studer, S. Burkhardt, H.-U. Blaser, *Chem. Commun.* (1999) 1727.
125. P.J. Collier, J.A. Iggo, R. Whyman, *J. Mol. Catal. A* 146 (1999) 149.
126. X. Zuo, H. Liu, J. Tian, *J. Mol. Catal. A* 157 (2000) 217.
127. W.-R. Huck, T. Mallat, A. Baiker, *J. Catal.* 193 (2000) 1.
128. M. Studer, S. Burkhardt, A.F. Indolese, H.-U. Blaser, *Chem. Commun.* (2000) 1327.
129. S.D. Jackson, S.R. Watson, G. Webb, P.B. Wells, N.C. Young, in *Catalysis of Organic Reactions*, M.E. Ford (Ed.), Marcel Dekker, Inc., New York, 2001, p. 477.
130. E. Toukoniitty, P. Mäki-Arvela, M. Kuzma, A. Villela, A.K. Neyestanaki, T. Salmi, R. Sjöholm, R. Leino, E. Laine, D. Yu. Murzin, *J. Catal.* 204 (2001) 281.
131. Y. Huang, J. Chen, R. Li, Y. Li, L.-e Min, X. Li, *J. Mol. Catal. A* 170 (2001) 143.
132. O.J. Sonderegger, T. Bürgi, A. Baiker, *J. Catal.* 215 (2003) 116.
133. N. Künzle, T. Mallat, A. Baiker, *Appl. Catal. A* 238 (2003) 251.
134. Y. Nitta, K. Kobiro, *Chem. Lett.* (1995) 165.
135. H.-U. Blaser, M. Müller, *Stud. Surf. Sci. Catal.* 59 (1991) 73.
136. H.-U. Blaser, *Tetrahedron Asymm.* 2 (1991) 843.
137. A. Baiker, *J. Mol. Catal. A: Chem.* 115 (1997) 473.
138. P.B. Wells, A.G. Wilkinson, *Top. Catal.* 5 (1998) 39.
139. B.M. Choudary, V.L.K. Valli, A. Durga Prasad, *J. Chem. Soc. Chem. Commun.* (1990) 1186.
140. B.M. Choudary, S. Shobha Rani, N. Narender, *Catal. Lett.* 19 (1993) 299.
141. T. Moriguchi, Y.G. Guo, S. Yamamoto, Y. Matsubara, M. Yoshihara, T. Maeshima, *Chem. Express* 7 (1992) 625.
142. K. Morihara, S. Kawasaki, M. Kofuji, T. Shimada, *Bull. Chem. Soc. Jpn.* 66 (1993) 906.
143. C. Cativiela, F. Figueras, J.M. Fraile, J.I. Garcia, J.A. Mayoral, E. Pires, A.J. Royo, in *Abstracts, Europacat-1*, Montpellier, 1993, p. 395.
144. S. Feast, D. Bethell, P.C. Bulman Page, F. King, C.H. Rochester, M.R.H. Siddiqui, D.J. Willock, G.J. Hutchings, *J. Chem. Soc. Chem. Commun.* (1995) 2409.
145. G. Sundarababu, M. Leibovitch, D.R. Corbin, J.R. Scheffer, V. Ramamurthy, *Chem. Commun.* (1996) 2159.
146. J.C. Bailar, *Catal. Rev.* 10 (1974) 17.
147. (a) K.T. Wan, M.E. Davis, *Nature* 370 (1994) 449; (b) W. Kahlen, H.H. Wagner, W.F. Hölderich, *Catal. Lett.* 54 (1998) 85.
148. S.K. Tanielyan, R.L. Augustine, in *Catalysis of Organic Reactions*, F.E. Herkes (Ed.), Marcel Dekker, Inc., New York, 1998, p. 101.
149. M.-Y. Yin, G.-L. Yuan, Y.-Q. Wu, M.-Y. Huang, Y.-Y. Jiang, *J. Mol. Catal. A* 147 (1999) 93.
150. F.M. de Rege, D.K. Morita, K.C. Ott, W. Tumas, R.D. Broene, in *Catalysis of Organic Reactions*, M.E. Ford (Ed.), Marcel Dekker, Inc., New York, 2001, p. 439.
151. R.L. Augustine, S.K. Tanielyan, S. Anderson, H. Yang, Y. Gao, in *Catalysis of Organic Reactions*, M.E. Ford (Ed.), Marcel Dekker, Inc., New York, 2001, p. 497.

152. J.A.M. Brandts, J.G. Donkervoort, C. Ansems, P.H. Berben, A. Gerlach, M.J. Burk, in *Catalysis of Organic Reactions*, M.E. Ford (Ed.), Marcel Dekker, Inc., New York, 2001, p. 573.
153. C. Baleizão, B. Gigante, H. Garcia, A. Corma, *J. Catal.* 215 (2003) 199.
154. H.M. Hultman, M. de Lang, M. Nowotny, I.W.C.E. Arends, U. Hanefeld, R.A. Sheldon, T. Maschmeyer, *J. Catal.* 217 (2003) 264.
155. B. Pugin, M. Muller, *Stud. Surf. Sci. Catal.* 78 (1993) 107.
156. M.-Y. Yin, G.-L. Yuan, Y.-Q. Wu, M.-Y. Huang, Y.-Y. Jiang, *J. Mol. Catal. A* 147 (1999) 89.
157. R. ter Halle, B. Colasson, E. Schulz, M. Spagnol, M. Lemaire, *Tetrahedron Lett.* 41 (2000) 643.
158. P.K. Dhal, B.B. De, S. Sivaram, *J. Mol. Catal. A* 177 (2001) 71.
159. K.T. Wan, M.E. Davis, *J. Catal.* 148 (1994) 1.
160. C.E. Song, E.J. Roh, S.-gi Lee, I.O. Kim, *Tetrahedron Asymm.* 6 (1995) 2687.
161. B.B. Lohray, E. Nandanan, V. Bhushan, *Tetrahedron Asymm.* 7 (1996) 2805.
162. J. Matsui, I.A. Nicholls, T. Takeuchi, *Tetrahedron Asymm.* 7 (1996) 1357.
163. C. Bolm, A. Gerlach, *Eur. J. Org. Chem.* (1998) 21.
164. P. Salvadori, D. Pini, A. Petri, *Syn. Lett.* (1999) 1181.
165. P. Salvadori, D. Pini, A. Petri, A. Mandoli, in *Chiral Catalyst Immobilization and Recycling*, D.E. De Vos, I.F.J. Vankelecom, P.A. Jacobs (Eds.), Wiley-VCH, Weinheim, 2000, p. 235.
166. C.T. Kresge, M.E. Leonowicz, W.J. Roth, J.C. Vartuli, J.S. Beck, *Nature* 359 (1992) 710.
167. B.F.G. Johnson, S.A. Raynor, D.S. Shephard, T. Maschmeyer, J.M. Thomas, G. Sankar, S. Bromley, R. Oldroyd, L. Gladden, M.D. Mantle, *J. Chem. Soc. Chem. Commun.* (1999) 1167.
168. M.T. Reetz, *Angew. Chem. Int. Ed. Engl.* 40 (2001) 284.
169. M.T. Reetz, *Angew. Chem. Int. Ed.* 41 (2002) 1335.
170. B. Archibald, O. Brummer, M. Devenney, S. Gorer, B. Jandeleit, T. Uno, W.H. Weinberg, T. Weskamp, in *Handbook of Combinatorial Chemistry*, K.C. Nicolaou, R. Hanko, W. Hartwig (Eds.), Wiley-VCH, Weinheim, 2002, p. 885.
171. A.H. Hoveyda, in *Handbook of Combinatorial Chemistry*, K.C. Nicolaou, R. Hanko, W. Hartwig (Eds.), Wiley-VCH, Weinheim, 2002, p. 991.
172. C. Markert, A. Pfaltz, *Angew. Chem. Int. Ed. Engl.* 43 (2004) 2498.
173. A. Tai, T. Kikukawa, T. Sugimura, Y. Inoue, S. Abe, T. Osawa, T. Harada, *Bull. Chem. Soc. Jpn.* 67 (1994) 2473.
174. A. López-Martinez, M.A. Keane, *J. Mol. Catal. A* 153 (2000) 257.
175. T. Osawa, T. Harada, A. Tai, *J. Mol. Catal.* 87 (1994) 333.
176. J. Barrault, M. Seffen, C. Forquy, R. Brouard, *Stud. Surf. Sci. Catal.* 41 (1988) 361.
177. J. Barrault, S. Brunet, N. Essayem, *J. Mol. Catal.* 77 (1992) 321.
178. L. H. Gross, R. Rys, *J. Org. Chem.* 39 (1974) 2429.
179. T. Harada, Y. Izumi, *Chem. Lett.* (1978) 1195.
180. G. Witmann, G. Göndös, M. Bartók, *Helv. Chim. Acta* 73 (1990) 635.
181. A. Hoek, W.M.H. Sachtler, *J. Catal.* 58 (1979) 276.
182. Y. Nitta, T. Imanaka, S. Teranishi, *J. Catal.* 96 (1985) 429.
183. T. Osawa, T. Harada, A. Tai, *J. Catal.* 121 (1990) 7.
184. T. Harada, S. Onaka, A. Tai, Y. Izumi, *Chem. Lett.* (1977) 1131.
185. Y. Nitta, F. Sekine, T. Imanaka, S. Teranishi, *Bull. Chem. Soc. Jpn.* 54 (1981) 980.
186. M. Bartók, in *Stereochemistry of Heterogeneous Metal Catalysts*, Wiley, New York, 1985, p. 511.
187. Y. Nitta, F. Sekine, T. Imanaka, S. Teranishi, *J. Catal.* 74 (1982) 382.
188. Y. Nitta, O. Yamanishi, F. Sekine, T. Imanaka, S. Teranishi, *J. Catal.* 79 (1983) 475.
189. Y. Izumi, *Angew. Chem. Int. Ed. Engl.* 10 (1971) 371.
190. A. Hoek, H.M. Woerde, W.M.H. Sachtler, *Stud. Surf. Sci. Catal.* 7 (1981) 376.
191. A. Bennett, S. Christie, M.A. Keane, R.D. Peacock, G. Webb, *Catal. Today* 10 (1991) 363.
192. L.J. Bostelaar, W.M.H. Sachtler, *J. Mol. Catal.* 27 (1984) 387.
193. T. Osawa, *Chem. Lett.* (1985) 1609.
194. G. Webb, in *Chiral Reactions in Heterogeneous Catalysis*, G. Jannes, V. Dubois (Eds.), Plenum Press, New York, 1995, p. 61.
195. W.M.H. Sachtler, in *Catalysis in Organic Reactions*, L. Augustine (Ed.), *Chem. Ind.*, Marcel Dekker, New York, 22 (1985) 189.
196. E.N. Lipyart, V.A. Pavlov, E.I. Klabunovskii, *Kinet. Katal.* 12 (1971) 1491.
197. T. Harada, M. Yamamoto, S. Onaka, M. Imaida, A. Tai, Y. Izumi, *Bull. Chem. Soc. Jpn.* 54 (1981) 2323.

198. T. Harada, T Osawa, in *Chiral Reactions in Heterogeneous Catalysis*, G. Jannes, V. Dubois (Eds.), Plenum Press, New York, 1995, p. 83.

199. T. Osawa, A. Tai, Y. Imachi, S. Takasaki, *Chiral Reactions in Heterogeneous Catalysis*, G. Jannes, V. Dubois (Eds.), Plenum Press, New York, 1995, p. 75.

200. M.A. Keane, G. Webb, *J. Catal.* 136 (1992) 1; M.A. Keane, *Catal. Lett.* 19 (1993) 197.

201. E.I. Klabunovskii, *Russ. J. Phys. Chem.* 41 (1970) 1370.

202. M.O. Lorenzo, S. Haq, T. Bertrams, P. Murray, R. Raval, C.J. Baddeley, *J. Phys. Chem. B* 103 (1999) 10661.

203. V. Humblot, M.O. Lorenzo, C.J. Baddeley, S. Haq, R. Raval, *J. Am. Chem. Soc.* 126 (2004) 6460.

204. S. De Feyter, A. Gesquiere, K. Wurst, D.B. Amabilino, J. Veciana, F.C. De Schryver, *Angew. Chem. Int. Ed.* 40 (2001) 3217.

205. J.V. Barth, J. Wechesser, G. Trimarchi, M. Vladimirova, A. De Vita, C. Cai, H. Brune, P. Günter, K. Kern, *J. Am. Chem. Soc.* 124 (2002) 7991.

206. G.P. Lopinski, D.J. Moffat, D.D. Wayner, R.A. Wolkow, *Nature* 392 (1998) 909.

207. S.M. Barlow, K.J. Kitching, S. Haq, N.V. Richardson, *Surf. Sci.* 401 (1998) 322.

208. J. Weckesser, J.V. Barth, C. Cai, B. Muller, K. Kern, *Surf. Sci.* 431 (1999) 168.

209. M. Schunack, L. Petersen, A. Kühlne, E. Laegsgaard, I. Stensgaard, I. Johannsen, F. Besenbacher, *Phys. Rev. Lett.* 86 (2001) 456.

210. R. Viswanathan, J.A. Zasadzinski, D.K. Schwartz, *Nature* 368 (1994) 440.

211. M. Böhringer, K. Morgenstern, W.-D. Schneider, R. Berndt, *Angew. Chem. Int. Ed.* 38 (1999) 821.

212. E.I. Klabunovskii, Y.I. Petrov, *Dokl. Acad. Nauk SSSR* 173 (1967) 1125.

213. A. Tai, T. Harada, Y. Hiraki, S. Murakami, *Bull. Chem. Soc. Jpn.* 56 (1983) 1414.

214. T. Osawa, T. Harada, *Bull. Chem. Soc. Jpn.* 57 (1984) 1618.

215. W.-R. Huck, T. Bürgi, T. Mallat, A. Baiker, *J. Catal.* 205 (2002) 213.

216. J.A. Slipszenko, S.P. Griffiths, P. Johnston, K.E. Simons, W.A.H. Vermeer, P.B. Wells, *J. Catal.* 179 (1998) 267.

217. Y. Nitta, K. Kobiro, *Chem. Lett.* (1996) 897.

218. K. Borszeky, T. Mallat, A. Baiker, *Tetrahedron Asymm.* 8 (1997) 3745.

219. K. Borszeky, T. Mallat, A. Baiker, *Catal. Lett.* 41 (1996) 199.

220. Q.-H. Xia, S.-C. Shen, J. Song, S. Kawi, K. Hidajat, *J. Catal.* 219 (2003) 74.

221. A. Solladie-Cavallo, F. Hoernel, M. Schmitt, F. Garin, *J. Mol. Catal.* 195 (2003) 181.

222. Y. Nitta, *Top. Catal.* 13 (2000) 179.

223. I. Kun, B. Török, K. Felföldi, M. Bartok, *Appl. Catal. A* 203 (2000) 71.

224. G.V. Smith, J. Cheng, R. Song, *Catal. Lett.* 45 (1997) 73.

225. M. Maris, T. Bürgi, T. Mallat, A. Baiker, *J. Catal.* 226 (2004) 393.

226. H.U. Blaser, H.P. Jalett, D.M. Monti, J.F. Reber, J.T. Wehrli, *Stud. Surf. Sci. Catal.* 41 (1988) 153.

227. K.E. Simons, A. Ibbotson, P. Johnston, H. Plum, P.B. Wells, *J. Catal.* 150 (1994) 321.

228. P.J. Collier, T. Goulding, J.A. Iggo, R. Whyman, in *Chiral Reactions in Heterogeneous Catalysis*, G. Jannes, V. Dubois (Eds.), Plenum Press, New York, 1995, p. 105.

229. W. Reschetilowski, U. Böhmer, *DMGK-Conference*, Germany, 1993, p. 275.

230. U. Böhmer, K. Morgenschweiss, W. Reschetilowski, *Catal. Today* 24 (1995) 195.

231. J.M. Bonello, F.J. Williams, R.M. Lambert, *J. Am. Chem. Soc.* 125 (2003) 2723.

232. H.U. Blaser, M. Müller, in *Abstracts*, *Europacat-1*, Montpellier, 1993, p. 406.

233. H.U. Blaser, H.P. Jalett, D.M. Monti, A. Baiker, J.T. Wehrli, *Stud. Surf. Sci. Catal.* 67 (1991) 147.

234. I.M. Sutherland, A. Ibbotson, R.B. Moyes, P.B. Wells, *J. Catal.* 125 (1990) 77.

235. J.L. Margitfalvi, M. Hegedus, *J. Catal.* 156 (1995) 175.

236. O. Schwalm, J. Weber, J. Margitfalvi, A. Baiker, *J. Mol. Struct.* 297 (1993) 285.

237. O. Schwalm, J. Weber, B. Minder, A. Baiker, *Int. J. Quant. Chem.* 52 (1994) 191.

238. M. Bartók, B. Török, K. Balázsik, T. Bartók, *Catal. Lett.* 73 (2001) 127.

239. G. Vayner, K. Houk, Y.-K. Sun, *J. Am. Chem. Soc.* 126 (2004) 199.

240. A. Marinas, T. Mallat, A. Baiker, *J. Catal.* 221 (2004) 666.

241. N. Bonalumi, T. Bürgi, A. Baiker, *J. Am. Chem. Soc.* 125 (2003) 13342.

242. A. Vargas, T. Bürgi, A. Baiker, *New J. Chem.* 26 (2002) 807.

243. A. Vargas, T. Bürgi, M. von Arx, R. Hess, A. Baiker, *J. Catal.* 20 (2002) 489.

244. A. Vargas, T. Bürgi, A. Baiker, *J. Catal.* 226 (2004) 69.

245. D. Ferri, T. Bürgi, A. Baiker, *Chem. Commun.* (2001) 1172.

246. D. Ferri, T. Bürgi, *J. Am. Chem. Soc.* 123 (2001) 12074.
247. D. Ferri, T. Bürgi, A. Baiker, *J. Chem. Soc. Perkin Trans.* 2 (2002) 437.
248. R. Selke, K. Häupke, H.W. Krause, *J. Mol. Catal. A* 56 (1989) 315.
249. H.-U. Blaser, B. Pugin, M. Studer, in *Chiral Catalyst Immobilization and Recycling*, D.E. De Vos, I.F.J. Vankelecom, P.A. Jacobs (Eds.), Wiley-VCH, Weinheim, 2000, p. 1.
250. V. Vetere, M.B. Faraoni, G.F. Santori, J. Podestá, M.L. Casella, O.A. Ferretti, *J. Catal.* 226 (2004) 457.
251. J. Margitfalvi, in *Proc. 8th Int. Congr. Catal.*, vol. 4, Verlag-Chemie, Berlin, 1984, p. 891.
252. J.P. Candy, O. Ferretti, J.P. Bournonville, A. El Mansour, J. Basset, G. Martino, *J. Catal.* 112 (1988) 210.
253. G.F. Santori, M.L. Casella, O.A. Ferretti, *J. Mol. Catal. A: Chem.* 186 (2002) 223.
254. B. Didillon, J.P. Candy, F. Lepeltier, O.A. Ferretti, J.M. Basset, *Stud. Surf. Sci. Catal.* 78 (1994) 203.
255. F. Joó, A. Kathó, in *Aqueous Phase Organometallic Catalysis: Concepts and Applications*, B. Cornils, W.A. Herrmann (Eds.), Wiley-VCH, Weinheim, 1998.
256. B.E. Hanson, in *Chiral Catalyst Immobilization and Recycling*, D.E. De Vos, I.F.J. Vankelecom, P.A. Jacobs (Eds.), Wiley-VCH, Weinheim, 2000, p. 81.
257. D.E. Bergbreiter, in *Chiral Catalyst Immobilization and Recycling*, D.E. De Vos, I.F.J. Vankelecom, P.A. Jacobs (Eds.), Wiley-VCH, Weinheim, 2000, p. 43.
258. D.J. Cram, F.A.A. Elhafez, *J. Am. Chem. Soc.* 74 (1952) 5828.
259. D.J. Cram, F.D. Greene, *J. Am. Chem. Soc.* 75 (1953) 6004.
260. A. Mengel, O. Reiser, *Chem. Rev.* 99 (1999) 1191.
261. M. Cherest, H. Felkin, N. Prudent, *Tetrahedron Lett.* 18 (1968) 2199.
262. N.T. Anh, *Top. Curr. Chem.* 88 (1980) 145.
263. K. Tomioka, M. Shindo, K. Koga, *Tetrahedron Lett.* 31 (1990) 1739.
264. K.J. Shea, E. Wada, *J. Am. Chem. Soc.* 104 (1982) 5715.
265. J.M. Brown, *Angew. Chem. Int. Ed. Engl.* 26 (1987) 190.
266. H.W. Thompson, E. McPherson, B.L. Lences, *J. Org. Chem.* 41 (1976) 2903.
267. A. Gypser, H.-D. Scharf, *Synthesis* (1996) 349.
268. K. Nasar, M. Besson, P. Gallezot, F. Fache, M. Lemaire, in *Chiral Reactions in Heterogeneous Catalysis*, G. Jannes, V. Dubois (Eds.), Plenum Press, New York, 1995 p. 141.
269. K. Harada, *Asymm. Syn.* 5 (1985) 345.
270. T. Munegumi, T. Maruyama, M. Takasaki, K. Harada, *Bull. Chem. Soc. Jpn.* 63 (1990) 1832.
271. (a) M. Besson, P. Gallezot, C. Pinel, S. Neto, in *Heterogeneous Catalysis and Fine Chemicals IV*, H.U. Blaser, A. Baiker, R. Prins (Eds.), *Stud. Surf. Sci. Catal.*, Elsevier Science B.V., The Netherlands, 108 (1997) 215; (b) M. Besson, B. Blanc, M. Champelet, P. Gallezot, K. Nasar, C. Pinel, *J. Catal.* 170 (1997) 254.
272. P.W. Collins, S.W. Djuric, *Chem. Rev.* 93 (1993) 1533.
273. F. Cocu, S. Coman, C. Ta nase, D. Macovei, V.I. Pârvulescu, in *Heterogeneous Catalysis and Fine Chemicals IV*, H.U. Blaser, A. Baiker, R. Prins (Eds.), *Stud. Surf. Sci. Catal.*, Elsevier Science B.V., The Netherlands, 108 (1997) 207.
274. S. Coman, F. Cocu, J.F. Roux, V.I. Pârvulescu, S. Kaliaguine, in *Mesoporous Molecular Sieves*, L. Bonneviot, F. Beland, C. Danumah, S. Giasson, S. Kaliaguine (Eds.), *Stud. Surf. Sci. Catal.*, Elsevier Science BV, The Netherlands, 117 (1998) 501.
275. S. Coman, F. Cocu, V.I. Pârvulescu, B. Tesche, H. Bonnemann, J.F. Roux, S. Kaliaguine, P.A. Jacobs, *J. Mol. Catal.* 146 (1999) 247.
276. M. De bruyn, S. Coman, R. Bota, V.I. Pârvulescu, D.E. De Vos, P.A. Jacobs, *Angew. Chem. Int. Ed. Engl.* 115 (2003) 5491.
277. R.L. Augustine, *J. Org. Chem.* 23 (1958) 1853.
278. R.L. Augustine, A.D. Broom, *J. Org. Chem.* 25 (1960) 802.
279. R.L. Augustine, *J. Org. Chem.* 28 (1963) 152.
280. I.N. Lisichkina, A.I. Vinogradova, B.O. Tserevitinov, M.B. Saporovskaya, V.K. Latov, V.M. Belikov, *Tetrahedron Asymm.* 2 (1990) 567.
281. C.F. de Graauw, K.A. Peters, H. van Bekkum, J. Huskens, *Synthesis* 10 (1994) 1007.
282. R.S. Downing, H. van Bekkum, R.A. Sheldon, *CATTECH* 2 (1997) 95.
283. E.J. Creyghton, S.D. Ganeshie, R.S. Downing, H. van Bekkum, *J. Mol. Catal.* 115 (1997) 457.
284. J.C. van der Waal, E.J. Creyghton, P.J. Kunkeler, K. Tan, H. van Bekkum, *Topics Catal.* 4 (1998) 261.
285. S. Coman, C. Bendic, M. Hillebrand, E. Angelescu, V.I. Pârvulescu, A. Petride, M. Banciu, in *Catalysis of Organic Reactions*, F.E. Herkes (Ed.), Marcel Dekker, Inc., New York, 1998, p. 169.

286. A.G.M. Barrett, F. Damiani, *J. Org. Chem.* 64 (1999) 1410.
287. M.E. Kuehne, F. Xu, *J. Org. Chem.* 63 (1998) 9427.
288. M.E. Kuehne, F. Xu, *J. Org. Chem.* 63 (1998) 9434.
289. S. Deng, B. Yu, Y. Lou, Y. Hui, *J. Org. Chem.* 64 (1999) 202.
290. S. Wenglowsky, L.S. Hegedus, *J. Am. Chem. Soc.* 120 (1998) 12468.
291. K. Fujiwara, A. Murai, M.Y. Yamashita, T. Yasumoto, *J. Am. Chem. Soc.* 120 (1998) 10770.
292. S.E. Denmark, J.A. Dixon, *J. Org. Chem.* 63 (1998) 6167.
293. S.E. Denmark, M. Seierstad, *J. Org. Chem.* 64 (1999) 1610.
294. J. Kaminska, M.A. Schwegler, A.J. Hoefnagel, H. van Bekkum, *Recl. Trav. Chim. Pays-Bas* 111 (1992) 432.
295. C.-J. Liu, W.-Y. Yu, S.-G. Li, C.-M. Che, *J. Org. Chem.* 63 (1998) 7364.
296. C.-J. Liu, S.-G. Li, W.-Q. Pang, C.-M. Che, *J. Chem. Soc. Chem. Commun.* (1997) 65.
297. J.T. Groves, T.E. Nemo, *J. Am. Chem. Soc.* 105 (1983) 5786.
298. A. Bhaumik, T. Tatsumi, *J. Catal.* 182 (1999) 349.
299. W. Adam, R. Kumar, T.I. Reddy, M. Renz, *Angew. Chem. Int. Ed. Engl.* 35 (1996) 880.
300. M. Dusi, T. Mallat, A. Baiker, *J. Mol. Catal.* 138 (1999) 15.

CHAPTER 14 QUESTIONS

Question 1

Which are the possible approaches for the preparation of optically active products by chemical transformation of optically inactive starting materials?

Question 2

State the difference between the kinetic resolution and the asymmetric synthesis?

Question 3

What kind of reactions are associated with asymmetric synthesis?

Question 4

Is the formation of an enantiomeric excess kinetically or thermodynamically controlled?

Question 5

Is the activation function the only criterion to select an enantioselective catalyst?

Question 6

Describe, in general, an enantioselective hydrogenation mechanism in the presence of a chiral-modified metal/support catalyst.

Question 7

Indicate the structural elements that are crucial for the functioning of the cinchona alkaloids as chiral modifiers.

Question 8

What are the most important ways to immobilize a homogeneous chiral complex onto a solid support?

Question 9

What are the main drawbacks of a heterogeneized chiral catalyst and how can they be solved?

Question 10

What are the main parameters influencing the diastereoselective reactions?

Index

A

Active catalysts generation conditioning, in colloidal nanoparticles catalysis, 75–76
Active structures design, catalysis regulation by, 244–248
  chemical tuning, 244–245
  ReO$_x$ clusters produced *in situ*, 246–248
Adsorption–desorption and thermal techniques, in heterogeneous catalysis, 7–11
  microcalorimetry, 10–11
  surface area and pore structure, 7–8
  temperature-programmed desorption and reaction, 8–9
  thermogravimetry and thermal analysis, 9–10
AES (Auger electron spectroscopy), in heterogeneous catalysts characterization, 20–21
Ag-based catalysts, 1,3-Butadiene epoxidation with, 405–407
Alkoxide sol–gel process, in TiO$_2$ photocatalysis, 436
Anion–cation pairing, in heterogeneous catalysts, 482–484
Anion-exchange materials intercalation, and POMs, 480–481
Asymmetric catalysis, by heterogeneous catalysts, 493–523
ATR (autothermal reforming) Stoichiometry, 201
Au model catalyst, 349–355
  Au/TiO$_2$ catalyst, 6
Autocompensation model, for metal oxides surface reconstruction, 45
Automotive catalysis, dynamic experiments in stage II screening for, 407–409

B

'Beads' in CombiChem approaches, 398–399
Beer–Lambert law, in photocatalysis, 428
Benzylation, in metal oxides catalysis, 53–54
Binary transition metal oxides, electrical and magnetic properties, 41
'Binary trees', 397
BJH method, in heterogeneous catalysts characterization, 7
Bronsted acidity/basicity, 48–51
1,3-Butadiene
  epoxidation, with Ag-based catalysts, 405–407
  hydrogenation, 52–53

C

Calcium hydroxyapatite (HAP) modified with Ti(IV) photocatalyst, 444
Catalysis/Catalysts, *see also individual entries*
  and chemical reaction engineering, 195–224, *see also* Chemical reaction engineering

colloidal nanoparticles in, *see* Colloidal nanoparticles catalysis
high-throughput experimentation and combinatorial approaches in, 373–421, *see also separate entry*
by metal oxides, *see* Metal oxides catalysis
Catalyst surfaces, structure and reaction control at, 229–254
  methanol selective oxidation regulation, on modified Mo (112) surface, 236–244, *see also separate entry*
  reaction intermediate and transition-state analogue for, 248–254, *see also separate entry*
  regulation, by active structures design, 244–248, *see also separate entry*
  regulation, by coadsorbed molecules, 231–244
  self-assisted dehydrogenation of ethanol on an Nb/SiO$_2$ catalyst, 231–233
  WGS reactions, *see separate entry*
Catalysts testing
  in gas–liquid reactions, 411–420
  in gas–liquid–solid reactions, 411–420
  in gas-phase reactions, 390–411
  in liquid–liquid reactions, 411–420
Catalytic activity vs. leach time, 150
Catalytic cracking, and zeolites, 110–113
Catalytic dewaxing, and zeolites, 116
Catalytic processes of hydrogen production for PEMFCs, 205–224
  contact time influence, 212–214
  feed gas composition influence, 211
  methanation reactions, 211–212
  nanostructured Cu$_x$Ce$_{1-x}$O$_{2-y}$ catalysts in, 214–224
  oxygen storage capacity influence, 209–211
  WGS reaction in, 206–214
Catalytic reaction mechanisms, 337–368
Catenoid, 266
CFD (computational fluid dynamics) modeling, in gas-phase applications, 403
Chemical reaction engineering, and catalysis, 195–224
  and heterogeneous catalysis, 197–199
  hydrogen production and cleaning, 199–224
  low-temperature fuel cells technology, 200–204, *see also separate entry*
Chiral auxiliaries, 521
Chiral-modified platinum hydrogenation catalyst and related systems, 510–517
  kinetic models, 512
  mechanistic investigations, 512–517
Chiral pocket, 514–515
Chloride–oxide exchange catalysis, 56–57
  chlorocarbons, 56
  Freons, 56–57
Chlorocarbons, 56
Cinchona-modified platinum catalysts, 510–517
Claisen–Schmidt condensation asymmetric epoxidation, 54–55

CO (carbon monoxide)
CO hydrogenation studies, on single-crystal surfaces, 338–340
CO methanation, poisons and promoters effects on, 339–340
CO oxidation, on single-crystal surfaces, 340
Colloidal nanoparticles catalysis, 63–86
active catalysts generation, conditioning, 75–76
applications, 74–85
colloidal Pd/C catalysts durability, 82
colloidal Pt/Ru catalysts, 84
heterogeneous reactions in, 74–85, *see also separate entry*
quasi-homogeneous reactions in, 74
reduction methods in, 66–73
stabilization modes in, 64–66
stable particle formation mechanism in, 64
Confinement concept, 519
Copolymer crosslinking by POMs, 479–480
Copper-catalyzed Suzuki cross-coupling reactions, 77
$Cu_{0.2}Ce_{0.8}O_{2-y}$ catalysts, oxygen storage capacity, 210

D

Dealumination of zeolites, 106
Dehydration and dehydrogenation, in metal oxides catalysis, 53
Dehydrochlorination, in metal oxides catalysis, 53
$DeNO_x$ technology, in automotive catalysis, 407–408
Diastereoselective catalysis, 519–523
in heterogeneous hydrogenation, 520
Dispersed systems physical chemistry, basic principles, 261–273
classic thermodynamic theory, limits, 266–267
curved interface, 264–265
flat interface, 262–264
Gibbsian classical thermodynamic theory, 261–262
surface curvature, 265–266
Young–Laplace equation, 267

E

Einstein law of photochemical equivalence, 428
EM (electron microscopy)
in heterogeneous catalysts characterization, 5–7
in microporous and mesoporous catalysts characterization, 130
Enantioselection
and diastereoselective syntheses, energy profiles, 498
enantioselective heterogeneous catalytic reactions, 493–523
enantioselective hydrogenation, 80
enantioselective synthesis vs. kinetic resolution, 497
enantioselective synthesis, reaction coordinate of, 496
mechanistic investigations and hypotheses, 504–507
ESR (electron spin resonance), in heterogeneous catalysts characterization, 18–19

F

Faujasite-type zeolites, 97–101
structure, 98
FCC (fluid catalytic cracking), and zeolites, 110–113
'Fine chemicals', in catalysts testing, 411
Fischer–Tropsch (FT) synthesis, in catalysis, 410
Fluorite structure, of metal oxide catalysts, 43
Fractal approach, for particles and pores modeling, 314–320
Franck–Condon principle, in photocatalysis, 428
Frenkel type defects, 46–47
Freons, 56–57
Friedel–Crafts reactions, in metal oxides catalysis, 53–54
FSM (folded sheet mesoporous) type catalysts, 119
Fuel cells
fuel cell catalysts, 83–85
primary fuel processing, for low-temperature fuel cells, 204–224
types, 204

G

Gas chromatography, in CombiChem approaches, in catalysis, 383
Gas-phase reactions, catalysts testing in, 390–411
gas–liquid reactions, 411–420
gas–liquid–solid reactions, 411–420
stage I testing, 396–402, *see also separate entry*
stage II testing, 402–411
Gas-to-liquid (GTL) technology, in catalysis, 410
Gibbs–Freundlich–Ostwald equation, in dispersed systems, 265
Gibbsian classical thermodynamic theory, 261–262

H

Heat transfer, in catalysts testing, 385
Heterogeneized homogeneous catalysts, 517–519
Heterogeneous asymmetric catalysis, 493–523
and stereodifferentiation, principles, 496–499
chiral-modified platinum hydrogenation catalyst and related systems, 510–517, *see also separate entry*
diastereoselective catalysis, 519–523
historical developments, 499–502
organic polymers, 519
tartaric acid modified me/support hydrogenation catalysts, 502–510, *see also separate entry*
Heterogeneous catalysis, *see also individual entries*
and chemical reaction engineering, 197–199
Heterogeneous catalysts, characterization, 1–26
adsorption–desorption and thermal techniques in, 7–11, *see also separate entry*
Auger electron spectroscopy for, 20–21
electron spin resonance (ESR) for, 18–19
infrared (IR) spectroscopy for, 12–13
low-energy ion scattering for, 21

model catalysts, 22–25, *see also separate entry*
nuclear magnetic resonance (NMR) spectroscopy for, 16–18
optical spectroscopies for, 12–19
Raman spectroscopy for, 13–15
secondary-ion mass spectroscopy for, 21–22
structural techniques for, 2–7, *see also separate entry*
surface-sensitive spectroscopies for, 19–22
ultraviolet photoelectron spectroscopy for, 19–20
ultraviolet–visible (UV–Vis) spectroscopy for, 15–16
x-ray photoelectron spectroscopy for, 19–20
Heterogeneous catalysts, surface Science studies, 337–368
bimetallic surfaces, 340–343
CO hydrogenation studies, on single-crystal surfaces, 338–340
CO oxidation, on single-crystal surfaces, 340
experimental techniques, 334
in situ studies, 355–368, *see also separate entry*
Heterogeneous catalysts, POMs based, 474–485
active carbon, 475–477
alkylammonium ions formation, 478–479
and anion-exchange materials intercalation, 480–481
anion–cation pairing, 482–484
copolymer crosslinking by, 479–480
covalent bond formation, 484
dispersion onto inert supports, 475–478
immobilization on surface-modified supports, 482–485
insoluble solid ionic materials formation, 478–480, *see also separate entry*
metal ions formation, 478
silica and MCM-41, 477–478
Heterogeneous diastereoselective synthesis, 521
Heterogeneous Fenton-type catalysts, 445
Heterogeneous photocatalysis, 427–453, *see also* Photocatalysis
kinetic studies in, 445–446
semiconductors in, 429, 444
Heterogeneous reactions, in colloidal nanoparticles catalysis, 74–85
fuel cell catalysts, 83–85
heterogeneous catalysts, 76–83
precursor concept, 74–75
High-throughput experimentation, *see* HTE
HJ (Haines jumps) mechanism, in texture genesis, 268–270
Homogeneous catalysts, POMs based, 465–474
Honda–Fujishima effect, in photocatalysis, 429
HTE (high-throughput experimentation) and CombiChem (combinatorial) approaches, in catalysis, 373–421
analytical techniques for screening, 382–385
classical statistical designs approaches, 377–378
definition and scope, 374–376
description-driven approaches, 376–377
experimental planning and data handling, 376–380
local optima search, techniques for, 378–380
refinery catalysis applications in, 409–411, *see also separate entry*
stage I and stage II screening, 380–382
synthetic approaches for, 385–390
HTs (hydrotalcitelike) compounds, 480–481
Hydrocracking process, 113–114

and zeolites, 113–114
Hydrogen and energy pathways sources, 200
Hydrogen fuel cell catalysts, 83–85
Hydrogen production and cleaning, 199–224
Hydrogenation, CO hydrogenation, *see under* CO
Hydrogen–Deuterium exchange reaction, 51–52
Hydrosols preparation, surfactants used for, 71

I

In situ studies, on heterogeneous catalysts, 355–368
elevated pressure XPS, 363–368
in situ STM, 363
vibrational spectroscopy, 355–362
Insoluble solid ionic materials formation, 478–480
alkylammonium ions, 478–479
copolymer crosslinking with POM, 479–480
metal ions, 478
IR (infrared spectroscopy)
in heterogeneous catalysts characterization, 12–13
in microporous and mesoporous catalysts characterization, 132–134
IR thermography, in CombiChem approaches, in catalysis, 383

K

Kelvin–Gibbs equation, in dispersed systems, 265
Klabunovskii model, for adsorbed reactant and adsorbed TA interaction, 507–508

L

Lacunary POMs, 474
Langmuir–Hinshelwood model for the photocatalytic reaction, 446
Leaching kinetics, in skeletal catalysis, 144–145, 148
LEIS (low-energy ion scattering), in heterogeneous catalysts characterization, 21
Lewis acidity/basicity, 48–51
Ligand accelerated catalysis, 498
Lipophilic nanostructured metal colloids, 74
Liquid–liquid reactions, catalysts testing in, 411–420
Liquid-phase catalysis
stage I screening for, 413–418
stage II screening for, 418–420
Liquid-phase oxidations, by POMs, 463–487, *see also* POMs
Low-temperature fuel cells technology, 200–204
autothermal reforming, 201–202
fuel cells and primary fuel processing for, 204–224
fuel reforming, 200–202
preferential oxidation of carbon monoxide, 202–204
steam reforming, 201
water-gas shift reaction, 202
LPG aromatization, and zeolites, 116–117

M

Mars and van Krevelen mechanism
  on metallic oxides, 54
  for selective CO oxidation, 224
MAS NMR (magic angle spinning nuclear magnetic
      resonance), 130–132
Mass spectrometric analysis, in CombiChem approaches,
      in catalysis, 383
MCM (mobile composition of matter) type catalysts,
      119–125
Mesoporous catalysts, 118–126
  ordered mesoporous silica (OMS) materials, 118–125,
      see also separate entry
  nonsiliceous ordered mesoporous materials, 125–126
metal adsorption modeling, in supported metal catalysts,
      174–177
metal oxides catalysis, 39–57
  classification, 41
  crystal structures, 42–44
  defect sites, 46–48
  insulators, 41–42
  properties, 40–44
  semiconductors, 42
  solid acids and bases, 48–51, *see separate entry*
  surface reconstruction, 44–46
  surface structures, 44–48
Metal oxides surfaces, reconstruction, 45
Metal surface reconstruction model, 45
Methanation reactions, 211–212
Methanol selective oxidation regulation, on modified Mo
      (112) surface, 236–244
  by extra oxygen atoms, 242–244
  reaction aspect, 236–240
  reaction scheme, 240–242
  steady-state methanol oxidation, reaction kinetics, 242
Microcalorimetry, in heterogeneous catalysts
      characterization, 10–11
Microporous and mesoporous catalysts, 95–134, *see also*
      Zeolites
  characterization, *see separate entry below*
  porosity in, 96
Microporous and mesoporous catalysts, characterization,
      126–134
  physisorption analysis, 128–130
  using BET equation, 128
  using BJH method, 129
  using electron microscopy, 130
  using infrared spectroscopy, 132–134
  using nonlocal density functional theory, 129
  using nuclear magnetic resonance spectroscopy,
      130–132
  x-ray diffraction, 127–128
Microwave radiation, and photocatalysis, 452
Mixed oxide and composites containing Titania, 441
Model catalysts, 22–25
  model-supported catalysts, characterization, 24
  single-crystal surfaces as, *see separate entry below*
Model catalysts, single-crystal surfaces as, 344–355
  amorphous SiO_2 films, 345–348
  crystalline SiO2 films, 348

highly defective TiO_x films, 348–349
  supported metal clusters, 349–355
  thin-film growth, 345–349
  TiO_2 thin films, 345
Molecular catalysts, synthetic approaches for, 386–390
Molecular imprinting, and catalyst surfaces reactions,
      248–252
Mordenite catalysts, 115–116
  structure, 103
Morpho-independent textural parameters, 280–290
  density and porosity, 280–283
  mean sizes of particles and pores, 290–293
  porosity properties, 284–289
  porous solids texture modeling, general problems,
      293–301, *see also separate entry*
  specific surface area, 289–290
  true, apparent, and bulk density, experimental
      techniques of measurements of, 283–284
MTG (methanol to gasoline) conversion, 117–118
MTO(methanol to olefins) conversion, 117–118

N

Nanopowders and nanostructured metal colloids, 69
Nanostructured $Cu_xCe_{1-x}O_{2-y}$ catalysts in PEMFCs,
      214–224
Nanostructured metal colloids formation via the salt
      reduction method, 65
Neodymium, as an ion modifier for TiO_2 photocatalysis, 439
Neural networks, in catalysis, 379
Nickel catalysts, modification, 503
NMR (nuclear magnetic resonance) spectroscopy
  in heterogeneous catalysts characterization, 16–18
  in microporous and mesoporous catalysts
      characterization, 130–132
n-Paraffins isomerization, and zeolites, 114–116

O

Ockam's razor, principles, 300
OCM (oxidative coupling of methane), 55–56
OMS (ordered mesoporous silica) materials, 118–125
  catalysis with, 123–125
  OMS-catalyzed reactions, 122–123
  surface modifications, 122–123
Optical spectroscopies, in heterogeneous catalysts
      characterization, 12–19
OR (Ostwald ripening), in texture genesis, 269–270
Organic polymers, 519
Organosols, 70–71, 74

P

PAFC (phosphoric acid fuel cells), catalysts for, 84
PEMFC (proton-exchange membrane fuel cells), 204–205
  catalysts for, 84

catalytic processes of hydrogen production for, 205–224, *see also separate entry*
cross section, 205
Percolation theory, 320–324
Peroxometalates, 472–474
Petrochemical processes, zeolites in, 109–117
    catalytic dewaxing, 116
    fluid catalytic cracking, 110–113
    hydrocracking, 113–114
    LPG aromatization, 116–117
    methanol to gasoline and methanol to olefins, 117–118
    n-paraffins isomerization, 114–116
Photocatalysis, 427–453, *see also* Heterogeneous photocatalysis
    and microwave radiation, 452
    combinatorial approaches in preparation and testing in, 446–447
    energy storage in, 450
    general principles, 430–433
    mixed oxide and composites containing Titania, 441
    monoliths containing Titania, 443
    noble metal deposited on Titania surfaces, 441–443
    O_2 role in, 448
    photocatalyst deactivation, 452
    photoinduced deposition of various metals onto semiconductor, 449–450
    reactions involving, 448–450
    solar photocatalysis, 450
    sonophotocatalysis, 450–452
    TiO_2 photocatalysts, 433–438, *see also separate entry*
    ZnO as, 443–444
Photoinduced deposition of various metals onto semiconductor, 449–450
Photothermal deflection spectroscopy, in CombiChem approaches, in catalysis, 383
pH shift modeling, in supported metal catalysts, 168–174
Physisorption analysis, in microporous and mesoporous catalysts characterization, 128–130
PM-IRAS, for *in situ* studies on heterogeneous catalysts, 358, 360
POMs (polyoxometalates) catalysts, 444–445
    copolymer crosslinking with, 479–480
    dispersion onto active carbon, 475–477
    dispersion onto inert supports, 475–478
    dispersion onto silica and MCM-41, 477–478
    heterogeneous catalysts based, 474–485
    heterogeneous catalysts with POMs based, 474–485, *see also separate entry*
    heterogenization, 476
    homogeneous catalysts with POMs based, 465–474
    immobilization on surface-modified supports, 482–485, *see also separate entry*
    in structural chemistry, 464
    intercalation into anion-exchange materials, 480–481
    lacunary POMs, 474
    liquid-phase oxidations by, 463–487
    mixed-addenda POMs, 465
    peroxometalates, 472–474
    POMs based homogeneous catalysts, 465–474
    porous POMs, 486
    transition-metal-substituted POMs, 465–472

Pores and particles arrangement, lateral and statistical models, 320–327
    backbone clusters in, 324
    problem of bonds, 321–322
    problem of sites, 322–324
    stochastic and other statistical models of long-range order, 324–327
Porous catalysts, 96–97, *see also* Zeolites
    structure characterization methods, 279
Porous solids texture modeling, general problems, 293–301
    classification, 294–299
    generalized models and systematic sets of models, 299–301
    morphology, 293–294
Positive asymmetric amplification, 498
POX (partial oxidation) stoichiometry, 201
Precursor concept, in colloidal nanoparticles catalysis, 74–75
PSs (porous systems) textures, *see* Texturology
Pt/SiO_2 catalysts, 178
    oxide PZC measurement, 179
    preparation methods, 177–179
    preparation methods, 180
    Pt tetraammine adsorption over silica, 177–185
    tuning finishing conditions to retain high dispersion, 182–185
    uptake–pH survey to identify optimal pH, 179–182
Pycnometric technique, for density measurement, 283–284
Pyridine, for surface acidity characterization, 51
PZC determination of silica, 179

Q

quasi-homogeneous reactions, in colloidal nanoparticles catalysis, 74

R

RA (radial anisotrophy) of PSs, 299–301
Raman spectroscopy, in heterogeneous catalysts characterization, 13–15
Raney catalysts, 141–142
Rayleight—Debae–Gans equation, 317
Reaction intermediate and transition-state analogue, for catalyst surfaces reactions, 248–254
    and reaction intermediate design, 252–254
    by molecular imprinting, 248–252
Refinery catalysis applications, in HTE, 409–411
    liquid–gas separation, 411
    pressure flow and pressure control, 410
Revised adsorption model development, for supported metal catalysts, 166–177
    metal adsorption modeling in, 174–177
    pH shift modeling in, 168–174
    qualitative discrimination of mechanisms, 166–168

Rh/C catalysts activity, 77
Rh/Pt/C and mixed Rh 1Pt /C catalysts, 79
Rock salt structure, of metal oxide catalysts, 43
RPA model, 176–177
Rutile structure, of metal oxide catalysts, 43

S

Sabatier principle, 198
Sachtler model, for adsorbed reactant and adsorbed TA
    interaction, 507–508
Schotty type defects, 46–47
SEA (strong electrostatic adsorption) approach, for
    supported metal catalysts preparation, 162,
    177–190
SERS (surface-enhanced Raman spectroscopy), in
    heterogeneous catalysts characterization,
    14–15
SIMS (secondary-ion mass spectroscopy), in heterogeneous
    catalysts characterization, 21–22
Single-bead reactors, 398–401
    'single-bead approach', in HTE, 398
Single-crystal surfaces
    as model catalysts, 344–355, *see also separate entry*
    CO hydrogenation studies on, 338–340
    CO oxidation on, 340
SiO_2
    amorphous films, in heterogeneous catalysis, 345–348
    and MCM-41 catalyst, 477–478
    crystalline films, 348
    silica gel texture formation, 273
Skeletal catalysts, 141–154
    advantages/disadvantages, 153–154
    applications, 151–153
    deactivation/aging of, 149–151
    leaching in, 144–145
    preparation, 142–147
    promoters in, 145–147
    reactions catalyzed by, 143, 146
    skeletal copper catalysts, 142–154
    skeletal nickel catalysts, 142–154
    skeletal nickel leaching, 148
    structures, 147–149
Solar photocatalysis, 450
Solid acids and bases, in metal oxides catalysis, 48–51
    benzylation, 53–54
    chloride–oxide exchange catalysis, 56–57, *see also*
        separate entry
    dehydration and dehydrogenation, 53
    dehydrochlorination, 53
    examples, 51–57
    hydrogen–deuterium exchange reaction, 51–52
    hydroxyl groups as Bronsted acids and Lewis bases, 50
    Lewis acidity/basicity and Bronsted acidity/basicity,
        spectroscopic methods of detecting, 50–51
    metal cations as Lewis acids, 49
    oxidation catalysis, 54–56
    oxidative coupling of methane, 55–56
    oxidative dehydrogenation, 54–55
    oxygen anions as Lewis bases, 49

Solid-state inorganic catalysts, synthetic approaches for,
    387–390
Sonophotocatalysis, 450–452
Stage I screening/testing
    alternative stage I screening concepts, 417–418
    and stage II screening, analytical techniques in, 382
    and stage II screening, distinction, 381
    for liquid-phase catalysis, 413–418
    of catalysts for gas-phase reactions, 396–402
    optimal use of, 401–402
    reactant distribution for stage I screening systems,
        396–398
Stage II screening/testing
    for liquid-phase catalysis, 418–420
    in automotive catalysis, 407–409
    in gas-phase applications, 402–411
Steam reforming, 201
Structural techniques, for heterogeneous catalysts
    characterization, 2–7
    electron microscopy (EM), 5–7
    transmission electron microscopy (TEM), 5
    x-ray absorption spectroscopy, 3–5
    x-ray diffraction, 2–3
Supported metal catalysts, scientific method of
    preparation, 161–190
    early work, 162–166
    extension of SEA to alumina and carbon, 185–187
    preparation, 75
    revised adsorption model for, development, 166–177,
        see also separate entry
Surface-capillary phenomena, and texture genesis,
    267–273
Surface-modified supports, POMs immobilization on,
    482–485
    anion–cation pairing, 482–484
    covalent bond formation, 484

T

Tartaric acid modified me/support hydrogenation
    catalysts, 502–510
    catalyst preparation process, 502–503
    mechanistic investigations and hypotheses for
        enantioselection, 504–507
    modification process, 503
    proposed mechanisms, 507–510
    substrate and hydrogenation parameters, 503–504
TEM (transmission electron microscopy), in
    heterogeneous catalysts characterization, 5
Texture genesis, mechanisms and processes, 267–273
    drying process in, 271–272
    fundamental mechanisms, 267–269
    fundamental processes, 269–273
    sol–gel technology in, 272
Texturology, 257–328
    characterization, absorption as primary instrument for,
        274–280
    disordered packings investigations, 311–314
    dispersed systems physical chemistry, basic principles,
        261–273

fundamental characteristics, 260
modeling particles and pores in a local arrangement, 301–314
morpho-independent textural parameters, 280–290, *see also separate entry*
ordered packings investigations, 306–311
particles and pores modeling, fractal approach, 314–320
pores and particles arrangement, lateral and statistical models, 320–327
Thermogravimetry and thermal analysis, in heterogeneous catalysts characterization, 9–10
Thiele modulus, in catalysts testing, 392–393
TiO$_2$ photocatalysts, 433–438
colloidal systems in, 435
modified Titania photocatalysts, 438–441
neodymium as an ion modifier for, 439
preparation procedures, 435–438
transition metal ions in, 440
Transition metal ions, in TiO$_2$ photocatalysis, 440
Tungsten-based catalyst library, 416

WGS reaction kinetics, 210
with or without methanation reaction, 212–213
Wurtzite structure, of metal oxide catalysts, 43

X

XPS (x-ray photoelectron spectroscopy)
for *in situ* studies on heterogeneous catalysts, 363–368
in heterogeneous catalysts characterization, 19–20
X-ray absorption spectroscopy, in heterogeneous catalysts characterization, 3–5
X-ray diffraction
for different manganese oxides before and after pretreatment in H$_2$, 3
for TiO$_2$ samples obtained by hydrothermal treatments at 80°C, 4
in heterogeneous catalysts characterization, 2–3
in microporous and mesoporous catalysts characterization, 127–128

U

UPS (ultraviolet photoelectron spectroscopy), in heterogeneous catalysts characterization, 19–20
UV–Vis (ultraviolet–visible) spectroscopy, in heterogeneous catalysts characterization, 15–16

V

Vanadia catalysts, 54–55
Vibrational spectroscopy, for *in situ* studies on heterogeneous catalysts, 355–362
Voronoi–Delaunay method, for corpuscular and sponge-like porous solids description, 301–306

W

Washburn equation, in pores and particles arrangement, 321
WGS (water-gas shift) reactions
CO conversion for, 208, 211
copper–ceria catalysts in, 212–213
in PEMFCs, 206–214
on Rh/CeO$_2$, 235–236
on ZnO mechanism, 234–235
reactant-promoted, 234–236

Z

Zeolites, 97–118
catalytic application of, 107–118
dealumination of, 106
description, 97–101
frameworks, 97
in catalytic processes, 101–103
in petrochemical processes, 109–117, *see also separate entry*
in TiO$_2$ photocatalysis, 436
industrial production, 104
mesoporous catalysts, 118–126, *see also separate entry*
metals and metal complexes in, 106–107
microporous and mesoporous catalysts, characterization, 126–134, *see also separate entry*
Mordenite zeolite, 102–103
pore diameters, 100
post-synthesis treatment and modification of, 105–107
production, 103–105
protonation of, 105–106
selectivity in, 108
surface area determination, by BET method, 101
Zeolite A, 103
Zeolite X and Zeolite Y, 101–102
Zeolite-catalyzed reactions toward organic products, 109–110
ZSM-5 zeolite, 102
ZnO, as photocatalyst, 443–444
ZSM-5 cracking catalyst
and zeolites, 116
structure, 102

9 780367 390815